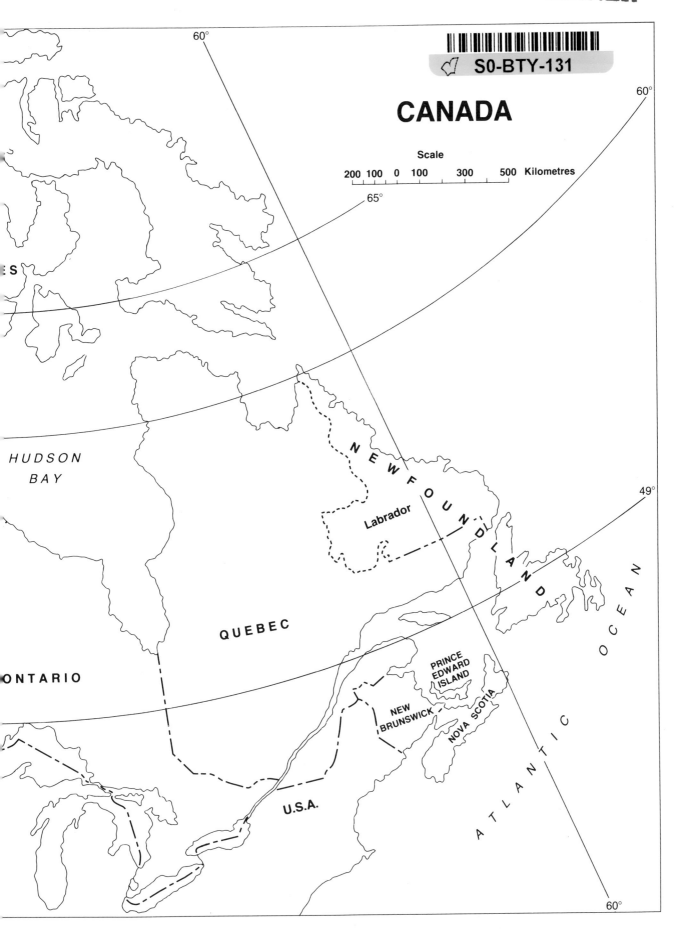

GREG BRENNER

S0-BTY-131

CANADA

Scale

200 100 0 100   300   500   Kilometres

60°

60°

65°

49°

60°

HUDSON BAY

NEWFOUNDLAND

Labrador

QUEBEC

ONTARIO

PRINCE EDWARD ISLAND

NEW BRUNSWICK

NOVA SCOTIA

ATLANTIC OCEAN

U.S.A.

ONE HUNDRED YEARS OF PROGRESS

The year 1986 is the centennial of the Research Branch, Agriculture Canada.

On 2 June 1886, *The Experimental Farm Station Act* received Royal Assent. The passage of this legislation marked the creation of the first five experimental farms located at Nappan, Nova Scotia; Ottawa, Ontario; Brandon, Manitoba; Indian Head, Saskatchewan (then called the North-West Territories); and Agassiz, British Columbia. From this beginning has grown the current system of over forty research establishments that stretch from St. John's West, Newfoundland, to Saanichton, British Columbia.

The original experimental farms were established to serve the farming community and assist the Canadian agricultural industry during its early development. Today, the Research Branch continues to search for new technology that will ensure the development and maintenance of a competitive agri-food industry.

Research programs focus on soil management, crop and animal productivity, protection and resource utilization, biotechnology, and food processing and quality.

COMPENDIUM OF PLANT DISEASE AND DECAY FUNGI IN CANADA 1960-1980

J. H. Ginns
Biosystematics Research Centre
Ottawa, Ontario

Research Branch
Agriculture Canada
Publication 1813   1986

© Minister of Supply and Services Canada 1986

Available in Canada through

Associated Bookstores
and other booksellers

or by mail from

Canadian Government Publishing Centre
Supply and Services Canada
Ottawa, Canada K1A 0S9

Catalogue No. A43-1813/1986 E Canada: $23.40
ISBN 0-660-12223-5 Other Countries: $28.10

Price subject to change without notice

Req. No. 01796-6-B967

Canadian Cataloguing in Publication Data

Ginns, J. H. (James Herbert)

    Compendium of plant disease and decay fungi in
Canada, 1960-1980

Introduction in English and French.
Companion volume to: An annotated index of plant
diseases in Canada / by I.L. Conners. 1967.

1. Plant diseases — Canada. 2. Fungi, Phyto-
pathogenic — Canada. I. Canada. Agriculture
Canada. II. Conners, Ibra L. (Ibra Lockwood). An
annotated index of plant diseases in Canada. III. Title.

SB605.C3G56 1986   632'.3'0971   C86-099203-9

About the author ...

James H. Ginns, whose major research interest lies in the biology of wood-decaying fungi, has been a research mycologist with Agriculture Canada in Ottawa since 1969. He has worked as a forest pathologist for the Canadian Forestry Service, Victoria, British Columbia; the International Paper Co., Bainbridge, Georgia; and the United States Forestry Service, Upper Darby, Pennsylvania. His field studies have been concentrated on northern forests from the Yukon to Newfoundland. He has published nearly 80 scientific papers and is an active member of several professional societies. He holds Ph.D. degrees from the State University of New York and Syracuse University and has received training in plant pathology at West Virginia University and in forestry at the University of Connecticut.

Note sur l'auteur ...

Spécialisé en biologie des champignons déprédateurs du bois, James Ginns est depuis 1969 mycologiste chercheur au ministère de l'Agriculture du Canada à Ottawa. Auparavant, il avait occupé la fonction de pathologiste forestier pour le Service canadien des forêts, à Victoria (C.-B.), de la société International Paper à Bainbridge (Georgie) et du Service américain des forêts, à Upper Darby (Pennsylvanie). Ses travaux sur le terrain se sont concentrés sur la forêt boréale, du Yukon à Terre-Neuve. Il a à son actif près de 80 publications scientifiques et est membre à part entière de plusieurs sociétés savantes. Il détient des doctorats des universités de New York et de Syracuse et a reçu sa formation en pathologie végétale à l'Université de Virginie occidentale et en foresterie à l'Université du Connecticut.

FOREWORD

This publication, Compendium of plant disease and decay fungi in Canada, by J. Ginns, covering the period 1960-1980 inclusive, is a notable achievement that meets a specific need. Its predecessor, An annotated index of plant diseases in Canada, by I. L. Conners, published in 1967, was a landmark in Canadian plant pathological literature. Covering more than a century of investigations these two companion volumes provide an invaluable record of the occurrences of diseases of cultivated and wild plants and their associated fungi. The provision of a ready access to Canadian plant pathological literature by host, pathogen, and geographical region is a most valuable reference for pathologists involved in disease control and in international plant quarantine.

This compendium is needed not only because of the many additions to our knowledge of pathogens and related fungi over the last 21 years, but also to correct and strengthen those parts of Mr. Conners' publication that are recognized as being inadequate. Fungal taxonomy changes, and the compendium gives an update of nomenclature and synonymy that reflects the current taxonomic status of the fungi involved.

Although much of the Canadian phytopathological literature of recent years has not contained accurate appraisals of losses caused by crop diseases, Dr. Ginns has included recorded estimates of losses as well as several opinions expressed about wood decays and other damage caused by resident fungi to harvested products of forestry and agriculture. Such statements will aid in identifying areas worthy of research undertakings.

Mature scientists never fail to be impressed by the potential usefulness of detailed observations recorded in the literature. Many great scientific discoveries have stemmed from research triggered by reflecting on previously recorded facts of unrevealed significance. Therefore, the information in this compendium should prove stimulative in unraveling probable interactions among fungal pathogens, resident fungi, and their environmental substrata. For this reason, the inclusion in this index of fungi found growing on dead parts of living plants, particularly forest trees, is warranted. Because one of the purposes of this index is to facilitate the identification of fungal specimens, there is ample justification for including perthophytes that may have an impact yet to be revealed to industrial mycology.

C. D. McKeen
Former Coordinator (Plant Pathology)
  and Leader of Crop Protection
  Coordination
Research Branch, Agriculture Canada
Central Experimental Farm,
Ottawa, Ontario

INTRODUCTION

This compendium is a summary of the records of fungi on cultivated and wild vascular plants in Canada that have been collected and published from 1960 to the end of 1980. Nearly 600 genera of host plants are included and over 3900 species of fungi are recognized. Many of these fungi such as stem rust of wheat (*Puccinia graminis* f. sp. *tritici*) and root rot of Douglas fir (*Phellinus weirii*) cause plant diseases that significantly decrease crop yield or reduce postharvest quality. Included are many fungi whose economic importance, whether significant or insignificant, has not been demonstrated either because the fungi are rare or because their impact has not yet been studied. Because the main purpose of this compendium is to facilitate the identification of pathological specimens, it was decided to include those fungi that occur in association with disease symptoms or on dead plant tissue rather than to restrict it to known pathogens. Often the relationship between the fungus and the substrate is not fully understood and in some cases fungi thought to be unimportant were later discovered to be very important. In addition, there are those pathogens that exist in the mycelial state in the live host plant and only produce the more easily identified fruiting bodies after the host has died. The correct identification and cataloguing of plant pathogens is the basis of plant pathology, from the formulation of plant quarantine regulations to breeding for disease resistance.

Losses caused by plant disease fungi are only partially documented. Losses caused by some fungi have not been determined, whereas losses caused by other diseases are relatively well known. Cereal diseases have been the subject of detailed analysis in Manitoba [672]. In 1971 losses from diseases in wheat, barley, and oats were 174.5, 115.7, and 24.7 thousand tonnes or 8.8%, 5.1%, and 2.2% of the potential production [672]. These figures include losses caused by bacteria, viruses, and so on, as well as the fungi. The use of disease-resistant varieties has been effective in limiting the losses caused by some diseases. For example, the 1979 estimated annual value of wheat stem resistance in Manitoba and southeastern Saskatchewan was $217 million [637].

In forestry, the disease losses are divided into decays, hypoxylon canker, miscellaneous diseases, and dwarf mistletoes, which caused between 1977 and 1981 an estimated annual depletion of 44.9 million cubic metres of wood. The annual loss from decays is estimated to be 25 million cubic metres, a volume equivalent to nearly one-fifth of the wood harvested annually in Canada. Most of the losses caused by decay result from the actual destruction of wood by stem decays; the remainder is due to growth reduction and tree mortality caused by root decays. Hypoxylon canker, through tree mortality, caused the loss of 11.2 million cubic metres of wood.

Records of disease losses that have come to my attention have been inserted in the text. Many of these reports appeared in the annual reports of the Research Branch, Agriculture Canada, or the annually issued Forest Insect and Disease Conditions in Canada from the Canadian Forestry Service.

Another purpose of this volume is to provide easy access to the recent literature on Canadian plant pathology by host and by geographical location. It is a summary of what has been accomplished in Canada from 1960 to the end of 1980.

This volume also represents a step toward the production of a flora of Canadian fungi. The most extensive summary of Canadian fungi to date is An annotated index of plant diseases in Canada by I. L. Conners. Published in 1967, Mr. Conners' book provides an account from the beginnings of Canadian plant pathology through to the end of 1959 with the occasional citation from the early 1960's. His commentary is an invaluable view of our history in plant pathology and mycology, and is a record of observations on disease epidemiology and the economic significance of plant diseases. In 1983 Gourley [603a] published an important paper, "An annotated index of the fungi of Nova Scotia."

This compendium is intended to be a companion volume to Mr. Conners' book. Unlike Conners' book, which included bacterial and viral diseases, physiological disorders, and mineral deficiencies, only plant disease and decay fungi are listed. Because of the recent increased volume of literature on these subjects, they have been excluded from this compendium. Furthermore, Conners included diseases from Alaska and Greenland, whereas this compendium deals only with Canadian reports. Despite the narrower scope and the briefer nature of the comments in the compendium, the explosion of plant pathology literature in the 21 years surveyed resulted in this volume being as large as Conners'. The format is similar to Conners', thereby facilitating consultation between the two volumes. The entries in the compendium are cross-referenced to Conners' book so the reader immediately knows whether Conners treated that fungus. Conners put a strong emphasis on the historical aspects, whereas the compendium provides a guide to the recent literature as well as an index of contemporary names for the fungi. The compendium concentrates on literature published from 1960 through to the end of 1980, but it also includes some earlier references not cited by Conners.

Only reports linking the name of the fungus, the geographic locality, and at least the generic name of the host plant are included. A few records where the linkage was unclear are included, based on verification from cited specimens present in the National Mycological Herbariam (DAOM). Some reports had to be excluded because the information was too vague, for example, if the host was simply reported as "on twigs." Also, some groups of fungi were excluded, such as the well-known mycorrhizal species that are beneficial to the growth of the host plant and provide a degree of disease resistance and epiphytic lichenized fungi.

The geographic localities cited for the fungus records are the provinces and territories, or subdivisions of them. These units and their abbreviations used in the text are, from northwest to east: Yukon Territory (Yukon); Northwest Territories (NWT) or its districts, Mackenzie (Mack), Keewatin (Keew), and Franklin (Frank); and the 10 provinces, British Columbia (BC), Alberta (Alta), Saskatchewan (Sask), Manitoba (Man), Ontario (Ont), Quebec (Que), New Brunswick (NB), Nova Scotia (NS), Prince Edward Island (PEI), and Newfoundland (Nfld). Some records cite

Labrador (Labr), the mainland portion of Newfoundland.

The entries in the compendium came primarily from monographs and journals that publish the results of scientific studies. The principal journals that were carefully reviewed were Canadian Field-Naturalist, Canadian Journal of Botany, Canadian Journal of Forest Research, Canadian Journal of Plant Pathology, Canadian Journal of Plant Science, Proceedings of the Canadian Phytopathological Society, Canadian Plant Disease Survey, Forestry Chronicle, Fungi Canadenses, Mycologia, Mycological Papers (Kew), Mycotaxon, Naturaliste canadien (Québec), Phytoprotection, Plant Disease Reporter and its successor Plant Disease, Studies in Mycology (Baarn), and Syesis.

A diligent effort has been made by the mycologists of the Biosystematics Research Centre to update the nomenclature and synonymy and to link anamorphic and holomorphic names for the fungi that adequately reflect current taxonomic status. Some names remain in limbo because

the revision of fungal groups is incomplete or they have not been considered in the recent taxonomic literature. Comments on obviously misplaced names are made in the text. Furthermore, considerable research effort was expended in the critical evaluation not only of the preferred name for a fungus but also of the validity of some reports, that is, had the fungus or the host been misidentified. Both corrections of misidentifications as well as questionable records are commented upon in the text.

The compendium is composed of four main sections: introduction, host index, bibliography, and fungal indexes. Within the host index the generic names of plants are arranged in alphabetical order. Only plant species recorded in the literature search are included. The fungi recorded on each plant genus are listed in alphabetical order. The authorities for the scientific names of the fungi are given only in the indexes, a space-saving arrangement.

ACKNOWLEDGMENTS

The gathering and organizing of the data in this volume was greatly facilitated by the conscientious and enthusiastic assistance of several people. Michael Sarazin compiled the majority of the citations and completed the initial revision of the raw data. He was succeeded by Sheila Thomson and Modra Kaufert, who completed the data compilation and dealt with revisions and corrections. Ms. Kaufert efficiently collated the final handwritten draft of the text, carefully proofread much of the manuscript, and dealt with numerous items in a most efficient manner. Joy Bowerman prepared the initial version of the paragraphs on the genera of host plants and provided valuable assistance in organizing the early stages of the project. Sylvie Séguin prepared most of the raw index and assisted in other ways. Mona Meredith efficiently typed the text and bibliography and Mary Ann Martin typed the index.

The Director of the Biosystematics Research Centre, Mr. G. A. Mulligan, has actively supported the project and allotted as much technical support as possible to the project. My colleagues in the Institute provided the necessary expertise to permit the taxonomic revision and updating of the fungus names. Because certain groups of fungi (that is, orders, and so on) appeared more frequently than others, some colleagues devoted a considerable amount of time to the taxonomic revisions, especially John Bissett, Michael Corlett, J. A. Parmelee, Robert Shoemaker, and George White. W. J. Cody, graciously answered many questions concerning the names and distribution of vascular plants.

The numerous discussions with Claire Lamoureux and Laurie Wilson of Research Program Services, as well as their efforts in formatting the manuscript, inserting revisions and making corrections, have made this volume a much improved product.

Due to the large number of entries the preparation of the genus index and the species index proved to be a muchmore complex, time consuming task than had been expected. The efforts of several individuals simplified my job in completing the indexes. John Bissett provided helpful suggestions on techniques, and Marilyn Olmstead, Louise Boardman and Doris Hyndford of Systems & Consulting Directorate expended considerable effort in developing a series of programs to manipulate the large data base and provide the output in the desired format.

AVANT-PROPOS

Embrassant la période 1960-1980, la publication Compendium des champignons des maladies des plantes et des pourritures du bois au Canada par J. Ginns est une remarquable réalisation qui répond à un besoin bien réel. Son prédécesseur, An Annotated Index of Plant Diseases in Canada, paru en 1967 sous la plume de I.L. Conners, a été un jalon de la bibliographie phytopathologique du Canada. Regroupant ainsi plus d'un siècle d'observations, les deux volumes constituent une source inestimable de renseignements sur la fréquence d'apparition des maladies des plantes cultivées ou sauvages ainsi que des champignons correspondants. Un accès rapide et facile à la bibliographie canadienne en ce domaine, par hôte, par pathogène et par région géographique, est un outil des plus précieux pour le pathologiste engagé dans les activités de lutte antiparasitaire ou les mesures de quarantaine phytosanitaire internationale.

Ce compendium se justifie non seulement à cause des nombreuses nouvelles connaissances sur les pathogènes et des champignons apparentés au cours des 21 dernières années, mais aussi pour corriger ou renforcer les parties de l'oeuvre de M. Conners qu'on jugeait non adéquates. La taxonomie des champignons évolue, aussi le compendium présente-t-il une mise à jour de la nomenclature et de la synonymie qui correspond à la situation taxonomique actuelle des champignons considérés.

Bien que dans l'ensemble la bibliographie phytopathologique récente du Canada ne renferme pas d'évaluations exactes des pertes causées par les maladies des cultures, le Dr Ginns a inséré les chiffres publiés des pertes ainsi que plusieurs avis exprimés sur les pourritures du bois et sur les autres dépradations causées aux produits de la forêt et de l'agriculture. Ces opinions permettront de mieux cerner les domaines appelant de nouvelles recherches.

Les chercheurs chevronnés sont toujours impressionnés par l'utilité éventuelle des observations détaillées qui sont consignées dans la bibliographie. Plus d'une grande découverte scientifique découle de recherches inspirées par des réflexions sur des faits déjà rapportés, mais dont on n'avait pas encore dégagé toute l'importance. L'information réunie dans ce Compendium devrait donc servir de stimulant pour l'élucidation des interactions possibles entre les champignons pathogènes, la mycoflore permanente et leur substrat écologique. C'est pourquoi l'auteur a jugé bon d'inclure dans ce recueil les champignons trouvés sur les parties nécrosées des plantes vivantes, en particulier les arbres forestiers. Comme l'un des buts du compendium est de faciliter l'identification des spécimens fongiques, il y a toute raison d'y inclure les perthophytes dont les incidences pour la mycologie industrielle restent encore à élucider.

C. D. McKeen
Ancien coordonnateur (pathologie des plantes) et chef du groupe de coordination de la protection des cultures,
Direction générale de la recherche, Agriculture Canada
Ferme expérimentale centrale
Ottawa, Ont.

INTRODUCTION

Ce compendium est le résumé des observations faites sur des champignons qui croissaient sur des plantes vasculaires cultivées et sauvages recueillies et publiées de 1960 à la fin de 1980. Il comporte près de 600 genres de plantes-hôtes et plus de 3 900 espèces de champignons reconnus. Un grand nombre de ces champignons, comme la rouille de la tige du blé (*Puccinia graminis* f. sp. *tritici*) et la pourriture des racines du Douglas (*Phellinus weirii*) causent des maladies qui abaissent de façon significative le rendement des cultures ou leur qualité postrécolte. On trouvera aussi nombre de champignons dont l'importance économique, réelle ou anodine, n'a pas été confirmée parce que le champignon en cause est rare, ou que ses effets n'ont pas encore été étudiées. Comme l'objet premier de cet ouvrage est de faciliter l'identification des prélèvements pathologiques, on a jugé bon d'y inclure les champignons qui existent en association avec les symptômes de maladies ou sur les tissus végétaux morts, plutôt que de se limiter aux pathogènes connus. Souvent on ne comprend pas complètement les rapports entre le champignon et son substrat et dans certains cas des champignons qu'on croyait négligeables se sont révélés très importants. Il y a aussi les pathogènes qui vivent au stade mycélien dans la plante-hôte vivante et ne produisent les fructifications plus facilement identifiables qu'après la mort de l'hôte. L'identification et le catalogage précis des pathogènes des plantes est le fondement de la phytopathologie, depuis l'établissement des règlements phytosanitaires jusqu'à la sélection axée sur la résistance aux maladies.

Les pertes imputables aux champignons phytopathogènes ne sont que partiellement répertoriées, celles causées par quelques espèces n'ont pas été établies du tout et pour d'autres maladies on n'en a qu'une idée incomplète. Les maladies des céréales ont fait l'objet d'analyses détaillées au Manitoba (672). En 1971, les pertes causées par les maladies du blé, de l'orge et de l'avoine s'établissaient, respectivement, à 174,5 115,7 et 24,7 milliers de tonnes, soit l'équivalent de 8,8, 5,1, et 2,2 % de la production potentielle (672). Ces chiffres comprennent les manques à produire causés par les bactéries, les virus et les organismes apparentés aussi bien que par les champignons. L'usage de variétés résistantes aux maladies a permis de limiter les pertes provoquées par certaines maladies. Ainsi, on attribue à la résistance à la rouille de la tige du blé une valeur de sauvegarde de quelque 217 millions de dollars en 1979 au Manitoba et dans le sud-est de la Saskatchewan (637).

Dans l'exploitation forestière, les déperditions causées par les maladies se divisent en pourritures, chancre à hypoxylon, maladies diverses et faux-gui qui ensemble ont occasionné entre 1977 et 1981 une perte annuelle approximative de 44,9 millions de mètres cubes de bois. Les pertes causées par les seules pourritures s'évaluent à 25 millions de mètres cubes par an, soit l'équivalent de près du cinquième de la récolte annuelle de bois au Canada. Pour la plupart, elles résultent de la destruction du bois par les champignons lignivores, le restant étant associé à une diminution de la croissance et à la mortalité des arbres causées par les pourridiés. Le chancre à hypoxylon, en causant la mort des arbres, est responsable d'une perte de 11,2 millions de mètres cubes de bois.

Les mentions des pertes qui ont été portées à mon attention ont été insérées dans le corps du texte. Elles provenaient pour une grande part de rapports annuels de la Direction générale de la recherche à Agriculture Canada ou de la livraison annuelle de Insectes et maladies des forêts au Canada publiée par le Service canadien des forêts.

Un autre but de cet ouvrage est de faciliter l'accès à la bibliographie récente de la pathologie végétale au Canada, par hôte et par emplacement géographique. Sous ce rapport, c'est le résumé de tout ce qui s'est fait au pays de 1960 à la fin de 1980.

Ce compendium constitue en outre une étape vers la constitution d'une flore mycologique du Canada. Le répertoire le plus complet jusqu'à ce jour des champignons du Canada est l'Index des maladies des plantes au Canada de I.L. Conners, paru en 1967. Cet ouvrage remonte aux débuts de la phytopathologie canadienne et s'étend jusqu'à la fin de 1959, complété de quelques références du début des années 1960. Les commentaires de Conners offrent un coup d'oeil inestimable sur l'histoire de notre pathologie et de notre mycologie végétales et constitue une mine d'observations sur l'épidémiologie et l'importance économique des maladies des plantes. En 1983, Gourley (603 a) publiait l'important recueil An Annotated Index of the Fungi of Nova Scotia.

Ce compendium a été conçu pour accompagner l'ouvrage de Conners, dont il se distingue toutefois en se limitant aux champignons des maladies des plantes et des pourritures du bois, alors que Conners étendait son enquête aux maladies bactériennes et virales, aux troubles physiologiques et aux carences minérales. Par suite de l'accroissement récent de la bibliographie touchant ces trois derniers domaines, l'auteur a cru bon de les exclure du compendium. En outre, Conners traitait aussi des maladies observées en Alaska et au Groenland, tandis que nous ne retenons ici que les observations signalées au Canada. Même en limitant le champ de nos observations, en réduisant les commentaires à l'essentiel, cet ouvrage s'avère aussi volumineux que celui de Conners à cause de la surabondance bibliographique qui s'est développée en phytopathologie au cours des 21 années durant lesquelles s'est déroulée cette étude. La présentation adoptée est la même que celle de Conners ce qui facilite d'autant la consultation des deux recueils et de plus les espèces traitées dans le compendium portent le cas échéant un renvoi à l'ouvrage de Conners. Alors que Conners accordait une grande place aux aspects historiques, le compendium fournit un guide de la bibliographie récente ainsi qu'un index des appellations contemporaines des champignons. Il se concentre sur la bibliographie parue entre 1960 et la fin de 1980, mais il ajoute quelques références antérieures non relevées par Conners.

Seules ont été consignées les mentions réunissant le nom du champignon, l'emplacement géographique et au moins le nom générique de la plante-hôte. On a toutefois inclus quelques mentions où cette associa tion n'était pas claire, mais seulement après vérification des prélèvements confirmés de l'Herbier mycologique national (DAOM). Quelques entrées ont dû être rejetées parce que trop vagues, par exemple l'hôte était décrit simplement par la mention << sur brindilles >>. On a également omis certains groupes de champignons, notamment les espèces mycorhiziennes bien connues qui favorisent la croissance de la plante-hôte et lui confèrent un certain degré de résistance aux maladies. On a aussi écarté les champignons lichénisés épiphytes.

Les emplacements géographiques associés aux mentions d'un champignon sont les provinces et les territoires, ou des circonscriptions plus réduites. Les unités adoptées et leurs abréviations sont, du nord-ouest au sudest, le Territoire du Yukon (Yukon), les Territoires du Nord-Ouest (NWT) ou ses districts: Mackenzie (Mack), Keewatin (Keew) et Franklin (Frank); les dix provinces: Colombie-Britannique (BC), Alberta (Alta), Saskatchewan (Sask), Manitoba (Man), Ontario (Ont), Québec (Que), Nouveau-Brunswick (NB), Nouvelle-Ecosse (NS), Ile-du-Prince-Edouard (PEI) et Terre-Neuve (Nfld). Certaines mentions sont faites du Labrador (Labr), la portion continentale de Terre-Neuve.

Les entrées du compendium proviennent essentiellement de monographies et de revues qui publient les résultats d'études scientifiques. Les principales revues scientifiques compulsées sont: Canadian Field Naturalist, Canadian Journal of Botany, Canadian Journal of Forest Research, Canadian Journal of Plant Pathology, Canadian Journal of Plant Science, Proceedings of the Canadian Phytopathological Society, Canadian Plant Disease Survey, Forestry Chronicle, Fungi Canadenses, Mycologia, Mycological Papers (Kew), Mycotaxon, Naturaliste canadien (Québec), Phytoprotection, Plant Disease Reporter et son successeur Plant Disease, Studies in Mycology (Baarn), et Syesis.

Les mycologistes du Centre de recherches biosystématiques ont abattu un gros travail de mise à jour de la nomenclature et de la synonymie et se sont attachés à rapprocher les noms anamorphes et holomorphes qui reflètent bien la situation taxonomique récente. Certains noms ne sont pas mentionnés parce que la révision des groupes de champignons est incomplète ou parce qu'ils n'ont pas été retenus dans les récentes publications taxonomiques. Le corps du texte présente des commentaires sur les erreurs évidentes de noms. En outre, on a consacré beaucoup de recherches à évaluer d'un oeil critique, non seulement le nom le plus commun de certains champignons, mais aussi la validité de certaines informations, c.-à-d. l'éventualité d'une mésidentification du champignon ou de son hôte. Les corrections des erreurs d'identification de même que les informations douteuses font l'objet de commentaires dans le corps du texte.

Ce compendium se compose de quatre grandes sections: introduction, index des hôtes, bibliographie et index des champignons. A l'intérieur de l'index des hôtes, les noms de genres des plantes sont rangés par ordre alphabétique et seules sont incluses les espèces relevées dans la recherche bibliographique. Les champignons énumérés pour chaque genre de plante sont donnés par ordre alphabétique. Par souci d'économie de place, les auteurs des appellations scientifiques des champignons ne sont mentionnés que dans les index.

REMERCIEMENTS

La collecte et l'organisation des données ont été grandement facilitées par la collaboration minutieuse et enthousiaste de nombreuses personnes. Michael Sarazin a colligé la plupart des citations et s'est chargé de la première lecture des données brutes. Il a ensuite cédé la place à Sheila Thomson et Modra Kaufert qui ont terminé la compilation des données et apporté les révisions et corrections nécessaires. Mme Kaufert s'est occupé de l'organisation du manuscrit définitif et de la correction d'une bonne partie du texte en plus de s'occuper de divers autres aspects de la publication. Joy Bowerman a rédigé la version première des paragraphes sur les genres des plantes-hôtes et a apporté une aide précieuse à l'organisation des premières étapes du projet. Sylvie Séguin a préparé la plus grosse partie de l'index brut, cependant que Mona Meredith se chargeait de la dactylographie du texte et de la bibliographie et Mary Ann Martin de celle de l'index.

G.A. Mulligan, directeur du Centre de recherches biosystématiques a apporté son soutien constant à ce projet. Mes collègues de l'Institut ont mis leur expérience à contribution pour les révisions taxonomiques et la mise à jour des appellations des champignons. Comme certains groupes de champignons (ordres et ainsi de suite) paraissaient plus fréquemment que d'autres, quelques collègues ont passé beaucoup de temps aux révisions taxonomiques, en particulier John Bissett, Michael Corlett, J.A. Parmelee, Robert Shoemaker et George White. Enfin, W.J. Cody a aimablement éclairci de nombreuses questions concernant les noms et la répartition géographique des plantes vasculaires.

Les nombreuses discussions avec Claire Lamoureux et Laurie Wilson du Service de programmes de recherche ainsi que leur collaboration à la révision et la correction du manuscrit ont grandement contribué à l'améloriation de cet ouvrage.

Vu le nombre élevé d'insertions, la préparation de l'index des genres et espèces s'est avérée plus fastidieuse que prévue. De nombreux collaborateurs en ont facilité la réalisation, John Bissett du Centre de recherche biosystématique pour ses suggestions d'ordre technique, Marilyn Olmstead, Louise Boardman et Doris Hyndford de la Direction des systèmes et services consultatifs pour les efforts a la mise au point de programmes destinés à la manipulation d'énormes banques de données et la présentation générale du travail.

ABIES Mill.   PINACEAE

1. *A. alba* Mill., silver fir, sapin
    argenté; imported from Europe.
2. *A. amabilis* (Dougl.) Forb., amabilis
    fir, sapin amabilis; native, confined to
    the coast forest region of BC, unknown
    in Queen Charlotte Islands.
3. *A. balsamea* (L.) Mill., balsam fir,
    sapin baumier; native, occurs widely
    from c. Alta-Nfld & Labr.
4. *A. concolor* (Gord.) Lindl., white fir,
    sapin concolore; imported from s. USA &
    Mexico.
5. *A. grandis* (D. Don ex Lamb.) Lindl.,
    grand fir, sapin de Vancouver; native,
    BC in s. coastal regions and s. interior
    wet belt.
6. *A. holophylla* Maxim., needle fir,
    sapin à aiguilles; imported from Asia.
7. *A. homolepis* Siebold & Zucc., Nikko
    fir, sapin de Nikko; imported from Japan.
8. *A. lasiocarpa* (Hook.) Nutt., alpine
    fir, sapin de l'ouest; native, in
    subalpine forest region of Yukon, BC &
    w. Alta.
9. *A. mariessi* M.T. Mast; imported from
    Japan.
10. *A. pinsapo* Boiss., spanish fir, sapin
    d'Espagne; imported from Spain.
11. *A. sacchalinensis* (Friedr. Schmidt)
    M.T. Mast, Sakhalin fir, sapin de
    Sakhaline; imported from Japan. 11a.
    A. s. var. *mayriana* Myiabe & Kudo;
    imported from Japan.
12. *A. veitchii* Lindl., Veitch fir, sapin
    de Veitch; imported from Japan.
13. Host species not named, i.e., reported
    as *Abies* sp.

Acanthostigma parasiticum: on 3 NB [1032].
Aecidium sp.: on 3 Que [282].
Aleurocystidiellum subcruentatum: on 5 BC
    [1012].
Aleurodiscus sp.: on 3 Ont [1341, 1827];
    on 8 NWT-Alta [1063].
A. abietis (C:5): on 3 Ont, Que, NS
    [964]; on 5 BC [1012]; on 13 BC [964].
    Lignicolous.
A. amorphus (C:5): on 2, 4, 5, 8, 10 BC
    [1012]; on 3 Sask, Ont, Que, NS [964];
    on 3 Ont [1341, 1827], Que [282, 969,
    1377, 1385, F61, F74, F76], NB, NS, PEI
    [1032], Nfld [1659, 1660, 1661]; on 5, 8
    BC [964]; on 8 Yukon [460] as DAOM
    118961, NWT-Alta [1063], Alta [950]; on
    13 NB [1032]. Apparently this fungus
    can invade only weakened and/or dying
    tissue, i.e., suppressed branches in the
    lower crown or recently up-rooted
    trees. Associated with a white rot.
    Reports from the Rocky Mountains and
    westward are probably *A. grantii* [576].
A. canadensis (C:5): on 3 Ont [1341], Que
    [969, 964]. Associated with a white rot.
A. laurentianus (C:5): on 3 Que [964].
    Lignicolous.
A. lividocoeruleus: on 3 Ont [964, 1341],
    Que [969]. Associated with a white rot.
A. penicillatus (C:5): on 5 BC [964].
    Associated with a white rot.

Allescheria terrestris: on 3 NB [1610].
Alternaria sp.: on purposely wounded 3
    Que [910].
A. alternata (*A. tenuis* auct.): on 3
    Que [454], Nfld [1661].
Alysidium resinae var. *resinae*: on 3
    Man [1760].
Amphinema byssoides (*Peniophora b.*)
    (C:8): on 3 Que [969], NS [1032]; on 8
    NWT-Alta [1063], BC [1012], Alta [950].
    On rotted wood and debris, probably
    mycorrhizal, reported to be associated
    with a white rot.
Amphisphaeria sp.: on 3 Que [F60].
A. juniperi (C:5): on 13 NS [1032].
Amylocorticium cebennense (*Corticium c.*):
    on 3 Ont [1939], NS [1032, 1939].
    Associated with a brown rot.
Amylostereum chailletii (*Stereum c.*)
    (C:10): white stringy rot, carie
    blanche filandreuse: on 2, 5, 8 BC
    [1012]; on 2 NB [1032]; on 3 Ont [95,
    1742, 1964, F74], Que [95, 456, 454,
    648, 969, 910 purposely wounded, 1672],
    NB [95, 1032, 1341, 1742, 1740], NS
    [1032], Nfld [95, 1659, 1661, F62]; on 8
    BC [1233, 1718, 1719]; on 13 BC [1233].
Anomoporia albolutescens (*Poria a.*) (C:9):
    on 8 BC [1012]. Associated with a brown
    rot.
Antennatula lumbricoidea (*Hyphosoma l.*, *H.
    abietis*) (C:7): sooty mold, fumagine:
    on 2 BC [1012].
Antrodia albida (*Coriolellus sepium*)
    (C:5): on 3 NB [1032]. Associated with
    a brown rot.
A. carbonica (*Poria c.*) (C:9): on 5 BC
    [1012]; on 13 Ont [1828]. Associated
    with a brown rot.
A. crustulina (*Poria c.*) (C:9): on 3 Ont
    [1341], NS [1032]; on 5, 8 BC [1012].
    Associated with a white rot.
A. gossypia (*Polyporus destructor* var.
    resupinatus) (C:9): on 5 BC [1012].
    Associated with a brown rot.
A. heteromorpha (*Coriolellus h.*, *Trametes
    h.*) (C:5): on 3 Ont [1341], NS [1032],
    Nfld [1659, 1660, 1661]; on 8 BC
    [1012]. Associated with a brown rot.
A. lenis (*Poria l.*): on 13 BC [1012].
    Associated with a white rot.
A. lindbladii (*Poria cinerascens*) (C:9):
    on 3 Ont [1341], Que [F71]; on 8 BC
    [1012]. Associated with a white rot.
A. serialis (*Trametes s.*): on 8 BC
    [1012]. Associated with a brown rot.
A. sordida (*Poria oleagina*): on 8
    NWT-Alta [1063], Alta [950]. Associated
    with a brown rot.
A. variiformis (*Coriolellus v.*, *Trametes
    v.*) (C:5): on 3 Ont [1341], NS [1032];
    on 8 BC [1012]. Associated with a brown
    rot.
A. xantha (*Poria x.*) (C:9): on 8 BC
    [1012, 1341]. Associated with a brown
    rot.
Anungitea fragilis: on 3 Man [1760].
Appendiculella pinicola (*Meliola p.*): on
    3 Nfld [1660, F73, F74].
Armillaria mellea (C:5): root-rot,
    pourridié-agaric: on 2, 8 BC [1012]; on

3 Alta, Sask, Man [F63] also Alta [950], Man [F62], Ont [1341, 1964, 1966, 1956, 1957, F61], Que [910, 1377, 1672, F65, F66, F69, F73, F76], NB [1032, 1742], NS [1032, F73, F74, F75], Nfld [780, 1659, 1660, 1661, 1663, F62, F64, F65, F66, F67, F73, F74] also Sask, Man [F67], Ont, Que, NB, NS [F71, F72] and Que, NB [F68, F74, F75]; on 3, 8, 13 NWT-Alta [1063]; on 5 BC [F75]; on 3, 6 Nfld [F69]; on 6, 7, 9, 11, 11a, 12 Nfld [1661, 1663]; on 8 BC [1397, 1718, 1730, F65]; on 13 BC [1397], Nfld [781, 1661, F69, F71]. Associated with a white rot.

Ascocalyx abietis (Godronia a.) (anamorph *Bothrodiscus berenice*) (C:5): twig blight, brûlure des rameaux: on 3 Sask, Man [F67, F68, F69], Man [F66], Ont [411, F60, F61], Que [1690], NB, NS, PEI [1032].

A. tenuisporus: on 8 Yukon [460, 658], BC [658, 1012].

Ascochyta sp.: on 8 BC [1012].

Ascocoryne sarcoides: on 3 Ont [1966], Que [910] purposely wounded, NB [1610]; on 5, 8 BC [1012].

Aspergillus sp.: on 3 Que [932] purposely wounded, NB [1610].

A. fumigatus: on 3 NB [1610].

Asterina sp.: on 2 BC [1615]; on 2, 5 BC [1012].

Asterodon ferruginosus (C:5): on 3 Ont [1964], NB [1032]; on 8 BC [1012]. Associated with a white rot.

Asterostroma cervicolor: on 3 NS [1032]. Associated with a white rot.

Athelia epiphylla: on 8 BC [1012]. Associated with a white rot.

A. neuhoffii: on 3 Que [969]. Associated with a white rot.

A. pellicularis (Corticium p.) (C:6): on 3 PEI [1032]; on 8 BC [1012]. Lignicolous. The characters to be associated with this name are uncertain.

Atichia glomerulosa: on 5 BC [1012].

Aureobasidium sp.: on 3 NB [1610]; on 8 BC [502, 1012].

A. pullulans (C:5): on 3 Que [1216], NB [1032, 1610].

Auricularia auricula (A. auricularis) (C:5): on 3 NWT-Alta [1063], Ont [423, 1341, 1827], Que [282, 423, 867], NB, NS [423, 1032], Nfld [1659, 1660], Labr [867]; on 8 Yukon [460], BC [423, 953, 1012, 1341], Alta [423, 950]. Lignicolous, on dead trees or logs.

Auriporia aurea (Poria a.) (C:9): on 8 BC [1012]. Associated with a brown rot.

Basidiodendron cinerea: on 3 Que [1016]. Lignicolous.

B. deminuta: on 13 Ont [1016]. Lignicolous.

B. nodosa: on 3 Que and on 13 Ont [1016]. Lignicolous.

B. subreniformis: on 13 Ont [1016]. Lignicolous.

Basidioradulum radula (Corticium hydnans, Radulum orbiculare) (C:6, 10): on 3 Que [969], NB [1032, 1742], NS [1032], Nfld [1659, 1660]. Associated with a white rot.

Battarrina sp.: on 5 BC [1012].

Bispora betulina: on 3 NB [1610].

Bjerkandera adusta (Polyporus a.) (C:9): on 3 Que [454]; on 8 BC [1012]. Associated with a white rot.

Bondarzewia montana (Polyporus m.) (C:9): on 5 BC [1012]. Associated with a white rot.

Bothrodiscus sp.: on 8 Yukon, BC [658].

B. berenice (B. pinicola) (holomorph *Ascocalyx abietis*) (C:5): on 3 Ont [1763, 1828, 1827, F68], Que [1763], Nfld [1659, 1660, 1661]; on 3, 4 Ont [1340].

Botryobasidium vagum (Pellicularia v.) (C:8): on 3 Ont [1341], Que [969], NS [1032]. Associated with a white rot.

Botryohypochnus isabellinus (Botryobasidium i.): on 3 Ont [1341]. Associated with a white rot.

Botryosphaeria abietina (C:5): on 3 Que [70]; on 5 BC [1012].

Brachysporiella sp.: on 3 NB [1610].

Byssochlamys sp.: on 3 NB [1610].

Caeoma sp.: on 8 Alta [1999].

Calcarisporium sp.: on 3 NB [1610].

C. abietis: on 3 Man [1760].

Caliciopsis pseudotsugae (C:5): on 2 BC [495, 1073, F64]; on 5 BC [493, 1012, F62].

Calocera viscosa (C:5): on 3 NS [1032]. Associated with a white rot.

Caloscypha fulgens (anamorph *Geniculodendron pyriforme*): on stored seeds of 5 BC [1749]. See *Picea*.

Camarosporium sp.: on 8 BC [F61].

C. strobilinum: on 8 BC [1012, F69].

Cellulosporium sp. (*Cytosporium* sp.): on 2 BC [1012]; on 3 Ont [1828].

Cenococcum graniforme: on 3 Ont [1964].

Cephalosporium sp.: on 3 Ont [1964], Que [910 purposely wounded, 1216, F65], NB [1032, 1610], Nfld [1661], e. Canada [95]; on 8 BC [1012].

Ceraceomyces borealis: on 2 BC [568]. Apparently associated with a brown rot.

C. serpens (Merulius s.): on 3 Ont [1341]; on 13 or *Picea* sp. Que [568]. Associated with a white rot.

C. sublaevis (Corticium microsporum): on 3 PEI [1032]. Lignicolous.

C. tessulatus: on 8 BC [1012]. Associated with a white rot.

Ceratocystis sp.: on 3 Nfld [1661], e. Canada [95]; on 8 BC [1012, F64, F65, F66].

C. aequivaginata: on 3 Man [1190].

C. albida: on 3 NB [1610]. Usually on *Ulmus*.

C. angusticollis: on 3 Man [1190], Ont [647, 1340].

C. bicolor: on 3 Ont [647, 1340].

C. brunneocrinita (C:5): on 3 Man [1190], Ont [647, 1340, 1983], Que [1983], NB [647, 1340].

C. concentrica: on 3 Man, Ont [1190].

C. conicicollis: on 3 Man, Ont [1190].

C. curvicollis: on 3 Man, Ont [1190].

C. dryocoetidis (anamorph *Verticicladiella d.*): on 8 BC [878, 1012, 1123, 1340, F64, F67].

C. leucocarpa: on 3 Man [1190].

C. magnifica: on 3 Ont [647, 1340].

C. minor: on 3 Man [1190], Ont [647, 1340].

C. minuta: on 3 Man [1190], Ont [647, 1340].

C. nigra: on 3 Ont [647, 1340].

C. ossiformis: on 3 Man [1190].

C. parva: on 3 Man, Ont [1190].

C. piceae: on 3 Ont [647, 1340], Que [454], NB [1610].

C. pilifera: on 3 Man [1190].

C. tubicollis: on 3 Man [1190].

Ceratosphaeria rhenana: on 3 Ont [1340].

Ceriporia tarda (Poria t.) (C:9): on 8 BC [1012]. Associated with a white rot.

Cerocorticium hiemale (Corticium h.): on 3 Ont [965]. Lignicolous.

C. sulphureo-isabellinum (Corticium s.) (C:6): on 3 Ont [741], NS [741, 1032]; on 8 BC [1012]. Associated with a white rot.

Chaetoderma luna: on 2 BC [1012]. Associated with a brown rot.

Chalara cylindrica: on 3 Man [1760].

C. fungorum: on 8 BC [1012].

Chlorociboria aeruginascens (Chlorosplenium aeruginosum): on 3 Nfld [1661].

C. aeruginosa: on 3 Nfld [1659, 1660].

Chondrostereum purpureum (Stereum p.) (C:10): on 3 Que [456, 454, 1672], NB [1032, 1742]; on 8 BC [1012]. The causal agent of silver leaf disease, associated with a white rot.

Chrysosporium sp.: on 3 NB [1610].

Ciboria rufofusca (C:5): on 2, 5 BC [663, 1012]; on 3 Ont, Que [663]; on 13 BC [663]. Restricted to cone scales.

Cirrenalia donnae: on 3 Man [1760].

Cladosporium sp.: on 3 NWT-Alta [1063], Que [910] purposely wounded, NB [1610].

C. cladosporioides: on 3 NB [1610].

C. herbarum (C:5): on 3 Que [1216]; on 8 Yukon [460].

Clavulicium macounii: on 8 BC [1012]. On well-rotted wood.

Collybia dryophila: on 3 NS [1032].

Coltricia perennis (Polyporus p.): on 3 NS [1032]. Mycorrhizal, on pine, probably on sandy soil under *Abies*.

Columnocystis abietina (Stereum a.) (C:10): brown cubical rot, carie brune cubique: on 2 BC [1012]; on 3 NS [1032]; on 8 BC [954, 1012], NWT-Alta [1063]; on 13 BC [1828].

Confertobasidium olivaceo-album (Corticium fuscostratum): on 3 Que [1672]. Lignicolous.

Coniophora arida var. *suffocata (C. suffocata)* (C:5): on 5 BC [1012]; on 13 NS [1032]. Associated with a brown rot.

C. puteana (C:5): brown cubical rot, carie brune cubique: on 3 Ont [1964, 1956, 1957, F71, F72, F74], Que [969, 909, 910 purposely wounded, 1672, F66], NB [1032, 1742], NS [1032], Nfld [1661]; on 8 BC [1012, 1718, 1719]. Sometimes causes a brown rot in living trees.

Coniothyrium faullii (holomorph *Leptosphaeria f.*) (C:5): on 3 Ont [362, 1827, F68, F69], Que [1691]; on 8 BC, Alta [2003].

C. fuckelii (holomorph *Leptosphaeria coniothyrium*): on 3 NB [1610].

Conoplea fusca: on 3 Sask [1760].

C. geniculata (C:5): on 3 Man [1760], Ont [784].

C. juniperi var. *robusta*: on 3 Sask [1760, F67], Man [1760].

Cordana pauciseptata (C:5): on 3 Que [F66], NB, NS [796].

Coriolellus sp.: on 3 Nfld [1659, 1660, 1661].

Coriolus hirsutus (Polyporus h.) (C:9): on 3 NB [1032]; on 5 BC [1012]. Associated with a white rot.

C. pubescens (Polyporus p.): on 3 NB [1032]. Associated with a white rot.

C. versicolor (Polyporus v.): on 2, 5 BC [1012]; on 3 NB [1032]. Associated with a white rot.

Corticium sp.: on 3 Que [456].

C. centrifugum (C:5): on 3 NS [1032]. Name of uncertain application [864]. Specimens and records should be reexamined.

Costantinella terrestre (Verticillium t.): on 3 NB [1610].

Crepidotus herbarum (C:6): on 3 NB [1032].

C. sphaerosporus (C:6): on 8 BC [1012]. Collection DAVFP 8197 is probably *C. regularis*, fide S.A. Redhead.

Crustoderma dryinum: on 5 BC [1012]. Associated with a brown rot.

Crustomyces pini-canadensis (Corticium p.) (C:6): on 3 Nfld [1659, 1660, 1661]. Lignicolous.

Cryptoporus volvatus (Polyporus v.) (C:9): red butt rot, carie rouge du pied: on dead or dying 2 BC [180]; on 3 NB [1032]; on 2, 5 BC [1012]. Associated with a white rot.

Cylindrobasidium corrugum (Corticium c.) (C:5): on 8 NWT-Alta [1063], BC [1012]. Apparently associated with a white rot.

C. evolvens (Corticium e., C. laeve) (C:6): on 3 Ont [1341], Que [454, 910 purposely wounded, 1672, F65], NB [1032, 1610]; on 8 BC [1012]; on 13 BC [1233]. Associated with a white rot.

Cylindrocarpon sp.: on 3 Que [910 purposely wounded, 1215, 1216, F65].

Cylindrocephalum sp.: on 3 Que [910] purposely wounded, NB [1610].

Cytospora sp.: canker, chancre: on 3 Sask, Man [F67], Ont [1828, F61], Que [910 purposely wounded, F63], NB, NS, PEI [1032]; on 5, 8 BC [1012]; on 8 Yukon [460].

C. abietis (C:6): canker, chancre: on 3 Ont [1340].

C. friesii (holomorph *Valsa f.*) (C:6): canker, chancre: on 3 Ont [1340, 1827], Que [F60, F72, F73], PEI [F62].

C. kunzei (holomorph *Leucostoma k.*): canker, chancre: on 3 Que [454].

C. pini: canker, chancre: on 3 NS [F61].

Cytosporella sp.: on 8 BC [1012].

Dacrymyces minor (D. deliquescens var.
  m.) (C:6): on 3 NS [1032].
  Associated with a brown rot.

D. palmatus (C:6): on 3 Que [282], NB, NS
  [1032], Nfld [1659, 1660]; on 5 BC
  [1012]. Reported as associated with a
  white rot but laboratory studies showed
  it to cause a brown rot.

D. stillatus (D. deliquescens, D. d. var.
  deliquescens): on 3 NS [1032], Nfld
  [1660]. Lignicolous.

Daedalea sp.: on 3 Nfld [1659, 1660,
  1661].

Darkera abietis: needle blight: on 3
  Man, Ont [1947]; on 8 Yukon [460], BC
  [1012].

Dasyscyphus sp.: on 8 Yukon [460]; on 13
  BC [1340].

Delphinella sp.: on 8 BC [F76].

D. abietis: on 8 BC [1012, F64, F68].

D. balsameae (Rehmiellopsis b.) (C:10):
  tip blight, brûlure des pousses: on 3
  Ont [1340, F63, F64], NB [F65], NS [F60,
  F65, F69] also NB, NS [F62, F66, F67,
  F68, F72] and NB, NS, PEI [1032, F61,
  F63, F64], Nfld [1659, 1660, 1661, F64,
  F65, F68, F76]; on 4 Que [F60]; on 8 BC
  [1012, F65, F66].

*Dendrothele incrustans (Aleurocorticium
  i.)*: on 5 BC [965, 1012]. Lignicolous.

Dendryphiopsis atra (holomorph
  Microthelia incrustans): on 3 Man
  [1760].

Dermea sp.: on 8 BC [499, 502].

D. balsamea (anamorph *Foveostroma
  abietinum*) (C:6): on 2, 8 BC [1012];
  on 3 Ont [499, 1340, 1422, F61], Que
  [1690, F66, F69, F72, F73, F74, F75,
  F76], NB [F62] and NB, NS, PEI [1032,
  F65], Nfld [1661, F64, F65].

D. piceina: on 3 Que [1683, 1690, F68];
  on 13 Nfld [1683].

D. pseudotsugae (anamorph *Foveostroma
  boycei*): on 5 BC [1012].

D. rhytidiformans (anamorph
  Gelatinosporium sp.): on 8 BC [522,
  954, 1012, F70].

D. tetrasperma (anamorph *Gelatinosporium
  lunaspora*): on 5, 8 BC [1012].

Dichomitus squalens (Polyporus anceps)
  (C:9): red rot, carie rouge
  rayonnante: on 3 NWT-Alta [1063], Que
  [F65], NS [1032]. Associated with a
  white pocket rot.

Dimerium balsamicola (Dimeriella b.): on
  13 BC [464].

Diplodina parasitica (C:6): on 3 PEI
  [1032].

Ditiola radicata (C:6): on 3 NS [1032].
  Lignicolous.

Echinodontium tinctorium (C:6): Indian
  paint fungus: on 2, 5, 8 BC [1012]; on
  8, 13 NWT-Alta [1063]; on 8 BC [953,
  954, 1718, 1719], Alta [839, 950, F61];
  on 13 BC [1341]. Causes a heart rot of
  living trees. Associated with a white
  rot.

Encoeliopsis sp.: on 8 BC [F71].

Endophragmia glanduliformis: on 3 Sask,
  Man [1760].

Epicoccum purpurascens (E. nigrum): on 3
  Que [454, 910 purposely wounded].

Epipolaeum abietis (Dimerosporium a.)
  (C:6): on 2, 5, 8 BC [1012]; on 8 BC
  [953], Alta [952, F67]; on 13 BC [464].

E. pseudotsugae: on 3 Ont [1340].

E. terrieri (Dimeriella t.): on 13 BC
  [464].

Eriosphaeria vermicularia (C:6): on 3 Que
  [70]; on 8 BC [73].

*Euantennaria rhododendri (Limacinia
  alaskensis)* (C:7): sooty mold,
  fumagine: on 2, 5, 8 BC [1012].

Exidia glandulosa: on 3 Nfld [1610,
  1660]. Lignicolous.

Exidiopsis macrospora: on 8 BC [1012].
  Lignicolous.

Femsjonia peziziformis: on 3 Que [1341].
  Lignicolous.

Fibricium rude (Peniophora greschikii)
  (C:8): on 3 NS [1032]. Associated with
  a white rot.

Fomitopsis cajanderi (Fomes c.): on 3 NB,
  NS [1032]; on 5 BC [1012]. Associated
  with a brown rot.

F. officinalis (Fomes o.) (C:6): on 2, 5
  BC [1012]; on 3 Que [1377]. Associated
  with a brown rot.

F. pinicola (Fomes p.) (C:6): red belt
  fungus, brown cubical rot, carie brune
  cubique: on 2, 5, 8 BC [1012]; on 3, 8,
  13 NWT-Alta [1063]; on 3 Ont [1341], Que
  [282], NB [1032, 1742], NS, PEI [1032],
  Nfld [1659, 1660, 1661], Labr [867]; on
  8 BC [1341]. A common destroyer of dead
  coniferous timber, this fungus also
  causes heart rot in live trees.

F. rosea (Fomes r.) (C:6): brown cubical
  rot, carie brune cubique: on 3 Ont
  [1341], NB [1032]; on 13 Sask [1031].

Foveostroma abietinum (Gelatinosporium a.)
  (holomorph *Dermea balsamea*): on 3 Ont
  [1340, 1828].

Fusarium sp.: on 3 NB [1610].

F. oxysporum: on 3 NB [1610].

F. solani (holomorph *Nectria
  haematococca*): on 3 NB [1610].

Fusicoccum abietinum (C:6): red flag,
  chancre des rameaux: on 3 Que [F68],
  NB, NS [1032, F61, F63, F64, F65, F66,
  F67, F70, F74] also NB [1422, F62], NS
  [F60, F71], PEI [F65, F67], Nfld [1659,
  1660, 1661, F64, F65, F67, F68, F73,
  F74].

Fusidium sp.: on purposely wounded 3 Que
  [910].

Ganoderma applanatum (C:6): white mottled
  rot, carie blanche madrée: on 2, 5 BC
  [1012].

G. lucidum (C:7): on 3 NB, NS [1032].
  Associated with a white rot.

G. oregonense (C:7): on 2, 5 BC [1012].
  Associated with a white rot.

Gelatinosporium griseo-lanatum: on 5 BC
  [517].

G. pinicola (Micropera p.): on 5 BC
  [1012].

Geomyces pannorum (Chrysosporium p.): on
  3 NB [226].

Gliocladium viride: on 3 NB [1610].

Gloeocystidiellum furfuraceum (Corticium f.) (C:5): on 3 Ont [968], Que [969]; on 13 NS [1032]. Lignicolous.

G. ochraceum: on 3 Que [969]. Lignicolous.

Gloeophyllum odoratum (Trametes o.): on 13 BC [1012]. Associated with a brown rot.

G. sepiarium (Lenzites s.) (C:7): brown cubical rot, carie brune cubique: on 2, 5, 8 BC [1012]; on 3 Ont [1152, 1341, 1964], Que [282], NB [1032, 1610], NS, PEI [1032], Nfld [1659, 1660, 1661]; on 8 BC [953]; on 8, 13 NWT-Alta [1063].

Gloeosporium sp.: on 2 BC [1012, F62].

Godronia sp.: on 3 Que [1422].

Graphium sp.: on purposely wounded 3 Que [910], NB [1610].

Gremmeniella abietina (Scleroderris lagerbergii): on 3 Que [F68].

Grovesiella abieticola (Scleroderris a.) (C:10): on 2 BC [F72]; on 3 Man [F67]; on 2, 5, 8 BC [1012]; on 8 NWT-Alta [1063], BC [954, F63, F64, F66].

G. grantii: on 5 BC [513, 1012].

Guepiniopsis buccina (G. torta): on 8 NWT-Alta [1063], BC [954], Alta [951]. Lignicolous.

G. chrysocomus: on 2 BC [1012]. Lignicolous.

Gymnopilus penetrans (Flammula p.): on 3 PEI [1032].

G. sapineus (Flammula s.): on 3 Ont [1341], NS [1032].

G. spectabilis (Pholiota s.) (C:8): on 3 NB, NS [1032]; on 5, 8 BC [1012].

Haematostereum rugosum (Stereum r.): on 8 NWT-Alta [1063]. Associated with a white rot.

H. sanguinolentum (Stereum s.) (C:10): red heart-rot, carie rouge du sapin: on 2, 3, 5, 8 BC [1012]; on 2 NB [1032]; on 3, 8, 13 NWT-Alta [1063], Man [F60], Ont [1341, 1742, 1827, 1956, 1957, 1964, F74], Que [282, 374, 453, 454, 456, 648, 969, 910, 1216, 1377, 1385, 1395, 1672, F66], NB [374, 1742], NS [374], Nfld [1659, 1660, 1661, F72], Labr [F73], e. Canada [95]; on 3 NB, NS, PEI [1032, F62]; on 8 BC [1233, 1341, 1718, 1719]; on 13 BC [1233], Alta [951]. Associated with a white rot.

Hapalopilus nidulans (Polyporus n.): on 3 NS [1032]. Associated with a white rot.

Harpographium sp.: on 3 NB [1610].

Helminthosporium sp.: on 8 BC [1012].

Helotium sp.: on 3 Nfld [1659, 1660].

H. immaculatum (Mycena gracilis): on needles (fallen?) 3 Que [1385].

H. resinicola (anamorph *Stilbella* sp.): on 2 BC [54, 1012, F69]. An Ascomycete which needs to be redisposed.

Hendersonia sp.: on 3 NB [1032]. The name *Hendersonia* has been dropped from usage in favor of *Stagonospora*. However, the proper generic disposition for this record is uncertain.

Henningsomyces candidus (Solenia polyporoidea): on 3 NS [1032]. Associated with a white rot.

Hericium abietis (C:7): on 2, 5, 8 BC [1012]; on 8, 13 BC [1341]. Associated with a white rot.

Herpotrichia juniperi (H. nigra) (C:7): brown felt blight, feutrage brun: on 2 BC [1012]; on 3 Que [1666]; on 8 NWT-Alta [1063], BC [1012, 1340], Alta [951, 952].

Heterobasidion annosum (Fomes a.) (C:6): fomes root rot, maladie du rond: on 2 BC [1143]; on 2, 5 BC [1012]; on 2, 8 BC [1916]. Associated with a white rot. See comment under *Phlebiopsis gigantea*.

Hormodendrum sp.: on 3 NB [1610]. Probably referrable to *Cladosporium* sp.

Humicola sp.: on 3 NB [1610].

Hyalopsora aspidiotus (C:7): needle rust, rouille des aiguilles: 0 I on 2, 5, 8 BC [1012, 2008]; on 3 Ont [1236], Que [1377, F72, F73, F74], NS [1074].

Hydropus marginellus (Omphalia rugosodisca): on 3 NS [1032].

Hymenochaete badioferruginea (C:7): on 3 NS [1032]. Lignicolous. The species of the Hymenochaetaceae which have been categorized all caused white rots and it is generally accepted that the, as yet, uncategorized species also cause white rots.

H. corrugata: on 3 Ont [1341]. Associated with a white rot.

H. rubiginosa (C:7): on 8 BC [1012]. Lignicolous but see above.

H. tabacina (C:7): on 3 Ont [1341], NS [1032], Nfld [1659, 1660, 1661]; on 8 NWT-Alta [1063], BC [1012]. Associated with a white rot.

H. tenuis (C:7): on 8 BC [1012]. Associated with a white rot.

Hyphoderma sp.: on 8 BC [1012].

H. clavigerum: on 3 Que [969]. Associated with a white rot.

H. praetermissum (H. tenue, Peniophora t.) (C:8): on 3 Que [969], NB, NS [1032]; on 8 BC [1012]; on 13 Que [198]. Associated with a white rot.

H. puberum (Peniophora p.): on 3 Que [969]. Associated with a white rot.

H. roseocremeum: on 3 Que [969]. Associated with a white rot.

H. setigerum (Peniophora aspera) (C:8): on 3 NB [1032, 1742], PEI [1032]; on 5 BC [1012]; on 3, 13 NS [1032]. Associated with a white rot.

Hyphodontia sp.: on 3 NB [1610].

H. alienata (Peniophora a.): on 3 Ont [1341]. Lignicolous.

H. alutacea (Odontia a.): on 3 NS [1032]; on 13 Ont [923]. Associated with a white rot.

H. arguta (Odontia a.): on 3 NS [1032]. Associated with a white rot.

H. barba-jovis (Odontia b.) (C:9): on 3 NS [1032]. Associated with a white rot.

H. breviseta: on 3, 13 NS [1032]. Associated with a white rot. See under *Odontia lactea*.

H. crustosa (Odontia c.) (C:8): on 3 NS [1032]. Associated with a white rot.

H. floccosa: on 13 Que [929]. Associated with a white rot.

H. pallidula: on 3 Que [969]. Associated with a white rot.

H. spathulata (Odontia s.) (C:8): on 3 Ont [1341], NS [1032]. Associated with a white rot.

H. subalutacea (Peniophora s.): on 3 NS [1032]. Associated with a white rot.

Hypholoma capnoides (Naematoloma c.) (C:7): on 5 BC [1012, F64]; on 8 BC [1012]. Associated with a white rot.

H. fasciculare (Naematoloma f.) (C:7): on 5 BC [1012]. Associated with a white rot.

Hypodermella sp.: on 3 Sask, Man [F64].

Hypoxylon diathrauston: on 8 BC [1012].

Inonotus dryadeus (Polyporus d.) (C:9): on 5 BC [1012]. Associated with a white rot.

I. subiculosus (Poria s.): on 3 NB, NS [1032]. Associated with a white rot.

Ischnoderma resinosum (Polyporus r.) (C:9): on 3 Ont [1341]; on 5 BC [1012]. Associated with a white rot.

Isthmiella abietis (Bifusella a.) (C:5): needle cast, rouge: on 8 Yukon [460], NWT-Alta [1063], BC [953, 1012, F68], Alta [950, 951, 952, F62].

I. crepidiformis: on 3 Ont [1827].

I. faullii (Bifusella f.) (C:5): needle blight, rouge: on 3 Ont [362, 1340, 1827, F67, F72], Que [282, 1377, 1672, F68, F75], NB [F66, F67, F76], NS [F66] also NB, NS, PEI [1032, F64, F65, F69], Nfld [1659, 1660, 1661, F68, F74, F75] and Ont, Nfld [F65, F69, F71, F73, F76].

I. quadrispora: on 8 BC [1012, 2003, F68, F76], Alta [839, 950, 2003, F68].

Junghuhnia collabens (Poria rixosa) (C:9): on 3 Ont [1004], NS [1032]; on 5, 8 BC [1004]; on 8 BC [1012]. Associated with a white rot.

J. luteoalba (Poria l.): on 8 BC [579, 1012]. Associated with a white rot.

Kirschsteiniella thujina (Amphisphaeria t.) (C:7): blue stain, bleuissure: on 3 Que [374, 910, 1395, 1394, F61, F66], Nfld [1661].

Lachnellula agassizii (Dasyscyphus a., Lachnella a.) (C:6): on 2, 5, 8 BC [1012]; on 3 Ont [398, 1340], Que [282, 398, 1385, 1690], NB, NS, PEI [1032], Nfld [666, 1659, 1660], Labr [666]; on 8 NWT-Alta [1063], BC [953].

L. arida (Dasyscyphus a., Lachnella resinaria) (C:6): on 3 Ont [1340]; on 3, 8 NWT-Alta [1063]; on 3 Que [282, 1690]; on 8 BC [1012], Alta [951]; on 13 Alta [644].

L. calyciformis (Dasyscyphus c.) (C:6): on 2 BC [1012]; on 4 PEI [1032].

L. ciliata: on 3 Que [1690]; on 5 BC [1012].

L. gallica: on 3 Que [398, 1690, F68].

L. occidentalis: on 5 BC [1012].

Laeticorticium minnsiae: on 8 BC [1012]. Lignicolous.

Laetiporus sulphureus (Polyporus s.) (C:9): brown cubical rot, carie brune cubique: on 2, 5, 8 BC [1012].

Laurilia sulcata (Stereum s., Echinodontium s.) (C:10): on 8 BC [1012, 1341, 1719]. Associated with a white rot.

Lentinellus ursinus: on 3 NB [1032]. Associated with a white rot.

Lenzites sp.: on 3 Nfld [1659, 1660, 1661].

Leptosphaeria faullii (C:7): on 3 Ont [362, 1827, F69], Que [362, 1691, F67]; on 8 BC [1012, 2003, F68], Alta [2003].

Leptosporomyces galzinii (Athelia g.): on 3 Que [969]. Associated with a white rot.

Leucogyrophana mollusca (Merulius m., M. fugax auct.*)*: on 3 NS [1032]; on 13 Ont [1341]. Associated with a brown rot.

L. pinastri: on 13 BC [581]. Associated with a brown rot.

L. romellii: on 3 Ont, Que, NS, PEI [571]. Associated with a brown rot.

Leucostoma kunzei (Valsa k.) (anamorph *Cytospora k.*) (C:11): on 3 Que [935, F72, F75], NB [1032, F63], NS [1032].

Libertella sp.: on 3 NB [1032].

Lirula abietis-concoloris (Hypodermella a.): on 2, 5, 8 BC [1012]; on 8 Yukon [460], NWT-Alta [1063], BC [1340], Alta [950, F61, F62].

L. mirabilis (Hypodermella m.) (C:7): needle cast, rouge: on 3, 8 NWT-Alta [1063]; on 3 Alta [53, F61, F68], Man [F66], Ont [362, 1340, 1827, F68, F69], Que [362, 1377, 1672, 1674, F61, F65, F66, F68, F74], NB [F66, F68, F76], NS [F68] and NB, NS, PEI [1032, F65, F69]; on 8 Mack [53, F63].

L. nervata (Hypodermella n.) (C:7): needle cast, rouge: on 3, 8 NWT-Alta [1063]; on 3 Alta [53, 950, F63, F69], Man [F66], Ont [1340, F63, F72], Que [1377, 1672, 1674, F65, F66, F68], NB [F63, F76], NS [F73], Nfld [1659, 1660, 1661, F73, F74] also Sask, Man [F67, F68], NB, NS [F64, F66, F68, F70] and NB, NS, PEI [1032, F65, F67, F69]; on 8 BC [954], Alta [950]; on 13 Alta [F65].

L. punctata (Hypodermella p.) (C:7): needle cast, rouge: on 2, 5, 8 BC [1012, F68]; on 3 Ont [1340], Que [1672, F65].

Lophium mytilinum (C:7): on 3 Que [70].

Lophodermium sp. (C:7): needle cast, rouge: on 3 Sask, Man [F63], Que [F73], NB [1032], Nfld [1659, 1660, 1661, F64, F65, F66, F68, F73, F74]; on 8 NWT-Alta [1063].

L. consociatum: on 2 BC [1012].

L. decorum: on 5, 8 BC [1012].

L. lacerum (C:7): needle cast, rouge: on 3 Alta [F69], Ont [1063, 1340], Que [1377, 1672]; on 5 BC [1012].

L. piceae (C:7): needle cast, rouge: on 3 Ont [1063, 1340], Que [1672], NB [1950]; on 8 BC [1012], Alta [950].

L. uncinatum: on 2, 8 BC [1012].

Lophomerum sp.: on 3 Que [F73].

L. autumnale (Lophodermium a.) (C:7): needle cast, rouge: on 2 BC [F72]; on 2, 8 BC [1012]; on 3 Alta, Ont, Que, NB, NS [1220]; on 3 Alta [F64], Que [1377,

NB, NS [1032]; on 3, 8 NWT-Alta [1063];
    on 8 Yukon [F69], BC [2003], Alta [952,
    F62].
Lophophacidium hyperboreum: on 3 Que
    [1692, 1674].
Lycoperdon perlatum (C:7): on 8 BC
    [1012]. Probably on rotted wood.
L. pyriforme: on 3 Ont [1341]. Probably
    on rotted wood.
Marasmiellus filopes: on 2 BC and on 3
    Ont, Que, NB, Nfld [1435].
Maurodothina farriae: on 2, 5 BC [1012,
    1358, F70]. Published as *"farrae"*.
Melampsora sp.: on 3 Que [F66].
M. abieti-capraearum (M. epitea race
    a.-c. (C:7): needle rust,
    rouille des aiguilles: 0 I on 2, 4, 5,
    8 BC [1012]; on 3 Alta [F63], Sask, Man
    [F68] and Man [F65, F67, F69], Ont
    [1236, 1827, F69], Que [1377, F73], Nfld
    [1660]; on 13 BC, Alta, Sask, Man [2008].
M. epitea: needle rust, rouille des
    aiguilles: on 3 Alta [53, F61], NS
    [1032]; on 3, 8 NWT-Alta [1063].
M. medusae (M. albertensis): needle rust,
    rouille des aiguilles: on 4, 5 BC
    [1012, 2002]; on 13 BC [F65].
M. occidentalis: on 4 BC [1012, 2002]; on
    13 BC [F65].
*Melampsorella caryophyllacearum (M.
    cerastii)* (C:7): witches' broom rust,
    rouille-balai de sorciere: on 2, 5, 8
    BC [1012]; on 3 NWT-Alta [1063] also
    Alta [53, F61, F63, F65], Man [F60, F64,
    F70, F72], Ont [1236, 1828, F62, F63,
    F64, F69, F76], Que [282, 1377, F70,
    F73], NS [F71, F75], Nfld [333, 1659,
    1660, 1661, F61, F62, F63, F64, F65,
    F66, F67, F71, F73, F74, F75, F76] and
    Sask, Man [F65, F67], Ont, Que [1341,
    F66, F67, F68, F71, F72, F74, F75], NB,
    NS, PEI [1032, F62, F63, F64, F65, F66,
    F67, F69]; on 8 Yukon [460, F61], BC
    [1341, 2007], Alta [644, F61]; on 13
    Yukon, BC, Alta, Sask, Man [2008] and
    Alta [952], PEI [337].
*Meruliopsis albostramineus (Ceraceomerulius
    a.)*: on 3 Que [568]. Apparently
    associated with a brown rot.
*M. taxicola (Merulius t., M. ravenelii,
    Poria t.)* (C:9): on 3 Alta, Sask, Ont
    [568] and Ont [1341, 1660]; on 8 BC
    [568, 951, 1012]. Associated with a
    white rot.
Merulius tremellosus: on 2 BC [1012].
    Associated with a white rot and produces
    the antibiotics, merulinic acid and
    merulidial.
Metacapnodium sp.: on 8 BC [1012].
Metulodontia nivea: on 8 BC [1012].
    Associated with a white rot.
Microascus trigonosporus: on 3 Ont [86].
Microlychnus epicorticis: on 2 BC [506,
    1012].
Micromphale foetidum (Marasmius f.): on
    13 Ont [1306].
M. perforans (Marasmius p.): on 3 Que
    [1385] on fallen needles.

Micropera sp.: on 8 BC [F61]. Probably
    Foveostroma or *Gelatinosporium*.
Microthelia biformis: on 3 Que [F67].
Microthyrium microscopicum: on 8 BC
    [1012,
    F69].
Milesia sp.: rust, rouille: on 3 Que
    [F69], NB, NS, PEI [1032] and NS [F61,
    F64, F71], Nfld [1659, 1660, 1661, F64,
    F65, F66, F67, F68, F69].
M. fructuosa (M. intermedia) (C:7):
    needle rust, rouille des aiguilles: 0 I
    on 3 Ont [1236], Que [1377, F61, F66,
    F72, F73, F74], NS [1032], Nfld [1660,
    1661, F73].
M. laeviuscula (Milesina l.) (C:7):
    needle rust, rouille des aiguilles: 0 I
    on 5 BC [1012, 2008].
M. marginalis (C:7): needle rust, rouille
    des aiguilles: 0 I on 3 Ont [1236], Que
    [1377].
M. pycnograndis (M. polypodophila) (C:7):
    witches' broom rust, rouille-balai de
    sorcière: 0 I on 3 Ont [1236], Que
    [1377], NS [1032].
Milesina sp.: rust, rouille: on 2 BC
    [1012]; on 2, 5 BC [F74].
Monochaetia abietis: on 5 BC [1012].
    Specimens should be restudied.
Mucor sp.: on purposely wounded 3 Que
    [910].
Mucronella calva (M. aggregata) (C:7): on
    3 NS [1032]. Associated with a white
    rot.
*Myceliophthora thermophila (Sporotrichum
    t.)*: on 3 NB [1610].
Mycena elegantula: on 3 NS [1032].
M. lilacifolia: on 3 Que [1385], on
    rotten wood.
M. rubromarginata: on 3 Que [1385], on
    dead branches on live trees.
Mytilidion sp.: on 3, 8 NWT-Alta [1063].
Naematoloma sp.: on 2 BC [1012]; on 8 BC
    [1719]. Specimens probably referable to
    Hypholoma.
Nectria sp.: on 3 Que [942, F66, F67,
    F69, F70], Nfld [1659, 1660, 1661].
N. fuckeliana: on 2, 5 BC [F65]; on 3 Que
    [943, 1685], e. Canada [95].
N. modesta (C:8): on 3 Que [70].
N. neomacrospora (N. fuckeliana var.
    macrospora, N. macrospora): on 2, 4
    BC [1012]; on 3 Que [F74]; on 13 BC
    [176].
Neournula nordmanensis: on 5 BC [1224].
*Nothophacidium phyllophilum (N.
    abietinellum, Dermatea p.)* (C:8):
    needle cast, rouge: on 3 NWT-Alta
    [1063], Sask [F68], Man [F65], Ont
    [1340, 1443, F62, F63]; also Ont, Que,
    NS [F64], Que [1674, F74] and NS [1032];
    on 3, 8 Alta [F66]; on 8 BC [1012].
Odontia lactea (C:8): on 2, 8 BC [1012].
    Specimens are typically of *Hyphodontia
    breviseta* but the type is a specimen of
    Resinicium bicolor (i.e., the name *O.
    lactea* is a synonym of *R. bicolor*).
O. stipata: on 3 NS [1032]. The name is
    of uncertain application. This specimen
    may be *Fibrodontia gossypina*.
Odonticium laxa (Odontia l.): on 3 NB
    [1032]. Lignicolous.

Odontotrema hemisphaericum: on 3 Que [1690].

Oedemium didymum (holomorph *Chaetosphaerella fusca*):   on 3 NB [1610].

Oidiodendron tenuissimum (C:8):   on 3 NB [82].

Onnia circinata (*Polyporus c.*, *P. tomentosus* var. *c.*) (C:9):   red butt rot, carie rouge alvéolaire du pied: on 3 Que [1390, 1377], NB [1032]; on 8 BC [1012].   Associated with a white pocket rot.

O. tomentosa (*Polyporus t.*, *P. t.* var. *tomentosa*) (C:9):   red butt rot, carie rouge alvéolaire du pied: on 2, 8 BC [1012]; on 3 Ont [F71, F72], NS [1032].   Associated with a white pocket rot.

Ostenia obducta (*Polyporus osseus*) (C:9):   on 3 Ont [1341]; on 8 BC [1012].   Associated with a brown rot.

Ostreola sessilis: on 8 BC [1012].

Oxyporus corticola (*Poria c.*) (C:9):   on 8 BC [1012].   Associated with a white rot.

Paecilomyces sp.:   on 3 NB [1610].

P. farinosus: on 8 BC [152].

P. varioti: on 3 NB [1610].

Panellus mitis (*Pleurotus m.*):   on 3 Nfld [1659, 1660].

P. serotinus (*Panus betulinus*, *Pleurotus s.*) (C:9):   on 2, 8 BC [1012].   See *Acer*.

P. violaceofulvus: on 3 Que [1109].

Panus rudis (C:8):   on 8 BC [1012].

Parmastomyces transmutans (*Polyporus subcartilagineus*):   on 8 BC [1012, 1718]; on 8, 13 BC [F64].   Associated with a brown rot.

Patinella punctiformis (C:8):   on 13 NS [1032].

Penicillium sp.:   on 3 Ont [1964, 1966], Que [454, 910 purposely wounded], NB [1032, 1610], e. Canada [95].

Peniophora cinerea: on 3 Ont [1341], Que [454].   Associated with a white rot.

P. incarnata (C:8):   on 5 BC [1012].   Lignicolous.

P. piceae (*P. separans*) (C:8):   on 3 NB [1742]; on 8 BC [1012].   Associated with a white rot.

P. pini (*Stereum p.*):   on 3 Ont [1941].   Associated with a white rot.

P. pithya (C:8):   on 8 NWT-Alta [1063].   Associated with a white rot.

P. pseudopini (C:8):   on 3 Ont [1941]; on 8 BC [1012].   Associated with a white rot.

P. septentrionalis (C:8):   on 8 BC [1012, 1719].   Associated with a white rot.

Peridermium sp.:   on 3 Que [F63].

P. balsameum (C:8):   rust, rouille: on 3 Que [F71].   Name of dubius application, refers to *Uredinopsis* spp.

P. holwayi (C:8):   rust, rouille: on 3, 8 Alta [53]; on 3, 8 NWT-Alta [1063];   on 8 Yukon [F61], Mack [53].

Pezicula livida (anamorph *Cryptosporiopsis abietina*):   on 3 Que [1690].

Phacidiopycnis balsamicola: on 8 BC [1012].

Phacidium sp.:   on 3 Que [F62].

P. abietis (C:8):   on 3 Ont [1340, 1660, 1827, F63, F68, F76], Que [1674, 1677, F64, F65, F66, F69]; on 3 Ont, Que [1444]; on 3 Que, NB [F75]; on 8 BC [953, 1012], Alta [952].

P. infestans (C:8):   snow blight, brûlure printanière: on 3 Que [1672].   Records are suspect and specimens should be restudied.   In North America several fungi have been labelled *P. infestans* and in Europe the fungus occurs on *Pinus*.

Phaeocryptopus nudus (*Adelopus balsamicola*) (C:8):   needle blight, rouge: on 2, 8 BC [1012]; on 3 NWT-Alta [1063], Alta [F64, F66], Ont [73, 1340, 1615], Que [70, 73, 1675], NB [73], NB, NS, PEI [1032], Nfld [1659, 1660, 1661]; on 8 BC [2003], Alta [F66].

Phaeolus schweinitzii (*Polyporus s.*) (C:9):   brown cubical rot, carie brune cubique: on 2, 8 BC [1012]; on 3 Ont [1341, F71, F72], NB, NS, PEI [1032].

Phanerochaete carnosa (*Peniophora c.*) (C:8):   on 2 BC [1012]; on 3 Que [969], NS [1032].   Associated with a white rot.

P. sanguinea (*Peniophora s.*) (C:8):   on 5 BC [1012].   Associated with a white rot but the wood often stained red.

P. sordida (*Peniophora cremea*):   on 3 NB [1032, 1742]; on 8 NWT-Alta [1063], Alta [950].   Associated with a white rot.

Phellinus chrysoloma (*Fomes pini* var. *abietis*):   on 3 Que [1377], Labr [867].   Associated with a white rot.

P. ferrugineofuscus (*Poria f.*) (C:9):   on 8 BC [1012].   Associated with a white rot.

P. ferruginosus (*Poria f.*) (C:9):   on 3 Ont [1341], NS [1032].   Associated with a white rot.

P. gilvus (*Polyporus g.*):   on 13 BC [579].   Associated with a white rot.

P. nigrolimitatus (*Fomes n.*) (C:6):   white pocket rot, carie blanche alvéolaire: on 8 BC [1012].

P. pini (*Fomes p.*) (C:6):   red ring rot, carie blanche alvéolaire: on 2, 5, 8 BC [1012]; on 3, 8 NWT-Alta [1063]; on 3 Alta [53], Ont [1341], NB, NS, PEI [1032], Nfld [1659, 1660, 1661]; on 8 Yukon [460], BC [953], Alta [950, 951, 952].   Associated with a white rot.

P. punctatus (*Poria p.*):   on 3 Ont [1341]; on 8 BC [1012].   Associated with a white rot.

P. robustus (*Fomes r.*) (C:6):   on 3 NB [1032].   Associated with a white rot.

P. tsugina (*Poria t.*) (C:9):   on 8 BC [1718].   Associated with a white rot.

P. viticola (*Fomes v.*):   on 8 BC [1012].   Associated with a white rot.

P. weirii: on 2, 5, 8 BC [1012]; on 5 BC [F75].   Associated with a root rot of live trees.

Phialocephala bactrospora: on 3 NB [1610].

Phialophora sp.:   on 3 Que [910] purposely wounded, NB [1610].

P. americana: on 3 NB [1610].

P. fastigiata: on 3 NB [1610].

P. heteromorpha: on 3 NB [1610].

P. lagerbergii: on 3 NB [1610].

P. melinii: on 3 NB [279, 1610], NS
   [1548]; on 8 BC [279, 1012].

Phlebia centrifuga (P. albida auct., *P.
   mellea)* (C:8): on 2, 5, 8 BC [1012];
   on 3 Ont [1341], NB [1032]; on 8 BC
   [1341]. Associated with a white rot.

P. livida (Corticium l.) (C:6): on 13 NS
   [1032]. Associated with a white rot.

P. phlebioides (Peniophora p.): on 3 NB
   [1032]. Associated with a white rot.

P. radiata (C:8): on 3 Alta [568]; on 5,
   8 BC [1012]; on 8 BC [568]. Associated
   with a white rot and produces the
   antibiotic merulinic acid.

P. rufa: on 3 Alta [568]. Associated
   with a white rot.

P. unica (Peniophora u.) (C:8): on 8 BC
   [1012]. Lignicolous.

Phlebiopsis gigantea (Peniophora g.)
   (C:8): on 3 Que [454], NB [1032, 1610];
   on 8 BC [1012, 1719]. Associated with a
   white rot. A biological control agent,
   arthrospores sprayed on recently cut
   stumps of coniferous trees effectively
   prevented colonization of those stumps
   by the root parasitic fungus
   Heterobasidion annosum.

Pholiota alnicola (Flammula a.) (C:6): on
   2, 8 BC [1012]; on 3 Ont [1964].
   Associated with a white rot.

P. aurivella (C:8): on 2, 5, 8 BC [1012,
   F65]; on 3 NB [1032]; on 5 BC [F60].
   Associated with a white rot.

P. limonella (P. squarrosa-adiposa)
   (C:8): on 5 BC [1012]. Associated with
   a white rot.

Phoma sp.: on 3 Ont [973], Que [1216].

Phomopsis sp. (C:8): canker and dieback,
   chancre et dépérissement: on 3 Que
   [F66].

Phyllosticta sp. *(Phyllostictina* sp.):
   on 3 Ont [1827, F68], NB, NS [1032].

Phyllotopsis nidulans (C:8): on NWT-Alta
   [1063], BC [1012]. Associated with a
   white rot.

*Piloderma bicolor (Athelia b., Corticium
   b.)* (C:5): on 3 Que [969], NS [1032];
   on 8 BC [1012]. Mycorrhizal on, at
   least, *Arctostaphylos*, *Arbutus* and
   Pseudotsuga. Typically this fungus
   fruits on rotted wood and duff.

Pithya vulgaris (C:9): on 2, 5, 8 BC
   [1012].

Pleurocybella porrigens (Pleurotus p.)
   (C:9): on 3 Ont [1341], NB, NS [1032];
   on 5 BC [1012]. Lignicolous.

Pleurotus sp.: on 2 BC [1012].

P. ostreatus (C:9): on 5 BC [1012].
   Associated with a white rot.

P. strigosus (Panus s.): on 3 NB [1032].
   Associated with a white rot.

Polyporus sp.: on 3 Nfld [1661].

P. badius (P. picipes): on 3 NB [1032];
   on 8 BC [1012]. Typically on fallen
   branches and stems. Associated with a
   white rot.

P. brumalis: on 3 NB [1032]. Typically
   on deciduous wood.

"P." destructor (C:9): on 3 NS [1032].
   This name is of uncertain application
   [1500]. Specimens and records should be
   reexamined.

P. hirtus (C:9): brown cubical rot, carie
   brune cubique: on 3 NB, NS [1032]; on
   5, 8 BC [1012]; on 8 BC [1718].

P. varius (P. elegans) (C:9): on 2, 8 BC
   [1012]. Associated with a white rot.

Poria sp.: on 3 NB, NS [1032].

P. rivulosa: on 2, 13 BC [1012].
   Associated with a white rot.

P. subacida (C:9): on 2, 5, 8 BC
   [1012]; on 3 Que [910, 1377], NB [1032,
   1742], NS [1032]; on 8 BC [1341].

P. vulgaris (C:9): on 3 NS [1032]. This
   name is of uncertain application.

Potebniamyces balsamicola (C:10): on 2,
   5, 8 BC [502, F68]; on 3 Que [502, 1673,
   F67, F75], NB [1673]; on 8 Yukon [460],
   BC [F67, F75]. Associated with tree
   dieback.

P. balsamicola var. *balsamicola*: on 3
   Que [504, 1690]; on 8 BC [504, 1012].

P. balsamicola var. *boycei*: on 2, 5 BC
   [504, 1012, F70].

Pragmopora abietina: on 13 Nfld [657].

Pseudophacidium piceae: on 3 Que [1685]
   by inoculation.

Pseudoplectania melaena: on 13 NS [1032].

Pseudoscypha abietis: on 3 Ont [1340,
   1447].

Pseudotomentella tristis (P. umbrina): on
   3 Ont [1341].

Ptychogaster sp.: on 3 NB [1610].

Pucciniastrum sp.: needle rust, rouille
   des aiguilles: on 3 Ont [1341, F64,
   F67, F68], Que [282, F76], NB [F75].

P. epilobii (P. abieti-chamaenerii)
   (C:10): needle rust, rouille des
   aiguilles: on 2 BC [F72]; on 3 Yukon
   [F71], Alta [53, F64, F66, F70, F72],
   Man [F69, F72, F73, F74], Ont [1236,
   F64, F65, F66, F67, F68, F71, F72, F76],
   Que [1546, F61, F70], NB [F61, F69, F73,
   F74, F76], NS [F66], PEI [1032, F64,
   F65], Nfld [1659, 1660, 1661, F69, F70,
   F73, F74, F76]; on 3, 8, 13 NWT-Alta
   [1063]; on 3 Sask, Man [F65, F67, F68,
   F70, F71]; on 3 Ont, Que [1341, F62,
   F63, F73, F74, F75]; on 3 NB, NS [1032,
   F62, F63, F64, F65, F68]; on 4, 5, 8 BC
   [1012]; on 8 Yukon [460, F61], BC [953,
   954, 2007, F64, F65, F69, F70, F71, F72,
   F73, F74], Alta [737, 950, F66, F71,
   F73, F74]; on 13 Yukon, BC, Alta, Sask,
   Man [2008]; also Alta [F65], Que [F70].

P. goeppertianum (C:10): needle rust,
   rouille des aiguilles: on 2 BC [1012,
   F67]; on 3 Yukon, Sask, Man, NS [F71];
   on 3 Alta [F64], Man [F60, F65, F69],
   Ont [1236, 1341], Que [1377, F60, F75],
   NB [1859, 1861, 1862, F69, F73, F74,
   F76], NS [1032, F66, F72], PEI [1032,
   1857], Nfld [1660, F72]; on 3 NB, NS
   [1857, 1858, F68]; on 8 Yukon [7, 460],
   NWT-Alta [1063], Mack [F66], BC [644,
   1012, F67, F76], Alta [737, 950, 951,
   952, F66, F71]; on 13 Yukon, NWT, BC,
   Alta, Sask, Man [2008] and Alta [F65],
   Nfld [333].

P. pustulatum: needle rust, rouille des
aiguilles: on 3 Alta [F60], Ont [1546],
Que [1377, F65], NS [1546], Nfld [1546,
F64, F65, F66, F67, F68]; on 8 BC [1546].

Pycnoporellus alboluteus (Polyporus a.)
(C:9): on 2, 8 BC [1012]; on 8 NWT-Alta
[1063], Alta [950]. Associated with a
brown rot.

P. fulgens (Polyporus fibrillosus) (C:9):
on 3 Ont [1179], NS, PEI [1032].
Associated with a brown rot.

Pyrenochaeta sp.: on 3 NB [1610].

Radulum pallidum: on 3 PEI [1032].
Lignicolous. The generic name is
applied to Ascomycetes but *R. pallidum*
is a Basidiomycete.

*Ramaricium albo-ochraceum (Corticium a.,
Trechispora a.)* (C:5): on 3 Ont [572,
1341]; on 13 Ont [970], NB [572, 970].
Lignicolous.

Rectipilus fasciculatus (Solenia f.): on
3 Ont [1341]. Lignicolous.

Resinicium bicolor (Odontia b.) (C:8):
white stringy rot, carie blanche
filandreuse: on 3 Ont [1956, 1957], Que
[909, 907a, 908, 1672, F65], NB, NS
[1032], Nfld [1661]; on 5 BC [1012,
F60]; on 8 BC [1012, 1718, 1719]; on 13
BC [140] in heartwood of live tree.

Retinocyclus abietis (C:10): on 3 Que
[374, 909, 910 purposely wounded, 1216,
1395, F60], Nfld [1661]; on 8 BC [1012].

Rhabdocline pseudotsugae: on 13 NWT-Alta
[1063]. Common on *Pseudotsuga
menziesii*.

Rhabdogloeopsis balsameae (Kabatiella b.)
(C:7): on 3 NB [1032], NS [1032, F61].

Rhinocladiella sp.: on 3 NB [1610].

R. atrovirens: on 3 NB [1610].

R. elatior (C:10): on 3 e. Canada [95];
on 8 BC [1012].

Rhizocalyx abietis (anamorph *Rhizothyrium
a.*): on 3 Que [1675, F67].

Rhizoctonia sp.: on 3 Que [910] purposely
wounded.

Rhizosphaera sp.: on 3 Man [F63].

R. pini (R. abietis) (C:10): needle
blight, brûlure des aiguilles: on 3 Man
[F65], Ont [1340, 1828], NB, NS, PEI
[1032], Nfld [1659, 1660, 1661, F73,
F74].

Rhizothyrium abietis (holomorph
Rhizocalyx abietis) (C:10): needle
cast, rouge: on 3 Ont [1827, F68], Que
[1675], NB, PEI [1032], Nfld [1659,
1660, 1661]; on 3 Ont, Que, NB, Nfld
[363]; on 8 BC [1012, 2003, F68], Alta
[F68].

Rhodotorula sp.: on 3 NB [1610].

Rigidoporus sp.: on 8 BC [1012].

Saccharomyces sp.: on 3 NB [1610].

Sarcodontia sp. *(Oxydontia* sp.): on 8
BC [1012].

*Sarcotrochila balsameae (Stegopezizella
b.)*: on 2, 5, 8 BC [1012]; on 3
NWT-Alta [1063]; on 3 Alta [53], Sask
[F73, F74], Ont [174, 1340, 1827, F63,
F68], Que [1674], NB, NS [1032]; on 3
Alta, Ont, Que [F65]; on 4, 8 BC [1444];
on 4, 13 Ont [1444]; on 8 BC [2003].

S. piniperda: on 3 Man [F65]. Typically
on *Picea*.

Schizophyllum commune (C:10): on 3 Ont
[1341], NB, NS [1032], Nfld [1661].
Associated with a white rot, presumably
saprophytic.

Schizopora paradoxa (Poria versipora)
(C:9): on 5 BC [1012]. Associated with
a white rot.

Scleroderris sp.: on 3 NB [1032]; on 8 BC
[F67].

Sclerophoma sp.: on 5, 8 BC [1012].

S. pithyophila (holomorph *Sydowia
polyspora*): on 3 Que [910] purposely
wounded; on 2 BC [F69].

Sclerotinia kerneri: on 1 Que [892]; on 3
Que [433, F72].

*Scoleconectria cucurbitula (Creonectria
c.)*: on 3 Sask, Man, NB, NS [F67]; on
3 Ont, NB [175, 1340], Que [1685], Nfld
[1661]; on 3 NB, NS [1032]; on 8 BC
[1012].

S. scolecospora (C:10): on 3 NWT-Alta
[1063], Alta [53].

Scorias spongiosa (C:10): on 13 NS
[1032]. Usually associated with insects
on plants.

Scutellinia scutellata (C:10): on 8 BC
[1012].

Scytalidium lignicola: on 3 NB [1610].

Scytinostroma galactinum (Corticium g.)
(C:6): white stringy rot, carie blanche
filandreuse: on 3 Ont [1341, 1964,
1956, 1957, F72], Que [910 purposely
wounded, 1395, 1672, F60, F66], NB
[1742], NB, NS, PEI [1032], Nfld [1661];
on 8 BC [1012, 1718]; on 13 BC [1233].

Seiridium abietinum: on 3 Nfld [1763]; on
3, 13 Nfld [1762].

Septoria sp.: on 8 BC [1012].

*Serpula himantioides (Merulius h., S.
lacrimans* var. *h.)* (C:7): brown
cubical butt rot, carie brune cubique:
on 2, 8 BC [1012]; on 3 NB [1032, 1742].

S. lacrimans: on 8 BC [1012]. Associated
with a brown rot in buildings.

Seuratia sp.: on 8 BC [1012].

S. millardetii: on live needles of 3 Ont,
Nfld, and on 5, 8 BC [1042].

Sirodothis sp. (holomorph *Tympanis*
sp.): on 5, 8 BC [1012].

Sistotrema brinkmannii (Trechispora b.)
(C:10): red heartrot, carie rouge du
coeur: on 2, 5, 8 BC [1012]; on 3 Que
[969], NB [1032, 1610, 1742]; on 8 BC
[966]. Associated with a white rot.
Airborne spores sometimes contaminate
agar cultures in the laboratory and in
one instance resulted in *S.
brinkmannii* being designated the
teleomorph for *Phymatotrichum
omnivorum*.

S. raduloides (Trechispora r.) (C:10):
red heartrot, carie rouge du coeur: on
3 NB [1032]; on 8 BC [1012]. Associated
with a white rot.

Sistotremastrum suecicum: on 3 Que
[969]. Associated with a white rot.

Skeletocutis amorpha (Polyporus a.)
(C:9): on 13 BC [1012]. Associated
with a white rot.
S. odora (Poria o.): on 2 BC [F71]; on 3
NB [1032]; on 8 BC [579, 1012].
S. stellae (Poria s.): on 2, 8 BC
[1012]. Associated with a white pin rot.
S. subincarnata (Poria s.) (C:9): on 2, 8
BC [1012]; on 3 NS [1032]. Presumably
associated with a white rot.
Spadicoides atra: on 3 Sask, Man [1760].
S. grovei: on 3 Man [1760].
Sphaerellopsis filum (Darluca f.)
(holomorph *Eudarluca caricis*) (C:6):
on 5 BC [F60]. Mycoparasitic on rust
pustules.
Sphaerobasidium minutum (Xenasma m.): on
3 Que [969]. Lignicolous.
Sphaeropsis sapinea: on 3 Que [1687] by
inoculation.
Sporidesmium achromaticum: on 3 Sask, Man
[1760].
S. coronatum: on 3 Man [1760].
S. harknessii: on 3 Sask, Man [1760].
S. larvatum: on 3 Sask [1760], Man [824,
1760].
*Stegonsporiopsis cenangioides (Myxocyclus
c.)*: on 3 Que [F70].
Stemonitis fusca: slime mould: on 3
NWT-Alta [1063]. Presumably on rotted
wood.
Stenocybe major: on 13 BC, Ont [1812].
Streptomyces sp.: on 3 NB [1610] on wood
chips.
Stereum sp.: on 3 Nfld [1661].
S. hirsutum: on 3 Que [282]. Associated
with a white rot.
S. ostrea (C:10): on 8 BC [1012].
Associated with a white rot.
Sterigmatobotrys macrocarpa: on 3 Sask,
Man [1760].
Strasseria geniculata: on 3 Que [1278].
Subulicystidium rallum: on 3 Ont [744].
Lignicolous.
Sydowia polyspora (anamorph *Sclerophoma
pithyophila*): on 2, 5, 8 BC [1012]; on
3 Que [1686, F68].
Taeniolella rudis: on 3 Ont [815].
Tarzetta bronca (Geopyxis b.): on 13 Nfld
[666], litter.
Thermoascus aurantiacus: on 3 NB [1610].
Thermomyces lanuginosus (Humicola l.): on
3 NB [1610].
Thyronectria sp.: on 3 Que [910]
purposely wounded.
T. balsamea: on 3 Alta, Man, NB [F62] and
Sask, Man [F63, F67], Ont [175, 1340,
1422, F61], Que [F66, F72, F73, F74,
F75], NB, NS, PEI [1032]; on 8 BC [1012].
Thysanophora penicillioides: on 2 BC
[1012]; on 3 Man [1760], Ont, Que, NB,
NS, Nfld [872].
Tiarosporella abietis: on 3 Alta, Ont and
on 8 Yukon, Alta [1947].
Tomentella atrorubra: on 3 NS [1032].
T. bresadolae: on 3 Ont [1341]; on 13 Ont
[928].

T. ellisii (T. microspora): on 3 Ont
[929, 1341, 1827].
T. fuscoferruginosa: on 3 Ont [929, 1341].
T. griseoviolacea: on 13 Ont [1828].
T. molybdaea: on 3 Ont [929, 1341].
T. nitellina: on 3 NS [1341].
T. pilosa: on 3 Ont [1828].
T. ruttneri: on 3 Ont [1341]; on 13 Ont
[928].
T. sublilacina: on 3 Ont [1341, 1828,
1827].
T. terrestris (T. umbrinella): on 3 NS
[1827].
T. violaceofusca (T. trachychaites): on 3
Ont [1341].
Toxosporium camptospermum: on 3 NS [1032].
Trametes sp.: on 3 Nfld [1661].
T. cervina (Polyporus biformis N. Amer.
auct.*)*: on 3 Nfld [1659, 1660,
1661]. Associated with a white rot.
Trechispora sp.: on 3 Que [454].
T. farinacea (Grandinia f.): on 3 Que
[969]. Associated with a white rot.
*T. mollusca (Poria candidissima, T.
onusta)* (C:9): on 2, 8 BC [1012]; on 3
NS [970, 1032], PEI [1032], Nfld [1659,
1660, 1661]. Associated with a white
rot.
T. vaga: on 2 BC [1012]. Mycorrhizal on,
at least, *Picea* and *Pinus*.
Tremella sp.: on 5 BC [1012]. Associated
with a white rot.
T. encephala: on 3 NS, PEI [1032].
Mycoparasitic on *Haematostereum
sanguinolentum*.
T. foliacea (C:10): on 3 NS [1032]; on 8
BC [1012]. Lignicolous.
*Trichaptum abietinus (Hirschioporus a.,
Polyporus a.)* (C:9): white pocket rot,
carie blanche de l'aubier: on 3, 8
NWT-Alta [1063]; on 3 Ont, Que, NB, Nfld
[95] also Ont [1341], Que [282, 1378]
and Ont, NB [1742], NS, NB, PEI [1032],
Nfld [1659, 1660, 1661], Labr [867]; on
5, 8 BC [1012]; on 8 BC [953]; on 13 Ont
[1827]. In Conners (p.9) delete "butt
rot of heartwood of 4 BC", Nobles [1180]
subsequently redetermined these cultures
to be *Resinicium bicolor*.
T. fuscoviolaceus (Hirschioporus f.): on
2 BC [1012]; on 3 Sask, Ont, Que
[1028]. Associated with a white rot.
T. laricinus (Hirschioporus l.): on 3 Que
[1028], Labr [867]; on 8 BC [1012].
Associated with a white rot.
Trichocladium canadense: on 3 NB [1610].
Trichoderma sp. *(Pachybasium sp.)*: on 3
Que [909, 910 purposely wounded], NB
[1610].
T. polysporum (T. sporulosum): on 3 NB
[1610].
T. viride: on 3 Ont [1964, 1966], Que
[454], NB [1610], e. Canada [95]. One
of the fungi being evaluated for its
ability, through the production of
cellulases, to increase the
digestibility of plant fiber (forages,
wood shavings, etc.) for animal feed.

And the fungus has been used in attempts
to biologically control soil-borne
pathogens.

Trichosphaeria parasitica (C:10): white
felt blight, feutrage blanc: on 3 Que
[1672, F61].

Troposporella fumosa: on 3 Sask, Man
[1760].

Tryblidiopsis pinastri: on 3 Ont [1340].

Tuberculina persicina (C:10): on 5 BC
[F60]. Mycoparasitic on rust fungi.

Tubulicrinis accedens: on 3 Que [969].
Associated with a white rot.

T. glebulosus (Peniophora gracillima)
(C:8): on 13 NS [1032]. Associated
with a white rot.

T. globisporus: on 8 BC [742].
Lignicolous.

T. hamatus (Peniophora h.) (C:8): on 3
Ont [1935]. Associated with a white rot.

T. subulatus: on 3 Que [969]. Associated
with a white rot.

Tulasnella violacea: on 3 NB [1032].
Lignicolous.

Tympanis sp.: on 3 Ont [F61], Que [909,
910 purposely wounded].

T. abietina (C:10): on 3 Ont [973, 1340],
Que [1688, F68], NB [1032].

T. confusa: on 3 Que [1688] by
inoculation.

T. hansbroughiana: on 3 Que [1688].

T. hypopodia (T. piceina): on 3 Que
[1688] by inoculation; on 13 Alta, Ont,
Que, NB [1221].

T. laricina (anamorph *Sirodothis* sp.):
on 3 Que [1688] by inoculation; on 8 BC
[1012]; on 13 BC, Ont, Que, NB, Nfld
[1221].

T. pithya (C:10): on 3 Que [1688] by
inoculation.

T. truncatula (T. pinastri) (C:11): on 3
Sask, Man [F67] and Man [F66, F69], Que
[1688, F68], NS [1032, 1925]; on 8 BC
[1012]; on 13 Ont, Que, Nfld [1221].

T. tsugae: on 3 Que [1688] by
inoculation; on 13 Que, NB [1221].

Tyromyces balsameus (Polyporus b.) (C:9):
brown cubical rot, carie brune cubique:
on 3 Ont [1341], Que [282, 913, 910
purposely wounded, 1377, F65], NB [1032,
1610], NS [1032], Nfld [1660, 1661]; on
8 BC [1012].

T. borealis (Polyporus b.) (C:9): white
mottled rot, carie blanche madrée: on 3
Que [1377], NB [1032]; on 8 BC [1012].

T. caesius (Polyporus c.) (C:9): on 3 NS
[1032]. Associated with a brown rot.

T. chioneus (Polyporus albellus) (C:9):
on 3 NB, NS [1032]. Associated with a
white rot.

T. fragilis (Polyporus f.): on 3 NS
[1032]. Associated with a brown rot.

T. guttulatus (Polyporus g.) (C:9): brown
cubical rot, carie brune cubique: on 3
Que [1390].

T. leucospongia (Polyporus l.) (C:9): on
8 BC [1012]. Associated with a brown
rot.

T. mappa (Poria m.): on 3 Que [F71].
Associated with a white rot.

T. mollis (Polyporus m.) (C:9): on 3 NB,
NS [1032], Nfld [1659, 1660, 1661].
Associated with a brown rot.

*T. sericeomollis (Polyporus s., Poria s.,
Poria asiatica)* (C:9): on 3 Ont
[1341], Que [1672], NB, NS [1032]; on 8
BC [1012, 1718]. Associated with a
brown rot.

T. stipticus (Polyporus immitis): on 3 NB
[1032, 1742]; on 5 BC [1012].
Associated with a brown rot.

T. subvermispora (Poria s.): on 8 BC
[579, 1012]. Apparently associated with
a white rot.

T. undosus (Polyporus u.) (C:9): on 2 BC
[1012]; on 3 Que [F71]; on fallen trunk
of 13 Que [373]. Associated with a
brown rot.

Uredinopsis sp.: rust, rouille: on 2, 5
BC [F74]; on 2, 8 BC [F71]; on 3 Ont
[1341], Que [F69], NB [1032], NS [F61,
F71, F73], Nfld [1661].

U. americana (U. mirabilis) (C:11):
needle rust, rouille des aiguilles: 0 I
on 3 Ont [1236], Que [1377], NS [1032].

U. arthurii (C:11): rust, rouille: on 3
Que [1377]; on 13 Que [1375]. The
report by Faull apparently was the basis
of the records in Conners and in
Pomerleau [1377].

U. ceratophora (C:11): rust, rouille: 0
I on 3 Ont [1236].

U. hashiokai (C:11): rust, rouille: 0 I
on 5, 8 BC [1012, 2008].

U. longimucronata (C:11): rust, rouille:
0 I on 2, 5, 8 BC [2008]; on 3 Ont
[1236], Que [1377].

U. longimucronata f. cyclosora (C:11):
rust, rouille: on 2, 5, 8 BC [1012,
2007].

U. osmundae (C:11): rust, rouille: 0 I
on 3 Ont [1236, 1341], Que [1377].

U. phegopteridis (C:11): rust, rouille:
0 I on 3 Ont [1236], Que [1377]; on 8 BC
[1012, 2008, 2007].

*U. pteridis (U. macrosperma, Peridermium
pseudobalsameum)*: rust, rouille: 0 I
on 2, 5, 8 BC [1012, 2008]; on 2, 5 BC
[1012]; on 5 BC [1341].

U. struthiopteridis (C:11): rust,
rouille: 0 I on 3 Ont [1236], Que
[1377], NB [F68]; on 8 BC [1012, 2008,
2007].

Valsa sp. (C:11): on 3 Sask, Man [F67],
Ont [F61], Que [F66]; on 5, 8 BC [1012].

V. abietis (C:11): on 3 Ont [1340, 1422,
F61, F69], Que [1689 by inoculation,
F76], NB, NS [1032, F62], PEI [1032].

V. friesii (anamorph *Cytospora dubyi*)
(C:11): canker, chancre cytosporéen:
on 3 Ont [1340], Que [1689, F68, F70,
F72, F74, F75, F76], NB, PEI [1032].

V. pini: on 3 Que [1689] by inoculation,
NS [1032]. Usually on *Pinus*.

Vararia granulosa (Grandinia g.) (C:11):
on 2, 8 BC [1012]; on 5 BC [557]; on 13
Ont [557]. Associated with a white rot.
V. investiens : on 3 Ont [557].
Associated with a white rot.
V. racemosa (C:11): on 5 BC [557]; on 8
BC [1012]. Associated with a white rot.
Verticicladiella brachiata : on 3 NB
[1610].
V. dryocoetidis (holomorph *Ceratocystis
d.*) : on 8 BC [878].
*Vesiculomyces citrinus (Corticium radiosum,
Gloeocystidiellum c.)* (C:6): on 3 NB,
NS [1032]; on 8 BC [1012]. Associated
with a white rot.
Virgella robusta : on 2 BC [1012].
Wolfiporia extensa (Poria cocos) : on 3 NB
[1032]. Causing a brown rot.
Wrightoporia lenta (Poria l.) (C:9): on 8
BC [1012]. Associated with a brown rot.
W. illudens (Polyporus abieticola)
(C:9): on 3 Que [F71]. Associated rot
uncertain.
Xeromphalina campanella (Omphalia c.)
(C:11): white stringy rot, carie
blanche filandreuse: on 3 Que [1672],
NB [1032, 1610, 1742]; on 8 BC [1012].
Zythia resinae (Pycnidiella r.) (holomorph
Biatorella r.) (C:11): on 3 Que [F60].

ABUTILON Mill.   MALVACEAE

1. *A. theophrasti* Medic., velvet-leaf;
naturalized from India, in waste ground,
vacant lots and cult. fields, BC,
Sask-Que, NS, PEI.

Verticillium dahliae : on 1 Ont [1087].
V. nigrescens : on 1 Ont [1087].

ACER L.   ACERACEAE

1. *A. circinatum* Pursh, vine maple,
érable circiné; native, along the
s. coast of BC and 2 small areas on
Vancouver Island.
2. *A. ginnala* Maxim., Amur maple, érable
de tartarie; imported from Asia.
3. *A. glabrum* Torr., Rocky Mountain
maple, érable nain; replaced in Canada
by *A. g.* var. *douglasii* (Hook.)
Dipp., Douglas maple, érable nain;
native, extending from isolated spots on
the Alaska-BC border throughout c. and
s. BC and Vancouver Island and into sw.
Alta.
4. *A. macrophyllum* Pursh, bigleaf maple,
érable à grandes feuilles; native, along
coast and on islands of s. BC.
5. *A. negundo* L., Manitoba maple, érable
à giguère; native, Sask-Man to isolated
spots in w. and s. Ont but frequently
cultivated and naturalized beyond its
range. 5a. *A. n.* var. *interius*
(Britt.) Sarg.; native, common in
prairie region.
6. *A. palmatum* Thunb., Japanese maple,
érable du Japon; imported from Asia.

6a. *A. p.* var. *atropurpureum*, a
cultivated variety with reddish leaves.
7. *A. pensylvanicum* L., striped maple,
bois barré; native, from the e. shore of
Lake Superior to Gaspé, Que and NS but
not in deciduous forest region s. of
Lake Simcoe, Ont.
8. *A. platanoides* L., Norway maple,
érable de Norvège; imported from Europe.
9. *A. pseudoplatanus* L., sycamore maple,
érable blanc; imported from Europe and
w. Asia.
10. *A. rubrum* L., red maple, érable rouge;
native, w. boundary of Ont - s. Nfld.
11. *A. saccharinum* L., silver maple,
érable argenté; native, s. Ont-s. NB.
11a. *A. s.* var. *laciniatum* (Carr.)
Pax, Wier maple.
12. *A. saccharum* Marsh., sugar maple,
érable à sucre; native, w. Ont-NS. 12a.
A. s. ssp. *nigrum* (Michx. f.)
Desmarais, black maple, érable noir;
native to s. Ont.
13. *A. spicatum* Lam., mountain maple,
plaine bâtarde; native, s. Sask-Nfld.
14. Host species not named, i.e., reported
as *Acer* sp.

Acarosporina microspora (Schizoxylon m.) :
on 12 Que [F68].
Acremonium boreale : on 5 BC, Alta, Sask
[1707].
Acrospermum cuneolum (Glyphium elatum)
(C:12): on 13 NB [1032]; on 14 BC [589].
Acrostalagmus sp.: on 12 Que [941].
Alatospora acuminata : on 4 BC [43].
Aleurodiscus sp.: on 4 BC [1012].
A. botryosus (C:12): on 13 Ont [964].
Lignicolous.
A. canadensis : on 14 Ont [964]. See
Abies .
A. cerussatus (C:12): on 14 Ont [964].
Lignicolous.
A. oakesii (C:12): on 10, 12 Ont [1341]
and 12 Ont [964]. Lignicolous.
Allescheriella crocea : on 5 Sask [1760].
Alternaria sp.: on 10, 12 Que [F60] and
12 Que [945, purposely wounded].
A. alternata (A. tenuis auct.) : on 12
Ont [92].
Antrodia crustulina (Poria c.) : on 10 NS
[1036]. See *Abies* .
A. malicola (Trametes m.) : on 12 Ont
[1341]. Associated with a brown rot.
Armillaria mellea (C:12): root rot,
pourridié-agaric: on 4 BC [6]; on 7,
10, 12 Que [1376]; on 10, 12 Ont [1341];
on 11 Ont [F64]; on 12 Ont [6, F63], Que
[912, F66, F68, F70, F73, F74]; on logs
of 12 Que [911]; on 14 Ont [F66], Que
[1377]. See *Abies* .
Arthrobotrys arthrobotryoides : on 12 Ont
[92].
Articulospora tetracladia : on 4 BC [43].
Ascocoryne sarcoides (Coryne s.) : on 12
Que [941].
Ascotremella faginea : on dead trunks of
14 Que [1385].

A. turbinata: on 14 Que [1340].
  Lignicolous.
Aspergillus fumigatus: on 14 NB [1610].
Athelia scutellare (Corticium s.) (C:12):
  on 12 Ont [1341]. Lignicolous.
Aureobasidium sp. *(Pullularia* sp.*)*:
  on 12 Que [F60].
A. apocryptum (Gloeosporium a., Kabatiella
  a., G. saccharinum) (C:13):
  anthracnose, anthracnose: on 2 Ont
  [338]; on 5, 8 NB [1032, 1340, F64]; on
  7 NB [F64], NS [F62] and NB, NS, PEI
  [1032]; on 8 Que [337], NS [1032, F63];
  on 10 Ont [336, 337, F75], Que [F74], NB
  [F60], NS [F62, F66], PEI [1340]; on 10,
  12 Que [F61, F65, F75], NB, NS [339,
  1032, F63, F64, F65], PEI [1032, F64];
  on 10, 13 Nfld [1659, 1660, 1661, F65,
  F66, F67, F68, F74]; on 11 Que [336],
  NS [1032]; on 12 Ont [339, F76], Que
  [F69, F72, F73, F76], NB, NS [1340,
  1594, F61, F67, F75] and NB [F60], NS
  [F62, F66, F68, F69, F72], PEI [337,
  F61, F63, F72]; on 12, 13 Nfld [F73]; on
  13 NB [1032, 1340, F64], NS [1032, F63],
  Nfld [F61, F62, F76]; on 14 Ont [338,
  F74], Que [338], NB [F66, F73], NS
  [336], Nfld [1661, 1594] and NB, PEI
  [336, 337, F62].
A. pullulans: on purposely wounded 12 Que
  [945].
Basidiodendron eyrei: on 14 BC [1016].
  Associated with a white rot.
Bertia moriformis: on 13 NS [1032].
Bispora betulina: on 12 Ont [92]; on 14
  Que [1755], NB [1610].
Bisporella citrina (Helotium c.): on 7 NB
  [1032].
Bjerkandera adusta (Polyporus a.) (C:15):
  on 10 Ont [1341]; on 10, 12 Que [1390];
  on 12 NB [1032]; on 14 NS [1032]. See
  Abies.
B. fumosa (Polyporus f.): on 4 BC
  [1012]. Associated with a white rot.
Bondarzewia berkeleyi (Polyporus b.)
  (C:15): on 4 BC [1012]. Associated
  with a white rot.
Botryobasidium pruinatum: on 4 BC
  [1012]. Associated with a white rot.
B. vagum: on 1 BC [1012]. See *Abies*.
Botryohypochnus isabellinus (Botryobasidium
  i.): on 14 Ont [1341].
Botrytis cinerea (holomorph *Botryotinia*
  fuckeliana): on 14 Sask [1760].
Brachysporium nigrum: on 14 Ont [827].
Byssocorticium atrovirens: on 14 Ont
  [1341, 1827].
Cacumisporium capitulatum: on 14 Ont, Que
  [586].
C. tenebrosum: on 14 Que [587].
Calocera cornea: on 4 BC [1012].
  Associated with a brown pocket rot.
Cantharellus sp.: on 12 Ont [1341].
Cephalosporium sp.: on 10 Ont [973]; on
  12 Ont [92, 1828], Que [941, 945].
Ceraceomyces serpens: on bark of 12 Ont
  [568]; on stick of 14 Ont [568]. See
  Abies.

Ceratocystis sp.: on 12 Ont [94, 92].
C. acericola: on 12 Ont [647, 1340].
C. coerulescens: on 12 Ont [647].
C. major: on 12 Ont [647].
C. piceae: on 10 Ont [647, 1340].
C. spinulosa: on 11 Ont [1340]; on 12 Ont
  [647].
C. tenella: on 12 Ont [1340].
Ceriporia reticulata (Poria r.) (C:16):
  on 4 BC [1012, F62]. Lignicolous.
C. spissa (Poria s.): on 12 Ont [1341].
  Associated with a white rot.
C. tarda (Poria semitincta): on 12 NS
  [1036]. See *Abies*.
Cerocorticium confluens (Corticium c.)
  (C:12): on 14 NS [1032]. Lignicolous.
Cerrena unicolor (Daedalea u.) (C:12):
  white spongy rot, carie blanche
  spongieuse: on 3 BC [1855]; on 7, 10,
  12, 13, 14 NB [1032]; on 7 NS [1855]; on
  7, 12, 14 NS [1032]; on 8, 12 Ont
  [1855]; on 8, 10, 12, 14 Ont [1341]; on
  10 Que [1376], PEI [1032]; on 12 Ont
  [1827], Que [1391 dead branch, 1376,
  1378]; on 14 Que [1377, 1855, F66], PEI
  [1032].
Chalara sp.: on 12 Que [941].
Cheimonophyllum candidissimus (Pleurotus
  c.): on 10 NS [1032]. Lignicolous.
Chlorociboria aeruginascens: on 4 BC
  [1012].
C. aeruginosa: on 12 Ont [1340].
Chondrostereum purpureum (Stereum p.)
  (C:16): on 10, 12, 14 NS [1032]; on 12
  Ont [92, 1341], Que [911 logs, 945]; on
  14 BC [1012], Ont [1341]. See *Abies*.
Chytridium ?confervae: on decaying leaves
  of 1 BC [44].
Ciboria acericola (C:12): on 14 Ont [662].
C. acerina (C:12): on 14 Ont, Que [662].
Cladosporium sp.: on 14 NB [1610].
C. herbarum: on 4 BC [1012].
C. humile (C:12): leaf spot, tache des
  feuilles: on 10 Ont [1340], Nfld [1661].
C. macrocarpum: on 5 Man [1760].
Climacodon septentrionalis (Hydnum s.,
  Steccherinum s.) (C: 13, 16): white
  spongy rot, carie blanche spongieuse:
  on 10, 11 Ont [1341]; on 12 Que [1377,
  1376], NB [1032, F65]; on 14 Ont [1341,
  1660, F62]. Often on wounds on live
  trees.
Clitocybe truncicola: on stumps or dead
  trunks of 14 Que [1385].
Coccomyces coronatus (C:12): on 12 Ont
  [1340]; on 14 NS [1032].
C. tumidus: on 14 NS [1609].
Comatricha suksdorfii (C:12): on 10 Que
  [F60].
Coniochaeta subcorticalis: on 12 Ont
  [1340].
C. velutina: on 12 Ont [94, 92, 1789].
Coniophora arida (C. betulae) (C:12): on
  12 Ont [1341]. Associated with a brown
  rot.

C. puteana: on 12, 14 Ont [1828].   See
   Abies.
Coniothyrium sp.:   on 10 Ont [973]; on 12
   Ont [92].
Conoplea olivacea (C. sphaerica) (C:12):
   on 14 Ont [784].
Coprinus micaceus: on 12 Ont [92].
Cordana pauciseptata (C:12):   on decaying
   wood of 14 Que [796].
Coriolopsis gallica (Trametes hispida)
   (C:16):   white ring rot, carie blanche
   annelée: on 14 NWT-Alta [1063].
Coriolus hirsutus (Polyporus h.) (C:15):
   white spongy rot, carie blanche
   spongieuse: on 3, 4 BC [1012]; on 10,
   12, 13, 14 Ont [1341], NS [1032]; on 12
   NB [1032].
C. pubescens (Polyporus p., P. velutinus)
   (C:15):   on 10, 12, 14 Ont [1341]; on 12
   Ont [1063]; on 14 Ont [1827].   See
   Abies.
C. versicolor (Polyporus v.) (C:15):
   white spongy rot, carie blanche
   spongieuse: on 4 BC [1012]; on 7, 10,
   12, 14 Ont [1341], NS [1032]; on 10 Que
   [1376], NB, NS, PEI [1032]; on 12 Que
   [911 logs, 912, 1390], NB, NS [1032]; on
   13 NS, PEI [1032]; on 14 Que [1377], NB,
   NS [1032].
Coronicium albo-glaucum (Xenasma a.) on
   12, 14 Ont [967].   Lignicolous.
Corticium sp.:   on 12 Ont [1341]; on 12
   logs Que [911].
Corynespora pruni: on 13 Man [1760].
Crepidotus applanatus: on 10 Ont [1306].
C. mollis (Agaricus fulvotomentosus)
   (C:12):   on 4 BC [1012]; on 14 NS
   [1032].   In these two references the
   name appeared as *Crepidotus
   fulvotomentosus* Peck, but Peck
   published it as *Agaricus (Crepidotus)
   fulvotomentosus*.
C. versutus: on 13 Ont [1341].
Cristulariella depraedans: on 1 BC
   [1427]; on 7, 10, 12, 13 Ont [1427]; on
   10 Ont [1827]; on 13 Que [F64].
*Crustomyces subabruptus (Cystostereum
   s.)*:   on 4 BC [1012].   Lignicolous.
Cryptendoxyla hypophloia: on 14 Ont
   [1040], under bark on dead tree.
Cryptodiaporthe acerinum (C:12):   on 7,
   10, 12 Ont [1340]; on 10 Que [F63]; on
   12, 14 Ont [1442].
C. myinda (C:12):   on 12 Ont [F60].
   Should be compared with *C. acerinum*.
C. petiolophila (C. densissima var.
   spicata) (C:12):   on 5, 10, 13 Ont
   [1340]; on petioles and small twigs of
   12, 13 Ont [79].
Cryptosporella sp.:   on 10 Ont [1340]; on
   12 Ont [1827].
Cryptosporiopsis sp.:   on 5 Sask, Man
   [F69] and Man [1768].
Cucurbidothis pithyophila: on 1 BC [1012,
   F67].   Unusual record, usually on insect
   on *Pinus*.
Cucurbitaria caraganae: on 5 Sask [F69].

*Cylindrobasidium evolvens (Corticium e., C.
   laeve)* (C:12):   on 4 BC [1012]; on 12
   Ont [92, 1341], Que [911, 945 purposely
   wounded], NB [1032]; on 14 Ont [1341],
   NB [1610], NS [1032].   Lignicolous.
Cylindrocarpon sp.:   on 12 Ont [92].
C. candidulum (holomorph *Nectria
   veuillotiana*):   on 14 Ont [176].
Cylindrocephalum sp.:   on 10 Ont [1340].
Cyphellopsis anomala: on 4 BC [1012]; on
   decorticated wood of 14 BC [1424].
Cyptotrama asprata (Collybia lacunosa)
   (C:12):   on 14 NS [1032, 1436].
   Associated with a white rot.
Cystostereum murraii (Stereum m.) (C:16):
   on 5, 12 Ont [1341]; on 10 Que [1377,
   1376]; on 12 Que [941]; on 10, 14 NS
   [1032].   Associated with a white rot.
Cytospora sp.:   on 4 BC [1012]; on 5, 7,
   8, 10, 12, 14 NB [1032]; on 7, 10, 14
   Ont [1340]; on 10 Ont [1827], NS, PEI
   [1032], Nfld [1659, 1661]; on 11 NS
   [1032]; on 12 Que [941, 945 purposely
   wounded], PEI [1032].
C. ambiens (holomorph *Valsa ambiens*)
   (C:12):   canker, chancre: on 8 Ont
   [333]; on 11 Ont [339].
C. annulata (C:12):   on 5 Sask [F65, F69];
   on 7 Ont [1340].
C. decipiens: on purposely wounded 12 Que
   [945].
Dacrymyces palmatus (C:12):   on 13 NS
   [1032].   See *Abies*.
Daedaleopsis confragosa (Daedalea c.)
   (C:12):   on 10, 12, 14 Ont [1341]; on 12
   NB, NS [1032]; on 14 PEI [1032].
   Associated with a white rot.
Daldinia concentrica: on 14 Que [1377].
   Lignicolous.
Dasyscyphus patula (Lachnum p.):   on 14
   Ont [1340].
Datronia mollis (Trametes m.) (C:16):   on
   4 BC [1012]; on 7, 10, 12, 14 NS [1032];
   on 10 Ont [1341]; on 12 Ont [1341,
   1660], NB [1032]; on 14 Ont [1827].
   Associated with a white rot.
Dendrophoma pulvis-pyrius: on 12 Que
   [F67].
*Dendrothele alliacea (Aleurocorticium a.,
   Aleurodiscus acerinus* var.
   alliaceus) (C:12):   on 12, 14 Ont
   [965].   Lignicolous.
D. candida (Aleurocorticium c.):   on 14
   Ont [965].   Lignicolous.
D. dryina (Aleurocorticium d.):   on 14
   Ont, Que [965].   Lignicolous.
D. griseo-cana (Aleurocorticium g.):   on
   10 Que [965]; on 14 Ont [965].
   Lignicolous.
D. incrustans (Aleurocorticium i.):   on 4
   BC [965].   Lignicolous.
D. maculata: on 14 Ont, Que [965].
   Lignicolous.
Dendryphiopsis atra (holomorph
   Microthelia incrustans):   on 5 Man
   [1760].
Dermea acerina (C:12):   on 10, 12 Ont
   [1234], Que [F67], NB, NS [1032].

Diaporthe acerina (C:12): on 13 Ont
    [1340, F60], NS [1032]; on 14 Ont [1827].
D. dubia (C:12): on 12 Ont [1340, F60],
    Que [F70]; on 14 Ont [1340].
Diatrype sp.: on 14 BC [1012].
D. macounii: on 12 Que [1377, 1376, 1378].
D. stigma (C:12): on 10, 12, 14 Ont
    [1340].
Diatrypella frostii (C:12): on 12 Ont
    [1827]; on 12, 14 Ont [1340].
Diplodia sp.: on 5 Ont [1340].
*Diplodina acerina (Cytodiplospora a.,
    Discella a., Fusicoccum obtulusum,
    Septomyxa tulasnei, Stagonospora
    collapsa)* (C:16): on 5 Ont [1340,
    F66], NB [1032, 1036, F61], NS [1036],
    PEI [1032, 1036]; on 10 Ont [1340]; on
    13 Man [F67]; on 14 NS [1032].
Doratomyces purpureofuscus: on 5 Sask,
    Ont [1760].
D. stemonitis: on 14 Sask [1760].
Dothidea sambuci: on 5 Man [F68].
Dothiorella sp.: on 10 Ont [1340].
Durella compressa: on 12 Ont [1340]. The
    name is of uncertain application.
    Specimens and records should be
    reexamined.
Endophragmiella bisbyi (Endophragmia b.):
    on dead bark of 13 Man [801, 1760].
E. collapsa: on decaying wood and bark of
    14 Que [803].
E. subolivacea: on 14 Ont, Que [804]. On
    fallen branches, particularly around
    stromata of *Diatrype* sp. and *Valsa
    ambiens*.
Epicoccum sp.: on purposely wounded 12
    Que [945].
E. purpurascens: on 5 Sask [1760].
Erinellina miniopsis: on 10 Ont [1340].
    The generic name is a synonym of
    Dasyscyphus, but this species may not
    be a *Dasyscyphus*.
Eutypa spinosa: on 12, 14 Ont [1340].
Eutypella parasitica (C:13): on 8 Ont
    [853]; on 10 Ont [1340, 1828, 1827,
    F60], Que [918, 1376]; on 10, 11, 12 Ont
    [F66]; on 12 Ont [1340, 1828, F60], Que
    [?918, 940, 1377, 1376, F67, F68, F71,
    F72, F74]; on 12 Ont, Que [F73, F75,
    F76]; on 14 Ont [F69, F74].
E. stellulata (C:13): canker, chancre
    eutypelléen: on 12 Ont [1340].
Exidia sp.: on 13 NB [1032].
E. glandulosa (E. spiculosa): on 12 Ont
    [1341]. Lignicolous.
E. nucleata (C:13): on 4 BC [1012]; on
    12, 14 Ont [1341]; on 14 NS [1032].
    Lignicolous.
E. recisa: on 13 Ont [1341]. Lignicolous.
Exidiopsis grisea (Sebacina g.): on 13 NB
    [1032, 1036]. Lignicolous.
Favolus alveolaris (C:13): on 12 Ont
    [1341]; on 14 NB [1032]. Associated
    with a white rot.
Fomes fomentarius (C:13): white mottled
    rot, carie blanche madrée: on 10, 12
    Ont [1341]; on 12 Ont [1827], Que [941,

1376], NB, NS [1032]; on 14 Que [1377],
    NS [1032].
Fomitopsis pinicola (Fomes p.) (C:13): on
    10, 14 NB [1032]; on 12, 14 Ont [1341];
    on 12 Que [911] logs. See *Abies*.
Fusarium sp.: on 4 BC [1012]; on 12 Ont
    [94, 92], Que [941, 945 purposely
    wounded].
F. solani (holomorph *Nectria
    haematococca*) (C:13): on 8 Ont [F61,
    F62].
Fusicoccum sp.: on 14 NS [1032].
Ganoderma applanatum (Fomes a.) (C:13):
    white mottled rot, carie blanche
    madrée: on 4 BC [1012]; on 5, 10, 11,
    12, 14 Ont [1341]; on 10 NS, PEI [1032];
    on 12 Ont [92, F63], Que [911 logs,
    1391, 1376, 1378], NB, NS [1032]; on 14
    Ont [1341] Que [1377], NB, NS [1032].
G. lucidum: on 10 NS [1032]; on 14 Ont
    [1341]. See *Abies*.
G. resinaceum (G. sessile): on 10 Ont
    [1341]. Associated with a white rot.
Gliocladium roseum: on 12 Ont [92].
G. viride: on 14 NB [1610].
Gloeocystidiellum karstenii: on 10 NB
    [1032]. Associated with a white rot.
G. leucoxanthum (Corticium l.) (C:12): on
    12 NS [1032]. Associated with a white
    rot.
G. porosum: on 14 Ont [1341]. Associated
    with a white rot.
Gloeophyllum sepiarium (Lenzites s.): on
    4 BC [1012]. See *Abies*.
Gloeoporus dichrous (Polyporus d.)
    (C:15): on 10, 12, 14 Ont [1341]; on 12
    Ont [92], Que [1390]; on 14 Ont [1660],
    Que [568]. Associated with a white rot.
Gloeosporium sp.: on 4 BC [1012]; on 7
    Ont [1827]; on 8, 10, 11, 12, 13, 14 Ont
    [1340]; on 10, 12 Que [F75, F76]; on 10
    NB [1032]; on 12 Ont [F73], Que [1377].
G. decolorans (C:13): on 10 Que [F61]; on
    12 NB [1032]. The application of the
    name is uncertain. Specimens and
    records should be reexamined.
G. oblongisporum: on living leaves of 11
    Ont [439]. A species of *Cylindrium*,
    cf. *C. aerugonosum* [20].
Gnomonia cerastis: on 14 BC [79].
Gomphus floccosus (Cantharellus f.): on
    12 NB [1032].
Graphium sp.: on 12 Que [941]; on 14 NB
    [1610].
G. giganteum (C:13): on 12 Que [F62].
Graphostroma platystoma: on 14 Ont, Que
    [1343].
Gymnopilus spectabilis (Pholiota s.)
    (C:15): on 12 Ont [585], Que [941], NB,
    NS [1032]. Associated with a white rot.
Gymnosporangium cornutum: on 13 Nfld
    [1661]. The host was probably mountain
    ash (*Sorbus*) not mountain maple
    (*Acer*).
Gyromitra infula (Helvella i.): on 14 Ont
    [1828].

Haematostereum rugosum (Stereum r.): on 10 NS and on 12 NB [1032]. See *Abies*.

Hapalopilus nidulans (Polyporus n.) (C:15): on 10 Ont [1341], NS [1032]. See *Abies*.

Hendersonia sp.: on 12 Ont [92]. See comment under *Abies*.

H. sarmentorum (C:13): on 7 NS [1032]. See comment under *Abies*. However, the proper generic disposition for this species name is uncertain.

Hericium americanum (H. coralloides Amer. auct.*):* on 12 Ont [115]; on live or dead trunks of 14 Que [1385].

H. coralloides (H. ramosum) (C:13): white spongy rot, carie blanche spongieuse: on 12 Ont [1341]; on dead trunks of 14 Que [1385].

H. erinaceus (C:13): white spongy rot, carie blanche spongieuse: on 12 Ont [92, 1341], NB [1032]; on 14 NB [1032]. These specimens are probably a form of *H. americanum*. Neither the recent North American monograph of *Hericium* [712] nor the recent Quebec mycoflora [1385] accepted this species as part of the Canadian mycoflora, despite earlier Canadian records.

Herpotrichia pezizula (C:13): on 13 NS [1032].

H. schiedermayeriana: on 13 Ont [1666].

Heterobasidion annosum (Fomes a.) (C:13): on 4 BC [1012]. See *Abies*.

Heterochaetella dubia (C:13): on 14 Ont [1015]. Lignicolous.

Hydnochaete olivaceum: on 10 NS [1032]. Associated with a white rot.

Hygrophorus lignicola: on 12 NS [146], on a recently fallen tree. Perhaps not distinct from *Pleurotus dryinus*.

Hymenochaete corrugata (C:13): on 5, 12, 13 Ont [1341]; on 7, 10 NB [1032]; on 14 NS [1032]. See *Abies*.

H. spreta: on 13 NS [1032]. Associated with a white rot.

H. tabacina (C:13): canker, chancre hyménochétéen: on 4 BC [1012]; on 5, 7, 10, 12, 13 Ont [1341]; on 7, 10, 12, 13 NB, NS [1032]; on 10, 12, 13 PEI [1032]; on 12 Ont [92, 1827], Que [911, 912, 1376], Nfld [1659, 1660, 1661]; on 14 Ont [1341], Que [1377], NS, PEI [1032]. See *Abies*

Hymenoscyphus calyculus (Helotium virgultorum) (C:13): on 4 BC [1012].

H. epiphyllus (Helotium e.) (C:13): on 14 NS [1032].

Hyphoderma heterocystidium (Peniophora h.) (C:14): on 10, 12 Ont [1341]. Lignicolous.

H. litschaueri (Corticium l.) (C:12): on 12 NB [1032]. Lignicolous.

H. mutatum (Peniophora m.) (C:14): on 10 Ont, Que [1341]. Associated with a white rot.

H. praetermissum (H. tenue): on 14 Ont [198]. See *Abies*

H. puberum (Peniophora p.): on 12 NB [1032]; on 14 Ont [1341]. See *Abies*.

H. sambuci (Peniophora s.) (C:14): on 12 Ont [1341, 1828]. Associated with a white rot.

H. setigerum (Peniophora aspera) (C:14): on 4 BC [1012]; on 5, 12 Ont [1341]; on 7 NB [1032]; on 12, 14 NS [1032]. See *Abies*.

Hyphodontia arguta: on 12 Ont [1341]. See *Abies*.

H. crustosa (Odontia c.) (C:14): on 10 NB, NS and on 14 NS [1032]. See *Abies*.

H. spathulata (Odontia s.): on 10 NB, NS and on 12 NB [1032]. See *Abies*.

Hypholoma fasciculare (Naematoloma f.) (C:14): on 4 BC [1012]. See *Abies*.

Hypochnicium analogum: on 4 BC [1012]; on 12 Ont [1341]. Associated with a white rot.

H. bombycinum (Corticium b.) (C:12): on 12 Ont [1341]; on 13 NS [1032]. Associated with a white rot.

H. vellereum (Corticium v.) (C:12): white spongy rot, carie blanche spongieuse: on 10 NB [1032]; on 12 Ont [92, 94, 156, 1341], Que [941, 911 on logs].

Hypocreopsis lichenoides (anamorph *Stromatocrea cerebriforme*): on 13, 14 Ont [230].

Hypoderma rufilabrum (C:13): on 13 Ont [1340, 1828, F60], NS [1032]; on 14 Ont [1340]. Lignicolous.

Hypoxylon sp.: on 10, 13 NB, NS [1032]; on 12 Ont [1827, F63], NS [1032]; on 14 NS, PEI [1032].

H. cohaerens (C:13): on 10, 12 Ont [1340].

H. deustum (Ustulina vulgaris) (C:13): brittle white heart-rot, carie blanche friable: on 4 BC [1012]; on 10, 12 Que [1376]; on 12 Ont [1340], Que [1377], NS [1032]; on 14 Que [F65].

H. fragiforme (C:13): on 10, 12 Ont [1340].

H. fuscum: on 5 Man [F66]; on 12 Ont [1340].

H. mammatum (H. blakei) (C:13): canker, chancre: on 2 Sask [1594, F67]; on 10, 12 Ont [1340], Que [1376]; on 12 Que [1377, 1378].

H. multiforme (C:13): on 7 Ont [1340]; on 13 Nfld [1659, 1660, 1661].

H. rubiginosum (C:13): on 12 Ont [1340]; on 13 NB, NS [1032]; on 14 Ont [1340], NB [1610], PEI [1032].

H. serpens: on 12 Ont [1340].

Hypsizygus ulmarius (Pleurotus u.) (C:15): on 12 Ont [1341]; on 14 Que [1385].

Hysterium pulicare (C:13): on 12 Ont [1340]; on 14 NB [1032].

Inonotus cuticularis (Polyporus c.) (C:15): on 12 Ont [92, 1341]; on 14 BC [1014]. Associated with a white rot.

I. glomeratus (Polyporus g.) (C:15): on 10, 12 Que [1377, 1376]; on 12 Ont [92, 585, 1341, 1827], Que [941, 1390 on dead

trunk], NB [1032]; on 14 Ont [1341], NS
[1032]. Associated with a white rot.
I. radiatus (Polyporus r.) (C:15): on 14
NS [1032]. Associated with a white rot.
I. obliquus (Poria o.): on 10 Que
[1376]. Associated with a white rot.
Irpex lacteus (Polyporus tulipiferae)
(C:15): white spongy rot, carie blanche
spongieuse: on 10, 12, 13 Ont [1341],
NB [1032]; on 10 NS, PEI [1032]; on 12
Ont [92], Que [911 logs, 912]; on 13 Que
[1390]; on 14 Ont [1341], PEI [1032].
Junghuhnia nitida (Poria attenuata, P.
eupora): on 7 NS [1032]; on 10 NB
[1032, 1036]; on 12 PEI [1032]; on 14
Ont [1004]. Associated with a white rot.
Kavinia himantia: on 14 Ont [1341].
Associated with a white rot.
Laeticorticium roseocarneum
(Dendrocorticium r., Stereum r.)
(C:16): on 14 Ont, NB [930], NS
[1032]. Lignicolous.
L. roseum (Aleurodiscus r.): on 10 NB
[1032]. Associated with a white rot.
L. violaceum (Dendrocorticium v.): on 12
Ont [930]. Lignicolous.
Laetiporus sulphureus (Polyporus casearius,
P. sulphureus): on 10, 12 Ont [1341];
on 12 Que [1389] living trunk; on 14 Ont
[1341, 1828], Que [1377]. See *Abies*.
Lasiosphaeria ovina: on 12 Ont [1340].
Leiosphaerella falcata: on small, dead
branches of 14 Que [144].
Lentinellus cochleatus: on 10 PEI
[1032]. Associated with a white rot.
Lenzites sp.: on 12 Ont [92].
L. betulina (C:14): on 10, 14 NS [1032];
on 12 Ont [1341, 1827], NB [1032]; on 14
Ont [1341], Que [1391] on stump.
Associated with a white rot.
Leptographium sp.: on 12 Ont [92].
Leucostoma sp.: on 10 NB [1032].
Libertella sp.: on 12 Ont [92], Que [941].
L. acerina (C:14): on 12 Que [853].
Licrostroma subgiganteum: on 14 Que
[965]. Lignicolous.
Lophiostoma pileatum: on 12 Que [234].
Lophiotricha viridicoma: on 14 Ont [1340].
Lycogala flavofuscum (C:14): slime mold:
on 12 Que [F62]. Probably saprophytic
on rotted wood.
Lycoperdon pyriforme: puff-ball: on 12
Ont [1341]. Probably saprophytic on
rotted wood.
L. umbrinum (C:14): puff-ball: on 12 Ont
[1341]. Probably saprophytic on rotted
wood.
Macrodiaporthe everhartii (Melanconis e.,
Cryptodiaporthe magnispora) (C:14): on
10, 12 Ont [1340, F60], on 10 Ont
[1827], NS [1032]; on 13 Ont [1340], NS
[1032].
Marasmiellus candidus (Marasmius c.): on
4 BC [1012].
Margarinomyces sp.: on 12 Ont [1340].
Marssonina sp.: on 12 Ont [1340].
Massaria sp.: on 12 Ont [1340].

M. inquinans (C:14): dieback,
dépérissement massarien: on 10 Ont
[1340]; on 12 Ont [1340, 1660, 1827,
F71]; on 13 NS [1032]; on 14 Ont [1340,
1626], Que [1377, 1626].
Melanconiella sp.: on 5 Sask [F65].
Melanconiopsis inquinans (holomorph
Massariovalsa sudans) (C:14): canker,
chancre mélanconiopsien: on 4 BC [1012].
Melanconium sp.: on 4 BC [1012].
Melanomma sp.: on 12 Ont [1340].
M. pulvis-pyrius: on 12 Ont [1828].
Melasmia acerina (holomorph *Rhytisma*
acerinum): on 14 Ont [1763].
Menispora glauco-nigra: on 14 Que [826].
M. tortuosa: on 14 Ont, Que [826].
Meruliopsis corium (Merulius c.) (C:14):
on 4 BC [1012, 1341]; on 14 BC [568].
Associated with a white rot.
Merulius tremellosus (C:14): on 4 BC
[568, 1012]; on 10, 12 NB [1032]; on 12
Ont [1341]; on 14 Ont [568, 1828], Que
[568], NB [1610], NS [1032]. See
Abies.
Microdiplodia sp.: on 10 Ont [1827].
Mollisia caespitica: on 14 NS [1032].
M. cinerea (C:14): on 14 Ont [1827].
M. stictella (C:14): on 13 NS [1032].
Mucor sp.: on 12 Que [941, 945 purposely
wounded].
Mycena cyaneobasis: on 14 Ont [1306],
gregarious on rotten wood.
M. meliigena (M. corticola Amer. auct.*)*
(C:14): on 12 Ont [1341], Que [F66].
M. subcaerulea (Hygrophorus miniatus): on
logs of 14 Ont [1306].
Mycoacia fuscoatra: on 10 Ont [1341].
Associated with a white rot.
M. uda: on 4 BC [1012]. Associated with
a white rot.
Mycopappus aceris (Cercosporella a.)
(C:12): on 4 BC [1012].
Myxosporium sp.: on 1, 7, 13 Ont [1340];
on 5 Ont [1827]. These reports may be
of *Discula acerina*.
Nectria sp.: on 4 BC [F65]; on 5, 11, 14
Que [1377]; on 7 Que [1376]; on 10 Ont
[1340]; on 10, 12 Que [1377, 1376], NS
[1032].
N. cinnabarina (C:14): dieback and coral
spot, dépérissement: on 4 BC [1012]; on
5 Alta [F66], Sask [F67, F69], Man
[F67], NB, PEI [1032], NS [336, 1032];
on 8 NB [1032, F63], PEI [1032], Nfld
[334, 336 dead tips of live branches,
338, 1594]; on 9 Nfld [1660, 1661]; on
10, 11, 12, 13 Ont [1340]; on 10 NS
[1032], Nfld [338, 1661, F65]; on 11a
Ont [338]; on 12 Que [1378], NB [1032,
F72]; on 13 Sask [F66], Que [1392]; on
14 Que [1377], NB, NS, PEI [1032], Nfld
[337, 1661, F64].
N. coccinea (C:14): on 1, 4 BC [1012].
N. episphaeria: on 14 Que [F65].
N. galligena (C:14): target canker,
chancre nectrien: on 9 Nfld [F75]; on
10 NB [1032]; on 12 Que [1009, F60,
F66], NS [1032, F64].

N. ?galligena: on 10 Nfld [1660].

N. modesta (C:14): on 14 Ont [1340, F62, F63].

Neobulgaria pura: on 14 Ont [569]. Lignicolous, on branches etc. on the ground.

Nidularia sp.: on 10 Ont [1340]. Lignicolous, on twigs, etc. and wood chips on the ground.

Nodulisporium sp.: on 10 Ont [973]; on 12 Ont [93, 92, 585], Que [941].

Nummularia repanda: on 14 Man, Ont [862].

Oedemium didymum (holomorph *Chaetosphaerella fusca*): on 5 Man [821].

Oidium sp.: powdery mildew, blanc: on 1 BC [1816]. Probably the anamorph of an *Uncinula* or *Phyllactinia*.

Ophiobolus rubellus: on 5 Ont [1620].

Ophiocordyceps clavulata (C:14): on 13 Ont [F61, F62]. Usually on insects.

Orbilia coccinella: on 10 Ont [1340]. This name has been applied to several different fungi.

Otthia hypoxylon (C:14): on 5 Sask [F67].

Oxyporus latemarginata (*Poria l.*): on 12 Ont [1341]. Associated with a white rot.

O. populinus (*Fomes connatus*) (C:13): white spongy rot, carie blanche spongieuse: on 7 Ont [1827], NS [1032]; on 10 Ont [1341], Que [1377, 1376, F67]; on 10, 12, 14 NB, NS, PEI [1032]; on 12 Ont [1341, F69], Que [941, 1377, 1376, 1378, F67]; on 14 BC [1014], Ont [1341, 1828, 1827], Que [1391 dead, 1385]. Often causing heart rot in live trees.

O. similis (*Poria s.*): on 13 NB [1036]. Associated with a white rot.

Pachyella clypeata (*Peziza c.*): on 14 Ont [1340].

Panellus serotinus (*Pleurotus s.*) (C:15): white spongy rot, carie blanche spongieuse: on 10, 12 NB [1032]; on 12 Que [1376]; on 12, 14 NS [1032].

Panus rudis (C:14): on 10, 12 Ont [1341]; on 14 NS [1032]. Associated with a white rot.

Penicillium sp.: on 14 NB [1610].

Peniophora sp.: on purposely wounded 12 Que [945]. Lignicolous.

P. cinerea (C:14): on 10 NB, NS [1032]; on 12 Ont [92], Que [911] logs, NS [1032]; on 12, 13 Ont [1341]; on 14 PEI [1032]. See *Abies*.

P. decorticans: on 4 BC [1012]. Lignicolous.

P. incarnata (C:14): on 4 BC [1012]; on 8, 14 NS [1032]. Lignicolous.

P. nuda: on 8, 14 NS [1032]. Lignicolous.

P. violaceolivida: on 14 NS [1032]. Lignicolous.

Pezicula acericola (C:14): on 7, 12, 13 Ont [1340]; on 12, 13 NS and on 13 NB [1032].

P. carnea (C:14): dieback, dépérissement péziculéen: on 7 Ont [1340]; on 10 NB [1032]; on 14 Que [1392].

Phaeoisaria sparsa: on 13 Sask, Man [1760].

Phanerochaete laevis (*Peniophora affinis*) (C:14): on 10 PEI [1032]; on 12 Ont [1341]. Associated with a white rot.

P. sordida (*Peniophora cremea*) (C:14): on 4 BC [1012]; on 13 PEI [1032]. See *Abies*.

P. tuberculata (*Corticium t.*) (C:12): on 12 Ont [1341]. Lignicolous.

P. velutina (*Peniophora v.*): on 14 Ont [1828]. Associated with a white rot.

Phellinus conchatus (*Fomes c.*): on 4 BC [1012]; on 14 Ont [1341]. Associated with a white rot.

P. ferreus (*Poria f.*) (C:15): white spongy rot, carie blanche spongieuse: on 4 BC [1012]; on 7, 10, 12, 14 NB [1032]; on 10, 12 NS, PEI [1032]; on 14 BC [558], NS [1032].

P. ferruginosus (*Poria f.*) (C:15): white spongy rot, carie blanche spongieuse: on 7 NB [1032]; on 10 PEI [1032]; on 12 Que [1377], NS [1032]; on 14 BC [558], NS [1032].

P. igniarius (*Fomes i.*) (C:13): white trunk rot, carie blanche du tronc: on 3 BC [1012]; on 7 Ont [1341], Que [1376]; on 7, 10, 11, 12, 13, 14 NB [1032]; on 10 Ont [973], Que [1391 dead, 1376]; on 10, 12, 14 NS [1032]; on 12 Ont [92, 585, 1341, 1827], Que [941, 1376, 1378]; on 14 BC [558], Ont [1341, 1827], Que [1377].

P. laevigatus (*Fomes igniarius* var. *l.*, *Poria l.*) (C:13): on 7, 10, 12 NB, NS [1032]; on 10 Ont [1341]; on 14 NS [1032]. Associated with a white rot.

P. punctatus (*Poria p.*) (C:15): white spongy rot, carie blanche spongieuse: on 7, 12 Ont [1341]; on 7, 13 NB [1032].

P. viticola (*Fomes v.*): on 10 NB, NS [1032]; on 12 PEI [1032]; on 14 BC [558], NB [1032]. See *Abies*.

Phialocephala canadensis (C:14): on 14 Que [877].

P. fusca (C:14): on 14 Que [877].

Phialophora sp.: on 12 Ont [93, 92], Que [941, 945 purposely wounded]; on 14 NB [1610].

P. americana: on 14 NB [1610].

P. botulispora: on 12 Ont [279].

P. fastigiata: on 12 Ont [92].

P. lagerbergii: on 12 Ont [279]; on 14 NB [1610].

P. melinii: on 10 NB, NS [279]; on 12 Ont [93, 94, 92, 279]; on 14 NB [1610].

P. verrucosa: on 14 Que [279].

Phibalis furfuracea (*Cenangium f.*, *Encoelia f.*): on 5 Mack [53, F65], NWT-Alta [1063].

Phlebia deflectens (*Corticium d.*): on 12 NS [1032]. Lignicolous.

P. radiata (C:15): on 4 BC [1012]; on 12 Que [568]; on 14 Ont [568, 1828], NS [1032]. See *Abies*.

P. rufa (Merulius r., M. pilosus): on 4
BC [1012]; on 12 Ont [1341, 1660]; on 14
BC, Ont [568]. See *Abies*.

Phleogena faginea (P. decorticata): on 12
Ont [1827]. Lignicolous.

Phloeospora sp.: on 10, 12 NB [1032].

Phloeospora platanoides: leaf spot, tache
des feuilles: on 8 NS [338].

Pholiota sp.: on 14 Ont [463].

P. adiposa: on 12 Ont [1341]; on 14 Que
[1377]. Associated with a white rot.

P. albocrenulata (C:15): on 12 NS [1032];
on stumps of 14 Que [1385]. Associated
with a white rot.

P. alnicola (Flammula a.): on 12 Ont
[92]. See *Abies*.

P. aurivella (C:15 as *P. adiposa*): on
12 Ont [92], Que [941, 949]; on trunks
of 14 Que [1385]. See *Abies*.

P. flammans (P. kauffmaniana): on 12 Ont
[1341]. Associated with a white rot.

P. squarrosoides: on 10, 12 Que [1377];
on 12 Que [1376]. Lignicolous.

Phoma sp.: on 12 Ont [92], Que [941, 945
purposely wounded].

P. aceris-negundinis (C:15): on 5 PEI
[1032].

P. fumosa (C:15): on 5 Sask [F67], Man
[1768, F65, F67].

Phomopsis sp.: on 4 BC [1012]; on 10 Ont
[1340]; on 11, 12 NB [1032]; on 12 Ont
[92, 1340], NS [1032].

P. platanoides: on 12 Ont [1340].

Phyllactinia sp.: powdery mildew, blanc:
on 5 BC [1012].

P. guttata (P. corylea) (C:15): powdery
mildew, blanc: on 14 Que [1377].

Phyllosticta sp.: on 2 Ont [1340].

P. minima (P. acericola) (C:15): leaf
spot, tache des feuilles: on 2, 10, 12,
13 Ont [1340]; on 10 Ont [F69, F73], Que
[F61, F62, F76], NB, NS, PEI [1032,
F65], Nfld [1659, F68, F69, F70, F71];
on 10, 12 Que [1377, F72], NB, NS, PEI
[F64, F66], Nfld [1660, 1661, F65, F74,
F76]; on 12 Ont [339, F76], Que [333,
F73, F74], NB [1032, F74]; on 13 NB
[1032], Nfld [F73].

P. minutella: on 12 Ont [195].

P. negundinis (C:15): leaf spot, tache
des feuilles: on 5 Que [F62].

Phyllotopsis nidulans (Claudopus n.): on
12 Ont [1341]. See *Abies*.

Piggotia negundinis (C:15): leaf spot,
tache des feuilles: on 5 Ont [440], Que
[1377].

Plagiostoma acerophilum (Gnomonia a.):
on 7, 10, 13 Ont, Que [79]; on 14 Que
[141]. On leaf petioles and main veins
of overwintered leaves.

Pleospora sp.: on 4 BC [1012].

Pleuroceras tenella: on 10, 14 Ont [79].
On leaf blades, veins and upper portions
of petioles.

Pleurothecium recurvatum: on 14 Ont [587].

Pleurotus ostreatus (C:15): on 4 BC
[1012]; on 10, 12 Ont [1341]; on 14 Que
[1377]. See *Abies*.

P. strigosus (Panus s.): on 14 Que
[1385], NS [1032]. See *Abies*.

P. subareolatus: on 4 BC [1012].
Associated with a white rot.

Plicatura crispa (P. faginea, Trogia c.)
(C:16): on 10 NB [1032]; on 12 Ont
[1341]; on 14 NB [1036]. Lignicolous.

P. nivea: on 7 Ont [563, 1341].
Associated with a white rot.

Pluteus admirabilis: on 12 Ont [1341].
Lignicolous.

Polyporus sp. (C:15): on logs of 12 Que
[911].

P. badius (P. picipes) (C:15): on 4 BC
[1012]; on 12 Ont [1341]. See *Abies*.

P. brumalis: on 12 Ont [1341], NB [1032];
on 14 Ont [1341, 1828]. Lignicolous.

P. lentus (P. fagicola): on old trunks of
14 Que [1385]. Lignicolous.

P. squamosus: on 5, 12 Que [1377]; on 12
Ont [1341]; on 14 Ont [1828], Que
[1389], NS [1032]. Associated with a
white rot.

P. varius (P. elegans) (C:15): on 7, 12,
13, 14 Ont [1341]; on 12, 14 NS [1032].
See *Abies*.

Poria sp.: on 1 BC [1012]; on 7, 10, 13
NB [1032]; on 10, 12 NS [1032].
Lignicolous.

P. fraxinea (Fomes f.) (C:13): white
spongy rot, carie blanche spongieuse:
on 14 Ont [1828].

P. subacida: on 14 NS [1032]. See
Abies.

*Prosthecium appendiculata (Melanconiella
a.)*: on 10 Que [F67].

P. stylosporum (Pseudovalsa s.) (C:16):
on 12, 13 NS [1032]; on 13 Man [F67],
Ont [1340, F63].

*Protoventuria vancouverensis (Antennularia
v.)* (C:16): on bark of dead tree of 14
BC [73].

Pseudospiropes nodosus: on 14 BC [816].

P. simplex: on 14 Ont, Que [817].

Pulveria porrecta: on 12 Ont [1042].

Pycnoporus cinnabarinus (Polyporus c.)
(C:16): on 10, 14 Ont [1181, 1341], NS
[1032]; on 12 Ont [1341], Que [1390], NB
[1032]. Associated with a white rot.

Pyrenochaeta sp.: on 12 Ont [1827].

Ramaria stricta: on 12 Ont [1341].

Rebentischia sp.: on 12 Que [F70].

Resinicium bicolor (Odontia b.): on 14 NS
[1032]. See *Abies*.

Rhinocladiella sp.: on 12 Ont [92, 585].

R. anceps: on 12 Ont [92].

Rhodotus palmatus: on 10 Que [F71].
Lignicolous.

Rhytisma acerinum (C:16): tar spot, tache
goudronneuse: on 1 NB [1032]; on 2 Man
[334]; on 5 Que [1377]; on 8, 10, 11, 14
Ont [1340]; on 10 Ont [425, 1827, F65,
F66, F67, F68], Que [1340, 1377, F67,
F76], NB [F69], NS [F67, F69] also NB,
NS, PEI [1032, F65, F70], Nfld [1659,
F74]; on 10, 13 Ont [F64, F70], Nfld
[1660, 1661]; on 11 Ont [336], Que [333,
338, 1377], NB [1032, F63], NS, PEI

[1032]; on 12 Ont [F67], Que [1377], NB,
NS [1032]; on 13 NB, PEI [1032], Nfld
[F61, F62, F63, F64, F65]; on 14 Ont
[F69], NB, NS [1032], Nfld [1661].

R. punctatum (C:16): speckled tar spot,
tache goudronneuse ponctuée: on 4 BC
[335, 1012, 1340, 1594, 1816]; on 5 Ont
[1660, 1828, 1827, F69], NB [1032]; on
5, 13 Ont [1340, F64, F65, F67, F68]; on
7 Ont [1340], Que [1377, F75]; on 7, 10,
12, 13 NB, NS [1032, F65], PEI [F65]; on
10 Que [F62]; on 11 Que [1377]; on 12
Ont [1340, F70], Que [1377, F72, F75];
on 13 Man [F68], Ont [425, F66, F70],
Que [F67, F75], Nfld [1659, 1660, 1661,
F61, F62, F64, F65, F68, F73, F74]; on
14 Alta [338], Ont [1827], Nfld [1661].

Rigidoporus nigrescens (Poria n.): on 14
Ont [1341]. Associated with a white rot.

Rutstroemia setulata (Ombrophila s.)
(C:16): on 13 Ont [1340, F60].

Sarcoscypha coccinea: on 14 NB [1036].
Apparently saprophytic on fallen
branches.

S. occidentalis: on fallen trunks of 14
Que [1385].

Schizophyllum commune (C:16): white
spongy rot, carie blanche spongieuse:
on 1, 4 BC [1012]; on 10, 12, 14 Ont
[1341], NB, NS [1032]; on 12 Que [565];
on 14 Ont [565].

Schizopora paradoxa (Poria versipora)
(C:16): on 4 BC [1012]. See *Abies*.

Scutellinia scutellata: on 12 NB [1032];
on 14 Ont [1340].

Scytinostroma galactinum: on 10 PEI
[1032]; on 12 Ont [1341]. See *Abies*.

Sebacina epigaea (C:16): on 14 NS
[1032]. Associated with a white rot.

Septobasidium sp.: on 13 Ont [1828].
Probably on insects on *Acer*.

Septogloeum sp.: on 5 Sask [F60].
Possibly a specimen of *Diplodina
acerina*.

Septoria sp.: on 10 Que [F61]; on 12 Ont
[1340].

*S. aceris (Cylindrosporium acerinum,
Phloeospora canadensis, P. aceris)*
(C:15): small leaf spot, tache petite
des feuilles: on 1, 3 BC [1012, 1816];
on 3 BC [644, 953, F73]; on 4 BC [1816];
on 7 Ont [195]; on 7, 10, 12 NB, NS, PEI
[1032, F64]; on 7, 10, 12, 13 Ont
[1340]; on 10 NB, NS, PEI [F65, F66]; on
12 Que [334, 339, F62, F72], NB, NS, PEI
[F66]; on 12a NB [1032]; on 13 Que [282,
F62], NB, NS [1032], Nfld [1661]; on 14
NB, NS, PEI [F66].

S. belonidium: on living leaves of 11 Ont
[439].

S. negundinis (C:16): leaf spot, tache
des feuilles: on 5 Sask [F65, F71], Man
[F68], Que [1377].

Sistotrema brinkmannii: on 12 Ont
[1827]. See *Abies*.

Spadicoides atra: on 14 Ont, Que [791].

Sphaeronaema acerinum (holomorph *Dermea
acerina*) (C:16): on 12 Ont, NS [1340].

Sphaeropsis sp.: on 12 Ont [92, 1340].

S. albescens (C:16): dieback,
dépérissement sphéropsien: on 5 Sask
[F65, F67], Man [F67].

S. clintonii (C:16): on 13 [1340, F60].

Sporidesmium foliculatum: on 14 Ont
[825], Que [1018].

Stachybotrys atra: on 12 Ont [92].

Stagonospora collapsa: on 10 Ont [1340].

Steccherinum sp.: on 12 Ont [1341].
Lignicolous.

S. ciliolatum (Odontia c.): on 12 PEI
[1032]; on 14 Ont [1828]. Associated
with a white rot.

S. fimbriatum (Odontia f.) (C:14): on 4
BC [1012]; on 14 Ont [1341]. Associated
with a white rot.

S. ochraceum (C:16): on 4 BC [1012]; on
14 Ont [1827]. Associated with a white
rot.

Stegonsporium acerinum: on 12 Ont [92,
147 bark and branches].

S. pyriforme (S. ovatum) (C:16): canker,
chancre stéganosporien: on 8, 10, 11,
12 Ont [1340]; on 8 NS [1032]; on 10 NB
[1032]; on 12 Ont [646, 1827, F64, F67,
F69, F72], Que [335, F69], NS [F74] and
NB, NS, PEI [1032]; on 14 Ont, Que
[1625, F60], NB, NS [1032, 1625] and Que
[F66].

Stereum sp.: on 12 Ont [1827].
Lignicolous.

S. complicatum (C:16): on 12 Ont [1341].
Lignicolous.

S. hirsutum (C:16): on 10 Ont [1341]; on
12 Ont [1340, 1827]; on 14 NS [1032].
See *Abies*.

S. ochraceo-flavum (C:16): on 4 BC
[1012]. Lignicolous.

S. ostrea (C:16): on 7 NB, NS [1032]; on
10, 11 Ont [1341]; on 12 NB, NS, PEI
[1032]; on 13 NB [1032]; on 14 Ont
[1341], NS [1032]. See *Abies*.

Stictis radiata (C:16): on 12, 13 NS
[1032].

*Stigmina negundinis (Coryneum n.,
Sciniatosporium n.)* (C:16): twig
blight or canker, brûlure des rameaux:
on 5 BC, Man [1066], Alta, Sask, Man,
Ont, NB [1127], Sask, Man [1760], Sask,
Man, Ont [F65, F67], NB, NS, PEI [1032];
also on 5 Ont [1340], NB [F73], NS
[F63]; on 5a Sask [1760].

Streptomyces sp.: on wood chips of 14 NB
[1610].

Stromatocrea cerebriforme (holomorph
Hypocreopsis lichenoides): on 13, 14
Que [230].

Taphrina sp.: on 10 NB, NS [F69]; on 14
PEI [1032].

T. darkeri (C:16): leaf blister, cloque
des feuilles: on 3 BC [1816, F60].

T. dearnessii (C:16): leaf blister,
cloque des feuilles: on 10 Que [F74,
F75, F76], NB, NS [1032, F65, F66] and
NB [F74]; on 12, 14 NB [1032].

T. letifera (C:16): leaf blister, cloque
des feuilles: on 7 NS [1032]; on 13 Ont
[1340], NB [1032, F65].

T. sacchari: leaf blister, cloque des
feuilles: on 12 Que [1377, F73].

Tetragoniomyces uliginosus (Tremella u.):
on rotting petioles of 4 BC [900].
Mycoparasitic on basidiomycetous
sclerotia perhaps belonging to a species
of *Thanatephorus* [1187].
Thelephora sp.: on 14 Ont [1828].
Thyrsidina sp.: on 12 Ont [1827].
Tomentella sp.: on 14 Ont [1341].
T. botryoides: on 14 Ont [1341, 1828, 1827].
T. bryophila (T. pallidofulva): on 14 Ont [1341, 1828, 1827].
T. cinerascens: on 14 Ont [1341].
T. crinalis: on 14 Ont [1828].
T. neobourdotii: on 14 Ont [1341, 1828].
T. olivascens: on 14 Ont [1341].
T. spongiosa: on 10 NS [1032].
T. sublilacina: on 14 Ont [1341, 1828].
T. umbrinospora: on 14 Ont [1341, 1828].
T. viridescens: on 14 Ont [1341].
Torula ligniperda: on 12 Que [1378]; on 14 Que [1377].
Trechispora farinacea (Cristella f.): on 14 Ont [1341]. See *Abies*.
T. vaga: on 4 BC [1012, 970]. See *Abies*.
Trematosphaeria sp.: on 10 Ont [1340].
Tremella lutescens: on 14 NS [1032]. Lignicolous.
T. mesenterica: on 13 NS [1032]. Associated with a white rot.
Trichaptum biformis (Polyporus pargamenus) (C:15): white spongy rot, carie blanche spongieuse: on 7, 10 NB, NS [1032]; on 10, 12 Ont [1341]; on 11, 12 14 PEI [1032]; on 12 Que [1390]; on 13 NB [1032]; on 14 NWT-Alta [1063], Que [1377].
Trichocladium canadense: on 12 Ont [93, 94, 92, 585], Que [941, 945]; on 14 NB [1610].
T. opacum: on 13 Sask [1760].
Trichoderma sp.: on 12 Que [941].
T. lignorum: on 12 Ont [585]. Probably a specimen of *T. viride*.
T. viride: on 12 Ont [92], Que [945] purposely wounded; on 14 NB [1610]. See *Abies*.
Trichothecium roseum: on 5 Ont [1827].
Tricladium splendens: on 4 BC [43].
Tubercularia ulmea: on 2 Sask, Man [F68]; on 5 Sask, Man [F67].
T. vulgaris (holomorph *Nectria cinnabarina*) (C:16): on 4 BC [1012]; on 5 NB, PEI [F64]; on 12 Ont [1340, 1828].
Tubulicrinis glebulosus (Peniophora gracillima) (C:14): on 14 NS [1032]. See *Abies*.
Tympanis acericola (C:16): on 13 NS [1032, 1925].
T. pulchella: on 7 Que [1221].
T. truncatula: on 14 Que [1221].
Typhula erythropus: on petioles and blades of fallen leaves of 4 BC [901].
Tyromyces caesius (Polyporus c.) (C:15): on 4 BC [1012]; on 14 Ont [1341]. See *Abies*.

T. chioneus (Polyporus albellus) (C:15): on 12 Ont [1341], NS [1032]; on 14 Ont [1341, F63]. See *Abies*.
T. galactinus (Polyporus g.): on 14 Ont [1341]. See *Betula*.
T. stipticus (Polyporus immitis): on 14 Ont [1341]. See *Abies*.
Uncinula bicornis (C:16): powdery mildew, blanc: on 4 BC [335, 1012, 1257].
U. circinata (C:16): powdery mildew, blanc: on 10, 11, 12, 13 Ont [1257]; on 10, 13 Ont [1340], NS [1032, F61]; on 10, 12, 13 Que [1257]; on 12 Que [1377].
Valdensia heterodoxa (Asterobolus gaultheriae) (holomorph *Valdensinia h.*): on 4 BC [1438].
Valsa sp.: on 1 BC [193]; on 10 Ont [1340]; on 12 NB [1032].
V. ambiens (V. sordida) (anamorph *Cytospora a.*): on 10 NB [1032]; on 12 Que [1377].
V. clavigera: on 1 BC [193].
V. etherialis (C:16): on 12 NS [1032].
Vararia effuscata (C:16): on 10 Ont [557]. Associated with a white rot.
V. investiens (C:16): on 12 Ont [1341]. See *Abies*.
V. pallescens: on 12 Ont [1341]. Associated with a white rot.
Varicosporium elodeae: on 4 BC [43].
Venturia acerina: on 10, 12 Ont [1340]; on 10 NB [1032].
Verticillium sp. (C:16): wilt, flétrissure: on 6a BC [1816]; on 8 Que [1377]; on 12 Ont [94, 92], Que [941]; on 14 Ont [97, 96, 337].
V. albo-atrum: wilt, flétrissure: on 8 Ont [336]; on 8, 10, 12 Ont [F64]; on 12 Ont [1827], Que [336, F63]; on 14 Ont [38, 333, F69], Que [F69], PEI [1032].
V. dahliae (C:16): wilt, flétrissure: on 3, 4 BC [1978]; on 8 BC [1595]; on 12 Ont [676], Que [F73]; on 14 BC [1012], Que [F69].
Volvariella bombycina: on 12 Que [1385] on wounds. Lignicolous.
Wardomyces hughesii: on 14 Que [717].
Xenasma praeteritum: on 14 Ont [967]. Lignicolous.
X. rallum: on 14 Ont [967]. Lignicolous.
X. tulasnelloideum: on 14 Ont [967]. Lignicolous.
Xylaria sp.: on 1 BC [1012]; on 12 Ont [1340]; on 14 NB [1032].
X. hypoxylon: on 4 BC [1012].
X. polymorpha: on 5, 14 Ont [1340]; on 12 Ont [1340, 1828]; on 14 Que [1377].

ACHILLEA L.   COMPOSITAE

1. *A. millefolium* L., common yarrow, achillée millefeuille; imported from Europe and w. Asia, natzd. as weed in all regions.
2. *A. nigrescens* (E. Mey.) Rydb. (*A. borealis* Bong.), northern yarrow, herbe à dindes; an arctic-alpine species, Yukon-Nfld.
3. *A. ptarmica* L., sneezewort yarrow, achillée ptarmique; imported from Europe

and Asia, natzd., BC, Alta, Ont, Que, NS, Nfld and Labr.

Erysiphe cichoracearum: powdery mildew, blanc: on 1 Ont, Que [1257].
Leptosphaeria doliolum (C:17): on 1 Que [70].
L. millefolii (C:17): on 1 Ont [760].
Ophiobolus anguillidus: on dead stems of 1 Ont [1620].
O. erythrosporus (C:17): on *A.* sp. Que [70].
O. mathieui: on 1 Que [1620].
Puccinia millefolii (C:17): rust, rouille: III on 1 Yukon [460], Que [1236]; on 3 NB [333, 1194].
Venturia centaureae (C:17): on overwintered leaves of 2 Labr [73].
Wentiomyces fimbriatus (Venturia f.) (C:17): on 1 Que [70, 73].

ACHLYS DC.   BERBERIDACEAE

1. *A. triphylla* (Smith) DC., vanilla leaf, achlyde; native, BC and grown in gardens across Canada.

Ascochyta achlyicola: leaf spot on 1 BC [1816].
Rhizoctonia sp.: on 1 BC [1012].
Stagonospora achlydis (Ascochyta a.) (C:17): on 1 BC [1012].

ACORUS L.   ARACEAE

1. *A. calamus* L., sweet flag, acorus roseau; native, s. Mack, BC-PEI.

Uromyces sparganii (C:17): rust, rouille: II III on 1 Ont [1236].
U. sparganii spp. *sparganii* (C:17): rust, rouille: on 1 Ont, Que, NS [1287].

ACTAEA L.   RANUNCULACEAE

1. *A. pachypoda* Ellis, white baneberry, actée à gros pédicelles; native, Ont-PEI.
2. *A. rubra* (Ait.) Willd., red baneberry, actée rouge; native, BC-Nfld. 2a. *A. r.* ssp. *arguta* (Nutt.) Hult.; native, Yukon, BC-Alta.

Puccinia recondita s. l. (C:18): rust, rouille: 0 I on 1 Ont [1236]; on 2 BC [1012], Que [1236]; on 2a BC [1816].

AESCULUS L.   HIPPOCASTANACEAE

1. *A. carnea* Hayne (*A. hippocastanum* x *A. pavia* L.), red horse chestnut.
2. *A. hippocastanum* L., horse-chestnut, marronnier d'Inde; imported from Balkan peninsula.

Bjerkandera fumosa (Polyporus f.) (C:18): on 2 PEI [1032]. See *Acer*.
Chondrostereum purpureum (Stereum p.) (C:18): on 2 BC [1012, 1816]. It often fruits on wounds on live trees. See *Abies*.

Cladosporium herbarum: on 2 BC [1012, F72].
Coriolus versicolor (Polyporus v.): on 2 BC [1012, 1816]. See *Abies*.
Cryptodiaporthe aesculi: on 2 NB [F65, 1032].
Cylindrobasidium evolvens (Corticium laeve): on 2 NS [1032]. Apparently saprophytic on dead branches and stems. See *Abies*.
Cytospora sp.: on 2 NB [1032].
Gloeosporium sp.: on 2 NS [1032].
Guignardia aesculi (anamorph *Phyllosticta sphaeropsoidea*) (C:18): leaf blotch: on 2 Ont [1340], Que [1377], NS [F60, F63, F64, F70, F71, F72, 337, 338, 339], Nfld [1661]; on 2 NB, NS [F65, F66]; on 2 NB, NS, PEI [335, 1032, F61, F68, F69, F73]; on 2 NS, PEI [F62, F67, F75, F76, 333, 334, 336].
Haematostereum rugosum (Stereum r.): on 2 NS [1032]. See *Abies*.
Hyphoderma mutatum (Peniophora m.): on 2 NS [1032]. Lignicolous, typically on logs.
Hypoxylon deustum (C:18): on 2 NS [1032].
Laxitextum bicolor: on 2 NS [1032]. Lignicolous, typically on logs.
Nectria cinnabarina (anamorph *Tubercularia vulgaris*) (C:18): on 2 BC [1012, 1816], Ont [336], Que [336, 1377], NB, NS [1032], Nfld [334, 1594].
N. coccinea var. *faginata* (C:18): on 2 NB [1032].
Oxyporus populinus (Fomes connatus): on 2 NS [1032]. It often fruits on wounds on live trees. See *Acer*.
Phyllosticta paviae: on 2 Ont [1340, F73, F76]. This name is of uncertain application. Specimens cited are probably referable to *P. sphaeropsoidea* (holomorph *Guignardia aesculi*).
Pleospora sp.: on 2 BC [1012].
Pleurotus sapidus: on 2 NB [1032]. Causes a white rot of wood. It often fruits on wounds on live trees.
Poria sp.: on 2 NS [1032].
Schizophyllum commune: on 2 NS [1032]. Apparently saprophytic on dead branches and stems. See *Abies*.
Sclerotinia sclerotiorum: on 2 NS [605].
Tubercularia sp.: on 2 BC [1012]; on *A.* sp. Ont [1340].
Tubercularia vulgaris (holomorph *Nectria cinnabarina*): on 2 Ont [1340].
Uncinula flexuosa (C:18): powdery mildew, blanc: on 1 Ont [1257]; on 2 NS [F65, 1032].

AGOSERIS Raf.   COMPOSITAE

1. *A. glauca* (Pursh) Raf.; native, BC-w. Ont.

Puccinia columbiensis (C:19): rust, rouille: III on 1 Alta [644].
P. hieracii (C:19): rust, rouille: II III on 1 NWT-Alta [644, 1063].

AGRIMONIA L.   ROSACEAE

1. *A. eupatoria* L., aigremoine eupatoire;
   imported from Europe, w. Asia and n.
   Africa.
2. *A. gryposepala* Wallr., agrimony,
   aigremoine à sépales crochus; native,
   BC, Ont, Que, NS, PEI.
3. *A. pubescens* Wallr.; native, s. Ont.
4. *A. striata* Michx., grooved agrimony,
   aigremoine strieé; native, BC-Nfld.

Gnomonia comari (Apiognomonia guttulata)
   (anamorph *Zythia fragariae*) (C:19):
   on *A.* sp. NS [79].
Pucciniastrum agrimoniae (C:19): rust,
   rouille: II III on 2 Ont [1828]; on 2,
   3, 4 Ont [1236], Que [282]; on *A.* sp.
   Alta, Man [2008].
Sphaerotheca macularis (C:19): powdery
   mildew, blanc: on 1 Ont [1257]; on 2, 3
   Ont, Que [1254, 1257]; on 4 Alta [1254],
   Ont, Que [1257].

x *AGROHORDEUM*   GRAMINEAE

1. *A. macounii* (Vasey) Lepage (*Elymus
   m.* Vasey), Macoun's wild rye; hybrid
   from *Agropyron trachycaulum* var.
   trachycaulum x *Hordeum jubatum*;
   native, Yukon, Mack, BC-Man.

*Pseudoseptoria stomaticola (Selenophoma
   donacis)* (C:107): on 1 BC [1728].

AGROPYRON Gaertn.   GRAMINEAE

   New scientific data has resulted in much
spirited discussion and several publications
on the number of genera and their
circumscriptions in this group of the
Gramineae. The genera *Agropyron* and
Elymus are treated herein in a restricted
sense and several segregate genera are
recognized. The scheme of Dewey, Crop
Science 23: 637-642, 1983, has been
adopted. Recognition of the genus *Leymus*
seems justified on cytogenetic grounds. All
species of *Leymus* appear to be built on
two basic genomes, the J and X. They are
distinctly different from the S and H
genomes of *Elymus* and from the C genome of
Agropyron.

   The recognition of segregate host plant
genera was done with the hope that the
occurrence of parasitic fungi only on
certain genera or certain species would
provide evidence supporting or denying the
generic limits of the Gramineae recognized
herein. The recognition that parasitic
fungi (esp. the rust fungi) have coevolved
with their host plants has shown that
certain species (plant and fungus) were
misclassified.

1. *A. cristatum* (L.) Gaertn., crested
   wheatgrass, chiendent à crête; imported
   from Eurasia, natzd., Yukon, BC-Que, NS.

2. *A. desertorum* (Fisch.) Schult. (*A.
   sibiricum* auct.), desert wheatgrass,
   chiendent désertique; imported from
   Eurasia, natzd., Mack, BC-Ont.
3 Host species not named, i.e., reported
   as *Agropyron* sp. Some of these
   records may refer to species of
   Elymus, *Elytrigia* or *Pascopyrum*.

Acremonium boreale: on 1 BC, Alta, Sask
   [1707].
Ascochyta agropyrina var. *nana*: on 3
   Sask [1414].
A. leptospora: on 3 Sask [1414].
*Belonioscypha culmicola (Belonidium
   vexatum)*: on 3 Ont [228].
*Bipolaris sorokiniana (Helminthosporium
   sativum)* (holomorph *Cochliobolus
   sativus*) (C:20): foot rot, piétin
   commun: on 1 Alta [120, 133], Sask
   [961].
Cheilaria agrostis (Septogloeum oxysporum)
   (C:21): char spot: on 1 Sask [1700].
Claviceps purpurea (C:20): ergot, ergot:
   on 1 BC [1816]; on 1, 2 Sask [338]; on 1
   Alta [133]; on 3 Alta, NB [333].
Drechslera tritici-repentis (holomorph
   Pyrenophora tritici-repentis) (C:20):
   on 3 Man [1611].
Phloeospora idahoensis: stem eye spot,
   tache ocellée de la tige: on 3 Alta
   [1708].
Physarum sp.: slime mold: on 1 Sask
   [338].
Puccinia coronata (C:21): rust, rouille:
   on 3 Ont [1236].
P. graminis f. sp. *secalis* (C:21 under
   Puccinia g.): rye stem rust: on 3
   Alta, Ont [635].
P. striiformis (C:21): stripe rust,
   rouille striée: on 1 BC [1816].
Pyrenophora tritici-repentis (anamorph
   Drechslera t.): leaf spot: on 3 Sask
   [773, 1369].
Rhynchosporium secalis (C:21): scald: on
   3 BC [333].
Sclerotinia borealis (C:21): snow mold,
   moisissure nivéale: on 2 BC [1816].
Septoria elymi (C:21): leaf spot, tache
   septorienne: on 3 Alta [333].

AGROSTIS L.   GRAMINEAE

1. *A. borealis* Hartm., northern agrostis,
   agrostis boréal; native, BC-Que, Nfld,
   Labr.
2. *A. canina* L., velvet bent, agrostide
   des chiens; imported from Europe,
   natzd., BC-Nfld.
3. *A. capillaris* L. (*A. tenuis* Sibth.),
   colonial bent, agrostide commun;
   imported from Eurasia and N. Africa,
   natzd., BC, Ont-NS.
4. *A. exarata* Trin., spike red top;
   native, Mack, BC-Sask.
5. *A. gigantea* Roth (*A. alba* Am.
   auct.), red top, foin follette; imported
   from Eurasia, natzd., Yukon, BC-NS.

6. *A. perennans* (Walt.) Tuckerm., autumn bent, agrostis pérennant; native, Ont-Que.
7. *A. scabra* Willd. (*A. hyemalis* (Walt.) BSP), hairgrass, foin fore; native, Yukon, BC-Nfld.
8. *A. stolonifera* L., creeping bent, agrostis stolonifère; imported from Eurasia, natzd., BC-Nfld. 8a. *A. s.* var. *palustris (A. palustris* Huds.*)*, agrostide de marais; natzd., BC-Nfld.
9. *A. thurberiana* Hitchc., Thurber red top; native, BC-Alta.
10. Host species not named, i.e., reported as *Agrostis* sp.

Acremonium boreale: on 8 BC, Alta, Sask [1707].
Bipolaris sorokiniana (holomorph *Cochliobolus sativus*): on 3, 8a Alta [120] by inoculation.
Cheilaria agrostis (Septogloeum oxysporum) (C:23): leaf blotch, tache foliaire: on 8 Que [333].
Claviceps purpurea (C:22): ergot, ergot: on 5 Ont [335].
Dilophospora alopecuri (C:22): on 3 NB [333, 1195].
Drechslera catenaria (C:22): on 5 Alta [1611].
D. erythrospila (C:22): on 3 Ont [1611].
D. fugax (C:22): on 3 Ont [1611].
D. phlei (C:22): on 5 Ont [1611].
D. tritici-repentis (C:22): on 5, 7 Ont [1611].
Erysiphe graminis (C:22): powdery mildew, blanc: on 3 BC [1849]; on 4 BC [1729]; on 10 BC [334, 1816, 1849].
Fusarium sp.: on 3, 10 BC [1849].
F. nivale (C:22): pink snow mold, moisissure des neiges: on 3 BC [1849]; on 8a Alta [956, 1594], Ont [533]; on 10 BC [1849], Man [1367]. Unsettled taxonomic and nomenclatoral problems concerning the name *F. nivale* led me to cite the name as it has tradionally appeared in the Canadian literature.
Gaeumannomyces graminis (Ophiobolus g.) (C:22): on 3 BC [1201].
G. graminis var. *tritici*: on 3 BC [1619].
Helminthosporium sp.: on 3, 10 BC [1849]. Probably referable to either *Drechslera* or *Bipolaris*.
Laetisaria fuciforme (Corticium f.): on 3, 10 BC [1849]. One of several species which cause pink disease of turf grasses.
Leptosphaerulina australis: on 3 BC [1201].
Low-temperature basidiomycete (C:22): snow mold, moisure des neiges: on 3, 8a Alta [333]. See comment under *Medicago*.
Mastigosporium rubricosum (C:22): eye spot, tache ocellée: on 8 NB [333, 1195].
Olpidium brassicae: in roots of 8a Ont [68].

Phyllachora graminis (C:22): on 3, 5 NB [1195]; on 10 NB, NS [333].
Pseudorobillarda agrostis (Robillarda a.): on 3 BC [1201].
Puccinia sp.: rust, rouille: on 3, 10 BC [1849].
P. coronata (C:22): rust, rouille: on 5 BC [1012]; II III on 8 Que [1236].
P. graminis (C:22): rust, rouille: on 5, 7 BC [1816], Ont [1236]; on 7 BC [1012]; on 8 NB [333, 1195].
P. graminis f. sp. *agrostidis* (C:22): rust, rouille: on 10 Alta [334].
P. poae-nemoralis (P. brachypodii var. *p.)* (C:22): rust, rouille: II III on 1 Que [1236].
P. praegracilis var. *praegracilis*: rust, rouille: on 9 BC, Alta [1578].
P. recondita s. 1. (C:22): rust, rouille: on 3, 7 BC [1816]; on 6 NB [333, 1195]; on 7 BC [1012].
Ramularia pusilla (C:23): leaf spot, tache des feuilles: on 3, 6, 8 NB [333, 1195].
Rhizoctonia solani (holomorph *Thanatephorus cucumeris*) (C:23): brown patch, plaque brune: on 3 BC [1849]; on 10 Man [1367].
Sclerotinia borealis (C:23): snow mold, moisissure des neiges: on 2 BC [1816]; on 10 Sask [1705].
S. homeocarpa (C:23): dollar spot, brûlure en plaques: on 10 BC [1849].
Septoria calamagrostidis (C:23): on 7 BC [1729].
S. triseti (C:23): leaf mottle: on 5 BC [1816].
Stagonospora avenae (Septoria a.) (holomorph *Phaeosphaeria avenaria*) (C:23): on 3 BC [1729].
Typhula sp. (C:23): snow mold, moisissure des neiges: on 8a Ont [533], Que [1594].
T. incarnata (C:23 under *T.* spp.): snow mold, moisissure des neiges: on 8a Ont [533].
T. ishikariensis var. *canadensis*: snow mold, moisissure des neiges: on 8a Ont [533].
T. ishikariensis var. *ishikariensis*: snow mold, moisissure des neiges: on 8a Ont [533].
Ustilago salvei (U. striiformis auct.) (C:23): stripe smut, charbon strié: on 5 BC [1816].

AILANTHUS Desf.   SIMARUBACEAE

1. *A. altissima* (Mill.) Swingle, tree of heaven, ailante; imported from China, spread from cult., Ont-Que.

Armillaria mellea (C:23): root rot, pourridié-agaric: on 1 BC [1816].
Micropera sp.: on *A.* sp. BC [1012].

ALCEA L.   MALVACEAE

Biennial or short-lived perennial herbs from the e. Mediterranean region and c.

Asia. Some species frequently grown as ornamentals.

1. *A. ficifolia* L. (*Althaea f.* (L.) Cav.); identity of species not established with any certainty.
2. *A. rosea* L. (*Althaea r.* (L.) Cav.), hollyhock, rose-trémière; imported from Asia Minor.

Puccinia malvacearum (C:30): rust, rouille: III on 1, 2 Ont [1236]; on 2 BC, Que [333, 334, 335, 336, 337, 338, 1281]; on 2 BC [339, 1012, 1816], Alta [335, 338], Ont [1341], NB [337, 339], NS [1594]; on 2 Alta, Sask, Ont, NS, PEI [1281], Man, NB, PEI [333], also NB, NS, PEI [334], NB, NS [1036]; on *A.* sp. NB [1194].

Sclerotinia sp.: on 1 Sask [420, 1138], Man [1138].

S. sclerotiorum (C:30): sclerotinia stem rot, pourriture sclérotique: on 2 BC [337], Sask [336].

ALLIUM L.   LILIACEAE

1. *A. acuminatum* Hook., Hooker's onion; native, BC.
2. *A. ampeloprasum* L., leek, poireau; imported from Eurasia and N. Africa.
3. *A. amplectens* Torr.; imported from s. Pacific coast of North America.
4. *A. cepa* L., onion, oignon; known only from cultivation.
5. *A. cernuum* Roth, nodding onion; native, BC-Sask.
6. *A. geyeri* S. Wats.; native, Alta.
7. *A. sativum* L., garlic, ail; imported, perhaps derived from an Asiatic species.
8. *A. schoenoprasum* L., chives, ciboulette; imported from Eurasia.
9. *A. textile* A. Nels. & Macbr., prairie onion; native, Alta-Ont.

Alternaria alternata (*A. tenuis* auct.): on 8 Ont [1340].

Alternaria porri (C:24): purple blotch, tache pourpre: on 4 BC [1202], Man [336], Ont [334, 1460, 1461], Que [333, 335, 338, 339, 348, 1654, 1639, 1641, 1647, 1652, 1653, 1595], and Ont, Que [1594].

Aspergillus niger (C:24): black mold, moisissure noire du bulbe: on 4 BC [1816].

Botrytis sp. (C:24): on 4 BC [1202, 1817, 1818], Man, NS [335], Ont [335, 1460, 1461, 1594], Que [335, 348, 1640].

B. allii (C:24): neck rot, pourriture du col: on 4 BC [335, 1816], Alta [339], Man [333], Ont [1450], Que [334, 338], NB [339], PEI [337], also BC, Alta, Ont [1594], BC, Sask, Man [336], BC, NS [334, 337, 338, 339].

B. cinerea (holomorph *Botryotinia fuckeliana*) (C:24): on 2 BC [1816]; on

4 Ont [337], Que [335, 1651, 1653, 1595], NB [338], Que, NS [334, 336, 337, 338].

B. squamosa (holomorph *Sclerotinia s.*) (C:24): on 4 Ont [1774], Que [338, 339, 348, 1650, 1654, 1671, 1641, 1651, 1647, 1652, 1653, 1594, 1786, 1787].

Colletotrichum circinans (C:24): on 4 Ont [333, 337], Que [333]; on 8 Ont [1340].

Doratomyces purpureofuscus: on *A.* sp. Ont [1148] on seeds.

Fusarium sp. (C:24): on 4 BC [1202, 1816], Ont [1460, 1461, 1594].

Fusarium spp. - *Pythium* spp. complex: on 4 Ont [1461, 1594].

F. acuminatum (C:24 under *F.* spp.): on 4 BC [1816].

F. oxysporum f. sp. *cepae* (C:24): on 4 BC, Man, NS [334]; on 4 BC [1595, 1816, 1979], Ont [339], Que [348, 1641, 1651, 1652], NS [333, 336]; also BC, Que [333, 335, 336, 338, 339], BC, Ont [337, 1594].

F. solani (holomorph *Nectria haematococca*) (C:24): on 4 BC [335, 336, 337, 1816, 1979].

Heterosporium allii (C:24): leaf blight, brûlure hétérosporienne: on 2, 4 BC [1816].

Penicillium sp. (C:24): on 4 BC [1594], Ont [1460].

Peronospora destructor (C:24): downy mildew, mildiou: on 4 BC [337, 1202, 1816, 1817, 1818], Ont [333, 1595], Que [334, 335, 1305, 1650, 1641, 1640, 1651, 1653] also BC, Que [333, 336, 338].

Puccinia allii: rust, rouille: on imported plants of 7 Que [1250]; on 8 Ont [1341].

P. blasdalei (C:24): rust, rouille: on 1, 3 BC [1545, 1816]; on *A.* sp. Alta [952].

P. granulispora (C:24): rust, rouille: on 5 BC [1012, 1816] and BC, Alta [644, 1545].

P. mixta (C:24): rust, rouille: on 4, 8 BC [1545, 1816]; on 8 BC, Ont [1556].

P. mutabilis (C:25): rust, rouille: on 6, 9 Alta [1545].

Pyrenochaeta terrestris (C:25): pink root, racine rose: on 4 BC [337, 339, 1202, 1979], Ont [335, 336], Que [338, 339, 1647, 1652].

Pythium sp.: on 4 BC [334, 1816], Ont [1595].

P. intermedium: on 4 Ont [868].

P. irregulare (C:25): on 4 Ont [868].

P. paroecandrum: on 4 Ont [868].

Sclerotium cepivorum (C:25): white rot, pourriture blanche: on 4 BC [337, 338, 339, 1199, 1202, 1594, 1816, 1845, 1843, 1979], Que [335, 338, 339, 348, 1641, 1647, 1652, 1653]; on 7 BC [1816], Que [333]; on *A.* sp. BC [1844].

Stemphylium botryosum (holomorph *Pleospora herbarum*) (C:25): on 2, 4 Ont [667].

Stilbella flavescens: on 4 Que [2].

Ulocladium consortiale (*Stemphylium c.*) (C:24): from seeds of 4 Ont [667].

Urocystis magica (U. cepulae) (C:25):
    smut, charbon: on 4 BC [334, 336, 337,
    1202, 1545, 1816, 1817, 1818, 1979], Man
    [333, 335], Ont [337, 1460, 1461], Que
    [345, 346, 344, 342, 348, 343, 1305,
    1654, 1639, 1647, 1652, 1653, 1788],
    also BC, Ont, Que [1594], BC, Que [333,
    338, 339].
Uromyces aemulus (C:25):   rust, rouille:
    on 1 BC [1545, 1816].

ALNUS B. Ehr.   BETULACEAE

1. *A. crispa* (Ait.) Pursh, American green
    alder, aulne vert d'Amérique; native,
    Yukon-Keew, BC-NS, Nfld.   la. *A. c.*
    var. *mollis* Fern. (*A. mollis*
    Fern.), silky alder, aulne soyeux;
    native, Ont-NS, Nfld.   1b. *A. c.* var.
    sinuata (Regel) Hult. (*A. sinuata*
    (Regel) Rydb.), Sitka alder, aulne de
    Sitka; native, Yukon, BC, Alta.
2. *A. incana* (L.) Moench, white alder,
    aulne blanc; imported from Eurasia.
3. *A. rubra* Bong. (*A. oregana* Nutt.),
    red alder, aulne de l'Orégon; native, BC.
4. *A. rugosa* (Du Roi) Spreng., speckled
    alder, verne; native, BC-Nfld.   4a. *A.
    r.* var. *americana* (Regel) Fern.;
    native, Sask-Nfld.   4b. *A. r.* var.
    occidentalis (Dippel) Hitchc. (*A.
    tenuifolia* Nutt.), mountain alder,
    aulne à feuilles minces; native, Yukon,
    Mack, BC, Alta.
5. Host species not named, i.e., reported
    as *Alnus* sp.

Acrodictys globulosa: on 5 Man [1760].
Alternaria sp.:   on 1b BC [1012].
Anthostoma melanotes (C:25):   on 5 NS
    [1032].
A. microsporum var. *exudans* (C:25):   on
    5 BC [1012].
Antrodia albida (Trametes sepium):   on 5
    BC [1012].   See *Abies*.
A. crassa (Poria c.):   on 5 Nfld [1660].
    Associated with a brown rot of wood.
Apiognomonia alniella var. *alniella*
    (C:25):   on 5 Que [79].
Armillaria mellea (C:25):   root rot,
    pourridié-agaric: on 3 BC [6, 1012].
    See *Abies*.
Ascocoryne constricta: on 3 BC [1012].
Aureobasidium pullulans (C:25):   on 3 BC
    [1012].
*Basidioradulum radula (Hyphoderma r.,
    Radulum orbiculare)* (C:28):   on 3 BC
    [1012]; on 5 NS [1032].   Typically on
    dead branches and stems.   See *Abies*.
Bispora effusa: on 5 Man [1755, 1760].
Bisporella citrina (Helotium c.) (C:27):
    on 3 BC [1012]; on 5 NS [1032].
Bjerkandera adusta (Polyporus a.) (C:28):
    on 4a NB [1032].   See *Abies*.
Botryobasidium vagum (Pellicularia v.)
    (C:27):   on 3 BC [1012]; on 5 NS
    [1032].   Typically on rotted logs.   See
    Abies.
Caliciopsis sp.:   on 3 BC [1012, F72].

Calocera cornea (C:25):   on 3 BC [1012].
    See *Acer*.
Calosphaeria cryptospora: on 5 Man [1340].
Ceraceomyces serpens: on log of 3 BC
    [568].   See *Abies*.
Cercoseptoria vermiformis: on 3 BC [1012,
    F66].   These specimens have been
    redetermined as *Mycopappus alni*.
Ceriporia purpurea (Poria p.) (C:28):   on
    3 BC [1012].   Associated with a white
    rot.
Cerrena unicolor (Daedalea u.) (C:26):
    white spongy rot, carie blanche
    spongieuse: on 1b Yukon [460, F62]; on
    3 BC [1012]; on 4 Yukon [460]; on 4b BC
    [953]; on 5 NWT-Alta [1063], Mack, Alta
    [53], NS [1032, 1855], PEI [1032].
Chalara aurea: on 5 Man [1760].
Chaetosphaeria pulviscula: on 5 BC [826],
    as the *Menispora* state.
*Chloroisciboria aeruginascens (Chlorosplenium
    aeruginosum)* (C:25):   on 3 BC [1012];
    on 4 Nfld [1661]; on 5 NS [1032].
C. aeruginosa (C:25):   on 4 Ont [1340],
    Nfld [1660]; on 5 NWT-Alta [1063], Ont
    [1340], Nfld [1659].
Chondrostereum purpureum (Stereum p.)
    (C:28):   on 3 BC [1012]; on 5 Ont
    [1341].   Often fruiting on wounds on
    live trees.   See *Abies*.
Chytridium confervae: on 3 BC [44].
Chytriomyces poculatus: from decaying
    stems of 3 BC [44].   Typically
    saprophytic in soil.
Ciboria acerina: on 5 Ont [662].
C. alni (C:25):   on seeds of 5 Ont [663].
C. amentacea (C:25):   on 5 Man [F67], Ont
    [662].   This fungus sensu Groves &
    Elliott [662] is apparently an unnamed
    species [1590, p. 153].
C. caucus (C. amentacea sensu Fr.):   on
    5 Ont [662, 1590], Que [662].
    Confirmation in [1590] of many
    identifications on *Alnus*, *Populus* and
    Salix cited in [662].
Clavariadelphus fistulosus var. *contortus
    (Pistillaria alnicola)*:   on roots of 5
    Que [1385].
Coniochaeta ligniaria (Rosellinia l.)
    (C:28):   on 1b BC [1012].
Coniophora olivacea (C:26):   brown cubical
    rot, carie brune cubique: on 3 BC
    [1012].   Typically on rotted logs,
    sometimes rotting live *Picea* [577].
C. puteana (C:26):   brown cubical rot,
    carie brune cubique: on 3 BC [1012].
    Probably saprophytic on logs but in
    other tree genera it causes a heart rot
    of live trees.
Coriolus hirsutus (Polyporus h.) (C:28):
    on 3 BC [1012, 1341]; on 4b BC [1012];
    on 5 Ont [1341].   See *Abies*.
C. pubescens (Polyporus p., P. velutinus)
    (C:28):   on 4 Ont [1341]; on 4a NB
    [1032]; on 5 NWT-Alta [1063], BC [1012],
    Alta [53], NB, NS [1032].   See *Abies*.
C. versicolor (Polyporus v.) (C:28):   on 3
    BC [1012]; on 4 Nfld [1659, 1660, 1661];
    on 5 NB, NS, PEI [1032].   See *Abies*.

Corticium sp.: on 5 NWT-Alta [1063], Mack [53], NS [1032].

Crepidotus mollis (C. fulvotomentosus, C. haerens) (C:26): on 3 BC [1012].

Cristella sp.: on 4 Ont [1828]. This may be a *Trechispora*.

Cryptocoryneum condensatum: on 5 BC [1012].

Cryptodiaporthe oxystoma (Valsa o.) (C:26, 29): on 5 NWT-Alta [1063], Mack [53, F61], Ont [1340].

C. salicella (anamorph *Diplodina microsperma*): on 5 BC [1012].

Cryptospora sp.: on 5 Ont [1340]. *Cryptospora* is a synonym of *Ophiovalsa* and this record may be of a *Ophiovalsa*.

C. alnicola (C:26): on 4, 5 Ont [1340]; on 5 NS [1032]. See above.

C. aurantiaca (C:26): on 5 Man [F66], Ont [1340], NB, NS [1032]. See above.

Cryptosporella paucispora (C:26): on 5 Ont [1340, 1440, F60, F65].

Cryptosporium neesii (C:26): on 3 BC [1012].

Cylindrobasidium evolvens (Corticium laeve) (C:26): on 3 BC [1012]; on 5 NS [1032]. Typically on dead branches and stems. See *Abies*.

Cylindrocarpon willkommii: on 3 BC [1012].

Cylindrosporium alni (C:26): on 3 BC [1012].

Cyphellopsis anomala (Solenia a.) (C:28): on 1a NS [1032]; on 1b, 3 BC [1012]; on 5 BC [1424]. Presumably saprophytic. Associated with a white rot.

Cytidia salicina: on 5 Ont [1341], Nfld [1659, 1660, 1661]. Lignicolous, typically on dead *Salix*.

Cytospora sp. (C:26): on 4 Ont [1827], Nfld [1660, 1661]; on 5 Nfld [1659].

C. occulta: on 5 Man [F65].

C. truncata: on 4 Ont [1340].

Cytosporina sp.: on 1b Yukon [460].

Dacrymyces capitatus (D. deliquescens var. ellisii): on 5 BC [1012]. Causes a brown rot.

D. palmatus: on 5 Nfld [1659, 1660]. See *Abies*.

Daedaleopsis confragosa (Daedalea c.) (C:26): on 3 BC [1012]; on 4a NB, NS [1032]; on 5 Ont [1341], NS [1032]. See *Acer*.

Daldinia concentrica (C:26): on 4, 5 Ont [1340]; on 5 NS [1032].

D. vernicosa (C:26): on 5 Ont [1340].

Dasyscyphus sp.: on 1 Nfld [1659, 1660].

Datronia mollis (Trametes m.): on 5 NS [1032]. See *Acer*.

D. scutellatus (Fomes s.) (C:26): on 1a, 4a NS [1032]; on 4, 5 Ont [1341]; on 4a, 5 PEI [1032]; on 5 BC [1012]. Typically on dead trees. Associated with a white rot.

Dendryphiopsis atra (holomorph *Microthelia incrustans*): on 4a Sask [1760]; on 5 Man, Sask [1760].

Dermea sp.: on 5 Ont [1340].

Diaporthe sp.: on 5 Man [F65], Que [F62].

Diatrype disciformis (C:26): on 5 Man, Sask [F65].

D. stigma (C:26): on 4a, 5 NS [1032].

Diatrypella sp.: on 5 Ont [1340].

D. betulina: on 4, 5 Ont [1340].

D. discoidea var. *alni* (C:26): on 4, 5 Ont [1340].

D. tocciaeana (C:24): on 4 Ont [1340]; on 5 NS [1032].

D. verrucaeformis: on 5 Ont [1340].

Didymosphaeria oregonensis (C:26): canker, chancre didymosphérien: on 1b, 3, 4b BC [1012]; on 5 NWT-Alta [1063], BC [2010] and Alta [F65].

Diplococcium lawrencei: on 5 Man [1760].

Eichleriella leveilliana (C:26): on 2 NS [1032]. Lignicolous, presumably saprophytic.

Endophragmiella angustispora: on 5 in association with *Chaetosphaeria* sp. BC [800].

E. collapsa (Endophragmia c.): on 5 Man [803, 1760].

E. pallescens: on 5 Ont [799].

Erysiphe aggregata (C:26): powdery mildew, blanc: on 1, 4 Ont, Que, NS [1257]; on 3 BC [1012, 1257]; on 4 NB, NS, PEI [F65] also NB [1257]; on 4a NB [1032]; on 5 BC [1257], Ont [1257, 1340, F66], NB, NS [F63] also NB, NS, PEI [1032, F64].

Eutypa flavovirescens (C:26): on 3 BC [1012].

E. milliaria: on 5 NS [1032].

Eutypella sp.: on 5 Ont [1340].

E. alnifraga (C:26): on 1b BC [1012]; on 5 NS [1032].

E. cerviculata (C:26): on 4 Ont and on 5 Sask, Man, Ont [1340].

E. stellulata (C:26): dieback, dépérissement eutypelléne: on 3 BC [1012].

Exidia candida: on 3 BC [1012]. Lignicolous, apparently saprophytic.

E. crenata: on 1b BC [1012]. Lignicolous, apparently saprophytic.

E. glandulosa (C:26): on 1a NS [1032]; on 1b, 4 Yukon [460]; on 4, 4a NB [1032], Nfld [1659, 1660]; on 5 BC [1012], NS, PEI [1032]. Lignicolous, apparently saprophytic.

E. glandulosa f. *populi*: on 5 NWT-Alta [1063], Mack [53].

E. spiculosa: on 5 Ont [1341]. Lignicolous, apparently saprophytic.

Exidiopsis grisea: on 3 BC [1012]. Lignicolous, apparently saprophytic.

E. macrospora: on 4a NS [1032]. Lignicolous, apparently saprophytic.

Favolus alveolaris: on 5 NWT-Alta [1063], Mack [53, F65]. See *Acer*.

Femsjonia peziziformis: on 5 Ont [1341]. Lignicolous, presumably saprophytic.

Fenestella fenestrata (F. princeps) (C:26): on 5 Ont [1340].

F. minor (C:26): on 5 NS [1032].

Fibricium rude (F. greschikii): on 3 BC [1012]. See *Abies*.

Flagelloscypha citrispora : on 5 BC
[1424]. Lignicolous, apparently
saprophytic.
Flagellospora sp.: on 3 BC [43].
Fomes fomentarius (C:26): on 5 BC [1012],
Ont [1341]. Typically on dead trunks.
See *Acer*.
Fomitopsis pinicola (Fomes p.) (C:26): on
3 BC [1012]. See *Abies*.
Galzinia incrustans : on 5 BC [1012].
Lignicolous, apparently saprophytic.
Ganoderma applanatum (C:26): white
mottled rot, carie blanche madrée: on 3
BC [1012]; on 5 Sask [F67].
Gibberidea alnea (C:26): on 5 NS [1032].
*Gloeocystidiellum leucoxanthum (Corticium
l.)* (C:26): on 1b Yukon [460]; on 4a
NB [1032]; on 5 Nfld [1659, 1660,
1661]. Lignicolous, apparently
saprophytic.
G. ochraceum (Corticium o.) : on 5
NWT-Alta [1063], Mack [53].
Lignicolous, apparently saprophytic.
G. porosum (Corticium p.) (C:26): on 3 BC
[1012]; on 5 NS [1032]. Lignicolous,
apparently saprophytic.
Gloeophyllum sepiarium (Lenzites s.)
(C:27): on 3 BC [1012]; on 5 NWT-Alta
[1063]. See *Abies*.
Gloeoporus dichrous (Polyporus d.)
(C:28): on 5 BC [638, 1012, 1341]. See
Acer.
G. pannocinctus (Poria p.) (C:28): on 3
BC [1012]. Causes a white rot of wood.
Gloeosporium sp.: on 5 NB [1032].
Glyphium corrugatum : on 1b BC [589].
Gnomonia gnomon : on 5 Ont [1340].
G. setacea (C:26): on 1b Yukon [79, 460,
F60], BC [1012]; on 2 Ont [79].
Gnomoniella sp.: on 5 NS [1032].
Godronia sp.: on 4 Ont [1828].
G. cassandrae (anamorph *Fusicoccum
putrefaciens*): on 4a Que [1678].
G. cassandrae f. *betulicola* : on 4 Que
[F68]; on 4a Que [1684]. Originally
described from two European specimens.
G. fuliginosa : on 4a Que [1684] by
inoculation.
Gonytrichum caesium (holomorph
Chaetosphaeria inaequalis) (C:27): on
5 BC [1012].
Guepiniopsis alpina : on 3 BC [1012].
Lignicolous, apparently saprophytic.
Haematostereum gausapatum : on 5 BC
[1012]. Associated with a white rot.
H. rugosum (Stereum r.) (C:28): on 5 BC
[1012], NS [1032], Nfld [1659, 1660,
1661]. See *Abies*.
Hapalopilus nidulans (Polyporus n.) : on 5
Mack [1110]. See *Abies*.
Haploporus odorus (Trametes o.) : on 5
NWT-Alta [1063], Mack [579]. Associated
with a white rot. Typically on *Salix*.
Helminthosporium velutinum (C:27): on 3,
5 BC [810]; on 3 BC [1012].
Hendersoniopsis thelebola (holomorph
Melanconis t.): on 5 twigs and
branches Ont [1763].
Heterobasidion annosum (Fomes a.) (C:26):
on 3 BC [1012, F61]. Causes a root rot

of live trees, typically on coniferous
species. See *Abies*.
*Hohenbuehelia atrocaerulea (Pleurotus
a.)* : on 1b Yukon [460].
H. petaloides (Pleurotus spathulatus)
(C:28): on 3 BC [1012].
*Hygrophoropsis aurantiaca (Clitocybe
a.)* : on 3 BC [1012]. Associated with
a brown rot of well rotted wood,
probably mycorrhizal.
Hymenochaete agglutinans (C:27): on 2 NS
[1032]. Associated with a white rot.
Apparently sterile mycelium of *H.
corrugata* occurring where a live branch
or stem rests against a dead one.
H. cinnamomea (C:27): on 5 BC [1012].
Associated with a white rot.
H. corrugata (C:27): on 4, 5 Ont [1341];
on 4a, 5 NS [1032]. See *Abies*.
H. tabacina (C:27): on 3 BC [1012]; on 4
Labr [867]; on 4a NB [1032]; on 4a, 5
Nfld [1659, 1660, 1661]; on 5 Ont
[1341], Que [282], PEI [1032]. See
Abies.
Hyphoderma praetermissum (H. tenue) : on 3
BC [1012]. Lignicolous, typically on
logs.
H. puberum : on 3 BC [1012]. Typically on
logs. See *Abies*.
H. setigerum (Peniophora aspera) (C:27):
on 3 BC [1012]; on 5 NS, PEI [1032].
Typically on logs. See *Abies*.
Hyphodontia crustosa (Odontia c.) (C:27):
on 3 BC [1012, 1341]; on 5 NS [1032].
Typically on dead branches. See *Abies*.
H. quercina (Radulum q.) (C:28): on 3 BC
[1012]; on 5 NWT-Alta [1063], Mack
[53]. Lignicolous, on dead wood.
Hypocrea gelatinosa : on 5 NS [1032].
H. rufa (C:27): on 5 NB, NS [1032].
Hypoxylon sp.: on 5 NWT-Alta [1063], Mack
[53], BC [1340], NS [1032].
H. coccineum : on 5 Sask [F65].
H. fuscum (C:27): on 1b BC [1012]; on 4
Ont [1340], Nfld [1661]; on 5 NWT-Alta
[1063], Alta [53], Man [F66], Ont
[1340], Que [282, 1385], NB, NS, PEI
[1032], Nfld [1659, 1660].
H. mammatum (C:27): on 1b BC [1012]; on
4, 5 Ont [1340], Nfld [1661]; on 5
NWT-Alta [1063], Alta [F62], Sask, Man
[F66], NS [1032], Nfld [1659, 1660].
H. multiforme (C:27): on 1 Que [282]; on
3 BC [1012]; on 4a NB [1032]; on 5 BC
[1340].
H. rubiginosum (C:27): on 3 BC [1012]; on
5 NWT-Alta [1063], Mack [53].
Inonotus radiatus (Polyporus r.) (C:28):
on 4 Ont [1341]; on 4a, 5 NB [1032]; on
5 NS [1032]. See *Abies*.
Irpex lacteus (Polyporus tulipiferae)
(C:28): on 3 BC [1012]; on 4a NB
[1032]; on 5 Ont [1341]. See *Acer*.
*Junghuhnia nitida (Poria attenuata, P.
eupora)* : on 5 NS [1032]. See *Acer*.
*J. subfimbriata (Poria radula sensu
Bres.)* : on 3 BC [579, 1012].
Associated with a white rot.
Kuehneromyces mutabilis (C:27): on 3 BC
[1012]. Associated with a white rot.

Laeticorticium lundellii: on 5 Mack
   [53]. Lignicolous.
L. roseum (Aleurodiscus r.):   on 5
   NWT-Alta [1063]. See *Acer*.
L. violaceum (Dendrocorticium v.):   on 5
   NS [930]. Lignicolous.
Leucostoma persoonii (Valsa leucostoma)
   (anamorph *Cytospora rubescens*):   on 5
   NWT-Alta [1063].
Libertella sp.:   on 5 Sask, Man [F67].
*Lindtneria leucobryophila (Trechispora
   l.)*:   on 5 BC [970].
Macrotyphula fistulosa: on 3 BC [1012].
Marasmius androsaceus: on 5 PEI [1032].
Melampsoridium hiratsukanum (C:27):   rust,
   rouille: on 1 Alta [F63].
Melanconis sp.:   on 5 Ont [1340].
M. alni (anamorph *Melanconium apiocarpon*,
   M. sphaeroideum) (C:27):   on 4, 5 Sask,
   Man [F65, F69].
M. alni var. *marginalis* (C:27):   on 1
   Que [70]; on 4 Ont [1340]; on 4b Alta
   [950]; on 5 Mack [53, F66], Ont [1340,
   F68], NS [1032].
M. thelebola (anamorph *Hendersoniopsis
   t.*) (C:27):   on 4, 5 Ont [1340]; on 5
   Man [F66], NS [1032].
Melanconium sp. (C:27):   on 5 Ont [1340].
M. apiocarpon (holomorph *Melanconis
   alni*) (C:27):   on 4, 5 Ont [1340]; on
   4a Man [1768].
M. sphaeroideum (holomorph *Melanconis
   alni*) (C:27):   on 3 BC [1012].
Melanomma pulvis-pyrius (C:27):   on 3 BC
   [1012]; on 5 Ont [1340].
Menispora ciliata: on 5 BC [826].
M. glauca: on 5 BC [826].
Merismodes fasciculatus (Cyphella f.)
   (C:26):   on 1a, 5 NS [1032]; on 4 Ont
   [1341]; on 5 Yukon [460], BC [1012],
   Sask [F66], Ont [1341, F60], Que [282].
   Lignicolous, on dead wood.
Meruliopsis ambiguus: on 3 BC [568]. On
   dead wood, associated with a white rot.
M. corium: on 3 BC [1012]. See *Acer*.
Merulius tremellosus: on dead 5 Ont
   [568]. See *Abies*.
Microsphaera penicillata (C:27):   powdery
   mildew, blanc: on 4 Ont [1257]; on 5
   Sask, Man [F67], NS [1032].
Mitrula borealis: on dead foliage of 4
   Ont [1428].
M. lunulatospora: on dead foliage of 4
   Ont [1428].
Mollisia cinerea (C:27):   on 5 NS [1032].
Mycena algeriensis: on trunks of 5 Ont
   [1385].
M. leaiana: on dead branches of 5 Que
   [1385].
M. roseipallens: on debris of 5 Ont
   [1385].
M. viridimarginata: on dead branches or
   stumps of 5 Que [1385].
Mycoacia aurea: on 3 BC [1012].
   Lignicolous.
M. uda: on 5 BC [1012]. Lignicolous.
Mycocalia denudata: on partly rotted wood
   of 5 BC [190].
Mycoleptodon dichroum (C:27):   on 3 BC
   [1012]. Lignicolous.

Mycorhynchus sp.:   on 3 BC [1012].
Mycosphaerella sp.:   on 1b Yukon [460,
   F60].
M. alnicola: on 5 NB [1032].
M. punctiformis (C:27):   on 1b BC [1012,
   F60]; on 4b BC [1012].
Myxofusicoccum alni: on 5 Man [F65].
Myxosporium sp.:   on 5 Ont [1340].
Naemospora alni (C:27):   on 4 Ont [1340].
Nectria sp.:   on 5 BC [F60].
N. cinnabarina (anamorph *Tubercularia
   vulgaris*) (C:27):   on 5 NWT-Alta
   [1063], Mack [53], Que [282].
N. coccinea: on 1 Nfld [1659, 1660, 1661].
N. episphaeria (C:27):   on 3 BC [1012].
   Probably on pyrenomycetous fungi on
   Alnus.
N. galligena: on 5 Ont [1340].
N. pithoides (C:27):   on 3 BC [1012].
Nipterella parksii (Belonidium p.)
   (C:25):   on 3 BC [1012].
Ophiovalsa femoralis (Cryptospora f.)
   (C:26):   on 4, 5 Ont [1340]; on 5 Sask
   [1340, F66], NB, NS [1032].
O. suffusa (Cryptospora s.) (C:26):   on 5
   Sask, Man [F65], Ont [1340].
Panellus rigens (Panus salicinus) (C:27):
   on 1b BC [1012].
P. serotinus (Pleurotus s.) (C:28):   on 3
   BC [1012, F66]. See *Acer*.
P. stipticus (Panus s.):   on 3 BC [1012];
   on 4a, 5 NB, NS [1032]; on 5 PEI [1032],
   Nfld [1659, 1660, 1661]. Associated
   with a white rot.
Panus conchatus (P. torulosus):   on 3 BC
   [1012]. Lignicolous.
P. rudis (C:27):   on 5 Ont [1341]. See
   Acer.
Passalora alni: on 1a Que [379].
P. bacilligera (C:27):   leaf spot,
   tavelure: on 1b BC [1012]; on 5 Que
   [F61].
Paxillus sp.:   on 3 BC [1012].
Pellidiscus pallidus: on 5 BC [1424].
Peniophora sp.:   on 1, 4 Nfld [1659, 1660,
   1661]. Lignicolous.
P. aurantiaca (C:27):   on 1, 4b, 5
   NWT-Alta [1063]; on 1b, 3 BC [1012]; on
   4, 5 Nfld [1659, 1660, 1661]; on 4a NB
   [1032], Nfld [1661]; on 5 Yukon [460] as
   a sight record, Mack [1110], Alta [53,
   950], NS, PEI [1032]. Lignicolous, on
   dead wood.
P. cinerea (C:27):   on 5 BC [1012]. See
   Abies.
P. erikssonii (C:27): on 4 Yukon [460]; on
   5 NWT-Alta [1063], Ont [1341]. On dead
   wood.
P. erumpens: on 4a NB [1032].
   Lignicolous.
P. incarnata (C:27):   on 3 BC [1012].
   Lignicolous.
P. pithya: on 5 NB [1032]. See *Abies*.
P. polygonia: on 5 Mack [1110].
   Associated with a white rot.
Pezicula alni (C:28):   on 1a, 5 NS [1032];
   on 3 BC [1012]; on 4, 5 Ont [1340]; on
   4a Ont [F60].
P. alnicola (C:28):   on 5 Ont [1340].

P. aurantiaca: on 1, 5 Ont [1340]; on 5 Ont [F63].

Phaeocalicium compressulum: on 5 BC, Ont [1812].

Phaeoisaria sparsa: on 5 Man [1760].

Phaeomarasmius erinaceus: on 5 Ont [1434].

P. rhombosporus: on fallen leaves of 4 Que [1041]; on 5 Ont [1041].

Phanerochaete laevis (Peniophora affinis) (C:27): on 5 NS [1032]. See *Acer*.

P. sordida (Peniophora cremea) (C:27): on 3 BC [1012]; on 5 NS [1032]. See *Abies*.

Phellinus ferreus (Poria f.) (C:28): on 3 BC [1012, 1341]; on 5 NWT-Alta [1063], Mack [1110], Ont [1341], Que [282], NB, NS, PEI [1032], Nfld [1659, 1660, 1661]. See *Acer*.

P. ferruginosus (Poria f.) (C:28): on 5 NWT-Alta [1063], Mack [53, 1110], BC [953]. See *Abies*.

P. igniarius (Fomes i.) (C:26): white trunk rot, carie blanche du tronc: on 3 BC [1012]; on 4b Yukon [F62, 460]; on 5 NWT-Alta [1063], NS [1032].

P. laevigatus (Poria l.) (C:28): on 1 Que [282]; on 3 BC [1012]. See *Acer*.

P. punctatus (Poria p.) (C:28): on 5 NS [1032]. See *Abies*.

Phibalis furfuracea (Cenangium f., Encoelia f.) (C:26): on 3 BC [1012]; on 4b Mack [664]; on 5 NWT-Alta [1063] and Alta [F62], Sask, Man [F67], Ont [1340, 1827], Que [867, 1385 on live branches], Labr [867].

Phlebia radiata (C:28): on 2 NB [1032]; on 3 BC [568, 1012]; on 4a NB, NS [1032]; on 5 BC [568], Ont [1341]. On logs and dead branches. See *Abies*.

P. rufa: on 5 BC [568]. On logs and dead branches. See *Abies*.

Pholiota aurivella: on 3 BC [1012]. See *Abies*.

P. destruens: on 5 Yukon [460], BC [1012]. Lignicolous.

P. tuberculosa: on 5 BC [1012]. This species not recognized in North America in Smith & Hesler's generic monograph.

Phomopsis sp.: on 5 Ont [1340].

Phragmoporthe conformis: on 5 BC [79].

Phragmotrichum rivoclarinum (P. karstenii): on 5 Man [F65].

Phyllactinia guttata (P. corylea) (C:28): powdery mildew, blanc: on 1 Yukon, Ont, Que, NS [1257]; on 1b Yukon [460]; on 2 Sask, Ont, Que, NS [1257]; on 3 BC [1012, 1257]; on 4 Yukon [460], Ont, NB, NS [1257]; on 4b Yukon [F62], NWT-Alta [1063], Mack [53]; on 5 NWT-Alta [1063], Mack [53, F65], Sask, Man [F66, F68], Ont [1340, F63], NB, NS [1032], Nfld [1659, 1660, 1661], also Man [F67].

Phyllosticta sp.: on 5 Que [F61].

P. alnea (C:28): on 5 Que [F62].

Pirex concentricus (Irpex owensii): on 3 BC [1012]. Lignicolous.

Plagiostoma alneum var. *alneum*: on 5 BC [79].

Platychora alni: on 1b Yukon [460]; on 4a Que [F67].

Pleospora sp.: on 5 Ont [1340].

Pleurothecium sp.: on 3 BC [587].

Pleurotus ostreatus (C:28): on 3 BC [1012]. See *Abies*.

Plicatura crispa (P. faginea, Trogia c.) (C:29): on 3 BC [1012]; on 4a NB [1032]; on 4b Yukon [460] as a sight record; on 5 NWT-Alta [1063], Alta [53], Ont [1341], NB, NS [1032], Nfld [1659, 1660]. On dead branches and stems.

P. nivea (Merulius n., Trogia alni) (C:27, 29): on 1b BC [1341]; on 2 Ont, Que, NS [563, 568] also NS [1032]; on 3 BC [563, 568, 1012]; on 4 Ont [563, 568, 1341]; on 4a Que, NS [563, 568], NB [563, 1032]; on 5 NWT-Alta [1063], Ont [1341], Que [1385], NS [1032] also BC, Alta, Ont [563, 568]. On dead stems and branches. See *Acer*.

Podosphaera clandestina: powdery mildew, blanc: on 5 Man [F68]. Possibly the host was misdetermined, typically on *Amelanchier*.

Polyporus badius (P. picipes) (C:28): on 1 NWT-Alta [1063]; on 3 BC [1012, 1341]; on 5 Ont [1341]. Typically on fallen branches. See *Abies*.

P. brumalis (C:28): on 4a, 5 NB [1032]. Typically on fallen branches.

P. varius (P. elegans) (C:28): on 1b BC [1012, 1341]; on 3 BC [1012]; on 4b NWT-Alta [1063]; on 5 Mack [1110], Ont [1341], Labr [867]. Typically on fallen branches. See *Abies*.

Poria sp.: on 5 PEI [1032], Nfld [1659, 1660, 1661]. Lignicolous.

P. ohiensis (Fomes o.): on 4 Ont [1828]. Associated with a white rot.

P. subacida (C:28): on 3 BC [1012]. See *Abies*.

Psathyrella griseifolia: on fallen twigs of 5 Que [1385].

Pseudospiropes nodosus (Helminthosporium n.): on 5 BC [816, 1012].

P. simplex: on 4a, 5 Sask, Man [817, 1760]; on 5 BC [817].

Punctularia strigosozonata (Phlebia s.): on 4 Ont [1341]. Apparently saprophytic. Associated with a white rot.

Pycnoporus cinnabarinus (Polyporus c.) (C:28): on 4b BC [1012, 1181]; on 5 NWT-Alta [1063]. See *Acer*.

Schizophyllum commune: on 4a NB [1032], Nfld [1659, 1660, 1661]; on 5 NWT-Alta [1063], Mack [53], PEI [1032]. On dead branches and stems. See *Abies*.

Schizopora paradoxa (Poria versipora) (C:28): on 3 BC [1012]; on 5 NWT-Alta [1063]. See *Abies*.

Scorias spongiosa (C:28): on 5 Ont [1340], NB, NS [1032] See comment under *Abies*.

Scytinostroma galactinum: on 3 BC [1012]. See *Abies*.

Septoria alni (C:28):   leaf spot, tache septorienne: on 1b BC [1012]; on 4a, 5 NB [1032].

S. alnicola: on 5 Man [F66].

S. alnifolia (C:28): on 5 BC [1012], Alta [F66], Sask [F65].

S. carolinensis: on 5 Man [F65].

Septotrullula bacilligera: on 5 Man [1760].

Sistotrema brinkmannii: on 3 BC [1012]. See *Abies*.

Spadicoides atra: on 5 Sask, Man [1760].

S. bina: on 5 BC [790], Sask [1760].

S. klotzschii: on 5 BC [766, 794].

Sphaerobolus stellatus (C:28): on 5 BC [1012]. On dead wood.

Sporidesmium foliculatum: on 5 BC [825].

S. hormiscioides: on 5 BC [823].

Steccherinum fimbriatum (Odontia f.): on 5 NB [1032]. Typically on rotted wood. See *Acer*.

S. ochraceum (C:28): on 3 BC [1012]; on 5 NS [1032]. Typically on rotted wood. See *Acer*.

Stemonitis fusca: slime mold: on 5 NWT-Alta [1063]. Probably on rotted wood.

Stenocybe pullatula: on 5 BC, Ont [1812].

Stereum complicatum (C:28): on 3 BC [1012]. Lignicolous.

S. hirsutum (C:28): on 3 BC [1012]; on 4a NS [1032]; on 5 NS, PEI [1032]. See *Abies*.

S. ochraceo-flavum: on 5 Nfld [1659, 1660, 1661]. Lignicolous.

S. ostrea (C:28): on 3 BC [1012]. See *Abies*.

Strickeria obducens (C:28): on 1 Que [70].

Stromatocyphella conglobata (Cyphella c.) (C:26): on 5 BC [1012], NB [1032]. Lignicolous.

Taeniolella alta (C:28): on 5 Ont, Que [814].

Taphrina sp.: on 1b BC [1012]; on 5 NS [1032].

T. amentorum (C:28): on 5 Sask, Man [F67, F68], also Man [F66].

T. japonica (C:28): on 3 BC [1012].

T. occidentalis (C:28): leaf blister, cloque des feuilles: on 3 BC [1012]; on 4 Yukon [460]; on 4b NWT-Alta [1063], BC [1012]; on 5 Que [F73].

T. robinsoniana (C:28): catkin blister, cloque des chatons: on 1 Ont, Que, NB, NS [1257]; on 2 NS [1925]; on 4 Ont [1340], NB, NS, PEI [F65], Nfld [1340, F65]; on 4 NB [1257]; on 4, 4a, 5 Nfld [1659, 1660, 1661]; on 4a Ont [F60], NB, NS, PEI [1032, F66, F67], Nfld [F66]; on 5 Ont [1340, F63], Que [282], NB, NS [1032, F63], PEI [1032], Nfld [F64], also NS [F64].

T. tosquinetii (C:29): leaf blister, cloque des feuilles: on 1a NS [F66].

Tectella patellaris (Panellus p.): on 5 Yukon [1108].

Tomentella coerulea (T. papillata): on 5 BC [929, 1012, 1341]. On rotted wood.

T. ellisii (T. ochracea, T. microspora): on 5 Ont [929, 1341, 1828]. Lignicolous, on rotted wood.

T. sublilacina: on 5 Ont [1341, 1827]. Lignicolous, on rotted wood.

T. ramosissima (T. fuliginea): on 5 BC [1341]. Lignicolous, on rotted wood.

Trechispora mollusca (Poria candidissima, T. onusta): on 3 BC [1012]; on 5 NS [1032]. On rotted wood. See *Abies*.

Tremella aurantia (C:29): on 3 BC [1012]. Lignicolous, apparently saprophytic.

T. mesenterica (C:29): on 1b BC [1012]; on 3 BC [42, 1012], Labr [867]. Lignicolous, apparently saprophytic.

Trichaptum biformis (Polyporus pargamenus): on 5 NB [1032]. See *Acer*.

Trichoderma viride: on 5 Man [1760]. See *Abies*.

Tubercularia vulgaris (holomorph Nectria cinnabarina) (C:29): on 3 BC [1012].

Tubulicrinis glebulosus (Peniophora gracillima) (C:27): on 5 NS [1032]. Lignicolous, apparently saprophytic. See *Abies*.

Tulasnella violea (C:29): on 5 NS [1032]. Lignicolous, apparently saprophytic.

Tympanis alnea (anamorph *Sirodothis inversa*) (C:29): on 1b Yukon [460]; on 3 BC [1012, F61]; on 4 Ont [1340]; on 4b, 5 NWT-Alta [1063]; on 5 BC, Alta, Ont, Que, NB, Nfld [1221] and Alta [53, 644, F62], Sask, Man [F65, F67, F69], Ont [1340, F60], NS [1032, 1925].

T. alnea var. *hysterioides (T. hysterioides)* (C:29): on 1b BC [1012]; on 5 BC, Ont, Que, NS, Nfld [1221], also NS [1925].

T. pseudoalnea: on 4, 5 Que [1221].

T. truncatula: on 5 NB [1221].

Typhula erythropus: on 3 BC [901].

Tyromyces chioneus (Polyporus albellus) (C:28): on 3 BC [1012]; on 5 NS [1032], Nfld [1659, 1660, 1661]. Typically on dead branches and logs. See *Abies*.

T. tephroleucus (Polyporus t.) (C:28): on 3 BC [1012]. Causes a white rot of wood.

Valsa sp.: on 3 BC [193, 1012]; on 5 Ont [1340].

V. diatrypoides (C:29): on 4 Ont [1340]; on 5 Ont [F62].

V. stenospora (C:29): on 5 NS [1032].

V. truncata (C:29): on 4, 5 Ont [1340]; on 5 NS [1032].

Valsaria moroides (C:29): on 4, 5 Ont [1340]; on 5 BC [1012], NS [1032].

Valsella furva: on 5 Man [F66].

Vararia effuscata (C:29): on 5 PEI [1032]. See *Acer*.

V. investiens: on 5 Ont [1341]. See *Abies*.

Varicosporium elodeae: on 3 BC [43].

Venturia alnea: on living and overwintered leaves of 5 Que [73].

Volucrispora aurantiaca: on 3 BC [43].

Xenasma rimicola: on 5 Ont [967]. Lignicolous, apparently saprophytic.

Xylaria sp. (C:29): on 3 BC [1012].
Zalerion maritimum: on a submerged leaf
    of 5 BC [249].

ALOPERCURUS L.   GRAMINEAE

1. *A. aequalis* Sobol., short-awned
    foxtail, vulpin à courtes arêtes;
    native, Yukon, Mack, Alta-Nfld. la. *A.*
    a. var. *aequalis (A. geniculatus*
    var. *aristulatus* (Michx.) Torr.*)*.
2. *A. alpinus* Sm., alpine foxtail,
    alopécure des Alpes; native,
    Yukon-Frank, mtns. of BC, Cypress hills
    in Sask, n. Man and Ont, Labr.
3. *A. pratensis* L. *(A. arundinaceus*
    Poir.), meadow foxtail, vulpin des prés;
    imported from Eurasia, natzd., BC, Alta,
    Man-Nfld. 3a. *A. p.* ssp. *pratensis*
    (A. seravchanicus).

Acremonium boreale: on 3 BC, Alta, Sask
    [1707].
Drechslera catenaria (C:29): on 3 Ont
    [1611].
Hysteropezizella diminuens: on 2 Frank
    [664].
Mastigosporium album (C:29): on 3 Nfld
    [335, 336].
Puccinia graminis (C:29): rust, rouille:
    II III on 1a, 3, 3a Ont [1236].
P. recondita s. 1. (C:29): rust,
    rouille: II III on 3 Ont [1236].
Sclerotinia borealis (C:29): on 3 BC
    [1816].

ALTHAEA L.   MALVACEAE

1. *A. armeniaca* Ten.; imported from
    Mediterranean region & se. USSR.

See also *ALCEA*.

Puccinia malvacearum (C:30): rust,
    rouille: on 1 Ont [1236, 1281].

ALYSSUM   CRUCIFERAE

See *AURINIA*

AMARANTHUS L.   AMARANTHACEAE

1. *A. graecizans* L., matweed, amarante
    parente; native, BC-Man and natzd., Ont,
    Que. Precise native area difficult to
    delimit.
2. *A. retroflexus* L., green amaranth,
    amarante réfléchie; imported from
    tropical America, natzd., Mack, BC-PEI.

Albugo bliti (C:30): white rust,
    albugine: on 2 BC [1816].
Peronospora amaranthi: downy mildew,
    mildiou: on *A.* sp. BC [1816].
Polymyxa betae f. sp. *amaranthi*: on 2
    Ont [60].
Pythium sp.: on 1 BC [1977].
P. intermedium: on 1 Ont [868].
P. irregulare: on 1 Ont [868].
P. undulatum: on 2 Ont [868].

Rhizoctonia solani (holomorph
    Thanatephorus cucumeris): on 2 Que
    [317].
Rhizophydium graminis: on 2 Ont [59].
Thielaviopsis basicola: on 2 Ont [546].
Verticillium dahliae: wilt, flétrissure
    verticillienne: on 1 BC [1087, 1816,
    1978]; on 2 BC [38, 1816].

AMARYLLIS L.   AMARYLLIDACEAE

    Bulbous plants native to tropical
America and S. Africa; cult. under glass,
ornamental.

Stagonospora curtisii (C:30): leaf
    scorch, grillure: on *A.* sp. Ont
    [334].

AMBROSIA L.   COMPOSITAE

1. *A. artemisiifolia* L., common ragweed,
    petite herbe à poux; native, Mack,
    BC-Nfld. Native area uncertain due to
    weedy habit.
2. *A. trifida* L., giant ragweed, grande
    herbe à poux; native, BC-PEI. Native
    area uncertain due to weedy habit.

Albugo tragopogonis (C:30): on 1 Que
    [714].
Erysiphe cichoracearum (C:31): powdery
    mildew, blanc: on 1 Ont, Que and on 2
    Man, Ont [1257].
Ophiobolus anguillidus: on 1, 2, *A.* sp.
    Ont [1620]; on 1 Ont [1340].
O. drechsleri: on 2 Ont [1620].
O. fulgidus: on 2 Ont [1620].
Puccinia xanthii (C:31): on 2 Man [1245,
    1261], Ont [1236, 1261].
Verticillium dahliae: wilt, flétrissure
    verticillienne: on 1 Ont [1079]; on 1,
    2 Ont [1087].
V. nigrescens: on 1 Ont [1087].

AMELANCHIER Medic.   ROSACEAE

1. *A. alnifolia* Nutt., saskatoon berry,
    amélanchier à feuilles d'aulne; native,
    Yukon, Mack, BC-w. Que. la. *A. a.*
    var. *cusickii* (Fern.) Hitchc. *(A.*
    cusickii Fern.); native, s. BC. lb.
    A. a. var. *semiintegrifolia* (Hook.)
    Hitchc. *(A. florida* Lindl.); pacific
    serviceberry, amélanchier de l'Ouest;
    native, BC.
2. *A. bartramiana* (Tausch) Roemer,
    mountain juneberry, amélanchier boréal;
    native, Ont-Labr.
3. *A. canadensis* (L.) Medic. *(A.*
    oblongifolia (T. & G.) Roemer),
    canadian serviceberry, shad bush,
    amélanchier du Canada; native, c.
    Ont-PEI.
4. *A. gaspensis* (Wieg.) Fern. & Weath;
    native, Ont-n. NB.
5. *A. huronensis* Wieg.; native, Ont-Que.
6. *A. intermedia* Spach., purple
    serviceberry, amélanchier pourpre;
    native, Ont-Nfld.

7. *A. laevis* Wieg., Allegheny
serviceberry, amélanchier glabre;
native, Ont-Nfld.
8. *A. ovalis* Medic. (*A. vulgaris*
Moench), garden serviceberry,
amélanchier commun; imported from Europe.
9. *A. sanguinea* (Pursh) DC., round-leaved
serviceberry, amélanchier sanguin;
native, Ont-Nfld.
10. *A. spicata* (Lam.) K. Koch (*A.
humilis* Wieg.), low shad bush,
amélanchier bas; native, Ont-Que.
11. *A. stolonifera* Wieg., running
serviceberry, amélanchier stolonifère;
native, Ont-Nfld.
12. *A. wiegandii* Nielsen, Wiegand's
serviceberry, amélanchier de Wiegand;
native, Ont-Nfld.
13. Host species not named, i.e., reported
as *Amelanchier* sp.

Aleurodiscus cerussatus (C:31): on 1 BC
[1012]; on 13 BC [964].
Apiosporina collinsii (C:31): black
leafcurl, enroulure noire: on 1, 13
NWT-Alta [1063]; on 1 BC, Alta, Sask,
Man, Ont [292] also BC [1012], Alta [53,
644, 880, F69], Man [F70]; on 1b BC
[1816]; on 3 Ont, NB [292]; on 3, 13 NB,
NS [1032]; on 6 NS [1032]; on 13 Alta
[F62, F71, F72, F73], Sask, Man [F63,
F64, F66, F67, F68], Ont [1340, 1827,
F73, F74], Nfld [1659, 1660, 1661].
A. morbosum (Dibotryon m.): black knot,
nodule noir: on 1 NWT-Alta [1063].
Host probably misdetermined. This
fungus occurs on *Prunus.*
Botryosphaeria obtusa (C:31): black rot,
pourriture noire: on 6 NS [1614].
Cladosporium herbarum: on 13 Man [1760].
Conoplea olivacea (C. sphaerica): on 13
Man [1760, F66].
*Cryptosporiopsis pruinosa (Sphaeronaema
p.)* (holomorph *Pezicula pruinosa*)
(C:32): on 13 Man [F67], Ont [1340,
1660, 1827].
Cylindrosporium sp. (C:31): leaf spot,
tache foliaire: on 1 BC [1012]; on 1b
BC [1816].
Dermea bicolor (C:31): on 13 Sask [F67],
Man [F66], Ont [1340].
Diaporthe sp.: on 13 Ont [1340].
D. tuberculosa (C:31): twig blight,
brûlure des rameaux: on 13 Ont [1340,
F60], NS [1032].
Diplocarpon mespili (Fabraea maculata)
(anamorph *Entomosporium mespili*)
(C:31): on 1b BC [1816]; on 3 NB
[1032]; on 13 Man [F67, F68], NB, NS
[1032], PEI [335, 1032].
Entomosporium mespili (E. maculatum)
(holomorph *Diplocarpon mespili*)
(C:31): leaf spot, tache des feuilles:
on 13 PEI [F62].
Fibricium rude (F. greschikii): on 1 BC
[1012]. See *Abies.*
Gymnosporangium sp.: rust, rouille: on 1
NWT-Alta [1063], BC [954], Ont [1341];
on 2, 7, 9 Nfld [1659, 1660, 1661]; on

13 Alta [338, 1594], Sask [F63], Man
[F63, F64], Ont [1341], NS [F66], Nfld
[337, 1661, 1595].
G. biseptatum: rust, rouille: on 13
NWT-Alta [1063].
G. clavariiforme (C:31): rust, rouille:
O I on 1 BC [1012, 1247] and BC, Alta,
Sask, Man [2008]; on 2 Que [1236], Nfld
[1262]; on 2, 4 Que [1262]; on 3 Man,
Ont, Que, NS, PEI [1240] also Ont [1236,
1247], NS [1032]; on 3, 6, 10, 11, 13
Ont [1262]; on 7, 11 NS [1032]; on 7, 12
NS [1262]; on 10 Ont [1236]; on 13 BC
[644, 953, 1247, 1262], Alta [951, 952,
1247], Sask [F67], Ont [1236], Que [339,
1594], NB [F64], NS [1032, F63, F64,
F65], PEI [1262], Nfld [333, 339].
G. clavipes (C:31): quince rust, rouille
du cognassier: O I on 1 NWT-Alta
[1063], BC, Alta, Man, Ont [1263] and BC
[644, 1012, 1341, 1247, 1816], Alta [53,
950], Man [335], Ont [1236, 1240, 1341];
on 1b, 3, 5, 7, 9, 10, 13 [1236]; on 2
Que, Nfld [1263]; on 3 Ont, Que [1240],
NB [1263]; on 3, 6, 13 NB [1032]; on 3,
10, 13 NS [1032]; on 6 NS [1240]; on 7
Nfld [1662]; on 13 BC, Alta, Sask, Man,
Ont [1247] in Ont by inoculation, NB,
NS, PEI [1263] and BC [1063], Alta [904,
F61], Man [F67], Ont [1341, 1827, 1828,
F68], Nfld [335, 1659, 1660, 1661] also
Alta, Sask [F71].
G. connersii: rust, rouille: O I on 13
Ont [1247] by inocluation.
G. cornutum: rust, rouille: O I on 13
Alta [951].
G. inconspicuum (C:31): rust, rouille: 0
I on 1 BC [1012, 1269, 1816, 2008]; on
13 BC [1247], Ont [1247] by inoculation.
G. nelsonii (G. corniculans) (C:31):
rust, rouille: O I on 1 Yukon, BC,
Alta, Sask, Man [1271, 2008] also
NWT-Alta [1063], Mack, Ont [1271], BC
[644, 1012], Alta [53, 644], Sask
[1839]; on 1, 9 Ont [1236]; on 1, 1b BC
[1816]; on 1a, 1b BC [1271]; on 5, 6, 7
Ont [1271]; on 13 Yukon, NWT, BC, Alta,
Sask, Man, Ont [1247] in Ont by
inoculation, also BC [953], Alta [952,
F61], Man [1240], Ont [1240, 1236] and
Sask, Man [F67].
G. nidus-avis (G. juvenescens) (C:31):
rust, rouille: on 1 NWT-Alta [1063],
Mack [53], BC [644, 1012] also BC, Alta,
Man [1272] and Alta [557, 952]; on 1, 3
Ont by inoculation [1247]; on 1, 1b BC
[1816]; on 5, 7 Ont [1272]; on 13 Mack,
BC, Alta, Sask, Man [2008] also BC
[953], Alta [F60], Ont [1236, 1341] also
BC, Alta, Man [1247] in BC by
inoculation, and Sask, PEI [1272], Man,
Ont, Que, PEI [1240].
Hymenochaete corrugata: on 13 Ont
[1341]. See *Abies.*
H. tabacina (C:31): on 1 BC [1012]; on 13
PEI [1032]. See *Abies.*
Irpex lacteus (Polyporus tulipiferae)
(C:32): on 13 Ont [1341]. See *Acer.*
Leucostoma cincta (Valsa c.) (C:32): on
13 Ont [1340, 1828], NS [1032].

Lophodermium hysterioides (C:31): on 1 BC
[1012]; on 13 NWT-Alta [1063], Alta
[F63].
L. tumidum (C:31): on 13 Man [F66].
Probably a specimen of *L. aucupariae*.
See [1609] page 96.
Massaria conspurcata: on 13 Man [F67],
Ont [1340]. Typically on *Prunus*.
M. pruni (C:32): on 13 NS [1032].
Typically on *Prunus*.
Meruliopsis corium: on 1 BC [1012]; on 1b
BC [568]; on 13 Sask [568]. See *Acer*.
Microsphaera penicillata (M. alni):
powdery mildew, blanc: on 13 Ont [1827].
Monilinia amelanchieris (C:32): blossom
blight or fruit rot, pourriture
sclérotique: on 1, 13 Alta [338]; on 13
BC [338].
Nectria sp.: on 13 Ont [1340].
N. cinnabarina (anamorph *Tubercularia
vulgaris*) (C:32): on 13 Man [F67].
Nummularia discincola: on 13 Ont [1340],
Que [862].
Oedemium atrum (holomorph
Chaetosphaerella *fusca*): on 13 Sask
[1760].
Peniophora cinerea (C:32): on 13 Ont
[1341]. See *Abies*.
Pezicula pruinosa (C:32): on 13 Ont
[1340], NS [1032].
Phaeomarasmius erinaceus: on 13 Man
[1434].
Phellinus ferreus: on 13 BC [558]. See
Acer.
P. ferruginosus (Poria f.): on 13 NS
[1032]. See *Abies*.
P. punctatus (Poria p.): on 13 Ont
[1341]. See *Abies*.
Phragmidium sp.: on 1 NWT-Alta [1063].
Probably a misdetermination of the host
because this rust genus is not known to
occur on plants of the tribe Pomoideae.
Phyllosticta innumerabilis (C:32): leaf
spot, tache foliaire: on 1 BC [1012];
on 13 Que [F60].
Podosphaera clandestina (C:32): powdery
mildew, blanc: on 1 Ont [1257]; on 13
Man [F68], Ont, Que [1257], NS [1032,
1257, F64], Nfld [1659, 1660, 1661].
Puccinia sparganioides: rust, rouille:
on 13 Nfld [1661]. 0 I known only on
Fraxinus, II III alternate to
Spartina. This record needs to be
confirmed.
*Skeletocutis nivea (Polyporus
semipileatus)* (C:32): on 13 NS
[1032]. Associated with a white rot.
Sporidesmium agassizii: on 13 Man [1760].
Stigmatolemma poriaeforme: on 1 BC
[1012]. Associated with a white rot.
Strickeria obducens (Teichospora o.): on
1 BC [1012].
Synchytrium aureum: on 13 NS [1032].
Tomentella pallidofulva: on 13 BC [1341].
Tubercularia vulgaris (holomorph *Nectria
cinnabarina*) (C:32): on 1 BC [1012];
on 13 Ont [1340].
Valsa sp. (C:32): on 1 BC [1012]; on 13
Ont [1340].

V. ceratosperma (V. ceratophora) (C:32):
on 13 Ont [1340, F65].

AMMOPHILA Host   GRAMINEAE

1. *A. arenaria* (L.) Link, european
beachgrass; imported from Europe,
natzd., BC.
2. *A. breviligulata* Fern., beachgrass,
ammophile à ligule courte; native,
Ont-Nfld.

Puccinia coronata (C:32): crown rust,
rouille couronnée: on 1 BC [1816].
P. graminis (C:32): rust, rouille: II
III on 2 Ont [1236].

AMPHICARPA Ell.   LEGUMINOSAE

1. *A. bracteata* (L.) Fern., hog peanut,
amphicarpée bractéolée; native, Sask-NS.

Erysiphe communis (E. polygoni) (C:33):
powdery mildew, blanc: on 1 Ont [1257].
Uromyces appendiculatus (U. phaseoli):
rust, rouille: 0 I II III on 1 Ont
[1236].

AMSINCKIA Lehm.   BORAGINACEAE

Rough-hairy North & South American
annuals. Commonly called Tarweed or
Fiddleneck.

1. *A. menziesii* (Lehm.) Nels. & Macbr.;
native, Yukon, BC.

Erysiphe cichoracearum: powdery mildew,
blanc: on 1 BC [1012].

ANAPHALIS DC.   COMPOSITAE

1. *A. margaritacea* (L.) Clarke, pearly
everlasting, immortelle; native, Mack,
BC-Nfld.

Leptosphaeria doliolum (C:33): on 1 Que
[70].
L. ogilviensis (C:33): on *A.* sp. Que
[70].
Niesslia exilis (N. pusilla) (C:33): on 1
Que [70].
Ophiobolus mathieui: on 1 BC [1620].
Pleospora ambigua (C:33): on 1 Que [70].
Uromyces amoenus (C:33): rust, rouille:
on 1 BC [1012, 1816], Ont [1341], Que
[282].

ANCHUSA L.   BORAGINACEAE

1. *A. azurea* Mill., Italian bugloss,
buglosse d'Italie; imported from
Mediterranean region.

Sclerotinia sclerotiorum (C:33): on 1 BC
[1816].

ANDROMEDA L.   ERICACEAE

1. *A. glaucophylla* Link, bog rosemary,
   andromède glauque; native, Man-Nfld.
2. *A. polifolia* L., bog rosemary,
   andromède à feuilles de Polium; native,
   Yukon-Keew, BC-n. Que, s. Labr.

Epipolaeum andromedae: on 1 Ont [73].
   The source of this record was Rehm
   collection 2017. The part at DAOM and
   presumably elsewhere has the host
   misdetermined. The plant is *Andromeda
   glaucophylla*, not *Cassandra*.
Exobasidium vaccinii (C:33): on 1 Que
   [282]; on 2 BC [1816]; on *A.* sp. Que
   [1385].
E. vaccinii var. *vaccinii*: on 2 BC
   [1012].
Gibbera andromedae: on living and
   languishing leaves of 1 Ont, Que [73].
   The Ont record, collected by Dearness,
   is reported on *A. polifolia* but the
   packet (Ellis & Ev., N.A.F. 2363) at
   DAOM contains *A. glaucophylla* Link
   (*A. polifolia* sensu Dearness).
Godronia cassandrae: on 1 Que [1678].
G. cassandrae var. *cassandrae*: on 1 Que
   [1525].
Mycosphaerella minor (C:33): on 1 Que
   [70].
M. vaccinii (C:33): on 1 Que [70].
Placuntium andromedae (Rhytisma a.)
   (C:33): on 1 NWT, NS, Nfld [1340]; on 2
   Mack [53], BC [1012, 1816]; on *A.* sp.
   Ont [1340].

ANDROPOGON L.   GRAMINEAE

1. *A. gerardii* Vitman, big bluestem,
   barbon de Gérard; native, Sask-Que.

See also *SCHIZACYRIUM*.

Puccinia andropogonis (C:33): rust,
   rouille: II III on 1 Ont [1236].
P. ellisiana (C:33): rust, rouille: II
   III on 1 Ont [1236].

ANEMONE L.   RANUNCULACEAE

1. *A. canadensis* L., Canada anemone,
   anémone du Canada; native, sw. Mack,
   BC-PEI.
2. *A. coronaria* L., poppy anemone,
   anémone couronnée; imported from
   Mediterranean region. 2a. *A. c.* var.
   St. Brigid; a cultivar.
3. *A. cylindrica* Gray, long-headed
   anemone, anémone cylindrique; native
   BC-sw. Que.
4. *A. drummondii* Wats., Drummond's
   anemone, anémone de Drummond; native,
   Yukon, nw. Mack, BC, sw. Alta.
5. *A. multifida* Poir, cut-leaf anemone,
   anémone multifide; native, Yukon-Keew,
   Alta-NB, Nfld. 5a. *A. m.* f.
   multifida (*A. globosa* (T. & G.)
   Nutt.); native, Yukon-Keew, BC-Alta,
   Ont-Que & Nfld.

6. *A. nemorosa* L., european wood anemony;
   imported from Europe.
7. *A. parviflora* Michx., small-flowered
   anemone, anémone à petites fleurs;
   native, Yukon-Labr.
8. *A. patens* L. var. *wolfgangiana*
   (Bess.) Koch (*A. ludoviciana* Nutt.),
   prairie crocus, crocus; native,
   Yukon-Mack, BC-Ont.
9. *A. quinquefolia* L., wood anemone,
   anémone des bois; native, BC-NS. 9a.
   A. q. var. *interior* Fern.; native,
   Alta-NB.
10. *A. riparia* Fern., riverbank anemone,
   anémone des rivages; native, BC-NS and
   Nfld.
11. *A. virginiana* L., Thimbleweed, anémone
   de Virginie; native, NS.

Aecidium ranunculacearum (C:34): on 6 Ont
   [437].
Erysiphe communis (E. polygoni): powdery
   mildew, blanc: on 1 Ont [1236], Que
   [1257]; on 11 Ont [1257].
Phloeospora anemones (C:34): on 5 BC
   [1012]; on 5a BC [1816].
Puccinia anemones-virginianae (C:34):
   rust, rouille: III on 3 Que and on 10,
   11 Ont [1236].
P. gigantispora (C:34): rust, rouille:
   on 4, 7 Mack [53]; on 5, 7 BC [1012]; on
   5a, 7 BC [1816]; on 5 Yukon [460].
P. magnusiana (C:34): rust, rouille: 0 I
   on 3 Ont [1236].
P. pulsatillae (C:34): rust, rouille: on
   7 BC [1012, 1816], Que [1236]; III on 8
   Sask [1839].
P. recondita s.l. (C:34): rust, rouille:
   on 3, 5, 9, 11 Ont [1236]; on 9, *A.*
   sp. Ont [1341]; on 9a Ont [1828].
Tranzschelia sp.: on 9 Ont [1828].
T. anemones (Urocystis a.): on 5 Yukon
   [460]; on 9 Ont [1236, 1828]; on 8, *A.*
   sp. Ont [1341]. Conners (p. 34) felt
   that Nannfeldt's concept was too broad,
   thus Conners used *T. fusca* Diet.
T. discolor: on 2 BC [1816]. See comment
   under *Prunus*.
T. pruni-spinosae (C:34): rust, rouille:
   on 9 Ont [1236].
T. pruni-spinosae var. *discolor*: on 2
   BC [2008]; on 2a BC [1012]. See comment
   under *Prunus*.

ANETHUM L.   UMBELLIFERAE

1. *A. graveolens* L., dill, aneth;
   imported from sw. Asia.

Alternaria sp.: rusty root: on 1 Ont
   [1708].
Cylindrocarpon sp.: rusty root: on 1 Ont
   [1708].
Olpidium brassicae: on 1 Ont [67].
Penicillium sp.: on 1 Ont [1708].
Phoma anethi (C:34): blight, brûlure: on
   1 Sask [337].
Sclerotinia sclerotiorum: stem rot,
   pourriture sclérotique: on 1 Sask
   [1140].

ANGELICA L.   UMBELLIFERAE

1. *A. arguta* Nutt.; native, BC-Alta.
2. *A. altopurpurea* L., alexanders, angélique noire-pourprée; native, Ont-Nfld.
3. *A. dawsonii* Wats.; moist or wet montane slopes, se. BC-sw. Alta.
4. *A. genuflexa* Nutt.; native, BC-Alta.
5. *A. lucida* L., sea coast angelica, céloplèvre brillanté; native, Yukon, BC, Ont-Labr.

Cercospora angelicae (C:35): on 3 BC [1551].
Cercosporidium angelicae: on 2 Ont [379].
C. depressum (Passalora d.): on 2 Ont [1551]; on 4 BC [379, 1551].
Pollaccia peucedani (Asperisporium p., Fusicladium p.): on 1 BC [1012, 1578], Alta [1578].
Puccinia angelicae (C:35): rust, rouille: on 4 BC, Alta [1551].
P. coelopleuri (C:35): rust, rouille: on 5 BC [1551].
Ramularia archangelicae (C:35): on 1, 5 BC [1551].
Stigmina sp.: shot hole: on 1 BC [1816].

ANTHOXANTHUM L.   GRAMINEAE

1. *A. odoratum* L., sweet vernal grass, flouve odorante; imported from Eurasia, natzd., BC, Ont-Nfld.

Cercosporidium graminis (Passalora g.) (C:36): brown stripe, strie brune: on 1 NB [333, 1195].
Claviceps purpurea (C:35): ergot, ergot: on 1 NB [333, 1195].
Puccinia graminis (C:35): rust, rouille: on 1 NB [333]; on *A.* sp. NB [1195].

ANTHYLLIS L.   LEGUMINOSAE

1. *A. vulneraria* L., kidney vetch, vulnéraire; imported from Europe, introduced in Ont, Que and NB. Sometimes grown as an ornamental or for forage.

Sclerotinia sp.: rot, pourriture: on 1 BC [1816].

ANTIRRHINUM L.   SCROPHULARIACEAE

1. *A. majus* L., common or large snapdragon, gueule de loup; perennial herb from the Mediterranean region, widely grown as an ornamental, in e. Canada, especially s. Ont and NS, escapes to disturbed sites and roadsides but hardly persistent.

Botrytis cinerea (holomorph Botryotinia fuckeliana) (C:36): gray mold, moisissure grise: on 1 BC [1816], Ont [163], in greenhouse NS [334, 336].
Oidium sp.: on 1 in greenhouse BC [1816].

Peronospora antirrhini (C:36): downy mildew, mildiou: on 1 Ont [333].
Phyllosticta antirrhini (C:36): leaf spot, tache foliaire: on 1 BC [1816].
Phytophthora cactorum: on 1 Ont [336].
Puccinia antirrhini (C:36): rust, rouille: on 1 BC [333, 1012, 1816], II III Ont [1236].
Pythium sp. (C:36): root rot, pourriture des racines: on 1 Alta [335].
Septoria antirrhini (C:36): leaf spot, tache septorienne: on 1 BC [1816].
Verticillium albo-atrum (C:36): verticillium wilt, verticillienne: on *A.* sp. BC [38].
V. dahliae (C:36): verticillium wilt, verticillienne: on 1 BC [1816].

APARGIDIUM Torr. & Gray   COMPOSITAE

1. *A. boreale* (Bong.) Torr. & Gray, perennial herb of sphagnum bogs and wet meadows at low to fairly high elevations; s. Alaska-BC.

Puccinia eriophori var. *apargidii*: rust, rouille: on 1 BC [1559].

APIUM L.   UMBELLIFERAE

1. *A. graveolens* L. var. *dulce* DC., celery, céleri; commonly cultivated and widely distributed.

Acremonium apii (Cephalosporium a.) (C:37): brown spot, tache brune: on 1 Man, Ont [336].
Alternaria dauci: blight, brûlure: on 1 Ont [1461, 1594].
Botryotinia fuckeliana (anamorph Botrytis cinerea): on *A.* sp. Ont [666a].
Cercospora apii (C:37): early blight, brûlure cercosporéene: on 1 BC [1816], Que [333, 334, 339, 1594, 1639, 1640, 1650, 1652, 1654], NS [338].
Fusarium spp. - *Stemphylium* spp. complex: on 1 Ont [1461].
Fusarium oxysporum f. sp. *conglutinans*: on 1 BC [1816]. Misidentified? This f. sp. attacks crucifers, whereas *F. oxysporum* f. sp. *apii* attacks *Apium*.
Olpidium brassicae: on 1 BC [1977], Ont [67].
Phyllosticta apii: leaf spot, tache foliaire: on 1 Que [333].
Pythium sp. (C:37): damping off, fonte: on 1 BC [1977], Ont [868], Que [333, 334, 1639].
P. irregulare: on 1 Ont [868].
P. sulcatum: on 1 Ont [868] by inoculation.
P. sylvaticum: on 1 Ont [868] by inoculation.
Rhizoctonia sp.: on 1 Que [1639].
R. solani (holomorph Thanatephorus cucumeris) (C:37): damping off, fonte des semis: on 1 Que [333, 334].

Sclerotinia sclerotiorum (C:37): drop or
   sclerotinia rot, pourriture
   sclérotique: on 1 Que [333, 334, 335,
   338, 339, 348, 1594, 1639, 1641, 1647,
   1651, 1652, 1653, 1654].
Septoria apii (C:37): late blight,
   brûlure septorienne: on 1 BC [335], Ont
   [335, 1461], Que [333, 334, 335, 339].
S. apiicola (C:37): late blight, brûlure
   septorienne: on 1 Que [348, 1651, 1652].
S. apii-graveolentis (C:37): late blight,
   brûlure septorienne: on 1 BC [1816],
   Que [1639, 1640, 1641, 1650].
Stemphylium botryosum (C:37): from seeds
   of 1 Ont [667].
Ulocladium consortiale (Stemphylium c.):
   from seeds of 1 Ont [667].

APOCYNUM L.   APOCYNACEAE

1. *A. androsaemifolium* L., spreading
   dogbane, herbe à la puce; native,
   Yukon-Mack, BC-Nfld.

Dearnessia apocyni: on 1 Ont [195],
   presumably Ont [1763].
Ophiobolus mathieui: on 1 BC [1620].
O. rubellus: on 1 Ont [1620].
Phyllosticta apocyni: on 1 Ont [195].
Puccinia seymouriana (C:38): rust,
   rouille: on 1 NS [1036].
Verticillium dahliae: on 1 BC [1978].

AQUILEGIA L.   RANUNCULACEAE

1. *A. flavescens* Wats., yellow columbine,
   ancolie jaune; native, BC-Alta.
2. *A. formosa* Fisch., wild columbine,
   ancolie gracieuse; native, Yukon,
   BC-Alta.
3. *A. vulgaris* L., european columbine,
   gants de Notre-Dame; imported from
   Europe, escaping to roadsides, fields
   and borders of woods, BC, Ont-Nfld.
4. Host species not named, i.e., reported
   as *Aquilegia* sp.

Ascochyta aquilegiae (C:38): leaf spot,
   tache ascochytique: on 4 BC [1816].
Erysiphe communis (E. polygoni) (C:38):
   powdery mildew, blanc: on 3 Ont, NS
   [1257]; on 4 BC [333, 334, 335, 337,
   338, 339, 1816], Alta [337].
Haplobasidion thalictri: leaf blotch,
   brûlure des feuilles: on 4 BC [1816].
Nodulosphaeria aucta: on 2 BC [72]. The
   host should be reconfirmed because the
   fungus is restricted to *Clematis recta*
   [761].
Puccinia recondita s.l. (C:38): rust,
   rouille: on 1 NWT-Alta [1063], BC
   [954]; on 1, 2 BC [1012]; on 2, 4 BC
   [1816].

ARABIS L.   CRUCIFERAE

1. *A. arenicola* (Richards.) Gelert.;
   native, Keew-Frank, Sask-Que, Labr.

2. *A. caucasica* Schlecht. (*A. albida*
   Stev.), wall rock cress, arabette de
   caucase; imported from Europe,
   occasionally spreading from cultivation.
3. *A. divaricarpa* Nels.; native,
   Yukon-Mack, BC-NB.
4. *A. drummondii* Gray (*A. confinis*
   Wats.); native, Yukon-Mack, BC-NS, Nfld,
   Labr.
5. *A. glabra* (L.) Bernh., tower-mustard,
   tourette; imported from Eurasia,
   introduced Yukon-Mack, BC-NS.
6. *A. hirsuta* (L.) Scop.; native,
   Yukon-Mack, BC-NS.
7. *A. holboellii* Hornem.; native,
   Yukon-Mack, BC-Ont.
8. *A. lemmonii* Wats.; native, Yukon,
   BC-Alta.
9. *A. lyallii* Wats.; native, Yukon,
   BC-Alta.
10. *A. lyrata* L.; native, Yukon-Mack,
    BC-Que.
11. *A. nuttallii* Robins.; native, BC-Alta.
12. *A. puberula* Nutt.; native, BC.

*Albugo candida (A. cruciferarum, Cystopus
   candidus)* (C:39): white rust,
   albugine: on 2 BC [1816]; on 5 Ont
   [195].
Cladosporium subsclerotioideum: on 5 Ont
   [195].
Macrosporium mycophilum: on 5 Ont [195].
Mycosphaerella tassiana var. *tassiana*
   (C:39): on 7 Sask [1327].
Peronospora parasitica (C:39): downy
   mildew, mildiou: on 2 BC [334, 1816];
   on 5 Ont [195].
Puccinia arabicola (C:39): rust,
   rouille: 0 I III on 4 Ont [1236].
P. holboellii (P. thlaspeos auct.)
   (C:39): rust, rouille: on 1 Ont, Que
   [1568] also Que [1341]; on 1, *A.* sp.
   Ont [1236]; on 3 Ont [1568]; on 3, 4, 8,
   9, 11 BC [1568]; on 4, 7, 9, 11 Alta
   [1568]; on 6, 8 BC [1012, 1816]; on 7
   Yukon, NWT, BC [1568]; on 10 Yukon, BC,
   Sask, Que [1568]; on *A.* sp. Alta
   [644].
P. monoica (C:39): rust, rouille: on 6,
   12 BC [1012, 1816]; on 10 Ont [1236].
Selenophoma sp.: on *A.* sp. [1323].

ARALIA L.   ARALIACEAE

1. *A. nudicaulis* L., wild sarsaparilla,
   salsepareille; native, sw. Mack, BC-Nfld.

See also *FATSIA*.

Ceratobasidium anceps (C:39): on 1 Ont
   [1341].
Gnomonia sp.: on 1 Ont [1340].
G. similisetacea: on 1 Ont [79].
Mundkurella mossii: on 1 Alta, Sask, Ont
   [1577].
Nyssopsora clavellosa (Triphragmium c.)
   (C:39): on 1 BC [1012, 1816], Ont
   [437,1660], Que [282]; on 1, *A.* sp.
   Ont [1236, 1341], NB [1032].

Trichometasphaeria gloeospora (C:39): on 1 Que [70].

ARAUCARIA Juss.   PINACEAE

1. *A. araucana* (Molina) K. Koch, monkey puzzle, cult. in sw. BC.

Varicosporium elodeae: on 1 BC [43].

ARBUTUS L.   ERICACEAE

1. *A. menziesii* Pursh, arbutus or madrona, madrono; native, sw. coastal BC.

Aleurodiscus macrocystidiatus (C:39): on 1 BC [964, 1012] Lignicolous.
Amylocorticum cebennense: on 1 BC [1939]. See *Abies*.
Antrodia lindbladii (Poria cinerascens) (C:40): on 1 BC [1012]. See *Abies*.
A. malicola (Trametes m.): on 1 BC [1012]. See *Acer*.
A. serialis (Poria ferox): on *A.* sp. BC [1014]. See *Abies*.
Armillaria mellea: on 1 BC [1012, F74]. Causes a root rot. See *Abies*.
Ascochyta hanseni (C:39): leaf spot, tache des feuilles: on 1 BC [1816].
Athelia scutellare (Corticium s.) (C:39): on 1 BC [1012]. Lignicolous.
Basidiodendron caesio-cinerea (Sebacina c.): on 1 BC [1012]. Lignicolous.
B. eyrei: on 1 BC [1016]. See *Acer*.
Capnodium sp.: on 1 BC [1012, F62].
C. walteri: on 1 BC [798]. Associated with *Trialeurodes merlini*, a scale insect.
Coccomyces arbutifolius: on 1 BC [1609].
C. quadratus (C:39): on 1 BC [1012].
Coriolus versicolor (Polyporus v.) (C:40): on 1 BC [1012]. See *Abies*.
Cytospora sp.: on 1 BC [1012]. Lignicolous.
Dacrymyces capitatus (D. deliquescens var. *ellisii)*: on 1 BC [1012]. Lignicolous. See *Alnus*.
Dendrothele incrustans (Aleurocorticium i.): on 1 BC [965]. Lignicolous.
Didymosporium arbuticola (C:39): on 1 BC [1012, F62].
Diplodia maculata (C:39): on 1 BC [1012].
Exobasidium vaccinii (C:39): on 1 BC [1012, 1816, F69].
Hendersonula toruloidea: on 1 BC [1012, F72].
Hymenochaete spreta (C:39): on 1 BC [1012]. See *Acer*.
H. tabacina (C:39): on 1 BC [1012]. See *Abies*.
Hyphoderma amoenum: on 1 BC [1012]. Lignicolous.
H. puberum: on 1 BC [1012]. See *Abies*.
Hyphodontia crustosa: on 1 BC [1012]. See *Abies*.
Hysterographium vulvatum (C:39): on 1 BC [1012].
Meruliopsis corium: on *A.* sp. BC [568]. See *Acer*.

Monochaetia desmazierii (C:39): on 1 BC [1816].
M. monochaeta: on 1 BC [1012].
Mycoacia uda: on 1 BC [1012]. Lignicolous.
Mycosphaerella arbuticola (C:39): leaf spot, tache des feuilles: on 1 BC [1816].
Panellus serotinus (Pleurotus s.) (C:40): on 1 BC [F66]. See *Acer*.
Peniophora cinerea (C:40): on 1 BC [1012]. See *Abies*.
P. incarnata (C:40): on 1 BC [1012]. Lignicolous.
Phellinus ferreus (Poria f.) (C:40): on 1 BC [1012]. See *Acer*.
P. igniarius (Fomes i.) (C:39): white trunk rot, carie blanche du tronc: on 1 BC [1012].
Phlebia centrifuga (P. albida auct.): on 1 BC [1012]. See *Abies*.
P. deflectens (Corticium d.) (C:39): on 1 BC [1012]. See *Acer*.
Phoma sp.: on 1 BC [1012].
Pleurotus subareolatus: on 1 BC [1012]. See *Acer*.
Poria stenospora (C:40): on 1 BC [1012]. Presumably a misdetermination because the "type" specimen is a collection of *Dichomitus squalens* [1501].
Pucciniastrum sparsum (C:40): leaf rust, rouille des feuilles: on 1 BC [1012, 2008].
Rhytisma arbuti (C:40): tar spot, tache goudronneuse: on 1 BC [1816].
Seimatosporium arbuti (C:40): on 1 BC [1012].
Skeletocutis nivea (Polyporus semipileatus): on 1 BC [1012]. See *Amelanchier*.
Stereum hirsutum (C:40): on 1 BC [1012]. See *Abies*.
S. ostrea (C:40): on 1 BC [1012]. See *Abies*.
Stictis radiata (C:40): on 1 BC [1012].
Trechispora vaga: on 1 BC [1012]. See *Abies*.
Trichaptum abietinus (Hirschioporus a., Polyporus a.) (C:40): on 1 BC [1012, 1816]. See *Abies*.
Valsa sp.: on 1 BC [1012].

ARCEUTHOBIUM M. Bieb.   LORANTHACEAE

Small parasitic plants of the northern hemisphere, often referred to as dwarf mistletoes. Parasitic on Pinaceae.

1. *A. americanum* Nutt. ex Engelm., pine mistletoe; native, BC-Ont, parasitic on pine, often *Pinus contorta* var. *latifolia*.
2. *A. campylopodium* Engelm.; native, BC, parasitic on branches of conifers.
3. *A. douglasii* Engelm.; native, BC, parasitic on *Pseudotsuga menziesii*.
4. *A. pusillum* Peck, dwarf mistletoe, petit gui; native, Sask-Nfld, parasitic on *Picea*, rarely *Larix* or *Pinus strobus*.

5. *A. tsugense* (Rosend.) G.N. Jones;
   native, BC, parasitic on *Tsuga*
   heterophylla and *Pinus contorta* var.
   contorta.
6. Host species not named, i.e., reported
   as *Arceuthobium* sp.

Alternaria alternata: on 4 Man [1760].
Ascoconidium tsugae: canker, chancre: on
   5 BC [521].
Botryosphaeria tsugae: canker, chancre:
   on 5 BC [521].
Cladosporium sp.: on 1 NWT-Alta [1063],
   Alta [952], Sask [F70].
Coccomyces heterophyllae: canker,
   chancre: on 5 BC [521].
Colletotrichum gloeosporioides (holomorph
   Glomerella cingulata): on 1 NWT-Alta
   [1063], BC, Alta, Sask [1155] also BC
   [953], Alta [951, 952, 1157, F66, F68],
   Sask [F70]; on 6 BC [F70], Alta [F65].
Cylindrocarpon gillii (*Septogloeum g.*):
   on 1 NWT-Alta [1063], BC, Alta, Sask
   [1155] also Alta [952, F61, F66], Sask
   [F70]; on 1, 3, 5 BC [1012]; on 1, 3, 6
   BC [F63]; on 3 BC [1063]; on 6 Alta
   [F65].
Discocainia treleasei: canker, chancre:
   on 5 BC [521].
Glomerella cingulata (anamorph
   Colletotrichum gloeosporioides): on 1
   BC [1012].
Pestalotia sp.: on 5 BC [1012].
Pestalotiopsis maculiformans (*Pestalotia*
   m.): on 2 BC [F70].
Phomopsis lokoyae: canker, chancre: on 5
   BC [521].
Sphaeropsis sapinea (*Diplodia pinea*): on
   1 BC [1012].
Wallrothiella arceuthobii: on 1 NWT-Alta
   [1063], BC, Alta, Sask [1155] also Alta,
   Sask, Man [F70] and BC [905, 953, 1012],
   Alta [53, 951, F61, F66], Ont [1828]; on
   3 BC [1012, 1063, F61]; on 6 Alta [F65,
   F67], Sask [F63].
Xenomeris abietis: canker, chancre: on 5
   BC [521].

ARCTAGROSTIS Griseb.   GRAMINEAE

1. *A. latifolia* (R. Br.) Griseb.; native,
   Yukon-n. BC, n. Alta, n. Sask, n. Man,
   n. Que, Labr.

Hysteropezizella lyngei: on 1 Frank [664].
Selenophoma drabae: on 1 Frank [1244]. A
   misdetermination.

ARCTIUM L.   COMPOSITAE

1. *A. lappa* L., great burdock, grande
   bardane; natzd., Man-NB.
2. *A. minus* (Hill) Bernh., common
   burdock, bardane mineur; imported from
   Eurasia, natzd., BC-Nfld.
3. Host species not named, i.e., reported
   as *Arctium* sp.

Crucibulum laeve (C. vulgare): on 3 Ont
   [1341]. Probably on dead stems.
Erysiphe cichoracearum: powdery mildew,
   blanc: on 2 Ont [1257].
Ophiobolus acuminatus: on 3 Ont [1620].
O. anguillidus: on old stems of 3 Ont
   [1620].
Pseudohelotium canadense: on dead stems
   of 1 Ont [440].
Puccinia bardanae (C:40): rust, rouille:
   II III on 2 BC, Man, Ont, Que, NB, NS
   [1558] and Ont [1236].
Sclerotinia sclerotiorum: on 2 Man [750].
Scopinella solani: on 2 Ont [1039].
Xylaria filiformis: on 3 Ont [1340].

ARCTOPHILA Rupr.   GRAMINEAE

   Marshy tundra and margins of tundra
pools; widespread arctic, perennial grasses.

1. *A. fulva* var. *pendulina* (Loest.)
   Holmb.

Pseudoseptoria everhartii (Selenophoma e.)
   (C:41): on 1 Yukon [1731].

ARCTOSTAPHYLOS Adans.   ERICACEAE

1. *A. alpina* (L.) Spreng., alpine
   bearberry, busserole alpine; native,
   Yukon-Frank, Alta-Labr. la. *A. a.*
   ssp. *rubra* (Rehd. & Wils.) Hulten (*A.*
   rubra (Rehd. & Wils.) Fern.), red
   bearberry, arctostaphyle à fruits
   éclartes; native, generally the same
   range although slightly more southern
   than the typical form.
2. *A. columbiana* Piper, hairy manzanita,
   arctostaphyle de la
   Colombie-Britannique; native, BC.
3. *A. uva-ursi* (L.) Spreng., bearberry,
   arctostaphyle raisin d'ours; native,
   Yukon-Keew, Alta-Nfld.
4. Host species not named, i.e., reported
   as *Arctostaphylos* sp.

Chaetonaevia nannfeldtii: on leaves of 4
   Que, Labr [73].
Chrysomyxa arctostaphyli (aecial state
   Peridermium coloradense) (C:41): III
   on 3 Yukon, Mack, BC, Alta, Sask, Man
   [2008] also Yukon [460], Mack [F64, F65,
   F66, F67], BC [953, 954, 1012, 1816,
   2007], Alta [53, 950, 951, 952, F62],
   Ont [1236, 1341, F62, F66], Que [F60,
   F64] also NWT-Alta [1063], Sask, Man
   [F67, F68]; on 3, 4 Mack [53]; on 4 Ont
   [F65]. The aecial stage causes witches'
   brooms on *Picea* spp.
Conoplea fusca: on 3 Man [1760].
Coccomyces arctostaphyli: on 3 Alta
   [1609].
Eupropolella arctostaphyli: on 3 Alta
   [644].
Exobasidium sp.: on 3 Alta [952].
E. vaccinii (C:41): red leaf, rouge: on
   3 BC [1816]; on 4 Que [1385].

E. vaccinii var. *arctostaphyli* (C:41):
on 2 BC [1816]; on 2, 3 BC [1012]; on 3
Alta [644].
E. vaccinii var. *vaccinii* (C:41): on 3
Yukon [460], BC [1012].
Gibbera grumiformis (C:41): on leaves of
1 Que [70, 73].
Leptosphaeria hyperborea (C:41): on 1 Que
[70].
Monochaetia phyllostictea: on 2 BC [1012].
Phloeospora sp.: on 3 Ont [1340].
Podosphaera myrtillina: on 1 Keew [1257].
Protoventuria alpina (*Antennularia a.*,
Gibbera petrakii) (C:41): on 1 Que
[70]; on living or dying leaves of 4
Que, Labr [73].
P. alpina var. *major* (*Antennularia a.*
var. *major*): on dead branches and
leaf bases of 1 Que [73].
Pucciniastrum sparsum (C:41): rust,
rouille: II III on 1 Yukon [F64], Mack
[F62], BC [1816], Alta [F61]; on 1a
Yukon [460], NWT [729], Mack [53], BC
[954, 1012], Alta [53, 729, 951, 952],
Ont [1236] and NWT-Alta [1063]; on 4
Yukon, Mack, BC, Alta, Man, Ont [2008].
Sphaceloma sp.: on 3 Alta [644].
Xenomeris raetica (C:41): on leaves of 1
Que [70, 73].

ARENARIA L.   CARYOPHYLLACEAE

1. *A. capillaris* Poir.; native,
Yukon-Mack, BC-Alta.
2. *A. lateriflora* L., blunt-leaved
sandwort, sabline latériflore; native,
Yukon-Mack, Alta-Nfld.
3. *A. obtusiloba* (Rydb.) Fern.; native,
Yukon-Mack, BC-Alta, e. Que, Nfld.
4. *A. peploides* L., seabeach sandwort,
sabline faux-péplus; native, Yukon-BC,
Ont-Labr.
5. *A. rossii* R. Br.; native, Yukon-Frank,
BC-Alta, Man, Que.
6. *A. rubella* (Wahl.) Sm.; native,
Yukon-Frank, BC-Alta, Ont-Que, Nfld.

Mollisia epitypha: on 5 Frank [664].
Mycosphaerella densa (C:41): on 5, 6
Frank [1586].
M. tassiana var. *tassiana* (C:41): on 3
Que [70].
Pleospora helvetica (C:42): on 3 Que [70].
Puccinia arenariae (C:42): rust,
rouille: on 1 BC, Alta [1578]; on 4
Man, Que [1236].
Ustilago violacea (C:42): smut, charbon:
on 2 Que [1390].

ARISAEMA Mart.   ARACEAE

1. *A. dracontium* (L.) Schott, dragon
root, ariséma dragon; native, Ont-Que.
2. *A. triphyllum* (L.) Schott (*A.
atrorubens* (Ait.) Blume), jack in the
pulpit, petit prêcheur; native, Man-NS.

Ramularia arisaemae: on leaves of 2 Ont
[440].

Uromyces ari-triphylli (*Aecidium caladii*)
(C:42): rust, rouille: 0 I II III on
1, 2 Ont [1236]; on 2 Ont [437, 1341,
1828, 1827].

ARMERIA Willd.   PLUMBAGINACEAE

1. *A. maritima* (Mill.) Willd., thrift,
gazon d'Espagne; native, Mack-BC, Sask,
e. Que, Nfld, Labr. 1a. *A. m.* var.
labradorica (Wallr.) Inversen; native,
Mack-Keew, e. Que, Nfld-Labr.

Mycosphaerella minor: on 1a Que [70].
Scleropleella hyperborea (*Leptosphaeria
h.*) (C:42): on 1a Que [70].
Uromyces armeriae ssp. *hudsonicus*
(C:42): rust, rouille: 0 I II III on
1a Que [1236].
U. armeriae ssp. *pacificus* (C:42):
rust, rouille: on 1 BC [1012].

ARNICA L.   COMPOSITAE

1. *A. alpina* (L.) Olin; native,
Yukon-Que, Nfld-Labr. 1a. *A. a.* ssp.
angustifolia (Vahl) Maguire; native,
Yukon-BC, Man, Nfld-Labr.
2. *A. amplexicaulis* Nutt.; native,
Yukon-Mack, BC-Alta.
3. *A. chamissonis* Less.; native,
Yukon-Mack, BC-Que.
4. *A. cordifolia* Hook., heartleaf arnica,
arnica à feuilles cordées; native,
Yukon-Mack, BC-Sask.
5. *A. latifolia* Bong., broadleaf arnica,
arnica à larges feuilles; native,
Yukon-Mack, BC-Alta.
6. *A. lessingii* Greene; native,
Yukon-Mack, BC.
7. *A. louiseana* Farr; native, Yukon-Mack,
BC-Alta, Que-Nfld. 7a. *A. l.* ssp.
frigida (Mey.) Maguire (*A. frigida*
Mey.); native, Yukon-Mack, BC.
8. *A. mollis* Hook., hairy arnica, arnica
soyeux; native, Yukon-Mack, BC-Alta,
Que-NB.
9. *A. rydbergii* Greene; native, BC-Alta.

Didymella delphinii (C:43): on 5 BC [296].
Entyloma arnicale (C:43): on 1 Alta
[644]; on 2, 3, 4, 5, 8, 9 BC [1583]; on
4 Yukon, Alta [1583].
Puccinia arnicalis (C:43): rust,
rouille: on 4, 8 BC, on 4 Alta, on 6
Yukon, on 7a Mack, all from [1582]; and
on *A.* sp. Yukon [720] as *P. hieracii*.
P. hieracii in the broad sense
encompassed *P. arnicalis* but Savile
[1582] elucidated the reasons for
treating the rust on *Arnica* as *P.
arnicalis*.
Sphaerotheca fuliginea (C:43): powdery
mildew, blanc: on 1a Frank [1586].

ARONIA Medic.   ROSACEAE

1. *A. melanocarpa* (Michx.) Elliott
(*Pyrus m.* (Michx.) Willd.), black
chokeberry, aronia noir; native, Ont-PEI.

2. *A. prunifolia* (Marsh) Rehd. (*A.
floribunda* (Lindl.) Spach), purple
chokeberry, aronie à feuilles de
prunier; native, NB-PEI.

Acrotheca dearnessiana: on 1 Ont [195,
1502].
Cercosporella pirina: on 1 Ont [195].
Gymnosporangium clavipes (C:43): rust,
rouille: 0 I on 1 Ont [1236, 1263]; on
2 NS [1263].
Podosphaera clandestina (C:43): powdery
mildew, blanc: on 1 Que, NS [1257].

ARRHENATHERUM Beauv.   GRAMINEAE

1. *A. elatius* (L.) Beauv., tall oat
grass, avoine élevée; imported from
Eurasia, natzd., BC, Ont-NS, Nfld.

Bipolaris sorokiniana (holomorph
Cochliobolus sativus): on 1 Alta
[120] by inoculation.
Cercosporidium graminis (Passalora g.)
(C:43): brown stripe, strie brune: on
1 BC [1816].
Drechslera tritici-repentis (C:43): on
A. sp. Alta [1611].
Puccinia graminis (C:43): stem rust,
rouille de la tige: on 1 BC [1816].
Sclerotinia borealis (C:43): snow mold,
moisissure nivéale: on 1 BC [1816].
Ustilago avenae (C:43): smut, charbon:
on 1 BC [1816].

ARTEMESIA L.   COMPOSITAE

1. *A. absinthium* L., wormwood; native,
Man-Nfld.
2. *A. biennis* Willd., biennial sagewort,
armoise bisannuelle; native to w. USA
and New Zealand, introduced, Mack,
BC-PEI.
3. *A. campestris* L., sagewort, wormwood,
aurone des champs; native, Yukon-Frank,
BC-NS, Nfld-Labr. 3a. *A. c.* ssp.
borealis (Pall.) Hall & Clements (*A.
borealis* Pall.); native, Yukon,
BC-Alta, Que, Nfld-Labr.
4. *A. cana* Pursh, silvery sagebrush,
armoise argentée; native, Yukon,
Alta-Man.
5. *A. dracunculus* L., tarragon, estragon;
native, Yukon, BC-Man, sometimes
introduced in the east.
6. *A. frigida* Willd., prairie sagewort,
armoise rustique; native, Yukon-Mack,
Alta-Man, introduced, Ont, Que and NS.
7. *A. herriotii* Rydb.; native, Alta-Sask.
8. *A. ludoviciana* Nutt., western mugwort,
armoise de Louisiane; native, BC-Ont and
introduced into Que, NB and PEI.
9. *A. tilesii* Ledeb.; native, Yukon-Keew,
ne. Man-ne. Ont.
10. *A. tridentata* Nutt., big sagebrush,
armoise tridentée; native, BC-Alta.

Crucibulum parvulum: on *A.* sp. Alta
[191].

Erysiphe cichoracearum (C:44): powdery
mildew, blanc: on 7 NWT-Alta [1063],
Mack [53].
Mycosphaerella minor (C:44): on 3 Que
[70].
Mycovellosiella ferruginea: on 1 Ont
[382], causing yellowish discolorations
of upper leaf surface.
Ophiobolus tanaceti: on 2 Sask [1620].
Puccinia atrofusca (C:44): rust,
rouille: on *A.* sp. NWT-Alta [1063].
P. millefolii (C:44): rust, rouille: on
9 Yukon [7].
P. tanaceti (P. ludovicianae) (C:44):
rust, rouille: on 4 NWT-Alta [1063]; on
5 BC [1816]; on 5, 6, 10 BC [1012]; on 8
Ont [1236]; on Yukon [460].
Wettsteinina mirabilis (C:44): on 3a Que
[70].

ARUNCUS Adans.   ROSACEAE

1. *A. dioicus* (Walt.) Fern. (*A.
sylvester* Kostel.), goat's-beard, barbe
de bouc commun; native, BC, introduced
in Que, NS.

Cristulariella depraedans: on 1 BC [1427].

ASARUM L.   ARISTOLOCHIACEAE

1. *A. caudatum* Lindl., western wild
ginger; native, BC.

Puccinia asarina (C:44): rust, rouille:
on 1 BC [1012, 1816].

ASCLEPIAS L.   ASCLEPIADACEAE

1. *A. incarnata* L., swamp milkweed,
asclépiade incarnate; native, Man-PEI.
2. *A. speciosa* Torr., showy milkweed,
belle asclépiade; native, BC-Man.
3. *A. syriaca* L., common milkweed, herbe
à coton; native, Man-PEI.
4. *A. tuberosa* L., butterfly-weed,
asclépiade tubéreuse; native, Ont-Que.

Ophiobolus rubellus: on 3 Ont [1620].
Sphaerotheca fuliginea: powdery mildew,
blanc: on 4 Ont [1257].
Uromyces asclepiadis (C:45): rust,
rouille: II III on 1, 3 Ont [1236].
Verticillium dahliae: on 2 BC [1978].

ASPARAGUS L.   LILIACEAE

1. *A. officinalis* L. (*A. o.* var.
altilis L.), garden asparagus,
asperge; imported from Europe, known as
a garden escape in all provinces except
Sask.

Fusarium sp. (C:45): crown rot,
pourriture fusarienne: on 1 BC [1816],
Ont [337, 1460, 1461, 1594], Que [1639].
F. oxysporum f. sp. *asparagi*: root rot,
flétrissure fusarienne: on 1 Que [333].
Phomopsis asparagi (Phoma a.): stem
canker, chancre phoméen: on 1 Ont [335].

Puccinia asparagi (C:45): rust, rouille:
on 1 BC [333, 1012, 1816], Sask [1839],
Ont [335, 1236, 1341].
Pythium sp.: on 1 Ont [1461].
Rhizoctonia crocorum (holomorph
Helicobasidium purpureum) (C:45):
violet root rot, rhizoctone violet: on
1 BC [1816].
R. solani (holomorph *Thanatephorus
cucumeris*) (C:45): rhizoctonia,
rhizoctone commun: on 1 Ont [1461].
Sclerotinia sclerotiorum: on 1 Sask
[1140].
Stemphylium botryosum (holomorph
Pleospora herbarum): (C:45): from
seeds of 1 Ont [667].

ASTER L.   COMPOSITEAE

1. *A. acuminatus* Michx., acuminate aster,
aster acuminé; native, Ont-Nfld.
2. *A. borealis* (Torr. & Gray) Provanchant
(*A. junciformis* Rydb.), rush aster,
aster jonciformeé; native, Yukon-Mack,
Man-PEI.
3. *A. campestris* Nutt.; native, BC-Alta.
4. *A. ciliolatus* Lindl. (*A. lindleyanus*
Torr. & Gray), ciliolate aster, aster
ciliolé; native, Mack, BC-NS.
5. *A. conspicuus* Lindl., showy aster,
aster remarquable; native, BC-Sask.
6. *A. cordifolius* L., blue wood aster,
aster à feuilles cordées; native, Man-NS.
7. *A. engelmannii* (Eat.) Gray; native,
BC-Alta.
8. *A. ericoides* L. (*A. multiflorus*
Ait.), heath aster, aster ericoide;
native, Yukon-Mack, Alta-Que.
9. *A. laevis* L. (*A. concinnus* Colla),
smooth aster, aster lisse; native,
Yukon, BC-Ont.
10. *A. lateriflorus* (L.) Britt., calico
aster, aster latériflore; native,
Man-PEI.
11. *A. lowrieanus* Porter, Lowrie's aster,
aster de Lowrie; native, Ont-Que.
12. *A. macrophyllus* L., bigleaf aster,
aster à grandes feuilles; native,
Man-PEI.
13. *A. novae-angliae* L., New England
aster, aster de la Nouvelle-Angleterre;
native, Man-Que, NS.
14. *A. novae-belgii* L., New Belgium aster,
aster de la Nouvelle-Belgique; native,
Ont-Nfld. 14a. *A. n.-b.* var. *belgii*
(*A. johannensis* Fern.), Lake St. John
aster, aster du lac St. Jean; native,
Ont-Nfld.
15. *A. occidentalis* (Nutt.) Torr. & Gray;
native, Mack, BC-Alta.
16. *A. pilosus* Willd.; native, Ont-Que.
17. *A. ptarmicoides* (Nees) Torr. & Gray,
upland white aster, aster faux-ptarmica;
native, Sask-NB.
18. *A. puniceus* L., purple-stemmed aster,
aster ponceau; native, Mack, Alta-NS,
Nfld.
19. *A. sagittifolius* Wedemeyer,
arrow-leaved aster, aster à feuilles
sagittées; native, Man-Ont.

20. *A. sedifolius* L.; imported from Europe.
21. *A. sibiricus* L.; native, Yukon-Keew,
BC-Alta.
22. *A. simplex* Willd., simple aster, aster
simple; native, Sask-NB, Nfld.
23. *A. subspicatus* Nees (*A. douglasii*
Lindl., *A. foliaceus* Lindl.); native,
BC-Sask, Ont-NS, Nfld-Labr. 22a. *A.
s.* var. *apricus* (Gray) Boivin;
native, BC-Alta.
24. *A. tradescanti* L., Tradescant aster,
aster de Tradescant; native, Ont-NS,
Nfld.
25. *A. umbellatus* Mill., flat-topped
aster, aster à ombelles; native,
Alta-Nfld.
26. *A. vimineus* Lam., native, Ont-Que.
27. Host species not named, i.e., reported
as *Aster* sp.

Aecidium compositarum: on 4 Ont [437].
Probably a specimen of *Uromyces
silphii*.
Basidiophora entospora (C:46): downy
mildew, mildiou: on 1, 23 BC [1816].
Cercosporella virgaureae: on 8 Man [1352].
*Coleosporium asterum (C. solidaginis, C.
tussilaginis)* (C:46): red rust,
rouille rouge: on 4, 5, 27 NWT-Alta
[1063], Mack [53]; on 4, 5, 7, 23, BC
[1012]; on 4, 5, 15, 23, 23a BC [1816];
on 4, 27 BC [953, 954], Alta [952]; on
4, 6, 9, 11, 12, 13, 14, 16, 17, 18, 20,
22, 23, 24, 25, 26, 27 Ont [1236]; on 5
Alta [53, 950, 1402]; on 7, 23, BC
[2007]; on 14 Que [282]; on 27 Alta [53,
644, 950, 951, 1402], Sask [F63], Man
[F68], Ont [F67], Que [282, 1377], NB
[333, 1032, 1194], NS [1036].
Entyloma compositarum (C:46): on 21 Yukon
[460].
Erysiphe cichoracearum (C:46): powdery
mildew, blanc: on 2, 4, 9, 10, 11, 12,
13, 14, 18, 19, 22, 24, 25 Ont and on
10, 12, 13, 18, 22, 24 Que [1257]; on 13
BC [1816]; on 27 Que [333].
E. communis: on 27 Que [333]. Probably a
specimen of *E. cichoracearum*.
Leptosphaeria doliolum (C:46): on 27 Que
[70].
Macrosporium florigenum: covering flowers
of 27 in gardens Ont [439].
Nectria pedicularis (C:46): on 27 Que
[70].
Ophiobolus erythrosporus: on 12 Ont
[1620]; on 27 Que [70, 1620].
O. fulgidus (C:46): on 27 Ont [1620].
Puccinia sp.: rust, rouille: on 3 Mack
[53].
P. asteris (P. a. var. *purpurascens)*
(C:46): brown rust, rouille brune: on
4, 5, 23 BC [1012, 1816]; on 5 Alta
[1839]; on 6, 9, 12, 13, 18, 27 Ont
[1236]; on 9 Alta [952]; on 9, 23, 27
Alta [644]; on 12 Ont [437, 1341], Que
[282], NB, NS [1032]; on 23 BC, Alta
[1578]; on 27 NWT-Alta [1063], Ont
[1341], NS [1032].
P. dioicae (C:46): rust, rouille: on 1,
4, 6, 12, 14a, 18, 19, 22, 27 Ont

[1236]; on 4, 15 BC [1012, 1816]; on 12
Ont [1341]; on 27 Alta [557, 644], NB,
NS [1036].

Ramularia asteris (C:46): blight, brûlure
ramularienne: on 27 Man [380].

Septoria atropurpurea (C:46): leaf spot,
tache septorienne: on 12 Ont [1340].

Sphaerotheca fuliginea: powdery mildew,
blanc: on 18 Que [1257].

Uromyces junci (C:46): rust, rouille: 0
I on 12 Ont [1236]; on 27 Ont [1341].

U. silphii (C:46): rust, rouille: 0 I on
12 Ont [1236].

ASTRAGALUS L.   LEGUMINOSAE

1. *A. aboriginum* Richards.; native,
    Yukon, BC-Sask, Que. la. *A.
    aboriginum* var. *richardsonii* (Sheld)
    Boivin.
2. *A. agrestis* Dougl.; native,
    Yukon-Mack, BC-Ont.
3. *A. alpinus* L., alpine milk vetch,
    astragale des alpes; native,
    Yukon-Frank, n. BC-Labr.
4. *A. cicer* L., mountain chick pea,
    astragale pois chiche; imported from
    Europe, natzd., s. Alta-Man.
5. *A. eucosmus* Robins.; native,
    Yukon-Frank, BC-NB.
6. *A. iskanderi* Lipsky; imported from
    Iraq.
7. *A. lentiginosus* Dougl.; native, BC.
8. *A. lotiflorus* Hook.; native, BC-Man.
9. *A. missouriensis* Nutt.; native,
    Alta-Man.
10. *A. pectinatus* Dougl.; native, Alta-Man.
11. *A. purshii* Dougl.; native, BC-Sask.
    11a. *A. purshii* var. *purshii*;
    native, Alta-Sask.
12. *A. robbinsii* (Oakes) Gray; native,
    Yukon-Mack, BC-Labr. 12a. *A. r.* var.
    minor (Hook.) Barneby (*A. r.* var.
    occidentalis S. Wats.); native,
    Yukon-Mack, BC-Alta.
13. *A. tenellus* Pursh; native, Yukon-Mack,
    BC-Man.
14. *A. umbellatus* Bunge; native,
    Yukon-Mack, BC.

Asteromella pichbaueri (C:47): on 3 Frank
[1543].

*Phloeospora serebrianikowii (Cylindrosporium
s.)*: on 3 Frank [1244].

Polystigma astragali: on 2 BC [1578]; on
A. sp. Alta [644].

Rhizoctonia leguminicola: on 4 Alta
[125]. Blackpatch, not previously known
in Canada, was found in 1977 naturally
infecting fields in Alta.

Septoria serebrianikowii: on 9 Alta [644].

Sphaerotheca fuliginea (C:47): powdery
mildew, blanc: on 3 Que [1257].

Uromyces lapponicus (C:47): rust,
rouille: on 1 BC [1012]; III on 10, 13
Alta [644].

U. lapponicus var. *lapponicus*: on 1,
12a Alta, on la Frank, and on 5 Que
[1589].

U. phacae-frigidae (C:47): rust,
rouille: on 14 Yukon, Mack [1588].

U. punctatus: rust, rouille: on 6 Sask
and on 7, 8, 11a BC [1587].

ATHYRIUM Roth   POLYPODIACEAE

1. *A. felix-femina* (L.) Roth, lady fern,
    fougère femelle; native, Yukon, BC,
    Man-Nfld. la. *A. f.-f.* var.
    michauxii (Spreng.) Farw. (*A.
    angustum* (Willd.) Presl); native,
    Man-Nfld.

Typhula athyrii: on la Que [1385].

Uredinopsis longimucronata (C:47): rust,
rouille: on 1 BC [1816, 2007, 2008],
Alta [2008], II III Ont [1236]; on la
Que [1377].

U. longimucronata f. *cyclosora* (C:47):
rust, rouille: on 1 BC [1012].

AUCUBA Thunb.   CORNACEAE

1. *A. japonica* Thunb., Japanese aucuba,
    aucuba de Japon; imported from the
    Himalayas and Japan. Evergreen shrubs,
    often grown for their ornamental foliage
    and colourful berries.

*Botrytis cinerea (holomorph Botryotinia
fuckeliana)*: on 1 BC [1816].

AURINIA Desv.   CRUCIFERAE

Biennial or perennial herbs native to
central and s. Europe and Turkey. Some
grown in rock gardens and borders. Until
recently put in the genus *Alyssum*.

1. *A. saxatile* (L.) Desv., golden tuft,
    corbeille d'or; imported from Europe and
    Turkey.

Peronospora parasitica: downy mildew,
mildiou: on 1 BC [1012].

AVENA L.   GRAMINEAE

1. *A. barbata* Pott ex Link, slender wild
    oats; imported from Mediterranean region.
2. *A. brevis* Roth, little oat, avoine
    courte; imported from Europe.
3. *A. fatua* L., wild oat, avoine sauvage;
    imported from Eurasia, natzd., BC-Nfld.
4. *A. longiglumis* Dur.; imported from
    Mediterranean region.
5. *A. sativa* L. (*A. chinensis* Link),
    oat, avoine; imported from Mediterranean
    region, commonly cultivated.
6. *A. sterilis* L. (*A. ludoviciana*
    Dur.), animated oat, avoine stérile;
    imported from Mediterranean region and
    c. Asia.
7. *A. strigosa* Schreb., lopside oat,
    avoine strigeuse; imported from Europe.
8. Host species not named, i.e., reported
    as *Avena* sp.

Absidia glauca: on seeds of 5 in storage
Sask, Man [1895].
A. orchidis: on seeds of 5 in storage
Sask, Man [1895].
Acremonium boreale: on 3, 5 BC, Sask, Man
[1707].
Alternaria alternata (A. tenuis s.l.*)*
(C:48): on seeds of 5 Sask, Man [1895]
and Man [1114] on overwintered plants.
Aspergillus sp.: on seeds of 5 Alta
[222], Man [1894].
A. flavus: on seeds of 5 Sask, Man [1895].
A. fumigatus: on seeds of 5 Sask, Man
[1895].
A. versicolor: on seeds of 5 Sask, Man
[1895].
*Bipolaris sorokiniana (Helminthosporium
sativum)* (holomorph *Cochliobolus
sativus*) (C:48): root rot, piétin
helminthosporien: on 5 Alta [223, 334,
335], Sask [961] by inoculation, Man
[1114], Ont [336, 252, 330], NS [595]
also Sask, Man [240, 1895 on seeds],
Alta, NB [333].
Cephalosporium acremonium: on
overwintered plants of 5 Man [1114].
This name has been applied to several
fungi and its original application is
uncertain. The most common fungus in
this aggregate has been named
Acremonium strictum [537] and the
references herein probably dealt with
that species.
Cercosporidium graminis (Passalora g.)
(C:49): brown stripe, strie brune: on
5 Alta, Ont, NB [334] also NB [333,
1196].
Cladosporium sp.: on seeds of 5 Sask, Man
[1895].
Cladosporium cladosporioides (C:48): on
overwintered plants of 5 Man [1114].
Claviceps sp.: on 3 Alta [1781].
C. purpurea (C:48): ergot, ergot: on 5
BC [1816], Alta [1594], Man [1364]; on
flowers of 8 Que [1385].
Colletotrichum dematium: anthracnose,
anthracnose: on 5 Alta [1595].
C. graminicola (C:48): anthracnose,
anthracnose: on 5 BC, Alta, Que [337]
also Alta [335, 339, 693], Man [672],
Que [334, 335], NB [1196] and Que, NB
[333].
Dendryphion nanum: on 5 Ont [819].
*Drechslera avenacea (Helminthosporium
avenae)* (holomorph *Pyrenophora
chaetomioides*) (C:48): leaf blotch or
stripe, rayure des feuilles: on 3 Sask,
Ont [1611]; on 5 Alta, Sask, Man, Ont,
Que, NB, NS, PEI [1611] also BC [1816],
Alta [223, 337, 1331, 1333, 1594], Sask
[335, 336], Man [339, 672, 1070], NB
[273, 1196], NS [337], Nfld [336] also
Alta, Nfld [333, 334, 338, 339], Sask,
Man [1071, 1895], Ont, Que, NS, PEI
[271] and NB, NS, PEI [333, 854]. Leaf
spots, caused by *D. avenacea, Septoria
avenae* f. sp. *avenae* and *Pseudomonas
coronafaciens*, the latter a bacterium,
reduced the yield in Man in 1971 by 0.7
million bushels [672].

D. teres (Helminthosporium t.) (holomorph
Pyrenophora t.): on seeds of 5 Sask,
Man [1895].
Epicoccum purpurascens (C:48): on 5 Man
[1114].
Erysiphe graminis (C:48): powdery mildew,
blanc: on 1 Ont [1255, 1257]; on 5 BC
[333, 335, 1594, 1816], Ont [338, 339,
1257], Que [335].
Fusarium sp. (C:49): root rot,
fusariose: on 5 BC [1816], Alta [223,
333, 334, 335, 1594], Sask [339], Man
[334], Que [337], NB, NS, PEI [854].
Fusarium spp. (C:49): including some
combinations of *F. acuminatum, F.
avenaceum, F. oxysporum, F. poae, F.
sacchari, F. sambucinum, F.
sporotrichioides, F. sulphureum, F.
tricinctum*, but not necessarily all
these taxa: on overwintered plants of 5
Man [1114].
F. culmorum (C:49): root rot, fusariose:
on 5 Sask, NS [337].
F. graminearum (holomorph *Gibberella
zeae*) (C:49): root rot, fusariose: on
5 PEI [338].
F. nivale: on 5 BC [1731]. See
Agrostis.
Gonatobotrys simplex: on overwintered
plants of 5 Man [1114].
Heterosporium avenae (C:49): leaf blight,
brûlure hétérosporienne: on 5 BC [1816].
Hormodendrum sp.: on seeds of 5 Sask, Man
[1895].
Mucor spp.: on overwintered plants of 5
Man [1114].
Penicillium spp.: on 5 Alta [222], Man
[1114, 1894] also Sask, Man [1895].
Phaeosphaeria avenaria f. sp. *avenaria
(Leptosphaeria a.* f. sp. *a.)*
(anamorph *Stagonospora avenae*)
(C:49): septoria leaf blotch or black
stem, septoriose: on 5 Alta, Sask, Ont,
PEI [1594] also Ont, Que, NB, PEI
[269, 1776] also Ont, Que, NS, PEI
[271], Nfld [269] and Ont [257, 270,
283], Que [264, 1297].
Puccinia coronata (C:49): crown rust,
rouille couronnée: on 3 Man [336]; II
III on 4, 7 Que [1236]; on 5 Sask, Man
[336, 337, 338], Ont [255, 258, 263,
334, 336, 641], Que [333, 1298], NB
[333, 335, 1196], PEI [336] also Man
[672, 1070], Ont, Que [271, 335, 337],
NS, PEI [333, 334]; on 5, 8 Ont [1236].
In 1971 crown rust in Man reduced the
yield by 0.7 million bushels [672].
P. coronata var. *avenae (P. coronata* f.
sp. *avenae)* (C:49): crown rust,
rouille couronnée: on 3 Sask [640],
Alta [339, 477, 639], Sask, Man, Ont,
Que [339, 472, 477, 481, 471, 639, 685,
686, 687, 689, 690, 691, 1533], NS, PEI
[639, 640, 692, 691, 854, 1533] also
Sask, Man [474, 480, 1071], Ont [257,
271, 260, 474, 473, 1594], Que [480], NB
[472, 639], NS [481, 685, 690, 1532],
PEI [472, 478] also Man, Ont, Que [476,
478, 479, 482, 640, 692, 1532] and Man

[473, 624, 1059, 1595]; on 8 Sask, Man,
Ont, Que [483] and Man [1536] by
inoculation. A severe outbreak in e.
Ont and w. Que in 1980 and 1981 caused a
29% reduction in kernel weights.

P. graminis (C:50): stem rust, rouille de
la tige: on 2, 3, 6, 7, 8 Ont [1236];
on 5 BC, Sask, Man, Ont, Que [336, 338],
NB [333] also BC, Que, NS, PEI [333,
334] and BC [1816], Alta [336], Sask
[337], Man [672], Ont [334, 335], Que
[335]; on 8 ?Que, on oat straw
originating in Montreal [1239].

P. graminis ssp. *graminis* var. *stakmanii*
(P. graminis f. sp. *avenae)* (C:50):
stem rust, rouille de la tige: on 3 Man
[614, 1053]; on 5 BC [639, 614, 1044,
1046, 1052, 1055, 1533], Alta [1048,
1052, 1053], Sask, Man [614, 642, 1044,
1047, 1051, 1052, 1053, 1055, 1533],
Ont, Que [614, 625, 639, 642, 1044,
1045, 1046, 1054, 1055, 1056, 1057,
1058, 1533, 1594], NB [1057, 1594], NS,
PEI [639, 640, 854, 1049] also Man [339,
618, 639, 640, 1045, 1046, 1048, 1049,
1050, 1054, 1056, 1058, 1059, 1071,
1594, 1595], Ont [255, 266, 339, 618,
641, 1047, 1048, 1050, 1052, 1592], Que
[339, 638], NS [1047, 1058, 1533]; on 7
BC [639]. The worst epidemic in two
decades caused serious crop losses in
Man and e. Sask in 1977. An estimated
530,000 ha of the 800,000 ha in the rust
area were affected, sustaining losses
from 5% to near total loss.

P. recondita: on 5 Ont [337]. Probably a
misdetermination of the fungus or host.

Pyrenophora chaetomioides (P. avenae)
(anamorph *Drechslera avenacea*)
(C:48): leaf blotch or stripe, rayure
des feuilles: on 5 Ont [253, 260, 270].

P. tritici-repentis (anamorph *Drechslera*
t.): on 5 Ont [1611].

Pythium spp.: on 5 Ont [261].

P. aristosporum: on 5 Ont [261].

P. arrhenomanes: on 5 Ont [261].

P. irregulare: on 5 Ont [261].

P. tardicrescens: on 5 Ont [261].

P. torulosum: on 5 Ont [261].

P. volutum (C:50): on 5 Ont [261].

Rhizopus sp.: on seeds of 5 Man, Sask
[1895].

Septoria sp.: on 5 Man [1059].

Stagonospora avenae (Septoria a.)
(holomorph *Phaeosphaeria avenaria*)
(C:49): on 5 BC, Ont, Que, NB, NS [639,
640, 1533], PEI [1533] also Man [1533],
Que [1298], NB, NS, PEI [272, 273, 854]
also Man, PEI, Nfld [639] and NB [1196];
on 8 Que [110, 281, 1294].

S. avenae f. sp. *avenae (Septoria a.* f.
sp. *a.)* (holomorph *Phaeosphaeria*
avenaria f. sp. *avenaria*) (C:49):
septoria leaf blotch or black stem,
septoriose: on 5 Alta, Man, NB, NS,
PEI, Nfld [333], Sask, Ont, PEI [334]
also Sask [1071], Man [335, 336, 672,
1070, 1071, 1533], Ont [255, 258, 263,
266, 335], Que [326, 327, 338, 1298], NS

[337], PEI [266], Nfld [335, 337] and
Sask, Ont [336, 337, 338, 339]; on 8
Ont, Que, NB [265] and Ont [254]. See
comment under *Drechslera avenacea*.

Streptomyces sp.: on seeds of 5 Sask, Man
[1895].

Thanatephorus cucumeris (Pellicularia
filamentosa, P. praticola) (anamorph
Rhizoctonia solani): on 5 Sask [1874].

Thielaviopsis basicola: on 5 Ont [546].

Ustilago sp.: smut, charbon: on 5 Sask,
Man [1071].

U. avenae (C:50): loose smut, charbon
nu: on 5 BC [1816], Alta [223], Sask
[333, 336], Man [1020, 1021, 1022, 1023,
1888], NB [1196], Nfld [334, 339] also
Alta, PEI [333, 334], Sask, Man [121,
334, 338, 1178], NB, NS [333] and Man,
NB [337, 339]; on 8 BC, Sask [1178].

U. kolleri (U. hordei sensu lat., *U.*
levis) (C:50): covered smut, charbon
couvert: on 5 BC [1816], Alta, Sask
[333, 334, 1594], Man [334, 339, 1020,
1021, 1022, 1023, 1178, 1888, 1889,
1890, 1892], NB, NS, PEI [333, 854] also
Sask [336, 338, 1178, 1838], NB [1196],
PEI, Nfld [334].

AXYRIS L.   CHENOPODIACEAE

1. *A. amaranthoides* L., russian pigweed,
ansérine de Russie; imported from
Eurasia, natzd., Mack, BC-PEI.

Mycosphaerella tassiana var. *tassiana*:
on 1 Sask [1327].

BACCHARIS L.   COMPOSITAE

1. *B. halimifolia* L., groundsel tree,
bacchante de Virginie; native to the e.
USA. Sometimes planted as an ornamental
for the more or less persistent foliage
or profuse flowers.

Meruliopsis corium: on 1 BC [568]. See
Acer.

BALSAMORHIZA Nutt.   COMPOSITAE

1. *B. sagittata* (Pursh) Nutt., balsam
root; native, BC-Alta.

Puccinia balsamorhizae (C:51): rust,
rouille: on 1 BC [644, 1012, 1241,
1256, 1816]; on *B.* sp. BC [1249].

BAMBUSA Schreber   GRAMINEAE

Stout, tall, clump-forming perennial
grasses native to tropical and subtropical
Asia with a few species in Mexico and South
America. Used as ornamentals, and in the
tropics as timber and sources of pulp.
Young stems of some species are edible.

Eurotium sp.: on *B.* sp. BC [1012].

BARTSIA L.   SCROPHULARIACEAE

1. *B. alpina* L., velvet bells; native,
   Frank, n. Man-Que, Nfld-Labr.

Melasmia mougeotii: on 1 Que [1554].
Ramularia bartsiae: on 1 Que [1554].

BECKMANNIA Host   GRAMINEAE

1. *B. erucaeformis* (L.) Host; imported
   from Eurasia.
2. *B. syzigachne* (Steud.) Fern., slough
   grass, beckmannie à écailles unies;
   native, Yukon-Mack, BC-Que, NS-PEI.

Acremonium boreale: on 2 BC, Alta, Sask
   [1707].
Drechslera tritici-repentis (holomorph
   Pyrenophora t.) (C:52): on 2 Alta
   [1611].
Erysiphe sp.: powdery mildew, blanc: on 2
   Yukon [460].
E. graminis (C:52): powdery mildew,
   blanc: on 1, 2 Sask [1257]; on 2 Sask
   [1255].
Helminthosporium sp.: on 2 BC [1728].
   Specimen was probably a *Drechslera* sp.
   or *Bipolaris* sp.
Physoderma beckmanniae: on 2 BC [1728].

BEGONIA L.   BEGONIACEAE

1. *B. tuberhybrida* Voss; hybrid tuberous
   begonia, used here as referring to
   horticultural tuberous-rooted begonias.

Botrytis cinerea (holomorph *Botryotinia
   fuckeliana*) (C:52): gray mold,
   moisissure grise: on 1 BC [1816], Ont
   [163], Que [337]; on *B.* sp. Alta
   [338], Ont [334].
Erysiphe cichoracearum (C:52): powdery
   mildew, blanc: on 1 Sask [339]; on *B.*
   sp. Alta [333, 1594], Ont [333, 1257,
   1806], Que [333, 337, 338, 1257].
E. communis (C:52): on *B.* sp. Ont [334,
   335], Que [335]. Probably misdetermined
   specimens of *E. cichoracearum*.
Oidium sp.: powdery mildew, blanc: on 1
   BC [1816]. Conidial state of
   Erysiphales.

BERBERIS L.   BERBERIDACEAE

1. *B.* x *emarginata* Willd.; a hybrid
   between *B. sibirica* Pall. and *B.
   vulgaris*.
2. *B. thunbergii* DC., Japanese barberry,
   épine-vinette du Japon; imported from
   Japan.
3. *B. vulgaris* L., barberry,
   épine-vinette commune; imported from
   Europe, natzd., BC, s. Man-PEI. 3a. *B.
   v. c*v. *macrophylla (B. macrophylla*
   Hort.).
4. *B. wilsoniae* Hemsl., Wilson barberry,
   vinettier de Lady Wilson; imported from
   China.

5. Host species not named, i.e., reported
   as *Berberis* sp.

See also *MAHONIA* and *MAHOBERBERIS*

Coccomyces dentatus: on 5 BC [1609].
Cucurbitaria berberidis: on 5 Ont [1827].
Dothidella berberidis (C:53): canker,
   chancre: on 4 BC [1816].
Dothiorella ?ribis: on 2 Que [333].
Kabatiella berberidis (Gloeosporium b.):
   on 3 Ont [336], Que [334]. Probably a
   species of *Aureobasidium* [376, p. 193].
Phyllactinia guttata: powdery mildew,
   blanc: on 3 Ont [1257].
Phyllosticta berberidis (C:53): leaf
   spot, tache foliaire: on 5 Que [335].
Puccinia graminis (C:53): rust, rouille:
   0 I on 1, 3, 3a Ont [1236]; on 3 Ont
   [335, 337], Que [276, 337], NB [333]; on
   5 BC [1816], Ont [1828].
P. graminis f. sp. *avenae*: on 3 Ont
   [641], Man [638] by inoculation.
P. graminis f. sp. *tritici*: I on 3 NB
   [1196].
Thyronectria lamyi: on 3 Ont [175].
Verticillium albo-atrum (C:53): wilt,
   flétrissure verticillienne: on 3 BC
   [336]; on 5 Ont [333].
V. dahliae: on 2 Que [337].

BETA L.   CHENOPODIACEAE

1. *B. vulgaris* L.; native to the coasts
   of Europe, often cultivated. 1a. table
   beet, betterave potagère. 1b. sugar
   beet, betterave à sucre. 1c. *B. v.*
   var. *cicla* L., swiss chard, poirée à
   carde.

Alternaria alternata (A. tenuis auct.)
   (C:54): leaf spot, tache
   alternarienna: on 1a Ont [1460, 1594],
   Que [1651]; on 1b Que [336]; on 1c NS
   [337].
Aphanomyces cochlioides (C:53): root rot,
   pourriture des racines: on 1a, 1b Que
   [312, 313, 314, 315, 317, 1594].
Ascochyta betae (C:53): on 1a BC [1816].
Botrytis cinerea (holomorph *Botryotinia
   fuckeliana*) (C:53): on 1a Ont [1460,
   1594].
Cercospora beticola (C:53): leaf spot,
   tache cercosporéenne: on 1, 1a, 1b BC
   [1816]; on 1 Ont, Que [333, 335] and Que
   [336, 1639], Nfld [335]; on 1a Ont
   [1460, 1594], NS [338, 1594]; on 1b Man
   [336], Ont [333, 334, 336, 339], Que
   [335].
*Colletotrichum coccodes (C.
   atramentarium)*: on 1a NS [593].
C. dematium: on 1a NS [593].
Erysiphe sp.: powdery mildew, blanc: on
   1b Alta [705].
Fusarium sp. (C:54): on 1a Ont [1460,
   1461, 1594, 1595]; on 1b Alta [1594].
F. oxysporum (C:54): on 1b Alta [335].
Peronospora farinosa (C:54): downy
   mildew, mildiou: on 1a BC [1816].

Phoma sp.: on 1a, 1b Alta [1594].

P. betae (holomorph *Pleospora betae*)
(C:54): black leg, jambe noire: on 1
Alta [336], NS [335, 336]; on 1, 1a, 1b,
1c BC [1816]; on 1b Alta [333, 334], Que
[338].

Polymyxa betae f. sp. *betae*: on 1 Ont
[60].

Pythium spp.: on 1a Ont [1460, 1461,
1595]; on 1b Alta [335, 1594].

P. aphanidermatum (C:54): root rot,
pourriture pythienne: on 1 BC [1594];
on 1a Ont [1813].

Ramularia beticola (C:54): leaf spot,
tache ramularienne: on 1, 1a, 1b, 1c BC
[1816]; on 1b BC [337, 339].

Rhizoctonia solani (holomorph
Thanatophorus cucumeris) (C:54):
damping off or seedling blight, fonte
des semis ou brûlure des plantules: on
1 Alta [243], Ont [336]; on 1a Ont
[1461, 1594]; on 1a, 1b Que [312, 313,
314, 315, 317, 1594]; on 1b Alta [1594],
Que [333, 338].

Rhizopus sp. (C:54): on 1 BC [1816]; on
1b Alta [1594].

Scopulariopsis canadensis: on 1 BC [1148].

Septoria betae (C:54): leaf spot, tache
septorienne: on 1, 1b, 1c BC [1816].

Stemphylium botryosum (holomorph
Pleospora herbarum) (C:54): on 1 BC
[1816]; on 1, 1c Ont [667].

Streptomyces scabies (C:54): scab, gale
commune: on 1 NS [333, 336]; on 1a Que
[337], NB [339], NS [338]; on 1b BC
[1816], Alta, Que, NS [334].

Ulocladium consortiale (*Stemphylium c.*):
on 1, 1c Ont [667] isolated from seeds.

Uromyces betae (C:54): rust, rouille: on
1, 1a, 1b, 1c BC [1816].

Verticillium albo-atrum (C:54): on 1 Que
[333].

V. nigrescens: on 1a, 1b Que [38, 392,
393]; on 1b Que [749].

BETULA L.   BETULACEAE

1. *B. alleghaniensis* Britt. (*B. lutea*
   Michx. f.), yellow birch, merisier;
   native, Ont-Nfld.
2. *B. borealis* Spach, northern birch,
   bouleau boréal; native, Que, Nfld-Labr.
3. *B. glandulosa* Michx., dwarf birch,
   bouleau glanduleux; native, Yukon-Keew,
   BC-n. NS, Nfld-Labr.
4. *B. lenta* L., cherry birch, bouleau
   acajou; native, s. Ont-s. Que.
5. *B. occidentalis* Hook., water birch,
   merisier rouge; native, Yukon-Mack,
   BC-Man.
6. *B. papyrifera* Marsh., white birch,
   bouleau à papier; native, Yukon-Mack,
   BC-Labr. 6a. *B. p.* var. *kenaica*
   (W.H. Evans) A. Henry, Kenai birch;
   native to Alaska. 6b. *B. p.* var.
   neoalaskana (Sarg.) Raup (*A.
   resinifera* (Reg.) Britt., *B. p.* var.
   humilis Fern.) alaska birch, bouleau
   de l'Alaska; native, Yukon-Mack, n.
   BC-nw. Ont.
7. *B. pendula* Roth (*B. alba* L., *B.
   alba* var. *pendula*), European birch,
   bouleau pleureur; imported from Europe
   and Asia Minor. 7a. *B. p.* cv.
   gracilis. 7b. *B. p.* cv. *purpurea*
   (*B. alba* var. *atro-purpurea* H.
   Jaeg.).
8. *B. populifolia* Marsh., gray birch,
   bouleau gris; native, e. Ont-PEI. 8a. x
   B. caerulea-grandis Blanch., blue
   birch; Que, NS and PEI.
9. *B. pumila* L., swamp or low birch,
   bouleau nain; native, Yukon-Mack,
   BC-Nfld. 9a. *B. p.* var.
   glandulifera Regel (*B. glandulifera*
   (Regel) Butler), swamp birch, bouleau de
   savane; native, Yukon-Mack, BC-Que, Labr.
10. Host species not named, i.e., reported
    as *Betula* sp.

Acrostalagmus sp.: on 1 Que [941].

Acrostaphylus sp.: on 6 Que [1634].

Agrocybe firma (*Naucoria f.*) (C:57): on
10 NS [1032].

Aleurodiscus canadensis: on 6 Ont
[1341]. See *Abies*.

Alternaria sp.: on 1 Que [945].

Amphinema byssoides: on 10 Que [969].
Typically on well rotted wood and duff.
See *Abies*.

Amphisphaeria sp.: on 1 Ont [1340, 1828].

Anthostoma sp.: on 10 NWT-Alta [1063],
Alta [53].

Antrodia crustulina (*Poria c.*): on 6 Mack
[1110]. See *Abies*.

A. lindbladii (*Poria cinerascens*): on 6
NS [1032]. See *Abies*.

A. serialis (*Trametes s.*): on 6 NWT-Alta
[1063]. See *Abies*.

Aporpium caryae (C:55): on 1 NS [1032].
Associated with a white rot.

Armillaria sp.: on 6 Que [F61].

A. mellea (C:55): root rot,
pourridié-agaric: on 1, 6 Ont [6,
1341]; on 1 Que [1376, F70], NS [1739];
on 6 BC [1012], Que [F68, F70], Nfld
[1659, 1660, 1661]; on 10 NS [1032].
See *Abies*.

Arthrographis cuboidea: on 10 Ont [1635].

Ascocoryne sarcoides (*Coryne s.*): on 1
Que [941]; on 10 Ont [1340].
Lignicolous, typically on dead wood.

Aspergillus fumigatus: on 10 NB [1610].

Asterodon ferruginosus (C:55): on 10 BC
[1012]. See *Abies*.

Asteroma microspermum (*Cylindrosporella m.*,
Gloeosporium betulae-luteae) (C:56):
leaf spot, tache des feuilles: on 1 Que
[F62], NB, NS, PEI [1032]; on 6 Que
[F68]; on 10 Que [334].

Asterosporium asterospermum (*A.
hoffmanii*): on 10 Man [F66].

A. betulinum: on 1 Ont [1340]; on 1 twigs
NB, NS [1763]; on 6 Ont [1340].

Athelia epiphylla (*Corticium e.*): on 6
Mack [53]. See *Abies*.

A. neuhoffii: on 10 Que [969]. See
Abies.

Atopospora betulina (*Euryachora b.*,
Rehmiellopsis b., *Rehmiodothis b.*)

(C:55): tar spot, tache goudronneuse:
on 3 Yukon [298, 460], Mack [53], Alta
[F63, F65], Ont [298, 1340], Que [70,
298]; on 3, 5, 6, 10 NWT-Alta [1063]; on
6 BC [298, 954], Sask, Man [F66, F67,
F68], Ont [1340, F68]; on 6 Sask [F71];
on 9 Sask, Man [F68]; on 9a Sask, Man
[F66, F67], Ont [298]; on 10 Yukon
[F69], Mack [53], Alta [53, 298], Ont
[298, 1340], Ont, Que, Nfld [73] on
leaves.
Aureobasidium pullulans: on 1 Ont [945].
Bactrodesmium betulicola: on 5, 6, 10
Sask and 6, 9a, 10 Man [1760].
Basidiodendron eyrei: on 1 Que [1016].
See *Acer*.
Basidioradulum radula (Radulum orbiculare)
(C:59): on 6 PEI [1032]. See *Abies*.
Bispora betulina: on 10 NB [1610].
Bisporella citrina (Helotium c.) (C:57):
on 6 Ont [1340].
Bjerkandera adusta (Polyporus a.) (C:58):
white mottled rot, carie blanche
madrée: on 1, 6, 10 Ont [1341], NB
[1032]; on 6 BC [1012]; on 10 NS [1032].
B. fumosa (Polyporus f.) (C:58): on 6 BC
[1012]. See *Acer*.
Botryobasidium subcoronatum: on 10 Que
[969]. Associated with a white rot.
B. vagum: on 1 Ont [1341]; on 10 Que
[969]. See *Abies*.
*Botryohypochnus isabellinus (Botryobasidium
i.)*: on 10 Ont [1341]. See *Abies*.
Botrytis cinerea (holomorph *Botryotinia
fuckeliana*): on 1 NB [1036].
Brachysporium obovatum: on 1 Que [828].
B. nigrum: on 10 BC [827].
Byssocorticium atrovirens: on 10 Ont
[1341]. Typically on rotted wood and
duff.
Cacumisporium capitulatum: on 6 Man
[1760], Ont [586].
Calocera cornea (C:55): on 6 Ont [1341];
on 10 NS [1032]. See *Acer*.
Camarographium sp.: on 6 Ont [1340].
Cephalosporium sp.: on 1 Que [941, 945].
Ceraceomyces borealis: on 6 Que and on 10
Ont [568]. See *Abies*.
C. serpens: on 1 NS and on 10 Ont [568].
See *Abies*.
Ceratocystis sp.: on 1 Ont [1828].
C. fimbriata: on 6 Man [1190].
C. introcitrina: on 6 Man [1190].
C. olivacea: on 6 Ont [647, 1340].
C. piceae: on 6 Ont [647, 1340].
C. pilifera: on 1 Ont [647], Que [1340].
Ceriporia tarda (Poria semitincta)
(C:59): on 10 NS [1032]. See *Abies*.
Cerocorticium confluens (Corticium c.):
on 6 Mack [1110]. Lignicolous.
Cerrena unicolor (Daedalea u.): on 1 Ont
[1341], Que [1376]; on 6 NWT-Alta
[1063], Mack [53, 1110], BC [1012], Alta
[53], Man [1855], Ont [1341], Que
[1855], PEI [1032], Nfld [1659, 1660,
1661], Labr [1855]; on 10 Yukon [460],
NWT-Alta [1063], Mack [53], BC [1341],
Alta [53], Ont [1341, 1828, 1855], NB,
NS, PEI [1032]. See *Acer*.
Chalara sp.: on 1 Que [941].

Chlorociboria aeruginosa: on 1, 6, 10 Ont
[1340].
Chlorosplenium aeruginosum (C:55): on 10
NB, NS [1032].
C. versiforme (Midotis v.): on 10 Ont
[1340].
Chondrostereum purpureum (Stereum p.)
(C:59): silver leaf, plomb: on 1, 6
Ont [1341]; on 1 Que [282, 945]; on 6
Yukon [460], BC [1012], NWT-Alta [1063],
NB [1032], Nfld [1659, 1660, 1661]; on 8
NB, NS [1032]; on 10 BC [1341], Que
[969]. See *Abies*.
Ciboria betulicola: on 10 Ont [429, 431,
662], Que [426, 662].
C. peckiana (Rutstroemia macrospora): on
1 Ont, Que [1340].
Cladosporium sp.: on 10 NB [1610].
Clavicorona pyxidata: on 1 Ont [408].
Associated with a white rot.
Climacodon septentrionalis (Hydnum s.):
on 1 Que [1377]. Typically on *Acer*,
often fruiting on wounds on live trees.
See *Acer*.
Coccomyces coronatus: on 6 Ont [1340].
Coltricia perennis (Polyporus p.): on 10
NB [1032]. See *Abies*.
Coniophora sp.: on 10 Ont [1341].
Associated with a brown rot.
C. puteana: on 10 Ont [1828]. Sometimes
causing a brown rot in live trees. See
Abies.
Coniothyrium sp.: on 1 Ont [770].
Conoplea fusca: on 6 Man [1760]; on 10
Sask [F66].
C. geniculata: on 6 Man [1760].
C. olivacea: on 6 Man [1760].
C. sphaerica (C:55): on 10 Que [784].
Cordana pauciseptata (C:55): on 6 Que
[796].
Coriolus hirsutus (Polyporus h.) (C:58):
white spongy rot, carie blanche
spongieuse: on 1, 6, 10 Ont [1341]; on
1 Que [1390]; on 1, 8, 10 NS [1032]; on
6 NWT-Alta [1063], BC [1012]; on 10 Alta
[426], Ont [830].
C. pubescens (Polyporus p., P. velutinus)
(C:59): white spongy rot, carie blanche
spongieuse: on 1, 6 Ont [1341]; on 1
NB, NS [1032]; on 5 BC [1341]; on 6, 10
NWT-Alta [1063]; on 6 BC [1012], Alta
[426], Nfld [1659, 1660, 1661]; on 8 NS
[1032]; on 10 Nfld [1659, 1660].
C. versicolor (Polyporus v.) (C:59):
white spongy rot, carie blanche
spongieuse: on 1, 6, 10 Ont [1341]; on
1 Que [1390, 1377], NB, NS, PEI [1032];
on 6 NWT-Alta [1063], BC [1012], NB, NS,
PEI [1032], Nfld [1661]; on 7 PEI
[1032]; on 8 Que [1390], NB, NS [1032];
on 10 NB, NS [1032], Nfld [1659, 1660].
C. zonatus (Polyporus z.): on 6 Alta
[53]; on 6, 8 NS [1032]. Associated
with a white rot.
Corticium sp.: on 1 NS [1032].
C. centrifugum: on 6 NWT-Alta [1063].
This name has been applied to several
fungi. See *Abies*.
Corynespora bramleyi: on 6 Ont [1340].
C. cespitosa: on 6, 10 Man [1760].

Crepidotus sp.: on 10 NWT-Alta [1063].
C. albescens (C. phaseoliformis): on 10 Ont [724].
Cryptendoxyla hypophloia: on 1 Que [1165] under bark of dead tree; on 10 Ont [1040].
Cryptocline betularum (Gloeosporium b.) (C:57): on 10 Que [333], NB [1032].
Cryptospora alnicola: on 10 Man [1340, F66].
Cryptosporiopsis sp.: on 1 NB [1032].
Cylindrobasidium evolvens (Corticium e., C. laeve) (C:56): on 1 Que [945]; on 10 NB [1610]. See *Abies*.
Cylindrosporella sp.: on 1 Que and on 9a Man [F67].
C. leptothyrioides: on 6 Sask [F65].
Cylindrosporium sp.: on 6 Ont [1340], NB [1032, F76], NS [1032].
C. betulae (C:56): leaf spot, tache des feuilles: on 6 Ont [282, F61], NB [1032, 1036], Nfld [1659, 1660, 1661, F65, F68, F73, F74].
Cyphellopsis anomala (Solenia a.) (C:59): on 6 Ont [1341]. Lignicolous, probably saprophytic.
Cystostereum murraii (Stereum m.) (C:59): on 1 Que [941, 1376, 1377], NB [1032], NS [1739]; on 1, 6, 8, 10 NS [1032]; on 1, 6, 10 Ont [1341]; on 10 Que [969], NB [1032], Nfld [1660]. See *Acer*.
Cytidia salicina: on 6 Ont [1341]. Lignicolous, typically on dead branches and stems, esp. of *Salix*.
Cytospora sp.: on 1 Ont [770], Que [941, 945]; on 6 NWT-Alta [1063]; on 7 Alta [338]; on 10 BC [1012].
Daedaleopsis confragosa (Daedalea c.) (C:56): white spongy rot, carie blanche spongieuse: on 1, 6, 10 Ont [1341]; on 1 Que [1391]; on 6 BC [1012]; on 8 Que [1391]; on 1, 10 NB, NS, PEI [1032].
Daldinia concentrica (C:56): on 1, 6 Ont [1340]; on 10 NS [1032].
D. vernicosa: on 6 Ont [1828].
Datronia mollis (Trametes m.): on 1 Ont [1341], Que [1391]; on 6 NWT-Alta [1063], BC [1012], Alta [53, 579]. See *Acer*.
Dendrothele microspora (Aleurocorticium m.): on 1 Que [965].
Dendryphiopsis atra: on 6 Man [1760].
Dermea molliuscula (Dermatea m.) (C:56): on 1 Ont [654], NB, NS [1032]; on 1, 6 Que [654, 1340]; on 1, 10 NS [654].
Diaporthe alleghaniensis: on 1, 4, 6 Ont [14]; on 1 Ont [9, F68], Que [9, 14], NB, NS [14]; on 1, 10 Ont [1827].
D. beckhausii: on 10 Man [1340, F66].
D. eres f. sp. *betulae*: on 1 Ont [771].
Diatrype stigma (C:56): on 1, 6 Ont [1340]; on 1 NB [1032]; on 10 Ont [1494], NS [1032].
Diatrypella sp.: on 3, 6 Ont [1340].
D. betulina (C:56): on 1, 6 Ont [1340]; on 1 NB [1032]; on 6 BC [1012], Ont [1827], Nfld [1659, 1660, 1661]; on 10 NS [1032].
D. decorata (C:56): canker, chancre diatrypelléen: on 3 NWT-Alta [1063].

D. discoidea (C:56): on 10 NS [1032].
D. favacea (C:56): on 1 Ont [1340], Que [F71]; on 6 Ont [1340, 1828, 1827, F68]; on 10 Ont [1340], NS [1032].
Dichomera sp.: on 6 Ont [1340].
Didymochora betulina: on 3 Ont [1340].
Discina korfii: on 1 Ont [567].
Discula betulina (Gloeosporium betulicola) (C:57): anthracnose, anthracnose: on 6 Que [F62], NB, PEI [1036]; on 10 PEI [F76].
Disculina betulina (Cryptosporium b., C. neesii var. *betulinum)* (C:56): on 6 Ont [1340]; on 7 Ont [1827]; on 10 Man [F66].
Durandiella seriata (Godronia s.): on 6 Sask [F65].
Endophragmiella biseptata: on 1 Que [802].
Epicoccum sp.: on 1 Que [945]; on 6 BC [1012].
E. purpurascens (E. nigrum): on 1 NB [1032, 1036]; on 6 Sask [1760].
Eutypella sp.: on 6, 10 NWT-Alta [1063]; on 10 Alta [951].
E. angulosa (C:56): on 1, 6 Ont [1340]; on 6, 10 NWT-Alta [1063]; on 6 BC [1012], Alta [53], Ont [1828]; on 10 Mack [53], Alta [F62].
E. parasitica: on 1 Que [1377].
Exidia sp.: on 3 Yukon [460].
E. candida: on 3 BC [1012]; on 10 NWT-Alta [1063], Mack [53]. Lignicolous, probably saprophytic.
E. glandulosa (E. spiculosa): on 1 Ont [1341]. Lignicolous, probably saprophytic.
Favolus alveolaris (C:56): on 1 NB [1032]; on 6 Ont [1341]. See *Acer*.
Femsjonia peziziformis (F. luteoalba): on 1 Ont [1341], NS [1032]; on 10 Que [1385] on dead wood, NS [1032]. Lignicolous, probably saprophytic.
Fenestella fenestrata (F. princeps) (C:56): on 10 Sask [1340].
Flammula sp.: on 6 BC [1012]. Specimen would probably be redetermined as *Gymnopilus* or *Pholiota*.
Flaviporus semisupinus (Polyporus s.) (C:59): on 1, 10 NS [1032]. Associated with a white rot.
Fomes sp.: on 6 Nfld [1660, 1661]. Lignicolous.
F. fomentarius (C:56): white mottled rot, carie blanche madrée: on 1 Ont [10, 771], Que [941, 1391, 1376]; on 1, 6, 10 Ont [1341]; on 1, 6 Nfld [1659, 1660, 1661]; on 5 BC [1341]; on 6 Yukon [460], Mack [F61], BC [953, 1012], Alta [53, F61], NB, NS, PEI [1032]; on 6, 10 NWT-Alta [1063], Mack [53]; on 6, 7 Que [282]; on 8 NB, NS [1032]; on 10 Que [1377, 1385 on dead trunk].
Fomitopsis pinicola (Fomes p.) (C:56): on 1, 6, 10 Ont [1341]; on 1 Que [1376, 1391], NB, NS [1032]; on 6 NWT-Alta [1063], BC [1012], NB, NS [1032], Nfld [1659, 1660, 1661]; on 8 NB [1032]; on 10 Que [1377], NS [1032]. See *Abies*.
F. rosea (Fomes r.): on 1 Ont [1341]. See *Abies*.

Fusarium sp.: on 1 Que [941, 945].
Fusicoccum sp.: on 1 Ont [1827], NB
    [1032].
F. betulinum: on 1 NB [1032].
Galzinia incrustans: on 6 BC [1012].
    Lignicolous, probably saprophytic.
Ganoderma applanatum (Fomes a.) (C:56):
    white mottled rot, carie blanche
    madrée: on 1, 6, 10 Ont [1341]; on 1
    Que [282, 1377], NB, NS, PEI [1032]; on
    6, 10 Mack [579]; on 6 BC [1012], Que
    [1634], NS, PEI [1032], Nfld [1659,
    1660, 1661]; on 8 PEI [1032]; on 10 NS
    [1032].
G. lucidum (C:56): on 1 Ont [1341], NB
    [1032]. See *Abies*.
Gelatinosporium sp.: on 1 Ont [770,
    1340]; on 6 NB [1032].
G. fulvum (holomorph *Dermea molliuscula*)
    (C:56): on 1 Ont [1340].
G. magnum (C:56): on 1 Ont [1340].
Gliocladium sp.: on 6 Que [1634].
G. viride: on 10 NB [1610].
Gloeophyllum sepiarium (Lenzites s.)
    (C:57): on 6 BC [1012], NWT-Alta
    [1063]. See *Abies*.
Gloeoporus dichrous (Polyporus d.)
    (C:58): on 1, 6 Ont [1341]; on 6 Yukon
    [460, 579], Mack [1110], BC [1012]; on
    6, 10 NWT-Alta [1063], Mack [53]. See
    Acer.
G. pannocinctus (Poria p.): on 6 BC
    [1012]. See *Alnus*.
Gloeosporium sp.: on 6 BC [1012], Ont
    [1340], Que [F75], Nfld [1661].
Gnomonia intermedia (C:57): on 3 Ont [70].
G. intermedia var. *intermedia*: on 10
    Ont, Que [79].
Godronia sp.: on 1, 6 Que [F76].
G. cassandrae: on 6, 8 Que [1678]. Very
    probably f. *betulicola*.
G. cassandrae f. *betulicola*: on 1, 6
    Que [1678, 1684, F68]; on 6, 8 Que
    [1680]; on 8 Que [1684, F68]. See
    comment under *Alnus*.
G. fuliginosa: inoculated on 1, 8 Que and
    on 6 Que [1684].
G. multispora: on 6, 7 Ont [656]; on 6
    Que [F70].
Graphium sp.: on 1 Que [941]; on 10 NB
    [1610].
Gymnopilus spectabilis (Pholiota s.)
    (C:58): on 1 Que [941], NB, NS [1032];
    on 6 BC [1012]. See *Acer*.
Gyromitra infula (C:57): on 10 NS [566,
    1032]. Probably on well-rotted wood.
Haematostereum guasapatum: on 10 PEI
    [1032]. See *Alnus*.
H. rugosum (Stereum r.): on 6 Nfld [1659,
    1660, 1661]; on 6, 8, 10 NS [1032].
    See *Abies*.
Haplographium delicatum (holomorph
    Hyaloscypha dematiicola): on 6 Man
    [1760].
Hapalopilus nidulans (Polyporus n.)
    (C:58): on 1, 6 Ont [1341]; on 1, 10 NS
    [1032]; on 6 Mack [1110], Ont [F63].
    See *Abies*.

Helicogloea lagerheimii (C:57): on 10 BC
    [1012]. Lignicolous.
Helminthosporium velutinum: on 6 Ont
    [810].
Hemimyriangium betulae: on 6 Sask, Man
    [F67, F69], Ont [1340, F67, F68]; on 9a
    Sask, Man [F67]; on 10 Sask, Man [F66],
    Ont [1448].
Hericium sp.: on 10 NWT-Alta [1063].
    Lignicolous.
H. americanum (H. coralloides N. Amer.
    auct.) (C:57): on 10 NB, NS, PEI
    [1032]. See *Acer*.
H. coralloides (H. laciniatum, H. ramosum)
    (C:57): white spongy rot, carie blanche
    spongieuse: on 1 NS [1032]; on 6
    NWT-Alta [1063]; on 10 BC [1012], Alta
    [712], Que [1385] on dead trunks.
H. erinaceus (C:57): on 1 NB [1032].
    Specimen probably a form of *H.
    americanum*. See comment under *Acer*.
Heterochaetella dubia: on 10 Ont [1015].
    Lignicolous.
Hohenbuehelia angustata: on rotted wood
    of 10 Que [1385]. Naematophagous.
Hyalopesotum introcitrina: on 6 Man
    [1842].
Hyaloscypha hyalina (Lachnella h.): on 1
    Ont [1340].
Hymenochaete agglutinans: on 1 NB
    [1032]. See *Alnus*.
H. badioferruginea (C:57): on 6, 10 NS
    [1032]. See *Abies*.
H. corrugata (C:57): on 1, 6 Ont [1341];
    on 10 NS [1032]. See *Abies*.
H. tabacina (C:57): on 1 Ont [1341]; on 6
    NS [1032], Nfld [1659, 1660, 1661]; on
    10 BC [1012], NB, NS [1032]. See
    Abies.
Hyphoderma argillaceum: on 10 Que [969].
    Associated with a white rot.
H. heterocystidium (Peniophora h.): on 6
    Ont [1341]. Lignicolous.
H. praetermissum (H. tenue): on 10 Que
    [969]. See *Abies*.
H. setigerum (Peniophora aspera) (C:58):
    on 1 NB, NS [1032]; on 6 BC [1012], PEI
    [1032], Nfld [1659, 1660, 1661]; on 10
    NS [1032]. See *Abies*.
Hyphodontia alienata: on 10 Que [969].
    Lignicolous.
H. arguta (Odontia a.): on 1 NB, PEI
    [1032]. See *Abies*.
H. breviseta: on 10 NS [1032]. See
    Abies.
H. crustosa (Odontia c.): on 10 NS
    [1032]. See *Abies*.
H. pallidula: on 10 Que [969]. See
    Abies.
H. spathulata (Odontia s.) (C:58): on 6
    BC [1012]. See *Abies*.
Hypholoma dispersum (Naematoloma d.): on
    6 BC [1012]. Lignicolous.
H. fasciculare (Naematoloma f.) (C:57):
    white spongy rot, carie blanche
    spongieuse: on 6 BC [1012].
Hypochnicium vellereum (Corticium v.): on
    1 Que [941]. See *Acer*.
Hypocrea citrina: on 6 NS [1032].
Hypoxylon sp.: on 1 Que [945]; on 6 BC
    [1012]; on 10 NB [1032].

H. deustum (Ustulina vulgaris) (C:57): brittle white rot, carie blanche friable: on 1 Ont [1340], NB [1032], NS [1032, 1739].

H. fuscum (C:57): on 6 Ont [1340]; on 10 Man [F66].

H. mammatum (H. pruinatum) (C:57): on 1, 6 Ont [1340]; on 1 NB [1032]; on 6 Sask [F68]; on 10 Sask, Man [F66].

H. mediterraneum: on 1 Ont [1828].

H. multiforme (C:57): on 1, 6 Ont [1340], Nfld [1659, 1660, 1661]; on 1 NB, NS, PEI [1032]; on 10 BC [1012], Que [282], PEI [1032].

H. rubiginosum: on 1 Ont [1340]; on 10 NB [1610].

Hysterium pulicare (C:57): on 1 Ont [1340]; on 10 NS [1032].

Hysterographium sp.: on 10 Que [282].

Hysteropatella minor (C:57): on 10 NS [1032].

Inonotus cuticularis (Polyporus c.) (C:58): white spongy rot, carie blanche spongieuse: on 1 Ont [1341], NB [1032]; on 6 BC [1012].

I. glomeratus (Polyporus g.): on 1 Que [941]; on 6 BC [1012]. See *Acer*.

I. obliquus (Poria o.) (C:59): white spongy rot, carie blanche spongieuse: on 1 Que [941, 948, 1377, 1376], NS [1739]; on 1, 6 Ont [1341], NB, NS, PEI [1032]; on 6 NWT-Alta [1063], Mack [53, 1110], BC [1012], Sask, Man [F67], Ont [F64], Que [1377, 1634]; on 6, 8 Que [F67]; on 10 NS [1032].

I. radiatus (Polyporus r.) (C:59): on 1, 6 Ont [1341]; on 1, 8, 10 NB [1032]; on 1, 6, 10 NS [1032]. See *Acer*.

Intextomyces contiguus: on 6 BC [1012]. Lignicolous.

Irpex lacteus (Polyporus tulipiferae) (C:59): on 1, 6 Ont [1341]; on 1, 8 Que [1390]; on 6 NWT-Alta [1063], Mack [1110], BC [1012], NB [1032]; on 10 NS [1032]. See *Acer*.

Ischnoderma resinosus (Polyporus r.) (C:59): on 1, 6 Ont [1341]. See *Abies*.

Junghuhnia nitida (Poria attenuata): on 1, 10 NS [1032]. See *Acer*.

Kabatiella apocrypta (Gloeosporium a.): on 6 Nfld [1660, 1661].

Kuehneromyces mutabilis (Pholiota m.) (C:58): on 10 NS [1032], on 1 NS [1658]. See *Alnus*.

K. lignicola: on 6 NS [1032]. Reported as *K. vernalis* an illegitmate name.

Laeticorticium violaceum (Dendrocorticium v.): on 1 Ont [930]. Lignicolous.

Laetiporus sulphureus (Polyporus s.): on 1, 10 NS [1032]; on 10 Que [1377]. See *Abies*.

Lasiosordaria lignicola (Bombardia l.): on 6 Ont [1340].

Laxitextum bicolor: on 6 BC [1012], Ont [1341]. Lignicolous.

Lenzites betulina (C:57): on 1, 6, 10 Ont [1341], PEI [1032]; on 5 BC [1341]; on 6 Yukon [460, 579], NWT-Alta [1063], BC [1012], NB [1032]; on 6b Sask [579]; on 8 Que [1391]; on 10 NB, NS [1032]. See *Acer*.

Leptosporomyces galzinii (Athelia g.): on 10 Que [969]. See *Abies*.

Libertella sp. (C:57): on 1 Ont [770], Que [941, F67], NB [1032]; on 1, 6 Ont [1340].

L. betulina (C:57): on 1 Ont [1340, 1828], Que [1340]; on 1, 10 NS [1032]; on 6 Sask, Man [F67]; on 10 NWT-Alta [1063].

Lophidium compressum var. *microscopicum* (C:57): on 10 NS [1032]. The species was transferred to *Platystomum*, but the variety has not been.

Mamianiella coryli var. *spiralis*: on 10 Que [79].

Marasmius rotula: on 10 NB [1036].

Marssonina betulae (C:57): leaf spot, tache des feuilles: on 6 NB [1032].

Massaria pruni (C:57): on 10 NS [1032].

Massarina sp.: on 8a Ont [1340].

Melampsoridium betulinum (C:57): leaf rust, rouille des feuilles: II III on 1 Que [F62], NS [1032]; on 3 Ont [1236 published under the name *Melampsorella*, 1341, F65]; on 3, 6 Yukon [F61], BC [1012]; on 3, 5, 6, 10 NWT-Alta [1063]; on 3, 6, 6a, 6b Yukon [460]; on 5 Alta [F64]; on 6 Alta [F63], Ont [1341, F65], Que [1377, F74], Nfld [1660, 1661, F71, F72, F74]; on 6a, 7 Yukon [F62]; on 6, 8 Que [F62, F69]; on 7 BC [333, F60]; on 7, 7a BC [1012]; on 8 Ont [1828], Que [F70, F71], NB [1032, F64], NS [1032]; on 9 Alta [F69], Sask, Man [F68]; on 10 Yukon, Mack, BC, Alta, Sask, Man [2008] and Mack [53], BC [336], Sask [F67], Que [335, 1377].

Melanconis sp. (C:57): on 1, 6 Ont [1340].

M. betulina: on 6 NB [1036]. Specimen should be reexamined.

M. decorahensis (anamorph *Melanconium subviridis*) (C:57): on 1, 6 Ont [1340]; on 6 Ont [F63].

M. nigrospora (C:57): on 1 Ont [770, 1340, F60, F63], Que [F67]; on 10 NS [1032].

M. stilbostoma (anamorph *Melanconium bicolor*) (C:57): twig blight, brûlure des rameaux: on 6 Man [1768, F65], Ont [1340, F60, F63], Que [F71]; on 10 Sask [1340], NS [1032, 1036], Nfld [79].

Melanconium sp. (C:57): on 6 BC [1012], Ont [1340].

M. bicolor (holomorph *Melanconis stilbostoma*) (C:57): on 6 Ont [770, 769, F67], NB [1032]; on 7 Ont [338]; on 6, 10 Ont [1340].

M. parvulum (C:57): twig blight, brûlure des rameaux: on 7 Que [1377].

M. subviridis (holomorph *Melanconis decorahensis*): on 6 Ont [1340].

Melanomma pulvis-pyrius (M. subsparsum) (C:57): on 10 NS [1032].

Menispora manitobaensis: on 6 Man [1760].

M. tortuosa: on 10 Que [826].

Meruliopsis hirtellus: on 10 Que [568]. Associated with a white rot.

Merulius sp.: on 1 Ont [1828].
M. tremellosus (C:57): on 1, 10 Ont
[568]; on 1 NB [1032]; on 5 BC [568]; on
6 BC [1012], Que [568]; on 10 NB [1610],
NS [568]. See *Abies*.
Microsphaera penicillata (C:57): powdery
mildew, blanc: on 1 NS [1032].
Mollisia benesuada (C:57): on 10 NS
[1032].
Monilia sitophila (holomorph *Neurospora
s.*) (C:57): on 10 NS [1032].
Monodictys sp.: on 6 Ont [973].
M. levis: on 6 Sask [1760].
M. paradoxa: on 6, 10 Man [1760].
Mucor sp.: on 1 Que [941, 945].
Mycena leaiana (C:57): on 1 Ont [1341,
F62].
Mycocalicium pallescens (C:57): on 10 NS
[1032].
Mycosphaerella maculiformis (C:57): on 6
Que [70].
Myxocyclus polycistis (*Stegonsporium
muricatum*) (C:59): on 5, 10 Sask, Man
[1762]; on 6 Sask [1762], Ont [1340],
Que [F71] and Sask, Man [F67]; on 10 NWT
[1762], Sask, Man [1763] on twigs and
bark.
Myxosporium sp.: on 6 Ont [1340].
Nectria sp. (C:57): on 1 Ont [1827], Que
[1376]; on 6 Ont [1340]; on 10 Que
[1377].
N. cinnabarina (C:57): on 6 Sask [F67];
on 6, 9a Man [F67]; on 10 Man [F66], Que
[1377], NB [1032].
N. coccinea var. *faginata* (C:58): on 10
NS [1032].
N. episphaeria (C:58): on 10 Ont [F62].
N. galligena (C:58): canker, chancre
nectrien: on 1 Que [948, 1005, 1010,
1011, F76], NB, NS, PEI [1032]; on 6 Ont
[1340]; on 10 Que [F60].
N. pithoides (C:58): on 10 NS [1032].
Neobulgaria pura (*Ascotremella
turbinata*): on 6 Que [569]; on 10 Ont
[569, 1340]. Lignicolous, presumably
saprophytic.
Nummularia discincola: on 10 Que [862].
Odontia crustula: on 1 NB [1032].
Lignicolous.
Ophiovalsa betulae (*Cryptospora b.*)
(C:56): on 6 Ont [1340, 1660, 1827]; on
6, 7 Ont [F68]; on 10 NS [1032].
Orbilia botulispora (*O. paradoxa*): on 10
Ont [1340].
O. coccinella: on 6 Yukon [664].
Oxyporus populinus (*Fomes connatus*)
(C:56): on 1 Ont [1341], NB, NS
[1032]. See *Acer*.
Panellus serotinus (*Pleurotus s.*, *Panus
betulinus*) (C:58): white spongy rot,
carie blanche spongieuse: on 1 NB, NS
[1032]; on 6 BC [1012]; on 10 BC [F66],
NS [1032], Nfld [1293].
P. stipticus (*Panus s.*) (C:58): on 1, 10
NB, NS [1032]; on 6 BC [1012], Ont
[1341], NB, PEI [1032], Nfld [1659,
1660, 1661]. See *Alnus*.
Panus conchatus (*P. torulosus*): on 10 Que
[1385].
P. laevis: on 6 Nfld [1659, 1660, 1661].

P. rudis (C:58): on 1, 6 Ont [1341]; on 6
Yukon [460, F64], BC [1012].
Penicillium sp.: on 10 NB [1610].
Peniophora sp.: on 1 Que [945], NS
[1032]; on 6 PEI [1032], Nfld [1659,
1660, 1661]; on 10 Ont [1828], NB [1032].
P. cinerea aggregate (C:58): on 1 Ont
[10].
P. cinerea (C:58): on 1, 6 Ont [1341]; on
10 NB, NS [1032]. See *Abies*.
Phaeoisaria sparsa: on 6 Man [1760].
Phaeomarasmius erinaceus (*Pholiota e.*):
on 6 Ont [1434]; on 10 Que [1385] on
twigs and branches.
Phanerochaete sanguinea (*Peniophora s.*):
on 1 NS [1032]. See *Abies*.
P. sordida (*Peniophora cremea*): on 10 NB
[1032]. See *Abies*.
Phellinus conchatus (*Fomes c.*): on 1 NS
[1032]; on 10 Ont [1828]. See *Acer*.
P. everhartii (*Fomes e.*) (C:56): white
spongy rot, carie blanche spongieuse:
on 6 PEI [1032].
P. ferreus (*Poria f.*): on 10 Ont [1827].
See *Acer*.
P. ferruginosus (*Poria f.*) (C:59): on 1
Que [1376]; on 5 NWT-Alta [1063]; on 6
BC [1012]. See *Abies*.
P. igniarius (*Fomes i.*) (C:56): white
trunk rot, carie blanche du tronc: on
1, 6, 10 Ont [1341]; on 1 Que [941,
1376], NS [1739]; on 1, 6 Nfld [1659,
1660, 1661]; on 1, 6, 8 NB, NS, PEI
[1032]; on 5 Alta [950]; on 5, 6, 10
NWT-Alta [1063]; on 6 Yukon [460, F61],
BC [1012], Alta [53], Man [F60], Ont
[973], Que [1634], Nfld [F68, F69], Labr
[867]; on 6, 10 Mack [53]; on 10 BC
[953], Que [282, 1377], NB, NS [1032].
P. laevigatus (*Fomes igniarius* var. *l.*,
Poria l., *Poria prunicola*) (C:59):
white spongy rot, carie blanche
spongieuse: on 1, 6 Ont [10, 1341]; on
1 Que [1377], NS [1739] and NB, NS, PEI
[1032]; on 5 BC [1341]; on 6 BC [1012];
on 6, 7 Que [282]; on 6, 8, 10 NS [1032].
P. nigricans: on 6 Labr [867].
Associated with a white rot.
P. punctatus (*Poria p.*): on 6 Ont
[1341]. See *Abies*.
P. robustus (*Fomes r.*) (C:56): white
spongy rot, carie blanche spongieuse:
on 6 Nfld [1659, 1660, 1661].
Phialophora sp. (C:58): on 1 Que [941,
945]; on 6 Ont [1827], Que [1634]; on 10
NB [1610].
P. americana: on 10 NB [1610].
P. botulispora: on 10 Ont [279].
P. fastigiata: on 1 Ont, NB, NS [279].
P. lagerbergii: on 1 Ont, NB [279]; on 10
NB [1610].
P. melinii: on 1 NB, NS [279]; on 10 NB
[1610].
P. verrucosa: on 1 NS [279]; on 10 Ont
[279].
Phibalis furfuracea (C:56): on 10 BC
[1012].
Phlebia livida (*Corticium l.*) (C:56): on
10 NS [1032]. See *Abies*.

P. martiana (Peniophora m.): on 1 Ont [1828]. Lignicolous.

P. radiata (C:58): on 1, 6 Ont [568, 1341]; on 1 Que [568], NB, NS [1032]; on 6 BC [1012], NB [1032]; on 10 BC, Ont, NS [568]. See *Abies*.

Phlebiopsis gigantea (Peniophora g.): on 10 NB [1032]. See *Abies*.

Pholiota sp.: on 1 Que [463]; on 6 NWT-Alta [1063].

P. adiposa (C:58): on 1, 6 Ont [1341]; on 1 NS [1739]. See *Acer*.

P. albocrenulata (C:58): on 10 NS [1032]. Lignicolous.

P. alnicola (Flammula a.) (C:56): yellow checked rot, carie jaune craquelée: on 1 NB, NS [1032]. See *Abies*.

P. aurivella (C:58): on 1 Que [941, 949], NB, NS [1032]; on 10 NS [1032]. See *Abies*.

P. lenta (Flammula l.) (C:56): on 10 NS [1032].

P. limonella (P. squarrosa-adiposa): on 1 Ont [1341], NS [1032]. See *Abies*.

P. lucifera: on 1 NS [1032]. Not recognized from North America in the Hesler & Smith monograph. Specimens should be reexamined.

P. lutea (C:58): on 10 NB [1032]. Probably a synonym of *Gymnopilus spectabilis*, according to Hesler & Smith.

P. marginella: on 6 Nfld [1659, 1660]. These records are probably based on specimens of *Kuehneromyces lignicola*. *Pholiota marginella* is a distinct species.

P. squarrosoides (C:58): on 1 Ont [1341], Que [1377]; on 6 BC [1012], NS [1032]. Lignicolous.

P. subsquarrosa (C:58): on 6 BC [1012]. Not recognized in North America by Hesler & Smith. Earlier they redetermined collections labelled *P. subsquarrosa* to be *P. subvelutipes*. Subsequently *P. subvelutipes* has been recognized as a synonym of *P. limonella*.

Phoma sp.: on 1 Ont [770], Que [941, 945].

Phomopsis sp.: on 1 Ont [769, 771, 1340], Que [9, F67, F76], NB [9, 1032], NS [9]; on 1, 6 Ont [9, 770]; on 10 Ont [1340].

Phyllactinia guttata (P. corylea) (C:58): powdery mildew, blanc: on 2, 5, 6, 10 Ont [1257]; on 6 Yukon [1257], Sask, Man [F66, F67], Ont [1340, 1827], Que [F72, F75] and Man [1257]; on 6c Yukon [460]; on 10 Que [1377].

Phyllosticta sp.: on 8 Que [F67]; on 10 Que [282].

P. betulae (C:58): leaf spot, tache des feuilles: on 10 NB [1032].

P. betulina: on 6 Que [F61].

Phyllotopsis nidulans (C:58): on 6 BC [1012]; on 8 NB, NS [1032]. Lignicolous.

Pilidium acerinum (Leptothyrium medium): on 6 Sask [F65].

Piloderma bicolor (Corticium b.): on 6 Ont [1827]. See *Abies*.

Piptoporus betulinus (Polyporus b.) (C:58): on 1, 6, 10 Ont [1341]; on 1 NB, NS [1032]; on 5 BC [1341]; on 6 Yukon, Mack [1110], BC [1012], Alta [53], Que [282], NB, NS, PEI [1032], Nfld [1659, 1660, 1661, F68], Labr [867]; on 6, 10 NWT-Alta [1063]; on 7 Que [282]; on 8 Que [1390], NB, NS [1032]; on 10 Alta [53, 1014], BC [1014], Que [1385], NB, NS, PEI [1032]. Associated with a brown rot of dead trees.

Plagiostoma alneum var. *betulinum*: on 3 BC [79].

P. campylostylum var. *campylostylum (Gnomonia c.)* (C:57): on 3 BC [1012]; on 10 Ont [79], Que [70, 79].

Pleomassaria siparia (C:58): on 6 Ont [1340].

Pleurothecium recurvatum: on 10 Ont [587].

Pleurotus sp.: on 5 NWT-Alta [1063].

P. ostreatus (C:58): white spongy rot, carie blanche spongieuse: on 1 Que [1341]; on 6 Yukon [460], Ont [1341], Labr [867].

P. sapidus (C:58): on 1, 10 NS [1032]; on 6 Ont [1341]. See *Aesculus*.

P. strigosus (Panus s.): on 10 Ont [1341]. See *Abies*.

Plicatura crispa (Trogia c.) (C:60): on 1 Ont, Que [1341]; on 6, 10 NWT-Alta [1063]; on 6 BC [1012], Ont [1341]; on 10 Ont [1341], NS [1032]. Lignicolous, typically on dead, fallen branches.

P. nivea: on 1 Ont [563], Que [563, 568]; on 10 Man [563, 568]. Typically on dead branches and stems. See *Acer*.

Pluteus cervinus (C:58): on 6 BC [1012]; on 10 Yukon, Mack [669]. Typically on rotted wood. Associated with a white rot.

P. leonius: on rotten wood of 10 Que [1385].

Polyporus arcularis (C:58): on 6 Nfld [1659, 1660, 1661]. Lignicolous, typically on fallen branches.

P. badius (P. picipes) (C:58): on 1 Ont [1341], NB [1032]; on 10 Ont [1827]. Lignicolous, typically on logs or fallen branches.

P. brumalis (C:58): on 1, 6 Ont [1341]; on 6 Nfld [1659, 1660, 1661]; on 8 Que [1389]; on 10 NB, NS [1032]. Lignicolous, typically on logs or fallen branches.

P. varius (P. elegans) (C:58): on 1, 6 Ont [1341]; on 1, 10 NS [1032]; on 3, 6 Mack [53]; on 6, 10 NWT-Alta [1063]. See *Abies*.

Poria sp.: on 1 NS and on 6 NB [1032].

P. amylohypha (P. elongata): on 1 NB [1032]. Associated with a white rot.

P. subacida (C:59): on 1 NB, NS [1032]; on 6 NS [1032]; on 10 BC [1012]. See *Abies*.

Prosthemium betulinum (holomorph *Pleomassaria siparia*): on 1 Que [F67]; on 6 Ont [1340].

Pseudospiropes simplex: on 6 Man [1760]; on 10 Ont [817].

Pseudotomentella tristis: on 10 Ont
   [1828].   Lignicolous, typically on
   rotted wood.
Pseudovalsa lanciformis (anamorph
   Coryneum brachyurum) (C:59): on 6
   Ont [1340].
P. spinifera: on 10 Ont [79].
Pycnoporus cinnabarinus (*Polyporus c.*,
   Trametes c.) (C:59):   on 1 Ont, Que
   [1181], NB, NS, PEI [1032] and Que
   [1385]; on 1, 6, 10 Ont [1341]; on 1, 8
   Que [1390]; on 5 BC [1012, 1341]; on 6
   NWT-Alta [1063], BC [1012, 1181], Alta
   [579], PEI [1032], Nfld [1659, 1660,
   1661]; on 10 BC, Ont [1181], NS [1032].
   On logs and dead branches. See *Acer*.
Ramaricium albo-ochraceum (*Corticium a.*)
   (C:55):   on 10 NS [1032]. Lignicolous.
Rectipilus fasciculatus (*Solenia f.*)
   (C:59):   on 10 NS [1032]. Lignicolous,
   apparently saprophytic.
Resinicium bicolor (*Odontia b.*) (C:58):
   white stringy rot, carie blanche
   filandreuse: on 1 NB [1032], NS [1739].
Schizophyllum commune (C:59):   on 1 Ont
   [1341], NB, NS [1032]; on 6 NWT-Alta
   [1063], BC [1012], Ont [1341], PEI
   [1032], Nfld [1659, 1660, 1661]; on 8 NB
   [1032]; on 10 NS, PEI [1032]. See
   Abies.
Schizopora paradoxa (*Poria versipora*):   on
   1 NB [1032]. See *Abies*.
Scutellinia erinaceus (*Patella setosa*):
   on 10 NS [1032].
S. scutellata (*Patella s.*):   on 10 Mack
   [666].
Scytinostroma galactinum (*Corticium g.*)
   (C:56):   white stringy rot, carie
   blanche filandreuse: on 1 NS [1032,
   1739]; on 6 PEI [1032].
Septogloeum sp.:   on 6 Ont [1827].
Septoria sp.:   on 6 Que [F74], NB, NS
   [F70]; on 10 Que [F63].
S. betulae (C:59):   leaf spot, tache des
   feuilles: on 1 Ont [1340, F67], NS
   [1032]; on 6 Ont [1340], Que [F62, F68,
   F73], NB, NS, PEI [1032], Nfld [1659,
   1660, 1661, F76]; on 7 PEI [336]; on 8
   NB [1032].
S. betulicola (C:59):   on 6 Mack [F65],
   Sask [F67, F71], Man [F67], Que [F61].
Septotrullula bacilligera: on 6 Sask, Man
   [1760].
Serpula lacrimans: on 10 Que [969].
   Specimen presumably was from the forest,
   in which case the fungus was probably
   S. himantioides.
Sistotrema brinkmannii (*Trechispora b.*)
   (C:60):   white stringy rot, carie
   blanche filandreuse: on 1 NB, NS
   [1032]; on 6 NB [1032]; on 10 Que [969].
S. hirschii: on 10 Que [969]. Lignicolous.
Sistotremastrum suecicum: on 10 Que
   [969]. See *Abies*.
Skeletocutis alutacea (*Poria a.*) (C:59):
   on 6 BC [1012]. Associated with a white
   rot.
Spadicoides atra: on 1 Ont [791]; on 6
   Alta [791], Man [1760].
S. bina: on 6, 10 Man [1760].

Sphaeropsis alnicola: on 1 Ont [1340].
Sporidesmium achromaticum: on 6 Sask, Man
   [1760].
S. anglicum: on 6 Man [1760].
Steccherinum sp.:   on 1 NB [1032].
S. ciliolatum (*Odontia c.*):   on 1 PEI and
   on 6 NB [1032]. See *Acer*.
S. ochraceum (C:59):   on 6 BC [1012]; on
   10 NS [1032]. See *Acer*.
Stegonsporium sp.:   on 10 Ont [1340].
   Probably *Myxocyclus* sp. as
   Stegonsporium is not known from
   Betula.
Stereum sp.:   on 6 NWT-Alta [1063].
S. complicatum: on 6 NWT-Alta [1063].
   Lignicolous.
S. hirsutum (C:59):   white sap rot, carie
   blanche de l'aubier: on 1, 6 Ont
   [1341]; on 5, 6 NWT-Alta [1063]; on 5
   Alta [950]; on 6 BC [1012], Alta [53],
   NB, NS, PEI [1032], Nfld [1659, 1660,
   1661]; on 10 BC [1341], Que [969], NB
   [1032].
S. ochraceo-flavum (C:59):   on 6 Ont
   [1341], NS [1032]. Lignicolous.
S. ostrea (C:59):   white crumbly rot,
   carie blanche friable: on 1 Ont [1341];
   on 1, 6 NB, NS [1032]; on 6 BC [1012],
   Nfld [1659, 1660, 1661]; on 10 Que [969].
Stictis radiata (C:59):   on 10 NS [1032].
Streptomyces sp.:   on wood chips of 10 NB
   [1610].
Stromatoscypha fimbriata (*Porotheleum f.*)
   (C:59):   on 1, 10 NS [1032]; on 10 Ont
   [1341]. Typically on rotted wood.
   Associated with a white rot.
Subulicystidium longisporum (*Peniophora
   l.*):   on 1 Ont [1827]; on 10 Que
   [969]. Associated with a white rot.
Taeniolella exilis: on 1, 6 Ont, Que
   [813].
Tapesia fusca (C:59):   on 1 Ont [1340]; on
   10 NS [1032].
Taphrina sp.:   on 1 NB, NS [1032, F69,
   F70]; on 1 NB [F68], Nfld [1659, 1660,
   1661]; on 6 Nfld [1661].
T. americana (C:59):   on 6 BC [1012, F64],
   Ont [1340].
T. bacteriosperma (C:59):   on 3 BC [1012,
   F63].
T. boycei (C:59):   leaf blister, cloque
   des feuilles: on 10 NWT-Alta [1063].
T. carnea (C:59):   leaf blister, cloque
   des feuilles: on 1 Que [F67, F69, F71,
   F75, F76], NS [F65, F74] and Que, NB
   [F65, F73, F74]; on 1, 6 Que [1377], NB,
   NS, PEI [1032], Nfld [1661]; on 3 BC
   [F62]; on 6 Nfld [1659, 1660]; on 6, 7
   Ont [1340]; on 10 Que [282], NS [F66].
T. flava (C:59):   on 1 Que [1375]; on 6 BC
   [1012, F69], Ont [1340], Que [F73], NB,
   NS [1032] and NB, PEI [F64]; on 8 Que
   [1377], NS, PEI [1032].
T. nana (C:59):   on 3 Yukon [460], BC
   [1012, F69]; on 6 Alta [F60]; on 10
   NWT-Alta [1063].
T. robinsoniana: on 1 Nfld [1661].
Titanella pelorospora: on 10 Que [80].
Tomentella sp. (C:60):   on 6 BC [1012]; on
   10 Ont [1341].

T. bryophila (T. pallidofulva) (C:60): on
1, 10 Ont [1341].

T. coerulea (T. papillata): on 10 Ont
[929, 1341, 1828].

T. crinalis (Caldesiella ferruginosa): on
1 NS [1032].

T. ellisii (T. microspora, T. ochracea):
on 1 Ont [1828]; on 5 BC [929, 1341].

T. ferruginea (T. fusca) (C:60): on 1
Nfld [1660]; on 10 NB, NS [1032].

T. fuscoferruginosa: on 10 Ont [929,
1341].

T. lateritia: on 10 Ont [1341].

T. neobourdotii: on 1 Ont [927, 1341].

T. ramosissima (T. fuliginea): on 10 Ont
[1341, 1828].

T. sublilacina: on 1, 10 Ont [1341,
1827]; on 6 Ont [1341]; on 10 Ont [1828].

T. terrestris (T. umbrinella): on 10 Ont
[1827].

*Tomentellastrum badium (Tomentella
fimbriata)*: on 1 Nfld [1660].

Torula ligniperda: on 1, 6 Que [1377].

Trametes cervina (Polyporus biformis N.
Amer. auct.*)*: on 6 Mack [1110], Sask
[579]; on 8 NS [1032]; on 10 NWT-Alta
[1063], Alta [579]. See *Abies*.

T. narymicus: on 10 Ont [1339]. Perhaps
a misidentification.

*Trechispora farinacea (Cristella f.,
Grandinia f.)*: on 6 Mack [1110]; on 10
Que [969]. See *Abies*.

T. mollusca (Poria candidissima): on 10
Ont [1341], Que [970]. See *Abies*.

T. vaga (Cristella sulphurea): on 10 Ont
[1341]. See *Abies*.

Tremella mesenterica: on 6 Yukon [460].

Trichaptum abietinus (Hirschioporus a.):
on 6 Que [1028]. See *Abies*.

*T. biformis (Polyporus pargamenus,
Hirschioporus p.)* (C:58): white spongy
rot, carie blanche spongieuse: on 1 Ont
[1341], Que [1377, 1390], NB, NS, PEI
[1032]; on 6, 10 NWT-Alta [1063]; on 6
Yukon [460], BC [1012], Ont [1341], NB,
NS, PEI [1032], Nfld [1659, 1660, 1661];
on 8 Que [1390], NS [1032]; on 10 Mack
[53], Alta, Man [1823], Que [282], NB,
NS [1032].

T. subchartaceum (Polyporus s.): on 10 BC
[1012]. Associated with a white rot.

Trichocladium canadense (C:60): on 1 Que
[941]; on 6 Que [945]; on 10 NB [1610].

Trichoderma sp.: on 1 Que [941].

T. viride: on 6 Que [945]; on 10 NB
[1610]. See *Abies*.

Trimmatostroma sp.: on 1 Ont [1827].

T. betulinum: on 5, 6, 10 Sask and on 6,
9a, 10 Man [1760].

Tubercularia sp.: on 1, 6 Ont [1340].

Tubulicrinis glebulosus (T. gracillimus):
on 10 Que [969]. Lignicolous.

Tulasnella violea: on 10 NS [1032].
Lignicolous, apparently saprophytic.

Tympanis alnea (anamorph *Sirodothis
inversa*) (C:60): on 5 NWT-Alta [1063],
Alta [950]; on 10 Alta [1221, F62], Ont,
Que [1221].

T. mutata: on 6 Sask, Man [F67], Ont
[1221].

Tyromyces borealis (Polyporus b.): on 10
NB, NS [1032]. See *Abies*.

T. caesius (Polyporus c.): on 10 PEI
[1032]. See *Abies*.

*T. chioneus (Polyporus albellus, Tyromyces
a.)* (C:58): white spongy rot, carie
blanche spongieuse: on 1, 6 Ont [1341];
on 1 Que [1390]; on 1, 10 NB, NS [1032];
on 1, 6 Nfld [1659, 1660, 1661]; on 6 BC
[1012], NB [1032]; on 7 Que [282]; on 8
NS [1032].

T. galactinus (Polyporus g.) (C:58):
white spongy rot, carie blanche
spongieuse: on 1, 6 NS [1032].

T. kmetii: on 10 Alta [575]. Associated
with a white rot.

T. minusculoides (Polyporus m.): on 10
Ont [1341, F67]. Lignicolous.

*T. sericeomollis (Polyporus s., Poria
s.)*: on 6 NS [1032, 1036]. See
Abies.

T. tephroleucus (Polyporus t.) (C:59): on
1 NS [1032]; on 6 BC [1012], Ont
[1341]. See *Alnus*.

Valsa sp.: on 6 Man [1340].

Valsaria sp.: on 1 Que [F76].

V. insitiva: on 1 Ont [1340, F67 as
Valsa i.].

Vararia effuscata (C:60): on 1 Ont
[1341], NB, NS [1032]. See *Acer*.

V. investiens: on 1 NB [1032]. See
Abies.

Velutarina rufo-olivacea: on 1 Ont [1827].

Venturia ditricha (C:60): on 3 Que [70];
on 6 Ont [1340]; on 10 Ont, Que [73] on
overwintered leaves and petioles.

Verticillium sp.: on 1 Que [941].

Wolfiporia extensa (Poria cocos) (C:57):
on 1 NB [1032], NS [1739]. Typically a
butt and trunk rot of large, mature
trees. Apparently not fruiting until
after the tree has died. Associated
with a brown rot.

Wrightoporia lenta (Poria l.): on 10 Ont
[1827]. See *Abies*.

Xenasma tulasnelloideum (Corticium t.):
on 1 NS [1032]. Lignicolous.

Xylaria polymorpha: on 1 Ont [1828].
Typically on buried or partially buried
wood.

Xylobolus frustulatus (Stereum f.): on 1
Ont [1341]. Lignicolous.

BIDENS L.   COMPOSITAE

1. *B. cernua* L., nodding beggar-ticks,
   bident penché; native, BC-PEI.
2. *B. connata* Muhlenb., beggar-ticks,
   bident connée; native, Ont-PEI.
3. *B. coronata* (L.) Britt.; native, Ont.
4. *B. discoidea* (Torr. & Gray) Britt.,
   discoid beggar-ticks, bident discoide;
   native, Ont-NS.
5. *B. frondosa* L., large-leaved
   beggar-ticks, bident feuillu; native,
   BC-Nfld.
6. *B. vulgata* Greene, common
   beggar-ticks, bident vulgairè; native,
   BC-NS.

Entyloma compositarum (C:60): smut,
    charbon: on 1 BC [1816].
Ophiobolus rubellus: on *B*. sp. Ont
    [1620].
Puccinia obtecta: rust, rouille: 0 III
    on *B*. sp. Ont [1236, 1559].
Sphaerotheca fuliginea (C:60): powdery
    mildew, blanc: on 1, 2, 3, 4, 5, 6 Ont
    and on 2, 5 Que and 5, 6 Man [1257].
S. macularis (C:60): powdery mildew,
    blanc: on 1 BC [1816].

BRASSICA L.   CRUCIFERAE

1. *B. hirta* Moench, white mustard,
   moutarde blanche; imported w. Asia and
   the Mediterranean region, escaped from
   cult., Yukon, BC-PEI.
2. *B. juncea* (L.) Czerniak, chinese
   mustard, moutarde joncée; imported from
   Eurasia, escaped from cult., Mack,
   BC-Nfld.
3. *B. kaber* (DC.) Wheeler, wild mustard,
   moutarde sauvage; probably native in
   Mediterranean region, natzd.,
   Yukon-Mack, BC-Nfld. The common phase
   in North America is *B. k.* var.
   pinnatifida (Stokes) Wheeler.
4. *B. napus* L., rape, colza; imported
   from Eurasia. 4a. *B. n.*
   napobrassica group (*B. napobrassica*
   Mill.), rutabaga, chou-navet.
5. *B. oleracea* L., native to Eurasia.
   5a. acephala group (*B. o.* var.
   acephala DC.), kale, chou frisé. 5b.
   botrytis group (*B. o.* var. *botrytis*
   L.), cauliflower, chou-fleur. 5c.
   capitata group (*B. o.* var. *capitata*
   L.), cabbage, chou. 5d. gemmifera group
   (*B. o.* var. *gemmifera* Zenker),
   brussel sprouts, chou de bruxelles. 5e.
   gongylodes group (*B. caulorapa* Pasq.),
   kohlrabi, chou-rave. 5f. italica group
   (*B. o.* var. *italica* Plenck),
   broccoli, chou-broccoli.
6. *B. rapa* L. (*B. campestris* L., *B.
   c.* var. *annua*), turnip rape, chou
   champêtre; imported from Eurasia. 6a.
   chinensis group (*B. chinensis* L.),
   pakchoi, chou de Chine. 6b. pekinensis
   group (*B. pekinensis* (Low.) Rupr.),
   pe-tsai, chou de Shanton. 6c. rapifera
   group, turnip, navette.
7. Host species not named, i.e., reported
   as *Brassica* sp. or mustard.

NB: *B. napus*, napobrassica group, and *B.
    rapa*, rapifera group, are often
    confused as they may both be referred to
    as turnip or as rape.

Absidia ramosa: on 4, 6 Sask [1113].
Acremonium boreale: on 4, 6 BC, Alta,
    Sask [1707].
Albugo sp.: on 4, 6 Sask [1867]; on 7
    (mustard) Sask [1871].
A. candida (A. cruciferarum) (C:61):
    white rust, albugine ou rouille

blanche: on 1, 3, 4 Sask [1321]; on 2
    Sask [1139, 1309]; on 3, 4 Sask [339,
    1320, 1594]; on 4 Alta [122, 335, 337,
    1320, 1595], Sask [333, 336, 1315 on
    seeds, 1868, 1871], Man [339, 1320,
    1317] and Alta, Sask [334, 338, 1317];
    on 4, 7 Alta [339, 1594]; on 4a Sask
    [1319]; on 5c BC [1816]; on 6 Alta [129,
    126, 1669, 1796], Sask [1309, 1323], Man
    [135, 1365, 1366], Que [1878] and Alta,
    Sask, Man [241 on seeds, 1317]; on 7
    Alta [338], Sask [336], Que [337]. By
    1977 this fungus was an important
    problem on brown mustard in Sask.
    However new lines carrying resistance
    were being developed.
Alternaria sp.: on 1, 2 Sask [1139]; on 4
    Alta [336], Sask [333, 334, 338, 1865,
    1868]; on 4, 6 Sask [1113, 1867], Man
    [135, 1113, 1366]; on 5c Que [1640]; on
    5f NB [333].
A. alternata (A. tenuis auct.*)* (C:62):
    on 2 Sask [1311]; on 2, 4, 6 Sask
    isolated from diseased roots [1312]; on
    3, 6 Sask isolated from hypertrophies
    caused by *Albugo cruciferarum* [1324];
    on 4 Sask [336, 1311], Man [1112]; on 6
    Alta [130, 1850, 1851].
A. brassicae (C:62): gray leaf spot,
    tache grise: on 1, 4 Sask [1321]; on 2,
    4, 6 Alta inoculated [1669]; on 4 Alta
    [122, 129, 339, 378, 1320, 1595, 1798,
    1831, 1832, 1833], Sask [336, 1323,
    1311, 1594, 1779, 1871] and Sask, Man
    [339, 1320]; on 4, 6 Alta, Sask, Man
    [1314, 1317 on seeds]; on 4, 7 Alta
    [1594]; on 4a NS [337]; on 5b Ont
    [1460]; on 5b, 5c Ont [1452, 1461,
    1594]; on 5c Que [334, 339], NB [334];
    on 5f, 6c BC [1816]; on 6 Alta [130,
    129, 378, 1795, 1850], Sask [1311, 1324
    isolated from hypertrophies caused by
    Albugo cruciferarum, 1779]; on 7 Sask
    [339].
A. brassicicola (C:62): black leaf spot,
    tache noire: on 4, 6 Man [1365]; on
    seeds of 5, 5c, 5e Sask [1313]; on 5b,
    5c, 5f BC [1816]; on 5c Que [334, 339],
    NB [334], NS [1370], Nfld [1594].
A. raphani: on 1, 4 Sask [1320]; on 3, 6
    Sask isolated from hypertrophies caused
    by *Albugo cruciferarum* [1324]; on 4
    Alta [129, 339, 378], Sask [336, 1311,
    1323, 1779, 1871], Man [339]; on 4, 6
    Alta, Sask, Man [1314, 1317 on seeds];
    on 4, 7 Sask [339]; on 5d Sask [1313];
    on 6 Alta [129, 1850], Sask [378, 1311,
    1779].
Ampelomyces quisqualis: on 4 Sask
    [1871]. Hyperparasitic on Erysiphales.
Apiospora montagnei: on 4, 6 Sask, Man
    [1113].
Aspergillus candidus: on 4, 6 Sask
    [1113]; on 4 Man [1112].
A. flavus: on 4, 6 Sask [1113].
A. fumigatus (C:62): on 4, 6 Sask [1113].
A. glaucus group: on 4 Man [1112].
A. niger: on 4, 6 Sask [1113].
A. versicolor: on 4 Man [1112]; on 4, 6
    Sask, Man [1113].

Botrytis sp.: on 4a, 5c Alta [1594].
B. cinerea (holomorph *Botryotinia
    fuckeliana*) (C:62): gray mold,
    moisissure grise: on 4 Alta, Sask, Man
    [1317]; on 4a NB, Nfld [339], NS [335];
    on 5b, 5f, 6c BC [1816]; on 5c NB [334],
    NS [1370]; on 6 Alta, Man [1317] and
    isolated from hypertrophies caused by
    Albugo cruciferarum Sask [1324].
Chaetomium indicum: on 4, 6 Sask [1113].
Cladosporium sp.: on 2, 4 Sask [1311]; on
    3, 6 Sask isolated from hypertrophies
    caused by *Albugo cruciferarum* [1324].
C. cladosporioides (C:62): on 4 Man
    [1112]; on 4, 6 Man [1113].
C. herbarum: on 4, 6 Man [1113].
C. variabile (Heterosporium v.) (C:62):
    on 5c BC [1816].
Cylindrocarpon sp.: on 4a Que [334].
Dendryphion nanum: on 6 Alta [819].
Emericella nidulans (anamorph *Aspergillus
    n.*): on 4, 6 Sask [1113].
Epicoccum sp.: on 6 Sask isolated from
    hypertrophies caused by *Albugo
    cruciferarum* [1324].
Erysiphe communis (E. polygoni) (C:62):
    powdery mildew, blanc: on 4 Sask [334,
    336, 1321, 1323, 1594, 1868, 1871]; on
    4a, 5a BC [1816]; on 6c Ont [1452], Que
    [1257].
Eurotium amstelodami: on 4, 6 Sask [1113].
E. chevalieri: on 4, 6 Sask [1113].
E. repens: on 4, 6 Sask [1113].
E. rubrum: on 4, 6 Sask [1113].
Fusarium sp. (C:62): on 1 Sask [1320]; on
    1, 2 Sask [1139]; on 4 Sask [333, 339,
    1321, 1323, 1594, 1865, 1867], Alta
    [122, 339]; on 4a Que, Nfld [334]; on 5b
    Ont [1458, 1460]; on 5b, 5c Ont [1452,
    1594]; on 5c Ont [1461]; on 5d NB [336];
    on 6 Alta [129], Sask [1867]; on 7 Sask
    [336, 1871].
F. acuminatum (C:62): on 4 Sask [1320,
    1871]; on 7 Sask [339].
F. avenaceum (C:62): on 4a NS [334].
F. equiseti (C:62): on 4 in greenhouse
    Sask [1871].
F. oxysporum (C:62): on 6 Alta [130].
F. oxysporum f. sp. *conglutinans*
    (C:62): yellows, jaunisse: on 5b, 5c
    Ont [1452, 1458, 1594]; on 5c Ont [1457,
    1460, 1461].
F. poae: on 7 Sask [339].
F. roseum: on 2 Sask [1312]; on 4 Alta,
    Sask, Man [1314, 1317]; on 4 Sask
    [1312]; on 6 Alta [130, listed as
    "Avenaceum, Culmorum, Gibbosum"], Sask
    [1312, 1324 isolated from hypertrophies
    caused by *Albugo cruciferarum*]; on 6
    Alta, Sask, Man [1314, 1317].
F. solani (holomorph *Nectria
    haematococca*) (C:62): on 4 in
    greenhouse Sask [1871].
Gliocladium sp.: on 6 Alta [130].
G. roseum: on 4 Sask [1321, 1594].
Gonatobotrys sp.: on 6 Sask [1324]
    isolated from hypertrophies caused by
    Albugo cruciferarum.

Helminthosporium sp.: on 6 Alta [130].
    Specimen was probably *Bipolaris* sp. or
    Drechslera sp.
Leptosphaeria maculans (anamorph
    Plenodomus lingam) (C:62): black leg,
    jambe noire: on 2, 4, 6 Sask [1312]; on
    seeds of 4, 6 Alta, Sask, Man [1325]; on
    4 Sask [1310, 1318, 1323, 1594, 1595],
    Man [1318]; on 4, 6 Sask [1316]. A
    severe localized outbreak occurred in
    ne. Sask in 1977. The estimated yield
    loss in one field was 20%. This
    outbreak was apparently due to the
    presence of a strain of the fungus which
    discharged ascospores much earlier in
    the season than the common strain which
    has been known on the Canadian Prairies
    since 1957.
Ligniera pilorum: on 5 Ont [60].
Mucor pusillus: on 4, 6 Sask [1113].
Mycosphaerella sp.: on 4, 6 Sask [1867].
M. brassicicola (C:62): ring spot, tache
    annulaire: on 1 Sask [1139, 1321]; on
    2, 4, 6 Sask [1311]; on 3, 4 Sask
    [1594]; on 4 Alta [122, 339, 1320], Sask
    [333, 334, 336, 337, 338, 1320, 1321,
    1323, 1865, 1868, 1871], Man [1320]; on
    4a Sask [1319]; on 5b, 5c, 5d, 5f BC
    [1816]; on 6 Alta [129], Sask [1866],
    Man [1365, 1366]; on 7 (rape) Sask
    [1872]. In a recent publication [1326]
    it has been contended that previous
    reports of *M. brassicicola* on *B.*
    sp. (rape) in the prairie provinces are
    incorrect. The pathogen in question may
    be *Pseudocercosporella capsellae*.
M. tassiana var. *tassiana*: on 4, 6
    Alta, Sask [1327].
Nectria inventa (anamorph *Verticillium
    tenerum*): on 4 Alta [1832, 1833].
Olpidium brassicae: on 5 Ont [65].
Oospora sp.: on 4a Que [334]. Probably a
    species of *Oidium*.
Paecilomyces varioti (C:62): from seeds
    of 4a BC [150].
Penicillium sp.: on 4, 6 Sask [1113]; on
    4a Que [334].
*P. aurantiogriseum (P. cyclopium, P.
    verrucosum* var. *c.)* (C:62): on 4 Man
    [1112]; on 4, 6 Sask, Man [1113].
P. brevicompactum: on 4, 6 Sask, Man
    [1113].
P. chrysogenum: on 4, 6 Sask, Man [1113].
P. glabrum (P. frequentans): on 4, 6 Sask
    [1113].
P. griseofulvum (P. patulum): on 4, 6
    Sask [1113].
P. piceum: on 4, 6 Sask [1113].
Peronospora sp.: on 4, 6 Sask [1867].
P. parasitica (C:62): downy mildew,
    mildiou: on 2 Sask [1139]; on 4 Alta
    [334, 338, 339, 1594], Sask [339, 1321];
    on 4 Alta, Sask, Man [1320]; on 4a BC
    [333, 1816], Sask [1319], Que [334], NS
    [333, 336], Nfld [335, 337]; on 4a Que,
    NS, Nfld [338, 339, 1594]; on 5a BC
    [1816]; on 5b BC [1816, 1817], Ont
    [336], NS [338]; on 5c Ont [1452], Que
    [333, 1639, 1594], NS [333, 338], Nfld
    [335]; on 5c, 5d, 5f BC [1816, 1817,

1818]; on 6 Sask isolated from hypertrophies caused by *Albugo cruciferarum* [1324], Man [1365, 1366]; on 6c Que [1640]; on 7 Alta [336, 339].

Phoma sp.: on 4, 6 Sask [1867].

P. glomerata: on 4, 6 Sask, Man [1113].

P. medicaginis: on 4, 6 Sask [920].

Phytophthora megasperma: on 4a BC [1816].

Plasmodiophora brassicae (C:62): clubroot, hernie: on 4 Nfld [339]; on 4a, 5a, 5b, 5c, 5f, 6b, 6c BC [1816]; on 4a Ont [333, 1594], Que [1654]; on 4a Que, NB, NS, PEI [40, 334, 336, 338]; on 4a Que, NB, NS [333, 1594]; on 4a Que, NB, PEI [335, 337, 339]; on 5, 5c Ont [1595]; on 5, 5a, 5b, 5c Que [245]; on 5b BC [465, 1817], Ont [338], Que [40], NB [339], NS [334, 338], PEI [333]; on 5b, 5c Ont [1452, 1460, 1594]; on 5c BC [40, 334, 335, 339], Ont [40, 1458, 1461], Que [334, 335, inoculated 340, 347, 348, 1641], NB [338, 339, 1594], NS [40, 334, 335, 338, 1594], PEI [40, 333], Nfld [337, 338]; on 5c Que, NB [40, 333, 337]; on 5c, 5d, 5f BC [1817, 1818]; on 5d NB [40, 336, 337]; on 5f NB [337]; on 5f, 6a, 6c Ont [1452]; on 6c Ont [1461], PEI [40]; on 7 Ont [1453], Que [341].

Plenodomus lingam (Phoma l.) (holomorph *Leptosphaeria maculans*) (C:62): black leg, jambe: on 1, 3, 4 Sask [1321]; on 3 Sask [1594]; on 4 Sask [334, 338, 339, 1320, 1868, 1871]; on 4a BC [333, 1816], Sask [1319], Que [334], NS [335], Nfld [333, 334]; on 5c BC [1816], Nfld [338]; on 5d Ont [334]; on 7 Sask [339].

Pseudocercosporella capsellae: white leaf spot and gray stem: on 1, 4, 4a, 5b, 5c, 6, 6c Sask [1326]; on 6 Alta [1669], Ont [1455].

Pythium sp.: on 5b Ont [1458, 1460, 1594]; on 5b, 5c Ont [1452].

P. irregulare: on 5c Ont [868] by inoculation.

P. polymastum: on 4, 6 Sask [1873].

P. ultimum (C:63): on 4a BC [1816].

Rhinocladiella mansonii: on 4, 6 Sask [1113].

Rhizoctonia sp.: on 1, 2 Sask [1139]; on 4, 6 Man [1113]; on 6 Alta [129].

R. solani (holomorph *Thanatephorus cucumeris*) (C:63): wire stem, tige noire: on 1 Sask [1320]; on 2, 4, 6 Sask [1312]; on 4 Sask [334]; on 4a Alta [335, 337], Ont [333], Que [334, 339], NS [333, 334, 335, 336, 1595], PEI [337], Nfld [335] and NB, NS, PEI [338, 339, 1594]; on 5b BC [1816, 1817], Alta [336, 338], Ont [1460, 1594], NS [336, 338]; on 5b, 5c Ont [1452, 1458]; on 5c BC [1816, 1818], NB [339], NS [334, 335, 338, 1594]; on 5d NB [336]; on 6 Alta [130]; on 6c BC [1816]; on 7 Alta [333], Sask [339]. In 1979 this fungus, the cause of damping-off, seedling blight and foot rot, was present in 95% of the Sask fields sampled. The seedling phase of the disease was so severe that many farmers resowed.

Rhizopus sp. (C:63): on 6 Sask isolated from hypertrophies caused by *Albugo cruciferarum* [1324].

R. nigricans: on 4a Que [334].

R. oryzae: on 4, 6 Sask [1113].

Sclerotinia sp.: on 1, 4, 6 Sask [420]; on 1, 5c, 7 Sask, Man [1138]; on 4, 6 Sask [1867]; on 7 Sask [1871].

S. sclerotiorum (C:63): sclerotinia rot, pourriture sclérotique: on 1 Sask [1319, 1320, 1321]; on 1, 2 Sask [1139]; on 3 Man [750]; on 4 BC [122], Alta [339, 892, 1317], Sask [333, 334, 336, 338, 339, 776, 1133, 1140, 1312, 1314, 1317, 1321, 1323, on seeds 1479, 1594, 1865, 1868, 1871], Man [338, 339, 1365, 1366]; on 4a BC [1816], Alta [337], Sask [1319], Ont [337], Que [334], NS [335]; on 5b BC [1816], Ont [1458], Que [334], PEI [333]; on 5b, 5c Ont [1452, 1460, 1594]; on 5c BC [1816], Alta [335, 339, 1594], Ont [335], Que [333, 335, 336, 339, 1639], NB [339, 1594], NS [1594], PEI [333]; on 6 Alta [129, 130, 1317], Sask [776, 1140, 1312, 1317], Man [750, 1365, 1366]; on 7 Alta [1736, 1971, 1972], Sask [336 mustard, 339 yellow mustard, 1134 rape]. The biological control agent, *Coniothyrium minitans* reduced stem rot by 50% under conditions favoring moderate disease development but the mycoparasite was ineffective under conditions highly favorable to disease development. The critical period for infection by *Sclerotinia* seems to be the mid- to late-flowering stage of the crop.

Scopulariopsis fusca: on 4, 6 Sask [1113].

Selenophoma sp.: on 6 Sask [1323].

Septoria sp.: on 4 Sask [1321].

Sphaerotheca macularis: powdery mildew, blanc: on 6c Ont [1257].

Stemphylium sp.: on 6 Alta [130], Sask isolated from hypertrophies caused by *Albugo cruciferarum* [1324].

S. botryosum (C:62): from seeds of 5c, 6c Ont [667].

Streptomyces scabies (C:63): scab, gale: on 4a NB [334, 1594], NB, NS [335] and NB, PEI [333], Nfld [339]. Causes scab of potato, but can infect other plants.

Talaromyces thermophilus: on 4, 6 Sask [1113].

Tetracladium setigerum: on 5 BC [43].

Thanatephorus cucumeris (Pellicularia praticola) (anamorph *Rhizoctonia solani*): on 4 Sask [338]; on 4a Sask [1319].

Trichoderma pseudokoningii: on 4 Sask and 4, 6 Man [1113].

Typhula umbrina (C:63): on 6c BC [1816].

Ulocladium consortiale (Alternaria c., Stemphylium c.) (C:62): from seeds of 5b, 5c Ont [667].

Verticillium sp.: on 4a Que [334].

V. nigrescens: on 5c Que [38, 392, 393].

Volucrispora aurantiaca: on 5 BC [43].

Wallemia sebi: on seeds of 4 Man [1112].

BRAYA Sternb. & Hoppe   CRUCIFERAE

1. *B. humilis* (C.A. Mey.) Robins.;
   native, Yukon-Frank, BC-Que.
2. *B. purpurascens* (R. Br.) Bunge;
   native, Mack-Frank, Que, Labr.
3. *B. thorild-wulfii* Ostenf.; native,
   Frank.

Sphaerotheca fuliginea (C:64): powdery
   mildew, blanc: on 1, 2, 3 Frank [1586].

BRIZA L.   GRAMINEAE

1. *B. minor* L., little quaking grass,
   petite brige; imported from Eurasia.

Puccinia graminis (C:64): rust, rouille:
   II III on 1 Ont [1236].

BROMUS L.   GRAMINEAE

1. *B. biebersteinii* Roem. & Schult.;
   imported from USSR.
2. *B. bromoideus* (Lej.) Crépin (*B.
   arduennensis* Dumont); imported from
   Europe.
3. *B. carinatus* Hook. & Arn. (*B.
   alutensis* Rydb.), California brome;
   native, BC.
4. *B. ciliatus* L., fringed brome, brome
   cilié; native, Yukon-Mack, BC-Labr.
5. *B. erectus* Huds. sensu lato (*B.
   syriacus* Boiss. & Blanche, *B.
   variegatus* auct. non Bieb.), upright
   brome-grass, brome dressé; imported from
   Europe, natzd., Yukon, BC.
6. *B. inermis* Leyss. complex (including
   B. ornans Kom. & *B. pumpellianus*
   Scribn.), smooth brome, brome inerme;
   native, Yukon-Mack, BC-Nfld, the typical
   form imported from Eurasia. 6a. *B.
   inermis* var. *tweedyi* (Scribn.) Hitchc.
7. *B. japonicus* Thunb. (*B. patulus*
   Mert. & Koch), Japanese brome, brome du
   Japon; imported from Eurasia.
8. *B. laevipes* Shear; imported from w.
   USA.
9. *B. lanceolatus* Roth (*B. macrostachys*
   Desf.); imported from the Mediterranean
   region.
10. *B. latiglumis* (Shear) A.S. Hitchc.;
    native, Sask-NB.
11. *B. mollis* L., soft chess, brome mou;
    imported from Europe, natzd., BC-Alta,
    Ont-NS.
12. *B. oxydon* Schrenk; imported from the
    USSR.
13. *B. paulsenii* Drob.; imported from the
    USSR. 13a. *B. p.* ssp. *angrenicus*
    (Drob.) Tzvel.
14. *B. purgans* L., laxative brome-grass,
    brome purgatif; native, Alta-Que.
15. *B. rigidus* Roth (*B. macrantherus*
    Hack.); imported from Europe.
16. *B. secalinus* L., cheat, brome des
    siegles; imported from Eurasia.
17. *B. sitchensis* Trin.; native, BC.
18. *B. squarrosus* L.; imported from
    Eurasia, natzd., BC, Man.

19. *B. sterilis* L., barren brome grass,
    brome stérile; imported from Eurasia,
    natzd., BC, Ont.
20. *B. tectorum* L., downy brome, brome des
    toits; imported from Eurasia, natzd.,
    Yukon, BC-NS.
21. *B. tyttholepis* (Nevski) Holub.;
    imported from the USSR.
22. *B. vulgaris* (Hook.) Shear; native,
    BC-Alta.
23. *B. willdenowii* Kunth (*B. unioloides*
    HBK), rescue grass, brome de Schrader;
    imported from South America.
24. *Bromus* hybrids.
25. Host species not named, i.e., reported
    as *Bromus* sp.

Acremonium boreale: on 6 BC, Alta, Sask
   [1707].
Alternaria alternata (A. tenuis auct.*)*
   (C:64): on 6 Sask [1711].
*Bipolaris sorokiniana (Helminthosporium
   sativum)* (holomorph *Cochliobolus
   sativus*) (C:64): leaf spot, tache des
   feuilles: on 6 Alta [120], Sask [961].
*Cercosporidium graminis (Passalora g.,
   Scolecotrichum g.)* (C:65): brown
   stripe, strie-brune: on 3, 6 BC [1816];
   on 17 BC [1728].
Claviceps sp.: on 4 BC and on 6 Alta
   [1781].
C. purpurea (C:64): ergot, ergot: on 3,
   6, 7, 9, 12, 16, 18 Sask [339]; on 6 BC
   [1816], Alta [334, 338], Sask [336, 337,
   1184], Ont, Que [335], NB [333, 1195].
*Colletotrichum graminicola (C.
   vermicularia)*: on wilted leaves of 14
   Ont [1502]; on 25 Que [335].
Curvularia geniculata: on 25 Sask [339].
Drechslera bromi (holomorph *Pyrenophora
   bromi*) (C:65): leaf blotch, tache des
   feuilles: on 1, 6, 8, 23 Sask [339]; on
   6 Alta [119, 121, 123, 124, 133, 336,
   338, 1594], Sask [336, 338, 1184], Que
   [32, 335, 339], NB [1195] and Alta, Man,
   Ont, PEI [1611]; on 25 BC [1816], Alta
   [337, 339], Sask [333, 335], NB [333]
   and Alta, Sask, Ont [1611].
D. tritici-repentis (holomorph
   Pyrenophora t.): on 6 Alta [1595].
*Drechslera verticillata (Podosporiella
   v.)* (holomorph *Pyrenophora
   semeniperda*): on 6 Sask [1594].
Erysiphe graminis (C:65): powdery mildew,
   blanc: on 6 Sask [338, 1184]; on 25
   Sask [336].
Fusarium sp. (C:65): on 6 Sask [339,
   1695].
F. poae (C:65): on 6 Alta [121, 123, 124].
Hendersonia crastophila: on 6 Sask
   [1631]. This name is of dubious
   application, possibly the same as
   Wojnowicia hirta. See comment under
   Hendersonia on *Abies*.
Leptosphaeria sp.: on 6 Que [334].
   Appeared to be associated with
   Stagonospora bromi.
L. luctuosa: on 6 Sask [1631].
Ligniera pilorum: on 6 Ont [60].

Metasphaeria bromigena (C:65): on 6 Yukon [1731].

Myxormia atroviridis: on overwintered culms of 6 Sask [1631].

Ophiobolus herpotrichus: on 6 Sask [1620, 1631].

Ovularia sp.: on 3 BC [1728].

Phaeosphaeria herpotrichoides (Leptosphaeria h.): on 6 Sask [1631].

Phloeospora idahoensis (holomorph *Didymella festucae*): on 6 Alta [1708].

Phyllachora graminis: tar spot, rayure goudronneuse: on 6 Que [335].

Phyllosticta sp.: on 17 BC [1728].

Pithomyces chartarum: on overwintered culms of 6 Sask [1631].

Polymyxa graminis: on 6 Ont [60, 194].

Pseudorobillarda agrostis (Robillarda a.): on 6 Sask [1201].

Pseudoseptoria bromigena (Selenophoma b., Septoria b.) (C:65): leaf spot, tache des feuilles: on 3 Yukon [1731], BC [1728]; on 6 BC [1696, 1728], Alta [119, 121, 123, 124, 133, 1594, 1697], Sask [333, 335, 339, 1184, 1694, 1695, 1698, 1709, 1710, 1763], Man [1763], Nfld [336] and Alta, Sask [337, 338, 1696]; on 13a Sask [339]; on 25 Alta [336], Sask [333, 336], Nfld [335].

Puccinia coronata (C:65): crown rust, rouille couronnée: II III on 6 Yukon [1731], Alta [644]; on 25 BC [1012].

P. coronata var. *bromi* (C:65): rust, rouille: III on 25 Sask [1839].

P. recondita s.l. (C:65): leaf rust, rouille des feuilles: on 1, 2 Sask [339]; on 3, 11 BC [1816]; II III on 4, 10, 14 Ont [1236]; on 6 Sask [339]; on 22 BC [1012].

P. striiformis (C:65): stripe rust, rouille striée: on 3, 17 BC [1816]. In the strict sense *P. striiformis* is a rust of Triticeae. It remains to be demonstrated whether it is biologically distinct from the rust on *Bromus* [1585].

Pyrenophora bromi (C:65): leaf blotch, tache helminthosporienne: on 1, 3, 6, 21, 24 Sask [1699]; on 5, 6, 15 Sask [1696]; on 6 BC [1696], Alta [1594, 1595, 1696, 1697], Sask [1594, 1631, 1694, 1695, 1697, 1709, 1710], Que [1611].

P. semeniperda (anamorph *Podosporiella verticillata*): on seeds of 25 Alta [1617].

Ramularia pusilla (C:65): on 6 Yukon [1731].

Rhizoctonia solani (holomorph *Thanatephorus cucumeris*): on 25 Ont [243].

Rhizophydium graminis: on 6, 7 Ont [59].

Rhynchosporium secalis (C:65): scald, échaudage: on 6 BC [1696], Alta [119, 121, 123, 124, 338, 1594, 1696], Sask [336, 338, 339, 1184, 1694, 1695, 1696, 1709], NB [333, 1195]; on 25 Sask [1699].

Sclerophthora macrospora: downy mildew, mildiou: on 6 Alta, Sask, Man [338]; on 6 Alta [336].

Sclerotinia borealis (C:65): snow mold, moisissure des neiges: on 5, 6 BC [1816].

Septoria sp.: on 6 Sask [339, 1695].

S. bromi (C:65): leaf spot, tache septorienne: on 6 Sask [1184, 1631, 1694], Que [335]; on 25 Sask [336].

Sporotrichum sp.: on 6 Sask [338, 339, 1694, 1695].

Stagonospora avenae (Septoria a.) (holomorph *Phaeosphaeria avenaria*): on 3 BC [1728].

S. bromi (C:65): leaf spot, tache stagonosporéenne: on 4 NB [333, 1195]; on 6 Que [334].

S. foliicola: on 6 Sask [1631].

Stemphylium botryosum: on 6 Sask [1711].

Trichothecium roseum: on 6 Sask [1711].

Ustilago bullata (C:65): smut, charbon: on 4 NB [333, 1195]; on 11, 17, 19, 20 BC [1816]; on 20 BC [335, 336, 339].

BROWALLIA L.   SOLANACEAE

Annual or perennial herbs of tropical America. Often grown in the greenhouse or in gardens for their flowers.

1. *B. americana* L.; native to tropical America.
2. *B. viscosa* HBK; native to Peru.

Verticillium dahliae: on 1 BC [1978]; on 2 BC [1816].

BUPLEURUM L.   UMBELLIFERAE

1. *B. ranuculoides* L. (*B. americanum* C. & R.), wet places or shallow water at low to fairly high elevations, Yukon-Mack, se. BC-sw. Alta.

Puccinia bupleuri (C:66): rust, rouille: on 1 Yukon [1551].

BUXUS L.   BUXACEAE

1. *B. sempervirens* L., common box, buis commun; native to Europe, n. Africa and w. Asia.

Macrophoma candollei (C:66): leaf blight, brûlure des feuilles: on 1 BC [1816].

Microthyrium macrosporum: on 1 Ont [1342].

Phomopsis stictica (C:66): on 1 BC [1816].

Phyllosticta sp.: on 1 BC [1816].

Volutella buxi (C:66): on *B.* sp. Ont [1828].

CAKILE Hill   CRUCIFERAE

1. *C. edentula* (Bigel.) Hook., sea rocket, caquillier édentulé; native, BC, Ont-Labr.

Peronospora cakile (C:66): downy mildew, mildiou: on 1 Que [282].

CALAMAGROSTIS Adans.   GRAMINEAE

1. *C. canadensis* (Michx.) Nutt.,
   Bluejoint, foin bleu; native, Yukon-Labr.
2. *C. inexpansa* Gray, contracted reed
   grass, calamagrostis contracté; native,
   Yukon-Keew, BC-Que, NS, Nfld.
3. *C. purpurascens* R. Br. 3a. *C. p.*
   var. *maltei* Polunin; Mack, Alta-Man,
   Ont-Que, Nfld.
4. *C. rubescens* Buckl., pine grass,
   calamagrostide rougissant; native,
   BC-Man.
5. *C. stricta* (Timm.) Koeler (*C.
   neglecta* (Ehrh.) Gaertn., Mey. &
   Scherb.); native, Mack-Labr.

Cheilaria agrostis (Septogloeum oxysporum)
   (C:67): on 1 Alta [333], Que [1730].
Claviceps sp.: on 2 Sask [1781].
C. purpurea (C. microcephala) (C:66):
   ergot, ergot: on 1 BC [1816], Alta
   [333, 334, 1594], Ont, Que [335], NB
   [333, 1195]; on 5 Que [282]; on flowers
   of *C.* sp. Que [1385].
Colletotrichum graminicola (C:66): on 1
   Ont [1729].
Cylindrosporium calamagrostidis: on 1 Que
   [335].
Drechslera catenaria (C:66): on 1 Ont
   [1611].
D. tritici-repentis (holomorph
   Pyrenophora t.) (C:66): on 1 Ont
   [1611].
Fusarium avenaceum (C:66): on 1 NB [333,
   1195].
Mastigosporium rubricosum (C:67): on 1
   Que [335], NB [333, 1195].
Phaeoseptoria sp.: on 2 Man [1729].
Pseudoseptoria everhartii (Selenophoma e.)
   (C:67): on 2 BC [1728]; on 3 Yukon
   [1731]; on 3a Frank [1729].
Puccinia coronata (C:67): crown rust,
   rouille couronnée: II III on 1 Ont
   [1236]; on 3 Yukon [1731]; on 4 BC
   [1012]; on *C.* sp. Que [335].
P. coronata f. sp. *calamagrostidis*
   (C:67): on 1 NB [333, 1195].
Stagonospora simplicior (C:67): on 2 Man
   [1729].
Ustilago salvei (U. striiformis auct.*)*
   (C:67): smut, charbon: on *C.* sp. Man
   [1594].

CALAMOVILFA (Gray) Hack.   GRAMINEAE

1. *C. longifolia* (Hook.) Scribn.,
   sand-reed; native, BC-Ont.

Claviceps sp.: on 1 Alta [1781].
Drechslera tritici-repentis (holomorph
   Pyrenophora t.) (C:67): on 1 Ont
   [1611].
Puccinia sporoboli (C:67): rust,
   rouille: II III on 1 Alta, Ont [1236].

CALENDULA L.   COMPOSITAE

1. *C. officinalis* L., pot-marigold, souci
   officinal; imported from Europe.

Entyloma polysporum: smut, charbon: on 1
   NS [337, 338]; on *C.* sp. NS [336].

CALLISTEPHUS Cass.   COMPOSITAE

1. *C. chinensis* (L.) Nees, china aster,
   reine-marguerite; native to China.

Botrytis cinerea (holomorph *Botryotinia
   fuckeliana*) (C:68): stem or flower
   blight, moisissure grise: on 1 BC
   [1816].
Coleosporium asterum (C:68): red rust,
   rouille rouge: II III on 1 Ont [1236],
   NS [334].
Fusarium oxysporum f. sp. *callistephi*
   (C:68): wilt, flétrissure fusarienne:
   on 1 BC [1816], Alta [338, 339], Sask
   [339], Man [334], NS [336].
Oidium sp.: powdery mildew, blanc: on 1
   BC [1816].
Phytophthora cryptogea (C:68): foot rot,
   mildiou du pied: on 1 BC [1816].
Verticillium dahliae: wilt, flétrissure:
   on 1 BC [1816, 1978].

CALOCEDRUS Kurz.   CUPRESSACEAE

   Monoecious, coniferous, evergreen trees
native to Taiwan, s. China, and Burma.

1. *C. decurrens* (Torr.) Florin
   (*Libocedrus d.* Torr.); imported from
   USA.

Lepteutypa cupressi: on 1 BC [1012].

CALOCHORTUS Pursh   LILIACEAE

1. *C. apiculatus* Baker; native, BC-Alta.
2. *C. macrocarpus* Dougl., sagebrush
   mariposa lily; native, BC.

Puccinia calochorti (C:68): rust,
   rouille: on 1 BC [1012]; on 1, 2 BC
   [1545, 1816]; on 2 BC [1586].

CALTHA L.   RANUNCULACEAE

1. *C. asarifolia* DC.; native, BC.
2. *C. leptosepala* DC., elkslip, populage
   à sépales étroits; native, Yukon,
   BC-Alta.
3. *C. palustris* L., cowslip, souci d'eau;
   native, Yukon-Frank, BC-Nfld.

Aecidium sp.: rust, rouille: on 2 BC
   [1012].
Botryotinia calthae (C:68): on 3 Ont, Que
   [718].
Erysiphe communis (E. polygoni) (C:68):
   powdery mildew, blanc: on 3 Alta, Ont
   [1257].
Pseudopeziza calthae (Fabraea rousseauana)
   (C:68): on 2 BC [1816].
Puccinia sp.: rust, rouille: on 2 BC
   [1012].
P. areolata (C:68): rust, rouille: on 1
   BC [1012]; on 1, 2 BC [1816].

P. calthae (C:68): rust, rouille: 0 I II
 III on 3 Ont [1236].
P. calthicola (C:68): rust, rouille: 0 I
 II III on 3 Ont [1236].
Verpatinia calthicola (C:68): on 3 Ont
 [662].

CAMASSIA Lindl.   LILIACEAE

1. *C. quamash* (Pursh) Greene, quamash;
 native, found in wet meadows, BC-sw.
 Alta.

Urocystis colchici (C:68): smut,
 charbon: on 1 BC [1012, 1545, 1816].

CAMELINA Crantz   CRUCIFERAE

1. *C. microcarpa* Andrz., small-seeded
 false flax, caméline à petits fruits;
 imported from Eurasia, introduced,
 BC-Nfld.

The following five fungi were isolated
from hypertrophies caused by *Albugo
cruciferarum* on 1 Sask [1324]:

Alternaria alternata.
A. brassicae.
Cladosporium sp.
Epicoccum sp.
Fusarium roseum.

CAMELLIA L.   THEACEAE

1. *C. japonica* L., common camellia,
 camélia du Japon; native to Japan, s.
 Korea and Taiwan.

Hendersonia sp.: on 1 BC [335, 1816].
 See comment under *Abies*.
Varicosporium elodeae: on 1 BC [43].

CAMPANULA L.   CAMPANULACEAE

1. *C. medium* L., canterbury bells,
 campanule carillon; imported from Europe.
2. *C. persicifolia* L., peach-leaved
 bellflower, campanule à feuilles de
 pêcher; native from Europe to ne. Asia.
3. *C. rapunculoides* L., creeping
 bellflower, campanule fausse-raiponce;
 native to Eurasia.
4. *C. rotundifolia* L., bluebell,
 campanule à feuilles rondes; native,
 Yukon-Labr, only as a garden escape in
 PEI.

Coleosporium campanulae (C:69): rust,
 rouille: II III on 1, 2 BC [2008]; on 2
 BC [1012, 1816]; on 2, 3, 4, *C.* sp.
 Ont [1236]; on 4 Ont [1828], Que [F66];
 on *C.* sp. Ont [1341].
Pleospora helvetica (C:69): on 4 Que [70].
P. phaeospora (C:69): on 4 Que [70].
Ramularia macrospora (C:69): leaf spot,
 tache ramularienne: on 1 BC [1816].
Sclerotinia sclerotiorum (C:69): stem
 rot, pourriture sclérotique: on 1 BC
 [1816]; on 3 Sask [1140].

CANNA L.   CANNACEAE

1. *C.* x *generalis* L.H. Bailey, common
 garden canna, balisier; introduced,
 herbs of tropical America.

Alternaria sp.: on 1 Ont [1808].
Myrothecium verrucaria: on 1 Ont [1808].

CAPSELLA Medic.   CRUCIFERAE

1. *C. bursa-pastoris* (L.) Medic.,
 shepherd's purse, tabouret; imported
 from Eurasia, natzd., Yukon-Mack,
 BC-Nfld.

Albugo candida (*A. cruciferarum*) (C:70):
 white rust, rouille blanche: on 1 BC
 [1816], Sask [1321, 1323, 1594].
Colletotrichum dematium: on 1 Sask [1323].
Ligniera pilorum: on 1 Ont [60].
Mycosphaerella brassicicola: ringspot,
 tache annulaire: on 1 Sask [1319, 1323].
Peronospora parasitica (C:70): downy
 mildew, mildiou: on 1 BC [1816], Sask
 [1320].
Pseudocercosporella capsellae: on 1 Sask
 [1326].
Pythium sp.: isolated from infected
 rootlets of 1 BC [1977].
P. polymastum: on 1 Sask [1873].
Rhizophydium graminis: on 1 Ont [59].
Verticillium dahliae: on 1 BC [38, 1816,
 1978].

CAPSICUM L.   SOLANACEAE

1. *C. annuum* L. var. *annuum* (*C.
 frutescens* in part), sweet or red
 pepper, piment; frequently cultivated.

Alternaria solani (C:70): early blight,
 brûlure alternarienne: on 1 Ont [1460,
 1594], Que [348].
Botrytis cinerea (holomorph *Botryotinia
 fuckeliana*) (C:70): gray mold,
 moisissure grise: on 1 BC [1816], Alta
 [339], NS [337]; on *C.* sp. NS [336].
Colletotrichum coccodes (C:70):
 anthracnose, anthracnose: on 1 Ont
 [334, 836].
Fusarium sp. (C:70): on 1 Ont [1460,
 1594].
Leiothecium ellipsoideum: on 1 BC [1012,
 1535].
Petriella sordida: on *C.* sp. Ont [85].
Phoma destructiva: phoma rot, pourriture
 phoméenne: on 1 Ont [1460, 1594].
Pythium sp. (C:70): damping-off, fonte
 des semis: on 1 Ont [868 by
 inoculation, 1460, 1594].
P. irregulare: on 1 Ont [868] by
 inoculation.
P. sulcatum: on 1 Ont [868] by
 inoculation.
P. sylvaticum: on 1 Ont [868] by
 inoculation.
P. ultimum (C:70): leak, pourriture
 aqueuse: on 1 BC [1816].

Rhizoctonia solani (holomorph
    Thanatephorus cucumeris) (C:70):   on 1
    Ont [334, 1460, 1594].
Sclerotinia sclerotiorum (C:70):   rot,
    pourriture sclérotique: on 1 BC [1816].
Stilbella flavescens: on 1 Que [2].   A
    symptomless parasite.
Ulocladium consortiale (Stemphylium c.):
    on 1 BC [1816].
Verticillium sp. (C:70):   wilt,
    flétrissure verticillienne: on 1 Ont
    [161, 1594]; on *C.* sp. Ont [96].
V. albo-atrum: wilt, flétrissure
    verticillienne: on 1 Ont [333].
V. dahliae: wilt, flétrissure
    verticillienne: on 1 BC [38, 337, 338,
    339, 1816, 1978, 1981], Ont [334, 337,
    338, 339, 461, 1087, 1460]; on *C.* sp.
    BC, Ont [335, 336].

CARAGANA Lam.   LEGUMINOSAE

1. *C. arborescens* Lam., Siberian pea
    tree, arbre aux pois; imported from
    Asia, escaped from cult., Alta-Man.
2. Host species not named, i.e., reported
    as *Caragana* sp.

Camarosporium caraganae (C:71):   on 2 Sask
    [1594, F67, F68].   A presumed anamorph
    of *Cucurbitaria caraganae*.
Cladosporium herbarum: on 2 Sask [1760].
C. macrocarpum: on 2 Sask [1760].
Cucurbitaria caraganae (C:71):   on 2 Sask,
    Man [1594, F67, F68], Sask [F68], Man [F63].
Cytospora sp.:   on 2 NWT-Alta [1063].
Fusarium solani (holomorph *Nectria
    haematococca*) (C:71):   crown rot or
    wilt, pourridié fusarien: on 1 Que
    [334].
Phyllosticta sp. (C:71):   on 2 Alta [337].
Rhizoctonia lilacina: on 1 Sask [1511].
Septoria caraganae (C:71):   leaf spot,
    tache septorienne: on 1 Sask, Man [339,
    F66, F70]; on 2 Alta [F66], Sask [1594,
    F65, F72, F73, F74], Sask, Man [F68,
    F69, F71], Man [F67].
Tubercularia ulmea: on 2 Sask [F67].
T. vulgaris (holomorph *Nectria
    cinnabarina*) (C:71):   on 2 Sask, Man
    [1594].

CARDAMINE L.   CRUCIFERAE

1. *C. bellidifolia* L.; native,
    Yukon-Alta, n. Que, Labr.
2. *C. digitata* Richards.; native, n.
    Yukon - c. Keew.
3. *C. occidentalis* (Wats.) Howell;
    native, BC.
4. *C. pratensis* L., cuckoo-flower,
    cardamine des prés; native, Yukon-Keew,
    Alta-Labr.

Botrytis cinerea (holomorph *Botryotinia
    fuckeliana*) (C:71):   on 1 Frank [664,
    1586].
Mycosphaerella pyrenaica (C:71):   on 1
    Frank [1586].

Peronospora parasitica (C:71):   downy
    mildew, mildiou: on 2 Yukon [460]; on 3
    BC [1816].
Puccinia cruciferarum (C:71):   rust,
    rouille: on 1 Frank [1239].
P. cruciferarum ssp. *borealis* (C:71):
    rust, rouille: on 1 Yukon, BC [1548].
P. cruciferarum ssp. *nearctica* (C:71):
    on 1 Frank [1586, 1548], Que [1548]; on
    1, 4 Keew [1548].
Synchytrium aureum (C:71):   on 4 Frank
    [1586].

CAREX L.   CYPERACEAE

1. *C. anthoxanthea* Presl; native,
    Yukon-Mack, BC.
2. *C. aquatilis* Wahlenb., aquatic sedge,
    carex aquatique; native, Yukon-Labr.
    2a. *C. a.* var. *altior* (Rydb.) Mack.
    (*C. substricta* (Kukenth.) Mack.),
    subnarrow sedge, carex subétroit;
    native, transcontinental. 2b. *C. a.*
    var. *stans* (Drej.) Boott; native,
    transcontinental.
3. *C. arcta* Boott, bear sedge, carex dru;
    native, Yukon, BC-NB.
4. *C. arctata* Boott, compressed sedge,
    carex comprimé; native, Ont-Nfld.
5. *C. atherodes* Spreng.; native,
    Yukon-Mack, BC-Que.
6. *C. atrata* L.; native, Yukon-Mack,
    BC-Alta, Ont-Que.
7. *C. atrofusca* Schkuhr; native,
    Yukon-Keew, Man-Que, Labr.
8. *C. backii* Boott, Back's sedge, carex
    de Back; native, BC-NB.
9. *C. bebbii* (Bailey) Fern., Bebb's
    sedge, carex de Bebb; native, Mack,
    BC-Nfld.
10. *C. bigelowii* All., Bigelow's sedge,
    carex de Bigelow; native, Mack-Keew,
    Sask-Man, Que, Nfld-Labr.
11. *C. brevior* (Dewey) Mack., short headed
    sedge, carex à têtes courtes; native,
    BC-Que.
12. *C. brunnescens* (Pers.) Poir, brownish
    sedge, carex brunâtre; native,
    Yukon-Keew, BC-NS, Nfld.
13. *C. canescens* L., silvery sedge, carex
    blanchâtre; native, Yukon-Keew, BC-Nfld.
14. *C. cephalophora* Muhl., oval-headed
    sedge, carex porte-tête; native,
    Ont-Que.
15. *C. chordorrhiza* L. f., creeping sedge,
    carex à long rhizome; native,
    Yukon-Frank, BC-Man, Que-NB, PEI-Nfld.
16. *C. colorata* Mack.; native, Man-Ont.
17. *C. conncinoides* Mack.; native, BC-Alta.
18. *C. crawfordii* Fern., Crawford's sedge,
    carex de Crawford; native, Yukon-Mack,
    BC-Nfld.
19. *C. crinita* Lam., fringed sedge, carex
    crépu; native, Man-Nfld.
20. *C. cristatella* Britt., crested sedge,
    carex accrêté; native, Alta-Que.
21. *C. debilis* Michx., flexuous sedge,
    carex flexueux; native, Ont-Nfld.
22. *C. demissa* Hornem.; native, Ont-Nfld.

23. *C. disperma* Dewey, two-seeded sedge, carex disperme; native, Yukon-Keew, BC-Nfld.
24. *C. eburnea* Boott, ivory sedge, carex ivoirin; native, Yukon-Mack, BC-Nfld.
25. *C. exilis* Dewey, starved sedge, carex maigre; native, Ont-NS, Nfld-Labr.
26. *C. festucaceae* Schkuhr; native, Ont.
27. *C. flava* L., yellow sedge, carex jaune; native, BC-Nfld.
28. *C. folliculata* L., folliculate sedge, carex folliculé; native, Ont-Nfld.
29. *C. geyeri* Boott; native, BC-Alta.
30. *C. gracillima* Schw., filiform sedge, carex filiforme; native, Man-Nfld.
31. *C. gynocrates* Wormsk., ridged sedge, carex à côtes; native, Yukon-Keew, BC-Labr.
32. *C. hirtifolia* Mack., hairy sedge, carex à feuilles poilues; native, Ont-NS.
33. *C. houghtonii* Torr., Houghton's sedge, carex de Houghton; native, Alta-NS.
34. *C. intumescens* Rudge, bladder sedge, carex gonflé; native, Man-NS.
35. *C. lacustris* Willd., lake sedge, carex lacustre; native, Alta-PEI.
36. *C. lasiocarpa* Ehrh. (*C. filiformis* auct. Amer.), villose sedge, carex à fruits tomenteux; native, Mack, BC-NS.
37. *C. laxa* Wahlenb.; native, Mack.
38. *C. laxiflora* Lam., distant-flowered sedge, carex laxiflore; native, Man-NS. 38a. *C. l.* var. *blanda* (Dewey) Boott (*C. blanda*), smooth sedge, carex lisse; native, Man-Que.
39. *C. lenticularis* Michx., lenticular sedge, carex lenticulaire; native, BC-NS, Nfld.
40. *C. leptalea* Wahlenb., bristle-stalked sedge, carex à tiges grêles; native, Yukon-Mack, BC-Nfld.
41. *C. limosa* L., mud sedge, carex des bourbiers; native, Yukon-Keew, Ont-Labr.
42. *C. livida* (Wahlenb.) Willd., livid sedge, carex livide; native, Yukon, BC-Nfld.
43. *C. lupuliformis* Sartwell; native, Ont.
44. *C. lupulina* Muhl., hop sedge, carex houblon; native, Ont-NS.
45. *C. lyngbyei* Hornem.; native, BC, Que, Labr.
46. *C. macrochaete* Mey.; native, Yukon, BC.
47. *C. maritima* Gunn; native, Yukon-Keew, Man, Nfld-Labr.
48. *C. mertensii* Prescott; native, Yukon, BC.
49. *C. misandra* R. Br.; native, Yukon-Keew, Alta, Labr.
50. *C. nardina* Fries; native, Yukon-Frank, BC-Alta, Que.
51. *C. obnupta* Bailey; native, BC.
52. *C. obtusata* Lilj.; native, Yukon-Mack, BC-Man.
53. *C. pauciflora* Lightf., pauciflorous sedge, carex pauciflore; native, Yukon, BC-Nfld.
54. *C. paupercula* Michx., stunted sedge, carex chétif; native, Yukon-Mack, BC-Nfld. 54a. *C. p.* var. *p.* (*C. magellanica* Lam.).

55. *C. pedunculata* Muhl., peduncled sedge, carex pédoncule; native, Man-Nfld.
56. *C. pensylvanica* Lam., Pennsylvania sedge, carex de Pensylvanie; native, BC-NS.
57. *C. petricosa* Dewey; native, Yukon-Mack, BC-Alta, Que, Nfld.
58. *C. phyllomanica* Boott; native, BC.
59. *C. plantaginea* Lam., plantain leaved sedge, carex plantain; native, Ont-NS.
60. *C. pluriflora* Hult.; native, BC.
61. *C. praegracilis* Boott; native, Yukon, BC-Ont.
62. *C. pseudo-cyperus* L., cyperus-like sedge, carex faux-souchet; native, Alta-Nfld.
63. *C. pyrenaica* Wahlenb.; native, Yukon-Mack, BC-Alta.
64. *C. retrorsa* Schw., retrorse sedge, carex réfléchi; native, Mack, BC-Nfld.
65. *C. rosea* Schkuhr, stellate sedge, carex roseau; native, Man-PEI.
66. *C. rostrata* Stokes, beaked sedge, carex rostré; native, Yukon-Keew, BC-Nfld. 66a. *C. r.* var. *utriculata* (Boott) Bailey; somewhat more southern in distribution.
67. *C. rupestris* Bellardi, rock sedge, carex des rochers; native, Yukon-Keew, Alta, Man, Que, Nfld-Labr.
68. *C. salina* Wahlenb. (*C. lanceata* Dew.), salt-marsh sedge, carex salin; native, Man-Labr.
69. *C. saxatilis* L.; native, Frank, Sask-Labr. 69a. *C. s.* var. *miliaris* (Michx.) Bailey (*C. miliaris* Michx.), miliary sedge, carex miliaire; native, Frank, Sask-NS, Nfld-Labr. 69b. *C. s.* var. *s.* (*C. s.* ssp. *laxa* (Trautv.) Kalela); native, Frank, n. Que.
70. *C. scabrata* Schw., rough sedge, carex scabre; native, Ont-PEI.
71. *C. scirpoidea* Michx., scirpoid sedge, carex faux-scirpe; native, Yukon-Alta, Man-Que, NS, Nfld-Labr.
72. *C. scoparia* Schkuhr, broom sedge, carex à balais; native, BC-Nfld.
73. *C. siccata* Dewey (*C. foenia* Willd.); native, Yukon-Mack, BC-Que.
74. *C. sitchensis* Prescott; native, BC.
75. *C. sparganioides* Muhl., bur-reed sedge, carex faux-rubanier; native, Ont-Que.
76. *C. spectabilis* Dewey; native, BC-Alta.
77. *C. sprengelii* Dewey, Sprengel's sedge, carex de Sprengel; native, BC-NB.
78. *C. stipata* Muhl., stipitate sedge, carex stipité; native, BC-Nfld.
79. *C. stricta* Lam., stiff sedge, carex raide; native, Man-NS. 79a. *C. s.* var. *strictior* (Dewey) Carey (*C. strictior* Dewey); native, Ont-Que, NS.
80. *C. tenera* Dewey, weak sedge, carex faible; native, BC-NS.
81. *C. tetanica* Schkuhr; native, Man-Ont.
82. *C. tribuloides* Wahlenb., blunt broom sedge, carex tribuloide; native, Ont-NS.
83. *C. trisperma* Dewey, three-fruited sedge, carex trisperme; native, Mack, BC-Nfld.

84. *C. vaginata* Tausch, sheathed sedge, carex engainé; native, Yukon-NB, Nfld-Labr.
85. *C. vesicaria* L., inflated sedge, carex vésiculeux; native, BC-Nfld. 85a. *C. v.* var. *monile* (Tuckerm.) Fern.; native, BC, Ont-NS.
86. *C. viridula* Michx., greenish sedge, carex verdâtre; native, Yukon-Mack, BC-Nfld.
87. *C. vulpinoidea* Michx., fox sedge, carex faux-vulpin; native, BC-Nfld.
88. Host species not named, i.e., reported as *Carex* sp.

Acremonium sp.: on 88 NWT [1969]. Probably saprophytic.
Anthracoidea altera: smut, charbon: on 49 Frank [1169].
A. americana: smut, charbon: on 66 Yukon, Que and on 85 Ont [1169].
A. atratae (C:73): smut, charbon: on 2b Frank [1586]; on 6 Yukon, BC [1169]; on 76 BC [1578].
A. bigelowii (Cintractia limosa var. *minor)* (C:73): smut, charbon: on 10 Que, Labr [1171].
A. caricis (Ustilago c., U. urceolorum) (C:73): smut, charbon: on fruits of 13, 73 Ont [437]; on 49 Frank [1372].
A. caricis-pauciflorae (C:73): smut, charbon: on 53 BC [1169].
A. caryophylleae (C:73): smut, charbon: on 52 Yukon [1169].
A. elynae (Cintractia carpophila var. *elynae)* (C:73): smut, charbon: on 61 BC [1816].
A. fischeri (Cintractia f.) (C:73): smut, charbon: on 13 Mack, BC [1168, 1169].
A. heterospora (Cintractia caricis var. *acutarum)* (C:73): smut, charbon: on 2 Yukon [1171], Ont [1341]; on 2a Que [282].
A. karii (Cintractia carpophila var. *carpophila* sensu Savile, pro parte*)* (C:73): smut, charbon; on 23 Yukon, Alta [460, 1570].
A. lasiocarpae (C:73): smut, charbon: on 69a Nfld and on 69b Que [1169].
A. laxae: smut, charbon: on 37 Mack [1169].
A. limosa (Cintractia limosa var. *limosa)* (C:73): smut, charbon: on 41 Que [282].
A. liroi: smut, charbon: on 68 Frank, Labr [1171].
A. misandrae (C:73): smut, charbon: on 7 Keew and on 57 BC [1169].
A. nardinae (A. elynae var. *nardinae)* (C:73): smut, charbon: on 50 Frank [1586].
A. paniceae (C:73): smut, charbon: on 84 "Yukon-Nfld" [1169], Alta [644].
A. rupestris (C:73): smut, charbon: on 67 Frank [1244].
A. scirpoideae (C:73): smut, charbon: on 71 Alta [644].
A. subinclusa (C:73): smut, charbon: on 69 Que and on 85 Ont [1169].

A. turfosa: smut, charbon: on 25 Que, NS, Labr and on 31 Que [1168, 1169].
Buergenerula biseptata: on 2 Yukon [76].
Cercospora caricis (C. caricina, C. microstigma): on leaves of 38 [1502], 65 [440] Ont.
Cladosporium cladosporioides: on 88 NWT [1969].
C. herbarum: on 88 BC [1012].
Clasterosporium caricinum (C:74): on 2 Yukon, Mack and on 2b Frank [1586].
Claviceps purpurea: ergot, ergot: on 73 Ont [1827]. Specimens should be compared with *C. grohii*.
Didymella glacialis (C:74): on 88 Que [70].
Galerina subbadipes: on rotted debris of 88 Que [1385].
Geomyces pannorum (Chrysosporium verrucosum): saprophytic on 88 NWT [1969].
Leptosphaeria sp. (C:74): on 4 Que [282].
Melanodothis caricis: on 1, 42, 45, 46, 53, 58, 60, 74, 86, 88 BC [11]; on 42 BC [1012]; on 53, 74 BC [13].
Melanotus caricicola: on 88 Ont [1437].
Mortierella sp.: saprophytic on 88 NWT [1969].
Mucor sp.: saprophytic on 88 NWT [1969].
Mycena cariciophila: on 88 NB [1432].
Mycosphaerella lineolata (C:74): on 88 Que [70].
M. tassiana var. *tassiana* (C:74): on 88 Que [70].
Myriosclerotinia caricis-ampullaceae: on 2 Man [666].
M. longisclerotialis: on 88 Man [666].
Neottiospora caricina: on old leaves of 88 Ont [1161, 1616].
Orphanomyces arcticus (C:75): smut, charbon: on 47 Frank [1562, 1586].
Paraphaeosphaeria michotii (Leptosphaeria folliculata) (C:74): on 28 Ont [1621].
Penicillium sp.: saprophytic on 88 NWT [1969].
Phaeosphaeria eustoma: on 88 Que [70].
P. herpotrichoides: on 88 Que [70].
Phialophora sp.: saprophytic on 88 NWT [1969].
Phoma herbarum: saprophytic on 88 NWT [1969].
Phomatospora therophila (C:74): on 88 Que [70].
Pleospora straminis: on leaves of 88 Que [142].
Puccinia sp.: rust, rouille: on 17 BC [1012].
P. atrofusca (C:75): rust, rouille: on 29 Alta [1578].
P. bolleyana (C:75): rust, rouille: II III on 33 Que [F60]; on 43, 44 Ont [1236].
P. caricina (C:75): rust, rouille: II III on 2, 10, 59 Que [1236]; on 4 Ont [1341]; on 4, 12, 13, 16, 18, 19, 21, 22, 24, 27, 30, 32, 34, 35, 38a, 39, 55, 62, 66a, 70, 79a, 81, 83, 85, 87, 88 Ont [1236]; on fossilized 5 Sask [394a]; on 48 BC [1012]; on 88 Alta [952], Ont [1828].

P. caricis-shepherdiae (C:75): rust,
rouille: II III on 10 Que [1236]; on 24
Ont [1236]; on 88 BC [953], Alta [951].
P. caricis-strictae: rust, rouille: on
51 BC [1012].
P. dioicae (C:75): rust, rouille: II III
on 8, 9, 11, 14, 15, 18, 20, 23, 26, 56,
65, 72, 73, 75, 77, 78, 80, 82, 87, 88
Ont [1236]; on 88 Alta [952].
P. karelica (C:75): rust, rouille: II
III on 54 Ont, Que [1236].
P. karelica ssp. *karelica*: rust,
rouille: II III on 60 BC [1550].
P. karelica ssp. *laurentina*: rust,
rouille: on 54a Ont, Que, NS, Nfld,
Labr [1550].
P. limosae (C:75): rust, rouille: II III
on 3, 19, 41, 85, Ont [1236]; on 41 BC
[1550].
P. microsora (C:75): rust, rouille: II
II III on 64, 85, 85a Ont [1236].
P. minutissima (C:75): rust, rouille: II
III on 2, 36, 40 Ont [1236].
P. uniporula (*P. caricis* var. *uniporula*)
(C:75): rust, rouille: on 4 Que [282].
Rutola graminis: on 44 Ont [332].
Rutstroemia paludosa (C:75): on 88 Ont
[429], Que [662].
Schizonella pusilla (*S. melanogramma*
auct.) (C:75): on 71 Alta [644].
Sclerotinia sp.: on 2b Frank [1586].
S. arctica (C:75): on 2 Keew and on 2b
Frank [428, 664].
Septoria caricis (C:75): on 84 NB [1036].
Thecaphora apicis (C:75): on 63 BC [1578].
Torula graminis (*T. caricina*): on dead
leaves of 44 Ont [440].
Trimmatostroma sp.: on 88 NWT [1969].
Uromyces perigynius (C:75): rust,
rouille: II III on 9, 20, 82 Ont [1236].
Wettsteinina macrotheca (C:75): on 88 Que
[70].
W. niesslii (C:75): on 88 Mack [70].

CARPINUS L.   BETULACEAE

1. *C. caroliniana* T. Walt. (*C.
    americana* Michx.), blue-beech, charme
    de Carolinie; native, Ont-Que. 1a. *C.
    c.* var. *virginiana* (Marsh.) Fern.;
    northern form of *C. caroliniana*.
2. Host species not named, i.e., reported
    as *Carpinus* sp.

Aschersonia carpinicola: on dead bark of
1 Ont [440]. Probably an insect.
Cheirospora botryospora: on 1 Ont [1340].
Conoplea globosa (C:75): on 1 Ont [784].
C. olivacea (*C. sphaerica*) (C:75): on 2
Ont, Que [784].
Coriolus versicolor (*Polyporus v.*): on 2
Ont [1827]. See *Abies*.
Coryneum carpinicola: on 1 Ont [1762,
1763]; on 2 Ont [1762].
Cryptosporiopsis fasciculata (holomorph
Pezicula carpinea): on 1 Ont [1340].
Diaporthe bakeri: on 2 Ont [1340].
Diatrype albopruinosa: on 1 Ont [1354].
Eutypella cerviculata: on 2 Ont [1340].

Ganoderma applanatum (C:75): on 2 NS
[1032]. See *Abies*.
Lenzites betulina: on 1 Ont [1341]. See
Acer.
Massarina eburnea: on 2 Ont [1340].
Melanconis chrysostroma var. *ellisii*
(*M. chrysostroma* var. *carpinigera*
(C:75): on 1, 2 Ont [1340]; on 1a Ont
[F60].
Nectria sp.: on 2 Ont [1828].
Pezicula carpinea (C:76): on 1, 2 Ont
[1340].
Phellinus igniarius (*Fomes i.*) (C:75):
white trunk rot, carie blanche du
tronc: on 1 Ont [1341], Que [1377].
Phyllactinia guttata: powdery mildew,
blanc: on 2 Ont [1257].
Pleomassaria carpini (C:76): on 1 Ont
[F63]; on 2 Ont [1340].
Robergea cubicularis: on 2 Ont [1608].
Rutstroemia bolaris (C:76): on 1a Ont
[F60]; on 2 Ont [1340].
Taphrina australis (*Gloeosporium
carpinicolum*) (C:75): on leaves of 1
Ont [440].
Teratosperma cornigerum (*Clasterosporium
pulcherrimum*): on 1, 2 Ont [808].

CARTHAMUS L.   COMPOSITAE

1. *C. tinctorius* L., safflower, safran
    bâtard; imported from Eurasia.

Alternaria carthami (C:76): leaf spot,
tache alternarienna: on 1 Sask [1139],
Man [335], Ont [334, 335, 336].
Botrytis cinerea (holomorph *Botryotinia
fuckeliana*) (C:76): gray mold,
moisissure grise: on 1 Alta, Man, Ont
[1505], Sask on seeds [1314], Ont [335,
336].
Fusarium sp. (C:76): on 1 Sask [1139],
Ont [335].
F. roseum: on seeds of 1 Sask [1314].
Puccinia carthami (C:76): rust, rouille:
on 1 Alta [1558], Sask [1139, 1558], Man
[1558, 1838], Ont [335, 336].
Sclerotinia sclerotiorum (C:76): head
rot, pourriture sclérotique: on seeds 1
Alta, Man [1505].

CARYA Nutt.   JUGLANDACEAE

1. *C. cordiformis* (Wangenh.) K. Koch (*C.
    amara* Raf.), bitternut, caryer à noix
    amères; native, Ont-Que.
2. *C. glabra* (Mill.) Sweet, pignut
    hickory, caryer à cochous; native, Ont.
3. *C. ovata* (Mill.) K. Koch, shagbark
    hickory, caryer à noix douces; native,
    Ont-Que.
4. Host species not named, i.e., reported
    as *Carya* sp.

Asteroma caryae (*Gloeosporium c.*,
Phyllosticta c.) (C:76): anthracnose,
anthracnose: on 3 Ont [1340], Que [F76].
Botryodiplodia sp.: on 1 Que [F64, F69].
Ceuthospora caryae: on 1 Ont [195].
Chaetomium sp.: on 1 Ont [1827].

Coniothyrium sp.: on 1 Ont [1827].
Conoplea globosa: on 2 Ont [1340].
C. sphaerica. (C:76): on 4 Que [784].
Coriolus versicolor (Polyporus v.): on 3
    Ont [1341]. See *Abies*.
*Dendrothele microsporum (Aleurocorticium
    m.)*: on 4 Ont [965]. Lignicolous,
    typically on dead branches.
Diplodia caryigena: on 1 Ont [195].
Endophragmiella cesatii: on 3 Ont [809].
Fenestella amorpha: on 3 Ont [1340].
Fusicoccum juglandis: on 1 Ont [195].
Ganoderma applanatum: on 4 Ont [1341].
    See *Abies*.
Gloeosporium sp.: on 4 Ont [1340].
Inonotus cuticularis (Polyporus c.): on 1
    Que [F71]. See *Acer*.
Melanconis sudans: on 3 Ont [1340].
Melanospora sp. *(Ceratostoma* sp.): on 1
    Ont [1340].
Microstroma juglandis (C:76): leaf spot,
    moisissure blanche: on 3 Ont [F60].
Mycosphaerella dendroides: on 2, 3 Ont
    [1340].
Phellinus conchatus (Fomes c.): on 3 Ont
    [1341]. See *Acer*.
P. igniarius (Fomes i.): on 1 Que [F71].
    See *Acer*.
Phomopsis sp.: on 1 Ont, Que [1340]; on 3
    Que [1377].
Phragmocephala stemphylioides: on 3 Ont
    [808].
Schizophyllum commune (C:76): white
    spongy rot, carie blanche spongieuse:
    on 1 Que [1377]; on 3 Ont [1341].
Stereum sp.: on 4 Ont [1341].
Typhula spathulata (Pistillaria s.): on
    dead branches of 4 Que [1385].

CASSIOPE D. Don   ERICACEAE

1. *C. lycopodioides* (Pall.) D. Don;
    native, BC.
2. *C. mertensiana* (Bong.) D. Don, white
    heather; native, BC-Alta.
3. *C. tetragona* (L.) D. Don,
    bell-heather, cassiopée tétragone;
    native, Yukon-Alta, n. Que, Labr. 3a.
    C. tetragona var. *saximontana*
    (Small) Hitchc.; native, Yukon-BC.

Discocainia arctica (Cenangium a.): on 3
    Frank [1244].
Exobasidium sp.: on 1 Alta [952].
E. vaccinii (C:77): on 2 BC [1012, 1816];
    on 3 Frank [1543], BC [1012].
E. vaccinii var. *vaccinii* (C:77): on 3
    Frank [1244, 1586]; on 3a Alta [644].
Gnomoniella hyparctica (C:77): on 3 Mack
    [764], Keew, Nfld [79].
Odontotrema cassiopes: on 3 Frank [764].
*Protoventuria latispora (Antennularia l.,
    Gibbera l.)*: on overwintered leaves of
    3 NWT [73]; on 3 Frank, Alta [764].
*Scleropleella hyperborea (Leptosphaeria h.,
    Leptosphaerulina h.)* (C:77): on 3 Yukon
    [460].

CASTANEA Mill.   FAGACEAE

1. *C. dentata* (Marsh.) Borkh., chestnut,
    châtaigner d'Amérique; native, s. Ont.
2. *C. sativa* Mill., European chestnut,
    châtaigner d'Europe; native in s.
    Europe, n. Africa and w. Asia.
3. Host species not named, i.e., reported
    as *Castanea* sp.

Cryphonectria parasitica (Endothia p.)
    (C:77): blight, brûlure du châtaigner:
    on 1 BC [1816], Ont [338, 1340, F61]; on
    2 BC [1012]; on 3 Ont [1340, F62, F64,
    F65].
Ganoderma applanatum: on 3 Ont [1341].
    See *Abies*.
Microsphaera penicillata: powdery mildew,
    blanc: on 1 Ont [1257].
Ophiovalsa cinctula (Cryptospora c.)
    (C:77): on 1 Ont [1340]; on 3 Ont [F60].
Phyllactinia guttata: powdery mildew,
    blanc: on 1 Ont [1257].
Pseudovalsa modonia (Melanconis m.): on 3
    Ont [1340].
Schizophyllum commune: white spongy rot,
    carie blanche spongieuse: on 2 BC
    [1012].
Tomentella cinerascens: on 1 Ont [1828].

CASTILLEJA Mutis   SCROPHLARIACEAE

1. *C. angustifolia* (Nutt.) G. Don;
    native, BC-Alta.
2. *C. coccinea* (L.) K. Spreng., scarlet
    painted cup, castilléjie éclarte;
    native, Sask-Ont.
3. *C. miniata* Dougl. ex Hook.; native,
    BC-Ont.
4. *C. occidentalis* Torr.; native, BC-Alta.
5. *C. raupii* Pennell; native, Yukon-Keew,
    BC-Ont.
6. *C. rhexifolia* Rydb.; native, BC-Alta.
7. *C. septentrionalis* Lindl., common
    yellow paint-brush, castilléjie
    septentrionale; native, Alta-NB, Nfld.
8. Host species not named, i.e., reported
    as *Castilleja* sp.

Cronartium sp.: rust, rouille: on 3, 8
    BC [1816]; on 6 Sask [1997, F61].
    Probably *Peridermium stalactiforme*.
C. coleosporioides (aecial state
    Peridermium stalactiforme) (C:77):
    rust, rouille: II III on 1, 3 BC [735];
    on 2 Ont [1828]; on 3 BC [2006
    inoculated], Alta [557, 1402, F61]; on
    3, 4, 5, 6, 7 Alta [735]; on 3, 4, 5, 7,
    8 NWT-Alta [1063]; on 3, 8 Alta [950,
    952]; on 4, 7 Alta [F64]; on 5 Mack
    [F63], Mack, Alta [53]; on 6 Sask [735];
    on 8 BC [953, 954], Alta [951, 1408].
C. coleosporioides f. *album*: rust,
    rouille: II III on 3 BC [1012, 2006],
    Alta [2008].
C. coleosporioides f. *coleosporioides*:
    rust, rouille: on 3 BC [1012].
C. ribicola: rust, rouille: on 3 Alta
    [734] by inoculation.
Ophiobolus niesslii: on 2 Man [1620].

Peridermium stalactiforme (holomorph
  Cronartium coleosporioides): rust,
  rouille: on 3 Alta [1398]; on 6 Sask
  [1997].
Ramularia castillejae: on 3 BC [1554].
Sphaerotheca fuliginea (C:77): powdery
  mildew, blanc: on 3 BC [1554].

CATALPA Scop.   BIGNONIACEAE

1. *C. bignonioides* Walt., common catalpa,
   catalpe commun; native in the USA.
2. *C. speciosa* Warder ex Engelm.,
   northern catalpa, catalpe élégant;
   native in the USA.

Alternaria sp.: on 1 Ont [1594].
A. catalpae (C:77): on 1 Ont [F61]; on
  C. sp. Ont [1340].
Botrytis cinerea (holomorph *Botryotinia
  fuckeliana*) (C:78): on 2 BC [1816].
Irpex lacteus (*Polyporus tulipiferae*): on
  C. sp. Ont [1341]. See *Acer*.
Verticillium albo-atrum (C:78): wilt,
  flétrissure: on 2 Que [336]; on *C.*
  sp. Ont [333, F64].
V. dahliae: wilt, flétrissure: on 2 BC
  [339, 1978].

CAULOPHYLLUM Michx.   BERBERIDACEAE

1. *C. thalictroides* (L.) Michx., blue
   cohosh, caulophylle faux-pigamon;
   native, Man-NS.

Erysiphe cichoracearum: powdery mildew,
  blanc: on 1 Ont [1257].
Streptotinia caulophylli (C:78): on 1 Ont
  [430], Que [427, 430].

CEANOTHUS L.   RHAMNACEAE

1. *C. sanguineus* Pursh, tea tree; native,
   BC.
2. *C. velutinus* Dougl., snow brush,
   céanothe velouté; native, BC-Alta.

Cylindrosporium ceanothi (C:78): on 1, 2
  BC [1816]; on 2 BC [1012].

CEDRUS Trew   PINACEAE

1. *C. deodara* (D. Don) G. Don, deodar
   cedar, cèdre de l'Himalaya; imported
   from the Himalayas.

Cytospora abietis: on 1 BC [1012].
Sydowia polyspora (anamorph *Sclerophoma
  pithyophila*): on *C.* sp. BC [1012].

CELASTRUS L.   CELASTRACEAE

1. *C. scandens* L., bittersweet, célastre
   grimpant; native, Sask-NB.

Phyllactinia guttata (C:78): powdery
  mildew, blanc: on 1 Man, Ont, Que
  [1257].
Septogloeum thomasianum: on 1 Ont [1767].

CELTIS L.   ULMACEAE

1. *C. occidentalis* L., hackberry, bois
   inconnu; native, Man-Que.

Polyporus lentus (*P. fagicola*): on *C.*
  sp. Que [1385]. Apparently saprophytic
  on logs.
Sphaerotheca phytoptophila: powdery
  mildew, blanc: on 1 Ont [1257].

CENTAUREA L.   COMPOSITAE

1. *C. cyanus* L., bachelor's buttons,
   bluet; imported from Europe, escaped
   from cult., BC-Alta, Ont-Nfld.
2. *C. diffusa* Lam., diffuse knapweed,
   centaurée diffuse; imported from
   Eurasia, introduced, BC-Alta.
3. *C. nigra* L., black knapweed, centaurée
   noire; imported from Europe, escaped
   from cult., BC, Ont.

Microsphaeropsis centaureae: on 2 BC
  [1130, 1922, 1923].
Puccinia centaureae var. *centaureae*:
  rust, rouille: on 3 NS [1558].
P. cyani (C:79): rust, rouille: 0 II III
  on 1 BC [1816], Ont [1236, 1557, 1558],
  NS [1557].
P. cyani var. *sublevis*: rust, rouille:
  on 1 BC [1558].
Sclerotinia sclerotiorum: on 2 BC [892,
  1130, 1922, 1923].
Septoria centaureicola var. *brevispora*
  (C:79): leaf spot, tache septorienne:
  on 1 BC [1816].

CEPHALANTHUS L.   RUBIACEAE

1. *C. occidentalis* L., buttonbush, bois
   noir; native, Ont-PEI.

Microsphaera penicillata: powdery mildew,
  blanc: on 1 Que [1257].
Puccinia seymouriana (C:79): rust,
  rouille: 0 I on 1 Ont [1341], Que
  [1236].

CERASTIUM L.   CARYOPHYLLACEAE

1. *C. alpinum* L.; native, Mack-Keew, n.
   Sask- n. Man, Labr.
2. *C. arvense* L., field chickweed,
   ceraiste des champs; native, Yukon-Mack,
   Alta-Nfld.
3. *C. beeringianum* C. & S.; native,
   Yukon-Frank, BC-Alta, Nfld-Labr.
4. *C. regelii* Ostenf.; native, Mack-Frank.
5. *C. vulgatum* L., common mouse-ear
   chickweed, céraiste vulgaire; imported
   from Eurasia, natzd., Yukon, BC-Nfld.

Isariopsis episphaeria (C:79): on 1
  Frank, Keew and on 3, 4 Frank [1586].
Melampsorella caryophyllacearum (*M.
  cerastii*) (C:79): rust, rouille: II
  III on 1 Que [1236]; on 2 NWT-Alta
  [1063], Alta [644, 950, 952]; on 2, 5 BC

[1012, 1816], Alta [F63, F67]; on *C.*
   sp. BC, Alta, Sask, Man [2008], Que
   [1377].
Mollisia affin. *epitypha*: on 4 Frank
   [664].
Mycosphaerella densa (C:79): on *C.* sp.
   Frank [1586].
Ophiobolus anguillidus: on 5 Ont [1620].
Puccinia arenariae (C:80): rust,
   rouille: on 1 Frank [1244], Man, Que
   [1236].

CERCIS L.   LEGUMINOSAE

1. *C. canadensis* L., redbud, bouton
   rouge; native to USA with one very old
   report from Ont, however, it is
   cultivated in Ont.

Verticillium dahliae: on 1 Ont [1811].

CHAENOMELES Lindl.   ROSACEAE

1. *C. japonica* (Thunb.) Lindl. ex Spach.,
   Japanese flowering quince, cognassier du
   Japon; imported from Japan.

Monilinia laxa: on 1 BC [334, 1816].
Podosphaera sp.: powdery mildew, blanc:
   on *C.* sp. BC [1816].

CHAMAECYPARIS Spach   PINACEAE

1. *C. lawsoniana* (A. Murr.) Parl., Lawson
   false cypress, cyprès de Lawson; native
   to the sw. USA. 1a. *C. l.* var.
   elwoodii; a cultivar.
2. *C. nootkatensis* (D. Don) Spach.,
   yellow cypress, cyprès jaune; native,
   BC. 2a. *C. n.* var. *pendula*; a
   cultivar.
3. *C. pisifera* Sieb. & Zucc., Sawara
   cedar; native to Japan. 3a. *C. p.*
   var. *filifera nana aurea*.

Armillaria mellea (C:80): root rot,
   pourridié-agaric: on 1 BC [1012]. See
   Abies.
Aspergillus sp.: on *C.* sp. BC [1012].
Aureobasidium pullulans: on 2 BC [1722].
Cytospora abietis: on 2 BC [507, 508,
   1012].
Gymnosporangium nootkatense (C:80): rust,
   rouille: II III on 2 BC [1012, 1247,
   2008].
Herpotrichia juniperi: on 2 BC [1012].
Kabatina thujae: on 2 BC [508]; on 2, 2a
   BC [1012].
Kirschsteiniella thujina: on 2 BC [1722].
Lepteutypa cupressi: on 1 BC [1012].
Lophodermium juniperinum (L. juniperi):
   on *C.* sp. BC [1012].
Nidularia sp.: on 2 BC [1012].
Pestalotia thujae: on 2 BC [508, 1012].
   An invalid name because it was published
   without a Latin description.
Pestalotiopsis funerea (Pestalotia f.)
   (C:80): on 2 BC [507, 508, 1012].
Phellinus weirii (Poria w.) (C:80): on 2
   BC [1012, F65]. See *Abies*.

Phoma sp.: on 1 BC [1012].
Phytophthora sp.: on 1 BC [25, 337]; on
   1a, 3a BC [23] by inoculation.
P. cinnamomi (C:80): on 1 BC [23, 25, 26,
   334, 337, 338, 1012]; on 1a BC [24,
   339]; on 3a BC [23] by inoculation.
P. lateralis (C:80): root and crown rot,
   pourridié phytophthoréen: on 1 BC [23,
   25, 333, 337, 1012]; on 1a BC [26].
Pleospora laricina: on 2 BC [508, 1012].
Pythium sp. (C:80): on 1 BC [1012].
Serpula himantioides: on 2 BC [1012].
   See *Abies*.
Seynesiella juniperi: on 2 BC [1012, F72].
Tyromyces sericeomollis (Polyporus s.):
   on 2 BC [1012]. See *Abies*.
Xeromphalina campanella (C:80): white
   stringy rot, carie blanche filandreuse:
   on 2 BC [1012].

CHAMAEDAPHNE Moench   ERICACEAE

1. *C. calyculata* (L.) Moench,
   leatherleaf, faux-bluets; native,
   Yukon-Keew, BC-Nfld.

Ascochyta cassandrae: on 1 Ont [1828].
   Apparently a *Discella*.
Chrysomyxa ledi (C:80): rust, rouille:
   on 1 Ont [1341].
C. ledi var. *cassandrae (C. cassandrae)*
   (C:80): rust, rouille: II III on 1
   Yukon, Mack, BC, Alta, Sask, Man [2008],
   NWT-Alta [1063], Mack, Alta [53], BC
   [1012, 1816], Alta [F64], Sask, Man
   [F67], Ont [1236], Que [1377]; on *C.*
   sp. Ont [F60].
Exobasidium vaccinii (C:80): on 1 Que
   [282], NS [1594]; on *C.* sp. Que [1385].
Gibbera cassandrae (C:80): on 1 Que [70,
   73].
Godronia cassandrae (anamorph *Fusicoccum
   putrefaciens*): on 1 Que [1678] by
   inoculation.
G. cassandrae f. *cassandrae*: on 1 Ont
   [656], Que [656, 1525].
Lophodermium orbiculare: on 1 Ont [1340,
   1827].
Marasmiellus paludosus: on 1 NB [1435].
Pseudophacidium ledi: on 1 Que [1685].
Psilachnum cassandrae: on 1 Ont, Que, NB
   [1630].
*Stigmatea pulchella (Gibbera p., Venturia
   p.)* (C:80): on 1 Ont [1340], Ont, Que,
   NB, NS, Nfld [73], NB [1032]; on *C.*
   sp. Ont [F60].

CHEIRANTHUS L.   CRUCIFERAE

1. *C. cheiri* L., wallflower, giroflée
   jaune; imported from s. Europe.

Alternaria brassicicola: on seeds of 1
   Sask [1313].
Botrytis cinerea (holomorph *Botryotinia
   fuckeliana*) (C:81): on 1 BC [1816].
Leptosphaeria maculans (anamorph
   Plenodomus lingam): on seeds of 1
   Sask [1325].

Peronospora parasitica (C:81): downy
mildew, mildiou: on 1 BC [1816].
Phytophthora megasperma (C:81): foot rot,
pourriture du pied: on 1 BC [1816].

CHELONE L.   SCROPHULARIACEAE

1. *C. glabra* L., turtlehead, galane
glabre; native, Man-Nfld.

Erysiphe galeopsidis (C:81): powdery
mildew, blanc: on 1 Ont, Que [1257].
Mollisia atrata: on 1 Ont [1340].

CHENOPODIUM L.   CHENOPODIACEAE

1. *C. album* L., lamb's quarters,
chou-gras; imported from Eurasia, a
common weed, Yukon-Mack, BC-Nfld.
2. *C. capitatum* (L.) Aschers.,
strawberry-blite; native, Yukon-Mack,
Sask-NS.

Acremonium boreale: on 1 BC, Alta, Sask
[1707].
Aphanomyces cochlioides: on 1 Que [317].
Peronospora chenopodii: downy mildew,
mildiou: on 2 Yukon [460].
P. farinosa (C:81): downy mildew,
mildiou: on 1 BC [1816].
Polymyxa betae: on 1 Ont [60].
P. betae f. sp. *betae*: on 1 Ont [60].
Pythium sp.: on 1 BC [1977].
Rhizoctonia solani (holomorph
Thanatephorus cucumeris): on 1 Que
[317].
Thielaviopsis basicola: on 1 Ont [546].
Verticillium dahliae: on 1 BC [38, 1816,
1978].

CHIMAPHILA Pursh   PYROLACEAE

1. *C. umbellata* (L.) W. Barton,
pipsissewa, chimaphile à ombelles;
native, BC-Nfld. 1a. *C. u.* var.
occidentalis (Rydb.) S.F. Blake;
native, BC-Ont.

Alysidiopsis pipsissewae: on 1a Man
[1760].
Mycosphaerella chimaphilae (C:81): on 1
Ont [1340].
Pucciniastrum pyrolae (C:81): rust,
rouille: II III on 1 Yukon, Mack, BC,
Alta, Sask, Man [2008], BC [1816]; on 1a
BC [1012].

CHOISYA Kunth   RUTACEAE

1. *C. ternata* HBK, Mexican Orange;
imported from Mexico.

Nectria cinnabarina (anamorph
Tubercularia vulgaris) (C:81): on 1
BC [1816].

CHRYSANTHEMUM L.   COMPOSITAE

1. *C. leucanthemum* L., daisy, marguerite;

imported from Eurasia, natzd., Yukon,
BC-Nfld.
2. *C. maximum* Ramond, Shasta daisy;
imported from Europe.
3. *C. x morifolium* Ramat. (C. *sinense*
Sabine), chrysanthemum, chrysanthème;
cutigen of Chinese origin.
4. Host species not named, i.e., reported
as *Chrysanthemum* sp.

Ascochyta chrysanthemi: on 4 Ont [338].
Botrytis cinerea (holomorph *Botryotinia
fuckeliana*) (C:82): gray mold,
moisissure grise: on 3 BC [1816]; on 4
Ont [333, 338, 339].
Erysiphe cichoracearum (C:82): powdery
mildew, blanc: on 3 BC [1816], Ont
[1257], NB [1194]; on 4 Alta, Que [337].
E. communis (C:82): powdery mildew,
blanc: on 4 Ont [334], NB [333].
Probably specimens of *E. cichoracearum*.
Fusarium oxysporum (C:82): wilt,
flétrissure: on 4 Que [333].
Ophiobolus mathieui: on 1 BC [1620].
Phyllosticta chrysanthemi: on leaves of 3
Ont [439].
Puccinia chrysanthemi (C:82): rust,
rouille: II on 3 BC [1816]; on 4 BC
[333, 1012], Ont [334, 339, 1236], Que
[1236].
Sclerotinia sclerotiorum (C:82): wilt or
dieback, flétrissure sclérotique: on 3
BC [1816].
Septoria chrysanthemi (C:82): leaf spot,
tache septorienne: on 2, 3 BC [1816];
on 4 NS [333].
S. chrysanthemella (*Cylindrosporium
chrysanthemi*) (C:82): killing deformed
plants of 3 in greenhouse Ont [439].
Stemphylium floridanum: on 4 Que [339].
Verticillium albo-atrum: wilt,
flétrissure verticilliene: on 3 Man
[38]; on 4 BC [337].
V. dahliae: wilt, flétrissure
verticilliene: on 3 BC [1816].

CHRYSOSPLENIUM L.   SAXIFRAGACEAE

1. *C. wrightii* Franch. & Sav.; native,
Yukon-Mack. Small semi-aquatic,
creeping or prostrate herbs.

Puccinia austroberingiana ssp.
austroberingiana: rust, rouille: on
1 Yukon [1560].

CICER L.   LEGUMINOSAE

1. *C. arietinum* L., chick pea, pois
chiche; imported from sw. Asia.

Ascochyta rabiei: on 1 Sask [1135].
Botrytis cinerea (holomorph *Botryotinia
fuckeliana*): on 1 Man [887].

CICHORIUM L.   COMPOSITAE

1. *C. intybus* L., chicory, chicorée;
native to Europe, n. Africa and w. Asia.

Puccinia hieracii (C:83):   rust, rouille:
    0 I II III on 1 BC [1012], Ont [1236].
Ramularia cichorii (C:83):   leaf spot,
    tache ramularienne: on 1 BC [1816].
Sclerotinia sclerotiorum (C:83):   drop,
    affaissement sclérotique: on 1 Ont
    [338].

CICUTA L.   UMBELLIFERAE

1. *C. bulbifera* L.; Swamps and wet
    thickets, BC-NS, Nfld.
2. *C. douglasii* (DC.) C. & R.,
    beaver-poison, carotte à Moreau; native,
    Mack-Alta.
3. *C. maculata* L., water hemlock,
    cicutaire maculée; native, Yukon-Mack,
    n. BC-PEI.

Puccinia cicutae (C:83):   rust, rouille:
    0 I II III on 2 BC [1816]; on 3 Ont
    [1236].
Uromyces americanus: rust, rouille: on 1
    Que [1559]. Teleomorph on *Scirpus* spp.
U. lineolatus ssp. *nearcticus*:   rust,
    rouille: on 3 Sask, NS [1559].
    Teleomorph on *Scirpus* spp.

CIMICIFUGA L.   RANUNCULACEAE

1. *C. racemosa* (L.) Nutt., black
    snakeroot; native, Ont.

Puccinia recondita s. l. (C:83):   rust,
    rouille: 0 I on 1 Ont [1236].

CINNA L.   GRAMINEAE

1. *C. arundinacea* L., stout woodreed,
    cinna roseau; native, Ont-Que.

Drechslera catenaria (C:83): on 1 Ont
    [1611].
Puccinia graminis (C:83):   rust, rouille:
    II III on 1 Ont [1236].

CIRCAEA L.   ONAGRACEAE

1. *C. alpina* L., alpine enchanter's
    nightshade, circée alpine; native, Mack,
    BC-Nfld.   1a. *C. alpina* var.
    pacifica (Aschers. & Magnus) Jones
    (*C. pacifica* A. & M.).
2. *C. quadrisculata* (Maxim.) Franch. &
    Sav., Lutetian enchanter's nightshade,
    circée de Lutèce; native, Man-NS, Nfld.
    The N. American plant is referable to
    C. quadrisculata var. *canadensis*
    (L.) Hara.

Puccinia circaeae (C:83):   rust, rouille:
    III on 1 BC [1012], Alta, Sask, Man, NS,
    Nfld [1546]; on 1, 1a BC [1546, 1816];
    on 1, 2 Ont [1236, 1546], Que [1546].

CIRSIUM Mill.   COMPOSITAE

1. *C. arvense* (L.) Scop., Canada thistle,
    chardon des champs; imported from
    Europe, natzd., Mack, BC-Nfld.   1a. *C.
    arvense* f. *albiflorum* (Rand. & Redf.)
    Hoffm.
2. *C. brevistylum* Cronq.; native, BC.
3. *C. drummondii* Torr. & Gray; native,
    Mack, BC-Ont.
4. *C. edule* Nutt.; native, BC.
5. *C. flodmanii* (Rydb.) Arthur, (*C.
    canescens* auth., not Nutt.), Flodman's
    thistle, chardon de Flodman; native,
    Alta-Ont.
6. *C. foliosum* DC. (*C. scariosum*
    Nutt.); native, Yukon-Mack, BC-Alta, e.
    Que.
7. *C. hookerianum* Nutt.; native, BC-Alta.
8. *C. undulatum* (Nutt.) Spreng.; native,
    BC-Ont.
9. *C. vulgare* (Savi) Ten. (*C.
    lanceolatum* Scop., Cnicus 1. L.),
    bull thistle, chardon vulgaire; imported
    from Eurasia, natzd., BC-Nfld.
10. Host species not named, i.e., reported
    as *Cirsium* sp.

Calyptella capula: on 1 BC [1424].
Lachnella villosa: on 1 BC [1424].
Mycosphaerella tassiana var. *tassiana*
    (C:84): on 10 Sask [1327].
Ophiobolus acuminatus (C:84): on 1, 9, 10
    Ont [1620]; on 9, 10 BC [1620]; on 10 NS
    [1620].
O. anguillidus: on 5 Ont [1340].
O. rubellus: on 1 Man, Ont [1620].
Puccinia calcitrapae (C:84):   rust,
    rouille: 0 II III on 8 BC [1012, 1816];
    on 9 Ont [1236]; on 10 BC [1816 as *C.
    breweri*].
P. cnici (C:84):   rust, rouille: 0 I II
    III on 1 Ont [1828]; on 2 BC [1558]; on
    9 BC [1012, 1816], BC, Alta, Sask, Ont,
    Que, NB, NS, Nfld [1558], Ont [1236].
P. cnici var. *cnici*: on 2 BC [1012].
P. inclusa: rust, rouille: on 8 BC,
    Alta, Sask [1558].
P. inclusa var. *boreohesperia*: on 3
    Alta [1558].
P. inclusa var. *flodmanii*: on 5 Sask,
    Man [1558].
P. laschii: rust, rouille: on 4, 8 BC
    [1558]; on 5 Alta, Sask, Man [1558]; on
    6 Alta [1558]. Conners (p. 84) cited
    specimens under *P. calcitrapae*, a name
    which is now applied in a narrower sense
    to a fungus on *Centaurea* in Europe.
P. laschii var. *laschii* (C:84 as *P.
    calcitrapae*):   rust, rouille: II III
    on 7 Alta [644].
P. punctiformis (C:84):   rust, rouille: 0
    II III on 1 BC [1816], BC, Alta, Ont,
    Que, NB, NS, Nfld [1558], Ont [1236,
    1341], Que [282]; on 1, 1a, BC [1012].
Sclerotinia sp.: on 1 Sask [420, 1138],
    Man [1138].

CITRULLUS Neck.   CUCURBITACEAE

1. *C. lanatus* (Thunb.) Matsum. & Nakai
    (*C. vulgaris* Schrad.), watermelon,
    melon d'eau; native to tropical and s.
    Africa.

Alternaria cucumerina (C:84): alternaria
rot, pourriture alternarienne: on 1 NS
[1594].
Colletotrichum orbiculare (C. lagenarium)
(C:84): anthracnose, anthracnose: on 1
NS [336].
Olpidium radicale: on 1 Ont [58].
Specimens probably were *O.
cucurbitacearum*.
Verticillium dahliae: on 1 BC [1978], Ont
[1087].

CLARKIA Pursh   ONAGRACEAE

1. *C. amoena* (Lehm.) A. Nels. & Macbr.;
   native, BC.
2. *C. pulchella* Pursh; native, BC.
3. *C. unguiculata* Lindl. (*C. elegans*
   Daigl.); native to California.

Peronospora arthuri (C:85): downy mildew,
   mildiou: on 2, 3 BC [1546]; on 3 BC
   [1816].
Pucciniastrum epilobii: rust, rouille:
   II III on 3 Ont [1236 from inoculum from
   Epilobium hirsutum]. Probably a
   specimen of *P. pustulatum*.
P. pustulatum (C:85): rust, rouille: on
   1 BC [1816], BC, Alta, Sask, Ont, Que,
   PEI [1546]; on 2, 3 Ont [1546].

CLAYTONIA L.   PORTULACACEAE

1. *C. caroliniana* Michx., spring beauty,
   claytonie de Caroline; native, BC,
   Yukon, Ont-Nfld.
2. *C. lanceolata* Pursh; native, BC-Sask.
3. *C. virginiana* L., spring beauty,
   claytonie de Virginie; native, Ont-Que.

Endophyllum lacus-regis (C:85): rust,
   rouille: 0 I III on 1 Ont [1236, 1827].
Puccinia claytoniicola (C:85): rust,
   rouille: on 2 BC [1012, 1341, 1816].
P. mariae-wilsoniae (C:85): rust,
   rouille: 0 I III on 1, 3 Ont [1236].
Uromyces claytoniae: rust, rouille: on
   C. sp. Que [1341].

CLEMATIS L.   RANUNCULACEAE

1. *C.* x *jackmanii* T. Moore (*C.
   lanuginosa* x *C. viticella*); a
   cultivar that originated in England.
2. *C. ligusticifolia* Nutt.,
   virgin's-bower, clématite à feuilles en
   coeur; native, BC-Man.
3. *C. occidentalis* (Hornem.) DC. 3a. *C.
   occidentalis* var. *occidentalis* (*C.
   verticillaris* DC.), purple clematis,
   clématite verticillé; native, Ont-NB.
4. *C. viorna* L., leather-flower; native,
   s. Ont.
5. *C. virginiana* L., devil's darning
   needle, clématite de Virginie; native,
   Man-PEI.

Cercospora squalidula (C:85): leaf spot,
   tache cercosporéene: on 2 BC [1816].

Didymaria clematidis (C:85): on 2 BC
   [1816].
Erysiphe communis (E. polygoni) (C:85):
   powdery mildew, blanc: on 1 BC [1816];
   on 5 Ont [1257].
Phyllosticta clematidis: on leaves of 4
   Ont [439].
Puccinia recondita s. l. (C:85): rust,
   rouille: 0 I on 1 BC [1012, 1816]; on
   3a, 5 Ont [1236]; on 5 Ont [1341], NB,
   NS [1032].
Ramularia saximontanensis: on *C.* sp.
   Alta [644].

CLINTONIA Raf.   LILIACEAE

1. *C. borealis* (Ait.) Raf., corn lily,
   clintonie boréale; native, Man-Nfld.
2. *C. uniflora* (Schult.) Kunth, queen's
   cup, clintonie à une fleur; native,
   BC-Alta.

Puccinia mesomajalis (P. mesomegala)
   (C:86): rust, rouille: III on 1 Ont
   [437, 1236, 1341], Ont, Que, Nfld
   [1545], Que [282]; on 2 BC [1012, 1816],
   BC, Alta [1545].

CLIVIA Lindl.   AMARYLLIDACEAE

1. *C. miniata* Regel, scarlet Kaffir lily,
   clivie vermillon; native to s. Africa.

*Colletotrichum gloeosporioides (C.
   himantophylli)* (holomorph *Glomerella
   cingulata*) (C:86): anthracnose,
   anthracnose: on 1 BC [1816].

COCHLEARIA L.   CRUCIFERAE

1. *C. officinalis* L. (*Draba corymbosa*
   R. Br.), scurvy grass, cuillerée; native
   along the coasts, Yukon-Labr, BC, n.
   Que, Nfld.

Mycosphaerella pyrenaica (C:86): on 1
   Frank [1586].
Peronospora parasitica (C:86): downy
   mildew, mildiou: on 1 Frank [1244].
Puccinia eutremae (C:86): rust, rouille:
   on 1 Mack, Frank, Keew [1563], Frank
   [1244].

COLCHICUM L.   LILIACEAE

1. *C. autumnale* L., autumn crocus,
   colchique; imported from Europe and n.
   Africa.

Urocystis colchici (C:86): smut,
   charbon: on 1 BC [1545, 1816 on
   imported corms], Ont [1545].

COLLINSIA Nutt.   SCROPHULARIACEAE

1. *C. parviflora* Dougl. ex Lindl.,
   blue-lips; native, BC-Ont.

Sphaerotheca fuliginea: powdery mildew,
   blanc: on 1 BC [1554].

COLLOMIA Nutt.   POLEMONIACEAE

1. *C. linearis* Nutt., narrow-leaved
    collomia, collomia à feuilles linéaires;
    native, BC-Ont, introduced, Yukon,
    Que-PEI.

Sphaerotheca macularis (C:86): powdery
    mildew, blanc: on 1 BC [1012, 1254],
    Sask [1254].

COMANDRA L.   SANTALACEAE

1. *C. umbellata* (L.) Nutt., bastard
    toadflax, comandre à ombelle; native,
    Yukon-Mack, BC-Nfld. 1a. *C. umbellata*
    var. *pallida* (DC.) Jones (*C. pallida*
    DC.), pale comandra, comandre pâle;
    native, Yukon-Mack, BC-Man. 1b. *C.*
    umbellata var. *umbellata* (*C.*
    richardsiana Fern.); transcontinental.

Cladosporium herbarum: on 1 Sask [1760].
Cronartium comandrae (C:86): comandra
    blister rust, rouille-tumeur: II III on
    1 Ont [1236], Que [1377]; on 1a NWT-Alta
    [1063], BC [1012, 1816], BC, Alta [644],
    BC, Alta, Sask, Man, Ont [735], Alta
    [53, 951, 1410, 1402, F61], Sask [F63],
    Ont [1236]; on 1a, 1b Sask, Man [1399];
    on 1b Sask, Man, Ont, Que [735]; on 1b,
    C. sp. Ont [1236, 1828]; on *C*. sp.
    Yukon, Mack, BC, Alta, Sask, Man [2008],
    Mack [F62], BC, Man [F67], Alta [950,
    951, 952], Sask [F68], Sask, Man [F64].
Puccinia andropogonis (C:87): rust,
    rouille: 0 I on 1, 1b Ont [1236].
P. comandrae (C:87): rust, rouille: on
    1a NWT-Alta [1063].

COMPTONIA L'Her.   MYRICACEAE

1. *C. peregrina* (L.) J. Coult.,
    sweet-fern, comptonie voyageuse; native,
    Ont-PEI. 1a. *C. peregrina* var.
    asplenifolia (L.) Fern. (*C.*
    asplenifolia L.).

Cronartium comptoniae (C:87): sweet-fern
    blister rust, rouille-tumeur: II III on
    1 Ont [652, 735, 1236, 1827, 1828, F70],
    Que [735], NB [1032], NB, NS [735, 1856,
    F68]; on 1a Que [1377].
Cryptodiaporthe aubertii var. *comptoniae*
    (anamorph *Uniseta flagellifera*): on
    1a Ont [1340].
Isariopsis dearnessii: on 1a Ont [195].

CONIMITELLA Rydb.   SAXIFRAGACEAE

1. *C. williamsii* (Eat.) Rydb.; native,
    Alta. A perennial, scapose herb.

Puccinia saxifragae var. *heucherarum*:
    rust, rouille: on 1 Alta [1578, 1580].

CONIOSELINUM Hoffm.   UMBELLIFERAE

1. *C. chinense* (L.) BSP (*C. pacificum*
    Wats., *C. gmelinii* C. & S.), hemlock

parsley; coastal BC, Ont-PEI and
    Nfld-Labr.

Cercospora selini-gmelini (C:87): on 1
    BC, Nfld [1551].
Plasmopara carlottae: downy mildew,
    mildiou: on 1 BC [1551].
Puccinia grenfelliana: rust, rouille: on
    1 Que, Nfld [1551].
P. ligustici (C:87): rust, rouille: on 1
    BC [1551].
Septoria levistici: on 1 BC, Que [1551].

CONRINGIA Lk.   CRUCIFERAE

1. *C. orientalis* (L.) Dumort,
    hare's-ear-mustard, conringia oriental;
    imported from Eurasia, introduced into
    all provinces.

Pseudocercosporella capsellae: on 1 Sask
    [1326].

CONSOLIDA DC.   RANUNCULACEAE

1. *C. orientalis* (J. Gay) Schrodinger
    (including most garden material
    marked *Delphinium ajacis* L.),
    larkspur, bec d'oiseau; native to n.
    Africa, s. Europe and w. Asia.

Erysiphe polygoni: powdery mildew,
    blanc: on 1 Sask, Ont [1257].

CONVALLARIA L.   LILIACEAE

1. *C. majalis* L., lily-of-the-valley,
    muguet; imported from Eurasia, sometimes
    a garden escape from Ont-PEI.

Botrytis sp.: on 1 BC [1816].
B. cinerea (holomorph *Botryotinia*
    fuckeliana) (C:87): gray mold,
    moisissure grise: on 1 BC, Ont [1545].
Gloeosporium convallariae (C:87):
    anthracnose, anthracnose: on 1 Ont
    [334]. The features to be associated
    with this species are uncertain as it
    has not been re-examined in modern times.
Mycosphaerella convallaria: on 1 Ont
    [1094].
Puccinia sessilis (C:87): rust, rouille:
    on 1 NS [333, 334].

CONVOLVULUS L.   CONVOLVULACEAE

1. *C. arvensis* L., field bindweed,
    liseron des champs; imported from
    Eurasia, natzd., BC-PEI.
2. *C. sepium* L., hedge bindweed, wild
    morning-glory, gloire du matin;
    cosmopolitan in temperate regions, very
    variable species, BC-Nfld.

Diaporthopsis sepium: on 2 Ont [79].
Gloeosporium sp.: on *C*. sp. Ont [1828].
Puccinia convolvuli (C:87): rust,
    rouille: 0 I II III on 2 Ont, Que
    [1236]; on *C*. sp. Ont [1341].

Septoria convolvuli (C:87): on 1 BC
[1816].

COPTIS Salisb.   RANUNCULACEAE

1. *C. trifolia* (L.) Salisb.; native, Asia
   and Alaska, BC. 1a. *C. trifolia* var.
   groenlandica (Oeder) Hult. (*C.*
   groenlandica (Oeder) Fern.),
   goldthread, savoyane; native, Mack,
   Keew, n. Alta-Nfld.

Lambertella copticola: on 1a Ont [1799].
Mycosphaerella coptis (anamorph *Septoria*
   coptidis) (C:87): on 1a Que [70].
Septoria coptidis (holomorph
   Mycosphaerella coptis) (C:88): on 1a
   Que [70].
Wettsteinina mirabilis (C:88): on 1a Que
   [70].

COREOPSIS L.   COMPOSITAE

1. *C. grandiflora* Hogg ex Sweet; imported
   from the USA, introduced from Ont-NB.

Sphaerotheca macularis: powdery mildew,
   blanc: on 1 BC [1816].

CORNUS L.   CORNACEAE

1. *C. alba* L., tatarian dogwood,
   cornouiller blanc; imported from
   Siberia, n. China and n. Korea. 1a. *C.*
   alba var. *elegantissima*; a form with
   the leaves edged with white.
2. *C. alternifolia* L. f.,
   alternate-leaved dogwood, cornouillier à
   feuilles alternes; native, Man-Nfld.
3. *C.* x *californica* C.A. Mey. (*C.*
   occidentalis x *C. sericea)*; native,
   BC.
4. *C. canadensis* L., bunchberry,
   quatre-temps; native, Yukon-Mack,
   BC-Nfld.
5. *C. nuttallii* Audub., Pacific dogwood,
   cornouiller de Pacifique; native, BC.
6. *C. occidentalis* (Torr. & Gray) Cav.
   (*C. pubescens* Nutt.); native, BC.
7. *C. racemosa* Lam. (*C. paniculata*
   L'Hér.); native, s. Ont.
8. *C. rugosa* Lam. (*C. circinata*
   L'Her.), roundleaf dogwood, bois de
   calumet; native, Man-NS.
9. *C. sericea* L. (*C. stolonifera*
   Michx.), red osier dogwood, hart rouge;
   native, Yukon-Mack, BC-Nfld.
10. Host species not named, i.e., reported
    as *Cornus* sp.

Ascochyta cornicola: on 9 Que [334].
Basidioradulum radula (*Hyphoderma r.)*: on
   9 BC [1012]. See *Abies*.
Botryotinia fuckeliana (anamorph *Botrytis*
   cinerea): on 5 BC [1012, 1816].
Conoplea olivacea (*C. sphaerica*) (C:88):
   on 10 Que [784].
Corynespora cambrensis: on 10 Man [1760].

Cryptodiaporthe corni (*Apioporthe c.)*
   (anamorph *Zythia aurantiaca*) (C:88):
   on 1a Ont [336]; on 2 NS [1032].
C. salicella: on 10 BC [1012]. Usual
   host is *Salix*. Specimen is probably
   C. corni.
Cryptosporiopsis cornina (*Phoma*
   paniculatae) (holomorph *Pezicula*
   corni) (C:88): on dead limbs of 7 Ont
   [440]; on 9 Sask, Man [F65]; on 10 Sask,
   Ont [1763].
Daedaleopsis confragosa (*Daedalea c.)*: on
   5 BC [1012]. See *Acer*.
Diaporthe sp.: on 5 BC [193].
D. eres (C:88): on 10 Ont [1340, F65].
Diplodia sp.: on 10 Ont [1340].
Diplodina macrospora (C:89): on dead
   limbs of 10 Nfld [441].
Elsinoe corni: on 10 Nfld [1661].
Exosporium occidentale: on 10 Sask [1760].
Fracchiaea callista: on 9 Que [F69].
Gloeosporium sp.: on 5 BC [334, 1816].
Glomerella cingulata: on 5 BC [1012].
Glomopsis corni (*Glomerularia c.*, as
   Gloeosporium (sic) *c.)* (C:88): leaf
   blight, brûlure des feuilles: on 4
   NWT-Alta [1063], BC [1816], Alta [53,
   644], Sask, Man, Ont, Que [1760], Ont
   [1340], NB, NS [1032].
Glyphium corrugatum: on 9 BC [589].
Helminthosporium velutinum: on 5 BC and
   on 9 Ont [810].
Hyphodontia crustosa: on 9 BC [1012].
   See *Abies*.
Hypoxylon fuscum: on 10 Sask [F67].
Libertella sp.: on 10 BC [1012].
Macrophoma sp.: on 10 NWT-Alta [1063].
Melanconium sp.: on 5 BC [1012].
Meruliopsis corium (*Merulius confluens*)
   (C:88): on 5 BC [568, 1012]. See
   Acer.
Microsphaera penicillata (C:88): powdery
   mildew, blanc: on 1 Ont and on 2 Ont,
   Que [1257].
Mollisia stictella (C:88): on 2 NS [1032].
Monilia corni: on 5 BC [1012].
Monilinia corni (C:88): leaf blight,
   brûlure sclérotique: on 5 BC [336,
   1816].
Myxosporium sp.: on 9 Ont [1340, 1827].
M. corni: on 10 Ont [1340].
M. nitidum (C:88): on 2, 10 NB [1032].
M. roumegueri: on 9 Sask [F65].
Nectria sp.: on 5 BC [1012].
N. cinnabarina (anamorph *Tubercularia*
   vulgaris): on 10 BC [1012, F74], Man
   [F67].
Odontia sp.: on 5 BC [1012].
Peniophora cinerea (C:88): on 5 BC
   [1012]. See *Abies*.
Pezicula corni (C:88): on 6 BC [1816]; on
   8, 10 Ont [1340]; on 9 BC [1012]; on 10
   Sask [F67], Que [F68].
Phellinus ferreus (*Poria f.)* (C:89): on
   5, 9 BC [1012]. See *Acer*.
P. igniarius (*Fomes i.)* (C:88): on 5 BC
   [1012]. Causes a heart rot of live
   trees. See *Acer*.
Phoma sp.: on 5 BC [1816].
Phomopsis sp.: on 10.Ont [1828].

Phyllactinia guttata (P. corylea) (C:89):
powdery mildew, blanc: on 1 Man, Ont
and on 2 Ont [1257]; on 3, 5, 6, 9 BC
[1816]; on 5 BC [1340]; on 5, 9 BC
[1012]; on 9 Mack, Ont, Que, NB [1257];
on 10 Alta [334], Sask, Man [F67].

Phytophthora sp.: on 5 BC [23].

P. cactorum (C:89): crown canker, mildiou
de collet: on 5 BC [337, 1012, 1816].

P. cinnamomi: on 5 BC [23] by inoculation.

Pseudomassaria corni (C:89): on 9 Man
[F66]; on 10 Man [F67], Ont [295, 1340,
F63].

P. foliicola (C:89): on 4 Que [71, 70].

Puccinia sp.: rust, rouille: on 9 BC
[1012].

P. porphyrogenita (C:89): rust, rouille:
III on 4 NWT-Alta [1063], Mack [53], BC
[1012, 1416, 1816], Ont [1236, 1341,
1828], Que [282], NB, NS, PEI [1032],
Nfld [1659, 1660, 1661].

Ramularia stolonifera: on 9 Sask, Man
[F66]; on 10 Sask, Man [F67].

Sarcinella heterospora: on 10 Ont [1340].

Sclerophoma ambigua: on 5 BC [524, 1012,
F72].

Seiridium corni (Pestalotia c.): on 9
Sask [1757], Man [1757, F66]; on 10
Sask, Man [1763, F67].

Septoria canadensis (C:89): on 4 Ont
[1340], Que [70], NS [1032].

S. cornicola (C:89): leaf spot, tache
septorienne: on 3 BC [1816]; on 4 NS
[1032]; on 5 BC [1416]; on 9 Sask [F65],
Que [1340]; on 10 Alta [338].

Sporidesmium achromaticum: on 10 Man
[1760].

Stereum sp.: on 5 BC [1012].

Stigmina cornicola: on 2, 10 Ont [1762].

Tremella lutescens (C:89): on 2 NS
[1032]. Apparently saprophytic on wood.

Tympanis fasciculata: on 8 Ont [1221].

Valsa sp.: on 9 BC [1012]; on 10 Alta
[644].

Valsaria sp.: on 10 Ont [1340].

Zythia aurantiaca (holomorph
Cryptodiaporthe corni) (C:89): on 10
Ont [1340, F73].

CORYLUS L.   BETULACEAE

1. *C. americana* Marsh., American
   hazelnut, noisetier d'Amérique; native,
   Man-Que.
2. *C. avellana* L., European filbert,
   avelinier; imported from Europe.
3. *C. cornuta* Marsh. (*C. rostrata*
   Ait.), beaked hazelnut, noisetier à long
   bec; native, BC-Nfld. 3a. *C. cornuta*
   var. *californica* (DC.) Sharp; native,
   BC.
4. Host species not named, i.e., reported
   as *Corylus* sp.

Allantoporthe decedens (Diaporthe d.)
(C:89): on 3 NS [1032]; on 4 Ont [79,
1340, F63].

Anisogramma anomala (Apioporthe a.)
(C:89): on 3 NS [1032]; on 4 Ont [1340,
1440, F60], NS [1440].

Cenangium fuckelii (C:89): on 3 NS [1032].

Conoplea geniculata (C:89): on 4 Ont
[784].

C. globosa: on 4 Man [1760].

C. olivacea (C. sphaerica) (C:89): on 4
Man [1760], Ont [784].

Cryptospora sp.: on 4 Ont [1340].

Cryptosporiopsis coryli (Catinula turgida)
(C:89): on 3 Ont [1340]; on 4 Ont
[1763].

Dacrymyces minor: on 4 Ont [1341]. See
Abies.

*Dendrothele maculata (Aleurocorticium
m.)*: on 3, 4 Ont [965]. See *Acer*.

Diaporthopsis sp.: on 3 NB [1032].

Diatrype albopruinosa (C:89): on 1, 3, 4
Ont, also on 3 Sask and on 4 Man [1354].

Diatrypella verrucaeformis (C:89): on 4
Ont [1340].

Diplodia sarmentorum: on 3 BC [1012].

Fenestella sp.: on 4 Ont [1340].

Fumago vagans: on 4 Man [1760].
Specimens are typically a mixture of
hyphae from several dematiaceous
hyphomycetes.

Gnomonia gnomon: on 4 Ont [79].

Helminthosporium velutinum: on 4 Ont
[810]. Probably mycoparasitic on
Anisogramma anomala.

Hymenochaete corrugata (C:89): on 3 Ont
[1341]. See *Abies*.

H. tabacina: on 3 BC [1012], Ont [1341];
on 3, 4 NB [1032]. See *Abies*.

Hyphodermella corrugata: on 4 BC [1012].
Lignicolous, typically on logs and
fallen branches.

Hyphodontia crustosa: on 4 BC [1012].
See *Abies*.

Hypocreopsis lichenoides: on 3 Ont [230].

Hypoxylon sp.: on 4 NB [1032].

H. fuscum (C:89): on 3 Man [F66], NS
[1032]; on 4 BC [1012], Ont [1340, 1828].

H. rubiginosum (C:89): on 3a BC [1012].

Inonotus radiatus (Polyporus r.) (C:90):
on 4 NS [1032]. See *Acer*.

Lenzites betulina (C:89): on 4 NS
[1032]. See *Acer*.

*Mamianiella coryli (Gnomoniella c.,
Mamiania c.)* (C:89): leaf spot, tache
ponctuée: on 3 Que [282], NB [1032,
F65, F66], NS [1032], Nfld [1659, 1660,
1661]; on 3a BC [1012, 1816]; on 4 Man
[F68], Ont [1340, 1827, F62], NB [1032,
F64], NS [1032].

M. coryli var. *coryli*: on 1, 3 Sask,
Man and on 3 BC, Nfld [79].

M. coryli var. *spiralis*: on 4 Alta,
Sask, Man, Ont, NB, NS [79].

Massaria lantanae: on 4 Ont [80].

M. inquinans (M. vomitoria): on 4 Ont
[1340]. Typically on *Acer*.

Melanomma pulvis-pyrius (C:89): on 3a BC
[1012].

Meruliopsis corium: on 3a BC [1012]. See
Acer.

Microsphaera penicillata: powdery mildew,
blanc: on 3 Ont [1257], NS [1032].

Ophiovalsa suffusa (Cryptospora s. var.
nuda) (C:89): on 3 NS [1032, 1340];
on 4 Ont [1340].

Peniophora cinerea (C:90): on 3 Ont
[1341], NB [1032]; on 3a BC [1012]. See
Abies.
P. incarnata: on 3 NS [1032]. See
Abies.
Pezicula corylina (C:90): on 3, 4 Ont
[1340], NS [1032]; on 4 Ont F63].
Phaeoisaria sparsa: on 4 Man [1760].
Phaeomarasmius erinaceus (Pholiota e.):
on 4 Man [1434], Que [1385].
Phanerochaete sordida (Peniophora cremea)
(C:90): on 4 BC [1012]. See *Abies*.
Phellinus ferreus (Poria f.) (C:90): on 4
BC [1012]. See *Acer*.
Phibalis furfuracea (Cenangium f.)
(C:89): on 4 Ont [1340].
Phomopsis revellens (C:90): canker,
chancre: on 2 BC [1816].
Phyllactinia guttata (C:90): powdery
mildew, blanc: on 2 BC [337]; on 3 Ont
[1257], NS [1257, 1032].
Piggotia coryli (Gloeosporium c.,
Monostichella c.) (C:89): leaf spot,
anthracnose: on 1 NS [1032]; on 3 Ont
[282, 1340, 1827], NB [1032]; on 3a BC
[1816]; on 4 Ont [1763], NB, NS [1032].
Pirex concentricus (Irpex owensii):
on 4 BC [1012]. On dead stems, etc.
See *Alnus*.
Poria sp.: on 3 BC [1012].
Pseudospiropes nodosus: on 2 BC [816].
Septoria corylina (C:90): leaf spot,
tache septorienne: on 3 NB, NS, PEI
[1032].
Sillia ferruginea: on 4 Ont [79].
Skeletocutis nivea (Polyporus
semipileatus) (C:90): on 4 BC [1012].
See *Acer*.
Stereum hirsutum (C:90): on 3 NS [1032].
See *Abies*
Tremella sp.: on 3a BC [1012]. Probably
saprophytic on wood. However, most
species are mycoparasites.
Trimmatostroma sp.: on 4 Ont [1827].
Tubercularia vulgaris (holomorph *Nectria*
cinnabarina) (C:90): on 4 BC [1012].

COSMOS Cav.   COMPOSITAE

1. *C. bipinnatus* Cav., common cosmos,
   cosmos bipenné; imported from Mexico.

Botrytis cinerea (holomorph *Botryotinia*
fuckeliana) (C:90): gray mold,
moisissure grise: on *C.* sp. NB [333,
1194].
Erysiphe cichoracearum: powdery mildew,
blanc: on 1 Ont [1257].
Sclerotinia sclerotiorum (C:90): on 1 BC
[1816].

COTINUS Duham.   ANACARDIACEAE

1. *C. obovatus* Raf., American smoke tree;
   imported from USA.

Botryodiplodia sp.: on 1 Ont [1340].
Verticillium albo-atrum: on *C.* sp. Ont
[F64].

COTONEASTER Ehrh.   ROSACEAE

1. *C. acutifolius* Turcz.; imported from
   n. China, introduced from Alta-Ont.
2. *C. franchetii* Bois.; imported from
   China and Burma.
3. *C. frigidus* Wallich ex Lindl.;
   imported from the Himalayas.
4. *C. horizontalis* Decne., rock
   cotoneaster, cotonéastre horizontal;
   imported from w. China.
5. *C. simonsii* Bak.; imported from the
   Himalayas.
6. Host species not named, i.e., reported
   as *Cotoneaster* sp.

Armillaria mellea: root rot,
pourridié-agaric: on 6 BC [1012].
Cytospora ambiens (C:90): on 5 BC [1816].
Gymnosporangium clavipes (C:90): quince
rust, rouille du cognassier: 0 I on 1
Man [1263].
Macrosporium sp.: on 6 NWT-Alta [1063].
Nectria cinnabarina (anamorph
Tubercularia vulgaris) (C:90): on 3
BC [1816]; on 6 NWT-Alta [1063], Ont
[1340].
Peniophora cinerea: on 6 Alta [338]. See
Abies.
Phyllosticta cotoneaster: on 6 Que [337].
P. sanguinea (C:90): leaf spot, tache
foliaire: on 1 Que [333].
Phytophthora cactorum (C:90): dark berry,
baie noire: on 2, 4 BC [1816].
Valsella sp.: on 6 NWT-Alta [1063].

CRATAEGUS L.   ROSACEAE

1. *C.* x *anomala* Sarg. (*C. intricata*
   Lange x *C. mollis* (Torr. & Gray)
   Scheele), abnormal hawthorn, aubépine
   anormale.
2. *C. beata* Sarg.; native, Ont.
3. *C. calopodendron* (J.F. Ehrb.) Medic.
   (*C. tomentosa* auct. non L.),
   blackthorn; native, Ont.
4. *C. chrysocarpa* Ashe, fireberry,
   aubépine à pommes dorées; native,
   Man-Nfld.
5. *C. coccinea* L. (*C. floribunda* K.
   Koch), scarlet hawthorne, épine
   écarlate; native, Ont-NS.
6. *C. columbiana* T.J. Howell, Columbia
   hawthorn, aubépine de Columbia; native,
   BC.
7. *C. crus-galli* L., cockspur hawthorn,
   ergot-de-coq; native, Ont-Que.
8. *C. curvisepala* Lindm. (*C. oxyacantha*
   L.), English hawthorn, épine blanche;
   imported from Eurasia.
9. *C. douglasii* Lindl., black hawthorn,
   aubépine de Douglas; native, BC-Man.
10. *C. fucosa* Sarg.; described from
    Massachusetts in 1903 but not found in
    any modern treatment.
11. *C. intricata* Lange; reported from s.
    Ont.
12. *C. laurentiana* Sarg., Laurentian
    hawthorn, aubépine laurentienne; native,
    Ont-Nfld.

13. *C. macracantha* Lodd. (*C. delicatibilis* Sarg.), scarlet haw, aubépine à épines longues; native, Ont.
14. *C. mollis* (Torr. & Gray) Scheele (*C. canadensis* Sarg.); thickets and open woods, Ont.
15. *C. monogyna* Jacq., English hawthorn, aubépine monogyne; imported from Eurasia, spread from cultivation, BC, Ont. 15a. *C. monogyna* ssp. *brevispina* (G. Kunze) Franco (*C. brevispina* G. Kunze); imported from Europe.
16. *C. pedicellata* Sarg. (*C. caesa* Ashe); native, Ont.
17. *C. pinnatiloba* Lange; a species from the Caucasus region not found in modern treatments.
18. *C. punctata* Jacq., dotted hawthorn, aubépine ponctuée; native, Ont-Que.
19. *C. roanensis* Ashe (*C. colorata* Sarg.); native, Ont.
20. *C. rotundifolia* Moench (*C. brunetiana* Sarg.); native, BC-NS.
21. *C. sanguinea* Pall., red-haw hawthorn, épine rouge; imported from Siberia.
22. *C. submollis* Sarg., velvety hawthorn, pommetier rouge; native, Ont-Que.
23. *C. succulenta* Link, fleshy hawthorn, aubépine succulente; native, Man-PEI.
24. Host species not named, i.e., reported as *Crataegus* sp.

Botryosphaeria obtusa (C:91): on 17 Ont [1614].
Cenangium crataegi: on 24 Ont [1340]. This specimen may be *Durandiella lenticellicola*. Groves [655] did not find any *Cenangium* on the type of *C. crataegi*.
Cerrena unicolor (Daedalea u.): on 24 NB [1032]. See *Acer*.
Chondrostereum purpureum (Stereum p.) (C:91): on 24 NB [1032, 1036]. Lignicolous, often on wounds on live stems. See *Abies*.
Conoplea olivacea (C. sphaerica) (C:91): on 24 Ont [784].
Corniculariella harpographoidea: on 24 Ont [401].
Cylindrosporium brevispina (C:91): on 15a BC [1816].
Dendrothele griseo-cana (Aleurocorticium g.): on 24 Ont [965]. Lignicolous, typically on dead branches.
Diaporthe eres (C:91): on 24 Ont [1340, F60].
Diplocarpon mespili (Fabraea maculata): leaf spot, entomosporiose: on 8 BC [333, 334, 1816]; on 24 BC [335, 338, 339, 1594], Ont [333, 335], NB [1032], NS [333].
Durandiella lenticellicola (C:91): on 24 Ont [1340].
Entomosporium mespili (E. maculatum, E. thuemenii) (C:91): leaf blight, brûlure entomosporienne: on 8 NB, NS [1032]; on 24 BC [1012], Ont [1340].
Gloeosporium sp.: on 24 Ont [336].

Gymnosporangium sp.: rust, rouille: on 6, 15a BC [337]; on 24 Sask, Man [F63], Ont [336], Que [336, 337].
G. betheli (G. tubulatum) (C:91): rust, rouille: 0 I on 4 Ont [1247 inoculated]; on 6, 9 BC [1012, 1816]; on 9 BC [644, 1267]; on 24 NWT-Alta [1063], BC [953, 1341, 1247, 1267, 1816], BC, Alta, Sask, Man [2008].
G. clavariiforme (C:91): rust, rouille: 0 I on 8 BC [1012, 1247, 1262, 1816], Ont [1236, 1262]; on 23 Alta, Que [339]; on 24 BC, Alta, Sask, Man [2008], Ont [1247 inoculated].
G. clavipes (C:91): quince rust, rouille du cognassier: 0 I on 2, 8, 10, 15, 16, 18, 23, 24 Ont [1236]; on 4 BC [1247]; on 6, 9 BC [1012, 1816]; on 8 NS [1032]; on 9 BC [644], Ont [1247]; on 18 Ont [1240]; on 24 Sask [F73], Sask, Ont, Que, NS [1263], Man [1247], Ont [1341, 1828, F68, F71, F72, F74], Que [333], NS, PEI [1032].
G. connersii: rust, rouille: 0 I on 4, 24 Alta, Sask, Man [1247, 1251]; on 9 Sask [1251]; on 11, 14, 22 inoculated, 24 Ont [1240]; on 20 Man, Ont, Que [1240], Que [1251]; on 24 Man [F67], Ont, Que [1251]; inoculated on 24 Ont [1247].
G. globosum (Roestelia lacerata) (C:91): hawthorn rust, rouille de l'aubépine: 0 I on 1, 2, 4, 7, 15 Ont [1268]; on 3 Ont [437]; on 3, 10, 16, 24 Ont [1236, 1268]; on 5, 8, 13, 18, 19, 20, 21 Ont [1236]; on 5, 24 Que [1268]; on 7, 24 Ont [1242]; on 8 Ont [334]; on 24 Ont [1341, 1828, F69, F71, F72, F74], Ont, Que [1240].
G. inconspicuum: rust, rouille: 0 I on 9 BC [1247].
Hendersonia discosioides: on leaves of 24 Ont [440]. The proper generic disposition for this species is uncertain. See *Abies*.
Inonotus radiatus (Polyporus r.): on 24 Ont [1341]. See *Acer*.
Leucostoma persoonii (anamorph *Cytospora rubescens*): on 24 NS [1032].
Micropera sp.: on 24 Ont [1827].
Monilinia johnsonii: on 24 NB [1032].
Nectria cinnabarina (anamorph *Tubercularia vulgaris*): on 24 NB [1032].
Phanerochaete laevis (Peniophora affinis): on 24 NS [1032]. See *Acer*.
Phellinus ferreus (Poria f.) (C:91): on 24 BC [1012]. See *Acer*.
P. ferruginosus (Poria f.): on 24 NS [1032]. See *Abies*.
Phyllactinia guttata (P. corylea) (C:91): powdery mildew, blanc: on 24 Alta [1594].
Phyllosticta rubra: on 24 Ont [1340].
Podosphaera clandestina (C:91): powdery mildew, blanc: on 7 Que [F60]; on 8 BC [1816]; on 24 BC [336], Alta, Ont, Que, NS [1257], NS [1032].
Schizoxylon compositum (C:91): on 24 Ont [1608].

Spilocaea sp.: on 24 Man [F68].
Valsa sp.: on 24 Ont [1827].
V. ceratosperma (C:91): on 24 NS [1032, F62].

CROCUS L.   IRIDACEAE

1. Cormous herbs native to Mediterranean region and sw. Asia; widely cultivated; host species not named, i.e., reported as *Crocus* sp.

Alternaria sp.: on 1 BC [1816].
Botrytis cinerea (holomorph *Botryotinia fuckeliana*) (C:92): gray mold, moisissure grise: on 1 BC [1816].
Heterosporium sp.: on 1 BC [1816].
Papulaspora sepedonioides: on 1 Man [1937].
Penicillium sp.: on 1 BC [1816].

CRYPTOGRAMMA R. Br.   POLYPODIACEAE

1. *C. crispa* (L.) R. Br., mountain parsley; native, Mack, BC-Ont. 1a. *C. crispa* var. *acrostichoides* (R. Br.) C.B. Clarke; native, Mack, BC-Ont.
2. *C. stelleri* (S.G. Gmel.) Prantl, slender cliff brake; native, Yukon-Nfld.

Hyalopsora cheilanthis (C:92): rust, rouille: II III on 2 Ont [1236], Que [282].
Milesia darkeri (Milesina d.) (C:92): rust, rouille: II III on 1a BC [1012, 2008], Alta [2008].

CRYPTOTAENIA DC.   UMBELLIFERAE

1. *C. canadensis* (L.) DC., honewort, cerfeuil sauvage; native, Man-NB.

Puccinia cryptotaeniae (C:92): rust, rouille: 0 III on 1 Ont [1236].

CUCUMIS L.   CUCURBITACEAE

1. *C. melo* L., melon, melon; native to w. Africa. 1a. *C. melo* Inodorus group (*C. melo* var. *inodorus* Naud.), honeydew melon, melon miel. 1b. *C. melo* Reticulatus group (*C. melo* var. *reticulatus* Ser.), muskmelon, cantaloup.
2. *C. sativus* L., cucumber, concombre; native to s. Asia. 2a. *C. sativus* var. *anglicus* Bailey, English forcing cucumber; grown in greenhouses.

Alternaria sp. (C:92): leaf spot, tache des feuilles: on 1b BC [1816]; on 2 Ont, NS [1594], NS [334, 336].
A. alternata (A. tenuis auct.*)* (C:92): on 2 BC [334, 1816, 1817, 1818].
A. cucumerina (C:92): leaf spot, tache alternarienne: on 1 Ont [333, 334, 335, 1458, 1460, 1594]; on 2 BC [338, 339, 1816, 1817, 1818], Ont [334, 1460, 1461, 1594], NB [333], NS [337, 338, 339, 1594], PEI [337].

Ascochyta cucumis (holomorph *Didymella bryoniae*) (C:92): on 2 BC [1816].
Botrytis cinerea (holomorph *Botryotinia fuckeliana*) (C:92): gray mold, moisissure grise: on 2 in greenhouse BC [1816], Alta [338], Ont [333, 335, 337], Ont, NB [336, 338, 339], NB [1594].
Cladosporium cucumerinum (C:92): scab, gale: on 1 Ont [1460, 1594], NS [335]; on 2 BC [336, 1816, 1817, 1818], BC, Que, NB, NS [337, 338, 339], Ont [338, 1460], Ont, Que, NB, NS [335, 336, 1594], Ont, Que, NB, PEI [333, 334], Que [321, 938, 939, 1639], NS [334].
Colletotrichum orbiculare (C. lagenarium) (C:92): anthracnose, anthracnose: on 1 Ont [333, 334, 335]; on 2 Ont [335], Que [333, 335, 1639].
Didymella bryoniae (Mycosphaerella citrullina) (anamorph *Ascochyta cucumis*) (C:93): stem blight, pourriture noire: on 2 BC [1416].
Erysiphe cichoracearum (C:92): powdery mildew, blanc: on 1 Ont [339, 1460, 1594]; on 2 Ont [333, 337, 338, 469, 1460, 1461, 1594], NB [339]. In greenhouses good control can be obtained by spraying with the mycoparasite *Ampelomyces quisqualis*.
E. communis: powdery mildew, blanc: on 1 Ont [335]; on 2 Ont, NS [334, 335, 336 in greenhouse]. Specimens probably were *E. cichoracearum*.
Fusarium sp. (C:93): on 1 Ont [1461, 1594], NS [336]; on 1b BC [1816]; on 2 BC [334, 338, 1594, 1817, 1818], Ont [1458, 1460, 1594].
F. oxysporum f. sp. *melonis* (C:93): wilt, flétrissure fusarienne: on 1 Ont [333, 336, 1088, 1932, 1934]; on 1a BC [1816]; on 2 Ont [1595].
Oidium sp.: powdery mildew, blanc: on 2 BC [1816].
Olpidium cucurbitacearum: on 2 Ont [64].
O. radicale: on 1, 2 Ont [58]. Specimens probably were *O. cucurbitacearum*.
Phomopsis cucurbitae (C:93): black rot, pourriture noire: in greenhouse on 2 BC [27].
P. sclerotioides: on 2a BC [1200].
Phyllosticta sp.: on 2 Que [335].
Pseudoperonospora cubensis (C:93): downy mildew, mildiou: on 2 Alta, Ont, Que [334].
Pythium sp. (C:93): damping-off, fonte des semis: on 1 Ont by inoculation [868], on 2 BC [1816], Ont [333, 1460, 1595].
P. debaryanum: damping-off, fonte des semis: in greenhouses and cold frames on 2 Que [336].
P. irregulare (C:93): damping-off, fonte des semis: on 1, 2 Ont [868] by inoculation.
P. sulcatum: on 1, 2 Ont [868] by inoculation.
P. ultimum (C:93): damping-off, fonte des semis: on 1 Ont [1085, 1086 inoculated]; on 1a, 1b BC [1816].

Rhizoctonia solani (holomorph
  Thanatephorus cucumeris) (C:93): foot
  rot, pourriture du pied: on 2 BC [334,
  1816], Ont [1460, 1594], PEI [338].
Sclerotinia sp.: on 2 Sask, Man [1138].
S. sclerotiorum (C:93): stem or fruit
  rot, pourriture sclérotique: on 2 BC
  [334, 337, 338, 1816], Alta [335], NS
  [336].
Septoria cucurbitacearum (C:93): leaf
  spot, tache septorienne: on 1 NS [336];
  on 2 NS [338].
Trichothecium roseum (C:93): foliage rot,
  moisissure rose: on 2 BC [1595], Ont
  [333].
Verticillium albo-atrum (C:93): on 2 NS
  [337].
V. dahliae: on 1 BC [1978], Ont [1087],
  Que [38, 392, 393]; on 1b BC [1816]; on
  2 BC [38, 1978].

CUCURBITA L.   CUCURBITACEAE

1. *C. ficifolia* Bouché, Malabar gourd,
    courge de Siam; native to tropical
    America.
2. *C. maxima* Duchesne, winter squash,
    courge; native to S. America.
3. *C. pepo* L., summer squash or pumpkin,
    citrouille; native to tropical America.
4. Host species not named, i.e., reported
    as *Cucurbita* sp.

Alternaria sp. (C:94): on 2, 3 BC [1816].
A. alternata (A. tenuis auct.*)* (C:94):
  on 2 NS [334, 1594].
A. cucumerina (C:94): alternaria spot,
  tache alternarienne: on 2 NS [337,
  1594].
Ascochyta cucumis (holomorph *Didymella
  bryoniae*) (C:94): leaf spot, tache
  ascochytique: on 3 BC [1816].
Botrytis sp.: on 2 NS [1594].
B. cinerea (holomorph *Botryotinia
  fuckeliana*) (C:94): gray mold,
  moisissure grise: on 2 NS [336] in
  storage; on 3 BC [1816].
Cladosporium cucumerinum (C:94): scab,
  gale: on 2 NS [335].
C. herbarum: on 4 NS [335].
Colletotrichum coccodes (C:94): on 2 NS
  [334].
C. orbiculare (C. lagenarium) (C:94): on
  2 NS [334].
*Didymella bryoniae (Mycosphaerella
  citrullina, M. melonis)* (anamorph
  Ascochyta cucumis) (C:94): black rot,
  pourriture noire: on 2 BC [334, 1816,
  1817, 1818], NS [336, 1594]; on 3, 4 NS
  [352].
Erysiphe cichoracearum (C:94): powdery
  mildew, blanc: on 2 BC, Ont [337]; on 3
  BC [337, 338].
E. communis: powdery mildew, blanc: on 2
  BC [334, 336], NS [334] in greenhouse;
  on 3 BC [334, 335, 336]; on 4 BC [335].
  Specimens probably were *E.
  cichoracearum*.
Fusarium sp. (C:94): on 2 NB [1594]; on 3
  Alta [335].

F. oxysporum (C:94): on 2 BC [1816], NS
  [1594].
F. roseum: on 2 NS [338, 1594].
F. solani f. sp. *cucurbitae*: on 3 Ont
  [334].
Oidium sp.: powdery mildew, blanc: on 2,
  3 BC [1816].
Olpidium radicale: on 2, 3 Ont [58].
Phomopsis sclerotioides: on 1 BC [1200].
Rhizopus sp. (C:94): on 2 NS [1594]; on 3
  Ont [1460, 1594].
Sclerotinia sp.: on 2 NB [1594]; on 2, 3
  Sask [420], Sask, Man [1138].
Sclerotinia sclerotiorum (C:94):
  sclerotinia rot, pourriture
  sclérotique: on 2 NB [339], NS [1594];
  on 3 BC [337].
Septoria cucurbitacearum (C:94): leaf
  spot, tache septorienne: on 2 BC
  [1816]; NS [333, 334, 336, 338, 1595].
Sphaerotheca fuliginea: powdery mildew,
  blanc: on 3 Ont [846].
Stemphylium botryosum: from seeds of 3
  Ont [667].
S. consortiale: from seeds on 2, 3 Ont
  [667].
Verticillium dahliae: on 2 BC [1816,
  1978]; on 4 BC [38, 335].

CUPRESSUS L.   PINACEAE

1. *C. macrocarpa* Hartweg, Monterey
    cypress, cyprès à gros fruits; native to
    California.

Ascocoryne sarcoides: on 1 BC [1012].
Coniophora puteana: on 1 BC [1012].
Lepteutypa cupressi: on 1 BC [1012].
Mytilidion sp.: on 1 BC [1012].
Phomopsis sp. (C:94): on 1 BC [1012].
Pithya cupressina: on *C.* sp. Ont [385].
Pleospora sp.: on *C.* sp. BC [1012].

CYCLAMEN L.   PRIMULACEAE

1. *C. persicum* Mill., florist's cyclamen,
    cyclamen; imported from e. Mediterranean
    region.

Botrytis cinerea (holomorph *Botryotinia
  fuckeliana*) (C:94): gray mold,
  moisissure grise: on 1 BC [333, 1816].
*Cylindrocarpon destructans (C.
  radicicola)*: on 1 Ont [336].

CYDONIA Mill.   ROSACEAE

1. *C. oblonga* Mill. (*C. vulgaris*
    Pers.), common quince, cognassier
    commun; imported from w. Asia.

Diplocarpon mespili (Fabraea maculata)
  (C:95): leaf blight, entomosporiose:
  on 1 BC [339, 1594, 1816].
Gymnosporangium clavariiforme (C:95):
  rust, rouille: on 1 NS [1262]. See
  note under *Pyrus*.
G. clavipes (C:95): quince rust, rouille
  du cognassier: 0 I on 1 Ont [1236].

CYNOGLOSSUM L.   BORAGINACEAE

Annual, biennial or perennial herbs of
the temperate zone, often cultivated.
Commonly known as Hound's Tongue or Beggar's
Lice because of the prickly seeds.

Ophiobolus rubellus: on *C.* sp. Ont
[1340].

CYNOSURUS L.   GRAMINEAE

1. *C. cristatus* L., crested dog's-tail,
   crételle commune; imported from Eurasia,
   introduced, BC, Ont-Nfld.

Cercosporidium graminis (Passalora g.)
(C:95): brown stripe, strie brune: on
1 BC [1816].

CYPERUS L.   CYPERACEAE

1. *C. esculentus* L., yellow nut sedge,
   amande de terre; native, s. Man, Ont-NS.
2. *C. strigosus* L., strigose cyperus,
   souchet hispide; native, Sask-Que.

Puccinia canaliculata (C:95): rust,
rouille: II III on 1, 2 Que [1236].

CYPRIPEDIUM L.   ORCHIDACEAE

1. *C. calceolus* L. var. *pubescens*
   (Willd.) Correll (*C. pubescens*
   Willd.), large yellow lady's-slipper,
   sabot de la vierge; native, Ont-NS.
2. *C. reginae* Walt. (*C. spectabile*
   Salisb.), showy lady's slipper, sabot de
   la vierge; native, Man-Nfld.

Cercospora cypripedii: on living leaves
of 1, 2 Ont [441].

CYSTOPTERIS Bernh.   POLYPODIACEAE

1. *C. bulbifera* (L.) Bernh., bulblet
   fern, cystoptéride bulbifère; native,
   Ont-Nfld.
2. *C. fragilis* (L.) Bernh., brittle fern,
   cystoptéride fragile; native,
   Yukon-Labr, BC, Man-NS, Nfld.

Hyalopsora polypodii (C:95): rust,
rouille: II III on 1 Ont [1236]; on 2
Yukon [7, 2008], BC [1012, 2008, F63],
Alta [644, 2008], Sask [2008], Ont
[1236].
Mycosphaerella filicinum (C:95): on 2 Que
[70].
Uredinopsis ceratophora (C:96): II^1
II^2 III on 1 Ont [1236].

CYTISUS L.   LEGUMINOSAE

1. *C.* x *beanii* Nichols: *C. ardoini*
   E. Fourn. x *C. purgans* (L.) Spach.
2. *C.* x *praecox* Bean: *C. multiflorus*
   (L'Hér. ex Ait.) Sweet x *C. purgans*.

3. *C. scoparius* (L.) Link, Scotch broom,
   genêt à balais; imported from Europe, a
   garden-escape, BC, NS and PEI.

Alternaria alternata: foliage blight: on
3 BC [1416].
Armillaria mellea: on 2 BC [1012].
Athelia scutellare (Corticium s.) (C:96):
on 3 BC [1012].
Coriolus versicolor (Polyporus v.)
(C:96): on 3 BC [1012].
*Cylindrobasidium evolvens (Corticium
laeve)*: on 3 BC [1012].
Diaporthe sp.: on 3 BC [1012].
Helminthosporium velutinum: on 3 BC [810,
1012, 1816].
Hyphoderma sambuci: on 3 BC [1012].
H. setigerum: on 3 BC [1012].
Lachnella alboviolascens (Cyphella a.):
on 3 BC [1424].
Mycoacia uda: on 3 BC [1012].
Pellidiscus pallidus: on 3 BC [1424].
Peniophora incarnata (C:96): on 3 BC
[1012].
Phomopsis sp.: on 3 BC [1012].
Pleiochaeta setosa: brown spots on living
leaves of 1 Ont [1345].
Pleospora sp.: on 3 BC [1012].
Scytinostroma galactinum: on 3 BC [1012].
Sebacina sp. (C:96): on 3 BC [1012].

DACTYLIS L.   GRAMINEAE

1. *D. glomerata* L., orchard grass,
   dactyle pelotonné; imported from
   Eurasia, introduced, BC-Nfld.

Acremonium boreale: on 1 BC, Alta, Sask
[1707].
Bipolaris sorokiniana (holomorph
Cochliobolus sativus): on 1 Alta
[120] by inoculation.
Cercosporidium graminis (Passalora g.)
(C:96): brown stripe, strie brune: on 1
BC [1816], Alta [333], Ont [335], NB
[333, 1195], Nfld [333].
Claviceps purpurea (C:96): ergot, ergot:
on 1 BC [337, 1816], Alta [133], Ont
[335], Que [337].
Colletotrichum graminicola: anthracnose,
anthracnose: on 1 Que [337].
Dinemasporium strigosum (D. graminum)
(holomorph *Phomatospora
dinemasporium*): on 1 BC [1012, 1816].
Erysiphe graminis (C:96): powdery mildew,
blanc: on 1 BC [1257, 1816], Que [32,
339].
Fusarium culmorum (C:96): foot rot, piétin
fusarien: on 1 BC [1816].
Lophodermium arundinaceum: on 1 BC [1012,
1816].
Mastigosporium rubricosum (C:96): purple
eye-spot, tache pourpre ocellée: on 1 BC
[1816], Que [282, 1729], NB [333, 1195],
PEI [336, 1974], Nfld [335, 1595].
Phyllosticta owensii (C:96): on 1 Que and
on *D.* sp. NB [333].
Pseudoseptoria donacis (Selenophoma d.):
on 1 Que [334].

Puccinia graminis (C:96): stem rust,
rouille de la tige: on 1 BC [1816], Ont
[1236].
*Ramaricium albo-ochraceum (Trechispora
a.)*: on 1 BC [970].
Rhynchosporium sp.: on *D.* sp. Ont [333].
R. orthosporum (C:96): scald, tache pâle:
on 1 Ont [333], NB [333, 1195].
Sclerophthora cryophila (C:96): downy
mildew, mildiou: on 1 BC [1816].
Sclerotinia borealis (C:96): snow mold,
moisissure nivéale: on 1 BC [1816].
Stagonospora arenaria (C:96): purple brown
spot, tache brun-pourpre: on 1 Que
[1729].
Uromyces dactylidis (C:96): leaf rust,
rouille des feuilles: on 1 BC [1012,
1816], Que [337].
Ustilago salvei (U. striiformis auct.)
(C:96): stripe smut, charbon strie: on 1
BC [1816].

DAHLIA Cav.   COMPOSITAE

1. Garden dahlia, dahlia; a cultigen
   probably derived from hybridization
   between *D. coccinea* Cav. and *D.
   pinnata* Cav. *(D. variabilis* (Willd.)
   Desf.)

Botrytis cinerea (holomorph *Botryotinia
fuckeliana*) (C:96): gray mold,
moisissure grise: on 1 BC [1816], Alta
[335], Que [337, 339].
Dendryphion nanum: on *D.* sp. BC [819].
Erysiphe cichoracearum (C:96): powdery
mildew, blanc: on 1 Ont [1257].
E. communis: powdery mildew, blanc: on 1
Ont [336]. Specimen probably *E.
cichoracearum*.
Fusarium spp. (C:96): on 1 Que [335].
Sclerotinia sclerotiorum (C:96):
sclerotinia rot, pourriture
sclerotique: on 1 BC [337].
Verticillium albo-atrum (C:96): wilt,
flétrissure verticillienne: on 1 Ont
[38, 333].

DANTHONIA DC.   GRAMINEAE

1. *D. spicata* (L.) Beauv., poverty-grass,
   danthonie à épi; native, BC-Nfld.

Atkinsonella hypoxylon (Balansia h.)
(anamorph *Ephelis borealis*) (C:97):
on *D.* sp. Que, NS [403], NS [21, 437].
Drechslera verticillata (Podosporiella v.)
(holomorph *Pyrenophora semeniperda*):
on 1 Ont [1729].
*Pseudoseptoria stomaticola (Selenophoma
donacis* var. *stomaticola)* (C:97):
on 1 Ont, Que [1729].
Pyrenophora semeniperda (anamorph
Drechslera verticillata): on 1 Ont
[1617].

DAPHNE L.   THYMELAEACEAE

1. *D. mezereum* L., daphne, daphné

mézéréon; imported from Eurasia, a
garden-escape, Ont-Nfld.

Botrytis cinerea (holomorph *Botryotinia
fuckeliana*) (C:97): twig blight,
brûlure des rameaux: on 1 BC [1816]; on
D. sp. BC [334].
Marssonina daphnes (Gloeosporium mezereum)
(C:97): leaf spot, anthracnose: on 1
BC [334, 1012, 1816], Ont [339]; on 1,
D. sp. BC [333].

DAUCUS L.   UMBELLIFERAE

1. *D. carota* L.; probably most records
   under this apply to *D. carota* var.
   sativa. 1a. *D. c.* var. *carota*,
   Queen Anne's lace, carotte sauvage;
   imported from Eurasia, introduced, BC,
   Sask-NS. 1b. *D. c.* var. *sativa* DC.,
   carrot, carotte; frequently cultivated.

Alternaria sp.: on 1 NS [989]; on 1b Ont
[871].
A. alternata: on 1 Ont [1708].
A. dauci (C:97): leaf blight, brûlure
alternarienne: on 1 BC [1816], Ont
[339, 921, 1461, 1594], Que [337, 339,
348, 1595, 1647, 1651, 1652, 1653], NB
[338, 339, 1594], NS [337, 338]; on 1b
Ont [335, 561], Que [333, 334, 335, 336,
1638, 1640, 1641, 1650, 1654, 1671], NS
[333, 334, 336, 339].
A. radicina (Stemphylium r.) (C:98):
black rot, pourriture noire: on 1 BC
[1816], NS [337, 339, 1594]; on 1b BC
[333, 334].
Botrytis cinerea (holomorph *Botryotinia
fuckeliana*) (C:97): gray mold,
moisissure grise: on 1 Alta [339,
1594], NB [337, 1594], NS [989]; on 1b
NS [333, 336].
Cercospora carotae (C:97): leaf spot,
brûlure cercosporéenne: on 1 BC [1816],
Ont [1461, 1594], Que [337, 338, 339,
348, 1595, 1647, 1651, 1652, 1653], NS
[339, 1594], Nfld [1594]; on 1b Ont
[335], Que [333, 334, 335, 336, 1638,
1639, 1640, 1641, 1650, 1654], NB [334],
NS [333, 334, 335, 336].
Chalara thielavioides (Chalaropsis t.):
dry rot, pourriture sèche: on 1 BC
[337, 1165], NS [1165] on carrots in
cold storage; on 1b BC [334].
*Colletotrichum coccodes (C.
atramentarium)*: on 1b NS [593]
inoculated.
C. dematium: on 1b NS [593] inoculated.
Cylindrocarpon destructans: on 1 Ont
[1708].
Cylindrosporium sp.: on 1b Ont [871].
Fusarium sp. (C:97): on 1 NS [989]; on 1b
Ont [871].
F. solani (holomorph *Nectria
haematococca*) (C:98): on 1 Ont [1629].
Geotrichum candidum: on 1 Alta [1594].
Gliocladium sp.: on 1 Ont [1708].
Melanospora papillata: from seeds of 1b
Man [1943].
Mucor sp.: on 1 Ont [1708].

Olpidium brassicae: tobacco necrosis
virus vector: on 1 Ont [67, 65]; on 1b
Ont, Que [67]; on 1b Ont [62, 871].
Penicillium sp.: on 1 Ont [1708], NS
[989]; on 1b Ont [871].
Phytophthora sp.: on stored roots of 1
Alta [1734].
P. megasperma (C:98): root rot, mildiou:
on 1 BC [1816].
Pythium sp.: on 1 Alta [1734]; on 1b Ont,
Que [67], Ont [868, 871].
P. afertile: on 1b Ont [868]; on *D.* sp.
BC [1977].
P. intermedium: on 1b Ont [868].
P. irregulare: on 1b Ont, Que [67], Ont
[868, 871]; on *D.* sp. BC [1977].
P. sulcatum: on 1b Ont [868, 871], Ont,
Que [67]; on 1b, *D.* sp. BC [1977].
P. sylvaticum: on 1 BC, Ont [1846]; on 1b
Ont [868]; on *D.* sp. BC [1977].
P. torulosum: on 1b Ont [868]; on *D.*
sp. BC [1977].
Rhizoctonia sp.: on 1 Que [348]; on 1b
Ont [871].
R. carotae: crater rot: on 1 Que [337].
R. crocorum (holomorph *Helicobasidium
purpureum*) (C:98): violet root rot,
rhizoctone violet: on 1 BC [1816].
Rhizopus sp. (C:98): on 1 NS [989].
Sclerotinia minor: on 1b Alta [892].
S. sclerotiorum (C:98): sclerotinia rot,
pourriture sclérotique: on 1 BC [1816],
Alta [337, 1594], Sask [338], Que
[1594], NB [337, 338, 1594], NS [338,
989, 1594]; on 1b Alta [335, 336, 1733],
Que [334, 1654], NB [336], NS [333, 334,
335, 336], PEI [334].
Stemphylium botryosum (C:97): from seeds
of 1b Ont [667].
Stilbella flavescens: on 1 Que [2].
Streptomyces scabies (C:98): scab, gale
commune: on 1b NS [334].
Thielaviopsis basicola: black root rot,
pourridié noir: on 1 BC [1416].
Ulocladium consortiale (*Stemphylium c.*):
from seeds of 1b Ont [667].

DECODON J.F. Gmel.   LYTHRACEAE

1. *D. verticillatus* (L.) Elliott, swamp
    loosestrife, décodon verticillé; native,
    Ont-Que.

Puccinia minutissima (C:98): rust,
rouille: 0 I on 1 Ont [1236].

DELPHINIUM L.   RANUNCULACEAE

1. *D.* x *cultorum* Voss; various garden
    plants of hybrid origin.
2. *D. elatum* L., bee larkspur,
    dauphinelle élevée; imported from
    Eurasia.

See also *CONSOLIDA*.

Botrytis cinerea (holomorph *Botryotinia
fuckeliana*) (C:98): gray mold,
moisissure grise: on 1 BC [1816].

Erysiphe communis (*E. polygoni*) (C:98):
powdery mildew, blanc: on 1 BC, Ont,
Que [1257]; on 2 Ont [1257]; on *D.* sp.
BC [333, 334], Alta [337, 339], NB [335].
Oidium sp.: powdery mildew, blanc: on 1
BC [1816].
Puccinia recondita s.l. (C:99): rust,
rouille: on *D.* sp. BC [1012, 1816].

DESCHAMPSIA Beauv.   GRAMINEAE

1. *D. atropurpurea* (Wahl.) Scheele;
    native, Yukon, BC-Alta, to n. Que, Labr,
    Nfld.
2. *D. caespitosa* (L.) Beauv., tufted hair
    grass, deschampsie cespiteuse; native,
    Mack-Labr, BC-Nfld. 2a. *D. c.* var.
    littoralis (Reut.) Richter; native, n.
    Mack-Frank.
3. *D. elongata* (Hook.) Munro, slender
    hair grass; native, Yukon, BC-Alta.
4. *D. flexuosa* (L.) Trin., common hair
    grass, deschampsie flexueuse; native,
    Ont-Nfld, introduced, BC.

Claviceps purpurea: ergot, ergot: on 4
Nfld [666].
Mycosphaerella lineolata (C:99): on 2a
Que [70].
Phaeosphaeria eustoma: on 2a Que [70].
Pseudoseptoria everhartii (*Selenophoma e.*)
(C:99): on 2 BC [1731].
P. stomaticola (*Selenophoma donacis* var.
stomaticola): on 3 BC [1728].
Puccinia coronata: on 2 Yukon [7].
P. poae-nemoralis (*P. poae-sudeticae* var.
airae) (C:99): rust, rouille: on 2
Yukon [1731].
P. praegracilis var. *connersii* (C:99):
rust, rouille: II III on 1 Que [1236].
Uromyces mysticus (*U. jacksonii*): rust,
rouille: on 3 BC [1012].

DESCURAINIA Webb & Berth.   CRUCIFERAE

1. *D. richardsonii* (Sweet) Schultz, gray
    tansy mustard, moutarde tanaisie grise;
    native, Yukon-Mack, Alta-Que. 1a. *D.
    r.* var. *richardsonii* (*Sisymbrium
    incisum* Gray var. *hartwegianum* AA.).
2. *D. sophia* (L.) Webb, flixweed, sagesse
    des chirurgiens; imported from Eurasia,
    introduced, Yukon, Mack-Frank, BC-Nfld.

Acremonium boreale: on 2 BC, Alta, Sask
[1707].
Albugo candida (*A. cruciferarum*) (C:100):
white rust, albugine: on 2 Sask [1323];
on *D.* sp. Sask [1321, 1594].
Alternaria brassicae: on 2 Sask [1323].
Colletotrichum dematium: on *D.* sp. Sask
[1319].
Mycosphaerella tassiana var. *tassiana*:
on 2 Sask [1327].
Selenophoma sp.: on 2 Sask [1323].
Verticillium dahliae: on 1a BC [1978].

DESMODIUM Desv.   LEGUMINOSAE

1. *D. canadense* (L.) DC., Canadian
   tick-trefoil, desmodie du Canada;
   native, Man-NS.
2. *D. glabellum* (Michx.) DC.; native to
   e. USA.
3. *D. paniculatum* (L.) DC.; native,
   Ont-Que.
4. *D. perplexum* Schub., perplexing
   tick-trefoil, desmodie perplexe; native
   to e. USA.

Microsphaera diffusa (C:100): powdery
   mildew, blanc: on 1 Ont, Que and on 3,
   4, *D.* sp. Ont [1257].
M. penicillata: powdery mildew, blanc:
   on 1 Ont [1257].
Uromyces hedysari-paniculati (C:100):
   rust, rouille: I II III on 1, 2 Ont
   [1236].

DIANTHUS L.   CARYOPHYLLACEAE

1. *D. barbatus* L., sweet william,
   jalousie; imported from Europe.
2. *D. caryophyllus* L., carnation,
   oeillet; perhaps native in the
   Mediterranean region.

Alternaria sp. (C:100): on *D.* sp. Alta,
   Ont, Que, NB [333].
A. dianthi (C:100): leaf spot or blight,
   brûlure alternarienne: on 1 NB [1194];
   on 2 BC [1816], Alta [333].
A. dianthicola (C:100): on 1 Ont, Que, NB
   [333].
Cladosporium elatum: on 2 BC [1816].
Fusarium equiseti (C:100): on 2 BC [1816].
F. oxysporium f. sp. *dianthi* (C:100):
   on 2 BC [1816].
F. poae (C:100): fusarium bud rot,
   pourriture fusarienne des boutons: on 2
   BC [1816].
Heteropatella valtellinensis (C:100):
   leaf rot, pourriture du feuillage: on 2
   BC [1816].
*Heterosporium echinulatum (Cladosporium
   e.)* (holomorph *Mycosphaerella
   dianthi*) (C:100): leaf spot, tache
   hétérosporienne: on 1, 2 BC [1816].
Phialophora cinerescens: on 2 Ont [165].
Rhizoctonia solani (holomorph
   Thanatephorus cucumeris) (C:100):
   stem rot, rhizoctone commun: on 2 Ont
   [335].
Uromyces dianthi (C:100): rust, rouille:
   on 2 BC [333, 1816], Ont, Que II III
   [1236], NS [334].
Verticillium albo-atrum: on 2 NS [334].

DIAPENSIA L.   DIAPENSIACEAE

1. *D. lapponica* L.; native, Mack, n. Man,
   east into Labr-Nfld.

Apiothyrium arcticum (C:101): on 1 Frank,
   Nfld [297].
Botryosphaeria diapensiae (Guignardia d.)
   (C:101): on 1 Que [74].

DICENTRA Bernh.   FUMARIACEAE

1. *D. spectabilis* (L.) Lem., bleeding
   heart, coeurs saignants; imported from
   Japan.

Botrytis cinerea (holomorph *Botryotinia
   fuckeliana*): gray mold, moisissure
   grise: on 1 Alta [337]; on *D.* sp.
   Alta [336].
Dendryphion sp.: on 1 NWT-Alta [1063].

DIERAMA K. Koch   IRIDACEAE

1. *D. pulcherrimum* (Hook. f.) Bak.;
   imported from s. Africa. Cultivated
   under glass in colder climates.

Cladosporium herbarum: on 1 BC [1012].

DIERVILLA Duham.   CAPRIFOLIACEAE

1. *D. lonicera* Mill., bush honeysuckle,
   herbe bleue; native, Sask-Nfld.

Godronia diervillae: on 1 Ont [656].
G. turbinata (C:101): on 1 Ont, NS [656].
Valdensia heterodoxa (holomorph
   Valdensinia h.): on 1 Que [1430].

DIGITALIS L.   SCROPHULARIACEAE

1. *D. lanata* J.F. Ehrh., Grecian
   foxglove, digitale laineuse; imported
   from Europe.
2. *D. purpurea* L., common foxglove,
   digitale pourprée; imported from Europe.

Colletotrichum fuscum (C:101):
   anthracnose, anthracnose: on 1 BC
   [1816].
Pythium sp. (C:101): stem and crown rot:
   on 1 BC [1816].
Ramularia variabilis (C:101): leaf spot,
   tache des feuilles: on 2 BC [1554,
   1816].
Septoria digitalis: septoria leaf spot,
   tache septorienne: on 1 BC [400].
Verticillium sp.: wilt: on 1 BC [1816].

DIRCA L.   THYMELAEACEAE

1. *D. palustris* L., leatherwood, bois de
   cuir; native, Ont-NS.

Aecidium hydnoideum (C:102): rust,
   rouille: 0 I on 1 Ont [1236, 1341,
   1827, 1828].
Phyllosticta dircae: on leaves of 1 Ont
   [439].

DISPORUM Salisb.   LILIACEAE

1. *D. hookeri* (Torr.) Nichols, fairy
   bells; native, BC-Alta. 1a. *D. h.*
   var. *oreganum* (S. Wats.) Q. Jones,
   Oregon fairy bells, dispore de l'Orégon;
   native, BC-Alta.
2. *D. trachycarpum* (Wats.) B. & H. (*D.
   majus* (Hook.) Britt.), white fairy

bells, dispore à fruit velu; native,
BC-Ont.

Cercospora streptopi: on 1a BC [1012,
    1547, 1816].
Septoria streptopodis (C:102): on 1a BC
    [976, 1547]; on 1a, 2 BC [1545, 1816];
    on *D.* sp. BC [1545].

DISTICHLIS Raf.   GRAMINEAE

1. *D. spicata* (L.) Greene; native, BC,
    NB, NS, PEI.
2. *D. stricta* (Torr.) Rydb.; native, BC,
    Mack, east into Man.

Dothidella aristidae: on 2 BC [1728].
Puccinia aristidae (C:102): rust,
    rouille: on 1 BC [1816].
Uromyces peckianus (C:102): rust,
    rouille: on 1 BC, NS, PEI [1578].

DRABA L.   CRUCIFERAE

1. *D. alpina* L., rock-cress draba, drave
    alpine; native, Yukon-Frank, into
    Man-Que. 1a. *D. a.* var. *nana* Hook.
    (*D. bellii* Holm); native, Yukon-Frank.
2. *D. arabisans* Michx., rock-cress draba,
    drave arabette; native, Ont, Que, Nfld.
3. *D. aurea* Vahl; native, n. BC, Yukon,
    Mack, and east into n. Que, Labr.
4. *D. cana* Rydb.; native, Alta, Yukon,
    Mack, east into NB.
5. *D. cinerea* Adams; native, Yukon-Frank,
    into BC, Sask, Ont, Que.
6. *D. fladnizensis* Wulfen, arctic draba;
    native, Yukon-Labr, into BC-n. Que. 6a.
    D. f. var. *heterotricha* (Lindbl.)
    Ball (*D. lactea* Adams); native,
    Yukon-Frank, n. BC, Man, Que.
7. *D. glabella* Pursh, glabrous draba,
    drave glabre; native, Yukon-Labr, Nfld,
    into BC, Man-NS.
8. *D. incana* L., twisted draba; native,
    Man-NB, PEI.
9. *D. incerta* Payson; native, Yukon,
    BC-Alta.
10. *D. longipes* Raup; native, Yukon, Mack,
    BC.
11. *D. nivalis* Lilj. (*D. lonchocarpa*
    Rydb. var. *l.*); native, Yukon-Labr,
    BC-Que, Nfld.
12. *D. norvegica* Gunn., Norwegian draba,
    drave de Norvège; native, Frank, Que,
    NS, Nfld, Labr.
13. *D. stenoloba* Ledeb.; native, c.
    Yukon-w. Mack, BC-sw. Alta.
14. *D. subcapitata* Simmons; native, Mack.

Mycosphaerella pyrenaica (C:103): on 1,
    1a, 6a, 14 Frank [1586].
M. tassiana (Sphaerella stellarianearum)
    (C:103): on 1 Keew [437].
Peronospora parasitica (C:103): downy
    mildew, mildiou: on 6a Frank [1543]; on
    D. sp. Frank [1244].
Pleospora hispida: on dead stems of *D.*
    sp. Keew [437].

Puccinia drabae (C:103): rust, rouille:
    on 1a, 5 Frank, on 6a, 7, 9 Keew, on 2
    Ont, on 2, 3, 4, 7, 8 Que, on 6a, 10,
    11, 13 Yukon, on 7 Man, on 8, 12 Nfld,
    on 9 Alta, on 9, 10 BC, and on 12 on NS,
    all from [1563]; and III on *D.* sp. Ont
    [1236].
P. drabicola: rust, rouille: on 5 Keew
    and on 9 BC [1563].
P. holboellii (P. thlaspeos auct.*)*
    (C:103): rust, rouille: on 4, *D.* sp.
    Alta [1568]; on 8 Que [1236, 1568].
Sphaerotheca fuliginea: powdery mildew,
    blanc: on 2 Ont [1257].

DRACAENA Vand. ex L.   AGAVACEAE

    Shrubby or tree-like plants in the old
world (1 sp. in America).

Phyllosticta dracaenae: leaf spot, tache
    des feuilles: on *D.* sp. Ont [339].

DRYAS L.   ROSACEAE

1. *D. drummondii* Richardson ex Hook.,
    yellow dryas, dryas de Drummond; native,
    Yukon-Mack, BC-Alta, Ont-Que, Nfld. 1a.
    D. d. var. *drummondii* (*D.* x
    lewinii Rouleau), BC-Sask, Ont-Que, w.
    Nfld.
2. *D. integrifolia* Vahl; native,
    Yukon-Frank, BC-Alta, Man-Nfld. 2a.
    D. i. f. *canescens* (Simmons) Fern.;
    arctic N. America, south to Alta, Que,
    Nfld.
3. *D. octopetala* ssp. *alaskensis*
    (Porsild) Hulten (*D. alaskensis*
    Porsild); native, Yukon-Mack.
4. Host species not named, i.e., reported
    as *Dryas* sp.

Cladosporium cladosporioides: from leaves
    of 4 NWT [1969].
*Geomyces pannorum (Chrysosporium
    verrucosum)*: from leaves of 4 NWT
    [1969].
Gnomonia dryadis (C:103): on 2 Labr [79].
G. sibbaldiae: on 2 Frank [79].
Gnomoniella vagans (C:103): on 2 Frank
    [79].
Isothea rhytismoides (C:104): on 1 Alta
    [644, 952]; on 2 Frank [1244, 1586].
Marasmius epidryas: on 2 Frank [1239]; on
    3 Yukon [1108]; on 4 Keew [1188] on dead
    parts, Alta [1920].
Pestalotia sp.: from leaves of 4 NWT
    [1969]. Specimen probably is
    Pestalotiopsis sp.
Phialophora sp.: from leaves of 4 NWT
    [1969].
Pseudomassaria islandica: on 1 Que, on 1,
    1a, 2, 2a Nfld and on 2 Frank, Que, all
    from [71].
P. minor: on 1 Que, Nfld and on 1a, 2, 2a
    Nfld, all from [71].
Synchytrium cupulatum (C:104): on 2 Frank
    [1586].
Trimmatostroma sp.: on 4 NWT [1969].

DRYOPTERIS Adans.   POLYPODIACEAE

1. *D. austriaca* (Jacq.) Woyn., spinulose wood fern, dryoptéride spinuleuse; native, Yukon-Mack, BC-Nfld.
2. *D. intermedia* (Willd.) Gray (*D. spinulosa* var. *i.* (Muhlenb.) C.V. Mort., *Thelypteris s.* var. *i.* (Muhl.) Nieuwl.); native, Ont-Nfld.
3. *D. marginalis* (L.) Gray (*Thelypteris m.* Nieuwl.), marginal shield-fern, dryoptéride marginale; native, Ont-NS.

See also *GYMNOCARPIUM* and *THELYPTERIS*.

Helicobasidium holospirum: on 2 NS [1032].
Milesia dilatata (C:104): rust, rouille: on 1 BC [1012, 1816, 2008].
M. fructuosa (M. intermedia) (C:104): rust, rouille: II III on 2 Ont [1236], Que [1377]; on *D.* sp. Que [F66, reported on "fern"].
M. marginalis (C:104): rust, rouille: II III on 3 Ont [1236], Que [1377].
Valdensia heterodoxa (Asterobolus gaultheriae) (holomorph *Valdensinia h.*): on 1 BC [1438].

DULICHIUM Pers.   CYPERACEAE

1. *D. arundinaceum* (L.) Britt., three-way sedge, dulichium roseau; native, BC, Man-Nfld.

Puccinia dioicae (C:104): rust, rouille: II III on 1 Ont [1236].

DUPONTIA R. Br.   GRAMINEAE

1. *D. fischeri* R. Br.; native, Yukon-Labr, south into n. Man.

Hysteropezizella diminuens: on 1 Frank [664].

ECHINOCHLOA Beauv.   GRAMINEAE

1. *E. crus-galli* (L.) Beauv., barnyard grass, pied-de-coq; native, BC-Nfld.

Drechslera dictyoides f. sp. *perenne*: on 1 Ont [1611].
Olpidium brassicae: on 1 BC [1977].
Pythium sp.: on 1 BC [1977].
Rhizoctonia solani (holomorph *Thanatephorus cucumeris*): on 1 Que [317].
Rhizophydium graminis: on 1 Ont [59].

ECHINOCYSTIS Torr. & Gray   CUCURBITACEAE

1. *E. lobata* Torr. & Gray, wild cucumber, concombre grimpant; native, Sask-NB.

Verticillium dahliae: on 1 BC [334, 1816, 1978].

ECHIUM L.   BORAGINACEAE

1. *E. vulgare* L., blueweed, vipérine; imported from Eurasia, introduced, BC-Nfld.

Ophiobolus cesatianus (Leptosphaeria c.): on 1 Ont [760]; on 1, *E.* sp. Ont [1620].
O. mathieui: on 1 Ont [1620].

ELEAGNUS L.   ELEAGNACEAE

1. *E. angustifolia* L., Russian olive, olivier de Bohême; imported from Eurasia, spreading from cultivation, BC-Alta, Man-Ont.
2. *E. commutata* Bernh., (*E. argentea* Pursh, non Moench), silverberry, chalef changeant; native, Yukon-Mack, BC-Que.

Camarosporium elaeagnellum: on 1 Sask [F68].
Cercosporidium manitobanum (Cercospora m.) (C:105): on 2 BC [1012], Sask, Man [1760].
Cladosporium macrocarpum: on 2 Sask [1760].
Nectria cinnabarina (anamorph *Tubercularia vulgaris*): on 1 Alta [F68].
Phomopsis arnoldiae (P. elaeagni): on 1 BC, Ont [15, 17], Que [17].
Puccinia caricis-shepherdiae (C:105): rust, rouille: on 1 BC [1816], Sask [337]; on 2 NWT-Alta [1063], BC [1012]; on *E.* sp. NWT-Alta [1063], Mack [53].
Verticillium albo-atrum: on 1 BC [337].
V. dahliae: on 1 BC [1978].

ELEOCHARIS R. Br.   CYPERACEAE

1. *E. palustris* (L.) R. & S., swamp spike rush, eléocharide des marais; native, Yukon-Mack, BC-Nfld.

Puccinia eleocharidis (C:106): rust, rouille: II III on 1 Man, Ont, Que [1559]; on 1, *E.* sp. [1236].

ELYMUS L.   GRAMINEAE

See discussion of generic circumscriptions under *Agropyron*.
1. *E. alaskanus* (Scribn. & Merr.) A. Löve ssp. *borealis* (Turcz.) A. & D. Löve (*Agropyron latiglume* (Scribn. & Sm.) Rydb.). This is taxonomically a difficult group. All fungus records herein were reported on *A. latiglume* in the Yukon.
2. *E. canadensis* L., Canada wild rye, élyme du Canada; native, Mack, BC-NS.
3. *E. fibrosus* (Schrenk) Tzvelev (*Agropyron f.* (Schrenk) Nevski); imported from USSR.
4. *E. glaucus* Buckl., blue wild rye, élyme glauque; native, BC-Ont. 4a. *E. g.* var. *virescens* (Piper) Bowden (*E. virescens* Piper); native, BC.

5. *E. hystrix* L. *(Hystrix patula
Moench)*; porcupine grass, hystrix
étalé; native, Man-NB.
6. *E. intermedia* (Host) Nevski ssp.
barbabulatus (Schur) Löve *(Agropyron
trichophorum* (Link) Richt.*)*; imported
from Eurasia, cultivated in test plots,
no known escapes.
7. *E. lanceolatus* (Scribn. & Sm.) Gould
(Agropyron dasystachyum (Hook.)
Scribn., *A. riparium* Scribn.*)*;
northern wheatgrass; native, BC-Ont.
7a. *E. l.* ssp. *albicans* (Scribn. &
Sm.) Dewey *(Agropyron a.* Scribn. &
Sm., *A. griffithsii* Scribn. & Sm.*)*;
native, BC-Sask. 7b. *E. l.* ssp.
yukonense (Scribn. & Merr.) Löve
(Agropyron y. Scribn. & Merr.*)*;
native, dry prairies, sandhills,
Yukon-n. BC.
8. *E. sibiricus* L., Siberian wild rye;
native, Mack, BC.
9. *E. subsecundus* (Link) A. & D. Löve
(Agropyron s. (Link) Hitchc.,
A. unilaterale Cass.*)*; native, c.
Yukon, BC-Nfld.
10. *E. trachycaulus* (Link) Gould
(Agropyron t. (Link) Malte*)*, slender
wheatgrass, chiendent à tiges rudes;
native, BC-Nfld.
11. *E. virginicus* L., Virginia wild rye,
élyme de Virginie; native, BC-Nfld.
12. Host species not named, i.e., reported
as *Elymus* sp. Some of these records
may be of species of *Leymus* or
Psathyrostachys.

See also x *Agrohordeum*.

*Bipolaris sorokiniana (Helminthosporium
sativum)* (holomorph *Cochliobolus
sativus*) (C:20, 106): foot rot, piétin
commun: on 2, 10 Alta [120] by
inoculation; on 8 Alta [133]; on 10 Sask
[961] by inoculation.
*Cercosporidium graminis (Passalora g.,
Scolecotrichum g.)* (C:106): brown
stripe, strié brune: on 4 BC [1816]; on
9 Yukon [1731].
Cheilaria agrostis (Septogloeum oxysporum)
(C:107): on 4 BC [1012, 1816].
Claviceps sp.: on 4, 7 BC and on 4a, 7a
Alta [1781].
C. purpurea (C:20, 106): ergot, ergot:
on 3, 8 Alta [133]; on 4 BC [1816]; on
12 BC [333], Ont [1340, 1660].
Drechslera erythrospila: on 12 Sask [338].
D. tritici-repentis (holomorph
Pyrenophora t.) (C:106): leaf blotch,
tache drechsleréenne: on 2 Ont and on
12 Alta, Man [1611].
Epicoccum sp.: on 4 BC [1012].
Erysiphe graminis (C:106): powdery
mildew, blanc: on 4a Sask [338]; on 5
Ont, Que [1257]; on 8 Alta [133].
Fusarium culmorum: on 12 Sask [337].
F. nivale: on 1 Yukon [1731]. See
Agrostis.

Leptosphaerulina sp. *(Pseudoplea
sp.)*: on 8 Alta [133].
Menispora ciliata: on 12 BC [826].
Pseudoseptoria donacis (Selenophoma d.)
(C:21): on 1 Yukon [1731].
P. obtusa (Selenophoma o.) (C:21): on 9
BC [1731].
Puccinia coronata (C:107): crown rust,
rouille couronée: II III on 2 Ont
[1236].
P. graminis (C:107): stem rust, rouille
de la tige: II III on 11 Ont [1236].
P. montanensis (C:21): leaf rust, rouille
brune: on 7a, 7, 9 Alta [350].
P. impatienti-elymi (P. recondita in
part*)* (C:107): leaf rust, rouille des
feuilles: on 4, 4a BC [1012]; on 8 Alta
[133]; on 10 BC [1816]; on 11 Ont
[1236]. This taxa has been segregated
from the *P. recondita* complex and is
recognized as a distinct species
[1561]. The name was published by
Arthur in Klebahn, Wirtsw. Rostp., p.
292, 1904.
P. striiformis (C:107): stripe rust,
rouille striée: on 4 BC [1012, 1816].
Rhizophydium graminis: on 2, 10 Ont [59].
Sclerotinia borealis (C:21, 107): snow
mold, moisissure nivéale: 2, 7, 8 BC
[1816].
Septoria sp. (C:107): on 6, 8 Alta [133].
Ustilago bullata (C:21): head smut,
charbon de l'épi: on 10 BC [1816].
U. salvei (U. striiformis auct.*)*
(C:21): stripe smut, charbon strié: on
10 BC [333, 1816].

ELYTRIGIA Desv.   GRAMINEAE

See discussion of generic circumscriptions
under *Agropyron*.

1. *E. intermedia* (Host) Nevski
(Agropyron i. (Host) Beauv.*)*,
intermediate wheatgrass, chiendent
intermédiaire; imported from Eurasia.
2. *E. repens* (L.) Nevski *(Agropyrum r.*
(L.) Beauv.*)*, quackgrass, chiendent
rampant; imported from Eurasia, natzd.,
Yukon-Mack, BC-Nfld.

Acremonium boreale: on 1 BC, Alta, Sask
[1707].
*Bipolaris sorokiniana (Helminthosporium
sativum)* (holomorph *Cochliobolus
sativus*) (C:20): foot rot, piétin
commun: on 1, 2 Alta by inoculation
[120]; on 1, 2 Sask but on 1 by
inoculation [961]; on 2 NS [595].
Claviceps purpurea (C:20): ergot, ergot:
on 2 Alta [333, 336], Ont [335], NB
[1195].
Dasyscyphus pudicellus: on 2 Ont [1340].
Drechslera tritici-repentis (holomorph
Pyrenophora t.) (C:20): on 2 Ont
[1611], NB [333, 1195].
Erysiphe graminis (C:20): powdery mildew,
blanc: on 1 Sask [333]; on 2 BC, Alta,
Sask, Man, Ont, Que, PEI [1257] and BC

[1816], Alta, Ont [1255], Que [282], NB [333, 1195], NS [335], PEI [336, 1974], Nfld [338].

Fusarium sp. (C:20): on 1 Sask [338].

F. avenaceum: on 2 NB [333, 1195].

Lagena radicicola (C:20): in roots of 2 Ont [68].

Ligniera pilorum: on 2 Ont [60].

Olpidium brassicae (C:20): in roots of 2 Ont [68].

Phyllachora graminis (C:21): tar spot: on 2 BC [1416], Ont [335, 1340, 1827], Que [337], NB [333, 1195], NS [1036].

Polymyxa graminis (C:21): on 2 Ont [68].

Pseudoseptoria donacis (Selenophoma d.) (C:21): on 2 Yukon [1731], BC [1728].

Puccinia coronata (C:21): rust, rouille: II III on 2 Ont [1236, 1341].

P. graminis (C:21): rust rouille: II III on 2 Ont [1236], Que, NS [337], NB [333, 1195].

P. graminis ssp. *graminis* var. *graminis (P. g.* f. sp. *tritici)* (C:21): on 2 Man [617].

P. graminis f. sp. *secalis* (C:21): rye stem rust: "overwintering" on 2 Man [616, 635].

P. montanensis (C:21): leaf rust, rouille brune: on 2 Sask [350].

P. recondita s. l. (C:21): rust, rouille: on 2 BC [1816], Ont [1236], Que [337], NB [333, 1195], NS [335].

P. striiformis (C:21): stripe rust, rouille striée: on 2 BC [1816].

Pythium irregulare: on 2 Ont [868].

P. ultimum: on 2 Ont [868].

Rhizoctonia solani (holomorph *Thanatephorus cucumeris*) (C:21): basal rot: on 1 Sask [338].

Rhizophydium graminis: on 2 Ont [59].

Rhynchosporium secalis (C:21): scald: on 2 BC [1816], NB [333, 1195].

Sclerophthora macrospora: downy mildew, mildiou: on 2 Man [338].

Sclerotinia borealis (C:21): snow mold, moisissure niveale: on 1 BC [1816].

Septoria elymi (C:21): leaf spot, tache septorienne: on 2 NB [333, 1195], NS [335], PEI [336, 1974].

Ustilago hypodytes (C:21): culm smut, charbon de la tige: on 2 BC [339, 1816].

U. spegazzinii: stem smut, charbon de la tige: on 2 BC [333, 334, 336, 337, 338]. These reports may be of *U. hypodytes*.

U. salvei (U. striiformis auct.) (C:21): stripe smut, charbon strié: on 2 NB [333, 1195].

Wojnowicia hirta (Hendersonia graminis): on 2 Que [282].

EMPETRUM L.   EMPETRACEAE

1. *E. nigrum* L., black crowberry, camarine noire; native, Yukon-Frank, BC-Nfld.

Botryosphaeria empetri (Physalospora e., P. crepiniana) (C:107): on 1 Frank, Que, Labr [74] on dead leaves, BC [1012].

B. hyperborea: on 1 Nlfd [74], at tip of leaf or forming a band across middle of leaf.

Chrysomyxa empetri (C:107): rust, rouille: II III on 1 Yukon, NWT, BC, Alta, Man [2008], Yukon [7, 460], NWT-Alta [1063], Mack [53, F66], Frank [1244], BC [1012, 1816], Alta [53, 950, 952], Sask [F66], Man, Que [1236], Nfld [1341] and Que [1377]; on 1, *E.* sp. BC [953, 954]; on *E.* sp. Mack [F64].

Cladosporium sp.: on 1 BC [1012].

Duplicaria empetri: on 1 Que [1411].

Euantennaria arctica (Limacinia a.): on 1 Que [70, 797].

Exobasidium empetri (C:107): on 1 BC [1816].

Herpotrichiella polyspora: on 1 Que [70].

Marasmiellus paludosus: on 1 NB [1433].

Mycosphaerella tassiana var. *tassiana* (C:107): on 1 Que [70].

Phaeangellina empetri: on 1 Nfld [1340].

Sphaeropezia empetri (C:107): on 1 BC [1012, 1816].

ENDYMION Dum.   LILIACEAE

Bulbous, scapose perennial herbs native to w. Europe and nw. Africa.

1. *E. hispanicus* (Mill.) Chouard (*Scilla h.* Mill.), Spanish bluebell; imported from Mediterranean region.
2. *E. non-scriptus* (L.) Garcke (*Scilla n.* (L.) Hoffmgg. & Link), harebell; imported from w. Europe.

Uromyces muscari (U. scillarum) (C:261): rust, rouille: on 1 BC [334]; on 1, 2 BC [1816]; on 2 BC [336, 1012].

U. muscari f. sp. *scillae* (C:261): rust, rouille: on 1 BC [1545].

EPIGAEA L.   ERICACEAE

1. *E. repens* L., mayflower, fleur de mai; native, Man-Nfld.

Cercospora epigaeae: on living leaves of 1 Ont [441].

Microsphaera vaccinii (M. penicillata var. *v.)* (C:107): powdery mildew, blanc: on 1 Ont, Que [1257].

EPILOBIUM L.   ONAGRACEAE

1. *E. alpinum* L.; native, Yukon-Mack, BC-Alta, Que, Nfld, Labr. 1a. *E. a.* var. *alpinum (E. anagallidifolium* Lam.); native, Yukon, BC-Alta, Que, Nfld, Labr. 1b. *E. a.* var. *clavatum* (Trel.) Hitchoc. (*E. clavatum* Trel.); native, BC-Alta. 1c. *E. a.* var. *gracillimum* (Trel.) Hitchc. (*E. oregonense* Haussk.); native, BC. 1d. *E. a.* var. *nutans* (Hornem.) Hook. (*E. hornemannii* Rchb.); native, Yukon, BC-Alta, Que, Nfld, Labr.

2. *E. angustifolium* L., fireweed, bouquets rouge; native, Yukon-Labr, BC-Nfld.
3. *E. coloratum* Biehler, purple-veined willow-herb, épilobe coloré; native, Ont-NS.
4. *E. davuricum* Fisch. (including *E. davuricum* var. *arcticum* (Sam.) Polunin); native, Yukon-Labr, BC, Man-Que.
5. *E. glandulosum* Lehm., glandular willow-herb, épilobe glanduleux; native, Yukon, BC-Que, NS-Nfld. 5a. *E. g.* var. *tenue* (Trel.) Hitchc. (*E. brevistylum* Barbey); native, BC.
6. *E. hirsutum* L., hairy willow-herb, épilobe hirsute; imported from Eurasia, introduced, Ont, Que and NS.
7. *E. latifolium* L., river beauty, épilobe à feuilles larges; native, Yukon-Labr, BC-Que, Nfld.
8. *E. leptophyllum* Raf., narrow-leaved willow-herb, épilobe à feuilles étroites; native, Mack, Alta-Nfld.
9. *E. luteum* Pursh; native, BC-Alta.
10. *E. palustre* L., swamp willow-herb, épilobe palustre; native, Yukon-Labr, BC-Nfld. 10a. *E. p.* var. *grammadophyllum* Haussk.
11. *E. paniculatum* Nutt.; native, BC-Man, introduced, Ont-Que.
12. *E. strictum* Muhl., narrow willow-herb, épilobe étroite; native, Ont-NS, PEI.
13. *E. watsonii* Barbey (*E. glandulosum* var. *adenocaulon* (Haussk.) Fern., *E. g.* var. *occidentale* (Trel.) Fern.); native, Yukon-Mack, BC-Nfld.
14. Host species not named, i.e., reported as *Epilobium* sp.

Botrytis cinerea (holomorph *Botryotinia fuckeliana*) (C:108): on 7 Frank [664].
Cercospora montana (Ramularia m.) (C:109): on 1d, 2, 11 BC, on 1d, 2, 5 Que, on 2 Yukon, Alta, Sask, Ont, on 3, 5 Ont and on 10 Keew, Man, all from [1546]; on 2 BC [1816], Alta [644].
Discostroma tosta (Clathridium t., Paradidymella t.) (anamorph *Seimatosporium passerinii*) (C:108): on 2 BC [1159], Que [70, 1159].
Ditopellopsis racemula: on 2 BC [79].
Doassansia epilobii (C:108): on 1, 1b, 1c, 1d BC and on 1d, 10 Que [1546].
Gloeosporium bowmanii: on leaves of 3 Ont [439]. A doubtful species of *Gloeosporium* [20].
Leptosphaeria doliolum (C:108): on 5 Que [70].
Mycosphaerella chamaenerii (anamorph *Ramularia c.*) (C:108): on 7 Frank, Keew, BC, Que [1546], Frank [1244].
M. minor (C:108): on 2 Que [70].
Nectria pedicularis (C:108): on 2 Que [70].
Pistillaria typhuloides (C:108): on 2 BC [1012].

Plasmopara latifolii (C:108): downy mildew, mildiou: on 7 Yukon, Keew, BC, Que [1546].
Psilocybe inquilina (P. acadiensis): on 14 NS [1693].
Puccinia dioicae (C:108): rust, rouille: 0 I on 2 BC [1012], Ont [1236, 1341].
P. epilobii (C:108): rust, rouille: on 10 Nfld [1341]; III on 10a Ont [1236].
P. epilobii ssp. *palustris* (C:108): on 10 Keew, BC, Ont, Nfld [1546].
P. gigantea (C:108): rust, rouille: on 2 NWT-Alta [1063], Mack [53], BC [1012, 1816], Alta, Man, Que [1546].
P. pulverulenta (C:108): rust, rouille: on 11 BC [1012, 1816]; on 14 NWT-Alta [1063].
P. scandica (C:108): rust, rouille: on 1 BC and on 1a, 1b BC, Alta [1546].
P. veratri (C:108): rust, rouille: on 1, 14 BC [1546]; on 7 BC [1012, 1816], Alta [1545]; on 9 BC [1545, 1816].
Pucciniastrum epilobii (P. abieti-chamaenerii) (C:108): rust, rouille: on 1, 2 BC [1816]; on 1d, 2, 5, 5a BC [1012]; on 2 Yukon, NWT, BC, Alta, Sask, Man [2008] also Yukon, Mack, BC, Alta, Sask, Ont, Que, Labr, Nfld [1546] and Yukon [460], NWT-Alta [1063], Mack [53], BC [2007], Alta [53, 560, 737, 1402, F65], Ont [1236, 1341, F70], Que [282, F60], NB, NS, PEI [1032], Nfld [1659, 1660, 1661]; II III on 3, 5, 6, 12 Ont [1236]; II III on 5 NWT-Alta [1063], BC [954], Alta [53, 950, 1402, F65], Que [F61]; on 7 Que [1546]; II III on 10 Que [1236].
P. pustulatum (C:109): rust, rouille: on 2 Que [1377]; on 3 Que and on 5 BC, Alta, Sask, Ont, Que, NS, PEI, Nfld [1546]; on 5, 5a BC [1816]; on 6 Ont, on 9 BC and on 10 BC, Sask, Man, Ont, Que, Nfld [1546].
Sphaerotheca epilobii: powdery mildew, blanc: on 5 BC [1012], BC, Ont, Que [1257]; on 8 Labr and on 10 Man, Labr, Nfld [1257].
S. macularis (C:109): powdery mildew, blanc: on 5 Que, on 8, 10 Labr and on 10 Man, Nfld [1546].
Sydowiella fenestrans (C:109): on 2 Que [70, 79].
Synchytrium ?aureum (C:109): on 1a Que, on 4, 10 Keew and on 7 Frank [1546].
Venturia adusta (Sphaerella effusa) (C:109): on overwintered leaves and stalks of 7 NWT, BC, Que [73].
V. maculiformis (C:109): on living and dead leaves of 14 Frank, BC, Que [73].

EQUISETUM L.   EQUISETACEAE

1. *E. arvense* L., common horsetail, queue de renard; native, Yukon-Labr, BC-Nfld.

Acremonium boreale: on 1 BC, Alta, Sask [1707].
Lachnella alboviolascens: on 1 BC [1424].
Thelephora terrestris: on *E.* sp. Ont [1828]. See *Pinus*.

Thielaviopsis basicola: on 1 Ont [546].
Verticillium dahliae: on 1 BC [1978].

ERANTHIS Salisb.   RANUNCULACEAE

1. *E. hyemalis* (L.) Salisb., winter
    aconite, éranthe; imported from Europe.

Urocystis eranthidis (C:109): smut,
    charbon: on 1 BC [1816].

ERECHTITES Raf.   COMPOSITAE

    Erect and coarse annual, odoriferous
herbs of America, Australia and New Zealand.

1. *E. hieracifolia* (L.) Raf., fireweed,
    crève-à-yeux; native, Ont-NS, PEI.

Sphaerotheca fuliginea: powdery mildew,
    blanc: on 1 Ont, Que [1257].

ERIGENIA Nutt.   UMBELLIFERAE

1. *E. bulbosa* (Michx.) Nutt.,
    harbinger-of-spring; native, Ont.

Puccinia erigeniae (C:109): rust,
    rouille: 0 I III on 1 Ont [1236].

ERIGERON L.   COMPOSITAE

1. *E. annuus* (L.) Pers., annual
    daisy-fleabane, vergerette annuelle;
    native, BC-Nfld.
2. *E. aureus* Greene; native, BC-Alta.
3. *E. canadensis* L., butterweed,
    vergerette du Canada; native, Mack,
    BC-Nfld.
4. *E. filifolius* Nutt.; native, BC.
5. *E. linearis* (Hook.) Piper; native, BC.
6. *E. philadelphicus* L., Philadelphia
    fleabane, vergerette de Philadelphie;
    native, Yukon-Mack, BC-Nfld.
7. *E. strigosus* Muhl., white-top,
    vergerette rude; native, BC-Nfld.   7a.
    E. s. var. *strigosus* f. *strigosus*
    (*E. ramosus* (Walt.) BSP); native,
    BC-Nfld.

Cercosporella virgaureae: on 1, *E.* sp.
    Que and on 1, 3, 6 Ont [1352]; on 3 Ont
    [380].
Coleosporium asterum (C. solidaginis)
    (C:110): rust, rouille: on *E.* sp.
    Que [1377].
Leptothyrium punctiforme: on 7 Ont [195].
Ophiobolus collapsus: on *E.* sp. Ont
    [1620].
Puccinia dioicae (C:110): rust, rouille:
    0 I on 6 Ont [1236].
P. grindeliae: rust, rouille: on 2, 4 BC
    [1578].
P. stipae (C:110): rust, rouille: 0 I on
    4 BC [644]; on 5 BC [1012, 1816].
Pythium sp.: on 1 Ont [868].
P. intermedium: on 1 Ont [868].
Septoria erigerontis (C:110): on 1, 7a
    Ont [195].

Sphaerotheca fulignea: powdery mildew,
    blanc: on 1, 3, 6 Ont and on 6 Que
    [1257].
Thielaviopsis basicola: on 3 Ont [546].

ERIOGONUM Michx.   POLYGONACEAE

1. *E. flavum* Nutt., yellow umbrella
    plant, eriogone jaune; native, BC-Man.
    1a. *E. f.* var. *flavum*; native,
    BC-Alta.
2. *E. heracleoides* Nutt.; native, BC.
3. *E. niveum* Dougl. ex Benth., snow
    eriogonum; native, BC.

Uromyces intricatus (C:110): rust,
    rouille: on 2 BC [1012, 1816].
U. intricatus var. *major*: rust,
    rouille: on 1a Sask and on 2 BC [1552].
U. intricatus var. *nivei*: rust,
    rouille: on 3 BC [1552].
U. intricatus var. *umbellati*: rust,
    rouille: on 2 BC [1552].

ERIOPHORUM L.   CYPERACEAE

1. *E. angustifolium* Honckeny (*E.
    polystachyon* auct. non L.),
    narrow-leaved cotton-grass, linaigrette
    à feuilles étroites; native, Yukon-Labr,
    BC-Nfld.
2. *E. chamissonis* Meyer, wet peat; Yukon,
    BC-Que, Nfld.
3. *E. scheuchzeri* Hoppe; native,
    Yukon-Labr, BC-Que, Nfld.
4. *E. tenellum* Nutt., filiform
    cotton-grass, linaigrette ténue; native,
    Ont-Nfld.
5. *E. vaginatum* L., dense cotton-grass,
    linaigrette dense; native, Yukon-Frank,
    BC-Nfld. 5a. *E. v.* ssp. *spissum*
    (Fern.) Hulten (*E. spissum* Fern.);
    native, Yukon-Frank, Alta-Sask, Ont-Nfld.
6. *E. virginicum* L., tawny bog-cotton,
    linaigrette de Virginie; native,
    Ont-Nfld.

Myriosclerotinia dennisii (Sclerotinia d.)
    (C:111): on 3 Frank [428, 664, 1243,
    1586].
M. vahliana (Sclerotinia v.) (C:111):
    on 1 Frank [1244], Nfld [428]; on 1, 3
    Mack [664], Frank [428, 664, 1239, 1244,
    1586]; on 5a, *E.* sp. Keew [428]; on
    E. sp. Frank, Keew, Nfld [666].
Puccinia angustata (C:111): rust,
    rouille: II III on 1 Sask, NS, Nfld and
    on 4 Ont [1559]; on 4, 6 Ont [1236].
P. eriophori: rust, rouille: on 2 BC and
    on 4, 6 Ont [1559].
P. eriophori var. *apargidii*: on 1 BC
    [1559].
Wettsteinina macrotheca (C:111): on 1 Que
    [70].
W. niesslii (C:111): on 1 Que [70].

ERYSIMUM L.   CRUCIFERAE

1. *E. asperum* (Nutt.) DC., western

wallflower; native, BC-Man, introduced, Ont-Que.
2. *E. cheiranthoides* L., wormseed mustard, vélar giroflée; imported from Eurasia, introduced, Yukon-Mack, BC-Nfld.
3. *E. inconspicuum* (Wats.) MacM., small flowered prairie rocket, vélar à petites fleurs; native, Yukon-Mack, BC-NS, Nfld. 3a. *E. i.* var. *inconspicuum* (*E. parviflorum* Nutt.); native, Yukon-Mack, BC-Ont.
4. *E. pallasii* (Pursh) Fern.; native, Yukon-Frank, Alta.

Mycosphaerella tassiana var. *tassiana*: on 1, 3a Sask [1327].
Puccinia holboellii (P. thlaspeos) (C:111): rust, rouille: on 4 Mack [1586], Frank [1239, 1568, 1586].
Pythium catenulatum: on 2 Ont [868].
P. irregulare: on 2 Ont [868].

ERYTHRONIUM L.   LILIACEAE

1. *E. americanum* Ker., trout lily, ail doux; native, Ont-NS.
2. *E. grandiflorum* Pursh, avalanche lily; native, BC-Alta.
3. *E. oregonum* Applegate, white fawn lily; native, BC.
4. *E. revolutum* J. Sm., pink fawn lily; native, BC.

Botrytis tulipae (C:111): blight, brûlure: on 2 BC [1816].
Ciborinia erythronii (C:111): on *E.* sp. Ont, Que [662].
Puccinia sessilis (C:111): rust, rouille: 0 I on 1 Ont [1236].
Urocystis erythronii (C:112): smut, charbon: on 1 Ont, Que [1545].
Uromyces heterodermus (C:112): rust, rouille: 0 III on 2 NWT-Alta [1063], Alta [644, 952, 1545]; on 2, 3, 4 BC [1545, 1816]; on 3 BC [1012].
Ustilago heufleri (C:112): smut, charbon: on 1 Ont, Que [1545]; on 3 BC [1545, 1816].

EUCALYPTUS L'Her.   MYRTACEAE

1. *E. rubida* H. Deane & Marden, candle-bark gum; imported from Australia.

Cytospora sp.: on 1 BC [1012].

EUONYMUS L.   CELASTRACEAE

1. *E. alata* (Thunb.) Siebold, winged spindle tree, fusain ailé; imported from e. Asia.

Diplodia ramulicola (C:112): twig blight, brûlure des rameaux: on 1 BC [1012, 1816].
Gloeosporium sp.: on 1 BC [1816].

EUPATORIUM L.   COMPOSITAE

1. *E. maculatum* L., joe-pye-weed, eupatoire maculée; native, BC-Nfld.
2. *E. perfoliatum* L., boneset, herbe à sonder; native, Man-PEI.
3. *E. purpureum* L., joe-pye-weed, eupatoire pourpre; native, BC-Nfld.
4. *E. rugosum* Houtt. (*E. ageratoides* L.), white snakeroot, eupatoire rugeuse; native, Sask-NS.

Aecidium ageratinae: rust, rouille: on 4 Ont, Que [1709].
Erysiphe cichoracearum (C:112): powdery mildew, blanc: on 1 Que and on 1, 2, 4 Ont [1257]; on 3 Ont [1828].
Puccinia eleocharidis (C:112): rust, rouille: 0 I on 2, 3 Ont [1236, 1559]; on 4 Ont [1236].
P. tenuis (C:112): rust, rouille: 0 I III on 4 Ont [1236], Que [351].

EUPHORBIA L.   EUPHORBIACEAE

1. *E. cyparissias* L., cypress spurge, euphorbe cyprès; imported from Eurasia, introduced, BC, Sask-Nfld.
2. *E. dentata* Michx.; imported from USA, introduced, s. Ont.
3. *E. esula* L., leafy spurge, euphorbe ésule; imported from Eurasia, introduced, BC-PEI.
4. *E. glyptosperma* Engelm.; native, BC-NB.
5. *E. humistrata* Engelm.; imported from e. USA, introduced, Ont-Que.
6. *E. nutans* Lag. (*E. maculata* auct. Amer., non L.); native, Ont-Que.
7. *E. peplus* L., petty spurge, euphorde des jardens; imported from Eurasia, introduced, BC, Sask-Nfld.
8. *E. pulcherrima* Willd. ex Klotzch, poinsettia, poinsettia; imported from Central America and Mexico.
9. *E. serpyllifolia* Pers., thyme-leaved spurge, euphorbe à feuilles de Serpolet; native, BC-Man, introduced, Ont-NB.
10. *E. vermiculata* Raf., hairy-stemmed spurge, euphorbe vermiculée; native, Ont-NS.

Alternaria alternata (A. tenuis auct.*)* (C:112): on 7 BC [1816].
Erysiphe communis (E. polygoni): powdery mildew, blanc: on 9 Ont [1236].
Melampsora sp.: rust, rouille: on 7 BC [1816].
M. euphorbiae (C:112): rust, rouille: 0 I II III on 1 Ont [1236].
M. ?euphorbiae-gerardianae (C:112): rust, rouille: on 7 BC [1012].
M. monticola (C:113): rust, rouille: on 7 BC [334, 1012, 1786].
Thielaviopsis basicola: on 8 Ont [337].
Uromyces euphorbiae (C:113): rust, rouille: 0 I II III on 2 Ont [1236]; on 4 BC [1012]; on 4, 5, 6, 9, 10, *E.* sp. Ont [1236].

U. pisi: rust, rouille: on 1 Ont
[1341]. Probably a specimen of *U.*
striatus.
U. striatus (C:113): rust, rouille: 0 I
on 1 Ont [1236]; on 1, 3 Ont [1237].

EUTREMA R. Br.   CRUCIFERAE

1. *E. edwardsii* R. Br.; native,
Yukon-Labr, n. BC- n. Alta, ne. Man, n.
Que.

Peronospora parasitica (C:113): downy
mildew, mildiou: on 1 Frank [1244].
Puccinia eutremae (C:113): rust,
rouille: on 1 Frank, Keew, Que [1563]
and Frank [1244].

FAGOPYRUM Mill.   POLYGONACEAE

1. *F. esculentum* Moench (*F. sagittatum*
Gilib.), buckwheat, sarrasin; imported
from Asia.

Alternaria sp.: on seeds of 1 Man [1119].
A. alternata: on leaves and seeds of 1
Man [2012].
Ascochyta fagopyri: on 1 Que [335].
Aspergillus candidus: on seeds of 1 in
storage Man [1119].
A. versicolor: on seeds of 1 in storage
Man [1119].
Bipolaris sorokiniana (holomorph
Cochliobolus sativus): on 1 Man
[2012].
Botrytis sp.: on 1 Man [1119].
B. cinerea (holomorph *Botryotinia*
fuckeliana): gray mold, moisissure
grise: on 1 Sask [1139, 1136], Que
[334, 335].
Cephalosporium sp.: on seeds of 1 Man
[1119].
Cladosporium sp.: on seeds of 1 Man
[1119].
Epicoccum sp.: on seeds of 1 Man [1119].
Fusarium sp.: on seeds of 1 Sask [1119,
1136], Man [1119].
Gonatobotrys sp.: on seeds of 1 Man
[1119].
Penicillium sp.: on seeds of 1 in storage
Man [1119].
Peronospora ducometi: downy mildew,
mildiou: on 1 Man [2013].
Rhizoctonia sp.: on 1 Sask [1136].
R. solani (holomorph *Thanatephorus*
cucumeris): on 1 NB [336].
Rhizopus sp.: on seeds of 1 Man [1119].
Streptomyces sp.: on seeds of 1 Man
[1119]. An Actinomycete.
Sclerotinia sclerotiorum: on 1 Sask
[1136].

FAGUS L.   FAGACEAE

1. *F. grandifolia* Ehrh., american beech,
hêtre américain; native, Ont-PEI.
2. *F. sylvatica* L., european beech, hêtre
européan; imported from Europe.
3. Host species not named, i.e., reported
as *Fagus* sp. Most reports were based

on specimens from the native forests
thus they would have been *F.*
grandifolia.

Agrocybe acericola (*Pholiota a.*) (C:115):
on 1 NS [1032].
Albatrellus peckianus (*Leucoporus p.*): on
1 Que [1385]. Apparently lignicolous on
dead roots.
Anisogramma sp. (*Apioporthe* sp.): on 1
Que [F64].
Anthostoma microsporum: on 3 Ont [1340].
Armillaria mella: root rot,
pourridié-agaric: on 1 Que [1376, 1377,
F66], NS [1032]. See *Abies*.
Ascocoryne sarcoides (*Coryne s.*) (C:114):
on 1 Que [1385] on dead trunks and
stumps, NS [1032].
Ascodichaena rugosa (*Dichaena faginea*)
(C:114): on 1 NS [1032].
Ascotremella faginea (C:113): on dead
trunks of 1 Que [1385]; on 3 Ont [1340].
Aspergillus fumigatus: on 3 NB [1610].
Asterosporium sp.: on 3 Ont [1340].
Probably *A. asterospermum*.
A. asterospermum (*A. hoffmanii*) (C:113):
on 1 Ont [1827, 1828, F63, F65], NS
[1032]; on 3 Ont [1340, 1763].
Basidioradulum radula (*Radulum*
orbiculare): on 1 NS [1036]. See
Abies.
Bispora betulina: on 3 NB [1610].
Bisporella citrina (*Helotium c.*): on 1 NB
[1036].
Bjerkandera adusta (*Polyporus a.*)
(C:115): white mottled rot, carie
blanche madrée: on 1 Que [1377], NB, NS
[1032].
B. fumosa (*Polyporus f.*): on 1 Que
[1376]. See *Acer*.
Botryobasidium pruinatum (*Pellicularia p.*)
(C:115): on 1 NS [1032]. Typically on
rotted wood.
Botryosphaeria quercuum (*B. fuliginosa*)
(C:113): on 1 NS [1032].
Botrytis cinerea (holomorph *Botryotinia*
fuckeliana): on 1 NB [1036].
Brachysporium nigrum: on decaying wood of
1 Ont, Que [827].
Calocera cornea (C:113): on 3 NS [1032].
Lignicolous, typically on logs. See
Acer.
Ceraceomyces serpens (*Merulius porinoides*)
(C:114): on 1 NS [1032]; on 3 Ont
[568]. See *Abies*.
Ceratocystis tenella: on 1 Ont [647]; on
3 Ont [1340].
Ceriporia tarda (*Poria semitincta*): on 1
NS [1036]. See *Abies*.
Ceriporiopsis subrufa (*Poria s.*): on
rotten log of 3 Ont [440].
Cerrena unicolor (*Daedalea u.*) (C:114):
on 1 Ont [1341], Que [1391, 1376], NB
[1032, 1741 associated with wood wasp
Tremex columba], NS, PEI [1032]. See
Acer.
Chaetosphaeria pulviscula (*Zignoella p.*)
(C:115): on 1 NS [1032].
Chalara cylindrosperma: on 3 Ont [1165].

Cheirospora botryospora (Thyrsidium b.)
(C:113): on 1 NS [1032].

*Chlorencoelia torta (Chlorociboria
rugipes)*: on 3 Ont [1340].

*Chlorociboria aeruginascens (Chlorosplenium
aeruginosum)* (C:113): on 3 NB, NS
[1032].

Chlorosplenium versiforme: on 3 Ont
[1340] as *Chloroscypha v.*, a
combination which has not been validly
published.

Chondrostereum purpureum (Stereum p.)
(C:115): silver leaf, plomb: on 1 NB
[1032, 1036]. Typically fruits on dead
wood. See *Abies*.

*Ciboria peckiana (Rutstroemia
macrospora)*: on 1 NB [1032].

Cladosporium sp.: on 3 NB [1610].

Clavaria sp.: on 1 NB [1032]. Probably
occurred on rotted wood.

*Climacodon septentrionalis (Hydnum s.,
Steccherinum s.)*: on 1 Que [1376]; on
3 Ont [1341]. Typically fruiting on
wounds on live trees. See *Acer*.

Clitocybe leptoloma (C:113): on 1 NS
[1032]. Lignicolous.

Coccomyces coronatus (C:114): on 1 NS
[1032]; on 3 NS [1609].

C. tumidus: on 3 NS [1609].

Coniophora puteana: on 1 NS [1032]. See
Abies.

Conoplea fusca: on 1 Ont [1340].

C. olivacea (C. sphaerica) (C:114): on 3
Ont, Que [784].

Coriolus hirsutus (Polyporus h.) (C:115):
white spongy rot, carie blanche
spongieuse: on 1 Que [1377], NB, NS
[1032]; on 1, 2 Ont [1341].

C. pubescens (Polyporus p., P. velutinus)
(C:115): on 1 Ont [1341], NB [1032,
1741], NS [1032]; on 3 Ont [1660]. See
Abies.

C. versicolor (Polyporus v.) (C:115):
white spongy rot, carie blanche
spongieuse: on 1 Que [1377], NB, NS,
PEI [1032] and NB [1741]; on 1, 3 Ont
[1341]; on 3 Ont [426].

Crepidotus applanatus (C:114): on 1 NB,
NS [1032]. Lignicolous.

C. stipitatus (C:114): on 1 NS [1032].
Lignicolous.

C. versutus (C:114): on 1 NS [1032].
Lignicolous.

Cryptodiaporthe galericulata (C:114): on
1 NS [1032]; on 3 Ont [1340].

Cudonia lutea (C:114): on 1 NS [1032].

Cylindrobasidium evolvens (Corticium e.):
on 3 NB [1610]. Typically fruiting on
dead branches and stems.

Cylindrocarpon sp.: on 1 NB [1032].

C. faginatum: on 1 NB, NS [1522].

Cyphellopsis anomala (Solenia a.)
(C:115): on 1 Ont [1341]. Lignicolous,
apparently saprophytic.

Daedalea quercina: on 1 NS [1036].
Associated with a brown rot.

Daedaleopsis confragosa (Daedalea c.)
(C:114): on 1 NB, NS [1032]. See
Acer.

Daldinia concentrica: on 1 NS [1032].
Typically on dead wood.

Dasyscyphus tricolor (Lachnella t.): on 3
NS [1032].

Datronia scutellatus (Fomes s.): on 3 Ont
[1341]. See *Alnus*.

*Dendrothele alliacea (Aleurocorticium
a.)*: on 1 Que [965]. Lignicolous.

Diaporthe fagi: on 1 Ont [F65]; on 3 Ont
[1340].

Diatrype sp.: on 3 Ont [1340].

D. albopruinosa: on 1 Ont [1354].

D. bullata: on 1 Ont [1828]; on 3 Ont
[1340].

D. stigma (C:114): on 1 NB [1032].

D. virescens (C:114): on 1 Ont, Que
[1355]; on 3 Ont [1340].

Diatrypella nigro-annulata (C:114): on 1
NS [1032].

*Discula umbrinella (D. quercina,
Gloeosporium fagicola)* (C:114): leaf
spot, anthracnose: on 1 Ont [1340], NB,
NS, PEI [1032, 1340, F64, F65] also NB,
NS [F63, F67, F75] and NB [F70, F76], NS
[F66, F74].

Dothiorella sp.: on 1 Ont [1827]; on 3
Ont [1828].

Eutypa spinosa (C:114): on 1 NS [1032].

Exidia glandulosa (C:114): on 1 NB, NS
[1032]. Lignicolous, typically on dead
wood.

Favolus alveolaris (C:114): on 1 NB, NS,
PEI [1032]; on 2 Ont [1341]. Often on
dead branches still attached to the
tree. See *Acer*.

Femsjonia peziziformis: on 1 NS [1032].
Lignicolous.

Flammulina velutipes: on 1 Que [1385] on
stumps and dead trunks, sometimes on
dead parts of live trees. Associated
with a white rot.

Flaviporus americana (Poria aestivale):
on 1 NS [1036]. Associated with a white
rot.

F. semisupinus (Polyporus s.): (C:115):
on 1 NS [1032]. See *Betula*.

Fomes fomentarius (C:114): white mottled
rot, carie blanche madrée: on 1 Que
[1377, 1376], NB, NS, PEI [1032]; on 3
Ont [1341].

Fomitopsis pinicola (Fomes p.) (C:114):
on 1 NB, NS [1032]. See *Abies*.

Fusarium avenaceum: on 1 NB [1036].

F. oxysporum var. *redolens (F.
redolens)*: on 1 NB [1036].

Fusicoccum sp.: on 1 Ont [1827].

Ganoderma applanatum (Fomes a.) (C:114):
white mottled rot, carie blanche
madrée: on 1 Que [1376, 1377], NB, NS,
PEI [1032]; on 2 BC [1012]; on 1, 3 Ont
[1341].

G. lucidum: on 1 PEI [1032]. See *Abies*.

Gliocladium viride: on 3 NB [1610].

Gliomastix musicola: on 1 Ont [399].

Gloeocystidiellum porosum (Corticium p.):
on 1 NB [1032, 1036]. Lignicolous,
probably saprophytic.

Gloeoporus dichrous (Polyporus d.): on 1
Ont [1063, 1341], NB, NS [1032]. See
Acer.

Gloeosporium sp.: on 1 NB, NS, PEI
[1032]; on 3 Ont [1340].

Gloniopsis curvatum (G. gerardiana):   on 3
Ont [1340].
Gonatorrhodiella highlei: on 1 NB, NS
[1032]. Specimens possibly
misdetermined, because the fungus
labelled *G. highlei* in North America
may not be the same as the original *G.
highlei*. In North America the fungus
seems to be mycoparasitic on *Nectria
coccinea* on *Fagus*.
Graphium sp.: on 3 NB [1610].
Haematostereum rugosum (Stereum r.): on 1
NS [1032]. See *Abies*.
H. sanguinolentum (Stereum s.): on 1 NS
[1032]. See *Abies*.
Hapalopilus nidulans (Polyporus n.)
(C:115): on 1 NS [1032]. See *Abies*.
Hericium sp.: on 1 Ont [1341], NB
[1032]. Lignicolous.
H. americanum (H. coralloides N. Amer.
auct.) (C:114): white spongy rot,
carie blanche spongieuse: on 1 Ont
[1341], Que [1385 on live or dead
trunks], NS [1032].
*H. coralloides (H. ramosum, Hydnum
caput-ursi)* (C:114): white spongy rot,
carie blanche spongieuse: on 1 Que
[1376, 1385], NS [1032].
H. erinaceus: on 1 Ont [1341], NS
[1032]. See comment under *Acer*.
Heterochaetella dubia (C:114): on 1, 3
Ont [1015]. Lignicolous, probably
saprophytic.
Hydnochaete olivaceum (C:114): on 1 NS
[1032]. Typically fruiting on dead
branches and small stems. See *Acer*.
Hymenochaete sp.: on 1 Ont [1341].
H. agglutinans: on 1 NB, NS, PEI [1032].
Apparently a sterile mycelial state of
H. corrugata. See *Alnus*.
H. badioferruginea (C:114): on 1 NS
[1032]. Lignicolous.
H. corrugata (C:114): on 1 Ont [1341],
NB, NS, PEI [1032]. See *Abies*.
H. corticolor (C:114): on 1 NS [1032].
Lignicolous but see *H. badioferruginea*
under *Abies*.
H. spreta: on 1 NS [1032]. See *Acer*.
H. tabacina (C:114): on 1 BC [1012], Que
[1376], NB, NS, PEI [1032]. See *Abies*.
Hymenoscyphus epiphyllus (Helotium e.)
(C:114): on 3 NS [1032].
Hyphoderma setigerum (Peniophora aspera)
(C:115): on 3 NS [1032]. See *Abies*.
Hyphodontia crustosa (Odontia c.)
(C:114): on 1 NS [1032]. See *Abies*.
H. papillosa: on 1 NS [1032]. Associated
with a white rot.
H. setulosa (Steccherinum s.): on 1 NS
[1032]. Lignicolous.
H. spathulata (Odontia s.): on 1 PEI
[1032]. Lignicolous.
Hypochnicium vellereum (Corticium v.)
(C:114): on 3 NS [1032]. See *Acer*.
Hypocrea patella (C:114): on 1 NS [1032].
H. rufa (C:114): on 1 NS [1032]; on 3 Ont
[1340].
Hypoxylon sp.: on 1 NB, NS [1032].
H. cohaerens (C:114): canker, chancre
hypoxylonien: on 1 Ont [F71], Que

[1376, 1377, 1385], NB, NS, PEI [1032];
on 1, 3 Ont [1340].
H. deustum (Ustulina vulgaris) (C:114):
brittle white heart rot, carie blanche
friable: on 1 Ont [1340], Que [1376,
1377], NB, NS [1032].
H. fragiforme (C:114): on 1 NB, NS
[1032]; on 1, 3 Ont [1340].
H. mammatum (H. blakei): on 1 Que [1376,
1377]; on 3 Ont [1340].
H. rubiginosum: on 1 Que [1340]; on 3 NB
[1610].
Hypsizygus ulmarius (Pleurotus u.): on 1
Que [1385]. See *Acer*.
Hysterographium mori (C:114): on 1 NS
[1032].
Hysteropatella minor (C:114): on 1 NS
[1032].
Inonotus cuticularis (Polyporus c.)
(C:115): on 1 Que [F71]. See *Acer*.
I. glomeratus (Polyporus g.) (C:115):
white spongy rot, carie blanche
spongieuse: on 1 Ont [1341], Que [1376,
1377], NS [1032].
I. obliquus (Poria o.): on 1 NB [1032].
Produces a sterile, clinker-like growth
on trunks of live trees. See *Acer*.
Irpex lacteus (Polyporus tulipiferae)
(C:115): on 1 Ont [1341], Que [1390],
NB, NS [1032]; on 3 Ont [1341, 1660].
Typically fruiting on dead branches.
See *Acer*.
*Junghuhnia nitida (Poria attenuata, P.
eupora)*: on 1 NS [1032, 1036]. See
Acer.
Kavinia alboviride (Mycoacia a.): on 3
Ont [1827]. Typically on rotted logs
and branches. Associated with a white
rot.
*Laeticorticium roseocarneum
(Dendrocorticium r., Stereum r.)*
(C:115): on 1 NS [1032]; on 3 NS
[930]. Lignicolous.
Lentaria micheneri: on dead trees of 1
Que [1385].
Lentinellus cochleatus: on 1 Que [1385],
on stumps or roots in NS [1032].
L. ursinus: on 1 Que [1385]. See *Abies*.
Lenzites sp.: on 1 Que [1376].
Lignicolous.
L. betulina (C:114): on 1 Ont [1341], NB,
NS [1032]. See *Acer*.
Libertella sp. (C:114): on 2, 3 Ont
[1340].
L. faginea (C:114): on 1 Ont [337], NB
[1032 misspelled as *L. faginata*]; on
3 Ont [1828, 1827].
Lopadostoma turgidum (C:114): on 1 Ont
[F62]; on 3 Ont [1340].
Lophiotricha viridicoma: on 3 Ont [1340].
Lycoperdon pyriforme: on 1 NS [1036].
Probably saprophytic on well-rotted wood.
L. subincarnatum (C:114): on 1 NS
[1032]. Probably saprophytic on
well-rotted wood.
Marasmius rotula: on 1 PEI [1032].
Lignicolous, probably saprophytic.
Menispora glauco-nigra: on 3 Ont, Que
[826].
M. tortuosa: on 3 Que [826].

Meruliopsis hirtellus: on 1 Ont [568].
    Apparently saprophytic. See *Betula*.
Merulius tremellosus (C:114): on 1 Ont
    [1341], Que [568], NS [1032]; on 3 NB
    [1610]. Occasionally associated with
    decay in live trees. See *Abies*.
Microsphaera penicillata (C:114): powdery
    mildew, blanc: on 1 NS [1032, F65].
Mollisia cinerea (C:114): on 3 NS [1032].
Mycena capillaris: on dead leaves of 1
    Ont [1385].
M. hemisphaerica (M. atroumbonata)
    (C:114): on 1 NS [1032].
M. leaiana: on 1 NB, NS [1032].
    Typically on logs.
M. roseipallens: on debris of 1 Ont
    [1385].
M. tenuiceps: on trunks of 1 Que [1385].
Mycoacia fuscoatra (Odontia f.): on 1 NS
    [1032]. Lignicolous.
M. uda: on 1 Ont [1341]. Lignicolous.
Mycosphaerella punctiformis (C:114): on 1
    Ont [F63], NS [1032]; on 1, 3 Ont [1340].
Myxosporium sp.: on 3 Ont [1340].
Nectria sp.: on 1 Que [1376, 1377], NB,
    NS [1032]; on 3 Nfld [1660].
N. cinnabarina (anamorph *Tubercularia
    vulgaris*): on 3 Ont [1340].
N. coccinea (C:114): on 1 NB, NS, PEI
    [1032].
N. coccinea var. *faginata* (C:114):
    beech canker, maladie corticale du
    hêtre: on 1 Que [F67, F71, F72, F73,
    F74, F75, F76], NB, NS, PEI [1032, F64,
    F65, F67, F69] also NB, NS [F61, F66]
    and NB [F68]; on 3 NB [1340].
N. episphaeria (C:114): on 1 Ont [F62].
    Usually mycoparasitic.
N. galligena (C:114): canker, chancre
    nectrien: on 1 Que [F65, F66], NB, NS
    [1032].
Neobulgaria pura (Ascotremella turbinata)
    (C:113): on dead trunks of 1 Que
    [1385]; on 3 NS [569].
Odontia sp.: on 3 NS [1032]. Lignicolous.
Omphalina epichysium (Omphalia e.)
    (C:115): on 1 NS [1032].
Orbilia auricolor (O. inflatula) (C:115):
    on 1 NS [1032].
Oxyporus populinus (Fomes connatus): on 1
    NS, PEI [1032]; on 3 Ont [1341]. See
    Acer.
Panellus serotinus (Pleurotus s.)
    (C:115): on 1 Que [1376], NB, NS
    [1032]. See *Acer*.
P. stipticus (Panus s.) (C:115): on 1 NB
    [1032, 1036], NS [1032]. See *Alnus*.
Panus conchatus (P. torulosus) (C:115):
    on 1 Que [1385], NS [1032]. Lignicolous.
P. rudis: on 1 NS [1032]. Lignicolous.
Paxillus involutus (C:115): on 1 NS
    [1032]. Mycorrhizal, probably fruiting
    on well-rotted wood.
Penicillium sp.: on 3 NB [1610].
Peniophora sp.: on 1 NB, PEI [1032].
P. cinerea (C:115): on 1 Ont [1341], NB
    [1032]. See *Abies*.
P. incarnata: on 1 NS [1032].
    Lignicolous.
P. nuda: on 1 NS [1032]. Lignicolous.

Phanerochaete laevis (Peniophora affinis)
    (C:115): on 1 NS [1032]. Lignicolous.
P. flavido-alba (Peniophora f.): on 1 NS
    [1032]. Lignicolous.
P. sanguinea (Peniophora s.) (C:115): on
    1 NS [1032]. See *Abies*.
P. sordida (Peniophora cremea) (C:115):
    on 1 NB [1032]. See *Abies*.
P. velutina (Peniophora v.) (C:115): on 1
    NS [1032]. Lignicolous.
Phellinus ferreus (Poria f.): on 1 NS
    [1032]. See *Acer*.
P. ferruginosus (Poria f.) (C:115): on 1
    Que [1376], NS [1032]. See *Abies*.
P. igniarius (Fomes i.) (C:114): white
    trunk rot, carie blanche du tronc: on 1
    Que [1376, 1377, 1391], NB, NS [1032];
    on 1, 3 Ont [1341].
P. laevigatus (Fomes igniarius var. *l.,
    Poria l.)*: on 1 Que [1376, 1377], NS,
    PEI [1032]. See *Acer*.
P. pini (Fomes p.) (C:114): on 1 NS
    [1032]. Rare on angiosperms. See
    Abies.
P. viticola (Fomes v.): on 1 NS [1032].
    See *Abies*.
Phialocephala bactrospora: on 1 NB [1032].
Phialophora sp.: on 3 NB [1610].
P. americana: on 3 NB [1610].
P. lagerbergii: on 3 NB [1610].
P. melinii: on 1 Ont, NB [279]; on 3 NB
    [1610].
Phlebia radiata (C:115): on 1 Que [568,
    1341], NB [568, 1032], NS [1032]. See
    Abies.
Phleogena faginea: on 1 Que [1385].
    Apparently saprophytic on dead wood.
Pholiota sp.: on 1 NS [1032].
    Lignicolous.
P. adiposa: on 1 Que [1376]. Probably a
    specimen of *P. aurivella*.
P. aurivella (C:115): on 1 Que [949], NS
    [1032]. See *Abies*.
P. malicola: on 1 NB [1032]. Lignicolous.
P. squarrosoides (C:115): on 1 Que
    [1376], NS [1032]. Lignicolous.
Phyllactinia guttata (P. corylea):
    powdery mildew, blanc: on 1 Ont [1257],
    Que [1377].
Phyllotopsis nidulans (C:115): on 1 NS
    [1032]. Lignicolous.
Pleurothecium recurvatum: on 1 Ont [587].
Pleurotus sp.: on 1 Que [1376].
    Lignicolous.
P. lignatilis (C:115): on 1 NS [1032].
    Associated with a brown heart rot of
    mature trees. Now in *Ossicaulis*.
P. ostreatus (C:115): on 1 NS [1032].
    Sometimes fruiting on wounds on live
    trees. See *Abies*.
P. sapidus: on 1 NB [1032]. Sometimes
    fruiting on wounds on live trees. See
    Aesculus.
P. strigosus (Panus s.): on 1 NS [1032].
    See *Abies*.
Plicatura crispa (P. faginea, Trogia c.)
    (C:115): on 1 Ont [1341], NB, NS [1032,
    1036]. Lignicolous, apparently
    saprophytic.

Pluteus leonius: on rotten wood of 1 Que [1385].
Polyporus sp.: on 1 NB [1036].
P. brumalis (C:115): on 1 NS [1032]. Typically fruiting on fallen branches and logs.
P. lentus (P. fagicola, P. squamosus var. *fagicola)*: on 1 Que [1385] on old trunks, NS [1032]. Often attached to buried logs or dead roots.
P. varius (P. elegans) (C:115): on 1 NS, PEI [1032]. Typically fruiting on fallen branches and logs. See *Abies*.
Poria sp.: on 1 NB, NS [1032].
P. amylohypha (P. elongata): on 1 NB [1036]. See *Betula*.
Propolis versicolor (P. faginea) (C:115): on 1 NS [1032].
Prosthemiella formosa (C:115): on 1 NS [1032].
Pseudospiropes simplex: on 1 Que [817].
Psilocybe silvatica: on 1 Ont, Que [1385]. Typically on rotted wood.
Pycnoporus cinnabarinus (Polyporus c.) (C:115): on 1 NB, NS [1032]; on 2 Ont [1341]. See *Acer*.
Quaternaria quaternata (C:115): on 1 Ont [F63]; on 3 Ont [1340].
Rosellinia conglobata var. *microtricha* (C:115): on 1 NS [1032].
Sarcoscypha coccinea (Plectania c.): on 3 Ont [1340]. Typically fruiting on wood buried under leaves.
Schizophyllum commune (C:115): on 1 BC [1012], Ont [1341], NB, NS [1032]. See *Abies*.
Scopuloides hydnoides (Odontia h.): on 1 NS [1032]. Associated with a white rot.
Solenia canadensis: on 1 NB [1032]; on 3 Ont [1827]. Lignicolous, probably saprophytic.
Spadicoides grovei: on decaying wood of 1 Que [793].
S. obovata: on decaying wood of 1 Que [792].
Stachylidium bicolor: on 1 NB [1032].
Steccherinum sp.: on 1 NB, NS [1032]. Lignicolous.
S. ochraceum (C:115): on 1 NS [1032]; on 3 Ont [1827]. Typically fruiting on logs and fallen branches. See *Acer*.
Stereum complicatum: on 1 Ont [1341]. Lignicolous.
S. hirsutum (C:115): on 1 Ont [1341], NS [1032]. See *Abies*.
S. ostrea (C:115). on 1 NB, NS [1032]. See *Abies*.
Streptomyces sp.: on wood chips of 3 NB [1610]. An Actinomycete.
Strickeria vilis (C:115): on 1 NS [1032].
Stromatoscypha fimbriata: on 1 NB, NS [1032]. Lignicolous, typically on rotten logs.
Tomentella sp.: on 1 Ont [1341].
T. cinerascens: on 1 Ont [1341].
T. subfusca (C:115): on 1 NS [1032].
T. sublilacina: on 1 Ont [1341, 1827, 1828].
Torula ligniperda: on 1 Que [1377].

Trametes sp.: on 1 Ont [1341]. Lignicolous.
T. cervina (Polyporus biformis N. Am. auct.*)*: on 1 Ont [1341], Que [1389]. See *Abies*.
Trechispora vaga (Corticium sulphureum): on 1 NS [1032]; on 3 NS [970]. See *Abies*.
Trematosphaeria melina: on 1 NB, NS [1032].
Tremella lutescens (C:115): on 3 NS [1032]. Lignicolous, probably saprophytic.
Trichaptum biformis (Polyporus pargamenus) (C:115): on 1 NB, NS, PEI [1032]. See *Acer*.
Trichocladium canadense (C:115): on 3 NB [1610].
Trichoderma viride: on 3 NB [1610]. See *Abies*.
Tubercularia sp.: on 1 NB [1032].
T. vulgaris (holomorph *Nectria cinnabarina*): on 1 Ont [1340].
Tyromyces caesius (Polyporus c.) (C:115): on 1 Ont [1341], NS [1032]. See *Abies*.
T. chioneus (Polyporus albellus) (C:115): on 1 NB, NS [1032]. See *Abies*.
Valsaria exasperans: on 1 Ont [1340].
Valsella sp.: on 3 Ont [1340].
Vararia investiens: on 1 Que [557]. See *Abies*.
Vesiculomyces citrinus (Gloeocystidiellum c.): on 1 NS [1032]. See *Abies*.
Volvariella bombycina: on 1 Que [1385]. Typically fruiting on wounds on live trees.
Xylaria castorea: on 1 Ont [1340], NS [1032].

FATSIA Decne. & Planch.   ARALIACEAE

1. *F. japonica* (Thunb.) Decne. & Planch. (*Aralia sieboldii* Hort. ex K. Koch); imported from Japan. Evergreen shrub, grown indoors for bold foliage effect.

Verticillium albo-atrum: on 1 BC [1816].

FESTUCA L.   GRAMINEAE

1. *F. altaica* Trin. (*F. scabrella* Torr.), rough fescue, fétuque scabre; native, Yukon-Mack, BC-Man.
2. *F. arundinacea* Schreb.; introduced, BC, Man-Que, NS.
3. *F. brachyphylla* Schultes; native, Yukon-Labr, BC, Man-Que, Nfld.
4. *F. idahoensis* Elmer; native, BC-Sask.
5. *F. occidentalis* Hook., western fescue; native, BC-Sask, Ont.
6. *F. ovina* L., sheep's fescue, fétuque ovine; native, Yukon-Frank, BC-Nfld.
7. *F. pratensis* Huds. (*F. elatior* L.), meadow fescue, fétuque élevée; imported from Eurasia, natzd., Yukon, BC-Alta, Man-Nfld.
8. *F. rubra* L., red fescue, fétuque rouge; native, Yukon-Labr, BC-Nfld. 8a. *F. r.* var. *commutata* Gaud.-Beaup.,

chewing fescue; native, Que-NS, Nfld.
8b. *F. r.* var. *rubra*; native,
Yukon-Keew, BC-Nfld, Labr.
9. *F. saximontana* Rydb.; native,
Yukon-Mack, BC-Que, Nfld, Labr.
10. *F. viridula* Vasey, green fescue;
native, BC.
11. Host species not named, i.e., reported
as *Festuca* sp.

Acremonium boreale: on 8 BC, Alta, Sask,
Man, Ont [1707].
Bipolaris sorokiniana (holomorph
Cochliobolus sativus) (C:116): root
rot, piétin: on 2, 6, 7, 8 Alta [120]
by inoculation.
Cercosporidium graminis (Passalora g.)
(C:116): brown stripe, strie brune: on
8 Alta [119, 121, 123, 133], NB [333,
1195].
Claviceps purpurea (C:116): ergot,
ergot: on 7 BC [1816], Alta [333]; on 8
NB [333, 1195].
Colletotrichum graminicola (C:116):
anthracnose, anthracnose: on 8 Sask
[336, 1185].
Dendryphion nanum: on 8 Alta [819].
Didymella festucae (anamorph *Phloeospora
idahoensis*): on 1 Yukon, Mack, on 1,
4, 6, 8 Alta, Sask, on 7 Alta, all from
[1714]; on 8, 9 Alta [1706]; on 8 BC
[1706], Alta [133].
Dilophospora alopecuri (C:116): twist,
torsion: on 7 NB [333, 1195].
Drechslera dictyoides (C:116): net
blotch, tache réticulée: on 2 Sask
[338]; on 7 BC [1816]; on 11 NB [333,
1195].
D. dictyoides f. sp. *perenne*: on 2 Ont,
on 7 Alta, Man, Ont and on 11 Alta
[1611].
Fusarium nivale (C:116): on 8 BC [1849],
Sask [1705]. See *Agrotis*.
F. poae (C:116): on 8 BC [1816]; on 11
BC, NB [333].
Helminthosporium spp.: on 8 BC [1849].
Hendersonia culmicola (C:116): on 8 NB
[333, 1195]. See comment under *Abies*.
Hysteropezizella lyngei: on 3 Frank [664].
Laetisaria fuciforme (Corticium f.): red
thread disease: on 8 BC [1849], Alta,
Sask [338]. Typically found on turf on
soils of low fertility, especially low
in nitrogen.
Phaeoseptoria festucae (C:116): on 8 NB
[333, 1195].
Phloeospora idahoensis (holomorph
Didymella festucae) (C:117): stem
eye spot, tache ocellée: on 1, 4 BC,
Alta, Sask [1703, 1704]; on 1, 2, 7, 8a,
8b Alta [1708]; on 8 Alta [119, 121,
123, 1595, 1715]; on 8b BC [1715], Alta
[1653]; on 11 BC [1703].
Phyllachora sp.: on 8 BC [1012].
P. silvatica (C:117): tar spot, rayue
goudronneuse: on 8 BC [1816].
*Pseudoseptoria stomaticola (Selenophoma
donacis* var. *s.)* (C:117): on 3 Yukon
[1731].

Puccinia coronata (C:117): crown rust,
rouille couronnée: II III on 7 Ont
[1236].
P. crandallii (C:117): leaf rust, rouille
des feuilles: on 8 BC [1012]; on 11
Alta [644] "associated but not rusted".
P. graminis (C:117): stem rust, rouille
de la tige: II III on 7 Ont [1236].
P. poae-nemoralis (C:117): rust,
rouille: on 7 BC [1816].
Ramularia pusilla (Ovularia p.) (C:117):
leaf spot, tache des feuilles: on 3
Yukon [1731]; on 8 Que [1729]; on 11 NB
[333, 1195].
Sclerotinia borealis (C:117): snow mold,
moisissure nivéale: on 7 BC [1816]; on
8 Sask [1705].
Spermospora subulata (C:117): on 8 NB
[333, 1195].
Stagonospora simplicior: on 8 Que [282].
Stemphylium botryosum (holomorph
Pleospora herbarum) (C:116): from
seed of 8 Ont [667].
Typhula sp. (C:117): snow mold,
moisissure nivéale: on 8 Man [1367].
T. incarnata (T. itoana) (C:117): on 8
Sask [338].

FICUS L.   MORACEAE

1. *F. carica* L., common fig, figuier;
imported from Mediterranean region.
2. *F. elastica* Roxb. ex Hornem., rubber
plant, caout chouc des jardins; imported
from India and Malaysia. 2a. *F.
elastica* cv. 'decora' (*F. decora*
Hort.).

Armillaria mellea: root rot,
pourridié-agaric: on 2 Alta [334].
Botrytis cinerea (holomorph *Botryotinia
fuckeliana*) (C:117): gray mold,
moisissure grise: on 1 BC [1816].
*Colletotrichum gloeosporioides
(Gloeosporium cingulatum)* (holomorph
Glomerella cingulata) (C:117):
anthracnose, anthracnose: on 2 Que
[333].
Glomerella cingulata: anthracnose,
anthracnose: on 2 BC [1816], Alta [337,
339], Ont [334], Que [334, 335]; on 2a
BC, Alta [334], the plants originally
from BC developed symptoms after arrival
in Alta.

FILIPENDULA Mill.   ROSACEAE

1. *F. rubra* (J. Hill) B.L. Robinson,
queen-of-the-prairie; imported from e.
USA, escaped from cultivation, Ont-Que,
NS, Nfld.
2. *F. ulmaria* (L.) Maxim.,
queen-of-the-meadow; imported from
Eurasia, escaped from cultivation,
Ont-Nfld.

Sphaerotheca macularis (C:117): powdery
mildew, blanc: on 1 Ont [1254]; on 1, 2
Ont [1257].

FORSYTHIA Vahl   OLEACEAE

Deciduous shrub; e. Asia and se. Europe;
cultivated, showy yellow flowers in
Spring.

Phomopsis sp.: on *F.* sp. NS [338].

FRAGARIA L.   ROSACEAE

1. *F.* x *ananassa* Duchesne (*F.
    chiloensis* var. *a.* Bailey). *F.
    chiloensis* x *F. virginiana*,
    cultivated strawberry, fraisier de
    jardin.
2. *F. chiloensis* Duchesne, beach
    strawberry; native, BC.
3. *F. virginiana* Duchesne, wild
    strawberry, fraisier des champs; native,
    Yukon-Mack, BC-Nfld, Labr.
4. Host species not named, i.e., reported
    as *Fragaria* sp.

Absidia sp.: on 1 NS [1027].
Acremoniella atra: on 4 NS [596].
Actinomucor sp.: on 1 in storage NS [980].
A. elegans (A. repens): on 4 NS [596].
Alternaria sp.: on 1 NS [980 in cold
    storage, 1027]; on 4 NS [596, 600].
Armillaria mellea (C:118): crown and root
    rot, pourridié-agaric: on 1 BC [335,
    1816].
Arthrobotrys sp.: on 4 NS [596].
Aspergillus sp.: on 4 NS [596].
Botrytis sp.: on 1 NB [611].
B. cinerea (holomorph *Botryotinia
    fuckeliana*) (C:118): gray mold,
    moisissure grise: on 1 BC [337, 486,
    488, 1594, 1816, 1817, 1818], Alta [259,
    333, 336], Man [259], Ont [259, 333,
    334], Que [111, 334, 389, 390], NB [333,
    337], NS [594, 980, 1027, 1595], PEI
    [333, 335, 336], Nfld [336] and NB, NS
    [259, 334, 335, 336, 338, 1594]; on 4
    Que [333], NS [592, 596, 600].
Calcarisporium arbuscula: on 4 NS [596].
Cephalosporium sp.: on 4 NS [596].
Chaetomium sp.: on 1 NS [1027]; on 4 NS
    [596].
C. anguipilium: on 4 NS [596].
    Originally described from dung.
C. aureum: on 4 NS [596].
C. cochliodes: on 4 NS [596].
C. funicolum: on 4 NS [596].
C. globosum (C:118): on 4 NS [596].
C. ochraceum: on 4 NS [596].
Cladosporium sp.: on 4 NS [596].
Coniothyrium sp.: on 4 NS [596].
C. fuckelii: on 4 NS [596].
Cylindrocarpon sp.: on 1 NB [611], NS
    [980, 993 in storage]; on 4 NS [596].
C. destructans (C. radicicola): on 1 in
    cold storage NS [981]; on 4 Ont [22].
Diachea leucopodia (C:118): slime mold:
    on 1 BC [333, 334, 1816]. Probably
    saprophytic.
Diplocarpon earliana (C:118): leaf
    scorch, tache pourpre: on 1 BC [1816],
    Ont [395, 396], Que [334, 335, 338], NB
    [337, 338, 339, 1594], NS [336, 1027],
    PEI [336]; on 4 Ont, Que [159], NS [592].

Discohainesia oenotherae (Pezizella o.)
    (C:118): tan rot, pourriture bistre:
    on 4 NS [600].
Doratomyces nanus: on 4 NS [596].
Epicoccum purpurascens (E. nigrum): on 4
    NS [596].
Fusarium sp. (C:118): on 1 NB [611], NS
    [980, 993, 981 in cold storage, 1027];
    on 4 Ont [96], NS [596].
Gliocladium sp.: on 1 in cold storage NS
    [980].
G. roseum: on 1 NS [1027]; on 4 NS [608,
    596].
Gliomastix sp.: on 4 NS [596].
G. murorum: on 4 NS [399].
Gloeosporium sp.: on 1 NB [611], NS [336,
    980, 1027]; on 4 NS [596].
Gnomonia comari (G. fruticola) (anamorph
    Zythia fragariae) (C:118): leaf
    blotch or petiole blight, brûlure du
    pêtiole: on 1 BC [1816], NS [333, 334,
    335, 336, 337, 338, 339, 1027, 1594]; on
    4 Que [79], NS [79, 592, 596, 598, 600],
    PEI [598].
Gonatobotrys simplex: on 4 NS [596].
Graphium sp.: on 4 NS [596].
Harknessia sp.: on 1 NS [980 in cold
    storage, 1027]; on 4 NS [596].
Hormodendrum sp.: on 1 NS [1027].
Macrophomina phaseolina (M. phaseoli): on
    4 NS [596].
Marssonina canadensis (C:118): on 3 BC
    [159, 160, 1012]; on 4 BC [1816].
Mucor sp.: on 4 NS [596].
Mycosphaerella fragariae (C:118): leaf
    spot, tache commune: on 1 Mack [334],
    BC [1816], Alta [334, 338], Sask [339],
    Ont [333, 335], Que [1594], NB, NS [333,
    334, 335, 336, 338, 339, 1594], PEI,
    Nfld [333, 334, 335, 336], and NS
    [1027], Nfld [337]; on 3 NB [1032]; on 4
    BC [1816], Alta [644], Ont, Que, NB
    [333], NS [592].
Myrothecium roridum: on 4 NS [596].
Penicillium sp.: on 1 NS [980, 1027]; on
    4 NS [596, 600].
Pestalotropsis sydowiana (Pestalotia s.):
    on 4 NS [596].
Peyronellaea sp.: on 4 NS [596].
Peziza ostracoderma: on 1 NS [1027].
Phomopsis obscurans (Dendrophoma o.)
    (C:118): leaf blight, brûlure des
    feuilles: on 1 Ont [333], Que [334],
    NB, NS [335, 336] and NS [333, 334,
    1027, 1594], PEI [336]; on 4 NS [592,
    596].
Phyllosticta sp.: on 1 Que [334].
P. fragaricola (C:118): leaf spot, tache
    foliaire: on 1 Que [335].
Phytophthora fragariae (C:118): red
    stele, stèle rouge: on 1 BC [334, 337,
    338, 339, 367, 368, 1594, 1816, 1817,
    1818], Alta [333], NB [1594], NS [333,
    334, 338, 339, 604, 1594, 1595], PEI
    [337, 1594]; on 2 BC [369]; on 4 BC
    [333]. The spread of red stele in NS
    during the late 1970s led to breeding
    programs which placed greater emphasis
    on field resistance.
Pyrenochaeta sp.: on 4 NS [596].

Ramularia tulasnei: on 4 BC [1012].
Rhizoctonia sp.: on 1 NB [334, 337, 611],
  NS [993].
R. *solani* (holomorph *Thanatephorus*
  cucumeris) (C:118): crown rot,
  rhizoctone commune: on 1 BC [337, 1816,
  1853], Que [336]; on 4 Que [333].
Rhizophydium graminis: on 3 Ont [59].
Rhizopus sp.: on 1 NS [980, 1027]; on 4
  Que [333].
R. *nigricans*: rhizopus rot, leak,
  moisissure chevelue: on 1 NB [336]; on
  4 NS [596, 600].
R. *stolonifer*: on 1 BC [1816].
Septoria sp.: on 1 NS [1027].
S. *aciculosa* (C:118): septoria leaf spot,
  tache septorienne: on 1 Que [334], NS
  [334, 335, 336, 337]; on 4 NS [333].
Sordaria sp.: on 1 NS [1027].
S. *fimicola*: on 4 NS [596].
S. *humana*: on 4 Ont [1340].
Sphaerotheca macularis (S. humuli)
  (C:118): powdery mildew, blanc: on 1
  BC [334, 339, 488, 489, 1594, 1816,
  1817, 1818], Alta [338], Ont [334,
  1257], Que [316, 320, 334, 337], NB
  [1594], NS, PEI [333, 334, 336, 338] and
  NS [335, 339, 603]; on 2 Ont [851]; on 4
  BC [333, 365], Que [333].
Spicara sp.: on 4 NS [596].
Sporotrichum sp.: on 1 in cold storage NS
  [981].
Stachybotrys atra: on 4 NS [596].
Trichocladium asperum: on 4 NS [596].
T. *opacum*: on 4 NS [596].
Trichoderma sp.: on 1 in cold storage NS
  [980]; on 4 NS [596].
Truncatella angusta (Pestalotia
  truncata): on 4 NS [596].
Typhula sp.: on plants of 1 in storage NS
  [981, 982, 1594].
Ulocladium sp.: on 4 NS [596].
Varicosporium elodeae: on 4 NS [596].
Verticillium sp. (C:118): wilt,
  flétrissure verticillienne: on 1 Ont
  [97, 161, 162], NS [1027]; on 4 Ont [96].
V. *albo-atrum*: wilt, flétrissure
  verticillienne: on 1 BC [1816], Ont
  [98], Que [335], NB, NS [335, 336] and
  NS [333, 334]; on 4 BC [1852], Ont [38].
V. *dahliae*: wilt, flétrissure
  verticillienne: on 1 BC [333, 334, 335,
  1816, 1978], Ont [337, 1296], Que [339,
  676], NB [338], NS [337, 338, 339, 997,
  1594]; on 2 Ont [1087]; on 4 BC, Ont, NB
  [38], NS [596, 608] and Ont [1295].
V. *nigrescens*: on 1 Que [392]; on 4 Que
  [38, 393].
Volutella sp.: on 4 NS [596].
Zygorhynchus sp.: on 4 NS [596].
Zythia fragariae (holomorph *Gnomonia*
  comari): on 4 NS [592].

FRAXINUS L.   OLEACEAE

1. F. *americana* L., white ash, frêne
   blanc; native, Ont-PEI.
2. F. *excelsior* L., european ash, grand
   frêne; imported from Europe and Asia
   Minor.

3. F. *latifolia* Benth., Oregon ash, frêne
   de l'Orégon; imported from w. USA.
4. F. *nigra* Marsh., black ash, frêne
   noire; native, Man-Nfld.
5. F. *pennsylvanica* Marsh., red ash,
   frêne rouge; native, Alta-PEI. 5a. F.
   p. var. *subintegerrima* (Vahl) Fern.,
   green ash, frêne vert; native, Sask-Que,
   planted, NS-PEI.
6. F. *velutina* Torr., velvet ash;
   imported from s. USA. 6a. F. *v.* var.
   glabra Reid, Arizona ash.
7. Host species not named, i.e., reported
   as *Fraxinus* sp.

Apiognomonia errabunda (anamorph *Discula*
  umbrinella): on 5a Sask [F66].
  Typically on *Fagus*.
Armillaria mellea: root rot,
  pourridié-agaric: on 1 Ont [F63]; on 4
  Que [F68]; on 7 Que [1377]. See *Abies*.
Basidiodendron deminuta: on 7 Ont
  [1016]. Lignicolous, probably
  saprophytic.
Bispora betulina: on 5 Man [1760].
Bjerkandera adusta (Polyporus a.): on 1
  Ont [1341]. See *Abies*.
Brachysporium nigrum: on 7 Que [827].
Ceratocystis crassivaginata: on 4 Ont
  [1190].
C. *pallidobrunnea*: on 4 Man [1190].
C. *spinulosa*: on 4 Man [1190].
Cerrena unicolor (Daedalea u.): on 4 NB
  [1032]. See *Acer*.
Chalara breviclavata: on ?7 Que [1165].
Ciboria acericola: on ?7 Ont [662].
Cladosporium sp.: on 7 NB [1036].
C. *acutum*: on fallen leaves of 7 Ont
  [440].
C. *macrocarpum*: on 5a Man [1760].
Coriolus versicolor (Polyporus v.): on 1,
  7 NB, NS [1032]; on 7 Ont [1341]. See
  Abies.
Corniculariella spina (Sphaerographium
  fraxini) (holomorph *Durandiella*
  fraxini) (C:120): on 1 Ont [1340,
  1763], Que [1763] on dead branches; on 7
  Ont [1828], NB [1032].
Cylindrosporium fraxini (holomorph
  Mycosphaerella effigurata): on 5a Man
  [F67].
Cylindrotrichum oligospermum: on 7 Man
  [1760].
Cystostereum murraii: on 7 NB [1032].
  See *Acer*.
Cytospora sp.: on 1 Ont [1340]; on 3 BC
  [1012]; on 1, 7 NB [1032].
C. *pruinosa (Cytophoma p., Dendrophoma p.)*
  (holomorph *Valsa cypri*) (C:120): on
  1, 4, 5, 7 Ont [1340]; on 1 NB [1032];
  on 7 Ont [1827].
Daedaleopsis confragosa (Daedalea c.): on
  1, 4 Ont [1341]. See *Acer*.
Dendrothele alliacea (Aleurocorticium
  a.): on 7 Ont, Que [965]. Lignicolous.
D. *griseo-cana (Aleurocorticium g.)*: on
  7 Ont [965]. Lignicolous.
D. *microspora (Aleurocorticium m.)*: on 7
  Ont [965]. Lignicolous.

Dendryphiopsis atra (holomorph
   Microthelia incrustans): on 5, 7 Man
   [1760].
Dermea tulasnei (C:120): on 1, 4 Ont
   [1340]; on 4 Que [F67].
Diatrype albopruinosa: on 7 Ont [1340].
D. stigma: on 7 Ont [1340].
Diplodia fraxini: on 5a Sask, Man [F67].
Discula umbrinella (*D. quercina,*
   Gloeosporium aridum, G. irregulare)
   (holomorph *Apiognomonia errubunda*)
   (C:120): anthracnose, anthracnose: on
   1 Ont [F75, F76], Que [F61], NB, NS
   [1032, 1340, F61, F62, F63, F64, F66,
   F75] and NS [335, 339, F67, F76]; on 4
   NB, NS [1032, F64] and NS [1340]; on 5
   Que [338, F65], PEI [1032]; on 6a BC
   [1416]; on 7 NB, NS [337, 1032, F65] and
   NS [338, 1594].
Dothiorella fraxinicola: on 7 Ont [1340].
Durandiella fraxini (anamorph
   Corniculariella spina) (C:120): on 1
   Ont [1340], Que [F70]; on 1, 5 NS [1032].
Eutypella stellulata: on 7 Ont [1340].
Fusarium oxysporum: on 1 Ont [F73].
Ganoderma applanatum (*Fomes a.*): white
   mottled rot, carie blanche madrée: on
   1, 4 Ont [1341]; on 2 NS [1032]; on 7
   Man [F67], Que [1377]. Sometimes
   causing decay in live trees.
Gloeoporus dichrous (*Polyporus d.*): on 4
   Ont [1341]. See *Acer*.
Gloeosporium sp.: on 1 Ont [1340], NS
   [1032].
Gloiodon strigosus: on 4 Que [F73].
   Associated with a white rot.
Haplographium delicatum (holomorph
   Hyaloscypha dematiicola): on 7 Sask,
   Man [1760].
Hericium coralloides (*H. ramosum*): on
   dead trunks of 7 Que [1385].
Heterobasidion annosum (*Fomes a.*): on 1
   Ont [1419]. Usually associated with
   root rot of live trees. See *Abies*.
Hymenochaete corrugata: on 4 Ont [1341].
   See *Abies*.
H. tabacina: on 4 Ont [1341]. See
   Abies.
Hypoxylon rubiginosum: on 4 Ont [1340].
H. vogesiacium (C:120): on 7 Ont [1340,
   F63].
Hypsizygus ulmarius (*Pleurotus u.*): on 1
   Ont [1341]. See *Acer*.
Hysterographium fraxini (C:120): on 1 Ont
   [1340, F60, F65]; on 7 Sask, Man [F66].
Irpex lacteus (*Polyporus tulipiferae*): on
   4 NB [1032]; on 7 Ont [1341]. See
   Acer.
Ischnoderma resinosum (*Polyporus r.*)
   (C:120): on 7 NB [1032]. See *Abies*.
Junghuhnia nitida (*Poria eupora*): on 7
   Ont [1004]. See *Acer*.
Laeticorticium violaceum (*Dendrocorticium*
   v.): on 4, 7 Ont [930]. Lignicolous.
Laetiporus sulphureus (*Polyporus s.*): on
   7 Ont [1341]. Often fruiting on wounds
   on live trees. See *Abies*.
Marssonina sp.: on 7 Que [F60].
Melanconium sp.: on 1 Ont [1340].

Merulius tremellosus: on old
   beaver-chewed stump of 4 Ont [568]. See
   Abies.
Mycosphaerella sp.: on 5a Man [F64]; on 7
   Nfld [1661].
M. effigurata (anamorphs *Asteromella*
   fraxini and *Cylindrosporium fraxini*)
   (C:120): leaf spot, tache des
   feuilles: on 1, 7 NS [1032]; on 4, 7 NB
   [1032].
M. fraxinicola: on 1 Que [1377], as *M.*
   fraxini.
M. punctiformis: on 7 Ont [1340].
Panellus stipticus (*Panus s.*): on 1 NS
   [1032]. See *Alnus*.
Phellinus conchatus (*Fomes c.*) (C:120):
   white spongy rot, carie blanche
   spongieuse: on 4 Ont [1660], Que [1376,
   1377]; on 4, 7 Ont [1341], NB [1032].
P. igniarius (*Fomes i.*) (C:120): white
   trunk rot, carie blanche du tronc: on 4
   Ont [1341], NB [1032]; on 7 Que [1377].
Phialophora melinii: on 4 Ont [279].
Phoma infossa (C:120): on 1, 4, 5 Ont
   [F61].
Phyllactinia guttata (*P. corylea*)
   (C:120): powdery mildew, blanc: on 1,
   4, 5, 7 Ont, on 4, 5 Que, on 5 Man
   [1257]; on 5a Sask, Man [F66, F67].
Phyllosticta sp.: on 5a Man [F60].
P. viridis: on 5a Sask, Man [F68]; on 7
   Ont [1340].
Piggotia sp.: on 7 NS [1036]. Probably
   Asteromella fraxini (holomorph
   Mycosphaerella effigurata).
Pleurotus ostreatus: on 7 Que [1377].
   Often fruiting on wounds on live trees.
   See *Abies*.
Polyporus badius (*P. picipes*): on 4 Ont
   [1341]. See *Abies*.
P. varius (*P. elegans*): on 4 Ont [1341].
   Typically fruiting on fallen branches
   and logs. See *Abies*.
Poria fraxinea (*Fomes f.*) (C:120): on 4
   Ont [1341]. See *Acer*.
P. fraxinophila (*Fomes f.*) (C:120): on 5
   Ont [1827]. Associated with a white rot.
Pseudospiropes simplex: on 5, 7 Man
   [1760]; on 7 Man [817].
Puccinia sparganioides (*P.*
   peridermiospora) (C:120): rust,
   rouille: 0 I on 1, 7 Ont [1236]; on 1
   Que [F74], NS, NB [1032, F62, F64, F65]
   and NS [335, 336, 338, 1594, F63, F66,
   F67, F68, F72, F73, F75]; on 1, 4 NS
   [F69, F70, F76]; on 4 NB [1032, F62], NS
   [1032]; on 5a Sask [2008], Man [2008,
   F64]; on 7 Que [1377], NB, NS, PEI
   [1032] and NS [334, 337, 339, F60, F61].
Rhabdospora sp.: on 5a Man [F67].
Scopuloides hydnoides (*Phlebia h.*): on 7
   NS [1032]. See *Fagus*.
Selenosporella cymbiformis: on 7 Man
   [1760].
Septoria sp.: on 1 Ont [1828].
Spadicoides bina: on 5 Man [1760].
Sphaeropsis nubilosa: on 5a Sask, Man
   [F68].

Spongipellis malicolus (Polyporus spumeus
    var. *malicola)*: on 1 Que [F71].
    Associated with a white rot.
Sporidesmium achromaticum: on 7 Man
    [1760].
Stictis radiata: on 4 Ont [1340].
Tomentella ruttneri: on 7 Ont [928, 1341].
Torula ligniperda: on 7 Que [1377].
Trametes cervina (Polyporus biformis N.
    Am. auct.*)*: on 4 Ont [1341]. See
    Abies.
Tubercularia sp.: on 7 NB [1032].
T. ulmea: on 5a Sask, Man [F68].
Valsa cypri (V. fraxinina) (C:120): on 4
    NB [1032].
Valsaria sp.: on 7 NWT-Alta [1063].
Xylohypha lignicola: on 5 Man [1760].

FRITILLARIA L.   LILIACEAE

1. *F. camschatcensis* (L.) Ker-Gawl,
    Kamchatka lily; native, Yukon, BC.
2. *F. lanceolata* Pursh, chocolate lily;
    native, BC.

Botrytis cinerea (holomorph Botryotinia
    fuckeliana) (C:121): gray mold,
    moisissure grise: on 1 BC [1545, 1816].
Phyllosticta fritillariae (C:121): on 2
    BC [1012].
Uromyces miurae (C:121): rust, rouille:
    on 1, 2 BC [1545, 1816]; on 2 BC [1012].

FUCHSIA L.   ONAGRADACEAE

1. *F.* x *hybrida* Hort. ex Jilm; commonly
    cultivated hybrid fuchsias, probably
    derived from *F. fulgeus* and *F.*
    magellanica.

Botrytis cinerea (holomorph Botryotinia
    fuckeliana) (C:121): gray mold,
    moisissure grise: on 1 Ont [163].
Poria sp.: on 7 BC [1012].
Pucciniastrum epilobii (C:121): rust,
    rouille: on 7 BC [1012, F73].
Verticillium albo-atrum (C:121): wilt,
    flétrissure verticillienne: on 7 BC
    [1816].

GAILLARDIA Foug.   COMPOSITAE

    Annual or perennial herbs, native,
mostly in far w. North America.

Entyloma sp.: smut, charbon: on *G.* sp.
    BC [1816].

GALEOPSIS L.   LABIATAE

1. *G. tetrahit* L., hemp nettle, ortie
    royale; imported from Eurasia,
    introduced, Mack, BC-Nfld, Labr.

Ophiobolus rubellus: on 1 Ont [1620].
Septoria galeopsidis (C:121): leaf spot,
    tache des feuilles: on 1 BC [1816].

GALIUM L.   RUBIACEAE

1. *G. aparine* L., cleavers, gratteron;
    native, BC-NS, Nfld.
2. *G. asprellum* Michx., rough bedstraw,
    gaillet piquant; native, Ont-Nfld.
3. *G. boreale* L., northern bedstraw,
    gaillet boréal; native, Yukon-Mack,
    BC-NS. 3a. *G. b.* var. *boreale (G.*
    septentrionale R. & S.); Yukon-Mack,
    BC-NB.
4. *G. palustre* L., marsh bedstraw,
    gaillet des marais; native, Ont-Nfld.
5. *G. trifidum* L., dyer's cleavers,
    tissavoyane rouge; native, Yukon-Keew,
    BC-Nfld, Labr.
6. *G. triflorum* Michx., sweet-scented
    bedstraw, gaillet odorant; native,
    Yukon-Mack, BC-Nfld, Labr.

Cercosporidium galii: on 6 Ont [379].
Erysiphe cichoracearum (C:121): powdery
    mildew, blanc: on 1, 6 Ont [1257].
Hainesia borealis (C:121): on 3 Yukon
    [460], BC [1012]; on 3a BC [1816].
Ophiobolus galii: on 2 Ont [1620].
Puccinia difformis: rust, rouille: on 1
    BC [1012]; on 3 NWT-Alta [1063]; on 6
    Mack [53].
P. punctata (C:122): rust, rouille: 0 I
    II III on 2, 4, *G.* sp. Ont [1236]; on
    5 NWT-Alta [1063].
P. punctata var. *troglodytes* (C:122):
    rust, rouille: II III on 6 BC [1012,
    1816], Ont [1236].
P. rubefaciens (C:122): rust, rouille:
    III on 3 Yukon [460], Mack, Alta [53],
    Alta [644], Ont [1236, 1341]; on 3, 5,
    G. sp. NWT-Alta [1063]; on 3, 6 BC
    [1012]; on 3a, 6 BC [1816]; on 6 Mack
    [53]; on *G.* sp. Alta [53].
Pucciniastrum guttatum (C:122): rust,
    rouille: II III on 6 BC [1012, 1816,
    2008], Ont [1236].
Sporonema punctiformis (Placosphaeria p.)
    (holomorph *Leptotrochila verrucosa)*
    (C:121): on 3 Alta [644, 951].

GAULTHERIA L.   ERICACEAE

1. *G. hispidula* (L.) Muhl. *(Chiogenes*
    h. (L.) Torr. & Gray), creeping
    snowberry, gaulthérie hispide; native,
    BC-Nfld, Labr.
2. *G. ovatifolia* Gray, Oregon
    wintergreen; native, BC.
3. *G. procumbens* L., wintergreen, thé des
    bois; native, Man-Nfld.
4. *G. shallon* Pursh, salal, salal;
    native, BC.

Asterina gaultheriae (Asterella g.)
    (C:122): on 3 Ont [1827].
Bulgaria melastoma (C:122): on 4 BC
    [1012].
Chrysomyxa chiogenis: rust, rouille: II
    III on 1 BC [1012, 1816, 2008], Ont
    [1236, 1341], Que [1377, F66]. The
    alternate host is *Picea* spp.

Coccomyces dentatus: on leaves of 4 BC
[832].
Coniothyrium sp.: on 4 BC [1012].
Dasyscyphus gaultheriae (Lachnella g.)
(C:122): on 4 BC [1816].
Gibbera gaultheriae (Venturia g.)
(C:122): on *G.* sp. BC, Ont [73], NS
[73, on living and dead leaves].
Hypoderma gaultheriae: on 4 BC [832 on
leaves, 1012 as *Lophodermium* sp.].
Apparently a secondary parasite
following *Valdensinia*.
Meliola sp. (C:122): on 2, 4 BC [1012,
1816].
Mycosphaerella gaultheriae (C:122): on 3
Ont [1340].
Pestalopeziza brunneo-pruinosa: on 4 BC
[1012, 1816].
Phacidium gaultheriae (C:122): on 4 BC
[1816].
Phellinus ferreus (Poria f.) (C:122): on
4 BC [1012, 1816]. Lignicolous,
associated with a white rot.
Phyllosticta gaultheriae (C:122): leaf
spot, tache des feuilles: on 4 BC
[1012, 1816].
Pucciniastrum goeppertianum: rust,
rouille: on *G.* sp. Alta [F65]. Host
plant probably misidentified. The rust
occurs on *Vaccinium*.
P. pyrolae: rust, rouille: on *G.* sp.
Mack [F66]. Host plant probably
misidentified.
*Valdensia heterodoxa (Asterobolus
gaultheriae)* (holomorph *Valdensinia
h.*): on 4 BC [1426, 1438].
Valdensinia heterodoxa (anamorph
Valdensia h.): on 4 BC [832, 1012,
1426, F76].
Varicosporium elodeae: on 4 BC [43].

GAYLUSSACIA HBK.   ERICACEAE

1. *G. baccata* (Wang.) K. Koch, black
huckleberry; native, Sask-Nfld.
2. *G. dumosa* (Andr.) Torr. & Gray, dwarf
huckleberry; native, NB-Nfld.

Exobasidium vaccinii (C:122): on 1 NS
[1036]; on *G.* sp. Que [1385].
Marasmiellus paludosus: on 2 NB [1433].

GAYOPHYTUM A. Juss.   ONAGRACEAE

Slender annuals native to the temperate
regions of w. North and South America.

1. *G. nuttallii* Torr. & Gray; native, BC.

Puccinia pulverulenta: rust, rouille: on
1 BC [1012, 1816].

GENTIANA L.   GENTIANACEAE

1. *G. andrewsii* Griseb., closed gentian,
gentiane close; native, Sask-Que.
2. *G. calycosa* Griseb., mountain bog
gentian; native, BC-Alta.
3. *G. glauca* Pall.; native, Yukon-w.
Mack, BC-sw. Alta.

4. *G. sceptrum* Griseb.; native, BC.

Phyllosticta sp.: on 4 BC [1012, 1816].
Puccinia gentianae (C:123): rust,
rouille: 0 II III on 1 Ont [1236].
Synchytrium sp. (C:123): on 4 BC [1012,
1816].
Uredo alaskana (C:123): on 3 Yukon [460].
Venturia atriseda (C:123): on
overwintered stalks of 2 BC [73].

GENTIANELLA Moench   GENTIANACEAE

1. *G. amarella* (L.) Borner (*Gentiana a.*
L.), felwort, gentiane amarelle; native,
Yukon-Mack, BC-Nfld, Labr.

Puccinia haleniae (C:123): rust,
rouille: III on 1 Ont [1236].

GEOCAULON Fern.   SANTALACEAE

1. *G. lividum* (Richards.) Fern.
(*Comandra livida* Richards.), northern
comandra, comandre du nord; native,
Yukon-Mack, n. Que-Labr, BC-NS, Nfld.

Cronartium comandrae (C:86): comandra
blister rust, rouille-tumeur: II III on
1 Yukon, Mack, BC, Alta, Sask, Man, Ont,
Que, NB [735], Yukon [460, F68], Mack,
Alta [53], Mack, Sask, Ont [1399], BC
[1012, 1816], Alta [F61, F69], Que
[1236, 1341], NB [F74]; on *G.* sp. Alta
[950, 951, 952].
Puccinia comandrae (C:87): rust,
rouille: III on 1 NWT-Alta [1063],
Mack, Alta [53], Ont [1236].

GERANIUM L.   GERANIACEAE

1. *G. erianthum* DC.; native, Yukon,
BC-Alta.
2. *G. maculatum* L., spotted cranesbill,
géranium maculé; native, Ont-Que.
3. *G. palmatum* Cav. (*G. anemonifolium*
L'Hér.); imported from Canary Islands.
4. *G. platypetalum* Fisch. & Mey.;
imported from Asia Minor.
5. *G. pratense* L., meadow cranesbill,
géranium des prés; imported from
Eurasia. This species is grown
mistakenly under *G. cinereus*, *G.
endressii*, *G. grandiflora*, *G.
platypetalum*, *G. sessiflorium* and *G.
sylvaticum*.
6. *G. richardsonii* Fisch. & Trautv. (*G.
albiflorum* Hook.), white geranium,
géranium de Richardson; native,
Yukon-Mack, BC-Sask.
7. *G. robertianum* L., herb Robert,
géranium à Robert; native, BC, Ont-Nfld.
8. *G. sylvaticum* L., wood cranesbill,
géranium des bois; imported from Eurasia.
9. *G. viscosissimum* Fisch. & Mey.;
native, BC-s. Sask.
10. Host species not named, i.e., reported
as *Geranium* sp.

Coleroa circinans (Venturia c.) (C:123):
on living leaves of 10 BC [73].
C. robertiani (Stigmatea r.) (C:123): on
7 Ont [1340], Ont, Que, NS [73, 304],
Que [305].
Erysiphe cichoracearum: powdery mildew,
blanc: on 2 Ont [1257].
Puccinia sp.: rust, rouille: on 10 Ont
[1341].
P. polygoni-amphibii (C:123): rust,
rouille: 0 I on 2 Ont [1236].
Sphaerotheca macularis (C:123): powdery
mildew, blanc: on 2 Ont [1257]; on 9 BC
[1254].
Uromyces geranii (C:123): rust, rouille:
0 I II III on 1 BC [1012, 1816]; on 3,
4, 5, 6, 8, 10 Ont [1236].
Verticillium dahliae: on 10 BC [38].

GERARDIA L.   SCROPHULARIACEAE

1. *G. flava* L. (*G. quercifolia* Pursh);
native, s. Ont.

Cercospora gerardiae: on leaves of 1 Ont
[439].

GEUM L.   ROSACEAE

1. *G. aleppicum* Jacq., yellow avens,
benoîte jaune; native, Yukon-Mack,
BC-Nfld. 1a. *G. a.* var. *strictum*
(Ait.) Fern.; native, BC-Que.
2. *G. calthifolium* Sm.; native, BC.
3. *G. canadense* Jacq., white avens,
benoîte blanche; native, Ont-NS.
4. *G. macrophyllum* Willd., large-leaved
avens, benoîte à grandes feuilles;
native, Yukon, BC-Nfld.
5. *G. quellyon* Sweet (*G. chilense*
Lindl.); imported from Chile.

Mycosphaerella caulicola (C:123): on 3
Que [70].
Peronospora potentillae: downy mildew,
mildiou: on 5 BC [1816].
Puccinia urbanis: rust, rouille: on 2 BC
[1564].
Sphaerotheca macularis (C:123): powdery
mildew, blanc: on 1a Ont [1257], Que
[1254]; on 3 Ont [1254, 1257]; on 4 BC
[1254].

GLADIOLUS L.   IRIDACEAE

1. *G.* x *hortulanus* L.H. Bailey,
gladiolus, glaieul; an inclusive group
for the prevailing cultivated kinds of
gladiolus to which no recognized
botanical specific name will now apply.
2. Host species not named, i.e., reported
as *Gladiolus* sp.

Botryotinia draytoni (anamorph *Botrytis
gladiolorum*) (C:124): core rot,
pourriture botrytique: on 1 BC [1816],
Alta [338], Que, NS [337]; on 2 NS [335].
Botrytis cinerea (holomorph *Botryotinia
fuckeliana*) (C:124): gray mold,

moisissure grise: on 1 BC [1816 in
greenhouse], Alta, NS [337]; on 2 NS
[335, 336].
Curvularia trifolii f. sp. *gladioli*
(C:124): leaf spot and corm rot,
curvulariose: on 1 Ont [337, 338], NS
[338]; on 2 Ont, Que [1807].
Fusarium sp.: on 1 Sask [1594]; on 2 PEI
[333].
F. oxysporum f. sp. *gladioli* (*F.
orthoceras* var. *g.*) (C:124): yellows
and corm rot, jaunisse fusarienne: on 1
BC [1816], Alta, Ont, Que, NS [337], Ont
[338, 339]; on 2 Man [334], Que [333,
336], NS [335, 336].
Papulaspora sepedonioides: on 1 BC [1937].
Penicillium sp.: on 1 BC [1816].
Phyllosticta gladioli: on 1 Que [339].
Septoria gladioli (C:124): corm hard rot
and leaf spot, septoriose: on 1 BC
[1816].
Stromatinia gladioli (anamorph *Sclerotium
g.*) (C:124): corm dry rot, and leaf
and stalk rot, pourriture sclérotique:
on 1 BC [1816], Que [338], NS [337,
338]; on 2 Ont, NS [333, 334, 336], Que
[336], NS [335].

GLEDITSIA L.   LEGUMINOSAE

1. *G. triacanthos* L., honey locust,
févier epineux; native, Ont.

Botrytis cinerea (holomorph *Botryotinia
fuckeliana*): gray mold, moisissure
grise: on 1 in nurseries BC [F76].
Camarosporium robiniae (C:124): on 1 Ont
[F63, F68, F69]; on 1, *G.* sp. Ont
[1340].
Lachnum sp. (C:124): on *G.* sp. BC
[1012].
Microsphaera penicillata (C:125): powdery
mildew, blanc: on 1 Ont [1257].
Phomopsis sp.: on 1 NS [1032].
Polyporus badius (P. picipes) (C:125): on
G. sp. BC [1012]. See *Abies*.

GLEHNIA F. Schmidt   UMBELLIFERAE

Low, somewhat fleshy Maritime herbs of
the Pacific shores of North America and e.
Asia.

1. *G. littoralis* Schmidt; native, BC.
The North American form is referable to
G. littoralis ssp. *leiocarpa*
(Mathias) Hult.

Ramularia glehniae: on 1 BC [1551].

GLOXINIA L'Hérit.   GESNERIACEAE

From tropical America, often cultivated,
in greenhouses, ornamental flower.

Chalara elegans: on *G.* sp. Ont [1165],
on tubers from Belgium. This name was
proposed for the phialoconidial morph of
Thielaviopsis basicola.

GLYCERIA R. Br.   GRAMINEAE

1. *G. borealis* (Nash) Batch., small
   floating manna-grass, glycérie boréal;
   native, Yukon, BC-Sask, Ont-Nfld, Labr.
2. *G. canadensis* (Michx.) Trin.,
   rattlesnake-grass, glycérie du Canada;
   native, Sask, Ont-Nfld.
3. *G. grandis* Wats., reed-meadow grass,
   glycérie géante; native, Yukon-Mack,
   BC-Nfld.
4. *G. striata* (Lam.) Hitchc., (*G.
   nervata* (Willd.) Trin.), fowl-meadow
   grass, glycérie striée; native,
   Yukon-Mack, BC-Nfld, Labr.

Claviceps sp.: on 1 Alta [1781].
C. purpurea (C:125): ergot, ergot: on 2
   Ont [1729].
Epichloë typhina (C:125): choke,
   quenouille: on 4, *G.* sp. Ont [889].
Passalora graminis (C:125): brown stripe,
   strie brune: on 3 NB [333, 1195].
Stagonospora avenae (Septoria a.)
   (holomorph *Phaeosphaeria avenaria*)
   (C:125): leaf spot, tache septorienne:
   on 2 Ont [1729]; on 2, 4 NB [333, 1195].
S. glycericola (C:125): on 3 Ont [1729].
Ustilago longissima (C:125): smut,
   charbon: on 3 NB [333, 1195].

GLYCINE L.   LEGUMINOSAE

1. *G. falcata* Benth.; imported from
   Australia.
2. *G. max* (L.) Merill, soybean, fève
   soja; imported from se. Asia.

Ascochyta sp. (C:125): leaf spot, tache
   des feuilles: on 2 BC [1816].
Candida sp.: on 2 in greenhouse Ont
   [1496].
Cephalosporium sp.: on 2 in greenhouse
   Ont [1496]
Cercospora sojina (C:125): frog-eye spot,
   tache ocellée: on 2 Ont [113]
   inoculated.
Chaetomium sp.: on 2 in greenhouse Ont
   [1496].
Cladosporium sp.: on 2 in greenhouse Ont
   [1496]
Colletotrichum truncatum (C. glycines)
   (C:125): anthracnose, anthracnose: on
   2 Ont [337].
Corynespora cassiicola (C:125): on 2 Ont
   [336, 337, 1592, 1596, 1599].
Cylindrocarpon sp.: on 2 Ont [1308, 1496].
Dendryphion nanum: on 2 Ont [819].
Diaporthe phaseolorum (C:125): on 2 Ont
   [333, 1597, 1899, 1901, 1909].
D. phaseolorum var. *caulivora* (C:125):
   stem canker, chancre de la tige: on 2
   Ont [336, 337].
D. phaseolorum var. *sojae* (C:125): pod
   and stem blight, brûlure phomopsienne:
   on 2 Ont [336, 337, 1592].
Doratomyces stemonitis: on 2 Ont [1148,
   1760].
Fusarium sp. (C:125): on 2 BC [1816], Ont
   [1308, 1496].

F. poae (C:125): on 2 Ont [169].
Gliocladium sp.: on 2 in greenhouse Ont
   [1496].
Helminthosporium sp.: on 2 in greenhouse
   Ont [1308].
Microsphaera diffusa: powdery mildew,
   blanc: on 1, 2 Ont [1257]; on 2 Ont
   [104].
Mucor sp.: on 2 in greenhouse Ont [1496]
Penicillium sp.: on 2 in greenhouse Ont
   [1308, 1496]
Periconia sp.: on 2 in greenhouse Ont
   [1308]
Peronospora manshurica (C:126): downy
   mildew, mildiou: on 2 BC [1816], Ont
   [104, 337].
Petriella sordida: on 2 Man [85].
Phialophora gregata (Cephalosporium g.)
   (C:125): brown stem rot, pourriture
   brune de la tige: on 2 Ont [337, 338,
   339].
Phoma sp.: on 2 in greenhouse Ont [1308,
   1496].
Phyllosticta glycinea (C:126): leaf spot,
   tache foliaire: on 2 BC [1816].
Phytophthora megasperma var. *sojae*
   (C:126): root and stalk rot, mildiou du
   pied: on 2 Ont [8, 206, 333, 334, 337,
   339, 404, 955, 1847].
Pullularia sp.: on 2 in greenhouse Ont
   [1308].
Pythium sp.: on 2 in greenhouse Ont
   [1308].
P. ultimum (C:126): stem and root rot,
   pourriture pythienne: on 2 Ont [725].
Rhizoctonia sp.: on 2 in greenhouse Ont
   [1496].
R. solani (holomorph *Thanatephorus
   cucumeris*) (C:126): damping off, fonte
   des semis: on 2 Ont [243, 335, 338].
Rhizopus sp.: on 2 in greenhouse Ont
   [1308, 1496].
Sclerotinia sclerotiorum (C:126): stem
   rot, pourridié sclérotique: on 2 Ont
   [338].
Septoria glycines (C:126): brown spot,
   tache brune: on 2 Ont [104, 113, 337,
   338, 339].
Trichoderma sp.: on 2 in greenhouse Ont
   [1496].

GNAPHALIUM L.   COMPOSITAE

1. *G. viscosum* HBK. (*G. macounii*
   Greene), poverty weed, gnaphale
   visqueuse; native, BC, Ont-PEI.

Puccinia investita (C:126): rust,
   rouille: 0 I III on 1 Ont [1236].

GOODYERA R. Br.   ORCHIDACEAE

1. *G. oblongifolia* Raf., giant
   rattlesnake plantain, goodyérie à
   feuilles oblongues; native, BC-NS.

Uredo goodyerae (C:126): rust, rouille:
   on 1 BC [1012, 1816].

GRINDELIA Willd.   COMPOSITAE

1. *G. integrifolia* DC.; native, BC.
2. *G. stricta* DC. (*G. integrifolia* var.
   macrophylla (Greene) Cronq.); native,
   BC.

Coleosporium asterum (C:127):   rust,
   rouille: II III on 1 BC [1816, 2007];
   on 2 BC [1012].

GYMNOCARPIUM Newm.   POLYPODIACEAE

1. *G. dryopteris* (L.) Newm. (*Dryopteris
   disjuncta* (Rupr.) Morton, *Thelypteris
   dryopteris* Slosson), oak fern,
   dryoptéride disjointe; native,
   Yukon-Mack, BC-NS, Nfld.

Hyalopsora aspidiotus (C:104):   rust,
   rouille: II III on 1 BC [1012, 1816,
   2008], Ont [1236, 2008], Que [282, 1341,
   1377, 2008, F61], NS [2008].
Uredinopsis phegopteridis (C:104):   rust,
   rouille: II III on 1 BC [1012, 1816,
   2007 by inoculation, 2008], Alta [2008],
   Ont [1236], Que [1377].

HABENARIA Willd.   ORCHIDACEAE

1. *H. dilatata* (Pursh) Hook., bog-candle,
   habénaire dilatée; native, Yukon-Mack,
   BC-Nfld, Labr.
2. *H. hyperborea* (L.) R. Br., northern
   green orchis, habénaire hyperboréale;
   native, Yukon-Mack, BC-Nfld, Labr.

Puccinia praegracilis (C:127):   rust,
   rouille: on 1, 2 BC [1012, 1816].
P. praegracilis var. *connersii* (C:127):
   rust, rouille: 0 I on 1 Que [1236].

HACKELIA Opiz   BORAGINACEAE

1. *H. virginiana* (L.) Johnston
   (*Echinospermum v.* (L.) Lehm.),
   beggar's lice, hackélia de Virginie;
   native, Ont-Que.

Erysiphe cichoracearum: powdery mildew,
   blanc: on 1 Ont, Que [1257].
Ophiobolus rubellus: on 1 Ont [1620].

HALENIA Borkh.   GENTIANACEAE

1. *H. deflexa* (Sm.) Griseb., spurred
   gentian, halénie défléchie; native,
   BC-NS, Nfld, Labr. 1a. *H. d.* var.
   brentoniana (Griseb.) Gray; native,
   Que, NS, Nfld, Labr.

Puccinia haleniae: rust, rouille: III on
   1 Ont [1236]; on 1a Nfld [1341].

HAMAMELIS L.   HAMAMELIDACEAE

1. *H. virginiana* L., witch-hazel, café du
   diable; native, Ont-NS.

Dermea hamamelidis (*Dermatea h.*) (C:128):
   on 1 Ont [654].
Gonatobotryum apiculatum: on 1 Ont, Que,
   NB, NS [879], NB [1032].
Massaria lantanae: on *H.* sp. Ont [80].
Pezicula hamamelidis (C:128):   on 1 Ont
   [1340, F63].
Phyllosticta hamamelidis: on 1 Ont [1340].
Podosphaera biuncinata: powdery mildew,
   blanc: on 1 Ont [1257].
Seimatosporium consocium (*Cryptostictis
   c.*):   on 1 Ont [1751].

HEBE Comm.   SCROPHULARIACEAE

1. *H. cupressoides* (Hook. f.) Cockayne &
   Allen; imported from New Zealand.

Rhizoctonia solani (holomorph
   Thanatephorus cucumeris):   on 1 BC
   [1816].
Thanatephorus cucumeris (*Pellicularia
   filamentosa*) (anamorph *Rhizoctonia
   solani*):   on 1 BC [334].

HEDEOMA Pers.   LABIATAE

   Annual or perennial herbs of North and
South America.

1. *H. hispida* Pursh; native, Alta-Que.

Erysiphe cichoracearum: powdery mildew,
   blanc: on 1 Ont [1257].

HEDERA L.   ARALIACEAE

1. *H. colchica* K. Koch, Colchis ivy,
   lierre de Perse; native from Caucasus to
   n. Iran.
2. *H. helix* L., english ivy, lierre
   d'Angleterre; imported from Europe, w.
   Asia and n. Africa.

Colletotrichum trichellum (*Amerosporium
   t.*) (C:128):   leaf spot, anthracnose:
   on 2 BC [1816].
Phyllosticta concentrica (holomorph
   Guignardia philoprina):   on 2 BC
   [1816].
Ramularia hedericola: on 1 BC [335, 1816].

HEDYSARUM L.   LEGUMINOSAE

1. *H. alpinum* L., alpine hedysarum,
   sainfoin alpin; native, Yukon-n. Que,
   BC-NB, Nfld. 1a. *H. a.* var.
   americanum Michx.; native, range of
   species.
2. *H. boreale* Nutt.; native, Yukon-n.
   Que, BC-Que. 2a. *H. b.* var.
   mackenzii (Richards.) Hitchc. f.
   mackenzii (*H. mackenzii* Richards.);
   native, Yukon-Keew, BC-Man.
3. *H. coronarium* L., sulla sweet-vetch,
   sainfoin d'Espagne; imported from Europe.
4. *H. sulphurescens* Rydb.; native,
   BC-Alta.

Diachora onobrychidis (anamorph
   Diachorella o.): on 1 Alta [644].
Diachorella onobrychidis f. *onobrychidis*
   (Placosphaeria o.) (holomorph *Diachora
   o.*) (C:128): on 3 Sask [1753].
Placosphaeria sp.: on 1 Alta [644].
   Specimens are probably *Diachorella
   onobrychidis*.
Uromyces hedysari-obscuri (C:128): rust,
   rouille: 0 I III on 1 Mack [53], Alta
   [951, 1839], Que [1236]; on 1, la Yukon
   [460]; on la Yukon [7]; on 1, 2a, 4,
   H. sp. NWT-Alta [1063]; on 1, 4 BC
   [1012]; on 1, 4, *H*. sp. Alta [644]; on
   la Que [282]; on la, 2a Ont [1236]; on
   2a, 4 BC [644]; on 4 BC [953].

HELENIUM L.   COMPOSITAE

1. *H. autumnale* L., sneezeweed, hélénie
   automnale; native, Mack, BC-Que.

Erysiphe cichoracearum: powdery mildew,
   blanc: on 1 Ont [1257].

HELIANTHUS L.   COMPOSITAE

1. *H. annuus* L., common sunflower,
   tournesol; native, BC-Man, introduced,
   Mack, Ont-PEI.
2. *H. debilis* Nutt.; imported from
   Florida. 2a. *H. d.* var.
   cucumerifolius (Torr. & Gray) Heiser,
   cucumber-leaved sunflower, soleil
   miniature; imported from Texas.
3. *H. decapetalus* L., ten-rayed
   sunflower, hélianthe à dix rayons;
   native, Ont-NB.
4. *H. divaricatus* L., divaricaté
   sunflower, hélianthe divariqué; native,
   Ont-Que.
5. *H. giganteus* L., gigantic sunflower,
   soleil géant; native, Ont-NS.
6. *H. kellermanii* Britt.; imported from
   e. USA.
7. *H. laetiflorus* Pers., beautiful
   sunflower, hélianthe à belles fleurs;
   native, Yukon, BC-Ont, introduced,
   Que-PEI. 7a. *H. l.* var.
   laetiflorus; introduced, BC, Ont-Que,
   Nfld. 7b. *H. l.* var. *rigidus*
   (Cass.) Fern.; native, BC-Man,
   introduced, Ont-Que, PEI. 7c. *H. l.*
   var. *subrhomboideus* (Rydb.) Fern.,
   rhombic-leaved sunflower, hélianthe
   subrhomboidal; native, Yukon, Alta-Man,
   introduced, Ont-NB.
8. *H. maximiliani* Schrab., Maximilian's
   sunflower, hélianthe de Maximilien;
   native, BC-Ont, introduced, Que, PEI.
9. *H. mollis* Lam., ashy sunflower;
   imported from e. USA.
10. *H. nuttalli* Torr. & Gray; native,
    BC-NS, Nfld. 10a. *H. n.* ssp.
    canadensis Long; native, BC-NS, Nfld.
    10b. *H. n.* ssp. *nuttalli*; native,
    Alta-Sask.
11. *H. petiolaris* Nutt.; native, Alta-Man,
    introduced, BC, Ont.

12. *H. strumosus* L. *(H. montanus* Wats.),
    strumose sunflower; native, Ont-NB.
13. *H. tuberosus* L., Jerusalem artichoke,
    topinambour; imported from s. USA and
    tropical America, a garden escape,
    Sask-PEI.
14. Host species not named, i.e., reported
    as *Helianthus* sp.

Alternaria sp.: on seedlings of 1 in
   greenhouse Man [1509].
A. alternata (A. tenuis auct.) (C:129):
   on 1 Sask [1139], Man [1760, 1509, from
   greenhouse grown seedlings].
Aspergillus sp.: on seedlings of 1 in
   greenhouse Man [1509].
*Bipolaris sorokiniana (Helminthosporium
   sativum)* (holomorph *Cochliobolus
   sativus*): on 1 Sask [1139].
Botrytis cinerea (holomorph *Botryotinia
   fuckeliana)* (C:129): gray mold,
   moisissure grise: on 1 Alta [1505], Man
   [748, 750, 1505, 1509], Ont [1505], Que
   [338], NS [335].
Cephalothecium sp.: on seedlings of 1 in
   greenhouse Man [1509].
Cladosporium sp.: on seedlings of 1 in
   greenhouse Man [1509].
C. macrocarpum: on 1 Man [1760].
Erysiphe cichoracearum (C:129): powdery
   mildew, blanc: on 1 Man, NB [333]; on
   1, 3, 7, 12, 13, 14 Ont, on 1, 13 Man,
   on 7, 13, 14 Que [1257]; on 3 BC [1816];
   on 8 Man [333]; on 14 NB [1194].
E. communis: on 3 BC [335]. Possibly a
   specimen of *E. cichoracearum.*
Fusarium sp. (C:129): on 1 Sask [1139
   from leaf spots and root rot, 1136].
Itersonilia sp.: on 1 Man [1727].
Mucor spinosus: on 1 Sask [1139].
Oidium sp.: powdery mildew, blanc: on 13
   BC [1816].
Ophiobolus rubellus: on 1, 14 Ont [1620].
Papulaspora sepedonioides: on 13 Ont
   [1937].
Penicillium sp.: on seedlings of 1 in
   greenhouse Man [1509].
Plasmopara halstedii (C:129): downy
   mildew, mildiou: on 1 Alta [337], Sask
   [1136], Man [335, 337, 339, 748, 750,
   752, 753, 754, 1420, 1509], Que [277,
   278, 335, 337, 588].
Puccinia helianthi (C:129): rust,
   rouille: 0 I II III on 1 BC [1506],
   Alta, Sask, Man [337, 1258, 1506], Sask
   [1136], Man [333, 335, 338, 339, 748,
   750, 752, 753, 754, 1420, 1509, 1595,
   1838], Ont [337], Que [840, 841
   inoculation]; on 1, 3, 5, 7b, 8, 12, 13
   Ont [1236, 1258]; on 1 Que, on 5 Ont,
   Man, on 10b BC, on 11 Alta [1258]; on
   2a, 3, 6, 7a, 7b, 7c, 8, 10a, 13 Ont, on
   9, 10b, 12 BC, Sask [1241]; on 4 Ont, on
   5, 7, 8, 10a, 11 Sask, on 10a Alta
   [1241, 1258]; on 8 Man [335, 1241,
   1258]; on 14 Man [1503], Ont [1236].
Rhizoctonia sp.: on 1 Sask [1136, 1139].
Rhizopus sp.: on 1 Sask [337, 1136,
   1139], Man [750, 754, 1509].
R. arrhizus: on 1 Sask [1139].

R. stolonifer: on 1 Man [1509].
Sclerotinia sp.: on 1 Sask [420, 1138],
    Man [1138].
S. minor: on 1 Ont [337, 892].
S. sclerotiorum (C:129): sclerotinia rot,
    flétrissure sclérotique: on 1 BC
    [1816], Alta [333, 336, 339, 1505,
    1594], Sask [336, 337, 892, 1136, 1139],
    Man [333, 335, 336, 337, 748, 750, 752,
    753, 754, 774, 775, 776, 777, 1505,
    1509, 1595], Ont [337, 1505], NB [1594];
    on 13 BC [1816]. In 1977 *Coniothyrium
    minitans* was used to control the rot
    and the result was an increased in yield
    of 24% in naturally infested fields in
    Man.
Septoria helianthi (C:129): leaf spot,
    tache septorienne: on 1 Man [335, 336,
    337, 748, 752, 753, 754, 1595].
Stemphylium sp.: on seedlings of 1 in
    greenhouse Man [1509].
Uromyces silphii: rust, rouille: 0 I on
    14 Ont [1236].
Verticillium albo-atrum (C129): leaf
    mottle, marbrure verticillienne: on 1
    Man [38, 333, 334, 335, 336, 337, 338,
    752, 753, 754, 1420, 1509], Que [38].
V. dahliae: on 1 BC [336, 1816, 1978],
    Man [339, 748, 750, 1595], Que [38, 392,
    393].
V. tenerum (V. lateritium) (holomorph
    Nectria inventa): on 1 Sask [1760].

HELIOPSIS Pers.   COMPOSITAE

1. *H. helianthoides* (L.) Sweet (*H.
    laevis* Pers.), ox-eye; native, s. Ont.

Septoria heliopsidis: on leaves of 1 Ont
    [440].

HELLEBORUS L.   RANUNCULACEAE

1. *H. niger* L., christmas rose, hellébore
    noire; imported from Europe.

Coniothyrium hellebori (C:130): black
    leaf spot, tache noire: on 1 BC [1012,
    1816].

HEPATICA Mill.   RANUNCULACEAE

1. *H. acutiloba* DC., sharp-lobed
    liverleaf, hépatique acutilobée; native,
    Ont-Que.

Tranzschelia pruni-spinosae (C:130):
    rust, rouille: 0 I on 1 Ont [1236].

HERACLEUM L.   UMBELLIFERAE

1. *H. lanatum* Michx. (*H. maximum*
    Bartr.), cow parsnip, berce très grande;
    native, Yukon-Mack, BC-Nfld, Labr.

Anaphysmene heraclei (Septoria h.): on
    H. sp. Alta [644].
Cladosporium macrocarpum: on *H.* sp.
    Sask [1760].

Epicoccum purpurascens: on *H.* sp. Sask
    [1760].
Leptosphaeria doliolum (C:130): on 1 Que
    [70].
Nodulosphaeria modesta: on 1 Que [70].
Ophiobolus rubellus (C:130): on 1 Que
    [70, 1620].
Phyllosticta heraclei: on leaves of 1 Ont
    [440]. Probably a species of
    Asteromella.
*Plagiosphaera umbelliferarum (Linocarpon
    u.)* (C:130): on 1 Que [70].
Pleospora helvetica (C:130): on 1 Que
    [70].
Ramularia heraclei (Cylindrosporium h.)
    (C:130): on 1 BC, Sask, Ont, Que
    [1551], a distinct fungus from later
    homonym *Cylindrosporium h.* Ell. & Ev.

HEUCHERA L.   SAXIFRAGACEAE

1. *H. chlorantha* Piper; native, BC.
2. *H. cylindrica* Dougl.; native,
    BC-Alta. 2a. *H. c.* var.
    cylindrica; native, BC-Alta. 2b. *H.
    c.* var. *glabella* (Torr. & Gray.)
    Wheelock (*H. c.* var. *orbicularis*
    (R., B. & L.) Calder & Savile, *H. c.*
    var. *septentrionalis* R., B. & L.);
    native, BC-Alta.
3. *H. glabra* Willd.; native, BC.
4. *H. micrantha* Dougl.; native, BC.
5. *H. richardsonii* R. Br.; native, Mack,
    BC-Ont.

Puccinia heucherae (C:131): rust,
    rouille: on 3 BC [1012, 1816]; on *H.*
    sp. Man [338].
P. heucherae var. *heucherae* (C:131):
    rust, rouille: on 1, 2a, 2b, 3, 4 BC
    [1560].
P. saxifragae (P. heucherae var. *s.)*
    (C:131): rust, rouille: III on 2 Alta
    [644]; on 5 Ont [1236].
P. saxifragae var. *heucherarum*: rust,
    rouille: on 5 BC, Man, Ont [1560].
P. saxifragae var. *holochloae*: rust,
    rouille: on 2, 2a, 2b BC, on 2b Alta
    [1560].

HIBISCUS L.   MALVACEAE

1. *H. trionum* L. flower-of-an-hour;
    Eurasian, locally introduced, in
    cultivation and waste ground, Sask-NS.

Verticillium dahliae (C:131): wilt,
    flétrissure verticillienne: on 1 Ont
    [1087].

HIERACEUM L.   COMPOSITAE

1. *H. albiflorum* Hook.; native, Mack,
    BC-Sask.
2. *H. aurantiacum* L., devil's
    paint-brush, épervière orangée; imported
    from Europe, natzd., BC-Alta, Man-Nfld.
3. *H. canadense* Michx., Canada hawkweed,
    épervière du Canada; native, BC-Sask,
    Ont-Nfld, Labr.

4. *H. pratense* Tausch, field hawkweed,
   épervière des prés; imported from
   Europe, natzd., BC, Ont-Nfld.
5. *H. scabrum* Michx., rough hawkweed,
   épervière scabre; native, Ont-PEI.
6. *H. scouleri* Hook.; native, BC.
7. *H. umbellatum* L., narrow-leaved
   hawkweed, accipitrine; native, BC-Ont.

Aecidium columbiense (C:131): rust,
   rouille: on 1 BC [1012, 1816].
Erysiphe cichoracearum (C:132): powdery
   mildew, blanc: on 2, 3 Ont [1257].
Leptosphaeria doliolum (C:132): on 4 Que
   [70].
Mycosphaerella caulicola (C:132): on 4
   Que [70].
Nectria pedicularis (C:132): on 4 Que
   [70].
Nodulosphaeria aquilana (C:132): on 4 Que
   [70].
Pleospora ambigua (C:132): on 4 Que [70].
Puccinia columbiensis (C:132): rust,
   rouille: on 6 BC [1012, 1816].
P. dioicae (C:132): rust, rouille: on
   H. sp. BC [1012].
P. fraseri (C:132): rust, rouille: III
   on 5 Ont [1236].
P. hieracii (C:132): rust, rouille: 0 I
   II III on 3, 5 Ont [1236]; on 7 NWT-Alta
   [1063], BC [1012]; on *H.* sp. Ont
   [1341], Que [282].
Trichometasphaeria gloeospora (C:132): on
   4 Que [70].

HIEROCHLOE R. Br.   GRAMINEAE

1. *H. odorata* (L.) Beauv., vanilla grass,
   foin d'odeur; native, Yukon-Mack, n.
   Que-Labr, BC-Nfld.

Colletotrichum graminicola (C:132): on 1
   Que [1729].
Claviceps purpurea (C:132): ergot,
   ergot: on 1 Nfld [666].
Erysiphe graminis: powdery mildew,
   blanc: on 1 Sask [1255, 1257].

HIPPEASTRUM Herb.   AMARYLLIDACEAE

   Herbs with tunicate bulbs, native to
tropical America and w. Africa. Grown as
house plants or outdoors.

Stagonospora curtisii: leaf scorch,
   grillure: on *H.* sp. BC [1816].

HOLCUS L.   GRAMINEAE

1. *H. lanatus* L., velvet grass, houlque
   laineuse; imported from Eurasia, natzd.,
   BC, Ont-NS, Nfld.

Dilophospora alopecuri (C:133): twist,
   torsion: on 1 BC [1816].
Entyloma dactylidis (C:133): leaf smut,
   charbon des feuilles: on 1 BC [1816].
Puccinia coronata (C:133): crown rust,
   rouille couronnée: on 1 BC [1012, 1816].

P. hordei (P. holcina): rust, rouille:
   on 1 BC [643, 1012].
P. recondita s. l. (C:133): rust,
   rouille: on 1 BC [1012, 1816].
Ustilago salvei (U. striiformis auct.)
   (C:133): stripe smut, charbon en
   stries: on 1 BC [1816].

HOLODISCUS Maxim.   ROSACEAE

1. *H. discolor* (Pursh) Maxim., ocean
   spray; native, BC.

Cylindrosporium spiraeicola (C:133): leaf
   spot, tache cylindrosporienne: on 1 BC
   [1816].
Datronia mollis (Trametes m.) (C:133): on
   1 BC [1012]. Lignicolous, associated
   with a white rot.
*Dendrothele incrustans (Aleurocorticium
   i.)*: on 1 BC [965, 1012]. Lignicolous.
Glyphium elatum: on 1 BC [589, 1012].
Hymenochaete rubiginosa (C:133): on 1 BC
   [1012]. Lignicolous associated with a
   white rot.
H. tabacina (C:133): on 1 BC [1012].
   Lignicolous, associated with a white rot.
Odontia sp.: on 1 BC [1012]. Lignicolous.
Peniophora cinerea (C:133): on 1 BC
   [1012]. Lignicolous.
P. incarnata: on 1 BC [1012].
   Lignicolous.
Phellinus ferreus (Poria f.) (C:133): on
   1 BC [1012]. Lignicolous, associated
   with a white rot.
Phyllactinia guttata (C:133): powdery
   mildew, blanc: on 1 BC [1816].
Stereum sp.: on 1 BC [1012]. Lignicolous.
Tulasnella violea: on 1 BC [1012].
   Lignicolous, probably saprophytic.

HORDEUM L.   GRAMINEAE

1. *H. brevisubulatum* (Trin.) Link,
   short-awned barley; imported from
   Eurasia.
2. *H. jubatum* L., squirrel-tail grass,
   queue d'écureuil; native, Yukon-Labr,
   BC-Nfld. 2a. *H. j.* ssp.
   breviaristatum Bowden (*H. nodosum*
   auct. Amer., non L.); native, BC-Sask,
   Nfld, Labr.
3. *H. vulgare* L., barley, orge; a
   commonly grown grain. The two major
   groups of cultivars are the two-rowed
   form, *H. distichon* and the six-rowed
   form, *H. hexastichon*.
4. *H. leporinum* x *vulgare*. An
   experimental hybrid which occurs
   infrequently as an escape.
5. Host species not named, i.e., reported
   as *Hordeum* sp.

Acremoniella sp.: from seeds of 3 PEI
   [1117, 1118]; on 5 PEI [1111].
Acremonium boreale: on 3 BC, Alta, Sask
   [1707].
Alternaria sp.: on 3 PEI [1117]; on 5 PEI
   [1111].

A. alternata (A. tenuis auct.*)* (C:133):
on 3 Man [335, 1114], Ont [1478], PEI
[1118].

Aspergillus sp.: on seeds of 3 Alta [222].

*Bipolaris sorokiniana (Helminthosporium
sativum)* (holomorph *Cochliobolus
sativus)* (C:133): spot blotch,
helminthosporiose: on 1 Alta [133]; on
2 Alta [48]; on 3 BC [1816], BC, Sask,
Man, Ont, Que [639, 1533], Alta [223,
1331, 1333, 1336, 1337], Alta, Sask, Man
[333, 334, 335, 336, 338, 339], Alta,
Man [1115], Sask [961], Sask, Man [337,
1594], Man [640, 672, 1070, 1114, 1887],
Ont [328, 329 on seeds, 331, 336, 339,
680], Ont, PEI [1115, 1594], Que [335,
416, 417], NB [273, 333, 639, 1196], NS
[272, 339], NS, PEI [273, 333, 337, 639,
1533], PEI [275, 334, 336, 338, 1025],
Nfld [335, 337, 1533]; on 4 Ont [256];
on 5 Alta, Man [197], Sask [697]. Leaf
spots caused by *B. sorokiniana,
Drechslera teres, Septoria passerinii,
Puccinia hordei, Rhynchosporium secalis*
and *Xanthomonas translucens,* the
latter a bacterium, were responsible for
a loss of 4.4 million bushels or 4.3% of
the potential yield in Man in 1971 [672].

Cephalosporium sp.: from seeds of 3 Man
[881], PEI [1117, 1118]; on 5 PEI [1111].

C. acremonium: on 3 Man [1114]. See
comment under *Avena.*

*Cercosporidium graminis (Scolecotrichum
g.)*: on 2 Yukon [1731].

Cladosporium sp.: from seeds of 3 PEI
[1117, 1118]; on 5 PEI [1111].

C. cladosporioides (C:134): on 3 Man
[1114].

Claviceps purpurea (C:134): ergot,
ergot: on 3 BC [1816], Alta [223, 334,
335, 338, 1333], Sask [334, 336, 338,
339], Man [335, 1364 inoculated], Que
[339], NB, NS [854], PEI [336, 854]; on
flowers of 5 Que [1385].

Cochliobolus sativus (anamorph *Bipolaris
sorokiniana*): spot blotch,
helminthosporiose: on 3 Alta [706,
1115, 1338, 1363, 1594], Sask [1115,
1115, 1338, 1594], Man [636, 1067, 1071,
1115, 1338, 1891, 1892 on seeds], Ont
[260, 261, 262, 263, 267, 1115, 1090,
1478 on seeds, 1594], Que [1412], NB
[854], NS [854, 1115, 1595], PEI [267,
854, 888, 1115, 1117, 1118 from seeds,
1893]; on 5 PEI [1111].

Colletotrichum dematium: anthracnose,
anthracnose: on 3 Alta [1595].

C. graminicola (C:134): anthracnose,
anthracnose: on 3 Alta [338, 693], Que
[333, 335].

Drechslera graminea (Helminthosporium g.)
(holomorph *Pyrenophora g.*) (C:134):
stripe, strie: on 3 BC [1816], PEI
[854]; on 5 Alta, Sask, Man, Ont, Que,
NS [1611].

D. teres (Helminthosporium t.) (holomorph
Pyrenophora t.) (C:134): net blotch,
rayure réticulée: on 3 BC, Alta, Sask
[335, 337], BC, Alta, Sask, Man, Ont,
Que, NB, NS, PEI [1611], Alta [223,

1330, 1332, 1333, 1335], Alta, Sask, Man
[333, 334, 336, 1887], Alta, Man, PEI
[1115], Man [335, 672, 881, 1068], NB
[333, 1196]; on 5 Alta, Man [197]. See
comment under *Bipolaris sorokiniana.*

D. tritici-repentis (holomorph
Pyrenophora t.) (C:134): on 3 Ont
[1611].

Epicoccum purpurascens (C:134): on 3 Man
[1114].

Erysiphe graminis (C:134): powdery
mildew, blanc: on 2 Yukon [1731], Mack
[1255, 1257]; on 3 BC [1012, 1255,
1816], BC, Alta, Sask, Man [333, 1257],
BC, Alta, Ont, Que [639], BC, Man, Ont
[335, 1533], BC, Ont, Que [334, 640],
Alta, Sask [339], Alta, Ont [336], Sask,
Ont [1594], Man [197, 1887], Ont [260,
261, 262, 338, 1115], Ont, Que, NS
[1257], Que [1255, 1533], NS [272], PEI
[333].

E. graminis f. sp. *hordei* (C:134):
powdery mildew, blanc: on 3 Ont [998].

Fusarium sp. (C:134): on 3 Alta [223,
339, 706, 1331, 1363], Alta, Sask [333,
334, 335, 336, 1594], Alta, Sask, Man,
PEI [337, 338], Man [1114], Ont [1090,
1091, 1594], NB [333], NB, PEI [273],
PEI [334, 336, 1117, 1118, 1594]; on 5
PEI [1111].

F. acuminatum (C:134): head blight,
brûlure de l'épi: on 3 Man [335, 1114].

F. avenaceum (C:134): head blight,
brûlure de l'épi: on 3 Man [1114].

F. culmorum (C:134): root rot and head
blight, fusariose: on 3 Alta [1336,
1337].

F. graminearum (holomorph *Gibberella
zeae*) (C:134): root rot and head
blight, fusariose: on 3 Alta [1337].
In 1980 the PEI crop was severely
damaged by the fungus and by
contamination with its mycotoxin,
vomitoxin.

F. nivale (C:134): on 2 Yukon, BC
[1731]. See *Agrostis.*

F. oxysporum (C:134): basal
discoloration, piétin fusarien: on 3
Man [1114].

F. poae (C:134): head blight, brûlure de
l'épi: on 3 Man [335, 1114].

F. sacchari: on 3 Man [1114].

F. sambucinum (C:134): head blight,
brûlure de l'épi: on 3 Man [1114].

F. sporotrichioides (C:134): head blight
and false scab, fausse gale: on 3 Man
[338, 1114].

F. sulphureum (holomorph *Gibberella
cyanogena*): on 3 Man [1114].

F. tricinctum: on 3 Man [1114].

Gaeumannomyces graminis (Ophiobolus g.)
(C:135): take-all, piétin-échaudage:
on 3 Alta [339].

Gonatobotrys simplex: on 3 Man [1114].

Heterosporium avenae (C:134): on 2 Yukon
[1731].

Mucor sp.: on 3 Ont [1090].

Olpidium brassicae (C:135): on 2 Ont [68].

Penicillium sp. (C:135): on 3 Alta [222], PEI [1117, 1118] from seeds; on 5 PEI [1111].

Polymyxa graminis (C:135): on 3 Ont [60, 68].

Pseudoseptoria stomaticola (Selenophoma donacis var. *s.)* (C:135): eye spot, tache ocellée: on 2 BC [1728]; on 3 NS, PEI [1534].

Puccinia graminis (C:135): stem rust, rouille de la tige: II III on 1 Alta [133]; on 2 Man [626], Ont [1236]; on 3 BC, Sask, Man, Ont [619, 630, 631], BC, Man, Ont [626, 627, 628, 632, 633, 639], BC, Man, Ont, Que [625, 634, 640, 1533], BC, Ont [623], Alta [1331], Alta, Sask, Ont [336, 621], Alta, Ont, Que [338], Sask, Ont [335, 629, 1594], Sask, Que [337, 629], Man [621, 629, 636, 1067, 1887], Man, Ont [622, 635], Man, NB, PEI [333], Ont [334, 339, 641, 1236], Que [619, 631, 635], NS [635, 640], PEI [334, 640, 854].

P. graminis ssp. *graminis* var. *graminis (P. graminis* f. sp. *tritici)* (C:135): rust, rouille: on 2 Sask [624], Sask, Man [625, 631], Man [639, 640, 1533]; on 3 Alta, Sask, Man, Ont, Que [620], Sask, Man [635], Man, Ont, PEI [639, 640].

P. graminis ssp. *graminis* var. *stakmanii (P. graminis* f. sp. *secalis)* (C:135): rust, rouille: on 2 Alta [635], Sask, Man [615, 625, 635], Man [617], Ont [615]; on 3 BC, Sask, Man, Ont [615], BC, Ont [335], Alta, Sask, Man, Ont, Que [620], Man [616]; on 5 Man, Ont, Que [635].

P. hordei (C:135): dwarf leaf rust, rouille naine des feuilles: II III on 3 BC [1816], BC, Man, Ont [639, 640], Alta, Ont [338], Sask, Ont [335, 336, 339], Man [197, 636, 672, 1887], Man, Ont, Que, NS [1533], Man, Que, NS, PEI, Nfld [333], Ont [261, 262, 263, 1236], Que [335, 337], Que, NB, NS, PEI [640], NB [1196], NS [337, 854], PEI [334, 639, 854], Nfld [339]. See comment under *Bipolaris sorokiniana*.

P. recondita s. l. (C:135): rust, rouille: on 1 Alta [133].

P. striiformis (C:135): stripe rust, rouille striée: II III on 2a BC [1816]; on 3 Alta [335].

Pyrenophora graminea (anamorph *Drechslera g.)* (C:134): stripe, strie: on 3 Alta, Sask, Man, NS [1793].

P. teres (anamorph *Drechslera t.)* (C:135): net blotch, rayure réticulée: on 3 BC, Sask, Man, Que [1533], Alta [334, 706, 1329, 1331, 1595], Alta, Sask, Man [338, 339, 639, 1791, 1792], Alta, Sask, Ont [1594], Sask, Man [1071], Man [196, 636, 640, 1067], Ont [338, 1611], NS, PEI [854], PEI [273, 275].

P. tritici-repentis (anamorph *Drechslera t.*): on 3 Ont [1611].

Pythium sp. (C:135): on 3 Alta [1594], Ont [261, 1089, 1090].

P. aristosporum: on 3 Ont [261].

P. arrhenomanes: on 3 Ont [261].

P. graminicola (C:135): browning root rot, piétin brun: on 3 Ont [1092].

P. tardicrescens: on 3 Alta [1595], Ont [261].

Ramularia pusilla (C:135): on 2 Yukon [1731].

Rhizophydium graminis: on 3, 5 Ont [59].

Rhizopus arrhizus (C:135): on overwintered plants of 3 Man [1114].

Rhynchosporium secalis (C:135): scald, tache pâle: on 3 BC [1816], BC, Alta, Sask [337], Alta [223, 639, 640, 706, 1115, 1331, 1333, 1533, 1595, 1667, 1668, 1928], Alta, Sask [334, 336, 1594], Alta, Sask, Man [333, 338, 339, 1887], Man [335, 672], Ont [261, 338, 1594], Que [334], NB, NS, PEI [854], NS [272, 273]; on 5 Alta [197]. See comment under *Bipolaris sorokiniana*.

Septoria sp.: on 1 Alta [133]; on 3 Alta [706, 1333].

S. passerinii (C:135): speckled leaf blotch, tache septorienne: on 3 BC, Sask, Man, Ont, Que [639], Alta [223, 1331], Alta, Sask, Man [333, 334, 336, 338], Alta, Man [335], Sask [1071], Sask, Man [339, 1887], Sask, NS [337], Man [636, 672, 1067], Man, Que, NS, PEI [1533], Ont [261], Que [640], NS, PEI [854]; on 5 Man [197]. See comment under *Bipolaris sorokiniana*.

Streptomyces spp.: on 3 PEI [1117, 1118].

Thanatephorus cucumeris (Pellicularia filamentosa, P. praticola) (anamorph *Rhizoctonia solani*): on 3 Sask [1874].

Urocystis agropyri (C:136): leaf smut, charbon des feuilles: on 2 Man [338]; on 2a BC [1816].

Uromyces mysticus (C:136): rust, rouille: on 2a BC [1816].

Ustilago bullata (C:136): head smut, charbon de l'epi: on 2 Alta [336]; on 5 Sask [333].

U. hordei (C:136): covered smut, charbon couvert: on 3 BC [1816], Alta [1331, 1333], Alta, Sask, Man [329, 333, 336, 1594, 1887], Sask, Man [337, 338, 339, 1071, 1803, 1804], Sask, Nfld [335], Man [491, 1021, 1022, 1023, 1067, 1070, 1106, 1888, 1889, 1890, 1891], NB [333, 1196], NS, PEI [854]; on 5 Man [653]. Losses due to barley smuts (*U. hordei*, *U. nigra* and *U. nuda*) in Man and Sask in 1981 were valued at 17.6 million dollars [1805].

U. kolleri: smut, charbon: on 3 Man [1891]. Probably a misdetermined specimen of *U. hordei*.

U. nigra (C:136): seedling-infecting or black loose smut, faux charbon nu: on 3 Alta, Sask, Man, Ont [334], Sask [1838], Sask, Man [338, 1071, 1594, 1803, 1804], Man [339, 491, 1070, 1067, 1106, 1887], NB [1196]. See comment under *U. hordei*.

U. nuda (C:136): true loose smut, charbon nu: on 3 BC [1816], Alta [223, 706, 1331, 1333], Alta, Sask, Man [333, 334,

336, 337, 339, 1594], Alta, Man [1107],
Sask [933, 1498], Sask, Man [1071, 1802,
1803, 1804, 1887], Sask, Ont, Que [335],
Man [491, 1070, 1067, 1106, 1892], Ont
[259, 333, 334, 336, 530, 1115, 1594],
Ont, Que, Nfld [339], NB [1196], NB, NS,
PEI [333, 854], NS, Nfld [337]. See
comment under *U. hordei*.

HOSTA Tratt.   LILIACEAE

Ornamental perennials, native of e. Asia.

Phyllosticta sp.: on *H.* sp. Que [335].

HOYA R. Br.   ASCLEPIADACEAE

1. *H. carnosa* (L. f.) R. Br., wax plant;
   imported from s. China and Australia.
   Grown as a house plant.

Colletotrichum gloeosporioides (holomorph
   Glomerella cingulata): anthracnose,
   anthracnose: on 1 BC [338].

HUMULUS L.   CANNABINACEAE

1. *H. lupulus* L., common hop, houblon
   commun; imported from Eurasia, natzd.,
   BC-Nfld.

Fumago vagans (C:137): sooty mold,
   moisissure charbonneuse: on 1 BC
   [1816]. See comment under *Corylus*.
Pseudoperonospora humuli (C:137): downy
   mildew, mildiou: on 1 BC [1816].
Pythium sp.: on 1 BC [1816].
Rhizoctonia solani (holomorph
   Thanatephorus cucumeris) (C:137): on
   1 BC [1816].
Sphaerotheca macularis (C:137): powdery
   mildew, blanc: on 1 Ont [1257, 1254].
Verticillium dahliae (C:137): wilt,
   flétrissure verticillienne: on 1 BC
   [1816, 1978]; on *H.* sp. BC [38].

HYACINTHUS L.   LILIACEAE

1. *H. orientalis* L., hyacinth, jacinthe;
   imported from Mediterranean region.

Dactylosporium macropus: on 1 BC [1816].
Uromyces muscari (C:137): rust, rouille:
   on 1 BC [1816].
U. muscari f. sp. *hyacinthi* (C:137):
   rust, rouille: on 1 BC [1545].

HYDRANGEA L.   SAXIFRAGACEAE

1. *H. macrophylla* (Thunb.) Ser.,
   big-leaved hydrangea, hortensia commun;
   imported from Japan.
2. *H. paniculata* Siebold, panicled
   hydrangea, hortensia en panicule;
   imported from China and Japan.
3. Host species not named, i.e., reported
   as *Hydrangea* sp.

Armillaria mellea: root rot,
   pourridié-agaric: on 3 BC [1012].

Botrytis cinerea (holomorph *Botryotinia
   fuckeliana*) (C:137): gray mold,
   moisissure grise: on 2 Sask [334] in
   greenhouse.
Erysiphe cichoracearum (C:137): powdery
   mildew, blanc: on 3 NB [333]. Based on
   DAOM collections, mildew on *Hydrangea*
   in Canada is known only in the conidial
   state [1257], thus it is questionable
   which teleomorph epithet is applicable.
E. communis (*E. polygoni*) (C:137):
   powdery mildew, blanc: on 1 Que [339];
   on 3 NB [334, 335, 338]. See above.
Nectria cinnabarina (anamorph
   Tubercularia vulgaris) (C:137):
   dieback or coral spot, deperissement
   nectrien: on 3 Que [337].
Oidium sp. (C:137): powdery mildew,
   blanc: on 3 BC [1816].
Schizopora paradoxa (Poria versipora): on
   3 BC [1012].
Sclerotinia sclerotiorum (C:138):
   sclerotinia stem rot, pourriture
   sclérotique: on 3 BC [1816].

HYDROCOTYLE L.   UMBELLIFERAE

1. *H. americana* L., marsh pennywort,
   hydrocotyle d'Amérique; native, Ont-Nfld.

Erysiphe communis (*E. polygoni*) (C:138):
   powdery mildew, blanc: on 1 Ont, Que
   [1257].

HYDROPHYLLUM L.   HYDROPHYLLACEAE

1. *H. canadense* L., Canadian waterleaf,
   hydrophylle du Canada; native, Ont-Que.
2. *H. tenuipes* Heller; native, BC.
3. *H. virginianum* L., John's cabbage,
   hydrophylle de Virginie; native, Man-Que.

Erysiphe cichoracearum: powdery mildew,
   blanc: on 1, 3 Ont [1257].
Puccinia hydrophylli (C:138): rust,
   rouille: III on 3 Ont [1236].
P. recondita s. l. (C:138): rust,
   rouille: on 2 BC [1012, 1816].
Septoria hydrophyli: on leaves of 3 Ont
   [440].

HYPERICUM L.   HYPERICACEAE

1. *H. boreale* (Britt.) Bickn., northern
   St. John's-wort, millepertuis boréal;
   native, Ont-Nfld.
2. *H. ellipticum* Hook., elliptic St.
   John's-wort, millepertuis elliptique;
   native, Ont-NS, Nfld.
3. *H. kalmianum* L., Kalm's St.
   John's-wort, millepertuis de Kalm;
   native, Ont-Que.
4. *H. mutilum* L., dwarf St. John's-wort,
   millepertuis nain; native, Man-NS.
5. *H. perforatum* L., common St.
   John's-wort, pertuisane; imported from
   Eurasia, natzd., BC, Ont-Nfld.
6. *H. punctatum* Lam., spotted St.
   John's-wort, millepertuis ponctué;
   native, Ont-Que, NS.

7. *H. spathulatum* (Spach) Steud., shrubby
   St. John's-wort, millepertuis prolifère;
   native, Ont.
8. *H. virginicum* L., marsh St.
   John's-wort, millepertuis de Virginie;
   native, Sask-Nfld.

Erysiphe communis (E. polygoni): powdery
   mildew, blanc: on 5 Ont [1257].
Uromyces sparganii (C:138): rust,
   rouille: 0 I on 8 Ont [1236, 1287], Que
   [1287].
U. triquetrus (C:138): rust, rouille: 0
   I II III on 1, 2, 3, 4, 6, 7, 8, *H.*
   sp. Ont [1236].

HYPOCHOERIS L.   COMPOSITAE

1. *H. radicata* L., spotted cat's-ear,
   porcelle enracinée; imported from
   Eurasia, introduced, BC, Sask, Ont-NS,
   Nfld.

Verticillium dahliae: on 1 BC [1978].

ILEX L.   AQUIFOLIACEAE

1. *I. aquifolium* L., English holly, houx
   commun; imported from Eurasia.
2. *I. verticillata* (L.) Gray,
   winterberry, apalanche; native, Ont-Nfld.

Boydia insculpta (C:139): canker,
   chancre: on 1 BC [29, 1816].
Ceuthospora phacidioides (C:139): on 1 BC
   [1012].
Diaporthe ilicis: on *I.* sp. Ont [1340].
Fumago vagans: sooty mold, moisissure
   charbonneuse: on 1 BC [1816]. See
   comment under *Corylus*.
Microsphaera penicillata (C:139): powdery
   mildew, blanc: on 2 Que, NB, NS [1257].
Phoma sp.: on 1 BC [1816].
Phytophthora ilicis (C:139): leaf and
   twig blight, brûlure: on 1 BC [1416,
   1816]; on *I.* sp. BC [334].
Trochila ilicis (C:139): on 1 BC [1012,
   1816].
Varicosporium elodeae: on 1 BC [43].

IMPATIENS L.   BALSAMINACEAE

1. *I. balsamina* L., garden balsam,
   balsamine cultivée; imported from
   Indonesia.
2. *I. capensis* Meerb., spotted
   touch-me-not, impatiente du Cap; native,
   Mack, BC-Nfld.
3. *I. noli-tangere* L., touch-me-not,
   herbe Ste.-Catharine; native, BC-Man.
4. *I. pallida* Nutt., pale touch-me-not,
   impatiente pâle; native, Ont-NS, Nfld.

Dasyscyphus sulphurea: on *I.* sp. Ont
   [1340].
Ophiobolus rubellus: on *I.* sp. Ont
   [1620].
Puccinia argentata (C:139): rust,
   rouille: on 3 Alta [53].

P. impatienti-elymi (P. recondita in
   part) (C:139): rust, rouille: 0 I on
   2, 4 Ont [1236]; on *I.* sp. Ont [1236,
   1828]. See comment under *Elymus*.
Stemphylium botryosum (holomorph
   Pleospora herbarum) (C:139): target
   spot, tache stemphylienne: on 1 BC
   [1816].

INULA L.   COMPOSITAE

1. *I. helenium* L., elecampane, inule
   aulnée; imported from Eurasia,
   introduced, BC, Ont-NS, Nfld.

Erysiphe cichoracearum (C:139): powdery
   mildew, blanc: on 1 Man, Ont [1257].

IPOMOEA L.   CONVOLVULACEAE

1. *I. purpurea* (L.) Roth, common morning
   glory, liseron pourpre; imported from
   tropical America, a garden escape,
   Ont-Que, NS.

Verticillium dahliae: on 1 BC [334,
   1978]; on *I.* sp. BC [1816].

IRIS L.   IRIDACEAE

1. *I. germanica* L., German iris, iris
   d'Allemagne; imported from Europe,
   widely grown in gardens. The aggregate
   species is probably a series of hybrids
   of uncertain parentage.
2. *I. versicolor* L., blue flag, clajeux;
   native, Man-Nfld, Labr.
3. *I. xiphium* L., Spanish iris, iris
   d'Espagne; imported from s. Europe.
4. Host species not named, i.e., reported
   as *Iris* sp.

Bipolaris iridis (Mystrosporium adustum)
   (C:140): ink disease, maladie d'encre:
   on 3 BC [1595, 1744]; on 4 BC [1816].
Botrytis cinerea (holomorph *Botryotinia
   fuckeliana*) (C:140): on 1 NB [1194];
   on 4 BC [1816], NB [333].
B. convoluta (holomorph *Botryotinia
   c.*): on 4 Ont [842].
*Cladosporium iridis (Heterosporium i., ?H.
   gracile)* (holomorph *Macrosphaerella
   macrospora*): on 3 BC [1595, 1744]; on
   4 BC [1816], Ont [1340], NB [1194].
*Macrosphaerella macrospora (Didymellina
   m.)* (anamorph *Cladosporium iridis*)
   (C:140): leaf spot, tache
   hétérosporienne: on 1 BC, Ont [337,
   338], NS [337]; on 4 BC, Man, Ont, Que
   [336], BC, NS [334], Que, NB, NS [335],
   NB, NS [333].
Mollisia iridis (C:140): on 2 Ont [1340].
Penicillium sp. (C:140): bulb rot,
   pourriture penicillienne: on 1 BC
   [339]; on 3 BC [1594].
Puccinia iridis (C:140): rust, rouille:
   II III on 2 Ont, Que [1236], NS [337].
P. sessilis (C:140): rust, rouille: 0 I
   on 2 Ont [1236].

Sclerotinia sclerotiorum (C:140):
   sclerotinia rot and wilt, flétrissure
   sclérotique: on 4 BC [1816].
Sclerotium tuliparum (C:140): gray bulb
   rot, pourriture grise des bulbs: on 4
   BC [1816].
Septoria breviuscula: on 2 Ont [195].
   The name is a later homonym of *S.
   breviuscula* Sacc. The specimens should
   be renamed.
Talaromyces luteus (Penicillium l.): on 4
   Ont [1340].
Venturia antherici: on overwintered
   leaves of 4 Nfld [73].

IVA L.   COMPOSITAE

1. *I. axillaris* Pursh, poverty-weed,
   herbe de pauvreté; native, BC-Man.

Puccinia intermixta (C:140): rust,
   rouille: on 1 Alta, Man [1245, 1280],
   Sask [1280].

IVESIA Torr. & Gray   ROSACEAE

1. *I. gordonii* (Hook.) Torr. & Gray.
   Perennial herbs in foothills of Rocky
   Mountains along Pacific coast.
   Sometimes grown in wild gardens in USA.

Phragmidium horkeliae: on 1 Alta [1579].

JUGLANS L.   JUGLANDACEAE

1. *J. cinerea* L., butternut, noyer
   tendre; native, Ont-NB.
2. *J. nigra* L., black walnut, noyer noir;
   native, s. Ont.
3. *J. regia* L., English walnut, noyer
   royal; imported from Eurasia.
4. Host species not named, i.e., reported
   as *Juglans* sp.

Aplosporella juglandis: on 2 Ont [1340].
   This may be an anamorph of
   Botryosphaeria obtusa.
Asterosporium asterospermum: on 4 Ont
   [1340].
Botryosphaeria obtusa (C:141): on 1 Ont
   [1614].
Cerrena unicolor (Daedalea u.): on 1 Que
   [1391]; on 2 Que [1855]. See *Acer*.
Coriolus versicolor (Polyporus v.)
   (C:141): on 4 BC [1012]. See *Abies*.
Cylindrocarpon sp.: on 2 Ont [1827].
Cytospora sp.: on 4 Ont [1340].
Diaporthe spiculosa: on 1 Ont [1340].
Fusarium lateritium (C:141): fusarium
   dieback, dépérissement fusarien: on 3
   BC [1816].
Ganoderma applanatum (Fomes a.): on 1 Que
   [1377]; on 1, 4 Ont [1341]. See *Abies*.
Gloeosporium sp.: on 2 Ont [1340].
Gnomonia caryae: on 1 Que [1377].
G. leptostyla (anamorph *Marssonina
   juglandis*) (C:141): leaf spot, tache
   des feuilles: on 1 NB, NS [1032]; on 4
   Ont [79, 1340], Que [337], NB [1032].

Irpex lacteus (Polyporus tulipiferae)
   (C:141): white spongy rot, carie
   blanche spongieuse: on 4 Ont [1341].
Marssonina juglandis (holomorph *Gnomonia
   leptostyla*) (C:141): leaf spot, tache
   des feuilles: on 1 BC [1816], Ont
   [1340], Que [F72, F73]; on 1, 2 Ont
   [1828, F76]; on 4 Ont [1340, F63].
Melanconis juglandis (anamorph
   Melanconium oblongum) (C:141):
   canker, chancre: on 1 Que [1377]; on 1,
   4 Ont [1340], NB [1032]; on 2 Ont [F64];
   on 4 Ont [337, F63].
Melanconium sp.: on 4 Ont [1340].
M. oblongum (holomorph *Melanconis
   juglandis*) (C:141): on 1, 4 Ont
   [1340]; on 2 Ont [1827].
Microstroma juglandis (C:141): white
   mold, moisissure blanche: on 1 Ont
   [339], Que [1377]; on 2 Ont [F74]; on 3
   BC [1816].
Peniophora cinerea: on 2 Ont [1341]. See
   Abies.
Phellinus igniarius (Fomes i.) (C:141):
   on 1 Que [1377]; on 1, 4 NB [1032]; on 4
   Ont [1341]. See *Acer*.
Phialophora richardsiae: on 1 Que [F71].
Phomopsis juglandina: on 3 Ont [F76].
Phyllactinia guttata (C:141): powdery
   mildew, blanc: on 3 BC [1816].
Trichocladium canadense: on 1 Que [F71].
Tubercularia vulgaris (holomorph *Nectria
   cinnabarina*): on 4 BC [1012].

JUNCUS L.   JUNCACEAE

1. *J. articulatus* L., jointed rush, jonc
   articulé; native, BC, Ont-Nfld.
2. *J. balticus* Willd., (*J. ater* Rydb.),
   baltic rush, jonc de la Baltique;
   native, Yukon-Labr, BC-Nfld.
3. *J. biglumis* L.; native, Yukon-Labr,
   BC-Alta, Man-Que.
4. *J. drummondii* Meyer; native,
   Yukon-Mack, BC-Alta.
5. *J. dudleyi* Wieg., Dudley's rush, jonc
   de Dudley; native, Yukon-Mack, BC-Nfld.
6. *J. effusus* L., soft rush, jonc épars;
   native, BC, Ont-Nfld. 6a. *J. e.* var.
   compactus Lej. & Court.; native,
   Ont-Nfld.
7. *J. ensifolius* Wikstr.; native, Yukon,
   BC-Sask.
8. *J. parryi* Engelm.; native, BC-Alta.
9. *J. tenuis* Willd., slender rush, jonc
   ténue; native, BC-Nfld, Labr.

Arthrinium cuspidatum: on 2 Sask [444].
Clathrospora heterospora var.
   simmonsii: on 4 BC [284].
Entorrhiza casparyana: smut, charbon: on
   1 BC [45].
Leptosphaeria sepalorum: on 8 BC [72].
Melanotus caricicola: on 6 Ont and on
   J. sp. Ont [1437].
Phyllachora therophila: on 6a NS [1032].
Platyspora pentamera (C:142): on 2 BC
   [285, 287].
Urocystis junci (C:142): smut, charbon:
   on 3 Frank [1586].

Uromyces junci-effusi (C:142): rust,
rouille: III on 6 Ont [1236]; on 7 BC
[1012].
U. silphii (C:142): rust, rouille: II
III on 5, 6, 9, *J.* sp. Ont [1236].

JUNIPERUS L.   CUPRESSACEAE

1. *J. chinensis* L., Chinese pyramid
   juniper, genévrier de Chine; imported
   from e. Asia. 1a. *J. c.* var.
   chinensis. 1b. *J. c.* var. *c.* cv.
   pfitzeriana. 1c. *J. c.* var. *c.* cv.
   plumosa.
2. *J. communis* L., common juniper,
   genévrier commun; native, Yukon-Keew,
   BC-Nfld, Labr. 2a. *J. c.* var.
   communis. 2b. *J. c.* var. *c.* cv.
   hibernica (*J. hibernica* Lodd. ex
   Loud.), Irish juniper. 2c. *J. c.* var.
   c. cv. suecica. 2d. *J. c.* var.
   depressa Pursh, dwarf juniper,
   genévrier nain; native,
   transcontinental. 2e. *J. c.* var.
   montana Ait.; native, transcontinental.
3. *J. conferta* Parl., shore juniper,
   genévrier des rovages; imported from
   Sakhalin and Japan.
4. *J. horizontalis* Moench, creeping
   juniper, savinier; native, Yukon-Mack,
   BC-Nfld.
5. *J. oxycedrus* L., prickly juniper,
   genévrier piquant; imported from Eurasia.
6. *J. rigida* Siebold & Zucc., needle
   juniper, genévrier rigide; imported from
   e. Asia.
7. *J. sabina* L., savin juniper, genévrier
   sabine; imported from Eurasia. 7a. *J.
   s.* cv. tamariscifolia. 7b. *J. s.* cv.
   variegata.
8. *J. scopulorum* Sarg., Rocky mountain
   juniper, genévrier des Rocheuses;
   native, BC-Alta.
9. *J. squamata* D. Don, single-seed
   juniper, genévrier écailleux; imported
   from Asia. 9a. *J. s.* var. *fargesii*
   Rehd. & E.H. Wils. 9b. *J. s.* var.
   squamatas cv. Meyeri. 9c. *J. s.* cv.
   prostrata.
10. *J. virginiana* L. (*J. fragans*
    Knight*)*, eastern red cedar, cèdre
    rouge; native, Ont-Que. 10a. *J. v.*
    var. *crebra* Fern. & Grisc.; the form
    found in Canada. 10b. *J. v.* cv.
    canaerti. 10c. *J. v.* cv.
    cinerascens. 10d. *J. v.* cv.
    elegantissima. 10e. *J. v.* cv.
    glauca. 10f. *J. v.* cv. pendula. 10g.
    J. v. cv. pendula viridis. 10h. *J.
    v.* cv. plumosa. 10i. *J. v.* cv.
    pyramidalis. 10j. *J. v.* cv. schottii.
11. Host species not named, i.e., reported
    as *Juniperus* sp.

Actidium nitidum: on 11 BC [1012].
Aleurodiscus lividocoeruleus: on 4 Man
[964]. See *Abies.*
Alternaria sp.: on 1 BC [1012].

Armillaria mellea: root rot,
pourridié-agaric: on 8 BC [1012, F74].
See *Abies*.
Ascochyta sp.: on 11 BC [1012].
Botryosphaeria sp.: on 7 Ont [675].
Botrytis sp.: on 7 BC [1012].
Cephalosporium sp.: on 4 BC [1012].
Ceuthospora phacidioides: on 11 Sask
[F65].
Chaetoscutula juniperi: on 2 Man [F67];
on 7b, 8 BC [1012].
*Chloroscypha sabinae (Kriegeria
juniperina)*: on 2 Yukon [460, 664],
NWT-Alta [1063], Mack [53], Mack, Alta
[F67], BC [1012]; on 4 Mack [F65], Sask
[F68].
Coriolus versicolor (Polyporus v.): on 10
Ont [1341]. Lignicolous. See *Abies*.
Corticium sp.: on 11 Ont [1827].
Crucibulum parvulum: on 4 Alta [191].
Lignicolous, probably saprophytic.
Cucurbitaria sp.: on 11 BC [1012].
Cytospora dubyi (holomorph *Valsa
friesii*) (C:142): on 11 BC [1012,
F73].
*Dendrothele griseo-cana (Aleurocorticium
g.)*: on 10 Ont [965]. Lignicolous.
D. nivosa (Aleurocorticium n.): on 10 Ont
[965, 1341]. Lignicolous.
Dothiorella sp.: on 7 Ont [675].
Epipolaeum sp.: on 2 BC [1012].
Gymnosporangium sp.: rust, rouille: on
2, 4, 8 NWT-Alta [1063]; on 10 Ont
[1341]; on 11 BC [954], Alta [1594].
G. aurantiacum (C:142): rust, rouille:
on 2d Que [1377]. The application of
this name is uncertain. Specimens from
North America fall within *G. cornutum*.
G. bermudianum: rust, rouille: I III on
4 Ont [1236]. These specimens have been
determined to be *Roestelia brucensis*.
G. betheli (G. tubulatum) (C:142): rust
gall, rouille-tumeur: III on 8 NWT-Alta
[1063], BC [338, 644, 953, 1012, 1247,
1267, 1816, F60]; on 10 BC [1816]; on 11
BC, Alta, Man, Ont, Que [2008].
G. clavariiforme (C:142): rust, rouille:
III on 2 NWT-Alta [1063], BC [953,
1816], BC, Alta [644], BC, Alta, Man,
Ont, Que, NB, NS, PEI [2008], Alta [335,
951, 952], Man [F67], NB, NS, PEI
[1032]; on 2b BC [339, 1816]; on 2b, 2c
BC [1012]; on 2b, 2d Ont [1236]; on 2c
BC [338], Que [1236], on 2d BC, Alta
[1247], BC, Alta, Ont, Que, NB, NS, PEI,
Nfld [1262]; on 4 NB [F64]; on 11 Ont,
Que, NB, NS, PEI, Nfld [1240].
G. clavipes (C:142): rust, rouille: III
on 2 Mack, BC, Sask, Man, Ont, Que, NB,
NS, PEI [2008], Mack [F68], BC [644,
1012, 1816], Alta [950], Man [F67], Ont
[F71, F72], NB, NS [1032]; on 2, 2c, 2d,
3, 5, 6, 7, 7a, 10 Ont [1236]; on 2, 2c,
7, 10 Ont [336]; on 2c Nfld [1662]; on
2d BC, Sask [1247, 1263], Alta [1247],
Man [1240], Ont, Que, NB, NS, PEI [1240,
1263]; on 4 Que, PEI [1240, 1263], NB
[F64]; on 4, 10 Ont [F74, F75]; on 10

Ont [339, 1240, 1263, F61, F68, F71]; on
10, 10b, 10e Ont [337]; on 10, 11 Ont
[1341]; on 11 NS [333].

G. connersii (C:142):   rust, rouille: III
on 4 Que [1240, 1251], NS [1251]; on 4,
8 Man [1247, 1251].

G. cornutum (C:142):   rust, rouille: III
on 2 Mack [53, F66], Mack, BC, Alta,
Sask, Ont, Nfld [2008], BC [1816], BC,
Alta [644], Alta [951], NB [1032]; on 2b
Mack, BC, Alta [1247, 1264], BC [1012],
Ont [1236], Ont, Que, NB, NS [1264]; on
4 NB [F64]; on 11 Alta [951, 952], Ont,
Que, NB, NS [1240].

G. fuscum (C:143):   rust, rouille: III on
1 BC [335, 1012]; on 1c, 11 BC [1816];
on 7 BC [181, 1253, 2000, F60], Que
[F61]; on 7, 7a, 7b, 9a, 9b, 9c BC [335,
1012, 1816]; on 11 BC [833, 1198, 1247,
1416, 2008].

G. gaeumannii: rust, rouille: on 2 Alta
[F68]; II III on 2d Alta [2008].

G. gaeumannii ssp. *albertense*:   rust,
rouille: II III on 2 Alta [644, 951,
952, F70, F71]; on 2d Alta [730, 1246,
1247, 1265].

G. globosum (C:143):   rust gall,
rouille-tumeur: III on 2d Que [1377];
on 4 NWT-Alta [1063], Ont [1268]; on 7,
10, 10b, 10c, 10d, 10f, 10g, 10h, 10i,
10j, 11 Ont [1236]; on 10 Ont [1242,
F68, F69, F72], Ont, Que [1240]; on 10,
10e, 11 Ont [337]; on 10, 11 Ont [1341];
on 10a Ont, Que [1268]; on 11 Ont [336,
F67].

G. haraeanum: on 1 BC [1012].   Occurs
only on imported *Juniper*.

G. inconspicuum (C:143):   rust, rouille:
III on 8 BC [1012, 1247, 1269, 1816,
2008].

G. juniperi-virginianae (C:143):
cedar-apple rust, rouille de Virginie:
III on 10 Ont [337, 339, 1236, 1341,
1595, 1839, F61]; on 10a Ont, Que
[1270]; on 10b Ont [336, 337]; on 10e
Ont [337]; on 11 Ont [338, 1240, 1341,
F64, F67, F74, F75], Que [338].

G. nelsonii (G. corniculans) (C:143):
rust gall, rouille-tumeur: III on 4
Yukon, Mack, Man, Ont [1271], Yukon [7],
Mack, BC, Alta, Sask, Man [1247], BC
[953], Alta [952, F66], Man [F63], Man,
Ont, NB [1240], Ont [1236, 1341]; on 4,
8 NWT-Alta [1063], BC [1012, 1816], BC,
Alta [644, 1271], Sask [1271]; on 8 Mack
[53], BC [1247], Alta [F61]; on 11
Yukon, Mack, BC, Alta, Sask, Man [2008].

G. nidus-avis (G. juvenescens) (C:143):
witches' broom, rouille-balai de
sorcière: III on 4 BC, Alta [644], BC,
Alta, Sask, Man [1247, 1272], Man [1240,
1838], Ont [1236, 1341], Ont, PEI [1240,
1272]; on 4, 8 NWT-Alta [1063], BC
[1012, 1816]; on 8 BC [644, 953, 1247,
1272], Sask [1594]; on 10 Ont [1236,
1240, 1272]; on 11 BC, Alta, Sask, Man,
PEI [2008], Alta [952].

G. tremelloides (C:143):   rust gall,
rouille-tumeur: III on 2 NWT-Alta

[1063], Mack [53, F66], Mack, BC, Alta
[2008], BC [644, 954, 1012, 1816], Alta
[950, 952]; on 2d BC [1247, 1266].

Hendersonia sp.:   on 2 BC [1012].   See
comment under *Abies*.

Herpotrichia juniperi (H. nigra) (C:143):
on 2 BC [1012, 1816], Alta [950, F66];
on 8 NWT-Alta [1063], Alta [950].

Holmiella sabina (Eutryblidiella s.)
(anamorph *Corniculariella* sp.):   on 4
Que [F66]; on 7 Ont [1344]; on 8 BC
[1012, 1344]; on 10 Ont [1344]; on 11
Alta [644], Ont [F66].

Hysterium acuminatum (C:143):   on 11 Sask
[F65].

Kabatina juniperi: on 1b, 11 BC [1012].

Lepteutypa cupressi: on 11 BC [1012].

Lophodermium chamaecyparisii: on 2 BC
[1012].

L. juniperinum (L. juniperi) (C:143):
needle cast, rouge: on 2 Alta [950,
951, 952, F64], Man [F66, F67, F68], Ont
[1828], Que [F65], NB, NS [1032];   on 2,
4 Yukon [460], NWT-Alta [1063], Mack
[53, F65, F67]; on 2b, 7, 7a, 8 BC
[1012]; on 2d Ont [1063]; on 4 NB [1032,
F64]; on 11 Sask [F65], Ont [1340, F73].

Microthyrium sp.:   on 8 BC [1012].

Mytilidion acicola: on 2 BC [1012].

M. decipiens (C:143):   on 2d, 2e Que [70].

Ojibwaya perpulchra: on 2, 11 Man and on
2d NS [1760].

Otthia sp.:   on 8 BC [1012].

Pestalotiopsis funerea (Pestalotia f.)
(C:143):   on 1b BC [1816]; on 10 BC
[507]; on 11 NS [1032].

Phlebia georgica (Rogersella eburnea):   on
11 Man [740].   Lignicolous.

Phoma sp. (C:143):   on 11 BC [1012, 1816].

Phomopsis juniperovora (C:143):   twig
blight, brûlere des rameaux: on 1b, 7a
BC [338]; on 7 BC [1012, F61, F72]; on
7, 7a BC [337]; on 7a BC [339]; on 8 Ont
[339]; on 10 Ont [F73, F74, F75]; on 10,
11 Ont [F72]; on 11 Ont [F64], Que [F61].

Pithya sp.:   on 10 BC [1012].

P. cupressina: on 1b, 4, 7 BC [1012].

Resinicium bicolor (Odontia b.):   on 11 NS
[1032].   Lignicolous.

Roestelia brucensis (C:143):   rust,
rouille: on 4 Ont [1240, 1273, 1279].
See under *Gymnosporangium bermudianum*.

Sclerophoma sp.:   on 8 BC [1012].

S. pithya: on 10 Ont [1827].

Seynesiella exigua: on dead needles of 2
Que [145].

S. juniperi (Stigmatea j.) (C:143):   on 2
Alta [F67], Man [F66]; on 8 BC [953,
1012, 1816, F68]; on 11 BC [73].

Stemphylium sp.:   on 7 BC [1012].

Stigmina deflectens: on 2 Man [1760].

*S. glomerulosa (Exosporium g.,
Sciniatosporium g.)* (C:142):   on 2 BC
[644, 953, 1012, 1127], Sask [1760,
F68], Ont [1340]; on 2, 2d, 11 Man
[1760]; on 2d Ont [1127].

*S. juniperina (Cercospora j., C. sequoiae
var. juniperi, Sciniatosporium j.)*
(C:142):   on 2 BC [1012, 1127, 1816],
Man [F65], Ont [747]; on 2, 2d Ont
[1127].

Valsa cenisia: on 10 Ont [F74].
V. friesii (anamorph *Cytospora dubyi*):
on 2d by inoculation, 4 Que [1689].

KALMIA L.   ERICACEAE

1. *K. angustifolia* L., sheep laurel,
crevard de mouton; native, Ont-Nfld,
Labr.
2. *K. polifolia* Wang. (*K. glauca* Ait.),
bog laurel, laurier des marais; native,
Yukon-Labr, BC-Nfld. 2a. *K. p.* var.
polifolia f. *polifolia* (*K. p.* var.
occidentalis Small); native,
Yukon-Keew, n. Que-Labr, BC-Alta,
Man-Nfld.

Dothidella kalmiae: on 1 Ont, NS [1340];
on 2a BC [1012, 1816].
Exobasidium vaccinii (C:143): on 2 Nfld
[1341]; on *K.* sp. Que [1385].
Gibbera kalmiae (C:143): on 1, 2 Que
[70]; on 2 Nfld [1340]; on living and
dead leaves, capsules and twigs of *K.*
sp. Que, Nfld [73].
Gibberidea rhododendri (*G. kalmiae*)
(C:143): on 1 Que [70, 763].
Godronia cassandrae (anamorph *Fusicoccum
putrefaciens*): on 1 Que [1678].
G. cassandrae f. *cassandrae*: on 1 Que
[1525].
Lophodermium exaridum (C:143): on 1 Ont
[1340, 1827].
Microsphaera vaccinii: powdery mildew,
blanc: on 2 Ont, Que [1257].
Mycosphaerella colorata (C:143): on 1 Ont
[1340], Que [70], NB [1032], Nfld [1659,
1660, 1661]; on *K.* sp. Ont [1340].
Phyllachora kalmiae (C:143): on 2 Que
[70].
Pseudomassaria erumpens: on 2 Que [71].
Tympanis alnea: on 1 Que [1221].

KOBRESIA Willd.   CYPERACEAE

1. *K. bellardii* (All.) Degl. (*K.
myosuroides* (Vill.) Fiori & Paol.);
native, Yukon-Labr, BC-Alta, Man.
2. *K. simpliscuiscula* (Wahl.) Mack.;
native, Yukon-Labr, BC-Alta, Man-Que,
Nfld.

Anthracoidea elynae (*Cintractia e.*, *C.
carpophila* var. *carpophila* sensu
Savile in part) (C:143): smut,
charbon: on 1 Yukon, Frank, BC, Que,
Nfld [906], Yukon [1542], Frank [1586],
Alta [644]; on 2 Frank, Keew, Man, Que
[1542].
A. elynae var. *elynae* (C:143): smut,
charbon: on 1 Frank [1244].
A. lindebergiae (*Cintractia l.*) (C:143):
smut, charbon: on 2 Frank, Keew, Man,
Que [906].
Schizonella elynae: on 1 Frank [1244].

KOELERIA Pers.   GRAMINEAE

1. *K. cristata* (L.) Pers. (*K. macrantha*
(Ledeb.) Spreng.), prairie June grass,

Koélérie à crète; native, Yukon-Mack,
BC-Que, Labr.

Ascochyta agropyrina var. *nana*: on *K.*
sp. Sask [1414].
Mycosphaerella tassiana (C:144): on 1
Alta, Sask [1362].
Platyspora permunda: on 1 Sask [1362].
Pseudoseptoria stomaticola (*Selenophoma
donacis* var. *s.*): on 1 BC [1728].
Pyrenophora tritici-repentis (anamorph
Drechslera t.): on *K.* sp. Sask
[1369].
Septoria andropogonis f. sp. *koeleriae*:
on 1 Alta, Sask [1362].
S. calamagrostidis f. sp. *koeleriae*: on
1 Sask [1362].
Stagonospora graminella: on 1 Sask [1362].

KRIGIA Schreb.   COMPOSITAE

Annual or perennial herbs with milky sap
native to North America. Commonly called
dwarf dandelion.

Puccinia hieracii: rust, rouille: on
K. sp. Ont [1341].

LABURNUM Medic.   LEGUMINOSAE

1. *L. anagyroides* Medic., golden-chain
laburnum, bois de lièvre; imported from
Europe.

Articulospora tetracladia: on *L.* sp. BC
[43].
Epicoccum purpurascens (*E. nigrum*): on
L. sp. BC [1012].
Nectria cinnabarina (anamorph
Tubercularia vulgaris): on 1 Nfld
[336].

LACTUCA L.   COMPOSITAE

1. *L. biennis* (Moench) Fern., tall blue
lettuce, laitue bisanmielle; native,
Yukon, BC-Nfld, Labr.
2. *L. canadensis* L., Canada lettuce,
laitue du Canada; native, Man-PEI.
3. *L. sativa* L., garden lettuce, laitue;
much cultivated, probably of asiatic
origin. 3a. *L. s.* var. *asparagina*
Bailey, celtuce; native of Eurasia.
4. *L. serriola* L. (*L. scariola* L.),
prickly lettuce, laitue piquante;
imported from Eurasia, natzd., BC-PEI.

Aecidium compositarum: on 2 Ont [437].
Probably a specimen of *Uromyces
silphii*.
Alternaria sp. (C:145): leaf spot, tache
des feuilles: on 3a BC [1816].
Botrytis cinerea (holomorph *Botryotinia
fuckeliana*) (C:145): gray mold,
moisissure grise: on 3 BC [1816], Ont
[1460, 1594], Que [336, 1594, 1654], NB
[337], NS [333, 334, 335, 336, 338,
1594], Nfld [333, 335].
Bremia lactucae (C:145): downy mildew,
mildiou: on 3 BC [1816], Que [333, 334,

335, 336, 338, 348, 1594, 1639, 1640,
1641, 1647, 1651, 1653, 1654], NS [334,
335, 336, 338, 1594].
Calyptella capula: on 4 BC [1424].
Cephalosporium sp.: on 3 Ont [202].
Erysiphe cichoracearum: powdery mildew,
blanc: on 1, 4 Ont [1257].
Fusarium sp.: on 3 Ont [1461, 1594].
F. oxysporium: on 3 Ont [202].
Marssonina panattoniana (C:145):
anthracnose, anthracnose: on 3 BC
[1816].
Olpidium sp.: on 3 Ont [333, 334, 335,
336]. Associated with tobacco necrosis
virus.
O. brassicae: on 3 BC [1977], Ont, Que
[67].
Ophiobolus anguillidus: on L. sp. Ont
[1340].
O. tanaceti: on 4 Ont [1620].
Penicillium sp.: on 3 Ont [202].
Puccinia dioicae (P. extensicola)
(C:145): rust, rouille: 0 I on 2, 3
Ont [1236]; on 3 Sask [339], NS [336];
on L. sp. Ont [1341].
P. minussensis (C:145): rust, rouille:
on 2 BC [1012].
Pythium sp.: on 3 BC [1977], Ont [335],
Que [868, 1594, 1654].
P. irregulare: on 3 Ont [868].
P. sulcatum: on 3 Ont [868] inoculated.
P. sylvaticum: on 3 Ont [868].
P. torulosum: on 3 Ont [868].
Rhizoctonia sp.: on 3 BC [1818].
Reported as *Rhizoctonia* complex, but
no indication was given as to the
composition of the complex.
Rhizoctonia solani (holomorph
Thanatephorus cucumeris) (C:145):
bottom rot, pourriture basale: on 3 BC
[339, 1594, 1817], Ont [336], Que [333,
334, 335, 338, 339, 348, 1594, 1639,
1640, 1647, 1653, 1654], NB [336, 337],
NS [334].
Sclerotinia sp.: on 3 Sask [420, 1138],
Man [1138].
S. sclerotiorum (C:145): drop,
affaissement sclérotique: on 3 BC [338,
339, 1594, 1816, 1817, 1818], Alta
[334], Ont [1460, 1461, 1594], Que [333,
334, 335, 336, 338, 1594, 1595, 1639,
1640, 1641, 1647, 1653, 1654, 1721], NB
[333, 339, 1594], NS [333, 335, 336,
337, 338, 1594], PEI [333].
Stemphylium botryosum (holomorph
Pleospora herbarum) (C:145): on
seeds of 3 Ont [667].
Ulocladium consortiale (Stemphylium c.):
on seeds of 3 Ont [667].
Verticillium albo-atrum: on seeds of 3
Ont [98].
V. dahliae (C:145): on 4 BC [38, 1816,
1978].

LAMIUM L.   LABIATAE

1. *L. amplexicaule* L., henbit, pain de
poule; imported from Eurasia, natzd.,
Mack, BC-NS, Nfld, Labr.

Peronospora lamii (C:146): downy mildew,
mildiou: on 1 BC [1816].
Verticillium dahliae: on 1 BC [1816,
1978].

LAPSANA L.   COMPOSITAE

1. *L. communis* L., nipplewort, herbe aux
mamelles; imported from Eurasia, natzd.,
BC, Man-Nfld.

Puccinia lapsanae (P. variabilis var.
lapsanae) (C:146): rust, rouille: 0
I II III on 1 BC [1012, 1816], Ont
[1236].
Ramularia lapsanae (C:146): on 1 BC
[1816].

LARIX Mill.   PINACEAE

1. *L. decidua* Mill., European larch,
mélèze d'Europe; imported from Europe.
2. *L. gmelinii* (Rupr.) Rupr., Dahurian
larch; imported from Siberia.
3. *L. laricina* (DuRoi) K. Koch, tamarack,
tamarac; native, Yukon-Keew, BC-Nfld,
Labr.
4. *L. leptolepis* (Siebold & Zucc.) Gord.,
Japanese larch, mélèze du Japon;
imported from Japan.
5. *L. lyallii* Parl., alpine larch, mélèze
de Lyall; native, BC-Alta.
6. *L. occidentalis* Nutt., western larch,
mélèze de l'ouest; native, BC-Alta.
7. *L. sibirica* Ledeb., Siberian larch,
mélèze de Siberie; imported from the
USSR.
8. Host species not named, i.e., reported
as *Larix* sp.

Acremonium boreale: on 3 BC, Alta, Sask
[1707].
Aleurodiscus spiniger (C:146): on 6 BC
[964]. Lignicolous.
Alternaria sp.: on 8 NB [1032].
Amylostereum chailletii: on 3 NB [1032];
on 6 BC [1012]. See *Abies*.
Antrodia serialis (Trametes s.): on 6 BC
[1012]. See *Abies*.
Armillaria mellea (C:146): root rot,
pourridié-agaric: on 3, 7 NWT-Alta
[1063]; on 3 Alta [53, F63], Ont [F62],
NB, NS [F71], Nfld [F65, F67]; on 3, 4
Nfld [1659, 1660, 1661, F69]; on 4 Nfld
[1663, F68]; on 4, 6 BC [1012, F62]; on
7 Alta [F65]; on 8 Ont [1341]. See
Abies.
Ascocalyx abietis (anamorph *Bothrodiscus
berenice*): on 3 Que [1690].
Ascotricha amphitricha: on 8 Sask [715].
Aureobasidium pullulans: on 6 BC [1012].
Botryosphaeria laricis (pycnidial anamorph
a form of *Sphaeropsis sapinea)*: on 3
Que, NS [1687].
Botrytis sp.: on 1 BC [1012].
B. cinerea (holomorph *Botryotinia
fuckeliana*): on 1 Que [F72].
Ceratocystis allantospora: on 3 Ont [647,
1340].

Chondrostereum purpureum (Stereum p.): on
3 NB [1032, 1036]. See *Abies*.
Cladosporium sp.: on 1 BC [1012].
Clitocybula familia: on dead trunks of 8
Que [1385].
Coccomyces cembrae (C:146): on 3 NS
[1032].
C. irretitus: on 3 NS [1609].
Coniophora puteana (C:146): on 3, 6 BC
[1012]. See *Picea*.
Coniothyrium sp.: on 6 BC [1012, F72].
Coriolus versicolor (Polyporus v.)
(C:147): on 3 NB, NS [1032]. See
Abies.
Corticium centrifugum: on 6 NWT-Alta
[1063], BC [953]. See *Abies*.
Cytospora sp.: on 1 Ont [1340, 1827]; on
3 Ont [1340], Que [F61], PEI [1032]; on
6 BC [1012].
C. curreyi: on 6 BC [953].
C. kunzei (holomorph *Leucostoma
kunzei*): on 3 Ont [1827].
Dacrymyces capitatus (D. ellisii): on 8
Que [282]. Lignicolous.
Dasyscyphus sp.: on 1 BC [1012]; on 3 Ont
[1340].
D. oblongospora (C:146): on 3 NB [1032].
No doubt a *Lachnellula*.
Dermea piceina: on 8 Ont [1340].
*Encoeliopsis laricina (Ascocalyx l.,
Crumenula l., Scleroderris l.)*
(anamorph *Brunchorstia laricina*): on
3 Que [F68, 1686, 1690]; on 6 BC [503,
660, 1012, 1827, F68, F69].
Endophragmia glanduliformis: on 3 Sask
[1760].
Epicoccum sp.: on 6 BC [1012].
Fomitopsis cajanderi (Fomes c.): on 3, 6
NWT-Alta [1063]. See *Abies*.
F. officinalis (Fomes o.) (C:147): on 6
BC [1012]. See *Abies*.
F. pinicola (Fomes p.) (C:147): brown
cubical rot, carie brune cubique: on 6
BC [1012].
F. rosea (Fomes r.): on 6 BC [1012]. See
Abies.
Fusicoccum sp.: on 3 Ont [1340]; on 3, 8
NB, NS [1032].
Gelatinosporium griseo-lanatum: on 1 BC
[517].
Gloeophyllum abietinum (Lenzites a.)
(C:147): on 8 NS [1032]. Rarely
recognized from North America. Perhaps
only a specimen of *G. sepiarium*.
G. sepiarium (Lenzites s.) (C:147): on 3
NWT-Alta [1063]. See *Abies*.
Godronia sp.: on 6 BC [F66].
Haematostereum sanguinolentum (Stereum s.)
(C:147): red heart-rot, carie rouge du
sapin: on 3 BC [F74], NB, NS, PEI
[F62]; on 3, 6 BC [1012]; on 3, 8 NB
[1032].
Hymenochaete tabacina: on 3 Ont [1341].
See *Abies*.
Hypochnicium bombycinum (Corticium b.):
on 8 NS [1032]. Lignicolous.
Hypodermella sp.: on 4 Nfld [1660, 1661].
H. laricis (C:147): needle cast, rouge:
on 3 Man [F66, F68], Ont [1340, 1828,

F62, F72], PEI [1032], Nfld [1659, 1660,
F64, F65, F66, F67, F68, F69, F70, F71,
F72, F73, F74, F75] also Man, NB, NS
[F73]; on 3, 4 Nfld [1661]; on 3, 8 NB
[1032]; on 5, 6 NWT-Alta [1063]; on 6 BC
[953, 1012, F60, F61, F62, F64, F66,
F70, F72, F74]; on 8 BC [F63, F65], Ont
[1340].
Junghuhnia collabens (Poria rixosa): on 3
Ont [1004]. See *Abies*.
Lachnellula arida: on 5 NWT-Alta [1063],
Alta [952].
L. flavovirens: on 6 BC [1012].
*L. occidentalis (Dasyscyphus o., Lachnella
o., Lachnellula hahniana)* (C:146-147):
on 3 Yukon [460], Ont [1828], Que [1690,
F68], NS [1032]; on 5 NWT-Alta [1063],
BC [1012]; on 6 Ont [398], BC [398,
1012]; on 8 Ont [1340].
L. suecica: on 3 Que [1690]; on 5 BC
[1012].
*L. willkommii (Dasyscyphus w., Lachnella
w.)*: on 3 Ont [1340, F67], Que [282].
Causes European Larch Canker.
Leucostoma curreyi (Valsa c.): on 3, 6
NWT-Alta [1063].
L. kunzei (anamorph *Cytospora k.*): on 1
Ont [1828], Que [935, 936, 1210, F66];
on 3 Que [F75]; on 3, 4 Que [935, 1210].
Libertella sp.: on 8 NB [1032].
Lophodermium laricinum (C:147): on 3, 5
NWT-Alta [1063]; on 3 Mack [53, F66], BC
[2005], Man [F66], Que [1377]; on 5 Alta
[F62]; on 5, 6 [1012, 2005, F68]; on 6
BC [953]; on 8 NS [1032].
Macrophoma sp.: on 1 BC [1012].
Melampsora sp.: rust, rouille: on 3, 5
NWT-Alta [1063]; on 3 Mack [53], NB
[F68], NS, PEI [F67, F68]; on 8 Mack,
Alta [F62]; Ont [1341, F69].
M. epitea: rust, rouille: on 3 NB [1032].
M. medusae (M. albertensis) (C:147):
needle rust, rouille des aiguilles: on
1, 4, 6 BC [2002] by inoculation; on 1,
3, 4, 6 BC [1012]; on 3, 8 NWT-Alta
[1063], Mack [53]; on 3 Alta [F64, F66],
Sask [F70], Ont [1236, F66], Que [1377,
F67], NB, NS [1032, F69], PEI [F69],
Nfld [1660, 1661, F71]; on 8 BC [F65],
Alta [952], Ont [1236], NB [1032].
M. occidentalis: rust, rouille: on 1, 4,
6 BC [1012, 2002 inoculated]; on 8 BC
[F65].
M. paradoxa (M. bigelowii) (C:147):
needle rust, rouille des aiguilles: 0 I
on 3 Yukon [460], Mack [53], BC [1012,
F73], Sask [F66], Man [F64, F66], Ont
[1236, 1236], Que [1377, F74], NB [F69,
F70], NS [1032, F70], PEI [1032, F64,
F69]; on 5 Alta [952, F69, F70]; on 6 BC
[1012, 2007]; on 8 Yukon, Mack, BC,
Alta, Sask, Man [2008] also Alta, Sask,
Man [F70], Nfld [1660] and Alta [F61].
Melampsoridium betulinum (C:147): on 3
Ont [1341], Que [1377].
Moellerodiscus advenulus: on 3 Ont [422].
Mytilidion sp.: on 3 Ont [1827].
M. gemmigenum: on 6 BC [1012].
Nectria fuckeliana: on 1, 3 Que [1685] by
inoculation.

Odontotrema hemisphaericum: on 3 Que
[1690].
Oidiodendron tenuissimum (C:147): on 3
Sask [82].
Pestalotia laricicola: on 3 Man [F66].
Pezicula livida (anamorph
Cryptosporiopsis abietina): on 3 Que
[1690].
Phaeolus schweinitzii (Polyporus s.)
(C:147): brown cubical rot, carie brune
cubique: on 3 Que [1377]; on 4 Ont
[F65]; on 3, 6 BC [1012].
Phellinus nigrolimitatus (Fomes n.)
(C:147): on 6 BC [1012]. See *Abies*.
P. pini (Fomes p.) (C:147): red ring rot,
carie blanche alvéolaire: on 3 Ont
[1341, F71], Que [1377]; on 5 NWT-Alta
[1063], Alta [952]; on 6 BC [1012,
1341]. See *Abies*.
P. weirii (Poria w.) (C:147): yellow ring
rot, carie jaune annelée: on 6 BC
[1012]. See *Abies*.
Phialophora fastigiata: on 3 Sask [279].
Pholiota alnicola (Flammula a.): on 3 BC
[1012]. See *Abies*.
P. aurivella: on 3 BC [1012]. See
Abies.
Physalospora laricis (C:147): on 3 NS
[1032].
Pleurostromella sp.: on 3, 8 Ont [1827].
Pluteus tomentosulus: on 8 Que [1385].
Lignicolous.
Poria subacida: on 3 BC [1012], Que
[1377]. See *Abies*.
Propolis rhodoleuca: on 8 Ont [1340].
Probably a form of *P. versicolor*.
Puccinia bolleyana: rust, rouille: on 3
Ont [F73]. A misdetermination, this
rust occurs on *Carex*.
Resinicium bicolor (Odontia b.) (C:147):
white stringy rot, carie blanche
filandreuse: on 2 BC [F62].
Sarcotrochila alpina: on 5 BC [1012,
2005]; on 6 BC [1012, F71].
Sclerophoma pithyophila: on 1 BC [F65,
F67]; on 3 Man [F67].
Scoleconectria cucurbitula: on 1 Ont
[1340]; on 3 Ont [F70].
Scytinostroma galactinum: on 6 BC
[1012]. See *Abies*.
Sirodothis sp.: on 3 Yukon [460].
Sistotrema brinkmannii (Trechispora b.):
on 8 NB [1032]. See *Abies*.
Sphaeropsis sapinea (Macrophoma s.)
(holomorph *Botryosphaeria laricis*):
on 1, 3 Que [1687] by inoculation.
Sporidesmium achromaticum: on 3 Sask
[1760].
Sydowia polyspora (anamorph *Sclerophoma
pithyophila*): on 1 BC [1012], Que
[F68]; on 1, 3 Que [1686].
Tomentella griseoumbrina: on 8 Ont [1341,
1828].
*Trichaptum abietinus (Hirschioporus a.,
Polyporus a.)* (C:147): pitted sap rot,
carie blanche du l'aubier: on 3, 6
NWT-Alta [1063]; on 6 BC [953, 1012].
T. laricinus (Hirschioporus l.): on 3 Que
[1028]. See *Abies*.
Trichothecium roseum: on 3 Man [1760].

Tryblidiopsis picea: on 1 Que [1690].
Tympanis abietina: on 1, 3 Que [1688], on
1 by inoculation.
T. alnea (anamorph *Sirodothis inversa*):
on 3 Que [1221]. An unusual report on a
coniferous host.
T. confusa: on 3 Que [1688] by
inoculation.
T. hansbroughiana: on 1, 3 Que [1688], on
1 by inoculation.
T. hypopodia (T. piceina): on 3 Que
[1688] by inoculation.
T. laricina (anamorph *Sirodothis* sp.)
(C:147): on 1 BC [1012], Que [1688] by
inoculation; on 3 Yukon [460], Sask
[F67], Ont [F66], Que [1688]; on 6 BC
[1012]; on 8 BC, Alta, Man, Que [1221].
T. pithya: on 3 Que [1688] by inoculation.
T. rhabdospora: on 8 Que [1221].
T. truncatula: on 3 Que [1688] by
inoculation; on 8 Ont, Que [1221].
T. tsugae: on 3 Que [1688] by inoculation.
Tyromyces placenta (Poria p.): on 3 Ont
[1341]. See *Pinus*.
T. sericeomollis: on 6 BC [1012]. See
Abies.
Valsa abietis: on 1, 3 Que [1689] by
inoculation; on 3 Yukon [460].
V. friesii (anamorph *Cytospora dubyi*):
on 1, 3 Que [1689], on 1 by inoculation.
Vararia granulosa: on 6 BC [557]. See
Abies.
V. racemosa: on 6 BC [557]. Lignicolous.

LATHYRUS L.   LEGUMINOSAE

1. *L. latifolius* L.; from Europe,
    garden-escape, BC, Ont-Que.
2. *L. nevadensis* Wats. (*L. nuttallii*
    Wats.); native, BC. The BC material is
    referable to ssp. *lanceolatus* (Howell)
    Hitchc. var. *pilosellus* (Peck) Hitchc.
3. *L. ochroleucus* Hook., pale vetching,
    gesse jaunâtre; native, Mack, BC-Que.
4. *L. odoratus* L., sweet pea, pois de
    senteur; imported from Italy.
5. *L. palustris* L., vetchling, gesse
    palustre; native, Yukon, BC-Nfld. 5a.
    L. p. var. *myrtifolius* (Muhl.) Gray;
    native, Ont-NS.
6. *L. sylvestris* L., everlasting pea,
    gesse des bois; imported from Eurasia, a
    garden escape BC, Ont-PEI.
7. Host species not named, i.e., reported
    as *Lathyrus* sp.

Acremonium boreale: on 7 BC, Alta, Sask
[1707].
Aphanomyces euteiches: root rot,
pourridié: on 4 Alta [334].
Ascochyta pinodella: on 7 Ont, Que [1898].
A. pisi (C:147): leaf spot, ascochytose:
on 5, 6 BC [1816].
Cladosporium macrocarpum: on 7 Man [1760].
Erysiphe communis (E. polygoni) (C:148):
powdery mildew, blanc: on 4 BC [1816],
Man [334], Que [333].
Fusarium sp. (C:148): root rot or wilt,
pourridié fusarien: on 4 Sask [339],
Man [335], Ont [1594], Nfld [334].

Microsphaera penicillata (M. alni)
(C:148): powdery mildew, blanc: on 3
Alta, Sask, Ont [1257]; on 5 Ont [1257];
on 7 Ont [1340].
M. penicillata var. *ludens*: on 1, 3, 4
Man [884]. On 3, 4 by inoculation.
These hosts may provide inoculum for
infection of cultivated fields of *Vicia
faba*.
Mycosphaerella tassiana (C:148): on 4 BC
[1816].
Peronospora lathyri-palustris (C:148):
downy mildew, mildiou: on 2 BC [1816].
Ramularia deusta (C:148): ramularia leaf
spot, tache ramularienne: on 4 BC [1816].
Septoria astragali (C:148): on 3 BC
[1816].
S. lathyri (C:148): on 7 BC [1012, 1816].
S. pisi: on 7 Ont, Que [1898].
Uromyces orobi: on 3 Ont [437]. Probably
a specimen of *U. viciae-fabae*.
U. viciae-fabae (U. fabae) (C:148): rust,
rouille: 0 I II III on 2, 5 BC [1012,
1816]; on 3, 5a Ont [1236]; on 7 Ont
[1341].
Verticillium dahliae: on 4 BC [336, 1816,
1978].

LAVANDULA L.   LABIATAE

1. *L. angustifolia* Mill., English
   lavender, lavande commune; imported from
   Mediterranean region. 1a. *L. a.* ssp.
   angustifolia (*L. officinalis* Chaix).

Septoria lavandulae (C:148): leaf spot,
tache des feuilles: on 1a BC [1816].

LEDUM L.   ERICACEAE

1. *L. glandulosum* Nutt., trapper's tea;
   native, BC-Alta.
2. *L. groenlandicum* Oed., Labrador tea,
   thé du Labrador; native, Yukon-Keew, n.
   Que-Labr, BC-Nfld. Some records on *L.
   glandulosum* may be included under this
   name.
3. *L. palustre* L.; native, Yukon-Labr,
   BC-Que. Represented in North America
   mostly by *L. p.* var. *decumbens* Ait.
   (*L. decumbens* (Ait.) Lodd. ex Steud.).
4. Host species not named, i.e., reported
   as *Ledum* sp.

Ampulliferina persimplex: on 2 Sask, Man
[1755, 1760].
Chrysomyxa ledi (C:148): rust, rouille:
II III on 1 BC [953], Alta [950, F66];
on 2 NWT-Alta [1063], Mack, Alta [F62],
BC [644], Alta [53, 644, 951], Sask
[F67, F68], Man [F67], Ont [1341, F66],
Que [1377, P67]; on 4 BC [953, 954, F72,
F73], Alta [951], Ont [F76].
C. ledi var. *ledi* (*C. l.* var.
glandulosi, C.l. var. *groenlandici*)
(C:148, 149): rust, rouille: II III on
1, 2, 3 BC [1012]; on 1, 2 BC [1816]; on
2 Ont [1236]; on 3 Yukon [460]; on 4
Yukon, NWT, BC, Alta, Sask, Man [2008].
The alternate host is *Picea* spp.

C. ledicola (C:149): rust, rouille de
l'épinette: II III on 1, 2 NWT-Alta
[1063]; on 2 Mack [F65, F66], Mack, Alta
[53, F62], BC [1816], Alta [560, 644,
1724], Sask, Man, Que [F67], Ont [1236,
1341, F63, F66], Que [282, 1341, 1377],
NB [F68], NB, NS, PEI [1032]; on 2, 3
Yukon [460], BC [953, 1012]; on 2, 4 BC
[954], Alta [951, 952]; on 3 Mack [53],
Frank [1244], Que [1236]; on 4 Yukon,
Mack, BC, Alta, Sask, Man [2008], BC
[953, F72, F73], Ont [F76].
C. woroninii (C:149): rust, rouille de
l'épinette: III on 2 Yukon [1402, F62,
F69], BC [1816]; on 2, 3 Yukon [460],
Mack [53, F66], BC [1012]; on 3 Que
[1236]; on 4 Yukon, Mack, BC [2008],
Yukon, BC [F60].
Coccomyces ledi: on 4 NWT, BC, Man [1609].
Elsinoe ledi (C:149): on 2 Ont [1340].
Encoeliopsis ledi: on 2 BC [644], Que
[660]; on 4 Ont [660].
Exobasidium sp.: on 2 NWT-Alta [1063],
Mack [53].
E. vaccinii (C:149): on 2 BC [1012].
E. vaccinii var. *vaccinii*: on 2 BC
[644, 1816].
E. vaccinii-uliginosi (C:149): on 4 Que
[1385].
Gibberidea rhododendri: on 2 Que [763].
Godronia cassandrae (anamorph *Fusicoccum
putrefaciens*): on 2 Que [1678].
G. cassandrae f. *cassandrae*: on 2 Que
[1525].
Lophodermium sphaeroides (C:149): on 2
Mack [53], Alta [F64], Man [F66], Ont
[1340]; on 3 NWT-Alta [1063].
Mycosphaerella sp.: on 2 Nfld [1660].
Protoventuria ledi (Antennularia l.): on
2 Que [73].
Pseudophacidium ledi: on 2 Que [1685].
Pycnostysanus azaleae: on 2 Sask [1760].
Rhytisma sp.: on 1 BC [1012].
Seimatosporium ledi: on 2 BC [1012] and
on living leaves [1359].
S. lichenicola (holomorph *Discostroma
corticola*): on 2 Que [1359].

LEERSIA Sw.   GRAMINEAE

1. *L. oryzoides* (L.) Swartz, rice
   cutgrass, léersie faux-riz; native, BC,
   s. Man-PEI.

Drechslera tritici-repentis (holomorph
Pyrenophora t.) (C:149): on 1 Ont
[1611].

LENS Mill.   LEGUMINOSAE

Herbs of the Mediterranean region and w.
Asia. One species is cultivated for its
seeds which are used as food and forage.

1. *L. culinaris* Medic., lentil, lentille;
   imported from sw. Asia.

Alternaria sp.: on 1 Sask [1095, 1097].
Ascochyta sp.: on 1 Sask [1141].
Botrytis sp.: on 1 Sask [1139].

B. *cinerea* (holomorph *Botryotinia*
    fuckeliana): on 1 Sask [1095, 1097].
Cladosporium sp.: on 1 Sask [1097].
Fusarium sp.: on 1 Sask [1095, 1139].
F. roseum: on 1 Sask [1097].
Rhizoctonia sp.: on 1 Sask [1095, 1097,
    1139].
Sclerotinia sp.: on 1 Sask [420, 1095,
    1138], Man [1138].
S. sclerotiorum: on 1 Sask [1097].

LEONTODON L.   COMPOSITAE

Low, stemless perennials native in
Eurasia but naturalized in North America.

Puccinia hieracii (C:149): rust,
    rouille: 0 I II III on *L.* sp. Ont
    [1236].

LEONURUS L.   LABIATAE

Biennial or perennial herbs of temperate
Europe and Asia.

1. *L. cardiaca* L., motherwort, agripaume
   cardiaque; imported from Eurasia,
   introduced, BC, Sask-PEI.

Ascochyta leonuri: on leaves of 1 Ont
    [440].
Sphaerotheca fuliginea: powdery mildew,
    blanc: on 1 Que [1257].

LEPIDIUM L.   CRUCIFERAE

1. *L. perfoliatum* L., clasping
   pepperweed; imported from Eurasia,
   introduced, BC-Sask, Ont-Que.

Albugo candida (*A. cruciferarum*) (C:149):
    white rust, albugine: on *L.* sp. Sask
    [1319, 1323].
Alternaria brassicae: on *L.* sp. Sask
    [339].
A. raphani: on *L.* sp. Sask [339].
Colletotrichum dematium: on *L.* sp. Sask
    [1319].
Verticillium dahliae: on 1 BC [1816,
    1978].

LEPTARRHENA R. Br.   SAXIFRAGACEAE

1. *L. pyrolifolia* (Don) Ser.; streambanks
   and wet meadows to moist alpine and
   subalpine slopes, Yukon, BC, sw. Alta.

Pseudomassaria occidentalis: on 1 BC [71].

LESPEDEZA Michx.   LEGUMINOSAE

1. *L. capitata* Michx.; native, s. Ont.
2. *L. hirta* (L.) Hornem.; native, s. Ont.
3. *L. intermedia* (Wats.) Britt.; native,
   s. Ont.

Cercospora latens (*C. lespedezae*): on
    leaves of 1 Ont [440].
Microsphaera diffusa: powdery mildew,
    blanc: on 1, 2, *L.* sp. Ont [1257].

Uromyces lespedezae-procumbentis (C:149):
    rust, rouille: II III on 1, 2, 3, *L.*
    sp. Ont [1236].

LEWISIA Pursh   PORTULACACEAE

1. *L. rediviva* Pursh, bitter root, racine
   amère; native, BC.

Uromyces unitus ssp. *unitus* (C:150):
    rust, rouille: on 1 BC [1012, 1816].

LEYMUS Hochst.   GRAMINEAE

See discussion of generic
circumscriptions under *Agropyron*. The
following species have been segregated from
Elymus.

1. *L. angustus* (Trin.) Pilger (*Elymus
   a.* Trin.), altai wild rye; imported
   from USSR.
2. *L. cinereus* (Scribn. & Merr.) A. Löve
   (*Elymus c.* Scribn. & Merr., *E.
   piperi* Bowden); native, BC-Sask.
3. *L. condensatus* (Presl) A. Löve
   (*Elymus c.* Presl), grant wild rye;
   native to California. Canadian reports
   are probably *L. cinereus*.
4. *L. inovatus* (Beal) Pilger (*Elymus i.*
   Beal), fuzzy wild rye, élyme; native,
   BC-Ont.
5. *L. mollis* (Trin.) Pilger (*Elymus m.*
   Trin.), sea lyme grass, siègle de mer;
   native, Yukon-Labr, BC-Nfld. 5a. *L.
   m.* ssp. *mollis* var. *mollis* f.
   mollis (*Elymus arenarius* var.
   villosus Mey.); native, Mack-Labr,
   BC-Nfld. 5b. *E. m.* ssp.
   villosissimus (Scribn.) ined. (*Elymus
   m.* ssp. *v.* (Scribn.) Löve); native,
   Keew-n. Que.

Bipolaris sorokiniana (*Helminthosporium
    sativum*) (holomorph *Cochliobolus
    sativus*) (C:106): on 1, 2 Alta [133].
Cladosporium herbarum: on 5 BC [1012].
Claviceps purpurea (C:106): ergot,
    ergot: on 4 Alta [1594]; on 5 BC
    [1729]; on 5a BC [1816], Man, Que [666].
Erysiphe graminis (C:106): powdery
    mildew, blanc: on 1, 2 Alta [133]; on 2
    BC [1257].
Lophodermium arundinaceum (C:106): on 5
    BC [1012, 1816], Keew [437].
L. arundinaceum var. *alpinum* (C:106):
    on 5b Frank [1543].
Pleospora affin. *herbarum* (C:107): on
    5b Frank [1543].
Pseudoseptoria stomaticola (*Selenophoma
    donacis*) (C:107): on 2 BC [1728].
Puccinia coronata (C:107): crown rust,
    rouille couronée: on 4 BC [1012], Alta
    [644, 1839].
P. montanensis (C:107): rust, rouille:
    on 2 BC [1012]; on 3 BC [1816].
P. impatienti-elymi (*P. recondita* in
    part) (C:107): leaf rust, rouille des
    feuilles: on 5 BC [1729]. See comment
    under *Elymus*.

Pyrenophora tritici-repentis: on 4 Sask
[1594].
Septoria infuscans: on 2 BC [1728].
Stagonospora avenae f. sp. *triticea*
    (*Septoria a.* f. sp. *triticea*)
    (holomorph *Phaeosphaeria avenaria* f.
    sp. *triticea*) (C:107): on 5 BC [1729].

LIATRIS Schreber   COMPOSITAE

1. *L. aspera* Michx.; native, sw. Ont.
2. *L. cylindracea* Michx.; native, s. Ont.

Puccinia liatridis (C:150): rust,
    rouille: 0 I on 1, 2 Ont [1236].

LIGUSTICUM L.   UMBELLIFERAE

1. *L. canbyi* C. & R.; native, BC.
2. *L. scothicum* L., Scotch lovage,
    livêche écossaise; native, BC, Ont-Nfld,
    Labr.

Puccinia ligustici (C:150): rust,
    rouille: III on 1 BC [1551]; on 2 Que
    [1236, 1551].
Ramularia terrae-novae: on 2 Nfld [1551].
Septoria levistici (C:150): on 2 NB, NS
    [1551].
Uromyces lineolatus ssp. *nearcticus*:
    rust, rouille: on 2 Que, PEI [1559].

LIGUSTRUM L.   OLEACEAE

1. *L. amurense* Carrière, Amur privet,
    troène de l'Amour; imported from n.
    China.
2. *L. vulgare* L., common privet, troène
    commun; imported from Europe, natzd.,
    BC, Ont, NS, Nfld.

Cercosporella howittii (C:150): leaf
    spot, tache des feuilles: on 2 BC
    [1816].
Glomerella cingulata (anamorph
    Colletotrichum gloeosporioides)
    (C:150): anthracnose, anthracnose: on
    1 Ont [337].
Microsphaera penicillata (C:150): powdery
    mildew, blanc: on 2 Ont [1257].
Nectria cinnabarina (anamorph
    Tubercularia vulgaris) (C:150): on
    L. sp. NB [1032].

LILIUM L.   LILIACEAE

1. *L. amabile* Palib.; imported from Korea.
2. *L. auratum* Lindl., golden-banded lily,
    lis doré du Japon; imported from Japan.
3. *L. canadense* L., Canada lily, lis du
    Canada; native, Ont-NS.
4. *L. candidum* L., Madonna lily, lis
    blanc; imported from Europe.
5. *L. columbianum* Hanson ex Bak. (*L.
    parviflorum* (Hook.) Holz.), Columbia
    lily; native, BC.
6. *L. davidii* Duchartre ex Elwes;
    imported from China. 6a. *L. d.* var.
    wilmottiae (E.H. Wils.) Raffill (*L.
    wilmottiae* E.H. Wils.).

7. *L. hansonii* Leichtl. ex D.D.T. Moore,
    Japanese Turk's Cap; imported from Japan
    and Korea.
8. *L. humboldtii* Roezl & Leichtl. ex
    Duchartre, Humboldt lily; imported from
    California.
9. *L. lancifolium* Thunb. (*L. tigronum*
    Ker-Gawl), tiger lily, lis tigré;
    imported from China, Korea and Japan.
10. *L. longiflorum* Thunb., Easter lily,
    lis à longues fleurs; imported from
    Japan.
11. *L. martagon* L., martagon lily, lis
    martagon; imported from Eurasia.
12. *L. michiganense* Farw., Michigan lily;
    native, Man-Ont.
13. *L. pardalinum* Kellogg, leopard lily;
    imported from w. USA.
14. *L. philadelphicum* L., wood lily, lis
    de Philadelphie; native, BC-Que. 14a.
    L. p. var. *andinum* (Nutt.)
    Ker-Gawl., western orange-cup lily;
    native, BC-Que.
15. *L. pumilum* Delile (*L. tenuifolium*
    Fisch.), coral lily; imported from
    Siberia and China.
16. *L. regale* E.H. Wils., regal lily, lis
    royal; imported from w. China.
17. *L. speciosum* Thunb., showy lily;
    imported from Japan.
18. *L. superbum* L., Turk's-cap lily,
    superbe; native, e. USA.
19. Host species not named, i.e., reported
    as *Lilium* sp.

Aecidium sp.: on 5 BC [1012, 1816].
Botrytis cinerea (holomorph *Botryotinia
    fuckeliana*) (C:151): gray mold,
    moisissure grise: on 16 NS [336]; on 19
    BC, Ont [1545], Sask [336].
B. elliptica (C:151): blight, brûlure
    botrytique: on 1, 6a, 15, 19 Sask
    [339]; on 4, 8, 10, 16, 17 BC [1816]; on
    9 Que [333]; on 16 NS [334, 336]; on 19
    Alta [339, 1594], Sask [336], Que, NS
    [335].
Cercosporella inconspicua (C:151): leaf
    spot, tache ovale: on 14a Alta [1545];
    on 19 Man [1545].
Colletotrichum dematium (*Vermicularia
    liliacearum*) (C:151): on 19 Sask
    [1712, 1713].
C. gloeosporioides (holomorph *Glomerella
    cingulata*): on 19 Sask [338].
Cylindrocarpon destructans (*C. radicicola*)
    (C:151): root rot, chancre des
    racines: on 19 Sask [338, 1712, 1713].
Fusarium sp. (C:151): on 19 Sask [1712].
F. oxysporum f. sp. *lilii* (C:151): on
    19 BC [1594, 1713].
Leptosphaeria lilii: on leaves of 18 Ont
    [439].
Mycosphaerella allicina: on 14 Alta
    [1880].
Penicillium sp.: on 19 Sask [1713].
Phyllosticta sp.: on 19 BC [1012].
P. lilii: on leaves of 18 Ont [439].
    Apparently a species of *Coniosporium*.

Rhizoctonia solani (holomorph
   Thanatephorus cucumeris): on 19 Sask
   [338, 1712], Ont [338].
Rhizopus sp.: on 2 BC [1816].
Uromyces holwayi (C:151): rust, rouille:
   0 I II III on 3, 18, 19 Ont [1236,
   1545]; on 5 BC [1012]; on 5, 10, 13 BC
   [1816]; on 5, 13, 14 BC [1545]; on 9 Ont
   [1341]; on 12 Ont [1236].

LIMONIUM Mill.   PLUMBAGINACEAE

1. *L. carolinianum* (Walt.) Britt. (*L.
   nashii* Small), sea lavender, lavande de
   mer; native, Que-Nfld.
2. *L. latifolium* (Sm.) O. Kuntze,
   wide-leaved sea lavender, slatice à
   larges feuilles; imported from Canary
   Islands.
3. *L. vulgare* Mill., Mediterranean sea
   lavender, saladell; imported from
   Mediterranean.

Uromyces limonii (C:151): rust, rouille:
   0 I II III on 1 BC [1816]; on 3 Ont
   [1236].
U. limonii-caroliniani (C:151): rust,
   rouille: on 2 Que [282].

LINARIA Mill.   SCROPHULARIACEAE

1. *L. vulgaris* Mill., butter-and-eggs,
   gueule de lion; imported from Eurasia,
   natzd., Mack, BC-Nfld.

Alternaria fallax (*Macrosporium f.*): on 1
   Ont [195].
Ophiobolus rubellus: on 1 Ont [1620].

LINNAEA Gronov.   CAPRIFOLIACEAE

1. *L. borealis* L., twinflower, linnée
   boréale; native, Yukon-Keew, n.
   Que-Labr, BC-Nfld. 1a. *L. b.* var.
   longiflora Torr. (*L. b.* var.
   americana (J. Forbes) Rehd.); range of
   species.

Ceramothyrium linnaeae: on 1a Ont [798].
Gnomonia linnaeae: on 1a Que [79].
Metacoleroa dickiei (*Gibbera d., Venturia
   d.*) (C:152): on 1 Yukon, BC, Alta,
   Man, Ont, Que [306], Yukon [460], BC
   [1012], Alta [644], Ont [1340]; on 1a
   Mack, Alta [53], BC, Ont, Que [73], Que
   [70].
Mycosphaerella linnaeae: on overwintered
   leaves of 1a Quebec [144].
Phylleutypa wittrockii: on 1 BC [644,
   1012]; on 1a BC [1816].
Thelephora terrestris: on 1 BC [1012].
   See *Pinus*.

LINUM L.   LINACEAE

1. *L. perenne* L., perennial flax, lin
   vivace; Yukon-Frank, BC-Que. 1a. *L.
   p.* var. *lewisii* (Pursh) Eat. & Wright
   (*L. lewisii* Pursh), wild blue flax,

lin de Lewis; range of species. 1b. *L.
   p.* var. *lewisii* f. *lepagei* (Boivin)
   Lepage (*L. lepagei* Boivin); native,
   Man-Ont.
2. *L. rigidum* Pursh, yellow flax; native,
   Alta-Man.
3. *L. usitatissimum* L., common flax, lin;
   imported from Eurasia, escaped from
   cultivation. Mack, BC-Nfld.
4. Host species not named, i.e., reported
   as *Linum* sp.

Alternaria alternata: on 3 Man [1114].
A. brassicae: on seeds of 3 Alta, Sask
   [1314].
A. linicola (C:152): brown stem blight,
   alternariose: on 3 Alta [1314] on
   seeds, Sask [336, 1314 on seeds, 1594];
   on 4 Sask [333].
A. raphani: on seeds of 3 Alta, Sask
   [1314].
Aspergillus sp.: on seeds of 3 Man [1894].
A. candidus: on 3 Man [1114].
A. glaucus group: on 3 Man [1114].
Aureobasidium lini (*Polyspora l.*)
   (C:153): browning and stem break,
   oxychromose: on 3 Alta [721], Sask
   [339, 1314].
Bipolaris sorokiniana (holomorph
   Cochliobolus sativus) (C:152): on 3
   Man [1114].
Botrytis cinerea (holomorph *Botryotinia
   fuckeliana*) (C:152): on seeds of 3
   Alta, Sask [1314].
Cephalosporium acremonium: on 3 Man
   [1114]. See comment under *Avena*.
Cladosporium cladosporioides (C:152): on
   3 Man [1114].
Diplogelasinospora princeps: on 4 Ont
   [208].
Embellisia chlamydospora (*Pseudostemphylium
   c.*): on 4 Sask [1618, 1657, 1760].
Epicoccum purpurascens (*E. nigrum*)
   (C:152): on 3 Man [1114].
Fusarium spp. (C:153): on 3 Man [1114];
   on 4 Sask [333, 1864].
F. acuminatum (C:153): on 3 Man [1114].
F. avenaceum (C:153): on 3 Man [1114].
F. oxysporum (C:153): on 3 Man [1114].
F. oxysporum f. sp. *lini* (C:153): wilt,
   flétrissure fusarienne: on 3 Alta [223,
   334], Sask [338], Ont [337], Que [337,
   870]; on 4 Sask, Que [975].
F. oxysporum var. *redolens* (C:153): on
   4 Man [333].
F. poae (C:153): on 3 Man [1114].
F. roseum: on seeds of 3 Alta, Sask
   [1314].
F. sacchari: on 3 Man [1114].
F. sambucinum: on 3 Man [1114].
F. sporotrichioides: on 3 Man [1114].
F. sulphureum: on 3 Man [1114].
F. tricinctum: on 3 Man [1114].
Gonatobotrys simplex (C:152): on 3 Man
   [1114].
Melampsora lini (C:153): rust, rouille:
   0 I II III on 1a Yukon [460], BC [226,
   1012]; on 1a, 3 BC [1816]; on 1b, 3 Ont
   [1236]; on 2, 4 NWT-Alta [1063]; on 3
   Alta [223, 640], Alta, Sask [334, 757],

Sask [336], Sask, Man [338, 2011], Man
[335, 338, 751, 757, 1595]; on 4 Mack
[53], Sask, Man [756], Man [755, 1838,
1840].
Microascus cirrosus: on 4 Ont [86].
Mucor sp.: on 3 Man [1114].
Paecilomyces variotii (C:152): on seeds
of 3 Ont [150].
Penicillium sp.: on 3 Man [1114] and on
seeds [1894].
Phoma medicaginis: on 3 Sask [920].
Pythium sp. (C:153): seedling blight and
root rot, brûlure des semis et pourridié
pythien: on 3 Alta, Man [338].
P. ultimum (C:154): seedling blight and
root rot, brûlure des semis et pourridié
pythien: on 4 Sask [333, 1864].
Rhizoctonia spp.: on 3 Alta [338, 1594],
Sask [1594], Man [338].
R. solani (*R. praticola*) (holomorph
Thanatephorus cucumeris) (C:154): on
3 BC, Alta [334], Sask [336], Que [337];
on 4 Sask [333, 1864], Man [333].
Rhizopus arrhizus: on 3 Man [1114].
Sclerotinia spp.: on 3 Sask [420, 1138],
Man [1138].
Selenophoma sp.: on 1a Sask [1323].
Possibly a specimen of *Selenophoma
linicola*.
Septoria linicola (holomorph
Mycosphaerella linorum): on 3 BC
[1816], Sask [337, 339], Man [335, 338],
Ont [336], Que [337]; on 4 Man [1507].
Stemphylium botryosum (holomorph
Pleospora herbarum) (C:154): on seeds
of 3 Ont [667].
Thanatephorus cucumeris (*Pellicularia
praticola*) (anamorph *Rhizoctonia
solani*) (C:153): seedling blight and
root rot, brûlure des plantules et
pourriture des racines: on 3 BC [1816],
Sask [1874].

LIQUIDAMBAR L.   HAMAMELIDACEAE

1. *L. styraciflua* L., sweet gum, copalme
d'Amérique; s. Ont.

Nectria cinnabarina (anamorph
Tubercularia vulgaris): on *L.* sp.
Ont [336].
Phomopsis sp.: on 1 BC [1012].

LIRIODENDRON L.   MAGNOLIACEAE

1. *L. tulipifera* L., tulip tree, tulipier
de Virginie; s. Ont.

Phyllactinia guttata: powdery mildew,
blanc: on 1 Ont [1257].
Trichothecium roseum (C:154): on 1 Ont
[1340]; on *L.* sp. Ont [F60].

LITHOPHRAGMA (Nutt.) Torr. & Gray
                                SAXIFRAGACEAE

1. *L. parviflora* (Hook.) Nutt.,
fringe-cup; native, BC-Alta.

Puccinia lithophragmae: rust, rouille:
on 1 BC [1560].

LLOYDIA Salisb. ex Reichb.   LILIACEAE

1. *L. serotina* (L.) Rchb., alpine lily;
native, Yukon-Mack, BC.

Puccinia kukkonenis (C:154): rust,
rouille: on 1 Yukon [1545].

LOBELIA L.   LOBELIACEAE

1. *L. siphilitica* L., blue cardinal
flower, cardinale bleue; native, Man-Ont.

Puccinia lobeliae (C:154): rust,
rouille: III on 1 Ont [1236].

LOLIUM L.   GRAMINEAE

1. *L. multiflorum* Lam., Italian rye
grass, ivraie d'Italie; imported from
Europe, natzd., Yukon, BC, Sask-Nfld.
2. *L. perenne* L., common darnel, ivraie
vivace; imported from Eurasia, natzd.,
BC-Nfld.

Bipolaris sorokiniana (holomorph
Cochliobolus sativus) (C:155): on 2
Alta [120] by inoculation.
Claviceps sp.: on 2 Ont [1781].
C. purpurea (C:155): ergot, ergot: on 1,
2 BC [1816]; on 2 NB [333, 1195].
Drechslera siccans (C:155): leaf blight,
brûlure drechsleréenne: on 1, 2, *L.*
sp. Ont [1611].
Ligniera pilorum: on 1 Ont [60].
Puccinia coronata (C:155): crown rust,
rouille couronnée: II III on 1, 2 BC
[1816]; on 2 Que [1236].
Ramularia pusilla (C:155): eye spot,
tache ocellée: on 1, 2 BC [1816].
Rhizophydium graminis: on 2 Ont [59].
Rhynchosporium secalis (C:155): scald,
tache pâle: on 2 BC [1816].
Sclerotinia borealis (C:155): snow mold,
moisissure nivéale: on 2 BC [1816].

LOMATIUM Raf.   UMBELLIFERAE

1. *L. ambiguum* (Nutt.) Coult. & Rose;
native, BC.
2. *L. brandegei* (Coult. & Rose) Macbr.;
native, BC.
3. *L. dissectum* (Nutt.) Mathias & Const.;
native, BC-Sask. 3a. *L. d.* var.
multifidum (Nutt.) Mathias & Const.
(*Leptotaenia multifida* Nutt.); BC-Alta.
4. *L. geyeri* (Wats.) Coult. & Rose;
native, BC-Man.
5. *L. macrocarpum* (Hook. & Arn.) Coult. &
Rose; native, BC-Man.
6. *L. martindalei* Coult. & Rose; native,
BC. Most Canadian material is referable
to *L. m.* var. *angustatum* Coult. &
Rose.
7. *L. triternatum* (Pursh) Coult. & Rose,
bush parsnip; native, BC-Alta.

8. *L. utriculatum* (Nutt.) Coult. & Rose,
   spring gold; native, BC.

Euryachora sp.: on 3 BC [1012].
Pollaccia peucedani (Asperisporium p.):
   on 2, 6 BC, on 5 Alta [1551].
Puccinia jonesii ssp. *asperior*
   (P. asperior) (C:155): rust,
   rouille: on 1, 3, 4, 5, 7, 8 BC [1551];
   on 3a BC [1816].
P. jonesii ssp. *jonesii*: rust,
   rouille: on 3a BC [1012]; on 3 Alta and
   on 3, 4, 5 BC [1551].
P. jonesii ssp.: rust, rouille: on 3, 5
   BC [1551]. Intermediate between ssp.
   jonesii and *asperior*.

LONICERA L.   CAPRIFOLIACEAE

1. *L. canadensis* Bartr., fly-honeysuckle,
   chèvrefeuille du Canada; native, Ont-PEI.
2. *L. ciliosa* (Pursh) Poir., western
   trumpet honeysuckle; native, BC.
3. *L. dioica* L., climbing honeysuckle,
   chèvrefeuille dioique; native, Mack,
   BC-Que. 3a. *L. dioica* var.
   glaucescens (Rydb.) Butters *(L.
   glaucescens* (Rydb.) Rydb.), orange
   honeysuckle, chèvrefeuille
   gris-bleuâtre; native, Mack, BC-Que.
4. *L. hirsuta* Eat., hairy honeysuckle,
   chèvrefeuille hirsute; native, Ont-Que.
5. *L. hispidula* Dougl. ex Torr. & Gray;
   native, BC.
6. *L. involucrata* (Richardson) Banks,
   black twinberry, chèvrefeuille
   involucrés; native, BC-NB.
7. *L. oblongifolia* (Goldie) Hook., swamp
   fly-honeysuckle, chèvrefeuille à
   feuilles oblongues; native, Sask-NS.
8. *L. periclymenum* L., woodbine,
   chèvrefeuille des bois; imported from
   Europe, a garden escape, BC, Ont, NS and
   Nfld.
9. *L. tatarica* L., Tatarian honeysuckle,
   chèvrefeuille de Tartarie; imported from
   Eurasia, a garden escape, Alta-NS.
10. *L. utahensis* S. Wats., red twinberry,
   chèvrefeuille de l'Utah; native, BC-Alta.
11. Host species not named, i.e., reported
   as *Lonicera* sp.

Ascochyta sp. (C:156): leaf and twig
   blight, brûlure des feuilles et
   ramilles: on 11 BC [333].
Botrytis cinerea (holomorph *Botryotinia
   fuckeliana*) (C:156): gray mold,
   moisissure grise: on 11 NB [333, 1032,
   1194].
Cercospora antipus (C:156): leaf spot,
   tache cercosporéenne: on 2 BC [1012,
   1816]; on 11 Sask [F66].
C. periclymeni: on 2 BC [1012].
Diaporthe eres: on 6 Ont [1340].
Diplodia lonicerae: on 9 Que [338].
Glomopsis lonicerae (Glomerularia l.)
   (holomorph *Herpobasidium deformans*):
   on 9 Ont, Que [1760]; on 11 NS [1594].
Herpobasidium deformans (anamorph
   Glomopsis lonicerae) (C:156): leaf

blight, brûlure des feuilles: on 1 NS
[1032]; on 9 Ont [333], Que [335, 336,
337, 338], NB [336]; on 11 NWT-Alta
[1063], Alta [F65], Que [334, 335, 339,
F65].
Hyponectria lonicerae (Guignardia l.): on
   5 BC [1012]; on 7 Ont [77].
Kabatia lonicerae: on 1, 7, 11 Ont
   [1340]; on 11 Ont [F65].
K. lonicerae var. *americana* (C:156):
   leaf spot, tache ponctuée: on 1 Que
   [282].
K. lonicerae var. *involucrata* (C:156):
   leaf spot, tache ponctuée: on 6 BC
   [1816], Alta [644].
K. lonicerae var. *periclymeni*: leaf
   spot, tache ponctuée: on 10 BC [1816].
K. mirabilis var. *oblongifoliae*
   (C:156): causing leaf lesions on 7 Ont
   [1763].
Lasiobotrys lonicerae: on 3a BC [644,
   953].
Lophium crenatum: on 6 Ont [1340]. The
   validity of this combination could not
   be confirmed.
Microsphaera penicillata (M. alni)
   (C:156): powdery mildew, blanc: on 1,
   3, 4, 11 Ont, Que [1257]; on 5 BC
   [1012]; on 5, 6, 8 BC [1816]; on 6, 9
   Ont [1257]; on 9, 11 Sask, Man [1257];
   on 11 BC [1257], Ont [1340], NB [1194].
M. penicillata var. *lonicerae* (C:156):
   powdery mildew, blanc: on 9 Man [884],
   Ont [333, 336], Que [336, 337]; on 11
   Ont [334], Que, NB [333].
Mycosphaerella minor (C:156): on 1 Que
   [70].
Mycovellosiella nopomingensis: on 3a, 11
   Sask, Man [1760].
Ophiobolus minor (C:156): on 6 BC [1620].
Phellinus ferreus (Poria f.) (C:156): on
   2 BC [1012, 1816]. See *Acer*.
Pseudospiropes saskatchewanensis: on 3a
   Sask [1760].
Stereum sp.: on 2 BC [1012]. Lignicolous.
Verticillium sp.: on 11 Ont [96, 97].
V. albo-atrum: on 11 Ont [38, 98].

LOTUS L.   LEGUMINOSAE

1. *L. corniculatus* L., birdsfoot-trefoil,
   patte d'oiseau; imported from Eurasia,
   introduced, BC-Alta, Man-NS, Nfld.
2. *L. pedunculatus* Cav. *(L. uliginosus*
   Schk.); introduced from Eurasia, s. BC,
   Sask, Ont-Que, NS.

Acremomium boreale: on 1 BC, Alta, Sask
   [1707].
Alternaria sp.: on 1 Alta [132].
Fusarium sp.: on 1 Alta [132].
Gliocladium sp.: on 1 Alta [132].
Papulaspora sp.: on 1 Alta [132].
Penicillium sp.: on 1 Alta [132].
Phoma sp. (C:156): leaf and stem spot,
   tache de la tige et des feuilles: on 1
   BC [1816], Alta [132].
P. sclerotioides (Plenodomus meliloti):
   on 1 Alta [128].
Pyrenochaeta sp.: on 1 Alta [132].

Rhizopus sp.: on 1 Alta [132].
Sclerotinia trifoliorum (C:156): wilt, fletrissure sclerotique: on 1 BC [1816], Que [63, 391]; on 2, *L.* sp. Que [63] by inoculation.
Stemphylium loti: stemphylium leaf spot, tache stemphylienne: on 1 Alta [132, 1594], Que [32, 339].
Verticillium albo-atrum: on 1 Que [36] by inoculation.
V. dahliae: on 1 Que [36] by inoculation.

LUNARIA L.   CRUCIFERAE

1. *L. annua* L. *(L. biennis Moench)*, honesty, lunaire; imported from Europe, a garden escape, BC, Man, Ont-Que, NS.

Helminthosporium lunariae (C:156): leaf spot, tache des feuilles: on 1 BC [1816].
Pythium polymastum: on 1 Sask [1873].
Septoria lunariae: on 1 Ont [439].

LUPINUS L.   LEGUMINOSAE

1. *L. albus* L., white lupine, lupin blanc; imported from Europe.
2. *L. arboreus* Sims, tree lupine; imported from California, introduced into sw. BC.
3. *L. arcticus* S. Wats.; native, Yukon-Frank, BC-Alta.
4. *L. nootkatensis* J. Donn, lupine, lupin; native, Yukon, BC-Alta, introduced, NS, Nfld.
5. *L. perennis* L., perennial lupine, lupin vivace; native, Ont.
6. *L. polyphyllus* Lindl., many-leaved lupine, lupin polyphylle; native, BC.
7. *L.* x *regalis* Bergermans *(L. polyphyllus* x *L. arboreus).*
8. *L. rivularis* Dougl.; native, BC.
9. Host species not named, i.e., reported as *Lupinus* sp.

Ascochyta sp. (C:157): leaf spot, tache des feuilles: on 9 BC [1816].
Botrytis cinerea (holomorph *Botryotinia fuckeliana*) (C:157): gray mold, moisissure grise: on 9 Que [335].
Cylindrosporium longisporum: on leaves of 5 Ont [439]. Apparently a species of *Cercospora*.
Erysiphe sp.: powdery mildew, blanc: on 7 BC [1012].
E. communis (E. polygoni) (C:157): powdery mildew, blanc: on 4, 9 BC [1816], on 4, 5, 6 BC [1257]; on 9 Yukon [460, 1257].
E. trifolii: powdery mildew, blanc: on 4 BC [1012]; on 5 Ont [1257].
Lachnella alboviolascens: on 2 BC [1424].
L. villosa: on 2 BC [1424].
Ovularia lupinicola (C:157): leaf spot or eye spot, tache ocellée: on 3, 9 BC [1816].
Peronospora trifoliorum (C:157): downy mildew, mildiou: on 9 BC [1816].
Thecaphora deformans: on 9 Alta [337].

Uromyces lupini (C:157): rust, rouille: on 8 BC [1012]; on 8, 9 BC [1816].
Verticillium albo-atrum: on 1 Que [36]; on 9 Que [1288].
V. dahliae: on 1 Que [36].

LUZULA DC.   JUNCACEAE

1. *L. acuminata* Raf., acuminate woodrush, luzule acuminée; native, Alta-Nfld.
2. *L. confusa* Lindeberg; native, Yukon-Labr, BC-Man, Que.
3. *L. nivalis* (Laest.) Beurl.; native, Yukon-Labr.
4. *L. parviflora* (Ehrh.) Desv., small-flowered woodrush, luzule parviflore; native, Yukon-Keew, n. Que-Labr, BC-Nfld. 4a. *L. p.* var. *melanocarpa* (Michx.) Buch *(L. piperi* (Cov.) Jones); native, essentially range of species. 4b. *L. p.* var. *parviflora*; native, range of species.

Botrytis cinerea (holomorph *Botryotinia fuckeliana*) (C:158): gray mold, moisissure grise: on 2 Frank [664, 1586].
Hysteropezizella pusilla (C:158): on 3 Frank [664].
Puccinia obscura (C:158): rust, rouille: II III on 1 Ont [1236]; on 4 BC [1012, 1816].
Thanatephorus cucumeris (Pellicularia filamentosa) (anamorph *Rhizoctonia solani*) (C:158): on 3 Frank [1586].
Ustilago vuijckii: smut, charbon: on 4a BC and on 4b Yukon, Alta [679].

LYCHNIS L.   CARYOPHYLLACEAE

1. *L. chalcedonica* L., Maltese cross, croix de Jérusalem; imported from Asia, a garden escape, BC, Sask-PEI.
2. *L. coronaria* (L.) Desr., rose campion, coquelourde des jardins; a garden escape, BC, Ont-Que.

Erysiphe cichoracearum: powdery mildew, blanc: on 1 NB [1036].
Phyllosticta lychnidis (C: 159): leaf spot, tache des feuilles: on 1 BC [1816]; on *L.* sp. Ont [339].
Septoria lychnidis (C:159): leaf spot, tache des feuilles: on 2 NB [333, 1194].

LYCIUM L.   SOLANACEAE

1. *L. chinense* Mill., Chinese matrimony vine, lyciet de Chine; imported from e. Asia.
2. *L. halimifolium* Mill., common matrimony vine, lyciet; imported from Eurasia, spreading from cultivation, BC-Sask, Ont, NS.

Oidium sp.: powdery mildew, blanc: on 1, 2 BC [1816]; on 2 BC [1594].
Puccinia tumidipes (C:159): rust, rouille: II III on 1, 2 Ont [1277]; on 2 Ont [1236].

LYCOPERSICON (LYCOPERSICUM) Mill.

SOLANACEAE

1. *L. cheesemanii* Riley; imported from
   the Galapagos Islands.
2. *L. glandulosum* Muller; imported from
   Peru.
3. *L. hirsutum* H.B.K.; imported from
   South America.
4. *L. lycopersicum* (L.) Karst.
   (*Lycopersicum esculentum* Mill.),
   tomato, tomate; imported from South
   America.
5. *L. peruvianum* (L.) Mill.; imported
   from Peru.
6. *L. pimpinellifolium* (Jusl.) Mill.,
   currant tomato; imported from Peru and
   Ecuador.
7. *L. esculentum* x *pimpinellifolium*,
   hybrid.

Alternaria sp.: on 4 NS [1594].
A. alternata (A. tenuis auct.*)* (C:159):
   on 4 BC [337, 338, 339, 1416, 1594], NS
   [713, 977, 978, 979, 985, 992].
A. porri f. sp. *solani*: on 4 Ont [99,
   100], Que [324].
A. solani (C:159): early blight, brûlure
   alternarienne: on 4 BC [337, 1416,
   1594, 1816], BC, Ont [1595], BC, Ont,
   Que, NB [336], BC, NB, NS [335, 338],
   Alta, Man [333], Alta, NB, NS [334], Ont
   [1460, 1461], Ont, NB, NS [333, 337,
   339, 1594], Que [338].
A. tomato (C:159): nailhead spot, tête de
   clou: on 4 BC [1594, 1816].
Botrytis cinerea (holomorph *Botryotinia
   fuckeliana*) (C:159): gray mold,
   moisissure grise: on 4 BC [1816], Alta
   [1594], Ont [1460, 1461], Ont, Que, NS
   [335], Ont, NB [339], Ont, NB, NS [334,
   336, 338, 1594], Ont, NS [333], NB, NS
   [337], NS [709, 713, 977, 978, 979, 985,
   992], PEI, Nfld [336].
Colletotrichum sp.: on 4 Ont [1460], NS
   [1594].
C. coccodes (C. atramentarium) (C:160):
   anthracnose, anthracnose: on 4 BC
   [1595, 1816], BC, Ont [336, 1594], BC,
   Ont, Que, NB [339], BC, NS [338], Ont,
   Que, NS [336], Ont, NS [333, 334], Que
   [324, 894], NS [335, 593, 713, 977, 978,
   979, 985, 992].
C. dematium (C:160): on 4 NS [593] by
   inoculation.
C. gloeosporioides (C. phomoides)
   (holomorph *Glomerella cingulata*)
   (C:160): on 4 BC, Que [337].
Fulvia fulva (Cladosporium f.) (C:159):
   leaf mold, moisissure olive: on 3 Ont
   [1292]; on 4 BC [1816, 1817, 1818], BC,
   Ont, NS [333, 338], BC, NS [335], Ont
   [187], Ont, NB [339], Ont, NS [336,
   337], Que [348], NS [334].
Fusarium sp. (C:160): on 4 Ont [1460,
   1461, 1594], NB [336, 1594], NS [977,
   985, 992, 1594].
F. oxysporum (C:160): wilt, flétrissure
   fusarienne: on 2, 3, 4, 5, 6, 7 Ont
   [855]; on 4 Ont [849], Que [232, 1800].

F. oxysporum f. sp. *lycopersici (F.
   lycopersici)* (C:160): wilt,
   flétrissure fusarienne: on 4 BC [28],
   Alta, Sask [335], Ont [334, 339, 1460,
   1594], NB [1594], NS [333].
F. oxysporum f. sp.
   radicis-lycopersici: on 4 Ont [848],
   Que [233].
Glomerella cingulata (anamorph
   Colletotrichum gloeosporioides): on 4
   Ont [838].
Mucor sp.: on 4 NS [977, 992].
Ophiobolus rubellus: on *L.* sp. Ont
   [1340].
Penicillium spp.: on 4 NS [713, 977, 978,
   992].
Phoma sp.: on 4 NS [979].
P. destructiva (C:160): fruit rot,
   pourriture phoméene: on 4 BC [1816], NS
   [336, 713, 977, 978, 985, 992].
P. lycopersici (Ascochyta l.) (holomorph
   Didymella l.): on 4 BC [1816].
P. terrestris (Pyrenochaeta t.) (C:161):
   pink root, racine rose: on 4 BC [334,
   1816]. Causes pink root of onions and
   reported from roots of other crops.
Phytophthora cactorum (C:160): fruit rot,
   mildiou polyphage des fruits: on 4 BC
   [1816].
P. infestans (C:160): late blight,
   mildiou: on 4 BC [333, 334, 335, 1816],
   Alta [339], Ont [333, 1459], Que [324,
   336, 337, 1594], NB [335, 336, 338,
   1594], NS [334, 335, 336, 337, 709, 713,
   978, 979, 985]; PEI [334, 337], Nfld
   [333].
P. parasitica (C:161): buckeye rot and
   stem rot, mildiou zoné: on 4 BC [336,
   1816], Ont [334, 338, 339].
Pleospora herbarum (anamorph *Stemphylium
   botryosum*): on 4 NS [597].
Preussia funiculata: on *L.* sp. Man
   [207].
Pythium sp. (C:161): damping off, fonte
   des semis: on 4 Ont [1460, 1461, 1594].
P. irregulare: on 4 Ont [868] by
   inoculation.
Rhizoctonia solani (holomorph
   Thanatephorus cucumeris) (C:161):
   damping-off and root rot, fonte des
   semis et rhizoctone commun: on 4 Alta
   [339], Ont [334, 336, 1460, 1461, 1594].
Sclerotinia sp.: on 4 Sask [420, 1138],
   Man [1138], NS [979, 992].
S. sclerotiorum (C:161): stem rot,
   pourriture sclérotique: on 4 BC [334,
   337, 338, 1816], Alta [336], NB [336,
   337, 1594], NS [333, 334, 335, 336, 337,
   338, 713, 977, 978], PEI [333].
Septoria lycopersici (C:161): leaf spot,
   tache septorienne: on 4 BC [1816], Man
   [333, 335], Ont [114, 333, 335, 339,
   1460, 1594], Que [324, 334, 336, 338].
Stemphylium botryosum: on 4 BC [1416].
S. botryosum f. sp. *lycopersici*: on 4
   NS [597].
S. solani (C:161): gray leaf spot, tache
   grise: on 4 Ont [334].
Stilbella flavescens: on 4 Que [2], a
   symptomless parasitic fungus.

Thanatephorus cucumeris (Pellicularia
 filamentosa) (anamorph *Rhizoctonia*
 solani) (C:160): on 4 Alta, Ont, NB
 [337].
Ulocladium consortiale (Stemphylium c.):
 on seeds of 4 Ont [667].
Verticillium sp. (C:161): wilt,
 flêtrissure verticillienne: on 4 BC
 [334, 338], Ont [96, 97, 161, 338], NS
 [985].
V. albo-atrum (C:161): wilt, flêtrissure
 verticillienne: on 4 Alta [335, 336],
 Ont [38, 98, 333, 335, 1087], Que [36,
 38, 246, 392, 393], NS [333, 334, 335,
 338].
V. dahliae (C:161): wilt, flêtrissure
 verticillienne: on 4 BC [38, 333, 335,
 336, 337, 339, 384, 1816, 1817, 1818,
 1978], Alta [335, 339], Ont [38, 334,
 335, 336, 337, 1087, 1460, 1594], Que
 [36, 38, 392, 393, 676, 1084], NS [339].
V. tenerum (V. lateritium) (holomorph
 Nectria inventa): wilt, flêtrissure
 verticillienne: on 4 Que [39] by
 inoculation.
V. nigrescens: wilt, flêtrissure
 verticillienne: on 4 Man [749], Ont
 [1087], Que [38, 392, 393].

LYCOPODIUM L.   LYCOPODIACEAE

1. *L. annotinum* L., bristly club moss,
 lycopode innovant; native, Yukon-Labr,
 BC-Nfld. 1a. *L. a.* var. *acerifolium*
 Fern.; native, transcontinental. 1b.
 L. a. var. *pungens* (La Pylaie)
 Desv.; native, Yukon-Labr, BC-Nfld.
2. *L. lucidulum* Michx., shining clubmoss,
 lycopode brillant; native, Man-Nfld.
3. *L. obscurum* L., groundpine, petits
 pins; native, BC-Nfld. 3a. *L. o.*
 var. *dendroideum* (Michx.) Eat.;
 native, BC-Nfld.

Conoplea geniculata: on *L.* sp. Sask
 [1760].
Dasyscyphus pteridis: on *L.* sp. Que
 [1392].
Epicoccum purpurascens: on 3a Man [1760].
Leptosphaeria lycopodina: on 1 Ont [1340,
 1827]; on 1b Que [70]; on *L.* sp. Ont,
 Nfld [760].
L. marciensis: on 1a, 1b Que [70].
Mycosphaerella lycopodii: on 1a, 1b Que
 [70].
Pseudomassaria lycopodina: on 1, 2 Ont
 [71].

LYCOPUS L.   LABIATAE

1. *L. americanus* Muhl., cut-leaf
 bugleweed, lycope d'Amérique; native,
 BC, Man-Que, NS-Nfld.
2. *L. asper* Greene; native, BC-Que.
3. *L. rubellus* Moench; native to e. USA.
4. *L. uniflorus* Michx., bugleweed, lycope
 uniflore; native, BC-Nfld, Labr.
5. *L. virginicus* L., bugleweed; native to
 e. USA.

Puccinia angustata (C:163): rust,
 rouille: 0 I on 1, 2, *M.* sp. Ont
 [1236]; on 2 Sask [1559]; on 3, 4, 5 Ont
 [1236, 1559]; on 4 Que, PEI [1559].

LYGODESMIA D. Don   COMPOSITAE

1. *L. juncea* (Pursh) Don, skeletonweed;
 native, BC-Man.

Puccinia stipae (C:163): rust, rouille:
 on 1 BC [1012, 1816].

LYSICHITON Schott.   ARACEAE

1. *L. americanum* Hult. & St.John, western
 skunk cabbage; native, BC.

Mitrula elegans: on 1 BC [1012].

LYSIMACHIA L.   PRIMULACEAE

1. *L. ciliata* L., fringed loosestrife,
 lysimaque ciliée; native, BC-PEI.
2. *L. thyrsiflora* L., tufted loosestrife,
 lysimaque thyrsiflore; native, Mack,
 Alta-PEI.

Clypeoporthella kriegeriana: on *L.* sp.
 Que [79].
Puccinia dayi (C:163): rust, rouille:
 III on 1 Ont [1236].
P. limosae (P. caricina var. *limosae)*
 (C:163): rust, rouille: 0 I on 2 BC
 [1550] only trace found, Ont [1236].
Uromyces acuminatus (C:163): rust,
 rouille: 0 I on 1 Ont [1236].

MADIA Molina   COMPOSITAE

1. *M. sativa* Molina, tarweed, madia
 cultivé; imported from w. USA and Chile,
 introduced, BC, Ont-Que, Nfld.

Coleosporium madiae (C:163): rust,
 rouille: on 1 BC [2008].

MAHOBERBERIS C.K. Schneider   BERBERIDACEAE

1. *M. neubertii* (Hort. ex Lem.) C.K.
 Schneid. var. *laxifolia (Berberis l.*
 Hort.); a hybrid between *Mahonia*
 aquifolium and *Berberis vulgaris*.

Puccinia graminis: rust, rouille: 0 I on
 1 Ont [1236].

MAHONIA Nutt.   BERBERIDACEAE

1. *M. aquifolium* (Pursh) Nutt. *(Berberis*
 a. Pursh), mahonia, mahonia à feuilles
 de houx; native, BC-Alta; introduced,
 Ont-Que.
2. *M. nervosa* (Pursh) Nutt. *(Berberis*
 n. Pursh*)*, Oregon grape, mahonia
 nervé; native, BC.
3. *M. repens* (Lindl.) G. Don *(Berberis*
 r. Lindl.*)*, creeping mahonia, mahonia
 rampant; native, BC-Alta.

Coccomyces coronatus: on 1, 2 BC [1816].

C. dentatus (C:164): on 1, 2 BC [1012].
Cumminsiella mirabilissima (C:164): rust,
    rouille: 0 I II III on 1 BC [1012,
    1415, 1416, 1816], Ont [1236]; on 2 Ont
    [336]; on 3 BC [560, 644, 1724]; on *M.*
    sp. Ont [1341, 1415].
Dennisiella babingtonii (Chaetothyrium
    b.): on 2 BC [1012].
Mycosphaerella sp.: on 2 BC [1012, 1816].
Phyllosticta sp.: on 2 Ont [337].
Puccinia brachypodii var. *arrhenatheri*:
    rust, rouille: on 1 BC [1012].
P. koeleriae (C:164): rust, rouille: on
    1 BC [1816].
P. pygmaea: rust, rouille: on 1 BC
    [1594].

MAIANTHEMUM Weber   LILIACEAE

1. *M. canadense* Desf., wild
    lily-of-the-valley, muguet; native,
    Mack, BC-Nfld, Labr. 1a. *M. c.* var.
    interius Fern.; native, Mack, BC-Ont.
2. *M. dilatatum* (Wood) Nels. & Macbr.,
    western maianthemum, maiantheme de
    l'ouest; native, BC.

Puccinia sp.: rust, rouille: on 1a
    NWT-Alta [1063].
Ramularia subsanguinea (C:164): on 1a BC
    [1545, 1816]; on 2 BC [1012, 1545, 1816].
Uromyces acuminatus var. *magnatus*
    (C:164): rust, rouille: on 1 NS [1032].

MALUS Mill.   ROSACEAE

1. *M. baccata* (L.) Borkh. *(Pyrus b.*
    L.), Siberian crabapple, pommetier
    microcarpe de Sibérie; imported from
    Asia.
2. *M. coronaria* (L.) Mill. *(Pyrus c.*
    L.), wild crabapple, pommier sauvage;
    native, Ont.
3. *M. floribunda* Siebold ex Van Houtte
    (Pyrus f. Siebold ex Kirsch*)*,
    Japanese crabapple, pommetier floribund;
    imported from Japan.
4. *M. fusca* (Raf.) C.K. Schneid. (*M.*
    diversifolia (Bong.) Roem., *M.*
    rivularis Dougl.), Pacific crabapple,
    pommier du Pacifique; native, BC.
5. *M. ioensis* (Wood) Britt. *(Pyrus i.*
    (Wood) L.H. Bailey), prairie crabapple,
    pommetier d'Iowa. 5a. *M. i.* cv.
    plena, Bechtel's crabapple.
6. *M. pumila* Mill. (*M. domestica*
    Borkh., *M. sylvestris* Mill., *Pyrus*
    malus L.), common apple, pommier;
    imported from Eurasia.
7. Host species not named, i.e., reported
    as *Malus* sp.

Acremonium kiliense: on 6 Que [319].
Alternaria sp.: on 6 BC [338], Ont
    [1182], Que [388], NB [1594], NS [995,
    1594].
A. alternata: on 6 Que [319], NS [1490].
A. mali (C:164): leaf spot and core rot,
    alternariose: on 6 Que [333], NB [1032].

Armillaria mellea (C:164): armillaria
    root rot, pourridié-agaric: on 6 NS
    [1594]; on 7 BC [1012]. See *Abies*.
Aureobasidium pullulans: on 6 Que [338].
Botryosphaeria obtusa (Physalospora o.)
    (C:164): black rot, pourriture noire:
    on 1 Sask [1614]; on 6 Man [337], Ont
    [338, 1614], Que [1614], NS [1490]; on 7
    Que [338], NS [1032].
Botryotinia fuckeliana (anamorph *Botrytis*
    cinerea): on 7 Ont [666a].
Botrytis sp.: on 6 NS [1490].
B. cinerea (holomorph *Botryotinia*
    fuckeliana) (C:164): gray mold,
    moisissure grise: on 6 BC [338, 339,
    1816], NS [995, 1488, 1490]; on 7 NB, NS
    [1032].
Butlerelfia eustacei: on 7 BC [1942].
    Causes a fish-eye spot on the fruit in
    storage.
Cerrena unicolor (Daedalea u.) (C:165):
    on 6 NB, NS [1032]. See *Acer*.
Chondrostereum purpureum (Stereum p.)
    (C:167): silver leaf, plomb: on 6 Que
    [969]; on 7 NB, NS, PEI [1032]. See
    Abies.
Cladosporium herbarum (C:165): on 7 NS
    [1032].
C. macrocarpum: on 7 Man [1760, F66].
Discostroma corticola (Clathridium c.)
    (anamorph *Seimatosporium*
    lichenicola): on 1 Man [1629].
Coniosporium mali: on 6, 7 BC [1816]; NS
    [1032].
Coriolus hirsutus (Polyporus h.) (C:167):
    on 7 BC [1012]. See *Abies*.
C. versicolor (Polyporus v.) (C:167): on
    6 BC [1012]; on 7 NS, PEI [1032]. See
    Abies.
Crepidotus fulvotomentosus: on 7 BC
    [1012].
Cryptosporiopsis malicorticis (Gloeosporium
    m., G. perennans) (holomorph *Pezicula*
    malicorticis): on 6 BC [337, 338, 339,
    1595, 1816], NS [338, 994, 995].
C. corticola (Myxosporium c.) (holomorph
    Pezicula c.) (C:166): surface bark
    canker, chancre de l'ecorse: on 6 BC
    [334].
Cylindrocarpon heteronema (C. mali): on 6
    BC [1816], NS [334].
C. pomi (holomorph *Mycosphaerella p.)*
    (C:165): fruit spot, tache des fruits:
    on 6 NS [1594].
Cytospora sp. (C:165): dieback and
    canker, chancre cytosporéen: on 6 Alta
    [338, 339], Que [333]; on 7 BC [1012],
    Ont [336, 1340, 1827], Que [336, F63].
C. ambiens (holomorph *Valsa a.)*: on 6
    NS [1490].
C. chrysosperma (holomorph *Valsa*
    sordida): on 7 Alta [F65].
Dendrothele griseo-cana (Aleurocorticium
    g.): on 6 Ont [965]. Lignicolous.
Exidia recisa: on 7 NS [570].
    Lignicolous, typically on dead limbs.
Flammulina velutipes (Collybia v.): on 7
    Ont [1150]. See *Fagus*.
Fomitopsis pinicola (Fomes p.): on 7 NS
    [1032]. See *Abies*.

Fusarium sp. (C:165): on 6 Que [388]; on
    7 NB [F72].
F. oxysporum: on 6 NS [1490].
F. roseum: on 6 NS [1490].
F. solani (holomorph *Nectria
    haematococca*): on 6 NS [1490].
Ganoderma applanatum (C:165): on 6 BC
    [1816]; on 7 NS [1032]. See *Acer*.
Gloeodes pomigena (C:165): sooty blotch,
    tache de suie: on 6 NS [337]; on 7 NB,
    NS, PEI [1032].
Gloeophyllum trabeum (Lenzites t.): on 7
    Ont [1341]. Associated with a brown rot.
Gloeosporium sp.: on 6 BC [333].
Glomerella cingulata (anamorph
    Colletotrichum gloeosporioides)
    (C:165): bitter rot, pourriture amère:
    on 6 BC [1816], NS [333]; on 7 NS [1032].
Gonatobotrys simplex (C:165): on 7 NB
    [1032].
Gymnopilus spectabilis (Pholiota s.)
    (C:166): on 7 NS [1032]. See *Acer*.
Gymnosporangium sp.: rust, rouille: on 7
    BC [1816], Ont [1341], Nfld [333].
G. clavipes (C:165): quince rust, rouille
    du cognassier: 0 I on 3, 6 Ont [1236];
    on 6 Que [337, 339, 1594], NS [1032]; on
    7 Sask [702], Sask, Que [1263], Ont
    [F71], Ont, Que, NS [866].
G. cornutum (C:166): rust, rouille: 0 I
    on 4 BC [1012, 1247, 1264, 1816].
G. globosum (C:166): hawthorn rust,
    rouille de l'aubépine: on 6 Ont, Que
    [333]; on 7 Man [335], Ont [1268].
G. juniperi-virginianae (C:166):
    cedar-apple rust, rouille de Virginie:
    0 I on 2 Ont [1248]; on 2, 5a, 6 Ont
    [1236, 1270]; on 5 Ont [336]; on 5a Que
    [1270]; on 6 Ont [865, 1839, F74, F75],
    Ont, Que [1595]; on 7 Ont [1340, 1341,
    F68].
G. nelsonii (C:166): rust, rouille; on 4
    BC [1271].
G. nootkatense (C:166): rust, rouille: 0
    I on 4 BC [1012, 1247, 1816, 2008].
Haematostereum rugosum (Stereum r.): on 7
    NS [1032, 1036]. See *Abies*.
Hymenochaete corrugata: on 7 NS [1032].
    See *Abies*.
Hyphodontia quercina (Radulum q.)
    (C:167): on 6 NS [1032]. Lignicolous.
Irpex lacteus (Polyporus tulipiferae)
    (C:167): on 6 NS [1032]; on 7 NS
    [1036]. See *Acer*.
Ischnoderma resinosum (Polyporus r.)
    (C:167): on 7 NS [1032]. See *Abies*.
Lenzites betulina (C:166): on 7 NS
    [1032]. See *Acer*.
Leptothyrium pomi (C:166): fly specks,
    moucheture: on 7 NB, NS, PEI [1032].
Leucostoma persoonii (Valsa leucostoma)
    (C:167): canker, chancre cytosporéen:
    on 6 Que [337].
Marssonina coronaria (M. coronariae): on
    2 Ont [1248, 1502].
Massaria excussa: on 7 Ont [1340]. This
    name has not been discussed in recent
    taxonomic treatments thus the features

to be associated with it are poorly
    known. A description can be found on
    page 750 in Ellis and Everhart [438].
Merulius tremellosus: on 6 NS [568]; on 7
    NS [1032]. See *Abies*.
*Microsphaeropsis olivacea (Coniothyrium
    o.)*: on 7 Man [F66].
Monilinia fructicola (C:166): brown rot,
    pourriture brune: on 6 Que [337], NS
    [1490].
M. laxa (C:166): brown rot, pourriture
    brune: on 6 NS [1490].
Mucor sp.: on 6 BC [1594].
Nectria cinnabarina (anamorph
    Tubercularia vulgaris) (C:166):
    dieback or coral spot, dépérissement
    nectrien: on 6 NB [337, 1594], NS [333,
    338, 339, 1594, 1595]; on 7 NWT-Alta
    [1063], Alta [337], Ont [1340], NB
    [1032].
Nectria coccinea var. *faginata* (C:166):
    on 7 NS [1032].
Nectria galligena (C:166): european
    canker, chancre européen: on 6 BC [335,
    338, 1416], NB [333, 336, 337, 338,
    339]; on 7 NB, NS [1032].
Neofabraea sp.: on 6 BC [1005]; on 7 BC
    [1012].
N. perennans (anamorph *Cryptosporiopsis
    p.*) (C:166): perennial canker, chance
    pérennant: on 6 BC [333, 334, 335, 336,
    338, 339, 1594, 1816]. Probably a
    species of *Pezicula*.
Oxyporus populinus (Fomes connatus)
    (C:165): on 7 NB, NS [1032]. See
    Acer.
Penicillium sp.: on 6 NB [333, 334], NS
    [995, 1488].
P. candidum (C:166): on 7 NB [1032]. The
    features to be associated with this name
    are uncertain, i.e., it is a nomen
    confusum.
P. expansum (C:166): blue-mold rot,
    moisissure bleue: on 6 BC [1816], NS
    [1490]; on 7 Ont [83, 1340], NS [1032].
Pezicula malicorticis (Neofabraea m.)
    (anamorph *Cryptosporiopsis m.*)
    (C:166): anthracnose, anthracnose: on
    6 BC [334, 336, 339, 1816], NS [336]; on
    7 Alta [336].
Phellinus conchatus (Fomes c.): on 7 Ont
    [1341]. See *Acer*.
P. igniarius (Fomes i.) (C:165): on 6 BC
    [1012], NB [1032]; on 6, 7 NS [1032].
    See *Acer*.
Phlyctaena vagabunda (Gloeosporium album)
    (holomorph *Pezicula alba*) (C:165):
    storage rot and canker, anthracnose: on
    6 NS [333, 337, 338, 994, 995, 1487,
    1488, 1492, 1594, 1595]; on 7 NB [1032].
Pholiota aurivella (C:166): on 7 NS
    [1032]. See *Abies*.
P. squarrosa (C:166): on 6 BC [1816]; on
    7 NS [1032]. See *Populus*.
Phoma pomi: on 7 NB [1032]. Because the
    name is a later homonym of *P. pomi*
    Schulzer, which is coincidentally a
    synonym of *P. macrostoma*, specimens
    should be redisposed.

*P. pomorum (Phyllosticta prunicola, P.
pyrina)* (C:166): on 7 Ont [1340], NS
[1032].
Phomopsis sp.: on 6 NS [1490]; on 7 Que
[F63].
P. mali (C:166): twig blight, brûlure
phomopsienne: on 7 BC [1816].
Phyllosticta solitaria: on 6 NB [336].
Phytophthora cactorum (C:166): crown rot,
pourridié du collet: on 6 BC [333, 334,
335, 336, 338, 339, 1076, 1594, 1595,
1816], NS [1076, 1490].
P. cambivora: on 6 BC [1816].
P. syringae: on 6 Alta [722], NS [1489].
Pleospora herbarum (anamorph *Stemphylium
botryosum*): on 6 BC [1007].
Pleurotus sp.: on 7 BC [1012].
P. dryinus (C:166): on 7 NS [1032].
Lignicolous.
Podosphaera clandestina (C:167): powdery
mildew, blanc: on 7 Ont [1257].
P. leucotricha (C:167): powdery mildew,
blanc: on 6 BC [333, 334, 335, 336,
337, 338, 339, 1077, 1594, 1816], Ont
[333, 337, 338, 339, 1257], Que [1257,
1594], NS [339, 1594]; on 7 BC [336],
Que [337, 339].
Polyporus varius (C:167): on 7 NB, NS
[1032]. See *Abies*.
Psathyrella candolleana (C:167): on 6 BC
[1816].
Rhizopus sp.: on 6 BC [333].
Sarcodontia setosa (Mycoacia s.): on 7
Ont [1341]. Lignicolous.
Schizophyllum commune (C:167): on 6 BC
[1816]; on 7 NWT-Alta [1063], NS
[1032]. See *Abies*.
Sclerotinia sp.: on 6 NS [995].
S. sclerotiorum (C:167): calyx-end rot,
pourriture sclérotique: on 6 BC [339],
Que [388], NB [339, 1594], NS [333, 334,
335, 336, 337].
Sphaeropsis malorum: on 6 BC [1816], NB
[336], NS [995]. The fungus from
Malus erroneously reported as *S.
malorum* is most likely the conidial
state of *Botryosphaeria obtusa*
[1614]. The conidial state does not
appear to have an anamorph name.
Further the combination *S. malorum*
Peck is illigetimate being a later
homonym and was never proposed by Peck.
Spilocaea pomi (holomorph *Venturia
inaequalis*): on 7 BC [1816], Ont [299,
1827, 1828, F68, F72].
Stemphylium sp.: on 6 BC [1594].
Stigmina negundinis (Sciniatosporium n.)
(C:167): on 6 BC [334, 1816]; on 7 BC,
Man [1127].
S. pallida: on 1, 7 Man [1762]; on 7 Man
[1760].
Stilbella flavescens: on 6 Que [2],
symptomless parasitic fungus.
Trichoderma sp.: on 6 NS [1490].
T. koningii (C:167): on 7 NS [1032].
Trichothecium roseum (C:167): pink mold
rot, moisissure rose: on 7 NB, NS, PEI
[1032].
Tubercularia vulgaris (holomorph *Nectria
cinnabarina*): on 7 Alta [F64].

Tympanis alnea (anamorph *Sirodothis
inversa*): on 7 Que, NB, NS [1221].
Tyromyces caesius (Polyporus c.): on 7 NS
[1032]. See *Abies*.
*Valdensia heterodoxa (Asterobolus
gaultheriae)* (holomorph *Valdensinia
h.*): on 4 BC [1438].
Valsa sp. (C:167): canker, chancre
cytosporéen: on 7 Ont [1827].
V. ambiens (V. sordida) (anamorph
Cytospora a.) (C:167): canker,
chancre cytosporéen: on 6 Que [337]; on
7 NWT-Alta [1063].
V. amphibola (C:167): canker, chancre
cytosporéen: on 7 Ont [1340, F63], NS
[1032].
Venturia inaequalis (anamorph *Spilocaea
pomi*) (C:167): scab, tavelure: on 1,
6 Man [289]; on 2, 6 Ont [289]; on 6 BC
[1075, 1595], BC, Alta, Que, NB [338,
1594], BC, Ont, Que, NB, NS [333, 334,
335, 337, 339], Man [336], Ont [338,
1182], Que [318, 322, 323, 325, 467,
1226], NB, NS [289, 336], NS [338, 1487,
1488, 1492, 1493], PEI [333, 334, 335],
Nfld [333, 336, 337, 339]; on 6, 7 BC
[1816], BC, Sask, Que [289], BC, Ont,
Que [336]; on 7 Alta [334], Ont [286,
310, 333, 1078], Que [335, 337, 339,
F62, F64, F65], Que, NB [289, 338], NB,
NS [1032, 1078], NB [1594], NS [309,
1491], cosmopolitan [73].
Xylobolus subpileatus (Stereum s.): on 6
NS [1036]; on 7 NS [1032]. Lignicolous.

MALVA L.   MALVACEAE

1. *M. alcea* L., hollyhock mallow, mauve
   alcée; imported from Europe, introduced,
   Ont-NS.
2. *M. moschata* L., musk mallow, mauve
   musquée; imported from Europe, a garden
   escape, BC, Man-Nfld.
3. *M. neglecta* Wallr., cheeses,
   fromagère; imported from Eurasia,
   introduced, BC-Alta, Man-Nfld.
4. *M. rotundifolia* L. *(M. pusilla
   Sm.)*, dwarf mallow, petite mauve;
   imported from Europe, introduced, BC,
   Sask-Que, PEI.
5. *M. sylvestris* L., high mallow, grande
   mauve; imported from Europe, introduced,
   BC-Alta, Ont-Que.
6. *M. verticillata* L., cluster mallow;
   imported from Asia, introduced, Alta-PEI.

Phomopsis malvacearum: on 4 BC [1340].
Puccinia malvacearum (C:169): rust,
rouille: III on 1, 3, *M.* sp. Man, on
2 NS, on 3 BC, Ont, Que, on 4 Sask, Ont,
NB, on 5, 6 Ont [1281]; on 1, 4, 5 Ont
[1236]; on 3, 4 BC [1816].
Thielaviopsis basicola: on 3 Ont [546].
Verticillium dahliae: on 4 BC [1816,
1978].

MARRUBIUM L.   LABIATAE

1. *M. vulgare* L., common horehound,
   masrube blaoc; imported from Eurasia,

introduced, BC, Sask, Ont-Que, NS.
Whitish-wooly, bitter, aromatic
perennial herb.

Ophiobolus erythrosporus: on 1 Ont [1620].

MATRICARIA L.   COMPOSITAE

1. *M. matricarioides* (Less.) C.L. Porter,
    pineapple weed, pomme de pré; imported
    from Eurasia, natzd., Mack, BC-Nfld,
    Labr.

Olpidium brassicae: on 1 BC [1977].
Pythium sp.: on 1 BC [1977].
Sphaerotheca fuliginea (C:169): powdery
    mildew, blanc: on 1 Que [1257].

MATTEUCIA Todaro   POLYPODIACEAE

1. *M. struthiopteris* (L.) Tod. *(Pteretis*
    nodulora (Michx.) Nieuwl.*)*, ostrich
    fern, matteuccie fougère-à-l'autruche;
    native, Mack, BC-Nfld.

Uredinopsis struthiopteridis (C:233):
    rust, rouille: II III on 1 BC [1012,
    1816, 2007, 2008], Alta, Man [2008], Ont
    [1236], Que [155, 282], NB [F68].
    Alternate hosts are *Abies balsamea* and
    A. lasiocarpa.

MATTHIOLA R. Br.   CRUCIFERAE

1. *M. incana* (L.) R. Br., common stock,
    matthiole blancheâtre; imported from
    Europe.

Alternaria raphani (C:169): leaf spot,
    tache alternarienne: on 1 Sask [1313].
Botrytis cinerea (holomorph *Botryotinia*
    fuckeliana) (C:169): gray mold,
    moisissure grise: on 1 BC [1816].
Fusarium solani (holomorph *Nectria*
    haematococca) (C:169): foot rot and
    wilt, pourridié fusarien: on 1 Que
    [333].
Oidium sp.: powdery mildew, blanc: on 1
    BC [1816].

MEDICAGO L.   LEGUMINOSAE

1. *M. falcata* L., yellow lucerne, luzerne
    jaune; imported from Eurasia,
    introduced, BC-Que, NS.
2. *M. lupulina* L., black medick,
    lupuline; imported from Eurasia, natzd.,
    BC-Nfld.
3. *M. sativa* L., alfalfa, luzerne;
    imported from Eurasia, escaped from
    cultivation, Mack, BC-PEI.
4. Host species not named, i.e., reported
    as *Medicago* sp.

Acremonium boreale: on 3 BC, Alta, Sask,
    Man, Ont [1707].
Alternaria sp.: on 3 Ont [244], Que [34].
Aphanomyces sp.: on 3 Ont [1093, 1825].
Ascochyta sp.: on 3 Que [34].
Aspergillus sp.: on 3 Ont [244], Que [34].

Aureobasidium sp. *(Pullularia* sp.*)*:
    on 3 Que [34].
Botrytis sp.: on 3 Ont [244], Que [34].
Cercospora zebrina (C:170): leaf spot,
    rayure nervale: on 3 Sask [336], Que
    [32, 339], PEI [1974].
Chaetomella sp.: on 3 Que [34].
Chaetomium sp.: on 3 Ont [244].
C. cochliodes: on 3 Que [34].
C. crispatum: on 3 Que [34].
C. dolichotrichum: on 3 Que [34].
C. globosum: on 3 Que [34].
Cladosporium sp. (C:170): on 3 Que [34].
Colletotrichum sp.: on 3 Que [34].
C. destructivum (C:170): anthracnose,
    anthracnose: on 3 Sask [336], Que
    [1472].
Coniothyrium sp.: on 3 Que [34].
Coprinus psychromorbidus: on 3 Alta
    [1820] as *Coprinus* sp. See [1439].
Cylindrocarpon sp. (C:170): on 3 Alta
    [333], Ont [244].
C. ehrenbergii (C:170): on 3 Ont [244],
    Que [34]. This name was considered to
    be a nomen nudum [247].
C. radicicola (C:170): on 3 Que [34].
Diplodia sp.: on 3 Que [34].
Epicoccum sp.: on 3 Que [34].
Fusarium sp. (C:170): root rot, pourridié
    fusarien: on 3 BC [339], Alta [333,
    336, 337, 338, 339, 1594], Que [490,
    1464, 1467, 1475], PEI [1594].
F. acuminatum (C:170): on 3 Ont [244].
F. avenaceum (C:170): on 3 Ont [241, 244].
F. oxysporum (C:170): on 3 Ont [241,
    244], Que [34, 1471].
F. oxysporum f. sp. *medicaginis*: on 3
    Que [1473] by inoculation.
F. oxysporum var. *redolens* (C:170): on
    3 Que [33] by inoculation.
F. roseum: on 3 BC [334, 716, 1816], Alta
    [333, 334, 335, 1595], Sask [334, 694],
    Que [34, 1471].
F. roseum "var. *acuminatum*": on 3 Que
    [1473]. This variety does not appear to
    have been validly published.
F. solani (holomorph *Nectria*
    haematococca) (C:170): on 3 Ont [241,
    244], Que [34, 1471].
F. tricinctum: on 3 Que [1471].
Helminthosporium sp.: on 3 Que [34].
    Probably a specimen of *Bipolaris*.
Hormodendrum sp.: on 3 Que [34].
Leptosphaeria pratensis (anamorph
    Stagonospora meliloti) (C:171): leaf
    spot, brûlure du melilot: on 3 BC
    [1816], Alta [119, 121, 124, 135].
Leptosphaerulina briosiana: on 3 Que [32,
    339, 1463, 1464, 1465, 1466, 1467, 1475,
    1745, 1746, 1747]. Conners [p. 171]
    listed "briosiana" as a synonym of
    "trifolii" but Graham and Luttrell
    [610] have shown that the two are
    distinct. In brief *L. briosiana*
    occurs on *Medicago* and rarely on
    Trifolium, whereas *L. trifolii*
    appears to be restricted to *Trifolium*.
L. trifolii *(Pseudoplea t.)* (C:171):
    pepper spot, tacheture noire: on 3 BC

[1816], Alta [119, 121, 123, 124]. The
specimens may be better named *L.
briosiana*.

*Leptotrochila medicaginis (Pseudopeziza
jonesii)* (anamorph *Sporonema
phacidioides*) (C:171): yellow leaf
blotch, tache jaune: on 3 BC [334, 339,
716, 1816], Alta [119, 121, 123, 124,
337, 338, 339, 1594, 1595], Sask [694,
695, 699], Que [32, 333, 339, 1475], NB
[333, 1195], PEI [335, 336, 1974].

Low-temperature basidiomycete (C:170):
basidiomycète frigophile: winter crown
rot or snow mold, pourridié hibernal:
on 3 BC [333], Alta [333, 336, 1919],
Sask [335, 336]. Some records may be of
Coprinus psychromorbidus [1439].

Macrosporium sp.: on 3 Ont [244].

Monilia sitophila (holomorph *Neurospora
s.*): on 3 Que [34].

Mucor sp.: on 3 Ont [244], Que [34].

Papularia sp.: on 3 Que [34]. The
specimens may be a species of
Arthrinium.

Penicillium sp.: on 3 Que [34].

Peronospora trifoliorum (P. aestivalis)
(C:171): downy mildew, mildiou: on 3
BC [333, 334, 716, 1816], Alta [119,
121, 123, 124, 134, 333, 337, 338, 339],
Sask [333, 334, 695], Ont [105], Que
[32, 1465, 1475].

Peziza sp.: on 3 Que [34].

P. ostracoderma: on 3 Que [34].

Phoma sp.: on 3 Que [34], PEI [336, 1974].

P. medicaginis (Ascochyta imperfecta)
(C:170): black stem, tige noire: on 3
BC [716, 1816], BC, Alta [339], BC,
Alta, Sask, Que [334], Alta [119, 121,
123, 124, 338], Alta, Sask [333, 335,
336, 1595], Alta, Sask, Man, Ont [696],
Alta, PEI [1594], Sask [694, 695, 699,
920], Ont [1340], Que [32, 1470, 1475],
NB [333, 1195]; on 4 BC, Alta, Sask,
Man, Ont [1102], Sask [1100, 1101].

P. medicaginis var. *medicaginis*: on 3
Que [1464, 1465, 1466, 1467, 1747].

P. sclerotioides (Plenodomus meliloti)
(C:171): brown root rot,
pourridié-plenodome: on 1 Yukon [1829];
on 3 Yukon [280, 1172, 1986], Alta
[1986]; on 4 Yukon [1173].

P. terrestris (Pyrenochaeta t.): on 3 Que
[34].

Physoderma alfalfae (C:171): crown wart,
tumeur noduleuse: on 3 BC [1816].

Phytophthora sp.: on 3 Ont [1093].

P. megasperma: on 3 Alta [240], Ont [61,
240, 244, 337].

Plenodomus sp.: on 3 Que [34]. Most
Plenodomus species are now placed in
Phoma.

Pleospora herbarum (anamorph *Stemphylium
botryosum*) (C:171): on 3 Ont [1340].

Pseudopeziza trifolii (P. medicaginis)
(C:171): common leaf spot, tache
commune: on 3 Alta [1595], Ont [102,
105, 1340], Que [32, 1464, 1465, 1466,
1467, 1475, 1747].

P. trifolii f. sp. *medicaginis-lupulinae*
(C:171): on 3 Sask, PEI [335].

P. trifolii f. sp. *medicaginis-sativae*
(C:171): common leaf spot, tache
commune: on 3 BC [334, 716, 1816], Alta
[119, 121, 123, 124, 333, 334, 337, 338,
339, 1594], Sask [334, 694, 695, 726,
1595], Que [333, 337, 339], NB [333,
1195], NS [337], PEI [333, 334, 336,
338, 1594, 1974].

Pythium sp.: on 3 Alta [1735], Ont [1093].

P. debaryanum: on 3 Ont [244], Que [33]
by inoculation.

P. hypogynum: on seedlings of 3 Alta
[1735].

P. irregulare: on 3 Ont [244].

P. paroecandrum: on seedlings of 3 Alta
[274].

P. sylvaticum: on seedlings of 3 Alta
[1735].

P. torulosum: on seedlings of 3 Alta
[1735].

P. ultimum: on seedlings of 3 Alta
[1735], Ont [244].

Rhizoctonia sp.: on 3 Que [34].

R. solani (holomorph *Thanatephorus
cucumeris*) (C:171): root and crown
rot, rhizoctone commun: on 3 BC [243,
334, 339, 716, 1816], Alta [243, 333,
334, 335, 336, 337, 338, 339, 1594,
1595], Sask [334, 694], Ont [244], Que
[33] by inoculation.

Rhizopus sp.: on 3 Ont [244], Que [34].

Sclerotinia sp.: on 3 Sask [420, 1138],
Man [1138], Que [34].

S. trifoliorum (C:171): sclerotinia wilt,
flétrissure sclerotique: on 2, 3 BC
[1816]; on 3 Que [63] by inoculation,
PEI [1594].

Sporonema phacidioides (holomorph
Leptotrochila medicaginis): on leaves
of 4 Sask, Man, Ont [1763].

*Stagonospora meliloti (Ascochyta
medicaginis)* (holomorph *Leptosphaeria
pratensis*): on 3 BC, Alta, Sask, Que
[339], Alta [338, 1594], Sask, NB [333],
Que [32], NB [1195].

Stemphylium sp.: on 3 Que [34].

S. botryosum (holomorph *Pleospora
herbarum*) (C:171): leaf spot, tache
stemphylienne: on 3 BC [1816], Sask,
NB, Nfld [333], Ont [102, 105, 334,
667], Que [32, 339, 1464, 1465, 1466,
1475, 1747], NB [1195].

Stilbella flavescens: on 4 Que [451].

Stysanus medius: on 3 Que [34].

Thielaviopsis basicola: on 2 Ont [546].

Trichoderma sp.: on 3 Ont [224].

T. album: on 3 Que [34]. *T. album* has
not been studied or redescribed in
modern times. The Canadian fungus is
probably *T. polysporum*.

T. viride (C:171): on 3 Que [33, 34].
See *Abies*.

Typhula trifolii (C:171): on 4 Ont [383].

*Ulocladium consortiale (Alternaria c.,
Stemphylium c.)* (C:170): on seeds of 3
Ont [667].

Uromyces striatus (C:172): rust,
rouille: II III on 1 Ont [1236]; on 2,

3 Ont [1236, 1237]; on 2 Que [1237]; on
3 Ont [1594].
Verticillium sp.: on 3 BC [1600, 1604,
1816].
V. albo-atrum: on 3 BC, Alta, Sask, Man,
Que [1605], Ont [37], Que [35, 36, 38].
V. dahliae: on 3 BC [1816], Que [36, 38].
V. tenerum (V. lateritium) (holomorph
Nectria inventa): on 3 Que [34].
V. nigrescens: on 3 Que [34, 39].
Zygorhynchus sp.: on 3 Que [34].

MELAMPYRUM L.   SCROPHULARIACEAE

1. *M. lineare* Desr., cow-wheat, mélampyre
linéaire; native, BC-Nfld. la. *M. l.*
var. *americanum* (Michx.) Beauv.;
native, Que-NS.

*Cronartium coleosporioides (C.
stalactiforme)* (C:172): rust,
rouille: II III on 1 NWT-Alta [1063],
BC [953], BC, Alta [F63], BC, Alta,
Sask, Man, Ont, Que, NB [735], Sask
[F70], Sask, Ont, Que [1554], Ont [1205,
1236], Ont, NB [F69, F70], Que [1377].
C. coleosporioides f. *coleosporioides*
(aecial state *Peridermium
stalactiforme*): rust, rouille: on 1 BC
[1012]. Causes perennial cankers on the
alternate hosts *Pinus banksiana* and
P. contorta.
Peridermium harknessii: rust, rouille:
on 1 Que [F60]. This rust does not
occur on *Melampyrum*. The specimens
were, no doubt a *Cronartium* sp.
P. stalactiforme (holomorph *Cronartium
coleosporioides* f. *coleosporioides*):
rust, rouille: on 1 Sask [1997], NB, NS
[1856].
Ramularia melampyri: on 1 Ont [1554]; on
leaves of la Ont [439].

MELICA L.   GRAMINEAE

1. *M. subulata* (Griseb.) Scribn.; native,
BC-Alta.

Erysiphe graminis (C:172): powdery
mildew, blanc: on 1 BC [1012, 1255,
1257, 1816].

MELILOTUS Mill.   LEGUMINOSAE

1. *M. alba* Desr., white sweet clover,
trèfle d'odeur blanc; imported from
Eurasia, natzd., Yukon-Mack, BC-Nfld,
Labr.
2. *M. indica* (L.) All., annual yellow
sweet clover; imported from Eurasia,
introduced, BC, Man, NS.
3. *M. officinalis* (L.) Pall., yellow
sweet clover, trèfle d'odeur jaune;
imported from Eurasia, natzd.,
Yukon-Mack, BC-Nfld.
4. Host species not named, i.e., reported
as *Melilotus* sp.

Acremonium boreale: on 3 BC, Alta, Sask
[1707].
Arthrobotrys oligospora: on 3 Que [30].
Ascochyta sp.: on 4 Alta [127].
A. viciae (A. caulicola, A. meliloti)
(C:172): on 1 BC [1816], Alta [131]; on
1, 3 Alta [119, 121, 123, 124]; on 3
Sask [131]; on 4 Sask [336, 694], Que
[335].
Cercospora zebrina: on 3, 4 Sask [336].
Cylindrocarpon sp. (C:172): root rot,
chancre des racines: on 4 Alta [333].
Dendryphion nanum: on 4 Man [819].
Fusarium sp. (C:172): root rot, pourridié
fusarien: on 3 Que [30]; on 4 Alta
[333], Sask [334].
F. acuminatum (C:172): root rot,
pourridié fusarien: on 4 Sask [336].
F. culmorum (C:173): root rot, pourridié
fusarien: on 4 Alta [338].
F. oxysporum (C:173): root rot, pourridié
fusarien: on 4 Sask [336].
F. oxysporum var. *redolens* (C:173):
root rot, pourridié fusarien: on 4 Alta
[338].
F. solani (holomorph *Nectria
haematococca*) (C:173): root rot,
pourridié fusarien: on 4 Sask [336].
Gliocladium sp.: on 3 Que [30].
Leptosphaeria pratensis (anamorph
Stagonospora meliloti) (C:173): leaf
spot, stem blight and root rot,
brûlure: on 1 BC [1816]; on 1, 3 Alta
[119, 121, 123, 124]; on 4 Sask [333].
Low-temperature basidiomycete: Conners
compilation (p. 172) from the Yukon is
an error.
Ophiobolus mathieui: on 4 Man [1620].
Penicillium sp.: on 3 Que [30].
Peronospora sp.: downy mildew, mildiou:
on 4 Alta [1594].
P. trifoliorum (C:173): downy mildew,
mildiou: on 1 BC [1816]; on 1, 3 Alta
[119, 121].
Phoma medicaginis (Ascochyta imperfecta)
(C:172): on 4 BC [333].
P. sclerotioides (Plenodomus meliloti)
(C:173): brown root rot,
pourridié-plénodome: on 1, 3 Alta
[128]; on 4 Alta [334, 335, 339, 1173,
1594].
Phytophthora cactorum (C:173): root rot,
pourridié phytophthoréen: on 4 Alta
[336, 337, 338].
Rhizopus sp.: on 3 Que [30].
Sclerotinia sp.: on 2 Sask [420, 1138],
Man [1138].
S. trifoliorum (C:173): sclerotinia wilt,
flétrissure sclérotique: on 1 BC [1816].
Selenophoma sp.: on 3 Sask [1323].
Septogloeum sp.: on 1 Ont [1340].
Stagonospora meliloti (holomorph
Leptosphaeria pratensis): on 4 Sask
[335].
Trichoderma sp.: on 3 Que [30].
Verticillium albo-atrum: on 1 Que [36].
V. dahliae: on 1 Que [36].

MENTHA L.   LABIATAE

1. *M. arvensis* L., field mint, menthe des
   champs; native, Yukon-Mack, BC-Nfld,
   Labr. la. *M. a.* var. *villosa* f.
   glabrata (Benth.) Stewart (*M. a.*
   var. *glabrata* (Benth.) Fern.);
   native. lb. *M. a.* var. *villosa* f.
   villosa (*M. a.* var. *canadensis*
   (L.) Brig., *M. canadensis* L.); native.
2. *M. piperita* L. (often considered a
   M. aquatica L. x *M. spicata* L.
   hybrid), peppermint, menthe pourée;
   imported from Europe, introduced, BC,
   Ont-PEI.
3. *M. spicata* L., spearmint, baume vert;
   imported from Eurasia, introduced, BC,
   Sask-PEI.

Erysiphe cichoracearum (C:174): powdery
   mildew, blanc: on 1 Ont, Que and on 1b
   Ont [1257].
Puccinia angustata (C:174): rust,
   rouille: 0 I on 1 Ont [1236]; on la
   Sask, Man, Ont [1559].
Puccinia menthae (C:174): rust, rouille:
   0 I II III on 1, 3 BC [1012]; on la, lb,
   3 Ont [1236]; on lb, 2, 3 BC [1816]; on
   M. sp. BC [338].
Verticillium albo-atrum (C:174): wilt,
   flétrissure verticillienne: on 2 Ont
   [335].

MENZIESIA Sm.   ERICACEAE

1. *M. ferruginea* Sm. (*M. f.* ssp.
   glabella (Gray) Calder & Taylor, *M.
   glabella* Gray), false azalea,
   menziézie ferrugineuse; native, Yukon,
   BC-Alta.

Cladosporium sp.: on 1 BC [953], Alta
   [950].
Exobasidium vaccinii (C:174): on 1 BC
   [1012, 1816].
Gibberidea rhododendri (*Melanomma r.*): on
   1 Alta [763], NWT-Alta [1063].
Godronia menziesiae: on 1 NWT-Alta
   [1063], Alta [656].
Melasmia menziesiae (teleomorph possibly a
   Rhytisma): on 1 NWT-Alta [1063], BC
   [954, 1012, 1816].
Rhytisma arbuti (C:174): on 1 NWT-Alta
   [1063], BC [644, 1012, 1816].
Valdensia heterodoxa (*Asterobolus
   gaultheriae*) (holomorph *Valdensinia
   h.*): on 1 BC [1438].

MERTENSIA Roth   BORAGINACEAE

1. *M. maritima* (L.) S.F. Gray,
   sea-lungwort; native, Yukon-Mack, Keew,
   BC coast, along the Hudson Bay-James Bay
   coasts of Man-Que, coasts of NB-NS,
   Nfld, Labr.
2. *M. paniculata* (Ait.) Don, bluebells,
   mertensia paniculé; native, Yukon-Mack,
   BC-Que.

Erysiphe cichoracearum (C:174): powdery
   mildew, blanc: on 2 Yukon [460], BC
   [1012, 1816], Ont [1257].
Puccinia mertensiae (P. hydrophylli ssp.
   mertensiae) (C:174): rust, rouille:
   on 1 Keew [719].

METASEQUOIA Hu & Cheng   TAXODIACEAE

1. *M. glyptostroboides* H.H. Hu & Cheng,
   dawn redwood; imported from Szechwan,
   China.

Diaporthe lokoyae (anamorph *Phomopsis
   l.*): on 1 BC [1012, F72].

MILIUM L.   GRAMINEAE

1. *M. effusum* L., diffuse millet grass,
   millet diffus; native, Man-NS, Nfld.

Puccinia graminis (C:174): rust,
   rouille: II III on 1 Ont [1236].

MIMULUS L.   SCROPHULARIACEAE

1. *M. lewisii* Pursh, red monkey-flower,
   mimule rose; native, BC-Alta.

Ramularia lewisii: on 1 BC [1554].

MITELLA L.   SAXIFRAGACEAE

1. *M. breweri* Gray; native, BC-Alta.
2. *M. diphylla* L., coolwort, mitrelle à
   deux feuilles; native, Ont-Que.
3. *M. nuda* L., mitrewort, mitrelle nue;
   native, Yukon-Keew, BC-Nfld, Labr.
4. *M. ovalis* Greene; native, BC.
5. *M. pentandra* Hook.; native, Yukon,
   BC-Alta.
6. *M. trifida* Graham; native, BC-Alta.

Puccinia austroberingiana ssp.
   austroberingiana (P. heucherae var.
   austroberingiana) (C:175): rust,
   rouille: on 1, 5 BC, Alta [1560]; on 5
   BC [1816], Alta [644]; on 6 BC [1560,
   1816].
P. heucherae (C:175): rust, rouille: on
   3 Mack [53], Ont [1341]; on 5 BC [1012].
P. heucherae var. *heucherae* (C:175):
   rust, rouille: III on 2 Ont [1236]; on
   3 BC [1012, 1239, 1816], Que [282]; on 4
   BC [1560, 1816].
P. heucherae var. *minor (P.
   congregata*): rust, rouille: on 2 Ont
   [1560]; on 3 Mack, BC, Alta, Sask, Man,
   Ont, Que, Nfld [1560], Ont [437].
P. saxifragae var. *mitellae*: rust,
   rouille: on 1, 5, 6 BC [1560].

MONARDA L.   LABIATAE

1. *M. didyma* L., bee-balm, monarde
   écarlate; imported from e. USA, a garden
   escape, Ont-Que.

2. *M. fistulosa* L., wild bergamot,
fistuleuse; native, BC-Que. 2a. *M. f.*
var. *menthaefolia* (Graham) Fern., wild
bergamot, menthe de cheval; native,
BC-Ont. 2b. *M. f.* var. *mollis* (L.)
Benth.; native, Sask-Ont.

Erysiphe cichoracearum: powdery mildew,
blanc: on 1, 2 Ont [1257].
Puccinia menthae (C:175): rust, rouille:
0 I II III on 1, 2, 2b Ont [1236]; on 2a
NWT-Alta [1063]; on 2b BC [1012, 1816].

MONESES Salisb.   PYROLACEAE

1. *M. uniflora* (L.) Gray, one-flowered
pyrola, monésès uniflore; native,
Yukon-Mack, BC-Nfld, Labr.

Chrysomyxa monesis (C:175): rust,
rouille: II III on 1 BC [1012, 1816,
2008]. The alternate host is *Picea
sitchensis*.
C. pirolata (*C. pyrolae*) (C:175): rust,
rouille: II III on 1 Yukon [460],
Yukon, BC [F62], Yukon, Mack, BC, Alta,
Sask, Man [2008], NWT-Alta [1063], BC
[1012, 1816], Alta [F65], Que [1377].
The alternate host is *Picea* spp.
Pucciniastrum pyrolae (C:175): rust,
rouille: II III on 1 Yukon, Mack, BC,
Alta, Sask, Man [2008], Yukon [460],
NWT-Alta [1063], Mack [53].

MORUS L.   MORACEAE

1. *M. alba* L., white mulberry, mûrier
blanc; imported from China, natzd., s.
BC, s. Ont - s. Que.

Diplodia mori: on 1 Ont [1340]. A later
homonym of *D. mori* Berk., published in
1847.
Diplodina acerina (*Septomyxa tulasnei*):
on *M.* sp. Que [336].
Fusarium lateritium f. sp. *mori* (*F.
lateritium* var. *mori*) (C:176):
dieback, dépérissement fusarien: on 1
BC [1816]; on *M.* sp. Ont [1340, F62].
Massaria zanthoxyli: on *M.* sp. Ont [80].
Nectria cinnabarina (anamorph
Tubercularia vulgaris): canker,
chancre: on 1 Que [336].
Pseudolachnea hispidula (*Dinemasporium
h.*): on 1 Ont [1340].

MUHLENBERGIA Schreb.   GRAMINEAE

1. *M. cuspidata* (Torr.) Rydb.; native,
Alta-Man.
2. *M. glomerata* (Willd.) Trin.,
agglomerated muhlenbergia, muhlenbergie,
agglomérée; native, Yukon-Mack, BC-Nfld.

Phyllachora vulgata (C:176): tar spot,
rayure goudronneuse: on 2 Ont [1340].
Puccinia schedonnardi (C:176): rust,
rouille: on 1 Man [1276].
Tilletia asperifolii (C:176): Duran &
Fischer [424] stated that *T.*

asperifolii "in 1947 was ... reported
in Canada (BC)" but the 1947 collection
in DAOM was not published until 1953
[466].
Uromyces minimus (C:176): rust, rouille:
II III on 2 Ont [1236].

MUSA L.   MUSACEAE

Banana family comprised of about 100
species from the tropics of both
hemispheres. Some are grown as indoor
ornamentals (e.g., the bird-of-paradise
flower).

Acremonium sp.: on *M.* sp. Ont [1601].
Blistum musae: on *M.* sp. Ont [1601].
This fungus developed on imported
bananas that had been incubated in a
moist chamber.
Chalara paradoxa: on *M.* sp. Ont [1165].
Verticillium sp.: on *M.* sp. Ont [1601].

MYOSOTIS L.   BORAGINACEAE

Soft-hairy annual or perennial herbs of
temperate regions, low; cultivated forms
mostly from Europe.

Oidium sp. (C:176): powdery mildew,
blanc: on *M.* sp. BC [339], Que [338].

MYRICA L.   MYRICACEAE

1. *M. gale* L., sweet gale, bois-sent-bon;
native, Yukon-Keew, n. Que-Labr, BC-Nfld.

Cronartium comptoniae (C:176): rust,
rouille: II III on 1 NWT-Alta [1063],
Mack [53, F65], Mack, BC, Alta, Sask,
Man, Ont, Que, NS [735], Mack, BC, Sask
[2008], Keew [732], BC [1012, 1122,
1816, 2006], Sask [F73], Ont [1236,
1341, 1828, F67, F69, F75], Que [1377],
NB, NS [1856], NS [F66].
Ramularia destructiva (C:176): on 1 BC
[1012, 1816], Ont [1340].
Uncinula adunca (*U. salicis*) (C:176):
powdery mildew, blanc: on 1 BC [1012,
1816]. These records were based on UBC
1889 and the specimen should be
restudied.

NARCISSUS L.   AMARYLLIDACEAE

1. *N. poeticus* L., poet's narcissus,
narcisse des poètes; imported from
Europe, a garden escape, Ont-NB, Nfld.
1a. *N. p.* var. *recurvis* (Haw.) Bak.
ex Burb.
2. *N. pseudonarcissus* L., common
daffodil, narcisse faux-narcisse;
imported from Europe, a garden escape,
Ont.
3. Host species not named, i.e., reported
as *Narcissus* sp.

Armillaria mellea (C:176): dry rot,
pourridié-agaric: on 3 BC [1816].

Botryotinia narcissicola (C:176): neck
    rot or smolder, feu du collet: on 2 BC
    [336]; on 3 BC [334, 337, 1816].
B. polyblastis (C:176): fire, feu: on 3
    BC [1816].
Ceratocystis narcissi: on 3 Man [1190].
Fusarium sp.: on 2 BC [336].
F. oxysporum f. sp. *narcissi* (C:177):
    basal rot, pourridié fusarien: on 3 BC
    [1816].
Heterosporium sp.: on 3 BC [1816].
Penicillium sp.: on 3 BC [1816].
Ramularia vallisumbrosae (C:177): white
    mold, moisissure blanche: on 3 BC [334,
    1816].
Sclerotium sp.: on 3 BC [334].
Stagonospora curtisii (C:177): leaf
    scorch, grillure: on 2 BC [336]; on 3
    BC [334, 337, 338, 1816].

NEMOPANTHUS Raf.   AQUIFOLIACEAE

1. *N. mucronatus* (L.) Trel., mountain
    holly, faux houx; native, Ont-Nfld.

Dermea peckiana (C:177): on 1 Ont [1340],
    Que [F69].
Durandiella nemopanthis (C:177): on 1 Ont
    [1340].
Godroniopsis nemopanthis (C:177): on 1
    Ont [1340].
Irpex lacteus (Polyporus tulipiferae): on
    1 Ont [1341]. See *Acer*.
Microsphaera penicillata (C:177): powdery
    mildew, blanc: on 1 Ont [1257].
Phellinus inermis: on 1 Ont [1235].
    Associated with a white rot.
Ramularia nemopanthis: on 1 Ont [1340].
Rhytisma prini (R. ilicis-canadensis)
    (C:177): on 1 Ont [1340], Que [F69].

NEPETA L.   LABIATAE

1. *N. cataria* L., catnip, herbe aux
    chats; imported from Eurasia,
    introduced, BC-Nfld.

Phialea cyathoidea (Helotium c.): on 1
    Ont [1340].

NESLIA Desv.   CRUCIFERAE

1. *N. paniculata* (L.) Desv., ball
    mustard, neslie paniculée; imported from
    Eurasia, introduced, Yukon-Mack, BC-Nfld.

*Pseudocercosporella capsellae
    (Cercosporella nesliae)* (C:178): on 1
    Alta [380], Sask [1326].

NICOTIANA L.   SOLANACEAE

1. *N. alata* Link & Otto, flowering
    tobacco; imported from South America.
2. *N. tabacum* L., tobacco, tabac;
    imported from tropical America.

Alternaria sp.: on 2 Ont [335, 336, 337,
    338, 339, 1290, 1594].

A. alternata (A. tenuis auct.) (C:178):
    on 2 Ont [548, 550].
A. longipes (C:178): brown spot, tache
    brune: on 2 Ont, NB [334].
Ascochyta daturae (A. nicotianae)
    (C:178): leaf spot, tache
    ascochytique: on *N.* sp. BC [1816].
Aspergillus sp.: on 2 Ont [550].
Botrytis cinerea (holomorph *Botryotinia
    fuckeliana*): on 2 Ont [548].
Fusarium sp.: on 2 Ont [338, 339, 1594].
F. tricinctum: on 2 Ont [548].
Myrothecium verrucaria: on leaves of
    seedlings of 2 Ont [549].
Penicillium sp.: on 2 Ont [550].
Pythium sp. (C:179): soft rot, pourriture
    pythienne: on 2 Ont [334, 335, 336,
    337, 338, 339, 549, 1290, 1594].
P. ultimum: on 2 Ont [550].
Rhizoctonia sp.: on 2 Ont [1290].
R. solani (holomorph *Thanatephorus
    cucumeris*) (C:179): sore shin, tige
    noire: on 2 Ont [334, 335, 336, 337,
    338, 339, 549, 550, 1594, 1595], NS, PEI
    [335].
Rhizopus sp.: on 2 Ont [338, 339, 1594].
R. arrhizus: on 2 Ont [545, 549].
R. circinans (R. reflexus): on 2 Ont
    [548].
Sclerotinia sclerotiorum: on 2 Ont [394,
    549].
Thanatephorus cucumeris (anamorph
    Rhizoctonia solani): on 2 Ont [550].
Thielaviopsis basicola (C:179): black
    root rot, pourridié noir: on 2 Ont
    [274, 334, 335, 337, 338, 339, 543, 544,
    546, 547 by inoculation, 549, 1290,
    1291, 1595], Que [1291], NS [1594]; on
    N. sp. Ont, Que [1397]. Losses from
    this disease were considerable in
    certain production areas of s. Ont
    between 1977-1981.
Verticillium dahliae: on 2 Que [676,
    1606, 1982 by inoculation].
V. nigrescens: on 2 Que [1607].

NIEREMBERGIA Ruiz & Pav.   SOLANACEAE

Perennial herbs and subshrubs, native to
Central and South America. In Canada grown
as indoor plants.

Botrytis cinerea (holomorph *Botryotinia
    fuckeliana*) (C:180): gray mold,
    moisissure grise: on *N.* sp. BC [1816].

NOTHOFAGUS Blume   FAGACEAE

Deciduous or evergreen, beech-like trees
or shrubs, native to temperate South
America, New Zealand, se. Australia, New
Caledonia and New Guinea.

Cucurbitaria sp.: on *N.* sp. BC [1012].
Valsa sp.: on *N.* sp. BC [1012].

NUPHAR Sm.   NYMPHAEACEAE

1. *N. advena* (Ait.) Ait. f. *(Nymphaea
    a.* Ait.); native, s. Ont.

Mycosphaerella pontederiae :   on 1 Ont
    [1340].
Ovularia nymphaerum :   on *N.* sp. BC
    [1816].

OEMLERIA Reichb.   ROSACEAE

1. *O. cerasiformis* (Torr. & Gray) Landon
    (*Osmaronia c.* (Torr. & Gray) Greene),
    osoberry; native, BC.

Cylindrobasidium evolvens (*Corticium
    laeve*) (C:182):   on 1 BC [1012].
Cylindrosporium nuttalliae (C:182):   on 1
    BC [1012, 1416, 1816]. This fungus is
    not a *Cylindrosporium*, but it has not
    been transferred to another genus.

OENANTHE L.   UMBELLIFERAE

1. *O. sarmentosa* Presl; native, BC.

Septoria oenanthis (C:180):   on 1 BC
    [1551, 1816].

OENOTHERA L.   ONAGRACEAE

1. *O. biennis* L., yellow evening
    primrose, mâche rouge; native, BC-Nfld.
    1a. *O. b.* var. *biennis* f. *biennis*
    (*O. victorinii* Gates & Catchside).
2. *O. nuttallii* Sweet, white evening
    primrose, onagre blanche; BC-Man,
    introduced, w. Ont.
3. *O. parviflora* L., small-flowered
    evening primrose, onagre parviflore;
    native, BC-Nfld.
4. *O. perennis* L.; introduced, BC, se.
    Man-NS, Nfld.

Erysiphe cichoracearum (C:181):   powdery
    mildew, blanc: on *O.* sp. BC [1816].
E. communis (*E. polygoni*) (C:181):
    powdery mildew, blanc: on 1 Ont, Que,
    NS [1257]; on 1, 3 Que [1546].
Gnomonia misella :   on 1 Ont [79].
Peronospora arthuri (C:181):   downy
    mildew, mildiou: on 1 Sask, Ont, NS,
    PEI [1546]; on 1a Que [1546].
Puccinia dioicae (C:181):   rust, rouille:
    0 I on 1 Ont [1236], NB [333, 1194].
Scopinella solani :   on 1 Ont [1039].
Septoria oenotherae (C:181):   leaf spot,
    tache septorienne: on 1 BC, Alta, Man,
    Ont, on 1, 2 Sask, on 1, 3, 4 Que, on 4
    Ont, NS, all from [1546]; on *O.* sp. BC
    [1816], NB [1032].
Uromyces plumbarius (C:181):   rust,
    rouille: on 3 NB [762].

ONCIDIUM Sw.   ORCHIDACEAE

1. *O. longipes* Lindl. & Paxt.; imported
    from Brazil.

Uredo behnickiana (C:181):   rust,
    rouille: II on 1 Ont [1236].

ONOBRYCHIS Mill.   LEGUMINOSAE

1. *O. viciifolia* Scop., sainfoin,
    esparcette; imported from Eurasia,
    introduced, BC, Ont.

Acremonium boreale :   on 1 BC, Alta, Sask
    [1707].
Ascochyta pisi (*A. orobi*) (C:181):   leaf
    and stem spot, tache ascochytique: on 1
    BC [1816].
Rhizoctonia leguminicola :   on 1 Alta [125].
Sclerotinia trifoliorum (C:181):
    sclerotinia wilt, flétrissure
    sclerotique: on 1 BC [1816].
Verticillium albo-atrum :   on 1 Que [36].
V. dahliae :   on 1 Que [36].

ONOCLEA L.   POLYPODIACEAE

1. *O. sensibilis* L., sensitive fern,
    onoclée sensible; native, Man-Nfld, Labr.

Gnomonia artospora :   on 1 Ont [79].
Uredinopsis americana (*U. mirabilis*)
    (C:181):   rust, rouille: II[1] II[2]
    III on 1 Ont [1236], Que [1377]. The
    alternate hosts are *Abies* species.

ONOSMODIUM Michx.   BORAGINACEAE

1. *O. molle* Michx., marble seed; native,
    Alta-Ont. 1a. *O. m.* var.
    hispidissimum (Mack.) Cronq. (*O.
    carolinianum* of most Canadian
    reports); native, Man-Ont.

Ophiobolus acuminatus :   on *O.* sp. Ont
    [1620], reported on *Osmodium.*
O. cesatianus :   on 1a Ont [1620].
O. niesslii :   on *O.* sp. Ont [1620].

OPLOPANAX (Torr. & Gray) Miq.   ARALIACEAE

1. *O. horridus* (Sm.) Miq., devil's club,
    bois piquant; native, Yukon, BC-Alta.

Ophiobolus mathieui :   on 1 BC [1620].

ORTHOCARPUS Nutt.   SCROPHULARIACEAE

1. *O. luteus* Nutt.; native, BC-Ont.

Cronartium coleosporioides (aecial state
    Peridermium stalactiforme):   rust,
    rouille: on 1 NWT-Alta [1063], Alta
    [735]. The alternate hosts are 2- and
    3-needle pines, e.g. *Pinus banksiana*
    and *P. contorta.*

ORYZOPSIS Michx.   GRAMINEAE

1. *O. asperifolia* Michx., rough mountain
    rice, oryopsis à feuilles rudes; native,
    Mack, BC-NS, Nfld.

Puccinia brachypodii-phoenicoides var.
    davisii :   rust, rouille: on 1 Ont
    [350].

P. pygmaea (C:182): rust, rouille: II
    III on 1 BC [1012, 1816]; on 1, *O.* sp.
    Ont [1236].

OSMORHIZA Raf.   UMBELLIFERAE

1. *O. chilensis* Hook. & Arn., Chile sweet
    cicely, osmorhize de Chile; native,
    BC-NS, Nfld.
2. *O. claytonii* (Michx.) C.B. Clarke,
    hairy sweet cicely, osmorhize de
    Clayton; native, Ont-Nfld.
3. *O. depauperata* Phil.; native, Mack,
    BC-NS, Nfld, Labr.
4. *O. longistylis* (Torr.) DC.,
    anise-root, osmorhize à long style;
    native, Alta-NS.
5. *O. purpurea* (Coult. & Rose) Suksdorf;
    native, BC-Alta.

Puccinia pimpinellae (C:182): rust,
    rouille: on 1, 5 BC [1012, 1816]; on 2,
    4, *O.* sp. Ont [1236]; on *O.* sp. BC,
    Alta [1551], Ont [1341].
P. pimpinellae ssp. *carlottae*: rust,
    rouille: on 1, 5 BC [1551].
P. pimpinellae ssp. *pimpinellae*: rust,
    rouille: on 1, 3 BC, Alta, on 2 Sask,
    on 5 BC [1551].
Ramularia reticulata: on 4 Ont [195].
Septoria aegopodii (*Phloeospora a.*)
    (C:182): on 1 BC [1816].
S. osmorrhizae (*Phloeospora o.*) (C:182):
    on 1, 3, 5 BC, on 2, *O.* sp. Ont, Que,
    on 4 Man, Ont [1551].

OSMUNDA L.   OSMUNDACEAE

1. *O. cinnamomea* L., cinnamon fern,
    osmonde cannelle; native, Ont-Nfld.
2. *O. claytoniana* L., interrupted fern,
    osmonde de Clayton; native, Man-Nfld,
    Labr.
3. *O. regalis* L., royal fern, osmonde
    royale; native, Ont-Nfld.

Mycosphaerella minor (C:182): on 3 Que
    [70].
Uredinopsis osmundae (C:182): rust,
    rouille: II III on 1, 2, 3 Ont [1236],
    Que [1377]; on 1 NS [1032, F65]; on 2
    Que [282], NB, NS [1032]; on *O.* sp.
    Ont [1236], NS, PEI [1032]. The
    alternate host is *Abies balsamea*.

OSTRYA Scop.   BETULACEAE

1. *O. virginiana* (P. Mill.) K. Koch,
    American hop hornbeam, bois de fer;
    native, Man-NS.
2. Host species not named, i.e., reported
    as *Ostrya* sp.

Aleurodiscus oakesii (C:183): on 1 Ont
    [964, 1341], Que [964, F67], NB [1032].
    Lignicolous.
Ceraceomyces serpens: on 1 Ont [568].
    Lignicolous.
Cerrena unicolor (*Daedalea u.*): on 1 Ont
    [1341]. See *Acer*.

Chondrostereum purpureum (*Stereum p.*): on
    1 Ont [1341]. See *Abies*.
Ciboria betulicola: on 2 Ont [429], Que
    [662].
Conoplea olivacea (*C. sphaerica*): on 2
    Ont [784].
Coriolus hirsutus (*Polyporus h.*): on 1
    Ont [1341]. See *Abies*.
C. versicolor (*Polyporus v.*): on 1 Ont
    [1341]. See *Abies*.
Cylindrosporium dearnessii (C:183): leaf
    spot, tache des feuilles: on 1 Ont
    [1340], NB, NS [1032]. Probably not a
    Cylindrosporium.
Daedaleopsis confragosa (*Daedalea c.*): on
    1 Ont [1341]. See *Acer*.
Dendrothele alliacea (*Aleurocorticium
    a.*): on 1 Ont [965]. Lignicolous.
D. griseo-cana (*Aleurocorticium g.*): on 1
    Ont, Que [965]. Lignicolous.
D. microspora (*Aleurocorticium m.*): on 1
    Ont [965]. Lignicolous.
Diatrype albopruinosa: on 1 Ont [1340,
    1354].
Diplodia sp.: on 2 Ont [1340].
Favolus alveolaris: on 1 Ont [1341]. See
    Acer.
Fomes fomentarius: on 1 Ont [1341]. See
    Acer.
Ganoderma applanatum (*Fomes a.*): on 1 Que
    [1391]. See *Acer*.
Gloeoporus dichrous (*Polyporus d.*): on 1
    Ont [1341]. See *Acer*.
Gnomonia ulmea: on 1 NB [1032]. The
    specimen may be *G. ostryae*.
Hendersonia ostryigena: on dead limbs of
    1 Ont [440]. The proper generic
    disposition for this species is
    uncertain. See *Abies*.
Hericium erinaceus: on 1 Ont [1341].
    Probably *H. americanum*. See comment
    under *Acer*.
Hymenochaete corrugata: on 1 Ont [1341].
    See *Abies*.
H. tabacina: on 1 Ont [1341], NB [1032].
    See *Abies*.
Hypoxylon cohaerens: on 1 Ont [1340].
H. rubiginosum: on 1 Ont [1340].
Melanconis ostryae (C:183): on 1 Ont, Que
    [1340].
Melanconium sp.: on 2 Ont [1340].
Microsphaera penicillata: powdery mildew,
    blanc: on 1 Que [1257].
Monostichella robergei (*Gloeosporium r.*):
    on 1, 2 Ont [1340].
Oxyporus populinus (*Fomes connatus*): on 1
    NB [1032]. See *Acer*.
Pezicula carpinea (anamorph
    Cryptosporiopsis fasciculata): on 1
    Ont [F62].
Phellinus igniarius (*Fomes i.*) (C:183):
    on 1 Ont [1341], Que [1341, 1377], NB
    [1032]. See *Acer*.
P. laevigatus (*Poria l.*): on 1 Ont
    [1341]. See *Acer*.
P. punctatus (*Poria p.*): on 1 Ont
    [1341]. See *Abies*.
Robergea cubicularis: on 2 Ont [1608].
Taphrina virginica (C:183): leaf blister,
    cloque des feuilles: on 1 Que [1377].

Tyromyces caesius (Polyporus c.): on 2
    Ont [1828]. See *Abies*.
Valsa sp.: on 2 Ont [1340].

OXALIS L.   OXALIDACEAE

1. *O. corniculata* L., creeping lady's
    sorrel, oxalide cornue; imported from
    Eurasia.
2. *O. corymbosa* DC.; imported from
    tropical America.
3. *O. montana* Raf.; common wood sorrel,
    oxalide de montagne; native, Ont-Nfld,
    Labr.
4. *O. rubra* St. Hil.; imported from
    Brazil.
5. *O. stricta* L., European wood sorrel,
    pain d'oiseau; native, Sask-PEI.

Microsphaera russellii (C:183): powdery
    mildew, blanc: on 1, 5 Ont, NS, on 3
    Ont, on 5 Que [1257].
Puccinia oxalidis (C:183): rust,
    rouille: II on 2, 4 Ont [1236].
P. sorghi: rust, rouille: on 1 Que
    [1103] by inoculation.

OXYRIA Hill   POLYGONACEAE

1. *O. digyna* (L.) Hill, mountain sorrel,
    oxyrie digyne; native, Yukon-Labr, Que,
    NS, Nfld.

Mycosphaerella oxyriae (C:183): on 1
    Yukon, Frank, Que [1586].
Puccinia oxyriae (C:183): rust, rouille:
    on 1 Frank, Keew [1586].
Ustilago vinosa (C:183): smut, charbon:
    on 1 Frank [1239, 1244, 1543, 1544,
    1586], Alta [644].
Venturia oxyriae (Coleroa o.) (C:183): on
    overwintered leaves and stalks of 1 BC
    [73].

OXYTROPIS DC.   LEGUMINOSAE

1. *O. campestris* (L.) DC.; native,
    Yukon-Labr, BC-NS, Nfld. la. *O. c.*
    var. *gracilis* (A. Nels.) Barneby;
    native, BC-Man. 1b. *O. c.* var.
    johannensis Fern., St. John's River
    oxytropis, oxytropis du fleuve St. Jean;
    native, Man, Que-NS, Nfld, Labr. 1c.
    O. c. var. *terrae-novae* (Fern.)
    Barneby.
2. *O. maydelliana* Trautv.; native,
    Yukon-Labr, BC, Que.
3. *O. nigrescens* (Pallas) Fisch.; native,
    Yukon-Frank.
4. *O. sericea* Nutt.; native, Yukon-Mack,
    BC-Man. Our plant is referable to *O.
    sericea* var. *spicata* (Hook.) Barneby.
5. *O. splendens* Dougl. ex Hook.; native,
    Yukon-Mack, BC-Ont.
6. *O. viscida* Nutt. ex Torr. & Gray;
    native, Mack-Frank, n. Que, BC-Alta.
    6a. *O. v.* var. *hudsonica* (Greene)
    Barneby; native, Mack-Frank, n. Que.

*Phloeospora serebrianikowii
    (Cylindrosporium s.)*: on 2 Frank
    [1244].
Phyllosticta sp.: on 2 Frank [1244].
Uromyces lapponicus (C:184): rust,
    rouille: on la Yukon [7]; on 2 Frank
    [1543].
U. lapponicus var. *oxytropis*: rust,
    rouille: on 1c Que, Nfld, on 2 Mack,
    Frank, on 2, 3 Yukon, on 2, 6a Keew, all
    from [1589].
U. punctatus: rust, rouille: on la BC,
    on 1b Que, on 4 BC, Man, on 5 Alta, all
    from [1587]; on *O.* sp. Que [282].

PAEONIA L.   RANUNCULACEAE

1. *P. lactiflora* Pallas, peony, pivoine;
    imported from e. Asia.

Botrytis paeoniae (C:184): botrytis
    blight, brûlure botrytique: on 1 BC,
    Alta, NS [336], Alta [335], Alta, Sask
    [333, 1594], Alta, Que [334, 337, 338,
    339], Man [334], Ont, Que, NB [333]; on
    P. sp. BC [1816], NB [1194].
Cercospora sp.: on *P.* sp. BC [1816].
Cladosporium paeoniae (C:184): leaf
    blotch, brûlure cladosporienne: on 1
    Ont [334]; on *P.* sp. BC [1816].
Rhizoctonia solani (holomorph
    Thanatephorus cucumeris): on 1 NB
    [335].
Sclerotinia sclerotiorum: on 1 Ont [336].
Septoria paeoniae (C:184): septoria leaf
    spot, tache septorienne: on 2 BC [1816].

PANAX L.   ARALIACEAE

1. *P. quinquefolius* L., ginseng; Ont-Que.
2. *P. trifolius* L., dwarf ginseng, petit
    ginseng; Ont-PEI.

Alternaria panax (C:185): leaf spot,
    brûlure alternarienne: on 1 BC [1816].
Puccinia araliae (C:185): rust, rouille:
    III on 2 Que [1236].
Ramularia sp. (C:185): disappearing rot,
    évanouissement: on 1 BC [1816].
Rhizoctonia solani (holomorph
    Thanatephorus cucumeris) (C:185):
    stem rot, rhizoctonie: on 1 BC [1816].

PANICUM L.   GRAMINEAE

1. *P. capillare* L., witch-grass,
    mousseline; native, BC-PEI.
2. *P. miliaceum* L., broom-corn millet,
    panic millet; imported from Asia,
    natzd., BC-Alta, Man-Que, NS, PEI.
3. *P. virgatum* L., switchgrass, panic
    raide; native, Sask-Que, NS.

Puccinia emaculata (C:185): rust,
    rouille: II III on 1 Ont [1236].
P. panici (C:185): rust, rouille: II III
    on 3 Ont [1236].
Ustilago destruens (C:185): smut,
    charbon: on 2 BC [1816].

PARNASSIA L.   SAXIFRAGACEAE

1. *P. fimbriata* Koenig; native,
    Yukon-Mack, BC-Alta.

Puccinia parnassiae: rust, rouille: on 1
    Yukon, BC, Alta [1565], BC [1012, 1816],
    Alta [952].
Synchytrium aureum (C:186): on 1 Alta
    [644].

PARRYA R. Br.   CRUCIFERAE

1. *P. nudicaulis* (L.) Boiss.; native,
    Yukon-Mack, BC.

Puccinia oudemansii (C:186): rust,
    rouille: on 1 Yukon, Frank [1563].

PARTHENOCISSUS Planch.   VITACEAE

1. *P. quinquefolia* (L.) Planch.
    (Ampelopsis q. (L.) Michx.),
    virginia creeper, vigne vierge; native,
    Ont-Que, a garden escape in Man, NB-Nfld.
2. *P. tricuspidata* (Sieb. & Zucc.)
    Planch., Boston ivy, vigne-vierge du
    Japon; imported from China and Japan.

Cladosporium herbarum: on 2 Que [1594].
Phyllosticta ampelicida (P. viticola)
    (holomorph *Guignardia bidwellii*): on
    2 Ont [334], Que [335].
Plasmopara viticola (C:186): downy
    mildew, mildiou: on 1 Man [335].
Septogloeum ampelopsidis: on 1 Ont [195].
Uncinula necator (C:186): powdery mildew,
    blanc: on 1 Alta [1594], Ont [334,
    1257], Que [1257].

PASCOPYRUM Löve   GRAMINEAE

    See discussion of generic
circumscriptions under *Agropyron*.

1. *P. smithii* (Rydb.) Löve *(Agropyrum
    s.* Rydb.), western wheatgrass or
    go-back grass, agropyre à tige bleue;
    native, s. prairies, Alta-Sask.

Puccinia montanensis (C:21): leaf rust,
    rouille brune: on 1 Man [350].

PASSIFLORA L.   PASSIFLORACEAE

    Vines climbing by tendrils, native to
the new and old world but principally to the
new world tropics. Often grown as
ornamentals. Commonly known as
passionflower (passiflore).

Alternaria passiflorae: on *P.* sp. BC
    [339].

PASTINACA L.   UMBELLIFERAE

1. *P. sativa* L., parsnip, panais;
    imported from Eurasia.

Alternaria sp.: on 1 Ont [1708].

A. radicina (Stemphylium r.): on 1 NS
    [337].
Botrytis sp.: on 1 Alta [1594].
Cercospora pastinacae (C:187): early
    blight, brûlure cercosporéenne: on 1 NB
    [333, 1594], NS [335, 336].
*Cylindrocarpon destructans (C.
    radicicola)*: on 1 NS [337].
Fusarium sp. (C:187): on 1 Ont [1708].
Itersonilia perplexans (C:187): canker,
    chancre: on 1 NB [1594], NS [334, 1594].
Mucor sp.: on 1 Ont [1708].
Ophiobolus rubellus: on 1 Ont [1620].
Phomopsis canadensis (C:187): stem
    blight, brûlure phomopsienne: on 1 Ont
    [195].
Pythium sp.: on 1 Ont [868, 1708].
P. irregulare: on 1 Ont [868] by
    inoculation.
P. sulcatum: on 1 Ont [868] by
    inoculation.
Ramularia pastinacae (Cercosporella p.)
    (C:187): leaf spot, tache
    ramularienne: on 1 BC [1816], Ont [333].
Rhizoctonia sp.: on 1 Ont [1708].
Rhizopus sp. (C:187): on 1 BC [1816].
Sclerotinia sclerotiorum (C:187):
    sclerotinia rot, pourriture
    sclérotique: on 1 BC [1816], Alta [337,
    1594].
Stemphylium botryosum (holomorph
    Pleospora herbarum) (C:187): on seeds
    of 1 Ont [667].
Streptomyces scabies (C:187): scab, gale
    commune: on 1 NS [336].
*Ulocladium consortiale (Alternaria c.,
    Stemphylium c.)* (C:187): on seeds of 1
    Ont [667].

PEDICULARIS L.   SCROPHULARIACEAE

1. *P. bracteosa* Benth.; native, BC-Alta.
2. *P. canadensis* L., common lousewort,
    pédiculaire du Canada; native, Man-Que.
3. *P. capitata* Adams; native,
    Yukon-Frank, BC-Alta.
4. *P. flammea* L.; native, Mack-Labr,
    Man-Que.
5. *P. labradorica* Wirsing, Labrador
    lousewort, pédiculaire du Labrador;
    native, Yukon-Labr, BC-Que.
6. *P. lanata* Cham. & Schlecht., woolly
    lousewort; native, Yukon-Frank, BC, Que.
7. *P. lanceolata* Michx., lanceolate
    lousewort, pédiculaire lancéolée;
    native, Man-Ont.
8. *P. langsdorfii* Fisch. ex Stev.;
    native, Yukon-Frank, BC-Alta. 8a. *P.
    l.* ssp. *arctica* (R. Br.) Pennell *(P.
    arctica* R. Br.).
9. *P. lapponica* L., Lapland lousewort,
    pediculaire de Laponie; native, Yukon,
    Man-Que, Labr.
10. *P. ornithorhynca* Benth.; native, BC.
11. *P. racemosa* Dougl.; native, BC.
12. *P. sudetica* Willd.; native,
    Yukon-Frank, BC, Man-Que.

Ascochyta pedicularidis (Diplodina p.)
    (C:187): on 4, 6 Keew and on 6, 8a

Frank [1554, 1586]; on 5 Alta, Que and
on 6 BC [1554].

Cronartium coleosporioides (aecial state
Peridermium stalactiforme): rust,
rouille: II III on 1 NWT-Alta [1063],
BC [735, 1554], Alta [735, 952]; on *P.*
sp. BC [953, 954], Alta [950, 951,
952]. The alternate hosts are 2- and
3-needle pines, e.g. *Pinus banksiana*
and *P. contorta.*

Mycosphaerella pedicularis (C:187): on 1,
10 BC, on 4, 6, 8a, 12 Frank, on 5 Alta,
Que, Nfld, on 9, 12 Keew, all from
[1554].

Peronospora pedicularis: on 3 Keew [1554].

Puccinia clintonii (C:188): rust,
rouille: III on 1 NWT-Alta [1063], BC
[560, 1816]; on 2 Ont, Que [1236]; on 3
Frank [1239].

P. clintonii var. *bracteosae*: rust,
rouille: on 1 BC [1012, 1553], Alta
[644, 1553].

P. clintonii var. *clintonii*: rust,
rouille: on 2 Ont, Que [1553].

P. clintonii var. *ornithorhynchae*:
rust, rouille: on 10 BC [1553].

P. clintonii var. *racemosae*: rust,
rouille: on 11 BC [1553].

P. helicalis (C:188): rust, rouille: on
3 Yukon, Frank [1553, 1586], Mack, Keew
[1586], Frank [1543].

P. pedicularis: rust, rouille: on 4
Frank [1244, 1554].

Ramularia obducens: on 1 BC [1012, 1816].

Sphaerotheca fuliginea (C:188): powdery
mildew, blanc: on 1 BC, Alta [1554],
Alta [644]; on 2 Ont [1554]; on 2, 7 Ont
[1257]; on 3 Yukon, Keew [1586], Keew
[1554]; on 3, 8a Frank [1554, 1586].

Thanatephorus cucumeris (Pellicularia
filamentosa) (anamorph *Rhizoctonia*
solani) (C:188): on 9 Frank [1586].

PELARGONIUM L'Her.   GERANIACEAE

1. *P.* x *domesticum* Bailey, *P.*
    angulosum var. *grandiflorum* x *P.*
    cucullatum, show geranium, géranium; a
    cultigen.
2. *P.* x *hortorum* Bailey, fish geranium;
    a cultigen.

Botrytis cinerea (holomorph *Botryotinia*
fuckeliana) (C:188): gray mold,
moisissure grise: on 1 Ont [337]; on 2
BC [1816]; on *P.* sp. BC [163, 333,
334, 338], Ont [334], Que [333, 334].

Fusarium sp. (C:188): stem rot, pourridié
fusarien: on *P.* sp. PEI [333].

Pythium sp.: on 2 BC [1816]; on *P.* sp.
Sask [335].

P. splendens: on 2 Ont [164] by
inoculation.

Verticillium sp.: on 2 BC [1816].

V. dahliae (C:188): verticillium wilt,
flétrissure verticillienne: on 2 BC
[336, 1816, 1978]; on *P.* sp. BC [334].

PELTANDRA Raf.   ARACEAE

1. *P. virginica* (L.) Schott & Endl.,
    tuckahoe, peltandre de Virginie; native,
    Ont-Que. Stemless, hardy perennial herb
    of swamps and margins of slow rivers in
    North America.

Uromyces ari-triphylli: rust, rouille: 0
I II III on 1 Ont [1236].

PENSTEMON Mitch.   SCROPHULARIACEAE

1. *P. confertus* Dougl., yellow
    beard-tongue, penstémon dense; native,
    BC-Alta.
2. *P. ellipticus* Coult. & Fisch.; native,
    BC-Alta.
3. *P. fruticosus* (Pursh) Greene; native,
    BC-Alta.
4. *P. grandiflorus* Nutt.; native, w. USA.
5. *P. hirsutus* (L.) Willd., hairy
    beard-tongue, penstémon hirsute; native,
    Ont-Que.
6. *P. ovatus* Dougl.; native, BC.
7. *P. procerus* Dougl.; native, Yukon,
    BC-Man. 7a. *P. p.* var. *procerus.*
8. *P. serrulatus* Menzies; native, BC-Alta.

Cercospora penstemonis (C:189): leaf
spot, tache cercosporéenne: on 6, 8 BC
[1816].

Puccinia andropogonis (C:189): rust,
rouille: 0 I on 5 Ont [1236]; on 7
Yukon [460].

P. dasantherae: rust, rouille: on 3 BC
[1012].

P. palmeri (C:189): rust, rouille: on 1,
2 BC [1816]; on 7 Yukon [7]; on 7a BC
[1012].

Ramularia nivosa (C:189): on 4 Man and on
8 BC [1554].

PEPEROMIA Ruiz & Pav.   PIPERACEAE

Mostly small, succulent herbs of wide
distribution in tropical and subtropical
areas.

Pythium spp.: on rooted cuttings of *P.*
sp. Ont [337].

PERIDERIDIA Reichb.   UMBELLIFERAE

1. *P. gairdneri* (Hook. & Arn.) Mathias,
    yampah; native, BC-Sask.

Cercosporidium punctiforme (Didymaria
platyspora): on 1 Alta [1551].
Puccinia ligustici: rust, rouille: on 1
BC [1551].

PETASITES Mill.   COMPOSITAE

1. *P. frigidus* (L.) Fries; native,
    Yukon-Labr, BC-Nfld. la. *P. f.* var.
    frigidus; native, Yukon-Frank,
    BC-Alta. 1b. *P. f.* var. *nivalis*
    (Greene) Cronq. *(P. nivalis* Greene,
    P. vitifolius Greene); native,

Yukon-Keew, BC-Que, Labr. 1c. *P. f.*
var. *palmatus* (Ait.) Cronq. (*P.*
palmatus (Ait.) Gray), palmate sweet
coltsfoot, pétasite palmé; native,
Yukon-Keew, BC-Nfld, Labr.
2. *P. sagittatus* (Baoks) Gray,
arrow-shaped sweet coltsfoot, pétasite
sagitté; native, Yukon-Keew, BC-Que,
Labr.
3. *P.* hybrids (*P. palmatus* x *P. nivalis;*
P. palmatus x *P. sagittatus*).

Puccinia conglomerata (*P. nardosmii*)
(C:189): rust, rouille, III on 1 Yukon
[1581]; on 1b, 1c Yukon, Ont, Que
[1581], Ont [1236]; on 1b, 1c, 2, 3 BC,
Alta [1581]; on 1c Yukon [7], NWT-Alta
[1063], BC [1012, 1816], Sask [1581],
Ont [437, 1341]; on 2 que [1236, 1581];
on *P.* sp. BC, Que [720].
P. poarum (C:189): rust, rouille: on 1
Yukon [460]; on 1c BC [1012, 1816].
Stagonospora petasitidis (C:189): on 1c
BC [1012, 1816].

PETROSELINUM Hoffm.   UMBELLIFERAE

1. *P. crispum* (P. Mill.) Mansf. (*P.*
hortense Hoffm.), parsley, persil;
imported from Europe, a garden escape,
BC, Ont, Nfld.

Olpidium brassicae: on 1 Ont [67].
Pythium sp.: on 1 Ont [868] by
inoculation.
P. irregulare: on 1 Ont [868] by
inoculation.
P. sulcatum: on 1 Ont [868] by
inoculation.
P. torulosum: on 1 Ont [868] by
inoculation.
Septoria petroselini: on 1 BC [1816].
Stemphylium botryosum (holomorph
Pleospora herbarum) (C:189): on seeds
of 1 Ont [667].
Ulocladium consortiale (*Stemphylium c.*):
on seeds of 1 Ont [667].

PETUNIA Juss.   SOLANACEAE

1. *P. x hybrida* Vilm., common petunia,
pétunia ou St. Joseph; a cultigen.

Botrytis cinerea (holomorph *Botryotinia*
fuckeliana) (C:190): gray mold,
moisissure grise: on 1 BC [1816], Ont,
que [334].
Erysiphe cichoracearum (C:190): powdery
mildew, blanc: on 1 Ont [1257].
Sclerotinia sclerotiorum (C:190):
sclerotinia rot and wilt, flétrissure
sclérotique: on 1 BC [338], Sask [337,
1140].

PHACELIA Juss.   HYDROPHYLLACEAE

1. *P. heterophylla* Pursh; native, BC(?).
Occurrence of this species in Canada is
doubtful.

Puccinia phaceliae: rust, rouille: on 1
Alta [1578]. Possibly specimens of *P.*
hastata Dougl.

PHALARIS L.   GRAMINEAE

1. *P. arundinacea* L., reed canary grass,
roseau; native, Yukon-Mack, BC-Nfld.
2. *P. canariensis* L., canary grass,
graines d'oiseau; imported from n.
Africa and the Canary Islands, natzd.,
Yukon-Mack, BC-PEI.

Bipolaris sorokiniana (holomorph
Cochliobolus sativus): on 2 Alta
[120].
Campanella subdendrophora: on 1 BC [1425].
Claviceps purpurea (C:190): ergot,
ergot: on 1 Alta [133].
Drechslera tritici-repentis (holomorph
Pyrenophora t.) (C:190): on 1 Ont
[1611].
Helminthosporium sp.: on 1 Sask [338].
Puccinia sessilis (C:190): leaf rust,
rouille des feuilles: II III on 1 Ont
[1236].
Rhynchosporium orthosporum: on *P.* sp.
Nfld [1595].
R. secalis (C:190): on 1 BC [1816]
Septoria sp.: on 1 Sask [338].
Stagonospora foliicola: on 1 Sask [1594].

PHASEOLUS L.   LEGUMINOSAE

1. *P. coccineus* L., scarlet runner bean,
haricot à rames; imported from tropical
America.
2. *P. vulgaris* L., green bean, kidney
bean, snapbean, fève; imported from
tropical America.
3. Host species not named, i.e., reported
as *Phaseolus* sp.

Alternaria sp.: on roots and/or
hypocotyls of 2 Ont [677].
A. alternata (*A. tenuis* auct.) (C:191):
on 2 NS [339].
Ascochyta sp.: on 2 NS [1594].
A. boltshauseri (C:190): leaf spot, tache
ascochytique: on 2 BC [338, 1816], Que
[338].
A. phaseolorum (C:190): leaf spot, tache
ascochytique: on 2 BC [1816].
Bipoloaris sorokiniana (holomorph
Cochliobolus sativus) (C:191): on 2
NB [336, 612], NS [595, 1594].
Botrytis cinerea (holomorph *Botryotinia*
fuckeliana) (C:190): gray mold,
moisissure grise: on 1 BC [337]; on 2
BC [1816, 1817, 1818], NB [333, 339],
NB, NS [334, 335, 336, 1594], PEI [334];
on 3 NB [338], NB, PEI [337].
Cochliobolus sativus (anamorph *Bipolaris*
sorokiniana): on roots and/or
hypocotyls of 2 Ont [677].
Colletotrichum lindemuthianum (C:191):
anthracnose, anthracnose: on 2 BC
[1816], Alta [334, 339], Ont [335, 339,
1775, 1836, 1837, 1906, 1907], Que
[335], NB [333, 334, 335, 336, 339,

1594], NS [333, 335], PEI [333, 334]; on 3 Ont, NB, NS, PEI [337], NB [338]. The Delta race of anthracnose was first observed in Ont in 1976. Subsequent surveys of Pedigreed seed fields revealed 30% infected in 1977 but by 1980 only 1 of 181 fields surveyed was infected. This was due to increased use of resistant bean varieties.

Corynespora cassiicola: on 2 Ont [1599].

Fusarium sp. (C:191): on 2 Alta, Que [1594], Ont [1460, 1461, 1777, 1594]; on 3 BC, Ont, PEI [338].

F. acuminatum (C:191): on 2 Alta [677].

F. avenaceum on 2 Ont [677].

F. culmorum (C:191): on 2 Ont [677].

F. equiseti (C:191): on 2 Ont [677].

F. graminearum (holomorph *Gibberella zeae*): on 2 Ont [677].

F. oxysporum (C:191): on 2 Ont [677].

F. poae (C:191): on 3 Ont [169].

F. solani (holomorph *Nectria haematococca*): on 2 Ont [677, 1775], Ont, NB [339].

F. solani f. sp. *phaseoli* (C:191): dry root rot, pourridié fusarien: on 2 BC [1816], Ont [250, 251, 334, 335, 336, 1776, 1905], Que [335].

F. tricinctum: on 2 Ont [677].

Gliocladium roseum: on 2 Ont [677].

Penicilliuum sp.: on 2 Ont [677].

Pythium sp. (C:191): damping off or root rot, fonte des semis ou pourridié pythien: on 2 BC [1816], Ont [677, 868].

P. irregulare: on 2 Ont [868] by inoculation

P. sulcatum: on 2 Ont [868] by inoculation.

P. sylvaticum: on 2 Ont [868].

P. torulosum: on 2 Ont [868] by inoculation.

Rhizoctonia solani (holomorph *Thanatephorus cucumeris*) (C:191): damping-off or stem rot, fonte des semis ou rhizoctonie: on 2 BC [1816], Ont [337, 339, 677, 1460, 1461, 1594, Que [214, 333, 336], NB [336, 339, 1594]; on 3 BC, Ont, PEI [338].

Rhizopus sp. on 2 Ont [677].

Sclerotinia sp.: on 2 Sask [420, 1138], Man [1138].

S. sclerotiorum (C:191) : stem rot, flétrissure sclerotique: on 1 BC [338]; on 2 BC [336, 1816, 1817, 1818], Alta [335, 772]. Sask [1319], Ont [250, 251, 334, 335, 336, 339, 671, 677, 1460, 1594, 1595, 1775, 1776, 1777, 1905, 1911], Que [335], NB [336, 339], NS [335, 339], PEI [333, 334]; on 3 BC, Ont, NB [337], Ont, NB, NS [338].

Stemphylium botryosum (holomorph *Pleospora herbarum*) (C:191): on seeds of 1 Ont [667].

Stilbella flavescens: on 2 Que [2], a symptomless parasitic fungus.

Trichoderma sp.: on 2 Ont [677].

Ulocladium consortiale (*Alternaria c.*, *Stemphylium c.*) (C:191): on seeds of 1, 2 Ont [667].

Uromyces appendiculatus (*U. phaseoli*, *U. p.* var. *phaseoli*, *U. p.* var. *typica*) (C:191): rust, rouille: 0 I II III on 1 BC [337, 338], Ont [1236]; on 2 BC [338, 1816], Ont [250, 251, 334, 335, 336, 1236, 1775, 1776, 1905]. Both reports of var. *typica* were based on the same specimens.

Verticillium nigrescens: on 3 Que [38, 392, 393].

PHILADELPHUS L.   SAXIFRAGACEAE

1. *P. lewisii* Pursh, mock orange, seringa de Lewis; native, BC-Alta.

Phyllactinia guttata (C:192): powdery mildew, blanc: on *P.* sp. BC [1816].

Septoria philadelphi (C:192): leaf spot, tache septorienne: on 1 BC [1816].

Verticillium albo-atrum: on *P.* sp. Que [337].

V. dahliae: on *P.* sp. Que [38].

PHIPPSIA (Trin.) R. Br.   GRAMINEAE

1. *P. algida* (Soland.) R. Br. (*P. concinna* (Fries) Lindeb.); native, Mack-Labr.

Hysteropezizella diminuens: on 1 Frank [664].

PHLEUM L.   GRAMINEAE

1. *P. pratense* L., timothy, mil; imported from Eurasia, natzd., Yukon-Mack, BC-Nfld. 1a. *P. p.* var. *nodosum* (L.) Huds. (*P. nodosum* L.).

2. Host species not named, i.e., reported as *Phleum* sp.

Acremonium boreale: on 1 BC, Alta, Sask [1707].

Belonioscypha culmicola (*Belonidium vexatum*): on 2 Ont [228].

Bipolaris sorokiniana (*Helminthosporium sativum*) (holomorph *Cochliobolus sativus*): on 1 Alta [120], Sask [961] both by inoculation; on 2 Sask [697].

Cercosporidium graminis (*Passalora g.*) (C:193): brown stripe, strie brune: on 1 BC [1816], Que [32], NB [333, 1195], PEI [1974]; on 2 Que [339].

Claviceps sp.: on 1 Alta [1781].

C. purpurea (C:192): ergot, ergot: on 1 Ont, Que [335]; on 2 BC [339], Alta [338].

Colletotrichum graminicola (C:192): anthracnose: on 2 Que [334].

Dendryphion nanum: on 1 Alta [819].

Drechslera phlei (C:192): on 1 Alta [119, 121, 123, 133, 1611], Sask [983, 1701 on seeds], Man [1611], Ont [1611, 1701 on seeds], 1729]; on 1a Sask [983].

Erysiphe graminis (C:192): powdery mildew, blanc: on 1 PEI [1974].

Heterosporium phlei (C:192): purple spot, tache pourpre: on 1 BC [1816], Alta [107, 119, 122, 123], Alta, Sask [1594],

Ont [1729], Que [282], Que, Nfld [334,
335], NB [1195], NB, NS, Nfld [333], PEI
[335, 1974], Nfld [337]; on 1, la Sask
[983]; on 2 Alta [338, 339], Sask, Nfld
[339].

Mycosphaerella lineolata (C:193): on 1
Que [70].

Olpidium brassicae: on 1 Ont [68].

Phaeosphaeria herpotrichoides: on 1 Que
[70].

Phloeospora idahoensis (holomorph
Didymella festucae): on 1 Alta [1708].

Phyllachora graminis (C:193): tar spot,
rayure goudronneuse: on 1 Que [334].

Pseudoseptoria donacis (Selenophoma d.)
(C:193): eye spot, tache ocellée: on 1
Nfld [336]; on 2 Que [339]. These
specimens on *Phleum* are probably *P.
stomaticola*.

P. stomaticola (Selenophoma donacis var.
s.) (C:193): eye spot, tache
ocellée: on 1 Nfld [1595].

Puccinia graminis (C:193): stem rust,
rouille de la tige: II III on 1 BC
[1012, 1816], Ont [1236], Que [337], NB
[333, 1195].

Rhynchosporina meinersii (C:193): on 1
Ont [1729].

Sclerophthora macrospora: downy mildew,
mildiou: on 1 Alta [338].

Sclerotinia borealis (C:193): snow mold,
moisissure des neiges: on 1 BC [1816].

Stemphylium botryosum (holomorph
Pleospora herbarum): on seeds of 1
Ont [667].

Ustilago salvei (U. striiformis auct.)
(C:193): stripe smut, charbon strié:
on 1 BC [1816], Que [336, 337], NB [333,
1195]; on 2 Ont [334].

PHLOX L.   POLEMONIACEAE

1. *P. divaricata* L., blue phlox, phlox
   divariqué; native, Ont-Que.
2. *P. drummondii* Hook., annual phlox;
   imported from Texas, a garden escape,
   Ont, NB.
3. *P. paniculata* L., perennial phlox;
   imported from e. USA, a garden escape,
   Ont-NS.
4. *P. subulata* L., moss pink; native,
   Ont., a garden escape, Ont-Que, NS.
5. Host species not named, i.e., reported
   as *Phlox* sp.

Colletotrichum dematium: on 5 Que [337].

Erysiphe cichoracearum (C:193): powdery
mildew, blanc: on 1, 2 Ont [1257]; on 3
BC [333, 1816], Ont [337, 1809], Que
[338]; on 3, 5 Ont, Que [1257]; on 5
Alta, Que [339], Alta, NS [1594], Ont
[333, 1340], Que, NS [333, 338], NB
[1194].

E. communis: powdery mildew, blanc: on 3
Alta, Ont [336]; on 5 Alta, Que [335],
Ont, Que, NS, PEI [334], Que [336].
Specimens may be *E. cichoracearum*.

Peronospora phlogina (C:193): downy
mildew, mildiou: on 3 Que [335]; on 4
BC [1816].

Phytophthora sp.: on 3 BC [1816].

Puccinia plumbaria (C:193): rust,
rouille: 0 I III on 1 Ont [1236].

Septoria divaricata (C:194): leaf spot,
tache septorienne: on 2, 3 BC [1816];
on 3 NB [333]; on 5 NB [333, 1194], NS
[334].

Verticillium dahliae: on 3 BC [334, 1978].

PHOTINIA Lindl.   ROSACEAE

1. *P. serrulata* Lindl.; imported from
   China.

Oidium sp.: powdery mildew, blanc: on 1
BC [335, 1816].

PHRAGMITES Adans.   GRAMINEAE

1. *P. australis* (Cav.) Trin. *(P.
   communis* (L.) Trin.*)*, reed, roseau;
   native, Mack, BC-PEI.

Puccinia magnusiana (C:194): rust,
rouille: II III on 1 Man [1341], Ont
[1236].

P. phragmitis (C:194): rust, rouille: II
III on 1 Man [1238], Ont [1236, 1238,
1341].

PHRYMA L.   PHRYMACEAE

1. *P. leptostachya* L., lopseed, phryme à
   épic grêles; native, Man-NB. A slender,
   glabrous perennial herb.

Ophiobolus anguillidus: on 1 Ont [1620].

PHYLLODOCE Salisb.   ERICACEAE

1. *P. caerulea* (L.) Bab., purple mountain
   heather, phyllodoce bleue; native,
   Mack-Labr, Man, Que, NS, Nfld.
2. *P. empetriformis* (Sm.) D. Don, pink
   mountain heather; native, Yukon-Mack,
   BC-Alta.

Exobasidium phyllodoces: on 2 BC [1012,
1578].

Herpotrichiella fusispora (C:194): on 1
Que [70].

Herpotrichia juniperi (H. nigra): on 2
Alta [952].

Physalospora hyperborea (C:194): on 1 Que
[70].

*Protoventuria latispora (Antennularia
l.)*: on overwintered leaves of 1 Que
[73].

PHYSALIS L.   SOLANACEAE

1. *P. heterophylla* Nees, clammy ground
   cherry, cerise de terre sauvage; native,
   Man-NS.
2. *P. virginiana* Mill.; native, Man-Que.

Aecidium physalidis (C:194): rust,
rouille: 0 I on 1 Ont [1236, 1282].

Puccinia physalidis (C:194): rust,
rouille: III on 1 Ont [1236], Ont, Que
[1283]; on 2 Man [1283].

PHYSOCARPUS Maxim.   ROSACEAE

1. *P. opulifolius* (L.) Maxim., ninebark,
bois à sept écorces; native, BC-Alta,
Ont-Que. Deciduous, spirea-like shrubs
with exfoliating bark.

Phellinus ferreus (Poria f.): on 1 BC
[1012]. See Acer.
Sphaerotheca macularis (S. humuli):
powdery mildew, blanc: on 1 Ont [1254,
1257, 1340].

PHYSOSTEGIA Benth.   LABIATAE

1. *P. parviflora* Nutt., small flowered
false dragonhead, physostégie à petites
fleurs; native, BC-Ont.
2. *P. virginiana* (L.) Benth., false
dragonhead, physostégie de Virginie;
native, Man-NS.

Puccinia physostegiae (C:195): rust,
rouille: on 1 BC [1012, 1816]; on 2 Ont
[1236].

PICEA A. Dietr.   PINACEAE

1. *P. abies* (L.) Karst., Norway spruce,
épinette de Norvège; imported from
Europe.
2. *P. engelmannii* Parry, Engelmann
spruce, épinette d'Engelmann; native,
BC-Alta.
3. *P. glauca* (Moench) Voss, *(P.
canadensis* (L.) BSP.), white spruce,
épinette blanche; native, Yukon-Keew, n.
Que-Labr, BC-Nfld. 3a. *P. g.* var.
albertiana (Brown) Sarg.; native,
Keew, BC-Sask.
4. *P. mariana* (Mill.) BSP., black spruce,
épinette noire; native, Yukon-Keew,
BC-Nfld, Labr.
5. *P. pungens* Engelm., Colorado blue
spruce, épinette de Colorado; imported
from w. USA. 5a. *P. pungens* cv.
kosteriana.
6. *P. rubens* Sarg. (*P. rubra* (Du Roi)
Link), red spruce, épinette rouge;
native, Ont-NS.
7. *P. sitchensis* (Bong.) Carr., Sitka
spruce, épinette de Sitka; native,
coastal-BC.
8. Host species not named, i.e., reported
as *Picea* sp.

Acanthostigma parasiticum: on 3 Que
[F62]; on 4 Que [F70].
*Aleurocystidiellum subcruentatum
(Aleurodiscus s.)* (C:195): twig
blight, brûlure des rameaux: on 3 BC
[964, 1012], Que [964, 969], NS [1032];
on 8 Ont, NS [964, 1060], Que [1060].
Aleurodiscus amorphus (C:195): on 3
NWT-Alta [1063], Alta [53, 964, 1060].
See *Abies*.

A. canadensis (C:195): on 3 Ont, Que
[964, 1060, 1341]; on 3 Ont [1827]; on 8
Ont, Que [1060]. See *Abies*.
A. fennicus (C:195): on 6 Que [964,
1060]. Lignicolous.
A. laurentianus (C:195): on 4 Que [964,
1060]. Lignicolous.
A. lividocoeruleus: on 3 Yukon, on 4 Ont,
on 8 BC, Alta, Ont, Que, Nfld, all from
[1060] and Ont [964], Que [969], NS
[1032]. See *Abies*.
A. penicillatus (C:175): on 3 Que [964,
1060]; on 7 BC [964, 1012]; on 8 BC
[1060]. See *Abies*.
A. piceinus (C:175): on 6 Ont [964], Que
[1060]; on 8 Ont [964, 1060].
Lignicolous.
Allescheria terrestris: on 8 BC [1723],
NB [1610].
Alternaria sp.: on 3, 5 Ont [1340]; on 4
Que [917]; on 5 Sask [1848].
A. alternata: on 8 Que [1610].
Amorphotheca resinae: on 7 BC [1012].
Amphinema byssoides (Peniophora b.)
(C:198): on 2, 3 NWT-Alta [1063]; on 2
BC [1060], Alta [950, 1060]; on 3 Yukon
[1060, 1110], Mack [53], BC [1012,
1060]; on 8 Que [969, 1060]. See
Abies.
Amylocorticium subincarnatum: on 3 BC
[1012, 1060]; on 3, 7 BC [1940]; on 8
Alta [1060], Ont [1940]. Associated
with a brown rot.
A. subsulphureum: on 8 Alta [1060].
Associated with a brown rot.
Amylostereum chailletii (Stereum c.)
(C:199): on 2, 8 NWT-Alta [1063] and
Alta [1062]; on 3, 7 BC [1012, 1062]; on
3 BC [1718], Ont [1062]; on 4 Que [917,
1062]; on 6 NB, NS [1032, 1062]; on 8
Alta [952], Que [969]. See *Abies*.
Anomoporia bombycina (Poria b.): on 8 BC
[1061], Alta [579, 1061], Ont [1341].
Associated with a brown rot.
A. myceliosa (Poria m.): on 7 BC [1061].
Associated with a white rot.
*Antrodia albida (Coriolellus sepium,
Trametes s.):* on 2 NWT-Alta [1063]; on
3 Yukon [1061]; on 8 Alta [579, 1061].
See *Abies*.
A. carbonica (Poria c.): on 7 BC [1012,
1061]. See *Abies*.
A. crustulina (Poria c.) (C:199): on 2
NWT-Alta [1063], BC [1012, 1061]; on 3
Yukon [1061, 1110], NWT-Alta [1061,
1063], Mack [53], BC [1012, 1061, 1341],
Ont [1341]; on 4 Nfld [1061], Labr
[867]; on 4, 6 NS [1032, 1061]; on 8 NS,
PEI [1032, 1061]. See *Abies*.
*A. heteromorpha (Coriolellus h., Trametes
h.)* (C:196): brown cubical rot, carie
brune cubique: on 2, 3, 8 NWT-Alta
[1063]; on 2 BC [954], Alta [950, 952,
1061]; on 2, 3, 7 BC [1012, 1061]; on 3
Yukon, NWT, Alta, NB, NS [1061] also
Yukon [1110], Mack [53], NB, NS [1032];
on 4 Alta, NS, Nfld [1061], Ont [1341],
NS [1032]; on 8 BC [1061], Alta [951],
Ont [1827], NS [1032, 1061].

A. *lindbladii (Poria cinerascens)*
(C:199): on 2, 8 Alta [579, 1061]; on 3
BC [1012]; on 8 NS [1032, 1061]. See
Abies.

A. *serialis (Coriolellus s., Trametes s.)*
(C:196): brown cubical rot, carie brune
cubique: on 2, 3, 4 BC, Alta [1061]; on
2, 3 NWT-Alta [1063], BC [1012]; on 3
Yukon, NWT, Sask, Que [1061], Mack [53];
on 4 NWT, Que, Nfld [1061], Mack [1110],
Ont [1341]; on 7 BC [1012, 1341] and BC,
Que [1061]; on 8 Alta [1061].

A. *sinuosa (Coriolellus s., Poria s.)*
(C:196): on 3 BC [1012, 1061]; on 4 NS
[1032, 1061]; on 7 BC [1012].
Associated with a brown rot.

A. *sordida (Poria oleagina)*: on 8 BC
[1061]. See *Abies*.

A. *variiformis (Coriolellus v., Trametes
v.)* (C:196): brown cubical rot, carie
brune cubique: on 2, 3 NWT-Alta [1063];
on 2 BC [1012, 1061, 1341]; on 3 Yukon,
BC, Alta, Sask, Man [1061]; on 3 BC
[1012], Sask, Man [579]; on 4, 8 Ont
[1341], Nfld [1659, 1660, 1661]; on 6
Ont, Que, NB, Nfld [1061] and NB [1032];
on 7 BC [1012]; on 8 NS [1032].

A. *xantha (Poria x.)* (C:199): brown
cubical rot, carie brune cubique: on 3
Yukon, NWT, BC, Alta, Nfld [1061] and
NWT-Alta [1063], Mack [53]; on 3, 4, 7
BC [1012]; on 4 NWT [1061], Mack [579],
Ont [1341]; on 7 BC [1061]; on 8 BC,
Alta [1061].

Aphanomyces euteiches: on 3 NS [1036].

*Appendiculella pinicola (Irenina p.,
Meliola p.)* (C:197): on 3 Nfld [1659,
1660, 1661, F73, F74]; on 7 BC [1012,
1358].

Armillaria mellea (C:195): root rot,
pourridié-agaric: on 1 BC [1012], Que
[1210, F64, F65, F76], NB, NS [F75],
Nfld [1659, 1660, 1661, 1663, F68,
F69]; on 2 Alta [1062, F65]; on 3
NWT-Alta [1063], Mack [53], BC [1012],
Alta [53, 950], Sask [1952], Sask-Man
[1951, F61], Ont [6, 1341, 1956, 1964,
1966, F62, F65], Que [914, 1212, 1222,
F64, F70, F74], NB [1032, F75], NS
[F75], Nfld [1659, 1660, 1661, 1663,
F65, F68, F69, F71] and BC, Alta, Sask,
Man, Ont, NB [1062], Ont, Que [F71, F72,
F76]; on 4 NWT-Alta [1063], Alta [F64],
Ont [90, 1062, 1341, 1956, 1958, 1964,
1966, F72], Que [F64, F65, F72], NB, NS
[F75], Nfld [1659, 1660, 1661, 1663,
F60, F64, F65, F67, F69, F73] and Ont,
Nfld [F63, F71, F74]; on 5 Man [1062],
NB, NS [F72]; on 6 Que [F67, F70], NB
[1032], NB, NS [1062, F71, F72, F74],
Nfld [1660, 1661, 1663, F69]; on 7 BC
[1012, 1062], Nfld [882, 1660, 1661,
1663, F68, F69, F73, F74,]; on 7 x 3
hybrid Nfld [1661]; on 8 Alta [F65, F68,
F71], Sask [1955], Ont [1341], Que
[1377], NB [1032, 1062], Nfld [1661,
F62]. See *Abies*.

Ascochyta piniperda (C:195): on 5 NB
[1032].

Ascocoryne cylichnium (Coryne c.): on 3
Que [1222].

A. *sarcoides (Coryne s.)* (C:196): on 3,
4, 8 NWT- Alta [1063]; on 3 Sask, Man
[1951], Ont [1966]; on 4 Ont [90, 1958,
1964, 1966], Que [917, 937]; on 8 Alta
[952], Que, NB, NS [1610].

Aspergillus sp.: on 8 BC [1012], NB
[1610].

A. *fumigatus*: on 8 BC [1723], NB, NS
[1610].

Asterodon ferruginosus (C:195): on 3
Yukon, BC [1062], BC [1012]. See
Abies.

Athelia decipiens: on 8 NS [1032]. See
Pinus.

A. *epiphylla (Corticium e.)*: on 3 Mack
[53]. See *Abies*.

A. *microspora*: on 8 Alta [1060].
Associated with a white rot.

A. *neuhoffii*: on 8 Alta, Que [1060], Que
[969]. See *Abies*.

A. *pellicularis* (C:196): on 8 Nfld [1659,
1660, 1661]. See *Abies*.

Atichia glomerulosa: on 4 Que [F69].

Aureobasidium sp.: on 8 NB [1610].

A. *pullulans (Pullularia p:)* (C:195): on
1 Que [1209, F61]; on 4 Que [457, 917,
937]; on 1, 4, 6 Que [F66]; on 8 Que, NB
[1610].

Auricularia auricula (A. auricularis)
(C:195): on 3 BC [1012], BC, NB [423]
and BC, Alta, NB [1062]; on 8 NB
[1032]. See *Abies*.

Bactrodesmium sp.: on 7 BC [1012].

B. *obliquum*: on 3 Sask, Man [1754]; on 4
Sask [1760].

*Basidioradulum radula (Corticium hydnans,
Hyphoderma r.)*: on 4 Ont [1060]; on 8
Que [969]. See *Abies*.

Bertia moriformis: on 8 Ont [1340].

Bispora betulina: on 8 NB [1610].

Bisporella citrina: on 3 BC [1012].

Bjerkandera adusta (Polyporus a.)
(C:198): on 3 Yukon [1061, 1110], BC
[1012]; on 4 Que [917], Nfld [1659,
1660, 1661]. See *Abies*.

B. *fumosa (Polyporus f.)*: on 3 Ont
[1341]. See *Acer*.

Bondarzewia berkeleyi (Polyporus b.)
(C:198): on 7 BC [1012, 1061]. See
Acer.

B. *montana (Polyporus m.)* (C:198): on 7
BC [1012, 1061]. See *Abies*.

Boreostereum radiatum: on 4 BC, Man
[1062]. Associated with a white rot.

Botryobasidium ansosum (Pellicularia a.)
(C:198): on 7 BC [1012, 1060].
Associated with a white rot.

B. *pruinatum*: on 8 Que [969, 1060]. See
Acer.

B. *subcoronatum (Pellicularia s.)*
(C:198): on 3 NWT-Alta [1063], Mack
[53], BC [1012, 1060]; on 8 Que [972,
1060], NS [1032]. See *Abies*.

B. *vagum (Corticium v., Pellicularia v.)*
(C:198): on 2 NWT-Alta [1063]; on 4 Que
[917]; on 8 Que [969, 1060], NB, NS
[1032]. See *Abies*.

*Botryohypochnus chordulatus (Pellicularia
   c.)*: on 3 NWT-Alta [1063], Mack [53].
   Lignicolous.
B. isabellinus (Pellicularia i.): on 3
   NWT-Alta [1063], Mack [53]. See *Abies*.
Botryosphaeria piceae: on 2, 3, 7 BC
   [1012, F65]; on 2, 7 BC [496, 507]; on 7
   BC [516].
Botrytis sp.: on 3, 7 BC [1012].
B. cinerea (holomorph *Botryotinia
   fuckeliana*): gray mold, moisissure
   grise: on 2 BC [1012, F67]; on 3 BC
   [F76], Ont [F72, F73, F74], Que [F67],
   on 4 Ont [F73], NB [1032, F64, F70].
Brachysporiella sp.: on 8 NB [1610].
Brevicellicium exile (Corticium e.): on 8
   Que [969]. Lignicolous.
Byssochlamys spp.: on 8 NB [1610].
*Byssocorticium terrestre (Byssoporia
   terrestris* var. *parksii, Poria t.)*:
   on 3, 8 Alta [1061] and 8 Alta [931].
   Mycorrhizal and associated with a brown
   rot.
Calcarisporium sp.: on 8 NB [1610].
Calocera viscosa: on 4 Nfld [1659,
   1660]. See *Abies*.
Caloscypha fulgens (anamorph
   Geniculodendron pyriforme): on stored
   seeds of 2, 3, 3 x 2 BC [1749]. The
   fungus spreads and kills dormant seeds
   during stratification (cold treatment)
   and in cool, moist seedbeds but is
   unable to infect germinating seeds
   [1225].
Camarosporium strobilinum: on 3 BC [1012,
   F76].
Candida sp.: on 1 Que [1211].
Capnobotrys neesii: on 8 NB [806].
Cenangium sp.: on 4 NWT-Alta [1063].
C. ferruginosum (C. abietis): on 6 Ont
   [F71].
Cenococcum graniforme: on 3, 4 Ont
   [1964]; on 4 Que [485].
Cephaloascus fragrans: on 8 Que [1610].
Cephalosporium sp.: on 1 Que [1211]; on 3
   NWT-Alta [1063]; on 4 Ont [90, 917,
   1964]; on 7 BC [1012]; on 8 Que, NB
   [1610].
C. terrestre: on 8 BC [1723]. This
   combination does not appear to have been
   validly published. Presumably it refers
   to the *Cephalosporium* anamorph of
   Allescheria terrestris.
Ceraceomyces borealis (Athelia b.): on 3
   BC [568, 1012]; on 3, 8 BC [1060]; on 4
   Alta [568, 1060]. See *Abies*.
*C. serpens (Byssomerulius s., Merulius
   s.)*: on 3 BC [1012], BC, Alta [568];
   on 8 Mack, Que [568], Alta [1060], Que
   [969]. See *Abies*.
C. sulphurinus (Phanerochaete s.): on 8
   Alta [1060]. Associated with a white
   rot.
C. tessulatus (Athelia t.): on 3 BC
   [1012, 1060]. See *Abies*.
Ceratocystis sp.: on 1, 3 Ont [1340]; on
   3 BC [F64], Que [F66].
C. aequivaginata: on 4 Man [1190].
C. albida: on 8 NB [1610].

C. allantospora: on 3 Ont [647, 1340]; on
   4 Man [1190].
C. angusticollis: on 4 Man [1190].
C. arborea: on 4 Man [1190].
C. bicolor: on 3 Ont [647, 1340].
C. cainii: on 4 Man [1190].
C. coerulescens: on 4 Man [1190].
C. coronata: on 4 Man, Ont [1190].
C. crassivaginata: on 4 Ont [647, 1190,
   1340].
C. crenulata: on 4 Man [1190].
C. dolominuta: on 4 Man [1190].
C. europhioides (C:195): on 3, 4 Ont
   [647, 1190, 1983]; on 4 Man [1190], Ont
   [1340].
C. huntii: on 4 Man [1190].
C. ips: on 3 Ont [647, 1190, 1340]; on 4,
   8 Man [1190].
C. leucocarpa: on 4 Man [1190].
C. minor: on 4 Man [1190], Ont [647,
   1340].
C. minuta: on 4 Man [1190], Ont [647,
   1340].
C. nigra: on 4 Ont [647, 1340].
C. ochracea: on 4 Ont [647, 1340].
C. olivacea: on 4 Man [1190], Ont [647,
   1190, 1340].
C. piceae: on 3, 4 Ont [647, 1340]; on 4
   Man [1190], Que [917]; on 8 Que, NB, NS
   [1610].
C. piceaperda: on 3 NS [1340].
C. pilifera: on 4 Man [1190].
C. pseudoeurophoides: on 4 Man [1190].
C. pseudonigra: on 4 Man [1190].
C. sagmatospora (C:195): on 3, 4 Ont
   [647, 1340]; on 4 Man, Ont [1190], Ont
   [1983].
C. stenoceras: on 8 Ont [647].
C. tenella: on 4 Ont [647].
C. tetropii: on 4 Ont [647, 1340].
C. tubicollis: on 4 Man [1190].
C. truncicola: on 3 Ont [647, 1340].
Ceriporia tarda (Poria t.) (C:199: on 3
   BC [1012, 1061]. See *Abies*.
Cerocorticium notabile (Corticium n.)
   (C:196): on 4 Ont [1060], Nfld [1659,
   1660, 1661]; on 8 Man [1060].
   Lignocolous.
Ceuthospora pithyophila: on 3 Sask [F66].
Chaetoderma luna (Peniophora l.) (C:198):
   on 2 NWT-Alta [1063]; on Yukon [1060];
   on 3, 7 BC [1012, 1060]; on 8 Alta
   [1060]. See *Abies*.
Chaetomium sp.: on 3 Ont [1340].
Chaetoscutula juniperi: on 4 Que [F66].
Chalara cylindrica: on 3 Sask and on 3,
   4, 8 Man [1760].
C. cylindrosperma: on 3, 4 Sask and on 3,
   8 Man [1760].
C. insignis: on 3 Man [1760].
Chondrostereum purpureum (Stereum p.)
   (C:199): on 3 BC [1012, 1062], Ont
   [1341]. See *Abies*.
Chrysomyxa sp.: rust, rouille: on 2, 3
   BC [F64]; on 3 NWT-Alta [1063], Mack
   [53], Sask [F75], Man [F62], Ont [1341],
   NS [F60, F61], Nfld [F61]; on 3, 4 Sask,
   Man [F63], Nfld [1660, 1661]; on 4 Sask
   [F75], Man [F62], Nfld [1659, F61, F63];
   on 5 Sask [F75], Man [F64]; on 6 NS

[F60]; on 8 Alta [F76], Sask [F73], Man [F76], Ont [1341], Nfld [1661, F62] and Alta, Man, NB, NS [F75], Sask, Man, NB [F60].

C. *arctostaphyli* (*Melampsorella caryophyllacearum* auct., *M. cerastii* auct.) (aecial state *Peridermium coloradense*) (C:195): witches'-broom rust, rouille-balai de sorciére: 0 I on 1 Alta [F63], Sask [F69]; on 1, 2, 3, 4, 8 NWT-Alta [1063]; on 2 BC [954], Alta [952]; on 2, 3, 4, 7 BC [1012]; on 3 Yukon [2007], Mack [53, F67], BC [953, 954, 2007, F74, F76], Alta [53, 644, 950, 951, 952], Sask [1341, F64, F65], Man [F71], Ont [F64, F66, F68, 1236], Que [F64, F68, F72, 1377], NS [F66], Nfld [F61, 1661] and Man, Ont [F63, F65, F70]; on 3, 4 Yukon [460]; on 4 Mack [53, F61, F63, F65, F67], BC [2007], Alta [53, 951, F70], Man [F63, F71], Ont [F62, F63, F65, F66, F69, F71, F73, 1236, 1341], Que [F64, F72, F76], NS [F62, F63, F66], Nfld [F65, F74, F75, F76, 1659, 1660, 1661] and Sask, Man, Ont [F64, F67, F70], NB, NS [1032, F64], Ont, Que [F68, F74, F75]; on 6 NB, NS [F64] and NS [1032, F62, F65]; on 7 Nfld [1660, F72]; on 8 Yukon, BC, Alta, Sask, Man [2008] and Yukon, BC [F61], Sask, Man [F72], Alta [F72], Man [F67, F73], Ont [F61, F72], NS [1032]. The brooms can cause economically significant losses through bole deformation, loss of increment and mortality.

C. *chiogenis* (C:195): needle rust, rouille des aiguilles: 0 I on 4 Ont [1236], Que [1377, F60, F66]; on 6 Que [F61]; on 8 Sask, Man [F67], Ont [2008], Que [1377]. The alternate host is *Gaultheria hispidula*.

C. *empetri* (C:195): rust, rouille: 0 I on 2 Alta [F64, F65]; on 3 Yukon [F76, 460], Mack [53, F64], BC [954, 1012], Alta [951, F65], Sask [F66], Man [1236], Que [1032, 1236, 1377, F60, F62, F66, F67], Nfld [1659, 1660, 1661, F73, F74]; on 3, 4, 5, 8 NWT-Alta [1063]; on 4 Alta [F62, F65], Man [F67]; on 6 Que [1377]; on 8 Yukon, Mack, BC [2008] and BC [951], Alta [952, F67], Nfld [1661]. The alternate host is *Empetrum nigrum*.

C. *ledi* (C:196): rust, rouille: 0 I on 2, 3, 4, 8 NWT-Alta [1063]; on 2, 3 BC [953, 1012]; on 2 Alta [951, F65]; on 3 Yukon [F62, F70, 460], Mack [53, F61, F62], BC [F72], Alta [53, F62], Sask [F71], Man [F67, F69], Ont [1827], NB, NS [1032], NS [F62], Nfld [1659, 1660, 1661, F67] and BC, Alta, Sask, Man [F73], NB, NS, PEI [F72], NB Nfld [F64, F65, F66]; on 3, 4 Alta [F65, F70], Alta, Ont [F60, F61, F64, F71], Sask, Man, Ont [F65, F68, F70], Ont [1236, F62, F63, F66, F67, F69], Que [F67, F69, F71, F72, F73], NS [F64, F67]; on 4 Mack [53, F61], BC [F73], Alta [F74], Sask [F71], Man [F69], Ont [1341], Que[282, 1377, F63, F65, F66, F70, F75], NB

[1032, F61, F76], PEI [1032], Nfld [1659, 1660, 1661, F64, F65, F66, F72] and Sask, Man [F67], Que, NB [F62, F64, F68, F74]; on 5 Ont [1236, 1341], NS [F65, F66]; on 6 Ont [1341], NB [F64] and NB, NS, PEI [1032]; on 4, 7 BC [1012, F72]; on 8 Yukon, Mack, Alta [F68], and Yukon [F72], BC [953, 954], Alta [950, F67, F69, F72, F74], Sask [F72], Man [F66], Ont [F72, F73, F74, F75, F76], Que [1377], NB, NS [1032], Nfld [1661].

C. *ledi* var. *cassandrae* (*C. cassandrae*) (C:196): rust, rouille: on 3, 4, 5, 6 Que [F60, 1377]; on 3 PEI [1032]; on 4, 6 Que [F61]; on 4 NS [1032]; on 5 NS [1032, F64]. The alternate host is *Chamaedaphne calyculata*.

C. *ledi* var. *ledi* (*C. ledi* var. *groenlandici*) (C:196): rust, rouille: on 3, 4, 5, 6 Que [F60]; on 4, 6 Que [F61]; on 8 Mack, BC, Alta, Sask, Man [2008]. The alternate hosts are *Ledum* spp.

C. *ledicola* (C:196): needle rust, rouille des aiguilles: 0 I on 2, 3, 4, 5, 8 NWT-Alta [1063]; on 2 BC [1012, F64], Alta [F65], PEI [1032]; on 2, 3 Alta [951, 952], NB, PEI [F69, F72]; on 3 Yukon [F70], Mack [53, F61], BC [953, 954, F64, F76], Alta [644], Sask [F71], Man [F69], Ont [F63], NB [F62, F65], NS [F70, F72] and Sask, Man [F65, F66, F67, F68, F70, F73]; on 3, 4 Yukon [460], BC [1012, F65, F72, F73], Alta [53, F60, F61, F65, F70, F71, F73], Ont [1236, 1341, 1828, F61, F65, F66, F70], Que [282, F62, F72], NS [F63, F67, F68, F71, F73], Nfld [1341, 1659, 1660, 1661, F64, F65, F66, F69, F73] and Alta, Ont, NB [F64], Ont, Que [F60, F67, F68, F69, F71]; on 3, 4, 6 NB, NS, PEI [1032]; on 4 Mack [53, F61, F62], Alta [F62, F74], Sask [F71], Man [F69], Sask, Man [F65, F66, F67, F68, F70, F73], Ont [F62], Que [1377, F61, F64, F65, F66, F70, F73, 75], Que, NB [F63, F74, F76], NS [F65, F66], Nfld [F67, F68, F70, F72, F74]; on 5 Ont [1236], Que [F61], NS [337, 1032]; on 5, 6 Que [F60]; on 6 NB [F64], NS [F63, F66, F73]; on 7 BC [1012, F72, F74], Nfld [1660]; on 8 Yukon, Mack, BC, Alta, Sask, Man [2008] also Yukon, Alta, Sask, Ont [F72], Yukon, Mack, Alta [F68] and BC [F63, F67], Alta [53, 951, 952, F67, F69, F74], Sask [339], Ont [F73, F74, F75, F76], NB [1032, F61], NS [1032, F62, F64], Nfld [337, 338]. The alternate hosts are *Ledum* spp.

C. *monesis* (C:196): cone rust, rouille des cônes: 0 I on 7 BC [1012, 2008]. The alternate host is *Moneses uniflora*.

C. *piperiana*: on 7 BC [1341]. The specimens should be reexamined because Ziller [2008] did not know of any Canadian specimens on *Picea*.

C. *pirolata* (*C. pyrolae*) (C:196): cone rust, rouille des cônes: 0 I on 2, 3, 4, 8 NWT-Alta [1063], on 2, 3, 4, 5, 7 BC [1012]; on 2 BC [954, F75]; on 2, 3

Alta [950, 951, 952, F64]; on 3 Yukon
[F62, F69, 460], Mack [53, F61], BC
[F62, F66, F70, F71, F75, F76], Alta
[F69, F71], Sask [F64], Man [F62, F63],
Ont [F60, F64, F65, F70, F72, F74, F75],
Que [282, F66], NB [F64], PEI [F66, F68]
and NB, NS [1032, F66, F68]; on 3, 4
Alta [53, F61, F70], Sask, Man [F65,
F67], Ont [1236, 1341, F67]; on 3, 4, 6
Que [1377]; on 4 Mack [53], Alta [951,
F64], Man [F63], Ont [F69], Que [F60],
NB, NS, PEI [F69] and NB, NS [1032]; on
4, 5 NB, NS, PEI [F66, F68]; on 5 BC
[F61], Alta [F67], NB, PEI [1032], NS
[F72]; on 8 Yukon, Mack, BC, Alta, Sask,
Man [2008] and Alta [F60, F62, F66, F67,
F68, F74], Sask [F74], Man [F75], Ont
[F62, F63, F68, F73], NB [1032].   The
alternate hosts are *Moneses* and
Pyrola spp.
C. *weirii* (C:196):   needle rust, rouille
    des aiguilles: III on 2, 3, 8 NWT-Alta
    [1063]; on 2 Alta [952], NB [F74]; on 2,
    3 BC [953, 1012]; on 3 Yukon [F69], Mack
    [53], BC [954, F63, F70, F74], Alta
    [F69, F70, F71], Sask [F64], Ont [1341,
    F67] and Sask, Man [F65, F66]; on 3, 8
    Alta [951, 952]; on 4 Ont [1236, 1341,
    F65], Que [F60, F64]; on 6 NS [1032]; on
    7 BC [1012, F71]; on 8 Yukon, Mack, BC,
    Alta, Sask, Man [2008] and BC [F65,
    F66], Alta [338, F66, F67, F68, F73].
C. *woroninii* (C:196):   rust, rouille: 0 I
    on 3 Yukon [F72, F74], Mack [53], BC
    [F74], Alta [53, F63], Que [1236]; on 3,
    4 Yukon [F69, F71, 460], NWT-Alta
    [1063], BC [1012, F76]; on 4 Mack [53,
    F63], BC [F75], Alta [F73]; on 8 Yukon,
    Mack, BC, Alta [2008] and Yukon [F66,
    F70], Mack [F66].   The alternate hosts
    are *Ledum* spp.
Chrysosporium sp.:   on 8 NB [1610].
C. *pruinosum*:   on 8 BC [1723].
Cladosporium sp.:   on 2, 3, 7 BC [1012];
    on 3 Ont [1828]; on 8 Yukon, Alta [F68],
    NB [1610].
C. *cladosporioides*:   on 8 NB [1610].
C. *macrocarpum*:   on 8 Man [1760].
C. *variabile*:   on 3, 4 Ont [F66]; on 4 Ont
    [1827].   Rarely found on *Picea*.
Clavulicium macounii: on 3 BC [1012].
    See *Abies*.
Colletotrichum gloeosporioides (holomorph
    Glomerella cingulata):   on 8 Alta
    [1155].
Collybia acervata: on 3 BC [1012, 1062].
    Associated with a white rot.
C. *dryophila*:   on 3 Ont [1341].   Probably
    occurred on well-rotted wood.
Columnocystis abietina (*Stereum a.*)
    (C:199):   on 2, 3, 8 NWT-Alta [1063]; on
    3 Yukon [1062], Mack [53]; on 7 BC
    [1012, 1062, 1341]; on 8 Que [969].   See
    Abies.
Comatospora suttonii: on 4 BC [1012,
    1360].
Confertobasidium olivaceo-album (*Athelia
    fuscostrata*, *Corticium f.*) (C:196):   on
    2 BC [1012, 1060]; on 8 NS [1032].
    Lignicolous.

Coniophora sp.:   on 3 NWT-Alta [1063],
    Mack [53].
C. *arida* (C:196):   on 8 Man [1062], Que
    [969, 1062].   See *Acer*.
C. *fusispora* (C:196):   on 8 NS [1032,
    1062].   Associated with a brown rot,
    sometimes in live trees.
C. *olivacea* (*Coniophorella o.*) (C:196):
    on 3 BC [1012, 1062]; on 4, 8 Alta
    [1062]; on 8 Man [1062].   Associated
    with a brown rot, sometimes in live
    trees.
C. *puteana* (C:196):   brown cubical rot,
    carie brune cubique: on 2 BC [1012,
    F67]; on 2, 3 BC, Alta, Man [1062]; on 3
    NWT-Alta [1063], Yukon [460], BC [1012,
    1718], Sask [1952], Sask-Man [1951],
    Ont, NB [1062]; on 3, 4 Ont [F71, F72];
    on 3a BC [1233]; on 4 Alta, Man, Ont,
    Que [1062] and Ont [90], Que [937], NB
    [1062]; on 4, 6, 8 NB [1032]; on 8 Sask
    [1955], Man [1062].   Sometimes causes a
    heart rot of live trees.
Coniothyrium sp.:   on 3 NWT-Alta [1063],
    BC [1012].
C. *fuckelii* (holomorph *Leptosphaeria
    coniothyrium*):   on 8 NB [1610].
Conoplea fusca: on 3 Man and on 3, 4, 8
    Sask [1760].
C. *geniculata*:   on 3, 4 Man and on 4 Sask
    [1760].
Cordana pauciseptata (C:196):   on 8 Que
    [1610].
Coriolus hirsutus (*Polyporus h.*):   on 3 BC
    [1012, 1341].   See *Abies*.
C. *pubescens* (*Polyporus p.*) (C:198):   on 3
    BC [1012, 1061].   See *Abies*.
C. *versicolor* (*Polyporus v.*) (C:199):   on
    8 NB [1032, 1061].   See *Abies*.
Corniculariella abietis (holomorph *Dermea
    grovesii*):   on 3 Ont [401].
Corticium sp.:   on 3 NWT-Alta [1063], Mack
    [53], on 7 BC [1012]; on 8 NS [1032].
C. *centrifugum*:   on 3, 4 NWT-Alta [1063].
    See *Abies*.
Costantinella terrestre (*Verticillium
    t.*):   on 8 NB [1610].
Crepidotus herbarum (C:196):   on 3 BC
    [1012].
Crinula caliciiformis (holomorph *Holwaya
    mucida*):   on 4 Man [1760].
Crustoderma dryinum (*Peniophora d.*)
    (C:198):   on 7 BC [1012, 1060]; on 8 Que
    [969, 1060].   See *Abies*.
Crustomyces pini-canadensis (*Radulum p.*):
    on 8 Ont [1341].
Cryptoporus volvatus (*Polyporus v.*)
    (C:199):   on 3 NWT, BC, Alta, Man, Ont,
    Que [1061] also Yukon [460], NWT-Alta
    [1063], Mack, Alta [53], BC [1012], Ont
    [1341]; on 7 BC [1012, 1061, 1341].   See
    Abies.
Cucurbidothis pithyophila: on 7 BC [1012,
    F64, F67].
Cylindrobasidium corrugum (*Corticium c.*):
    on 3 BC [1012].   See *Abies*.
C. *evolvens* (*Corticium e.*, *C. laeve*)
    (C:196):   white spongy rot, carie
    blanche spongieuse: on 3 BC [1012]; on

4 Que [917]; on 8 Que, NB [1610], Nfld
[1659, 1660, 1661].
Cylindrocarpon sp.: on 3 BC [1012]; on 4
NB [F72].
C. roseum: on 2, 7 BC [507].
Cylindrocephalum sp.: on 8 NB [1610].
Cylindrocladium floridanum: on 3, 4 Ont
[F75]; on 4 Ont [F74, F76].
C. scoparium: on 3 Que [F65].
Cylindrosporium sp.: on 3, 8 NB [1032].
Cystostereum murraii: on 8 Que [969,
1062]. See *Acer*.
Cytospora sp. (C:196): on 1 Que [1208,
1209, F70]; on 1, 3 Que [1210, 1211,
F61, F63, F64, F66]; on 1, 4, 8 Que
[F62]; on 2 BC [1012]; on 3 Yukon [460],
Que [915]; on 3, 5 Que [F65]; on 4 Que
[917, 937, 943, 947, F66, F70]; on 6 Ont
[1340], Que [F66]; on 8 Que [F71].
C. curreyi (C:197): on 2 Alta [952, F61].
C. kunzei (holomorph *Leucostoma k.*)
(C:197): canker, chancre cytosporéen:
on 1, 3 Ont [1340, 1594, F67]; on 3 Ont
[1738, 1828]; on 3, 5 Ont [F65]; on 5
Que [337, 338], Nfld [F75]; on 8 Alta
[952], Ont [1340].
C. kunzei var. *piceae*: on 1 Que [1672];
on 3 Ont [863].
Dacrymyces sp.: on 3 NWT-Alta [1063].
D. capitatus (D. ellisii): on 8 Que [282,
1062]. See *Alnus*.
D. palmatus: on 3 Yukon [460], NWT-Alta
[1063], BC [1012, 1062], Alta [53]; on 4
Alta [1062]; on 8 Que [282]. See
Abies.
D. stillatus (D. deliquescens): on 3
NWT-Alta [1063], Alta [53]. Lignicolous.
Dacryobolus sudans (Odontia s.) (C:198):
on 3 BC [1012, 1060]. Associated with a
brown rot.
Darkera parca: on 3 Alta [1947].
Dasyscyphus sp. (C:197): on 8 Que [282].
D. ellisianus (C:197): on 7 BC [1012].
Datronia mollis: on 3 Alta [1061]. See
Acer.
Dendrothele sp.: on 8 BC [1012].
Dermea grovesii: on 3 Ont [1340, 1448].
D. piceina (C:197): on 3 Ont [1683]; on 4
Ont [1340], Que [1683, F68].
Dichomera gemmicola: on 2, 3, 7 BC [526,
1012, F72]; on 3 Sask [526]; on buds of
8 BC, Sask [1763].
Dichomitus squalens (Polyporus anceps)
(C:198): on 3 Yukon [460], NWT-Alta
[1063], Mack [53], Que [F65], NS [1032],
NWT, Sask, Man, NS [1061]; on 4 Ont
[1341], NB [1032, 1061], Nfld [1061]; on
8 Ont [1061]. See *Abies*.
Dimerium balsamicola (Dimeriella b.): on
4 Que [F67].
Discocainia treleasei (Atropellis t.)
(C:195): on 7 Mack, Frank, BC [1446],
and BC [1012, F66].
Echinodontium tinctorium (C:197): brown
stringy rot, carie brune filandreuse:
on 2, 3, 7 BC [1062]; on 3 BC [1012,
1718]. See *Abies*.
Endophragmia glanduliformis: on 3 Sask,
Man [1760].

Epicoccum purpurascens (E. nigrum): on 3a
Alta [746]; on 4 Que [917].
Epipolaeum tsugae: on 3 BC [1012]. An
atypical host for this fungus.
Eriosphaeria vermicularia: on 3 NB [F66];
on 4 PEI [1032, F65].
*Euantennaria rhododendri (Limacinia
alaskensis)*: on 2 BC [1012, F64].
Exidia pinicola: on 3 Yukon [460].
Lignicolous. Previously known only from
the type specimen which was on branches
of *Pinus* in New York State.
E. saccharina: on 8 Ont [1341].
Lignicolous.
Exidiopsis calcea (Sebacina c.) (C:199):
on 3 Yukon, BC [1062], NWT-Alta [1063],
Mack, Alta [53]; on 4 NS [1032, 1062];
on 8 Man [1062], NS, PEI [1032, 1062].
Associated with a white rot.
E. macrospora: on 3 BC [1012, 1062].
Lignicolous.
Flagelloscypha citrispora: on 8 BC
[1424]. Lignicolous.
*Flaviporus overholtsii (Polyporus
canadensis)*: The name used in Conners
should be corrected.
F. semisupinus (Poria romellii): on 3 NWT
and on 8 Alta [1061]. See *Betula*.
Fomes sp.: on 4 Nfld [1660, 1661].
*Fomitopsis cajanderi (Fomes c., F.
subroseus)* (C:197): brown cubical rot,
carie brune cubique: on 2, 3, 8
NWT-Alta [1063]; on 2 BC, Alta [1061];
on 3 NWT, BC, Alta, Sask, Man [1061],
NB, NS [1032, 1061] and Mack [53, F67],
BC [1012], Alta [53], Man [1031]; on 4
Alta [1175], Ont [90, 1341] and Alta,
Ont [1061]; on 4, 6 NB, NS [1032, 1061];
on 8 Ont [1341], NB, NS, PEI [1032,
1061].
F. officinalis (Fomes o.) (C:197): brown
cubical rot, carie brune cubique: on 2,
7 BC [1061]; on 7 BC [1012].
F. pinicola (Fomes p.) (C:197): brown
cubical rot, carie brune cubique: on 2,
3, 4, 8 NWT-Alta [1063]; on 2 BC [1012,
1061]; on 3 Yukon, NWT, BC, Alta, Sask,
Man, Ont, Que, NS [1061] and Yukon [460,
1110], Mack, Alta [53], BC [1012], Ont
[1341], NS [1032], Nfld [1659, 1660,
1661]; on 3a BC [1233]; on 4 NWT, Ont,
Que, NB, NS [1061], Nfld [1659, 1660,
1661] and Mack [53], Ont [1150, 1341],
NB, NS [1032]; on 6 Que [1391], NB, NS
[1032, 1061]; on 7 BC [1012, 1061, 1149,
1341]; on 8 NWT, BC, Alta, Sask, Man,
Ont, Que, NB, NS, PEI [1061] and Mack
[53], BC [953, 954], Alta [1828], Ont
[1150, 1341, 1377], Que [282, 1061], NB,
NS, PEI [1032].
F. rosea (Fomes r.) (C:197): brown
cubical rot, carie brune cubique: on 2,
3, 8 NWT-Alta [1063]; on 2, 3 BC [1061];
on 3 Yukon, NWT, Sask, Ont, Que [1061],
Mack [53], BC [1012], Sask [F63]; on 3,
8 Ont [1031, 1341]; on 3, 4, 6 NB [1032,
1061]; on 4 Ont [1341], NS [1061]; on 8
NB, NS, PEI [1032, 1061].
Foveostroma sp. *(Micropera* sp.): on 3
BC [1012].

Fusarium sp.: on 3 Ont [1340], NS [1032];
on 3a Alta [746]; on 4 Que [F65], NB
[F72]; on 5 Yukon [460], Sask [1848]; on
6 Ont [1827]; on 8 Que [1377], NB [1610].

F. lateritium: on 3 Ont [1827].

F. oxysporum: on 3 NS [1885]; on 3a Alta
[746]; on 6 NB [1885]; on 7 BC [F69]; on
8 NB [1610].

F. solani (holomorph *Nectria
haematococca*): on 7 BC [F69]; on 8 NB
[1610].

Fusicoccum sp.: on 3 NB, NS [1032].

Fusidium sp.: on 4 Que [917].

Ganoderma applanatum (C:197): white
mottled rot, carie blanche madrée: on 3,
7 BC [1012, 1061]; on 3, 8 NWT-Alta
[1063]; on 4 Nfld [1660, 1661]; on 8 NS
[1032, 1061].

G. lucidum: on 8 NB, NS [1032]. See
Abies.

G. oregonense (C:197): on 7 Bc [1061].
See *Abies*.

G. tsugae (C:197): on 8 NS [1061].
Associated with a white rot.

Geniculodendron pyriforme (holomorph
Caloscypha fulgens): on seeds of 3, 7
BC [1225]. See *C. fulgens*.

Geotrichum sp.: on 3 NWT-Alta [1063].

Gliocladium roseum: on 1 NB [1036].

G. viride: on 8 Que, NB [1610].

Gloeocystidiellum furfuraceum (*Corticium
f.*): on 8 Que [969, 1060], NS [1032].
Lignicolous.

G. ochraceum (*Corticium o.*): on 2
NWT-Alta [1063]; on 3 Mack [53]. See
Abies.

Gloeophyllum odoratum (*Osmoporus o.*,
Trametes o.) (C:200): on 3 Yukon
[1061, 1110]; on 3, 7 BC [1012]; on 3,
4, 8 NWT-Alta [1063]; on 8 NB, NS [1032,
1061]. See *Abies*.

G. protractum: on 8 Labr [867]. Perhaps
a specimen of *G. odoratus*.

G. sepiarium (*Lenzites s.*) (C:197): brown
cubical rot, carie brune cubique: on 2,
3, 8 NWT-Alta [1063]; on 3 Yukon, NWT,
BC, Alta, Ont [1061], Yukon [460], Mack,
Alta [53], Ont [1341, F74]; on 3, 8 NB,
NS, PEI [1032, 1061]; on 3, 4, 7 BC
[1012]; on 4 NWT, BC, Alta, Ont, NB
[1061] and Mack [53, F66], Ont [1341,
F74], Que [917, 937], NB, NS [1032],
Nfld [1659, 1660, 1661]; on 6 NS [1032,
1061]; on 7 BC [1061]; on 8 Alta [1061],
Ont [1341], Que [282, 867], NB [1610],
Nfld [1061, 1659, 1660, 1661], Labr
[867].

Graphium sp.: on 1 Que [F65]; on 8 Que,
NB [1610].

Gremmeniella abietina (*Scleroderris
lagerbergii*) (anamorph *Brunchorstia
pinea*): on 1, 6 Que [1676] by
inoculation; on 3 Ont [1828]; on 3, 4
Que [1676, 1682, 1686, F68].

Gymnopilus spectabilis (*Pholiota s.*): on
3 BC [1012]. See *Acer*.

Haematostereum rugosum (*Stereum r.*): on 3
NWT-Alta [1063]. See *Abies*.

H. sanguinolentum (*Stereum s.*) (C:199):
red heart rot, carie rouge du sapin: on
2, 3, 4, 8 NWT-Alta [1063]; on 2 BC
[F74]; on 2, 3, 7 BC [1012]; on 3 Yukon,
BC, Alta, Man, Ont [1062] and Mack [53],
BC [1718], Alta [648], Sask [1951,
1952], Man [1951], Ont [1341]; on 3, 6
NB, NS [1032, 1062]; on 3a BC [1233]; on
4 Mack [53], Ont [90, 1341], Que [917,
937], NB [1032] and Man, Ont, Que, NB
[1062]; on 6 Que [1062]; on 7 BC [1062];
on 8 BC [1233], Alta [951], Man [1062],
Ont [1341], Que [969, 1377], NB, NS, PEI
[1032, 1062, F62], Nfld [1659, 1660,
1661]. See *Abies*.

Hansenula holstii: on 3, 6 Que [1967].

Harpographium sp.: on 8 NB [1610].

Helicobasidium corticioides: on 8 Alta
[1062]. Lignicolous.

Helicogloea lagerheimii: on 4 Ont
[1341]. Lignicolous.

Helotium immaculatum (*Mycena gracilis*):
on needles (? fallen) of 8 Que [1385].

H. resinicola (anamorph *Stilbella* sp.):
on 3, 7 BC [1012, F69]; on resin of 3 BC
[54]. An Ascomycete which needs to be
redisposed. Perhaps a *Bisporella*.

Henningsomyces candidus (*Solenia
polyporoidea*): on 8 NS [1032]. See
Abies.

Hericium abietis (C:197): on 7 BC [1012,
1062]. See *Abies*.

H. coralloides (*H. laciniatum*, *H. ramosum*)
(C:197): on 2 BC [954]; on 3 NWT-Alta
[1063], Alta [53, F62]. One report
[F62] was based on a specimen on
Populus, not *Picea* [712]. The
remaining reports should be reexamined
as *Picea* is an atypical host for this
species. See *Acer*.

Herpotrichia coulteri: on 2 BC [1012,
F69].

H. juniperi (*H. nigra*) (C:197): brown
felt blight, feutrage brune: on 2
NWT-Alta [1063], BC [954], Alta [951,
952]; on 2, 3, 7 BC [1012]; on 3 NS
[1032]; on 4 Que [1666], NS [1340, 1666].

Heterobasidion annosum (*Fomes a.*,
Fomitopsis a.) (C:197): fomes root
rot, maladie de rond: on 2, 7 BC
[1916]; on 3, 7 BC [1012, 1061]; on 7 BC
[1143, 1341]. See *Abies*.

Heyderia abietis: on 8 Que [1385] on
fallen needles, Labr [869].

Hormodendrum sp.: on 8 NS [1610].

Humicola sp.: on 8 NB [1610].

Hyaloscypha leuconica: on 4 Ont [1340].

Hymenochaete cinnamomea (C:197): on 3
Yukon [1062]; on 8 BC [1012, 1062]. See
Alnus.

H. fuliginosa (C:197): on 3 BC [1012,
1062]. Associated with a white rot.

H. tabacina (C:197): on 2, 3 NWT-Alta
[1063]; on 2 Alta [950, 1062]; on 3 Mack
[53]; on 8 NS [1032, 1062], Nfld [1659,
1660, 1661]. See *Abies*.

Hyphoderma argillaceum (*Peniophora a.*):
on 8 Ont [1341], Que [969]. See
Betula.

H. clavigerum: on 8 Que [1060]. See
  Abies.
H. karstenii (Peniophora nivea): on 3 NWT
  [1341]. Lignicolous.
H. polonense: on 8 Que [1060].
  Associated with a white rot.
*H. praetermissum (H. tenue, Peniophora
  t.)*: on 3 BC [1012]; on 8 Que [969,
  1060], NS [1032]. See *Abies*.
H. puberum (Peniophora p.): on 3 NWT-Alta
  [1063], Mack [53]; on 8 Que [969,
  1060]. See *Abies*.
H. resinosum: on 7 BC [1012, 1060].
  Associated with a white rot.
H. sambuci (Hyphodontia s.): on 8 Que
  [969, 1060]. See *Acer*.
H. setigerum (Peniophora aspera): on 3 BC
  [1060]; on 4 Ont [90]. See *Abies*.
H. terricola: on 3 Alta [1060].
  Lignicolous.
Hyphodontia sp.: on 8 NB [1610].
  Lignicolous.
H. abieticola: on 4 Nfld [1060].
  Associated with a white rot.
H. alienata: on 8 Que [969, 1060].
  Lignicolous.
H. alutacea (Odontia a.): on 8 Alta, Ont
  [923, 1060], NS [1032]. See *Abies*.
H. alutaria: on 8 Man [1060].
H. aspera: on 2, 3, 7 BC [1060]; on 3 BC
  [1012]; on 8 Que [1060]. See *Pinus*.
H. barba-jovis: on 8 Que [1060]. See
  Abies.
H. breviseta: on 4 Nfld [1060]; on 8 BC,
  Alta [1060]. See *Abies*.
H. crustosa (Odontia c.): on 4 BC [1060];
  on 8 NS [1032]. See *Abies*.
H. floccosa: on 8 Alta [1060], Que [969,
  1060]. See *Abies*.
H. pallidula (Peniophora p.) (C:198): on
  3 BC [1012, 1060]; on 8 Ont [1341], Que
  [969, 1060]. See *Abies*.
H. subalutacea (Peniophora s.): on 2
  NWT-Alta [1063]; on 8 Que [969]. See
  Abies.
Hypholoma capnoides (Naematoloma c.)
  (C:198): on 3 BC [1062]; on 3, 7 BC
  [1012]. See *Abies*.
Hypochnicium geogenium (Corticium g.)
  (C:196): on 3 NWT-Alta [1063], Mack
  [53]. Lignicolous.
Hypoxylon sp.: on 4 Ont [90].
Ischnoderma resinosum (Polyporus r.)
  (C:198): on 2 NWT-Alta [1063]; on 3 BC
  [1012, 1061]; on 8 NS [1061, 1032]. See
  Abies.
Isthmiella crepidiformis (Bifusella c.)
  (C:195): needle cast, rouge: on 1 Ont
  [1063]; on 2, 3, 4 NWT-Alta [1063], BC
  [1012, F68]; on 2 Alta [952, 1063, F65,
  F66]; on 3 Mack [53], Alta [53, 951,
  F61], Man [F67], Que [F74], Nfld [F69,
  F70]; on 3, 4 Yukon, Mack, Alta [F67,
  F68], Sask [F66, F67], Nfld [1660, 1661,
  F68, F73] and Yukon [460], Alta [F65,
  F66]; on 4 Mack [53, F63], Alta [951],
  Sask [F68], Man [F67], Ont [1340, 1828,
  F61, F64, F66, F69, F70, F72,F76], Que

[F67, F74, F75, F76], Nfld [1659, F74,
  F75]; on 6 Yukon, Mack [F66], NB [1032,
  F65]; on 8 Alta [951], Nfld [1661].
I. faullii: on 4 Ont [1828].
Junghuhnia collabens (Poria rixosa)
  (C:199): on 2, 3 BC [1012]; on 3 BC,
  Ont [1004, 1061]. See *Abies*.
J. luteoalba (Poria l.): on 8 Que [1004,
  1061, 1341]. See *Abies*.
*Kavinia alboviride (Mycoacia a., Oxydontia
  a.)*: on 6 Que [F67]; on 8 BC [1012].
  Typically on well-rotted wood. See
  Fagus.
Kirschsteiniella thujina: on 3 Que [F64];
  on 4 Que [937].
Kuehneromyces vernalis: on 3 BC [1012].
  See comment under *Betula*.
Lachnellula agassizii (Dasyscyphus a.)
  (C:197): on 7 BC [1012]; on 8 NWT-Alta
  [1063], Alta [951], Que [282, 1032,
  1385].
L. arida: on 8 NWT-Alta [1063].
L. chrysophthalma (C:197): on 2 NWT-Alta
  [1063], Alta [F62].
L. occidentalis: on 3, 4 Que [1690, F68].
L. resinaria: on 3 Ont [1340], Que [F70].
L. suecica: on 2, 3 BC [1012]; on 3, 4
  Que [1690].
*Laeticorticium lundellii (Dendrocorticium
  l.)*: on 3 NWT [1341], Mack [53], Alta
  [930]. Lignicolous.
L. piceinum (Dendrocorticium p.): on 8
  Ont [930]. Lignicolous.
L. violaceum (Dendrocorticium v.): on 8
  Ont [930]. Lignicolous.
Laetiporus sulphureus (Polyporus s.)
  (C:199): brown cubical rot, carie brune
  cubique: on 2, 3, 7 BC [1012, 1061]; on
  7 BC [1149].
*Laurilia sulcata (Stereum s., Echinodontium
  s.)* (C:199): on 2, 8 NWT-Alta [1063];
  on 2, 3, 8 Alta [1062]; on 3 BC [1012,
  1062, 1341]. See *Abies*.
Lentinus kauffmanii (C:197): on 7 BC
  [1012, 1062]. Associated with a brown
  rot. Now in *Neolentinus*.
Lenzites sp.: on 4 Nfld [1660, 1661].
L. betulina: on 3 NWT-Alta [1063]. See
  Acer.
Leptomelanconium piceae: on 4 Sask, Man,
  Ont [1763, 1764].
*Leptosporomyces galzinii (Athelia g.,
  Corticium g.)*: on 3 Yukon [1060],
  NWT-Alta [1063], Mack [53]; on 8 Que
  [969, 1060]. See *Abies*.
*Leucogyrophana mollusca (Merulius fugax
  auct.)* (C:198): on 3 NWT-Alta [1063],
  Mack [53]; on 8 Alta [1060], NS [1032].
  See *Abies*.
Leucostoma curreyi (Valsa c.): on 2
  NWT-Alta [1063].
L. kunzei (Valsa k.) (anamorph *Cytospora
  k.*): on 1, 3 Ont [F68, F72]; on 2 BC
  [1012]; on 3 Ont [F71], Que [F72]; on 3,
  5, 8 NB [1032]; on 4 Que [457, F75]; on
  8 NWT-Alta [1063], Ont [F69, F74, F75].
Lirula brevispora: on 2, 3 BC [1012,
  2004, F68]; on 2 BC [F69]. On needles.
*L. macrospora (Lophodermium filiforme, L.
  macrospora)* (C:197): needle cast,

rouge: on 2, 3, 4, 8 NWT-Alta [1063];
on 2, 3, 4, 7 BC [1012]; on 2 BC [954],
Alta [951, 952, F62, F65, F66]; on 3, 4
Yukon [460, F68], Alta [F65, F66, F67,
F68], Nfld [1659, 1660, 1661, F68]; on 3
Mack [53, F62, F67, F68], BC [953, F68,
F74, F75], Alta [53, 951], Que [1377],
NB [1032], NS [F73], PEI [1032, F69],
Nfld [F65]; on 3, 4, 6 NB, NS [F64,
F65]; on 4 Mack [53, F63, F67, F68],
Alta [F64], Man [F65], NB, NS, PEI
[1032], Nfld [F69, F70, F73, F74]; on 4,
6 PEI [F65]; on 6 NB [1032, F63, F70],
NS [1032, F60, F62]; on 7 BC [F73, F74];
on 8 Alta [F64], NB [1032].

Lophium mytilinum (C:197): on 3 NWT-Alta
[1063], Alta [53]; on 4, 8 NS [1032].

Lophodermium sp.: on 3, 4 Que [F73]; on
3, 6 NB [1032]; on 6 NS [1032]; on 8
Nfld [1660, 1661].

L. piceae (C:197): on 1 Ont [1063]; on 3
Alta [53], Ont [1340, 1827, F70, F76],
NS [1032]; on 3, 4 NWT-Alta [1063], Sask
[F69], Man [F66, F69], Que [1377, F74];
on 3, 7 BC [1012]; on 4 Mack [53, F66],
Sask [F67], Ont [1340], Que [F66], NB
[1032]; on 5 Ont [F73]; on 7 BC [1340];
on 8 Alta [F66].

Lophomerum sp.: on 3, 4 Que [F73].

L. darkeri: on 3 Yukon, BC, Alta [F68]
and Yukon [460], BC [1012], Que [1220,
F66], NB [1032, 1220].

L. septatum (*Lophodermium s.*) (C:197): on
3 Que [F63].

Lophophacidium hyperboreum (C:197): on 1,
4, 5 Que [1692, F69]; on 3, 4 Yukon
[F67]; on 2, 3, 4 Alta [F67]; on 3 BC
[1012, F76], Alta [F65]; on 3, 4 Ont
[F76]; on 3, 6 Que [1692, F67]; on 4, 6
Que [1444]; on 5 Que [F67].

Lycogala epidendrum: on 3 NWT-Alta
[1063], Mack, Alta [53]. Slime mold,
probably on rotted wood.

Lycoperdon perlatum: on 3 Ont [1827].
Probably on rotted wood.

Marasmiellus filopes: on 8 BC, Ont, Que,
NB [1435].

Marasmius androsaceus: on 8 NS [1032].

Mariannaea elegans: on 3 Sask [153].

Martinina panamaensis f. *tetraspora*
(*M. tetraspora*): on 4 Que [421, 436].

Melampsora medusae (*M. albertensis*):
needle rust, rouille des aiguilles: on
7 BC [1012, 2002], the latter record by
inoculation; on 8 BC [F65].

M. occidentalis: rust, rouille: on 7 BC
[1012, 2002], the latter record by
inoculation; on 8 BC [F65].

Melanconium sp.: on 3 Ont [1828].

Melanotus hartii: on mine timbers of 8
Ont [4]. Associated with a white rot.

Meruliopsis albostramineus (*Ceraceomerulius
a.*): on 8 Que [568]. See *Abies*.

Meruliopsis ambiguus (*Merulius a.*)
(C:197): on 3 Yukon [568], NWT-Alta
[1063], BC [568, 1012]; on 8 Man [568].
See *Alnus*.

M. taxicola (*Merulius t., Poria t.*)
(C:199): on 2, 3 NWT-Alta [1063], Alta
[568]; on 3 BC [1012, 1341], Sask, Ont

[568]; on 4 Ont [1341]; on 6, 8 Que
[568].

Merulius sp.: on 3 NWT-Alta [1063], Mack
[53].

M. bellus: on 3 NWT-Alta [1063], Mack
[53]. Probably specimens of
Ceraceomyces serpens.

M. incarnatus (*Byssomerulius i.*): on 8
Alta [1060]. Probably a specimen of
Meruliopsis ambiguus. Significantly
beyond the host and geographic range of
Merulius incarnatus.

Micraspis acicola (anamorph *Periperidium
a.*) (C:198): on 4 Ont [360], Nfld
[1340]; on 6 NS [1340].

Microcallis sp.: on 4 Que [F66].

Microlychnus epicorticis: on 7 BC [506,
1012].

Monilia geophila: on 1 [1211], 3, 4 [F64]
and 4 [937] Que.

Mucor sp.: on 8 Que [1610].

Mucronella calva (*M. aggregata*): on 2 BC
[1062]. See *Abies*.

Myceliophthora thermophila (*Sporotrichum
t.*): on 8 BC [1723], NB [1610].

Mycena rubromarginata: on 8 Que [1385].

Mytilidion sp.: on 3 NWT-Alta [1063].

M. karstenii: on 3 Man [F66].

M. tortile: on 3 NWT-Alta [1063].

Myxotrichum setosum: on 3 Sask [1635].

Nectria sp.: on 2 BC [1012]; on 3 Ont
[1340]; on 5 Nfld [F75].

N. fuckeliana: on 1, 3, 4 Que [1685], on
1 by inoculation; on 1, 3 Que [F66]; on
3 BC [1012]; on 8 NB [175].

Odontia sp.: on 4 Ont [1341]; on 7 BC
[1012].

O. lactea (C:198): on 3 BC [1012]. See
comment under *Abies*.

Odonticium romellii (*Odontia r.*): on 2
NWT-Alta [1063]. See *Pinus*.

Odontotrema hemisphaericum: on 3, 4 Que
[1690].

Oedemium didymum (holomorph
Chaetosphaerella fusca): on 8 NB
[1610].

Oedocephalum sp.: on 7 BC [1149].

Oidium microspermum: on 8 NWT-Alta
[1063], Alta [952].

Onnia circinata (*Inonotus c., Polyporus c.,
P. tomentosus* var. *c.*) (C:199): red
butt rot, carie rouge alvéolaire du
pied: on 1 Que [1210]; on 2 Alta
[1062]; on 2, 3 Alta [1341]; on 3 Sask
[1062], Man [579], Que [F65], NB [1032,
1062]; on 3, 7 BC [1012, 1062]; on 3, 4
Ont [1062, 1341], Que [F71]; on 4 Man
[1062]; on 6 NB, NS [1032, 1062]; on 8
Sask [1962], Que [1377]. See *Abies*.

O. tomentosa (*Inonotus t., Polyporus t.,
Polyporus t.* var. *tomentosus,
Xanthochrous t.*) (C:199): red butt
rot, carie rouge alvéolaire du pied: on
1, 3 Que [F64]; on 2, 3 NWT-Alta [1063],
BC [1012, 1062]; on 3 Alta [1062, F62],
Sask [158, 1952, 1954 by inoculation,
F60, F70], Ont [338, 1341, 1738, 1966],
Que [913, 915, 1210, 1212, 1222, F62,
F69, F76], NB [1062], NS [1032, F72],
Nfld [1659, 1660, 1661] and Sask, Man

[1062, 1951, 1959, 1965, F61, F63], Ont,
Que [F65, F66, F67]; on 3, 4 Sask [1953,
1960, 1962], Ont [1062, 1828, 1956, F71,
F72]; on 3, 6 Que [F70]; on 4 Alta, Man,
Que, NB [1062], Sask [1961], Ont [90,
1958, F73], Que [F67, F73], Nfld [1661],
Labr [F73]; on 6 NS [1062]; on 7 BC
[1062]; on 8 Sask [1955, F60], Man
[F60], Ont [F68, F69], Que [1385], NB,
NS, PEI [1032]. See *Abies*.

Osteina obducta (Polyporus osseus)
(C:198): on 7 BC [1012, 1061, 1341].
See *Abies*.

Ostreola consociata (C:198): on 3 Que
[359].

Oxyporus similis (Poria s.): on 7 BC
[1061]. See *Populus*.

Pachnocybe ferruginea: on 3 Que [F69].

Paecilomyces sp.: on 8 NB [1610].

P. varioti: on 8 NB [1610].

Panellus serotinus (Pleurotus s.)
(C:198): on 3 BC [1012, 1062]; on 8 BC
[F66]. See *Acer*.

Panus rudis: on 4 Ont [1341]. See *Acer*.

*Parmastomyces transmutans (Polyporus
subcartilagineus, Tyromyces
kravtzevianus)* (C:199): on 3 BC
[1718], Sask [579, 1061]; on 4 Que
[1061]; on 8 Sask, Man [579], NB
[1061]. See *Abies*.

Paullicorticium delicatissium: on 8 Que
[969, 1060]. Lignicolous.

P. niveo-cremeum (Corticium n.): on 2
NWT-Alta [1063]. Lignicolous.

P. pearsonii: on 8 Que [969, 1060].
Associated with a white rot.

Penicillium sp.: on 3, 4 Ont [1964,
1966], Que [917]; on 8 Que, NB, NS
[1610].

P. aurantiogriseum (P. cyclopium): on
seedlings of 3a in cold storage Alta
[746].

Peniophora sp.: on 2, 3, 8 NWT-Alta
[1063]; on 3 Mack [53].

P. cinerea: on 4 Que [917]; on 8 Alta
[1060]. See *Abies*.

"P." coccineo-fulva: on 8 Que [969]. The
type specimen is effete. The name has
been applied to a *Phanerochaete* which
may not be different from *P. velutina*.

P. piceae (P. separans) (C:198): on 2, 3
NWT-Alta [1063], Alta [1060]; on 2 Alta
[951]; on 3 BC [1012, 1060]; on 8 Nfld
[1060, 1659, 1660, 1661]. See *Abies*.

P. pithya (C:198): on 3 Ont [1060]; on 4
NB [1032, 1060]; on 8 NWT-Alta [1063],
Que, NS [1060], NB, NS [1032]. See
Abies.

P. pseudopini (C:198): on 3 BC [1060],
Alta [1941]; on 3, 4 Alta, Man [1060];
on 4 Que [917]. See *Abies*.

P. septentrionalis (C:198): red heart
rot, carie rouge du coeur: on 2, 3, 4
NWT-Alta [1063]; on 2, 3, 8 Alta [1060];
on 3 Mack [53], BC [1012, 1060], Sask
[1060], Ont [1341]; on 3a BC [1233]; on
4 Ont [90, 1060], Que [917, 937, 1060],
Nfld [1659, 1660, 1661]; on 6 NS [1060];
on 6, 8 NS [1032]. See *Abies*.

P. violaceolivida: on 2 NWT-Alta [1063].
Lignicolous.

Peridermium coloradense (telial state
Chrysomyxa arctostaphyli):
witches'-broom rust, rouille-balai de
sorcière: on 8 Nfld [333]. Perennial
and systemic, causes large witches'
brooms.

Periperidium acicola (holomorph *Micraspis
a.*) (C:198): on 4 Ont [360, 362, 1763].

Pestalotia sp.: on 1 Ont [1340].

Peyronelia rudis: on 3 Sask [1760].

Pezicula livida (anamorph
Cryptosporiopsis abietina): on 3 Que
[1690].

Peziza sylvestris: on 3 Ont [1828].

Pezizella sp.: on 7 BC [1012].

P. chapmanii: on 3 BC [1012, 1946].

Phacidium sp.: on 1, 6, 8 Que [F62]; on 4
Nfld [1661, F65].

P. abietis: on 1, 3, 4, 8 Que [F64]; on
1, 3, 5, 6 Que [1677].

P. infestans (C:198): snow blight,
brûlure printanière: on 4, 8 Nfld
[1661]; on 8 Que [1377]. Records are
suspect, see note under *Abies*.

Phaeocryptopus nudus: on 5, 7 BC [1012].

Phaeolus schweinitzii (Polyporus s.)
(C:198): brown cubical rot, carie brune
cubique: on 2, 3 NWT-Alta [1063]; on 2
BC [954]; on 3 NWT, BC, Man, Ont, NB, NS
[1061] and Mack [53, 579], BC [1012],
Ont [1341, 1828, F65, F67, F68, F72],
NB, NS [1032], Nfld [1659]; on 3, 4 Ofld
[1660, 1661]; on 4 Ont [1061, F68, F71,
F72]; on 7 BC [1012, 1061, 1149]; on 8
Que [282, 1377], NB, NS, PEI [1032,
1061].

*Phaeostalagmus tenuissimus (Verticillium
t.)*: on 4 Man [1760].

Phanerochaete carnosa (Peniophora c.)
(C:198): on 4 Nfld [1060]; on 8 NS
[1032].

P. sanguinea (Peniophora s.): on 2, 3
NWT-Alta [1063]; on 8 Que [969, 1060],
NS [1032]. See *Abies*.

*P. sordida (Peniophora cremea,
Phanerochaete c.)* (C:198): on 3
NWT-Alta [1063]; on 3 Mack [53], BC
[1012, 1060]. See *Abies*.

P. velutina (Peniophora v.): on 3
NWT-Alta [1063], Mack [53]; on 8 Alta
[1060], Que [969, 1060]. See *Acer*.

Phellinus chrysoloma (Fomes pini var.
abietis) (C:197): red ring rot, carie
blanche alvéolaire: on 4 Labr [867]; on
8 NS [1036]. See *Abies*.

P. ferreus: on 3 BC [1062]. See *Acer*.

P. ferrugineofuscus (Poria f.) (C:199):
on 3 NWT, BC, Ont [1062] and NWT-Alta
[1063], Mack [53, 579]; on 8 Alta
[1062]. See *Abies*.

P. ferruginosus (Poria f.) (C:199): on 3
BC [1012, 1062]; on 8 Alta [1062]. See
Abies.

P. nigrolimitatus (Fomes n.) (C:197):
white pocket rot, carie blanche
alvéolaire: on 2 NWT-Alta [1063], BC
[953]; on 3 Yukon [1110, 1062]; on 3, 7
BC [1012]; on 3, 7, 8 BC [1062]; on 7 BC
[1341].

P. pini (Fomes p., Trametes p.) (C:197):
red ring rot, carie blanche alvéolaire:
on 1 Ont [1341]; on 2, 3, 4, 8 NWT-Alta
[1063]; on 2 BC, Alta, Sask, Man, Ont,
Que, OB, NS [1062] and BC [953, 954,
1012, 1341], Alta [950, 951, 952, F72],
Que [1377]; on 3 Yukon [F62], Mack [53,
F60, F67], BC [1012, 1341], Alta [53],
Sask [932, 1952], Ont [1341, 1966, F66],
Que [913, 914, 1377], NB, NS [1032],
Labr [F73] and Sask, Man [1951, F63]; on
3, 4 Yukon [460], Nfld [1660, 1661,
F72]; on 4 Yukon [F62], Mack [53, 1110,
F60], BC [1012, F74], Ont [90], Que
[937, 938, 1377, 1391, F67], NB, NS
[1032, 1062], Nfld [1659], Labr [F73]
and Yukon, NWT, Alta, Sask, Man, Ont,
Que [1062]; on 5, 6 Que [1377]; on 6 Que
[1062, F70], NB, NS [1032, 1062]; on 7
BC [1012, 1062, 1149, 1341]; on 8 Alta
[950, 951], Man [1062], NB, NS, PEI
[1032, 1062], Nfld [1659, 1660, 1661].
See *Abies*.

P. punctatus (Poria p.): on 3 BC [1012].
See *Abies*.

P. tsugina (Poria t.): on 3 BC [1718].
See *Abies*.

P. viticola (Fomes v.): on 2 BC [1062];
on 3 Yukon [1062], NWT-Alta [1063], Mack
[53]. See *Abies*.

P. weirii: on 2, 7 BC [1062]; on 7 BC
[1012]. Causes a root and butt rot of
live trees. See *Abies*.

Phialea strobilina: on 4 Ont [1340].

Phialocephala bactrospora: on 8 NB, NS
[1610].

P. fusca (C:198): on 3 Ont [877].

P. repens: on 1 Que [F65].

Phialophora sp.: on 1 Que [F65]; on 4 Ont
[90]; on 8 Que, NB [1610].

P. alba: on 4 Ont [90].

P. americana: on 8 NB [1610].

P. botulispora: on 3 Ont [279].

P. fastigiata: on 2 BC [279, 1012]; on 4
Sask [279]; on 8 Que, NB [1610].

P. heteromorpha: on 8 NB [1610].

P. lagerbergii: on 8 Ont [279], NB [1610].

P. melinii: on 4, 6 NB [279]; on 8 Sask
[279], Que, NB [1610].

Phlebia centrifuga (P. albida auct.)
(C:198): on 3 BC [1012]. See *Abies*.

P. livida: on 8 Alta [1060], Que [969].
See *Abies*.

P. phlebioides: on 8 Que [969]. See
Abies.

P. radiata: on 3 NWT-Alta [1063]. See
Abies.

P. segregata (Peniophora livida) (C:198):
on 8 Que [969]. Lignicolous.

P. subserialis: on 8 Que [969, 1060].
See *Pinus*.

Phlebiopsis gigantea (Peniophora g.,
Phlebia g.) (C:198): white sap rot,
carie blanche de l'aubier: on 3
NWT-Alta [1063], BC [1012, 1060], Sask
[1951, 1952], Man [1951], NB [1032]; on
4 Que [917]; on 6 NS [1032]; on 8 Que
[969, 1060, 1610], NB [1610], Nfld
[1659, 1660, 1661]. See *Abies*.

Pholiota adiposa: on 2, 3 BC [1062],
Sask, Man [1951]. See *Acer*.

P. alnicola (Flammula a.) (C:197): yellow
checked rot, carie jaune craquelée: on
2 BC [F66]; on 2, 3, 4 BC [1012]; on 3
NWT-Alta [1063], BC [F60], Sask, Man
[F61] and BC, Alta, Man [1062]; on 3, 3a
Alta [387]; on 4 Ont [90, 1062, 1964,
F74]; on 8 Sask [1062]. See *Abies*.

P. aurivella (C:198): on 2, 3 BC [1062];
on 3 BC [1012]. See *Abies*.

P. limonella (P. squarrosa-adiposa)
(C:198): on 8 BC [1012, 1062]. See
Abies.

Phoma sp.: on 3 Ont [973]; on 3a Alta
[746]; on 7 BC [1012].

P. glomerata: on 3a Alta [746].

Phomopsis sp.: on 5 Ont [1827]; on 7 BC
[1012].

Phragmothyriella sp.: on 2 NWT-Alta
[1063]. Possibly a specimen of
Schizothyrium sp.

Phragmotrichum chailletii: on 3 BC
[1012]; on 4 Ont [1340].

Phyllotopsis nidulans: on 8 Nfld [1659,
1660]. See *Abies*.

Phytophthora cactorum: on 5 Sask [1848].

Piloderma bicolor (Athelia b., Corticium
b.) (C:196): on 3 NWT-Alta [1063],
Mack [53], BC [1012, 1060]; on 8 Alta
[1060], Que [969, 1060], NS [1032]. On
well-rotted wood or on fallen needles,
twigs, etc. See *Abies*.

P. byssinum (Athelia b.): on 3 NWT
[1060];
on 8 Alta [1060]. On well-rotted wood
or on fallen needles, twigs, etc.
Associated with a white rot.
Mycorrhizal on, at least, *Pseudotsuga*.

Piloporia albobrunnea (Poria a.): on 2
NWT-Alta [1063]; on 3 Yukon [1061].
Associated with a brown rot.

Pleospora sp.: on 2, 3 BC [1012]; on 6 NS
[1032].

Pleurostoma candollei (C:198): on 1 Que
[F62].

Pleurostromella sp.: on 3 Ont [1827].

Pleurotus ostreatus (C:198): white spongy
rot, carie blanche spongieuse: on 3 BC
[1012, 1062]; on 4 NWT [1062].

Plicatura crispa (Trogia c.) (C:200): on
3 BC [1012]. Lignicolous.

Polyporus sp.: on 4 Nfld [1660, 1661].

P. badius (P. picipes) (C:198): on 3 Nfld
[1659, 1660, 1661]; on 7 BC [1012,
1061]. See *Abies*.

P. hirtus (C:198): on 3 BC [1012, 1061,
1718]; on 6 Nfld [1061]. Uncommon,
typically attached to buried wood. See
Abies.

P. varius (P. elegans) (C:198): on 3 BC
[1012, 1061], Alta [1061]; on 8 NWT
[1061], Mack [579]. See *Abies*.

Poria sp.: on 3 NWT-Alta [1063], Nfld
[1659]; on 3, 4 Nfld [1660, 1661]; on 8
NB, NS [1032].

P. alpina: on 3 Alta [1061]. Associated
with a brown rot. Now in *Antrodia*.

P. byssina: on 3 Mack [579]; on 8 Alta
[579]. These records are probably
collections of *Flaviporus*
semisupinus. Lignicolous.

P. rivulosa: on 3 BC [1012]; on 3, 7 BC
   [1061]. See *Abies*.
P. sitchensis (C:199): on 4 Alta [1061].
   Associated with a brown rot.
Pragmopora bacillifera: on 8 Ont [657].
Propolis leonis (C:199): on 8 NS [1032].
P. rhodoleuca: on 4 Ont [1340].
P. versicolor: on 3 Ont [1340].
Pseudohiatula conigenoides: on 1 Que
   [1385]. Possibly specimens of
   Baeospora sp. or *Strobilurus* sp.
   because *P. conigenoides* occurs on
   inflorescences of *Magnolia* sp.
Pseudohydnum gelatinosum: on 8 Que
   [282]. Probably on well-rotted wood.
Pseudomerulius aureus (*Merulius a.*,
   Plicatura a.) (C:197): on 3 BC [568,
   1012, 1060]; on 8 Man [568]. See
   Pinus.
Pseudophacidium garmanii: on 2, 3 BC
   [520].
P. piceae: on 1, 6 Que [1685], on 6 by
   inoculation; on 3, 4 Que [1685, 1686,
   F68, F69]; on 4 Que [F65].
Pseudotomentella flavovirens: on 4 Ont
   [1341].
P. humicola: on 4 Ont [926, 1062, 1341].
P. tristis (*P. umbrina*): on 8 Ont [1341].
Ptychogaster sp.: on 8 NB [1610].
Pucciniastrum sp.: rust, rouille: on 3
   Yukon, Alta [F68], Mack [53], Sask, Man
   [F69, F73], Que [F67] and Man [F74].
P. americanum (C:199): needle rust,
   rouille des aiguilles: 0 I on 2 BC
   [1012, F62]; on 3 BC [1012, F72], Ont
   [1236], Que [1377, F60, F65], NB, PEI
   [1032]; on 8 BC, Man [2008].
P. arcticum (C:199): rust, rouille: 0 I
   on 3 Alta [2008], Ont [1828], Que [1377].
P. sparsum: rust, rouille: 0 I on 3
   Yukon [F64]; on 3 Yukon, Mack [729,
   F69]; on 3, 4 Yukon [460]; on 3, 5 Alta
   [729] by inoculation; on 4 Yukon [F64];
   on 5 Yukon, Mack [F69]; on 8 Yukon, NWT
   [2008], Alta [952].
Punctularia strigosozonata: on 8 Que
   [969, 1062]. See *Alnus*.
Pycnoporellus alboluteus (*Phaeolus a.*,
   Polyporus a.) (C:198): on 2, 3, 8
   NWT-Alta [1063]; on 2, 3, 7 BC [1012,
   1061]; on 2 BC [954], Alta [950]; on 7
   BC [1179, 1341]; on 8 Alta [1061]. See
   Abies.
P. fulgens (*Phaeolus fibrillosus*, *Polyporus
   f.*) (C:198): on 3 BC [1179]; on 4 Ont
   [1341]; on 3, 7 BC [1012, 1061]; on 8
   Nfld [1659, 1660, 1661]. See *Abies*.
Pycnoporus cinnabarinus (C:199): on 3 NS
   [1061, 1032]. Host appeared as *P.
   engelmannii*, presumably in error, in
   [1061].
Pyrenochaeta sp.: on 8 NB, NS [1610].
Pythium sp. (C:199): on 2 BC [1012]; on 7
   BC [F70]; on 8 Alta [337], NB [1885].
P. debaryanum: on 4 Que [F65]; on 5 Sask
   [1848]; on 8 Que [1377].
P. intermedium: on 2, 3a Alta [745].
P. sylvaticum: on 2 Alta and on 3 Sask
   [1846].

P. ultimum: on 3 BC [1012, F69]; on 3a
   Alta [746]; on 5 Sask [1848].
Ramaricium albo-ochraceum (*Trechispora
   a.*): on 4 Ont [572, 970]. Lignicolous.
Repetobasidium mirificum: on 8 Que [969,
   1060]. Lignicolous.
R. vile: on 8 Que [969, 1060].
   Lignicolous.
Resinicium bicolor (*Odontia b.*) (C:198):
   white stringy rot, carie blanche
   filandreuse: on 3 BC [1012, 1718],
   Sask, Man [1951] and BC, Sask [1060]; on
   4 Ont [90, 1341, 1964], Que [917], NB,
   NS [1032] and Ont, NB [1060]; on 8 NS
   [1032].
R. chiricahuaensis: on 8 Alta [1060].
   Associated with a white rot.
R. furfuraceum: on 8 Que [1060].
   Associated with a white rot.
Resupinatus applicatus (*R. atropellitus*):
   on 8 Que [282]. Lignicolous.
Retinocyclus abietis (C:199): leader
   dieback, dépérissement de la flèche: on
   1 Que [1209]; on 1, 3 Que [1211]; on 1,
   4, 6 Que [F66]; on 2 Alta [950]; on 2,
   3, 7 BC [1012]; on 3 Mack [F61], Alta
   [53], Ont [1340, 1827, F66]; on 3, 4
   Mack [53]; on 3, 4, 8 NWT-Alta [1063];
   on 4 Alta [F61], Ont [90], Que [457,
   937, 917]; on 8 Sask [F67].
R. olivaceus: on 3 Que [F71]; on 7 BC
   [1012].
Rhinocladiella sp.: on 8 NB [1610].
R. atrovirens (C:199): on 8 NB [1610].
R. elatior: on 8 Que [1610].
R. mansonii: on 2, 3 BC [1012].
Rhizina undulata: on 3, 7 BC [1012]; on 4
   Sask [F70], Ont [F72]; on 8 Ont [F74].
   Causes root rot in seedlings, esp. when
   planted on recently burnt sites. Also
   parasitizes trees adjacent to burnt
   sites.
Rhizoctonia solani (*R. dichotoma*, *R.
   praticola*) (holomorph *Thanatephorus
   cucumeris*): on 3 Sask [1511]; on 5
   Sask [1510]; on 6 NB [1885, F73], Sask
   [1848]; on 8 Que [1377].
Rhizopus nigricans: on 8 BC [1012].
Rhizosphaera kalkhoffii (C:199): needle
   cast, rouge: on 2, 8 Que [1377]; on 3
   Ont [F74, F76], NB [1032]; on 5 Ont
   [F73, F74, F75]; on 8 Que [1763].
R. pini: on 4 Man [F67].
Rhizothyrium abietis (holomorph
   Rhizocalyx a.): on 6 NB [1032, F65];
   on 8 Ont [1610].
Rhodotorula sp.: on 8 NB [1610].
Rigidoporus nigrescens (*Poria n.*)
   (C:199): on 7 BC [1012, 1061]. See
   Acer.
Rileya piceae: on 7 BC [516].
Rutstroemia bulgarioides (*Chlorosplenium
   b.*, *Chlorociboria strobilina*): on 4
   Mack [664], Ont [1340]; on 8 NS [1032].
Saccardinula sp.: on 4 Ont [1340].
Saccharomyces sp.: on 8 Que, NB [1610].
Sarcotrochila piniperda (*Naevia p.*)
   (C:199): on 3 NWT-Alta [1063] and Alta
   [F64, F68], Sask [F74], Man [F67], Ont
   [1340], NS [F66]; on 3, 4 Sask, Man

[F68]; on 7 BC [1012]; on 8 Alta [F66],
Ont [1444].
Schizophyllum commune (C:199): on 3 NB,
on 6, 8 NS and on 8 PEI [1032, 1062].
See *Abies*.
Sclerophoma sp.: on 3 BC [1012].
S. pithya: on 4 Ont [1827].
S. pithyophila (holomorph *Sydowia
polyspora*) (C:199): on 3, 4 Sask, Man
[F67]; on 5 Sask, Man and on 8 Ont [F68].
Scoleconectria cucurbitula: on 1, 3, 4
Que [1685] by inoculation.
Scopuloides hydnoides (*Odontia h.*, *Phlebia
h.*): on 3 NWT-Alta [1063], Mack [53];
on 8 Alta [1060]. See *Fagus*.
Scytalidium sp.: on 3 Que [354].
S. lignicola: on 3 Que [353, F69]; on 4
Man [1760], Ont [90]; on 8 NB, NS [1610].
Scytinostroma galactinum (*Corticium g.*)
(C:196): white stringy rot, carie
blanche filandreuse: on 1 Que [F65]; on
2, 3 BC [1062]; on 3 BC [1012, 1718],
Sask, Man [1951], Que [914], NB [1032,
1062]; on 3, 4 Ont [1062, 1341, 1956];
on 3, 6 Ont [F72]; on 3a, 8 BC [1233];
on 4 Man, Que, NB [1062] and Ont [90,
1964, 1958, F74], NB [1032]; on 6 Que
[1062]; on 8 Alta [1062], Que [969], NS
[1032, 1062], Nfld [1659, 1660, 1661].
Sebacina sp.: on 3 NWT-Alta [1063], Mack
[53].
S. incrustans: on 3 NWT-Alta [1063], Mack
[53]. Lignicolous.
Seimatosporium sp. (*Sciniatosporium
sp.*): on 3 Man [F66].
Septonema fasciculare: on 3 Sask, Man,
Que [1760].
Septotrullula sp.: on 3 BC [1012].
Serpula himantioides (*Merulius h.*, *S.
lacrimans* var. *h.*) (C:198): brown
cubical rot, carie brune cubique: on 3
BC, Alta, Man, Ont, NB [1062] and BC
[1012], NB [1032]; on 4 Ont, Que [1062]
and Ont [90].
S. incrassata (*Poria i.*): on 8 NB [1032,
1061]. Associated with a brown rot.
S. lacrimans (*S. l.* var. *lacrimans*): on
8 Ont [1341], NS [1032]. Associated
with a brown rot of timbers in
buildings. Not found in the forest.
Sirococcus conigenus (*S. strobilinus*,
Discella s.): on 1 Que [F70, F71]; on
2, 3, 7 BC [1012]; on 3 BC [F71, F72,
F76], Que [F70], NS [F75] and BC, Ont,
NB, PEI [F74]; on 3, 4 Que [1690, F68],
NB, PEI [1037]; on 4 Ont [1340, F67],
Que [F71, F72, F73, F74, F75], NB
[1886], PEI [1886, F74]; on 5 Ont [F73,
F74, F75, F76], Que [F68]; on 8 BC [18].
Sirodothis sp. (holomorph *Tympanis*
sp.): on 3 BC [1012].
Sistotrema brinkmannii (*Grandinia b.*,
Trechispora b.) (C:200): white stringy
rot, carie blanche filandreuse: on 3
NWT-Alta [1063], Mack [53], BC [1012];
on 4 Que [917]; on 6 NB [1060]; on 8 Que
[969, 1060], NB [1610], NS [1032, 1060].
S. diademiferum: on 8 Que [969, 1060].
Lignicolous.

S. raduloides (*Trechispora r.*) (C:200):
on 3 BC [1012]; on 6 NB [1032, 1060]; on
8 Que [1610]. See *Abies*.
Sistotremastrum suecicum: on 8 Que [969,
1060]. See *Abies*.
Skeletocutis amorpha (*Polyporus a.*)
(C:198): on 2, 3 BC [1012, 1061]. See
Abies.
S. nivea (*Polyporus semipileatus*, *Tyromyces
s.*): on 8 Ont [1061], Nfld [1659,
1660, 1661]. See *Amelanchier*.
S. odora (*Poria o.*): on 8 Yukon [1061].
S. stellae (*Poria s.*): on 3, 7 BC [1012];
on 4 Nfld [1061]; on 8 BC, Alta [1061].
See *Abies*.
S. subincarnata (*Poria s.*) (C:199): on 2,
3, 4, 8 Alta [1061]; on 2, 8 Alta [579];
on 3 BC [1012]; on 3, 7 BC [1061]; on 8
NS [1032, 1061]. Associated with a
white rot.
Spadicoides atra: on 3 Man [1760].
Sphaerobasidium minutum (*Xenasma m.*): on
8 Que [969, 1060]. See *Abies*.
S. subinvisible: on 8 Que [969, 1060].
Lignicolous.
Sphaeropsis ellisii (C:199): on 3 Ont
[1340]; on 5a Que [337].
S. sapinea (*Macrophoma s.*) (holomorph
Botryosphaeria laricis): on 1, 3, 5, 6
Que [1687].
Sporidesmium achromaticum: on 3 Sask
[1760]; on 3, 4 Man [1760].
Sporobolomyces sp.: on 8 Que [1610].
Steccherinum ciliolatum (*Hydnum c.*): on 8
Que [556, 1062]. See *Acer*.
Stemonitis axifera: slime mold: on 2
NWT-Alta [1063], Alta [950].
Stereum ostrea (*S. fasciatum*): on 4 Que
[917]. See *Abies*.
Sterigmatobotrys macrocarpa: on 8 Man
[1760].
Stigmina lautii: on 3 BC [1012, F73],
Sask [1760]; on 3, 4 Man [1760].
S. verrucosa (*Sciniatosporium v.*): on 2
BC [1012, F69]; on 3 BC [1127]; on 3, 4
BC [1012, 1760].
Strasseria geniculata: on 3 Que [1278].
Streptomyces sp.: on 8 NB [1610].
Strobilurus occidentalis: on 3, 7 BC,
Alta and on 8 Alta [1431].
Subulicystidium longisporum: on 4 BC
[1061]. See *Betula*.
Suillosporium cystidiatum: on 8 Ont
[1062], Que [969, 1062]. Lignicolous.
Sydowia polyspora (anamorph *Sclerophoma
pithyophila*): on 1, 3, 4, 6 Que [1686]
on 1 and 6 by inoculation; on 2, 3, 3 X
2, 7 BC [1012].
Taeniolella rudis: on 3 Que [815].
Talaromyces emersonii: on 8 BC [1723].
Reported as "*Byssochlamys emersonii*
Stolk-Apinis," a name which apparently
has not been validly published; see in
[1743] under "Materials examined".
Thanatephorus cucumeris (anamorph
Rhizoctonia solani): on 3 Que [1945].
Thelephora sp.: on 8 Ont [1828].
T. penicillata (*T. fimbriata*): on 3 Sask
[F63].
T. terrestris: on 4 Que [F76]; on 8 NS
[1032]. See *Pinus*.

Thermoascus aurantiacus: on 8 BC [1723],
NB [1610].
Thermomyces lanuginosus (Humicola l.): on
8 BC [1723], NB [1610].
Therrya sp. *(Coccophacidium* sp.*)*: on
3 BC [1012].
T. piceae: on 3 BC [519].
Thysanophora penicillioides: on 3 Man
[1760], NB [872]; on 3, 4, 8 Ont [872];
on 6 NS [872].
Tiarosporella parca: on 2, 3 Alta [1947].
Tomentella sp.: on 8 NB [1032].
T. bryophila: on 8 Man [1062].
T. calcicola: on 8 Alta [1062].
T. ramosissima (T. fuliginea): on 4, 8
Nfld [1062]; on 8 Ont [1062], Que [1062,
1341].
T. rubiginosa (C:199): on 3 Man [1062].
T. sublilacina: on 4 Nfld [1062]; on 8
Ont [1062, 1341].
T. violaceofusca: on 8 Man [1062].
*Tomentellastrum badium (T. floridanum,
Tomentella atroviolacea)*: on 3 Yukon
[1062]; on 8 Alta [1341].
Tomentellina fibrosa: on 3 Yukon [1062].
*Trechispora cohaerens (Corticium confine,
T. confine, T. submicrospora, Corticium
s.)*: on 8 Que [969, 1060].
Lignicolous.
T. farinacea (Grandinia f.): on 8 Alta
[1060], Que [969, 1060]. See *Abies*.
*T. mollusca (Poria candidissima,
Trechispora c., P. mollusca)* (C:199):
on 4 Nfld [1659, 1660, 1661]; on 7 BC
[1061]; on 8 Ont [970], Que [1060]. See
Abies.
T. pallidoaurantiaca: on 4 Alta [559,
1060]. See *Pinus*.
T. stellulata (Corticium s.): on 8 Que
[969, 1060]. Lignicolous.
T. vaga (Corticium sulphureum) (C:196):
on 3 NWT-Alta [1063], Mack [53], Nfld
[1060]; on 8 Man [970]. See *Abies*.
Tremella lutescens: on 4 Nfld [1659,
1660]. Lignicolous.
T. mesenterica: on 3 BC [1012, 1062].
See *Acer*.
*Trichaptum abietinum (Hirschioporus a.,
Polyporus a.)* (C:198): white pocket
rot, carie blanche de l'aubier: on 2,
3, 4 NWT-Alta [1063]; on 2 BC [953],
Alta [952]; on 3 Yukon [460, 1110], Mack
[53, 1028], BC [1012, 1341], Alta [53],
Ont [1341] and Yukon, NWT, BC, Alta, Ont
[1061]; on 3, 4, 8 NB, NS, PEI [1032,
1061]; on 4 NWT, Ont, Que [1061] and
Mack [53, 1110], BC [1012], Sask [1028],
Ont [1341], Nfld [1061]; on 4, 8 Nfld
[1659, 1660, 1661]; on 5 NB [1032]; on 6
NB, NS [1032, 1061]; on 7 BC [1012,
1061]; on 8 BC, Ont, Que [1061], Ont
[1827], Labr [867].
T. laricinus (Hirschioporus l.): on 2 BC
[1012]; on 4 Sask [1028], Labr [867].
See *Abies*.
Trichocladium canadense: on 8 NB [1610].
Trichoderma sp. *(Pachybasium* sp.*)*: on
3 Ont [F66]; on 8 NB [1610].
T. polysporum (T. sporulosum): on 8 NB
[1610].

T. viride: on 3, 4 Ont [1964, 1966]; on 4
Sask, Man [1760]. See *Abies*.
Tricholomopsis decora: on 3 BC [1062].
Associated with a white rot.
Trichosphaeria parasitica (C:200): on 3,
6 Que [F61].
Trichothecium roseum: on 2 NWT-Alta
[1063], Alta [950].
Triposporium elegans: on 4 Man [1760].
Tritirachium hydnicola: on 3 NWT-Alta
[1063].
Tryblidiopsis sp.: on 4 Que [F70].
T. picea: on 3, 4 Que [1690].
T. pinastri (C:200): on 2, 3, 7 BC [1012,
F63]; on 3, 4 Yukon [F63], Sask, Man
[F69]; on 3 Yukon [460], Sask [1340],
Ont [1340, 1827, 1828, F62], Que [F65],
PEI [1032]; on 4 BC [1012, F62], Ont
[1340], PEI [F62].
Tubercularia sp.: on 6 NB [1032].
T. ulmea: on 5 Sask [F69].
Tubulicrinis accedens: on 8 Que [969,
1060]. See *Abies*.
T. angustus (Peniophora a.): on 4 Ont
[1935]; on 8 Que [969, 1060].
Lignicolous.
T. calothrix: on 8 Que [969, 1060]. See
Pinus.
T. chaetophorus: on 8 Que [969, 1060].
Associated with a white rot.
*T. glebulosus (Peniophora gracillimus,
Tubulicrinis g.)*: on 3 NWT-Alta
[1063], Mack [53], BC [1012]; on 8 Que
[969, 1060]. See *Abies*.
T. hamatus: on 8 Que [969, 1060]. See
Abies.
T. juniperinus: on 8 Que [969].
Lignicolous.
T. medius (Peniophora m.): on 3 Yukon
[1110]. Lignicolous.
T. propinquus: on 3 Man [1060]; on 8 Que
[969, 1060]. Lignicolous.
T. subulatus: on 8 Que [969, 1060]. See
Abies.
Tylospora asterophora: on 8 Que [969,
1060]. Lignicolous.
T. fibrillosa (Hypochnus f.): on 8 Que
[924, 969, 1060]. Lignicolous.
Tympanis sp.: on 1 Que [1211]; on 4 Que
[917].
T. abietina: on 1, 3, 4, 6 Que [1688] on
1, 3 and 6 by inocluation.
T. confusa: on 3, 4 Que [1688] by
inoculation.
T. hansbroughiana: on 3, 4, 6 Que [1688]
on 6 by inoculation.
T. hypopodia (T. piceina) (C;200): on 1,
3, 4, 6 Que [1688]; on 3, 4 Que [F68];
on 4 Ont [90]; on 8 Que, NB [1221].
T. laricina (anamorph *Sirodothis* sp.):
on 3 BC [1012]; on 3, 4 Que [1688] by
inoculation; on 8 Que, NB [1221].
T. pithya: on 3, 4 Que [1688] by
inoculation.
T. truncatula (T. pinastri): on 3 Man
[F66]; on 3, 4 Que [1688] by inoculation.
T. tsugae: on 3, 4 Que [1688] by
inoculation; on 8 Ont [1221].
*Tyromyces balsameus (Polyporus b., P.
cutifractus)* (C:198): brown cubical

rot, carie brune cubique: on 3 BC,
Alta, Ont, NB [1061] and NB [1032]; on
3, 7 BC [1012]; on 4 Ont [90, 1061,
1341], Que [937]; on 7 BC [1061, 1341];
on 8 Que [1377], NB [1610], NS [1032,
1061].

T. *borealis (Abortiporus b., Polyporus b.)*
(C:98): white mottled rot, carie
blanche madrée: on 2, 3, 7 BC [1061];
on 3,7 BC [1012]; on 3, 6, 8 NB [1032,
1061]; on 6, 8 NS [1061]; on 8 Ont
[1377], NS [1032], Nfld [1659, 1660,
1661]. Associated with a white rot.

T. *caesius (Polyporus c.)* (C:198): on 3
Yukon, NWT, BC, Alta [1061]; on 3 BC
[1012]. See *Abies*.

T. *canadensis* (C:198): on 8 Ont [1013].
Associated with a white stringy rot.

T. *chioneus (Polyporus albellus, Tyromyces
a.)*: on 4 NB [1032, 1061]. See
Abies.

T. *guttulatus (Polyporus g.)* (C:198): on
7 BC [1012, 1061]. See *Abies*.

T. *lapponicus (Amylocystis l., Polyporus
l.)* (C:198): on 2, 3, 8 NWT-Alta
[1063]; on 2, 3, 7, 8 BC [1061]; on 2, 3
Alta [1061]; on 2 Alta [579]; on 3 Yukon
[1110]; on 3, 7 BC [1012]; on 7 BC
[1341]; on 8 Yukon [460]. See *Pinus*.

T. *mollis (Polyporus m.)*: on 3 NWT-Alta
[1063], Alta [579, 1061]. See *Abies*.

T. *placenta (Poria p., P. monticola)*
(C:199): on 3 NWT-Alta [1063]; on 7 BC
[1012, 1061, 1341]. See *Pinus*.

T. *ptychogaster*: on 8 NS [1635].
Lignicolous.

T. *sericeomollis (Polyporus s., Poria s.,
P. asiatica)* (C:199): on 2, 3, 7 BC
[1061]; on 3 BC [1718], Sask, Man
[1951]; on 3, 7 BC [1012]; on 6 NB
[1032, 1061]; See *Abies*.

T. *stipticus (Polyporus immitis, Tyromyces
i.)* (C:198): brown cubical rot, carie
brune cubique: on 3 NWT [1061], Mack
[579], Sask [579, 1061]; on 8 Que
[1341], NB [1032, 1061]. See *Abies*.

T. *subvermispora (Poria s.)*: on 7 BC
[1012]. See *Abies*.

T. *undosus (Polyporus u.)* (C:199): on 7
BC [1012, 1061]. See *Abies*.

*Uthatobasidium ochraceum (Botryobasidium
flavescens)*: on 3 BC [1060].
Lignicolous.

Valsa sp.: on 1, 3, 4, 6 Que [1223]; on
1, 3 Que [F61]; on 3 Yukon [460].

V. *abietis*: on 1, 3, 4, 5, 6 Que [1689];
on 3 BC [1012]; on 3, 4 Que [F68]; on 4
Que [F69].

V. *friesii* (anamorph *Cytospora dubyi*):
on 1, 3, 4, 6 Que [1689] on 1, 6 by
inoculation; on 3, 4 Que [F68].

V. *kunzei* var. *piceae*: on 1 Ont [863];
on 3 Sask [F65].

V. *pini*: on 1, 3, 4, 6 Que [1689] by
inoculation.

Vararia granulosa (Grandinia g.) (C:197,
200): on 2, 3 BC [1012, 1062]; on 8
Alta [557], Ont [1341]. See *Abies*.

V. *investiens*: on 8 Que [969, 1062], Nfld
[1659, 1660]. See *Abies*.

Verticicladiella abietina (C:200): on 1
Que [F65]; on 2 BC [1012, 874]; on 8 Ont
[874].

V. *brachiata* (C:200): on 4 NB [874]; on 8
NB [1610].

Verticillium sp.: on 4 Ont [90]; on 8 Ont
[1827].

*Vesiculomyces citrinus (Corticium radiosum,
Gloeocystidiellum c.)*: on 2 BC [1012];
on 8 NWT-Alta [1063], Alta [951], NS
[1032]. See *Abies*.

Xenasma gaspesicum: on 8 Que [969,
1060]. Lignicolous.

X. *grisellum*: on 8 Que [969, 1060].
Lignicolous.

X. *rallum*: on 7 BC [1060]. Lignicolous.

X. *rimicola*: on 8 Que [969, 1060].
Lignicolous.

X. *tulasnelloideum (Corticium t.)*: on 3
NWT-Alta [1063]. Lignicolous.

Xeromphalina campanella (C:200): white
stringy rot, carie blanche filandreuse:
on 2, 3 BC [1012]; on 2, 3, 4 Alta
[1062]; on 3, 8 NWT-Alta [1063]; on 3, 7
BC [1062]; on 3 Ont, NB [1062]; on 4 Ont
[90, 1062], Que [1062]; on 6 NB, NS
[1062]; on 8 NB [1610].

Zythia resinae (Pycnidiella r.) (holomorph
Biatorella r.): on 4 Que [937].

PINUS L.   PINACEAE

1. P. *albicaulis* Engelm., whitebark pine,
pin albicaule; native, BC-Alta.
2. P. *banksiana* Lamb. (P. *divaricata*
(Ait.) Dum.), Jack pine, pin gris;
native, Mack, Alta-PEI.
3. P. *cembra* L., Swiss stone pine, pin
des rochers de Suisse; imported from
Eurasia.
4. P. *contorta* var. *contorta* Dougl.,
shore pine, pin tordu; native, coastal
BC. 4a. P. *c.* var. *latifolia*
Engelm. (P. *murrayana* Grev. &
Balf.), lodgepole pine, pin lodgepole;
native, s. central Yukon-sw. Mack, BC-w.
Alta and Cypress Hills of se. Alta-sw.
Sask. Planted beyond native range.
5. P. *echinata* P. Mill., short-leaf pine,
pin jaune; imported from USA.
6. P. *flexilis* James, limber pine, pin
souple; native, BC-Alta.
7. P. *jeffreyi* Grev. & Balf.; imported
from sw. USA.
8. P. *lambertiana* Dougl., sugar pine, pin
à sucre; imported from w. USA.
9. P. *monticola* Dougl., western white
pine, pin blanc de l'Ouest; native,
BC-Alta.
10. P. *mugo* Turra (P. *montana* Mill.)
Swiss mountain pine, pin de montagne;
imported from Europe.
11. P. *muricata* D. Don, bishop pine, pin
épineux; imported from California.
12. P. *nigra* Arnold (including vars.
austriaca, *poirentianer* and
calabrica), Austrian pine, pin noir;
imported from Europe.
13. P. *palustris* P. Mill., longleaf pine,
pin à longues feuilles; imported from
se. USA.

14. *P. peuce* Griseb., Balkan pine, pin de Balkans; imported from the Balkans.
15. *P. pinaster* Ait., cluster pine, pin de Bordeaux; imported from the Mediterranean.
16. *P. ponderosa* Dougl., ponderosa pine, pin ponderosa; native, BC.
17. *P. radiata* D. Don, Monterrey pine, pin de Monterrey; imported from California.
18. *P. resinosa* Ait., red pine, pin rouge; native, sw. Man-Nfld.
19. *P. rigida* Mill., pitch pine, pin des corbeaux; native, Ont-Que. Along St. Lawrence river between Lake Ontario and Montréal.
20. *P. strobus* L., eastern white pine, pin blanc; native, sw. Man-Nfld.
21. *P. sylvestris* L., Scots pine, pin sylvestre; imported from Eurasia.
22. *P. taeda* L., loblolly pine, pin à encens; imported from se. USA.
23. *P. hybrids*; *P. contorta* x *banksiana*; *P. echinata* x *taeda*; *P. murrayana* x *banksiana*.
24. Host species not named, i.e., reported as *Pinus* sp.

Aleurodiscus lividocoeruleus (Corticium l., Gloeocystidiellum l.) (C:202): on 4 NWT-Alta [1063], Alta [53]; on 24 Ont [964]. See *Abies*.
A. penicillatus (C:200): on 4, 9, 16, 21 BC [1012]; on 4, 16, 21 BC [964]. See *Abies*.
Allescheria terrestris: on 24 BC [1723].
Alternaria sp.: on 2 Ont [91]; on 18 Que [1748]; on 20 Ont [974]; on 21 Sask [1848].
A. alternata (A. tenuis auct.*)* (C:200): on 2 Ont [F70].
Amphinema byssoides: on 2 Ont [1341]; on 4 Alta [1341, 1828]. See *Abies*.
Amylocorticium canadense: on 24 Ont, Que [1938]. Associated with a brown rot.
A. cebennense (Corticium c.) (C:201): on 9 BC [1012, 1939]; on 24 Ont [1939]. See *Abies*.
A. subincarnatum: on 24 Ont [1940]. See *Picea*.
A. subsulphureum: on 20 NS [1032]. See *Picea*.
Amylostereum chailletii (Stereum c.) (C:204): white stringy rot, carie blanche filandreuse: on 4 NWT-Alta [1063]; on 4a Alta [999, 1485]; on 9 BC [1012].
Anomoporia bombycina (Poria b.): on 4 Alta [579]. See *Picea*.
A. myceliosa (Poria m.): on 2 Ont [1341]; on 4 NWT-Alta [1063]. See *Picea*.
Anthostomella sp.: on 4 BC [1012].
Antrodia albida (Trametes sepium): on 4 NWT-Alta [1063], Alta [53]. See *Alnus*.
A. crassa (Poria c.) (C:204): on 9 BC [1012]. See *Alnus*.
A. crustulina (Poria c.): on 18 Ont [1341]. See *Abies*.
A. gossypia (Poria g.): on 2 Man [579]. See *Abies*.

A. heteromorpha (Coriolellus h., Trametes h.) (C:201): on 2, 4 NWT-Alta [1063]; on 4 BC [953]; on 9, 16 BC [1012]. See *Abies*.
A. lenis (Poria l.) (C:204): on 9 BC [1341]; on 16 BC [1012]; on 18 Ont [1341]. See *Abies*.
A. lindbladii (Poria cinerascens) (C:204): on 16 BC [1012]. See *Abies*.
A. serialis (Coriolellus s., Trametes s.) (C:201): on 2 Ont [1341]; on 4 NWT-Alta [1063], Alta [952]; on 24 NS [1032]. See *Abies*.
A. sinuosa (Coriolellus s., Poria s., P. vaporaria) (C:201): on 4 NWT-Alta [1063]; on 9, 16 BC [1012]; on 20 Ont [1963]. See *Picea*.
A. variiformis (Coriolellus v., Trametes v.) (C:201): on 4 NWT-Alta [1063], Alta [952]; on 9 BC [1012]. See *Abies*.
A. xantha (Poria x.) (C:204): on 4 NWT-Alta [1063]; on 9 BC [1012, 1341]; on 18, 20, 24 Ont [1341]; on 20 Ont [1827]. See *Abies*.
Aposphaeria sp.: on 4 BC [1012].
Arcyria nutans: on 4 NWT-Alta [1063].
Armillaria sp.: on 2 Que [F61]. Probably *Armillaria mellea*.
Armillaria mellea (C:201): root rot, pourridié-agaric: on 2 Mack [53], Alta, Sask, Man, Ont [F61] also Mack, Ont [F63, F64], Man [F62], Ont, Que [F62, F65, F68, F70, F72] and Ont [1827]; on 2, 4, 6 NWT-Alta [1063]; on 2, 10, 18, 20, 21 Que [1377]; on 2, 18 Ont [F66], Que [F74]; on 2, 18, 20 Ont [1341]; on 2, 18, 20, 21 Ont [1964, 1966, F73]; on 2, 18, 21 Que [F76]; on 2, 20 Ont [6]; on 2, 21 Que, NB, NS [F75]; on 4 Alta [950, 952, F60, F63, F64, F65, F66], Que [F66]; on 4a Alta [57, F71, F73]; on 4, 9, 10, 11, 15, 16, 17, 18, 21 BC [1012]; on 6 Alta [F62]; on 9 BC [1124]; on 11, 17 BC [F62]; on 15 BC [F63]; on 18 Ont [834, F64, F65], Que [F62, F73, F75], NB, NS [F74], NS [F70], Nfld [1663, F69]; on 18, 20 Ont [F62, F63], Nfld [1661]; on 18, 21 Que [F68, F69]; on 20 Ont [F74], NB, NS [F72, F74], NB [1032, F62], Nfld [1659, 1660]; on 20, 21 Ont [F62]; on 23 Alta [F69]; on 24 Alta [F68], Ont [F64, F69]. See *Abies*.
Arthrobotrys arthrobotryoides: on 2 Ont [88].
Ascocoryne cylichnium (Coryne c.): on 20 Que [F67].
A. sarcoides (Coryne s.) (C:201): on 2, 4 NWT-Alta [1063]; on 2 Ont [88, 89, 91]; on 2, 18, 20 Ont [1964, 1966]; on 4 Alta [182].
Aspergillus fumigatus: on 24 BC [1723].
Asterodon ferruginosus: on dead wood of 20 Que [1385]. See *Abies*.
Athelia decipiens (Corticium d.): on 2, 4 NWT-Alta [1063]. Associated with a white rot.
A. epiphylla (Corticium e.): on 2 Mack [53]. See *Abies*.
Atropellis sp.: on 15 BC [1012].

A. pinicola (C:201): canker, chancre
    atropellien: on 4, 9 BC [1012]; on 24
    BC [1446].
A. piniphila (C:201): canker, chancre
    atropellien: on 2, 4a Alta [F71]; on 2,
    4, 23 Alta [F69]; on 4, 16 BC [1012]; on
    4, 23 Alta [F68]; on 4, 24 NWT-Alta
    [1063]; on 4 BC, Alta [767], BC [953,
    F67, F68, F70], Alta [950, 951, 952,
    F60, F61, F63, F65], Sask [F65, F67]; on
    4a BC [F75], Alta [644, 768, 1340, 1485,
    F72, F73, F75]; on 23 Alta [F69]; on 24
    BC [F73], Alta [F62, F66].
A. tingens (C:201): on 9 BC [1012]; on 20
    NS [1032].
Aulographina pinorum: on 18 Ont [1340].
Aureobasidium sp. *(Pullularia* sp.*)*:
    on 20 Ont [974]. Possibly the
    Hormonema state of *Sydowia polyspora*.
A. pullulans (Pullularia p.) (C:201): on
    2, 18 Ont [1340]; on 2, 18, 20, 21 Ont
    [F67]; on 2, 21 Ont [F69]; on 2 Ont
    [88]; on 4 NWT-Alta [1063], Alta [182],
    Que [F67]; on 10, 18, 21 Que [F62]; on
    18 NB [1032]; on 20 Ont [1827]; on 21 NS
    [1032].
*Auriscalpium vulgare (Hydnum
    auriscalpium)*: on 2 Mack [1110], NS
    [1307]; on 4 BC [1012]; on 21 Ont
    [1341]; on 24 Que [1385]. On fallen
    cones.
*Basidiodendron caesio-cinerea (Sebacina
    c.)*: on 2 NWT-Alta [1063]; on 20 Que
    [1016]. Lignicolous.
B. eyrei (Sebacina e.): on 4 NWT-Alta
    [1063], Alta [950]. See *Acer*.
B. nodosa: on 24 Ont [1016].
B. pini: on 20 Ont [1016].
Basidioradulum radula (Radulum orbiculare)
    (C:204): on 24 NS [1032]. See *Abies*.
Biatorella sp.: on 1 BC [1012].
B. resinae (anamorph *Zythia r.*)
    (C:201): on 2, 4, 24 NWT-Alta [1063];
    on 2, 20 Que [1690]; on 2, 21 Ont
    [1340]; on 2 Alta [53, F62], Ont [F60];
    on 4, 9, 16 BC [1012]; on 4 Alta [951];
    on 20 Que [1685].
Bifusella pini (Hypoderma p.) (C:203): on
    6 NWT-Alta [1063], Alta [F63].
B. linearis (C:201): needle cast, rouge:
    on 1, 9 BC [1012]; on 1 Alta [F68]; on 6
    NWT-Alta [1063], Alta [950, F62, F66,
    F70]; on 20 Ont [1340], NB, NS [1032]
    and NS [F62].
B. saccata: on 1 BC [1012, F68].
Bjerkandera adusta (Polyporus a.): on 2
    Ont [1964]; on 18 Ont [1341]. See
    Abies.
Blyttiomyces aureus: saprophytic on
    pollen of 21 BC [177].
Boletopsis griseus (Polyporus g.): on 24
    NS [1032]. Typically on ground, perhaps
    mycorrhizal.
Botryobasidium subcoronatum: on 20 Ont
    [1341, 1660]. See *Betula*.
B. vagum (Pellicularia v.) (C:203): on 2,
    4 NWT-Alta [1063]; on 2 Alta [53], Ont
    [1341]; on 9 BC [1012]. See *Abies*.
*Botryohypochnus isabellinus (Botryobasidium
    i.)*: on 24 Ont [1341]. See *Abies*.

Botrytis sp.: on 21 NWT-Alta [1063].
B. cinerea (holomorph *Botryotinia
    fuckeliana*): on 2 Ont [F74, F76]; on
    4a BC [F76]; on 11 BC [1012, F66]; on
    18, 21 Ont [F73].
Bulliardella sp.: on 4 NWT-Alta [1063].
    Probably a specimen of *Actidium*.
Calcarisporium sp.: on 4 NWT-Alta [1063].
Caliciopsis pinea (C:201): canker,
    chancre caliciopsien: on 9 BC [495,
    1012, F62]; on 20 Ont [493, 1340, F60],
    NB, NS [1032].
C. pseudotsugae (C:201): on 21 BC [493,
    1012, F62].
Calocera cornea: on 20 NS [1032]. See
    Acer.
Candida sp.: on 18 Que [1748].
Capnodium sp.: on 20 Ont [1340].
C. pini: on 20 Ont [1340, F63].
Catinella nigro-olivacea: on 18 Ont
    [1340].
Cenangium sp.: on 20 Ont [974] by
    inoculation.
C. acuum (C:201): twig blight, brûlure
    des rameaux: on 2 Ont [1827]; on 9 BC
    [1012]; on 20, 21 NB [1032]; on 20 Ont
    [974, 1340], Que [1377].
C. atropurpureum (C:201): twig blight,
    brûlure des rameaux: on 2 Ont [1340,
    1828, F67]; on 18, 21 Ont [F62].
C. ferruginosum (C. abietis) (C:201):
    twig blight, brûlure des rameaux: on 2,
    4a, 18, 20, 21 Que [1690]; on 2, 12, 18
    Ont [F68]; on 2, 18, 20, 21 Ont [1340],
    Que [F68]; on 2, 18, 21 Ont [F71]; on 2,
    20, 21 Ont [F69]; on 4, 9, 16 BC [1012];
    on 18, 20 Ont [1828]; on 18, 21 Ont
    [F72, F74, F76]; on 18 Ont [1827, F75];
    on 21 Ont [F73, F75].
C. pithyum: on 9 BC [F65]. Perhaps a
    misdetermined specimen of *C.
    ferruginosum*.
Cenococcum graniforme: on 2, 18, 20 Ont
    [1964].
Cephaloascus fragrans: on 20 Ont [108].
Cephalosporium sp.: on 2, 18, 20 Ont
    [1964]; on 2 Ont [88]; on 4 NWT-Alta
    [1063], Alta [182]; on 18 Que [1748].
C. terrestre: on 24 BC [1723]. See note
    under *Picea*.
Ceraceomyces serpens (Merulius s.)
    (C:203): on 3 Alta [568]; on 16 BC
    [568, 1012, 1341]. See *Abies*.
C. tessulatus (Corticium t.): on 2
    NWT-Alta [1063], Alta [53]. See *Abies*.
Ceratobasidium fibulatum: on 2 Ont
    [1835]. The host and locality are
    uncertain and the name is not validly
    published [578]. The record should be
    disregarded.
Ceratocystis sp.: on 4 NWT-Alta [1063];
    on 4a BC [1481].
C. aequivaginata: on 2, 18 Man [1190].
C. alba: on 2 Man [1190].
C. allantospora: on 2 Man [1190]; on 18,
    20 Ont [647, 1340].
C. angusticollis: on 2, 18 Man [1190]; on
    18, 20 Ont [647]; on 18 Ont [1340].
C. columnaris: on 2 Man [1190].

C. coronata: on 2 Man and on 2, 18, 20
   Ont [1190].
C. crenulata: on 2 Man [1190].
C. deltoideospora: on 2, 18 Man and on
   18, 20 Ont [1190].
C. dolominuta: on 2 Man, Ont [1190]; on
   18 Ont [647, 1340].
C. europhioides (C:201): on 2 Man and on
   2, 18 Ont [1190]; on 18, 20 Ont [1340];
   on 18, 20, 21 Ont [647, 1983].
C. falcata (C:201): on 20 Ont [1340,
   1983].
C. huntii (C:201): on 2 Man [1190]; on 4
   NWT-Alta [1063], BC [1340]; on 4a BC
   [647, 1484]; on 18 Ont [647, 1340].
C. ips: on 2 Man [1190]; on 2, 18, 20 Ont
   [647, 1190, 1340]; on 21 Ont [647].
C. leucocarpa: on 2 Man [1190].
C. longispora: on 2 Man [1190].
C. major: on 18 Que [F71].
C. minima: on 2, 18 Man and on 2, 20, 21
   Ont [1190].
C. minor: on 2, 18, Man [1190]; on 2, 20,
   21 Ont [647, 1340]; on 2 Que [F68]; on 4
   NWT-Alta [1063], Alta [951]; on 4a BC
   [1481]; on 18 Ont [647].
C. minuta: on 2 Man [1190]; on 2, 18, 20
   Ont [647, 1340]; on 4a BC [1481]; on 21
   Ont [647].
C. montia: on 4a BC [1450, 1481], Alta
   [1482].
C. nigra: on 18, 20, 21 Ont [647, 1340].
C. olivacea: on 21 Ont [647, 1190, 1340].
C. perfecta: on 2 Man [1190].
C. piceae: on 2 Man [1190].
C. pilifera: on 2 Man [1190]; on 2, 18,
   20 Ont [647, 1340]; on 18 Que [F70].
C. sagmatospora: on 2 Man [1190], Ont
   [647]; on 2, 20 Ont [1340]; on 18, 20
   Ont [647, 1190, 1983].
C. spinifera: on 2 Man [1190].
C. stenoceras: on 20, 24 Ont [647, 1340].
C. tenella: on 2 Man [1190]; on 2, 18, 20
   Ont [647, 1340].
C. truncicola: on 18, 20 Ont [647, 1340].
C. tubicollis: on 2 Man and on 18 Ont
   [1190].
Ceriosporopsis halima: on wood of 9 in
   water BC [782].
Ceriporia purpurea (Poria p.) (C:204): on
   4 NWT-Alta [1063]. See *Alnus*.
C. reticulata (Poria r.): on 4 NWT-Alta
   [1063]. See *Acer*.
C. rhodella (Poria r.): on 20 Ont
   [1341]. Probably associated with a
   white rot.
Ceuthospora sp.: on 4 NWT-Alta [1063].
Chaetoderma luna (Peniophora l.) (C:203):
   on 4 NWT-Alta [1063]; on 9 BC [1012].
   See *Abies*.
Chaetophiophoma sp.: on 4 NWT-Alta [1063].
Chalara cylindrica: on 2 Man [1760].
Cheiromycella microscopica: on 2 Man
   [1760].
Cirrenalia macrocephala: on wood of 9 in
   water BC [782].
Cladosporium sp.: on 18 Ont [1828].
C. herbarum (C:201): on 2 Ont [88]; on 4
   BC [1012], Alta [182, 952].
C. macrocarpum: on 2 Man [1760].

C. variabile: on 2 Ont [1828, F70].
Coccomyces strobi (C:201): on 18 Ont
   [F70]; on 20 Ont, Que [F67], NB, NS
   [1032] also Ont [1340], Que [1685, F61],
   NB [F62], Ont, NB [1609] and Ont, NB, NS
   [1441].
Coleosporium sp.: on 2 Ont [F62]; on 18
   NB [1032]; on 21 NS [1032], Nfld [1660];
   on 24 NB, NS [F75].
C. asterum (C. solidaginis) (C:201):
   needle rust, rouille des aiguilles: on
   2 Mack [53], Sask, Man [F67, F68, F69],
   Man [F62, F63, F64, F74], Man, Ont
   [F73], Ont [F75], Ont, NS [F71], Ont,
   Nfld [F72], Que [1206], NB, NS [1032,
   F64, F65], NS [F68]; on 2, 4 Alta [53,
   F64, F65]; on 2, 4, 21 NWT-Alta [1063],
   Alta [F62]; on 2, 18 Ont [1236, F60,
   F62, F63, F65, F70], Ont, Que [F61, F66,
   F67, F74], Que [1377, F68, F69, F71,
   F73, F75]; on 2, 18, 24 Ont [1341]; on
   2, 18, 20, 21 Nfld [1660]; on 4 BC [953,
   954, 2007, F64], BC, Alta [F66], Alta
   [950, 952, F61, F63, F67]; on 4a BC
   [1012], BC, Alta [F72], Alta [F73], Ont
   [1828], NB [1032]; on 12 NB [F71], PEI
   [F76]; on 18 Ont [338, F73], Ont, Que
   [F72, F76], Que [1341, F70], Que, PEI
   [F64], NB, NS [F61, F64, F71, F72], NB,
   NS, PEI [1032], NB [F65], NS, PEI [F74],
   Nfld [1661, F63]; on 18, 20, 21 Nfld
   [F73]; on 21 NS [1032, F65], NB, NS
   [F63]; on 24 Mack [2008], BC, Alta,
   Sask, Man [2008, F70], Alta [951, 952],
   Man [839], Ont [F64], Ont, NB, NS [F68,
   F69], Que [338].
C. pinicola (C:201): needle rust, rouille
   des aiguilles: on 2 Ont, Que [1341],
   Ont [1445, F63, F64, F65, F66], NB [F73].
C. viburni: on 2 Que [1206, F66, F67],
   NB, NS [F74].
Colletotrichum gloeosporioides (holomorph
   Glomerella cingulata): on 2 Alta,
   Sask and on 4a BC, Alta [1155].
Columnocystis abietinum (Stereum a.): on
   4 NWT-Alta [1063]. See *Abies*.
Comatricha nigra: slime mold: on 4
   NWT-Alta [1063]. Presumably on
   well-rotted wood.
*Confertobasidium olivaceo-album (Athelia
   fuscostrata, Corticium f.)* (C:201):
   on 2 Ont [88]; on 2, 20 Ont [1341]; on 4
   NWT-Alta [1063]. Lignicolous.
Coniochaeta ligniaria (Rosellinia l.): on
   4 NWT-Alta [1063].
C. malacotricha: on 20 Ont [1340].
Coniophora sp.: on 20 Ont [1341], NB
   [1032].
C. arida (C. betulae) (C:201): on 4
   NWT-Alta [1063]; on 16 BC [1012]; on 20
   Ont [1828]. See *Acer*.
C. arida var. *suffocata (C. suffocata)*
   (C:201): on 4 NWT-Alta [1063]. See
   Abies.
C. olivacea: on 24 Ont [1341]. See
   Alnus.
C. puteana (C. kalmiae) (C:201): brown
   cubical rot, carie brune cubique: on 2
   Ont [88]; on 4 NWT-Alta [1063]; on 4, 16
   BC [1012]; on 4a Alta [999, 1000, 1002,

1485]; on 18 Ont [1341]; on 20 Ont
[1963], NB [1032, F62]. See *Abies*.
Coniothecium sp.: on 16 BC [1012].
Coniothyrium sp.: on 4 NWT-Alta [1063].
Connersia rilstonii (Pseudoeurotium r.):
  on wood of 9 in water BC [782].
Conoplea elegantula: on 18 Man [1760].
C. geniculata (C:201): on 4 Alta [784].
Coriolus pubescens (Polyporus p.): on 3
  NWT-Alta [1063]. See *Abies*.
Corticium sp.: on 4 NWT-Alta [1063]; on
  20 NB [1032].
C. centrifugum (C:201): on 2, 4, 24
  NWT-Alta [1063]; on 4 BC [953], Alta
  [950, 951]. See *Abies*.
Coryne sp.: on 4a Alta [1485]; on 20 Ont
  [1063].
Costantinella micheneri: on 4 NWT-Alta
  [1063].
Cristinia helvetica (Grandinia h.): on 4
  NWT-Alta [1063]. Associated with a
  white rot.
Cronartium sp. *(Peridermium* sp.)
  (C:202): rust, rouille: on 2 Mack
  [53], Ont [1828, F61], Que [F63]; on 2,
  16, 21 Ont [1341, F64]; on 2, 21 Ont
  [F62, F63, F67]; on 4, 16, 21 Ont
  [1236]; on 7 BC [1012]; on 21 Que
  [1341]; on 24 BC [F67].
C. coleosporioides (C. stalactiforme nom.
  nud.) (aecial stage *Peridermium
  stalactiforme)* (C:201): rust,
  rouille: 0 I on 2 Alta, Sask, Man, Ont,
  Que [735], Ont [1828, F69, F76], Ont,
  Que [F75], Que [1205, F65, F67], NS
  [F64], Nfld [1660, F74]; on 2, 4 NB, NS
  [1554, F66]; on 2, 4, 21, 24 NWT-Alta
  [1063]; on 2, 4a, 21 NB, NS [1032]; on
  2, 10 Que [1377]; on 2, 21 Ont [1341],
  Que [339, F66]; on 4 Mack, BC, Alta, Man
  [735], Mack, Alta [53], BC, Alta [2006];
  on 4, 16 BC [1341]; on 4, 21 BC [337],
  Ont [1236]; on 4a BC [F75, F76], BC,
  Alta [F73], Alta [1400, F72]; on 5 NB
  [F66], NS [735, F66]; on 10 Que [735];
  on 16, 21 BC [735]; on 21 Man [735]; on
  23, 24 Alta [F69]; on 24 Yukon, Mack
  [F69], BC [F72], Alta, Sask [F70], Alta
  [339], Alta, Ont [F71], Ont [F74].
C. coleosporioides f. *album*: rust,
  rouille: on 4 Alta [2006]; on 4a BC,
  Alta [1406].
C. coleosporioides f. *coleosporioides*:
  rust, rouille: on 4 BC [1012]; on 24
  Mack, BC, Alta, Sask, Man [2008].
C. coleosporioides "var.
  stalactiforme": rust, rouille: on 2
  Que [946]. Apparently referring to the
  aecial state, *Peridermium
  stalactiforme*. The combination with
  C. coleosporioides does not appear to
  have been published.
C. comandrae (C:202): gall rust,
  rouille-tumeur: 0 I on 2 Yukon, Mack
  [F72], Mack, Alta [53], Mack, Ont, Que,
  NB [735], Sask, Man, Ont [F64], Sask,
  Man, Que [F63], Sask, Que [F66], Man,
  Ont [F67], Ont [1236, 1341, F69, F76],
  Ont, Que [F75], Que [1377, F65, F68]; on
  2, 4 Yukon, Alta, Sask [F70]; on 2, 4,

21 NWT-Alta [1063], Alta, Sask [735]; on
2, 21 Sask [F67], Man [735]; on 2, 24
Ont [F70]; on 4 Yukon [460, 735, F68],
BC [1341, F68], Alta [F66]; on 4, 16 BC
[1012, 735]; on 4, 24 Alta [F65]; on 4a
BC [F76]; on 10 Alta [735, F72]; on 21
Alta [F64]; on 24 Yukon [F69], Yukon,
Alta, Sask, Man [2008, F71], NWT-Alta
[1063], Mack [F65], Mack, BC [2008],
Alta [F68], Ont, Que [F71].
C. comptoniae (C:202): rust, rouille: 0
  I on 2 NWT [732, 1407], Mack, Alta
  [F72], Sask, Man [735], Sask, Ont [F73],
  Ont [650, 651, 652, 1236, 1827, 1828,
  F64, F65, F66, F69], Ont, Que [F63, F67,
  F68, F70, F75, F76], Que [1377, F61,
  F71, F74], NB [F68, F74], NS [F75]; on
  2, 4 NWT-Alta [1063], Mack, Alta [53,
  735, F66], Alta [732]; on 2, 4, 5 NS
  [F68]; on 2, 4, 5, 21 NS [735]; on 2, 4,
  10, 21 Que, NB [735]; on 2, 4a Ont
  [F72], NB, NS [1856]; on 2, 10 NB
  [1032]; on 2, 12, 18, 21 Ont [735]; on
  2, 18, 21 Ont [1341]; on 4 BC [1341],
  Que [F62], NS [1032]; on 4, 7, 11, 15,
  16, 17 BC [735]; on 4, 11, 16, 17 BC
  [1012]; on 4a BC [1122, F75], Que
  [1210], NS [F71, F76]; on 5 NS [1856];
  on 10 Que, NB [1410]; on 11, 15, 17 BC
  [F61, F62]; on 11, 16, 17 BC [2006]; on
  11, 17 BC [1122, F63]; on 23 Mack [53];
  on 24 Mack, BC, Alta, Sask, Man [2008],
  Mack, Sask, Ont [F70], Mack [F69], BC
  [F64], Ont [F71, F74].
C. quercuum (C:202): gall rust,
  rouille-tumeur: 0 I on 2 Ont [337],
  Ont, Que [1377], NB [F61, F64]; on 2, 10
  NB [F65]; on 2, 18, 21 Ont [1236]; on 2,
  21 Ont [735, F68], NB, NS [1032]; on 4,
  4a, 5, 10 NB [1032]; on 5 NS [F64]; on
  21 Ont [339], Que [1377], NS [334]; on
  24 Ont [F74], NB, NS [F66].
C. ribicola (C:202): white pine blister
  rust, rouille vésiculeuse du pin
  blanc: 0 I on 1 BC [953], Alta [734,
  839, 951, F68, F70, F72, F73, F75]; on
  1, 3, 6, 8, 9, 20 BC [735]; on 1, 6
  NWT-Alta [1063], Alta [735, 950, F63];
  on 1, 8, 9, 20 BC [1012]; on 3, 20 Man
  [735, F64]; on 6 BC [954], Alta [F60,
  F61, F62]; on 8 BC [F61]; on 8, 20 Ont
  [735]; on 9 BC [734, 1341, F76], Man
  [F73]; on 20 Sask [735], Man [333, F62,
  F63], Ont [186, 1299, 1341, 1839, F62,
  F72, F75], Ont, Que, NB, NS, Nfld [F63,
  F65, F66, F67], Ont, Que, Nfld [F68,
  F69, F70], Ont, NB, NS, Nfld [F76], Que
  [337, 944, 1210, 1377, 1386, 1387, 1388,
  1685, F64, F71, F72, F73], Que, NB, NS,
  Nfld [735, F74], NB [336], NB, NS [1032,
  F75], NS, PEI [334], PEI [735], Nfld
  [1659, 1660, 1661, F64, F73]; on 24 BC
  [F72, F73], BC, Alta, Man [2008], Ont
  [338, F71, F74], Que [F75, F76].
Crumenulopsis sp.: on 4 BC [1012].
C. sororia (Crumenula s.): on 4 BC [1340].
Cryptoporus volvatus (Polyporus v.)
  (C:204): on 2 Que [F71]; on 2, 4
  NWT-Alta [1063]; on 2, 18, 20 Ont
  [1341]; on 4 Alta [951]; on 4, 16 BC

[180, 1012]; on 18 Ont [1827, 1828].
See *Abies*.

*Cucurbidothis pithyophila (Cucurbitaria
p.)* (C:202): stem girdle, chancre
cucurbidothien: on 6 Alta [F68]; on 9
BC [F67]; on 20 Ont [1340, F67], NS
[1032], Nfld [1659, 1660, 1661].

C. pithyophila var. *cembrae*: on 8 BC
[1012].

C. pithyophila var. *pithyophila*: on 8
BC [1012].

Cyclaneusma niveum (Naemacyclus n.)
(C:203): on 2 NB [F73]; on 2, 21 Que
[1377]; on 4, 10, 15, 16, 21 BC [1012];
on 6 NWT-Alta [1063], Alta [952, F65];
on 21 BC [F68], Man [F66], Ont [1340,
1828], Que [F67, F69].

*Cylindrobasidium evolvens (Corticium
laeve)* (C:201): on 4 NWT-Alta [1063].
See *Abies*.

Cylindrocarpon sp.: on 18 NB [F72], NS
[1885, F68]; on 20 [1032].

Cyptotrama asprata: on 20 Ont, Que
[1436]. See *Acer*.

Cytospora sp. (C:202): on 2 Que [946]; on
4 NWT-Alta [1063], Que [F67]; on 20 NB,
NS [1032].

C. kunzei (holomorph *Leucostoma k.*): on
20 Ont [F70], Que [337].

C. pini (C:202): on 20 Ont [1827, 1828,
F70, F71], NS [F61].

Cytosporella sp.: on 18 Ont [1828].

Dacrymyces sp.: on 18 Ont [1341].

D. capitatus (D. stipitatus) (C:202): on
4 NWT-Alta [1063], Alta [952]. See
Alnus.

D. punctiformis: on 4 NWT-Alta [1063]; on
20 Ont [1828]. Lignicolous.

*D. stillatus (D. abietinus, D.
deliquescens, D. d.* var.
deliquescens): on 2 Mack [53]; on 2,
4 NWT-Alta [1063]; on 2, 18, 20, 21, 24
Ont [1341]; on 20 NS [1032]. See
Abies.

Dacryobolus karstenii: on 9 BC [1012].
Associated with a brown rot.

Dacryonaema rufum: on ?24 BC [192].
Lignicolous.

Dactylium sp.: on 4 NWT-Alta [1063].

Dasyscyphus acuum: on 2 Ont [F73]; on 20
Ont [1340].

D. pulverulentus: on 4 NWT-Alta [1063].

Davisomycella ampla (Hypodermella a.)
(C:203): needle cast, rouge: on 2
Sask, Man, Ont, Que [F65, F66, F68],
Man, Ont [F67, F69], Ont [1063, 1340,
1827, F61, F63, F64, F71, F72, F73, F74,
F76], Que [1377, F60, F61, F67, F75,
F76]; NB [F62, F63, F66, F69, F70], NB,
NS [1032, F64, F65, F73, F74]; on 4
Yukon [460], BC [1012, 2009], Alta [950,
951], Sask [F66], Que [F67]; on 4a Alta,
Sask [F74]; on 23 Alta [F70]; on 24 Alta
[F69].

D. fragilis: on 2 Ont [364].

D. medusa (Hypodermella m.) (C:203): on 2
Ont [1340, F60, F62]; on 16 BC [1012].

D. montana (Hypodermella m.) (C:203): on
2 Mack [53]; on 2, 4, 24 NWT-Alta
[1063]; on 4 BC [1012], Alta [F60, F65,
F66], Sask [F66]; on 4, 23 Alta [F61].

*Dendrothele griseo-cana (Aleurocorticium
g.)*: on 20 Ont [965]. Lignicolous.

Dendryphiopsis atra (holomorph
Microthelia incrustans): on 2 Man
[1760].

Dermea balsamea: on 9 BC [1012].

D. pinicola (C:202): on 20 Que [1685,
1690, F65], NB [1032], NS [F64].

Dichomitus squalens (Polyporus anceps)
(C:204): on 2 Mack [53, 1110]; on 2, 4
NWT-Alta [1063]; on 2, 20, 21 Ont
[1341]; on 4, 16 BC [1012]; on 18, 24 NS
[1032]; on 19 Que [F65]. See *Abies*.

Didymium iridis: on 4 NWT-Alta [1063].

D. melanospermum: on 4 NWT-Alta [1063].

Didymosphaeria sp.: on 16 BC [1012].

Discina ancilis (D. perlata): on 4, 16 BC
[1012, F68].

Discosia strobilina (D. pini): on 2 Ont
[358].

Ditiola radicata (C:202): on 4 NWT-Alta
[1063]; on 24 NS [1032]. Lignicolous.

Dothiorella sp.: on 20 NS [1032].

Dothistroma sp.: on 21 Que [F69].

D. septospora var. *septospora (D. pini)*
(holomorph *Scirrhia pini*): on 4 Alta
[F67]; on 4, 11, 12, 15, 16, 17, 23 BC
[1232, F64]; on 9 BC [1232]; on 11, 17
BC [F65]; on 12 Ont [F73], Nfld [338,
F65]; on 12, 18 Nfld [1661]; on 15, 17
BC [1063]; on 21 Alta [338]; on 24 BC
[338, 523, F66].

D. septospora var. *lineare (D. pini* var.
l.) (holomorph *Scirrhia pini*): on 4
Sask [F66]; on needles of 24 BC [1763].

Elytroderma deformans (C:202): needle
cast, rouge: on 2 Alta [53, F63], Sask
[F67], Man [F65], Ont [1340, 1377, 1827,
1828, F62, F68], NB [1032, F65]; on 2, 4
Yukon [F68]; on 2, 4, 24 NWT-Alta
[1063], Alta [F70]; on 4 Yukon [460], BC
[953], Alta [644, 839, 950, 951, 952,
F64, F65, F67], Alta, Sask [F66]; on 4,
16 BC [F65, F68, F70]; on 4, 23 Alta
[F60, F69]; on 4a Alta [F72]; on 4a, 16
BC [1012, F71]; on 16 BC [F60, F61, F62,
F63, F64, F66, F67, F69, F74, F75]; on
16, 24 BC [F73]; on 24 Alta [F71].

Endocronartium sp.: on 21 Ont [1828].

E. harknessii (Peridermium h.) (C:201):
gall rust, rouille-tumeur: on 2 Mack,
Alta [F60], Mack, Alta, Man, Ont [F72],
Alta [F67], Alta, Sask, Man, Que [F73],
Sask [1997, F65], Sask, Man [F63, F64,
F67], Man [224, F62], Ont [F69, F75,
F76], Que [733, 946, F65, F70, F71,
F76], NB [733, 1863, F70], NB, NS [F69];
on 2, 4 Yukon, Alta, Man [F70], Mack,
Alta [53], Mack, Sask [735], Alta [739,
1830], Que [F68]; on 2, 4, 10, 21, Que
[F67]; on 2, 4, 18, 21 Alta, NB [735];

on 2, 4, 21 Que, NS [F68], NB, NS [F68];
on 2, 10 NB [F74]; on 2, 21 Man, Ont
[735, F68], Ont [1828, F70], Que [F60,
F66, F72, F74, F75], NB, NS [F67]; on 2,
4a Alta [1514, 1834]; on 2, 4a, 21 Que
[1210]; on 4 Yukon [460, 735], BC [954,
F69], Alta [950, 951, 952, F63]; on 4,
11, 12, 15, 16, 17 BC [735, 1012]; on 4,
16 BC [F66]; on 4, 16, 21 BC [2006]; on
4, 21 Alta [F64]; on 4, 24 BC [F68]; on
4a Mack, BC, Alta [F72], BC [738, 1409,
F73, F74, F75, F76], Alta [733, 1400,
1402, 1408]; on 10 BC, NB [1410]; on 10,
21, 23 BC [1012]; on 12 BC [F62]; on 15,
16, 21 BC [F61]; on 18 BC [735]; on 21
Man [F61, F69, F74], NB [F71], NS [F73,
F75]; on 23 Mack [53], Alta [F60, F69];
on 24 Yukon, NWT, BC, Alta, Sask, Man,
Ont, Que, NB, NS [2008], Yukon, Mack,
Alta [F69], Alta [57, F65, F68], Ont
[F71, F74].
Endophragmia nannfeldtii: on 2 Man [1760].
Enerthenema papillatum: on 4 NWT-Alta
    [1063].
Epicoccum sp.: on 18 Ont [1827].
E. purpurascens: on 4a Alta [746].
Erinellina rhaphidospora (C:202): on 9 BC
    [1012].
Europhium sp.: on 4a BC [1450, 1481].
E. aureum: on 4 BC [1483].
E. clavigerum: on 4, 16 BC [1483]; on 4a
    BC [1449, 1632], Alta [1633].
E. trinacriforme (C:202): on 9 BC [1012,
    1483].
Exidia saccharina (C:202): on 4 NWT-Alta
    [1063]. Lignicolous.
Fibricium rude (Peniophora greschikii):
    on 2 NWT-Alta [1063]. See *Abies*.
*Fomitopsis cajanderi (Fomes c., F.
    subroseus)* (C:202): on 2 Mack [1110],
    Mack, Alta [53], Ont [88], on 2, 4
    NWT-Alta [1063]; on 2, 24 Ont [1341]; on
    4 BC [1012]; on 4a Alta [953, 1485]; on
    20, 24 NB [1032]; on 21 Alta [F66]. See
    Abies.
F. officinalis (Fomes o.) (C:202): on 16
    BC [1012]. See *Abies*.
F. pinicola (Fomes p.) (C:202): on 2 Mack
    [53]; on 2, 4 NWT-Alta [1063]; on 2, 18,
    20 Ont [1341]; on 4, 9, 16 BC [1012]; on
    4a Alta [1485]; on 20 NB [1032]; on 21
    Que [F72]; on 24 Que [1377], NB, NS, PEI
    [1032]. See *Abies*.
F. rosea (Fomes r.): on 2, 18, 20 Ont
    [1341]; on 4 NWT-Alta [1063]. See
    Abies.
Foveostroma abietinum (Gelatinosporium a.)
    (holomorph *Dermea balsamae*): on 20
    Ont [1828].
Fumago sp.: on 2 Ont [91]; on 4 NWT-Alta
    [1063]; on 20 Ont [1828].
Fusarium sp. (C:202): on 4 BC [1012]; on
    4a Alta [746]; on 12 NB [F64]; on 18 Que
    [1748], NB [F72], NS [F68]; on 20 NB, NS
    [1032]; on 21 Sask [1848].
F. acuminatum: on 18 Ont [1340].
F. decemcellulare: on 18, 20 Ont [1828];
    on 18 Ont [F70]. This is a tropical
    species which, previously, was recorded
    only as far north as Oklahoma.

F. oxysporum (C:202): on 4a Alta [746];
    on 18 NB, NS [1885]; on 21 NS [1885].
F. solani (holomorph *Nectria
    haematococca*): on 18 Ont [1827].
Fusicoccum sp.: on 18 Ont [1827].
F. baccillare: on 4 Sask [F66].
Fusidium sp.: on 2 Ont [91].
Ganoderma applanatum (C:202): on 21 BC
    [1012, F60]. See *Abies*.
G. lucidum: on 20 NS [1032]. See *Abies*.
Gelatinosporium sp.: on 20 Ont [1340],
    Que [1685, 1690], NS [1032].
*G. pinicola (Cryptosporium p., Micropera
    p.)*: on 6 Que [F67]; on 9 BC [1012].
Geotrichum candidum: on 4 NWT-Alta [1063].
Gliocladium sp.: on 18 Ont [1827].
G. viride: on 2 Ont [830].
*Gloeocoryneum cinereum (Coryneum c.,
    Leptomelanconium c.)*: on 2 Sask, Man
    [F65]; on 4 NWT-Alta [1063], BC [839,
    953], Alta [952, 1128, F64, F65, F66];
    on 4, 16 BC [1012, 1128]; on 4a, 16 BC
    [F72].
*Gloeocystidiellum furfuraceum (Corticium
    f.)* (C:201): on 4 BC [968]; on 9 BC
    [1012]; on 24 Ont [1341]. Lignicolous.
Gloeophyllum odoratum (Trametes o.)
    (C:205): brown cubical rot, carie brune
    cubique: on 2, 4 NWT-Alta [1063]; on 4
    BC [1012]; on 18, 20, 24 Ont [1341]; on
    24 Alta [1341].
G. protractum (Trametes p.): on 2 Alta
    [53]. Perhaps a specimen of *G.
    odoratum*.
G. sepiarium (Lenzites s.) (C:203): on 2
    Mack [1110], Alta [53], NB [1032]; on 2,
    4 NWT-Alta [1063]; on 2, 20, 24 Ont
    [1341]; on 4, 9, 16 BC [1012]; on 4a
    Alta [999, 1000, 1485]; on 20 Ont
    [1963]; on 20, 24 NS [1032]. See
    Abies.
Gloeoporus dichrous (Polyporus d.)
    (C:204): on 2 Ont [1341]; on 4, 16 BC
    [1012]; on 24 NS [1032]. See *Acer*.
Glomerella cingulata (anamorph
    Colletotrichum gloeosporioides): on
    4a BC [F71].
Gorgoniceps ontariensis (C:202): on 18
    Ont [1340].
Grandinia sp.: on 2 NWT-Alta [1063].
Graphium sp.: on 18 Ont [1340].
*Gremmeniella abietina (Scleroderris
    lagerbergii)* (anamorph *Brunchorstia
    pinea*): on 1, 4, 16 BC [1012]; on 1,
    16 BC [731, F75]; on 2 Que [942, F71],
    NB [414, F74, F75]; on 2, 4, 18 Que
    [414]; on 2, 4, 18, 20, 21 Que [F68]; on
    2, 4a, 18, 20, 21 Ont [1828], Que
    [1682]; on 2, 12, 18 Ont [414, F72], NB
    [1063]; on 2, 18 Ont [1827, F65, F68,
    F69, F70, F71, F73, F76], Que [1218,
    1594, F67, F74], NB [F72, F73]; on 2,
    18, 20, 21 Ont [1418, F66]; on 2, 18, 21
    Ont [1340, F74], Que [943, F69, F70,
    F72, F76]; on 4 Alta [412]; on 4a BC
    [F76], Alta [731, F74]; on 12 Ont
    [1828], Nfld [1738]; on 18 Ont [409,
    410, 411, 413, 415, 1417, 1660, 1676],
    Que [916, 1384, F75], NS [F72]; on 18,
    20, 21 NS [F73]; on 18, 21 Ont [F67], NB

[F76], NS [F74]; on 20 Ont [414, 415, 1340], Que [F73], NB [1033]; on 21 Ont [F72], NB, Que [F71], NB [1033, 1035]; on 24 Ont [1594], Que [F75]. The European race was first discovered in Canada in Que in 1978. Subsequent surveys detected it in other pine plantations in Que, in two forest nurseries in NB and on ornamental pines in Nfld.

Griseoporia carbonaria (Trametes c.): on 2 Man [579]. On charred wood, associated with a brown rot.

Guepiniopsis buccina (G. torta): on 4 NWT-Alta [1063]. Lignicolous.

G. minuta: on 4 BC [1012].

Gyromitra esculenta: on 4 NWT-Alta [1063]. Presumably on rotted wood.

Gyrostroma sp.: on 4 NWT-Alta [1063].

Haematostereum sanguinolentum (Stereum s.) (C:205): on 2 Ont [88], NB, NS [F62]; on 2, 4 NWT-Alta [1063], Alta [53]; on 2, 18 PEI [F62]; on 2, 20 NB [1032]; on 4, 9, 16 BC [1012]; on 4a Alta [999, 1000, 1485]; on 9 BC [1341]; on 18, 20 PEI [1032]; on 18, 20, 24 Ont [1341]; on 20 Ont [1963], Que [1377]. See *Abies*.

Halonectria milfordensis: on wood of 9 in water BC [782].

Hansenula capsulata: on 4a BC [1481], Alta [1482]. Probably from frass in insect galleries in wood.

H. holstii: on 4a BC [1012, 1481], Alta [1482]. Probably from frass in insect galleries in wood.

Hapalopilus nidulans (Polyporus n.): on 4a NB [1032]. See *Abies*.

H. salmonicola (Poria s.): on 2 Mack [1110]; on 20 Ont [1341]. Associated with a brown rot.

Hemiphacidium longisporum: on 4 BC [1012, 2009].

H. planum: on 1, 9 BC [1012].

Hendersonia sp.: on 2 Ont [1340]; on 4 NWT-Alta [1063]; on 9 BC [1012]. See comment under *Abies*.

H. acuum: on 4 NWT-Alta [1063]. See comment under *Abies*.

H. pinicola (C:202): on 2 Man [F67], Man, Ont [F66], Que [1063], NB, NS [F75]; on 4 Yukon [460], Yukon, BC [F68], Yukon, Alta [F64], NWT-Alta [1063], BC [953, 1012, 2009], Alta [951, 952, F67]; on 4a BC [F73]; on 10 NB [F74]; on 24 Alta [F73]. See comment under *Abies*.

Herpotrichia coulteri (Neopeckia c.) (C:203): on 1 Alta [F61]; on 1, 4 BC [1012]; on 1, 6 NWT-Alta [1063]; on 4 BC [F69], Alta [F66]. Brown felt blight develops on foliage under snow at high elevations.

H. juniperi: on 4 BC [1012]. See above.

Heterobasidion annosum (Fomes a.) (C:202): fomes root rot, maladie du rond: on 2 Ont [F72]; on 4 BC [1916]; on 2, 18 Ont [338, 1341, F65, F66, F67, F69]; on 18 Ont [1462, 1828, F63, F64, F73]; on 18, 20, 21 Ont [F68]; on 20 Ont [1341, F69]; on 21 Ont [F67, F69]; on 24 Ont [649]. See *Abies*.

Heteroconium chaetospira (Septonema c.) (C:204): on 4 NWT-Alta [1063], Alta [182].

Heyderia abietis: on fallen needles of 24 Que [1385].

Hohenbuehelia mastrucata (Pleurotus m.): on 4 NWT-Alta [1063].

Hormodendrum sp.: on 4 Alta [1032].

Hormonema dematioides: on 4 NWT-Alta [1063].

Humicola alopallonella: on wood of 9 in water BC [782].

Hyaloscypha stevensonii: on 20 Ont [1340].

Hyphoderma argillaceum (Peniophora a.): on 20 Ont [1341, 1660]. See *Betula*.

H. setigerum (Peniophora aspera) (C:203): on 4 NWT-Alta [1063]; on 20 NB [1032]. See *Abies*.

H. ?tsugae (Corticium t.): on 2 NWT-Alta [1063]. Lignicolous.

Hyphodontia arguta: on 4 NWT-Alta [1063]. See *Abies*.

H. aspera: on 18 Ont [1341]. Associated with a white rot.

H. barba-jovis: on 2 Ont [1341]. See *Abies*.

H. breviseta: on 20 NB [1032]. See *Abies*.

H. crustosa (Odontia c.): on 2 NWT-Alta [1063]. See *Abies*.

H. hastata (Peniophora h.) (C:203): on 4 NWT-Alta [1063]; on 16 BC [1012, 1341]. Lignicolous.

H. pallidula (Peniophora p.) (C:203): on 20 NB [1032, F62]. See *Abies*.

H. spathulata (Odontia s.): on 20 Ont [1341, 1660], NB [1032]. See *Abies*.

H. subalutacea (Peniophora s.): on 4 NWT-Alta [1063]. See *Abies*.

Hypochnopsis mustialaensis (Coniophora m.): on 20 Ont [1341]. Typically on rotted wood.

Hypocrea gelatinosa: on 2 Ont [830].

H. rufa: on 4 NWT-Alta [1063].

Hypodermella sp.: on 4 NWT-Alta [1063].

Irpex lacteus (Polyporus tulipiferae): on 4 NWT-Alta [1063]. See *Acer*.

Ischnoderma resinosum (Polyporus r.) (C:204): on 9 BC [1012]. See *Abies*.

Junghuhnia collabens (Poria rixosa): on 2 Ont [1341]; on 4 NWT-Alta [1063], BC [1012]. See *Abies*.

J. luteoalba (Poria l.): on 2, 4 Alta [579, 1004]; on 18 Ont [1004]. See *Abies*.

Kabatiella sp.: on 4 NWT-Alta [1063].

Kuehneromyces vernalis: on 20 NB, NS [1032]. See comment under *Betula*.

Lachnellula agassizii (Dasyscyphus a.) (C:202): on 1, 4, 9, 17 BC [1012]; on 2 Ont [1340, 1828]; on 4 NWT-Alta [1063], Alta [952]; on 20 Que [1685, 1690], NB, NS, PEI [1032]; on bark and dead wood of 24 Que [1385].

L. arida: on 1, 4 BC [1012]; on 2 Que [1690]; on 4 NWT-Alta [1063], Alta [952].

L. calyciformis (Dasyscyphus c.): on 2 Ont [1340]; on 21 Que [F66].

L. chrysophthalma (C:203): on 2 Ont [F63]; on 2, 21 Ont [1340]; on 4 NWT-Alta [1063].

L. flavovirens: on 1, 4 BC [1012, F68].

L. fuscosanguinea (Lachnella f.) (C:203):
on 4 BC [1012]; on 6 Alta [950].

L. occidentalis (L. hahniana):   on 2 Mack,
Alta [53], Que [F68, F70]; on 2, 4
NWT-Alta [1063], Que [1690].

L. pini (Dasyscyphus p.):   on 1 BC [F65];
on 1, 6 NWT-Alta [1063]; on 1, 9 BC
[1012].

L. suecica: on 2 Que [F70]; on 2, 21 Que
[1690]; on 4, 9 BC [1012]; on 21 Ont
[F65].

*Laeticorticium lundellii (Dendrocorticium
l.)*:   on 4 Alta [930]. Lignicolous.

L. minnsiae (Aleurodiscus m.) (C:200):   on
4 NWT-Alta [1063]; on 16 BC [930].
Lignicolous.

L. pini (Aleurodiscus p.) (C:200):   on 2
NWT-Alta [1063]; on 20 Ont [930, 1341],
NB [1032]; on 24 Que [969]. Lignicolous.

Lecanosticta sp.:   on 6 Alta [950].

L. acicola (holomorph *Mycosphaerella
dearnessii*):   on 2, 4a Man [934]; on
needles of 24 Man [1763].

Lentinus lepideus (C:203):   brown cubical
rot, carie brune cubique: on 9 BC
[1012]. Now in *Neolentinus*.

Leocarpus fragilis: on 4 NWT-Alta [1063].

Leptographium sp. (C:203):   on 4 NWT-Alta
[1063], Alta [952]; on 4a BC [1481],
Alta [1482]; on 18 Ont [1827, 1828].

L. lundbergii: on 4 NWT-Alta [1063].

Leptostroma strobicola (holomorph
Meloderma desmazerii):   on 20 Ont
[1120].

Leptostromella sp.:   on 2 Ont [F74].

Leptomelanconium allescheri: on brown
needles of 9 BC [1764].

*Leucogyrophana mollusca (Merulius m., M.
fugax* auct.*)* (C:203):   on 2 Mack
[1110]; on 4 NWT-Alta [1063]; on 9 BC
[1341]; on 9, 16 BC [1012]; on 24 BC,
Ont [571]. See *Abies*.

L. pinastri: on 20 Que [571]; on 24 Ont
[571], Que [581]. See *Abies*.

L. pulverulenta: on old stump of 24 NS
[571]. Associated with a brown rot.

L. romellii: on 9 BC [571]. See *Abies*.

Leucostoma sp.:   on 6 Que [F67].

L. kunzei (Valsa k.) (anamorph *Cytospora
k.*) (C:205):   on 2, 20 Que [F72, F73];
on 20 Ont [1340, 1828, F70], Que [1685],
NB [F62], NB, NS [1032].

Libertella sp.:   on 9 BC [1012].

Linodochium hyalinum: on 2 Sask, Man
[1760]; on 4a Sask [1760]; on 9 BC
[1012].

Lophium mytilinum (C:203):   on 2 Man
[F66], Ont [F60]; on 2, 20 Ont [1340];
on 4 NWT-Alta [1063]; on 4, 9 BC [1012].

Lophodermella sp.:   on 9 BC [1012].

L. arcuata: on 1 BC [1012, F68], Alta
[F72].

L. concolor (Hypodermella c.) (C:203):
needle cast, rouge: on 2 Ont [1063,
1377], Que [F65, F66]; on 2, 4 Yukon,
Alta [F68]; on 2, 24 Mack [F69]; on 4
Yukon [460], NWT-Alta [1063], Mack, Alta
[53, F66], BC [839, 953, 1340, 2009,
F68, F69], BC, Alta [F67], Alta [950,

951, 952, F63, F64, F65]; on 4a BC, Alta
[F73], BC, NB [F76], Alta [F74], Alta,
Sask [F72]; on 4a, 21 BC [1012]; on 23
BC [F70]; on 23, 24 Alta [F69].

L. montivaga (Hypodermella m.) (C:203):
needle cast, rouge: on 4 Yukon [460,
F68], NWT-Alta [1063], BC [953], Alta
[952, F60, F63, F65]; on 4a BC [F76]; on
4a, 23 BC [1012]; on 23, 24 Alta [F70];
on 24 Yukon [F69].

Lophodermium sp.:   on 2, 18 NB [1032]; on
1, 16 BC [F63]; on 20 NB, NS [1032]; on
21 Nfld [F70]; on 21, 24 Nfld [1660,
1661].

L. molitoris: on 2 Sask, Ont [1120].

L. nitens (C:203):   needle cast, rouge:
on 1 Mack, Alta [F67], Alta [952, F74];
on 1, 6 Alta [951]; on 1, 9 BC [1012];
on 2, 20 Ont [1340]; on 6 NWT-Alta
[1063]; on 9 BC [922]; on 18, 20 Ont
[F75]; on 20 Ont [974, 1063, 1120,
1131], Ont, Que, NB [922], Que [1377],
NB, NS [1032], Nfld [1659, 1660, 1661].

L. pinastri (C:203):   needle cast, rouge:
on 2 Mack, Alta [53, F67], Mack, Sask
[F65], Mack, Man [F63], Man [F64, F69],
Ont [1132, 1637], NB [922, 1885, F73];
on 2, 4 NWT-Alta [1063], Alta [F62],
Sask, Man [F66]; on 2, 4, 7, 9, 10, 11,
12, 15, 16, 17, 18, 21, 24 BC [1012]; on
2, 9, 10, 23 Que [F67]; on 2, 16 Alta
[922]; on 2, 18 Ont [F69], Ont, Nfld
[F76], Que [F75], NS [F73]; on 2, 18, 20
Ont [1340], NB [1032]; on 2, 18, 20, 21
Ont [F73], Que [1377], NB, NS, PEI
[F64], NS [1032], Nfld [1660, 1661]; on
3 Man [F68]; on 4 BC [953], Alta [951,
952, F60, F64, F65, F66], Sask [F68]; on
4a, 18, 20 PEI [1032]; on 9 BC [922];
12, 18 Ont [1828]; on 17 BC [F70]; on 18
Ont [922, 1827, F70, F72], Ont, Que, NS
[F74], Que [F68], NS [1885, F71], PEI
[F62], Nfld [F69]; on 18, 21 BC [F64],
Que [922, F76]; on 20 Ont [974], Nfld
[1659]; on 21 BC [1140, 1416], NS [F63],
Nfld [F73, F74, F75]; on 21, 24 BC
[F68]; on 24 Yukon, Alta [F69], Nfld
[1661].

Lophophacidium hyperboreum: on 20 Que
[1692] by inoculation.

Lulworthia medusa: on wood of 9 in water
BC [782].

Marasmiellus filopes: on 4 BC [1435]; on
20 Ont [1435].

Marasmius sp.:   on 4 NWT-Alta [1063].

Melampsora medusae (M. albertensis)
(C:203):   rust, rouille: on 2, 4, 16,
17, 24 BC [F62]; on 2, 4a, 8, 16, 17 BC
[2002] by inoculation; on 2, 4a, 8, 16,
17, 21 BC [1012]; on 2, 4a, 16, 18, 21
BC [1125] by inoculation; on 4 BC [F69];
on 16 BC [F63]; on 24 BC [F65]. See *M.
pinitorqua* below.

M. occidentalis (C:203):   rust, rouille:
on 4, 16, 24 BC [F62]; on 4a, 8, 9, 16,
17, BC [1012, 2002 by inoculation]; on
4a, 16 BC [1125] by inoculation; on 24
BC [F65].

M. pinitorqua (C:203):   rust, rouille: on
10, 16, 21 BC [2001] by inoculation; on

16 BC [F60, F61]. Shown to be
misdeterminations [1125]. The correct
identification is *M. medusae*.

Melanconium sp.: on 1, 9 BC [1012].

Meliola sp.: on 4 NWT-Alta [1063].

Meloderma desmazierii (Hypoderma d.)
(C:202): needle cast, rouge: on 2, 18,
20, 21 Que [1377]; on 9 BC [1012]; on 20
Ont [974, 1063, 1827, F68], Que [F67],
NB [F63], NB, NS [1032].

*Meruliopsis albostramineus (Ceraceomerulius
a.)*: on 9 BC [568, 1012]. See *Abies*.

M. ambiguus (Merulius a.) (C:203): on 2
Ont, Que [568], Que [F68]; on 2, 4
NWT-Alta [1063]; on 4, 16 BC [568,
1012]; on 24 Alta [568]. See *Alnus*.

M. corium (Merulius c.): on 2 Ont [1341];
on 4 BC [1012]. See *Acer*.

M. taxicola (Poria t.) (C:204): on 2 Alta
[53], Que [568]; on 2, 4 NWT-Alta
[1063]; on 4 BC [1012]; on 24 Alta
[568]. See *Abies*.

Merulius sp.: on 4 NWT-Alta [1063]; on 20
Ont [1341].

Micropera sp.: on 6 Que [F67]; on 20 Ont
[1340]. These specimens are probably a
species of *Foveostroma*.

Mollisia cinerea: on 4 NWT-Alta [1063].

M. fallax: on 21 Ont [1340].

Monilia sp.: on 20 Ont [1827].

Monocillium sp.: on 2 Ont [91].

M. nordinii (Cephalosporium n.): on 4
NWT-Alta [1063], Alta [183].

Monodictys pelagica: on wood of 9 in
water BC [782].

Mortierella ramanniana (Mucor r.): on 2
Ont [91].

Mucronella calva (M. aggregata): on 4
NWT-Alta [1063]. See *Abies*.

*Myceliophthora thermophila (Sporotrichum
t.)*: on 24 BC [1723].

Mycena alcaliniformis: on carpet of
needles of 20 Ont [1385].

M. psammicola: on needles (?fallen) of 24
Ont [1385].

M. strobilinoides: on needles (?fallen)
of 24 Que [1385].

M. subincarnata: on needles (?fallen) of
20 Ont [1385].

Mycosphaerella sp.: on 20 Ont [1828].

*M. dearnessii (Scirrhia acicola, Systremma
a.)* (anamorph *Lecanosticta a.*): on
2, 4 Man [F65, F66]; on 2, 4a Man [338];
on 4 Man [F68].

M. pinicola: on 2 Man [F66]. Presumably
referring to *Sphaerella pinicola*. The
combination in *Mycosphaerella*
apparently has not been published.

M. tassiana: on 4 NWT-Alta [1063].

Mytilidion sp.: on 1 BC [1012]; on 2
NWT-Alta [1063], Mack [53].

M. gemmigenum: on 4 BC [1012], Sask [F66].

Myxosporella sp.: on 4 NWT-Alta [1063].

Naematoloma sp.: on 4 BC [1012]. The
specimen is probably referrable to
Hypholoma.

Nectria sp.: on 4 BC [1012]; on 20 NB, NS
[1032].

N. fuckeliana: on 2 Que [1525] by
inoculation; on 4a, 18, 20, 21 Que
[1685] by inoculation; on 17 BC [F65].

N. neomacrospora (N. fuckeliana var.
macrospora): on 9 BC [F71]; on 9, 17
BC [1012]; on 24 BC [176].

N. peziza: on 20 Ont [1340].

N. sanguinea (C:203): on 24 NS [1032].

Neournula nordmanensis: on 9 BC [1224].

Nodulisporium sp.: on 20 Ont [974] by
inoculation.

Odontia sp.: on 9 BC [1012]; on 20 Ont
[1341].

O. crustula: on 2 NWT-Alta [1063].
Lignicolous.

Odonticium romellii (Odontia r.): on 2
NWT-Alta [1063], Alta [53, 1341].
Apparently associated with a white rot.

Odontotrema hemisphaericum: on 2, 21 Que
[1690].

Oidiodendron tenuissimum (C:203): on 4
Alta [82].

Oidium corticale: on 4 NWT-Alta [1063].

*Onnia circinata (Polyporus c., P.
tomentosus* var. *c.)* (C:204): red
butt rot, carie rouge alvéolaire du
pied: on 4, 9 BC [1012]; on 20 Nfld
[1661]; on 24 Que [1377]. See *Abies*.

O. tomentosa (Polyporus t., P. t. var.
tomentosus) (C:204): red butt rot,
carie rouge alvéolaire du pied: on 2
Ont [88, 1964]; on 4, 16 BC [1012]; on
4a Alta [1002, 1485]; on 18 Que [F69];
on 20 Ont [1963]; on 20, 21 Ont [1828].
See *Abies*.

Orbilia botulispora (O. paradoxa): on 20
Ont [1340].

O. pannorum: on 20 Ont [1340]. Doubtful
record, specimens should be restudied.

O. picea: on 20 Ont [1340]. Doubtful
record, specimens should be restudied.

Ostreola sessilis: on 20 Que [359].

Paecilomyces varioti (C:203): on 2 Ont
[830]; on 4 NWT-Alta [1063], Alta [182,
150].

*Parmastomyces transmutans (Polyporus
subcartilagineus)*: on 2 Alta [579].
See *Abies*.

Peltosphaeria canadensis: on 12, 15, 21
BC [1012].

Penicillium sp.: on 2 Ont [830]; on 2,
18, 20, 21 Ont [1964, 1966]; on 18 Que
[1748].

P. aurantiogriseum (P. cyclopium): on 4a
Alta [746].

P. glabrum (P. frequentans) (C:203): on 4
Alta [182].

Peniophora sp.: on 2 Ont [830]; on 4 Alta
[952]; on 4, 24 NWT-Alta [1063]; on 20
Ont [1341].

P. cinerea (C:203): on 20 Ont [1963].
See *Abies*.

P. incarnata (C:203): on 4 NWT-Alta
[1063]. Lignicolous.

P. piceae (P. separans) (C:204): on 4
NWT-Alta [1063]. See *Abies*.

P. pini (Sterellum p., Stereum p.): on 4
Alta [182]; on 24 Que [969]. See
Abies.

P. pini ssp. *duplex (P. duplex)*
(C:203): on 2 Ont [1341, F67]; on 2,
18, 21 Ont [1941]; on 20 Ont [1827].

P. pseudopini (C:204): on 2 Mack, Alta
[53, F60], Alta [F61], Ont [88, 91, 89],
on 2, 4, 24 NWT-Alta [1063]; on 2, 18,
20, 21, 24 Ont [1941]; on 4 BC [1012,
1941], Alta [53, F60]; on 4a Alta [1002,
999, 1485, 1941]; on 20 Ont [1341], Que
[1941]; on 23 Alta [F60]. See *Abies*.

P. septentrionalis (C:204): red heart
rot, carie rouge du coeur: on 2, 20 Que
[F64]; on 4 NWT-Alta [1063], BC [1012],
Alta [952]. See *Abies*.

P. violaceolivida: on 2 Alta [53]; on 2,
4 NWT-Alta [1063]. Lignicolous.

Peridermium stalactiforme (holomorph
Cronartium coleosporioides) (C:202):
stem rust, rouille du tronc: on 2 Sask
[1997], Sask, Man [F64], Que [F68]; on 4
BC [F68], BC, Alta [F67], Alta [F60,
F62, F66], Man [F69]; on 4a Alta [738,
1398, 1409]; on 23 Alta [F61]; on 24
Alta [57, F68], Que [F71].

Pestalotia sp.: on 12 Ont [1340].
Probably referable to *Truncatella* or
Pestalotiopsis.

Pestalotiopsis funerea (*Pestalotia f.*):
on 2 Que [F72, F73]; on 2, 18 Que
[F76]; on 18 NS [F71].

Pezicula livida (anamorph
Cryptosporiopsis abietina): on 2 Ont
[1827]; on 2, 4, 21 Que [1690]; on 2, 18
Ont [1340].

Pezizella chapmanii: on 4, 16 BC [1012,
1946].

Phacidiopycnis sp.: on 4 BC [1012].

P. pseudotsugae (*Phomopsis strobi*)
(holomorph *Potebniamyces coniferarum*)
(C:204): on 4 Sask [F65]; on 20 Ont
[1828, F70], NB [F63], NB, NS [1032]; on
20, 21 Que [1679].

Phacidium abietis: on 18 Que [F75]; on 20
Que [1677] by inoculation.

Phaeolus schweinitzii (*Polyporus s.*)
(C:204): brown cubical rot, carie brune
cubique: on 2 Ont [88]; on 4, 9, 16 BC
[1012].

Phaeosclera dematioides: on 4 Alta [1636].

Phaeoseptoria contortae: on 4 Alta [950,
1285, F70].

Phanerochaete laevis (*Peniophora
affinis*): on 4 NWT-Alta [1063]. See
Acer.

P. burtii (*Peniophora b.*): on 20 NB
[1032]. Lignicolous.

P. sanguinea (*Peniophora s.*): on 2, 4
NWT-Alta [1063]; on 18 Ont [1341]. See
Abies.

P. velutina: on 2 Alta [53]. See *Acer*.

Phellinus pini (*Fomes p.*) (C:202): red
ring rot, carie blanche alvéolaire: on
1 Alta [951, F63]; on 1, 2, 4, 24
NWT-Alta [1063]; on 2 Mack [F60], Mack,
Alta [53], Ont [88, 91]; on 2, 10, 18,
20, 21 Que [1377]; on 2, 18, 20 Ont
[1341]; on 4 Alta [950, 952, 1485]; on
4, 9, 16 BC [1012]; on 4a Alta [1001,
1002]; on 20 NB [1032]. See *Abies*.

P. viticola (*Fomes v.*): on 4 NWT-Alta
[1063]. See *Abies*.

P. weirii: on 4, 9, 16 BC [1012]. See
Abies.

Phialocephala fusca (C:204): on 20, 24
Ont [877].

Phialophora sp.: on 2 Ont [91]; on 20 Ont
[974]; on 24 Ont [1827].

P. botulispora: on 2, 24 Ont [279].

P. fastigiata: on 2, 20, 24 Ont [279].

P. heteromorpha (C:204): on railroad tie
of 2 Ont [1918].

P. lagerbergii: on 2 Ont [279].

P. melinii: on 2 Ont [279].

P. richardsiae: on 24 Ont [279].

P. verrucosa: on 2 Ont [279].

Phlebia sp.: on 20 Ont [1341].

P. centrifuga (*P. albida* aucts.): on 2
Ont [1341]. See *Abies*.

P. flavoferruginea (*Peniophora f.*): on 20
Ont [1341]. Lignicolous.

P. phlebioides (*Peniophora p.*) (C:204):
on 2 Ont [91]; on 4, 16 BC [1012]; on 4,
24 NWT-Alta [1063]; on 4a Alta [999,
1000, 1485]; on 20 Ont [1963]. See
Abies.

P. radiata (*P. merismoides*) (C:204): on 4
NWT-Alta [1063], BC [1012], BC, Alta
[568]; on 4a Alta [1485]. See *Abies*.

P. subserialis (*Peniophora s.*): on 2 Mack
[1110]. Associated with a white rot.

Phlebiopsis gigantea (*Peniophora g.*)
(C:203): white sap rot, carie blanche
de l'aubier: on 2 Ont [88, 1963], NB
[1032]; on 2, 18, 20 Ont [1341]; on 4
NWT-Alta [1063]; on 4, 9, 16 BC [1012];
on 4a Alta [999, 1485]; on 24 Ont
[1827]. See *Abies*.

Pholiota alnicola (*Flammula a.*) (C:202):
yellow checked rot, carie jaune
craquelée: on 2 Ont [88]; on 4 NWT-Alta
[1063], BC [1012]; on 4a Alta [999,
1002, 1485]. See *Abies*.

Phoma sp.: on 2 Ont [91]; on 4 NWT-Alta
[1063], Alta [952]; on 4a Alta [746]; on
wood of 9 in water BC [782]; on 18 Ont
[1827], Que [1748]; on 20 Ont [973,
974], NB [1032].

P. glomerata (C:204): on 4a Alta [746].

Phomopsis sp.: on 4a, 17 BC [1012]; on
18, 20 Ont [1827]; on 20 Ont [1828], Que
[1377], NB, NS [1032].

Phytophthora cactorum: on 21 Sask [1848].

Phycomyces blakesleeanus: on 4 BC [1012].

Pichia pini: on 4a BC [1012, 1481], Alta
[1482].

Piloderma bicolor (*Corticium b.*): on 4
NWT-Alta [1063]. See *Abies*.

Piloporia albobrunnea (*Poria a.*) (C:204):
on 4 BC [1012], NWT-Alta [1063]. See
Picea.

Platygloea arrhytidiae: on 4 NWT-Alta
[1063].

Pleurocybella porrigens (*Pleurotus p.*)
(C:204): on 24 NB [1032]. Lignicolous.

Pleurotus atrocaeruleus var. *griseus*:
on 4 NWT-Alta [1063]. A species of
Hohenbuehelia.

Pluteus atromarginatus: on stumps of 20
Que [1385].

P. fuliginosus: on 20 Que [1385].

Polyporus sp.: on 20 Nfld [1659, 1660,
1661].

Poria sp.: on 4 NWT-Alta [1063].

P. byssina: on 4 NWT-Alta [1063]. These
records are probably collections of
Flaviporus semisupina.

P. subacida (C:204): on 2 NWT-Alta
[1063], Alta [53], Ont [88]; on 2, 18,
20 Ont [1341]; on 4, 9, 21 BC [F62]; on
24 Que [1377]. See *Abies*.

P. tenuis var. *pulchella*: on 16 BC
[1012]. Associated with a white rot.

Potebniamyces coniferarum (anamorph
Phacidiopycnis pseudotsugae): on 2
Que [F75]; on 2, 20, 21 Que [F68]; on
2, 21 Que [1690]; on 20 Que [1685]; on
21 Que [1679]; on 20, 21 Que [F69].

Pragmopora amphibola: on 2, 21 Que
[1690].

P. pithya (anamorph *Pragmopycnis p.*):
on 20 Que [1690, F68].

Propolis sp.: on 16 BC [1012].

P. leonis (C:204): on 4 Alta [644].

P. rhodoleuca (C:204): on 2, 21 Ont
[1340, F60]. A form on pine cones that
is doubtfully distinct from *P.
versicolor*.

P. rhodoleuca var. *strobilina*: on 4
NWT-Alta [1063], Alta [951]. See above.

Pseudocenangium succineum (holomorph
Phialea fumosella): on needles of
24 NB [1763].

Pseudohelotium pineti (*Belonium p.)*: on 4
NWT-Alta [1063].

Pseudohiatula conigenoides: on cones of
20 Que [1385]. Probably a specimen of
Baeospora myosura. See comment under
Picea.

Pseudomerulius aureus (*Merulius a.)*
(C:203): on 2, 20, 24 Ont [568]; on 4
NWT-Alta [1063], Alta [568]; on 20 Que
[568]; on 24 NB, NS [568], NS [1032].
Associated with a brown rot.

Pseudophacidium piceae: on 21 Que [1685].

Pseudotomentella atrofusca: on 20 BC
[1828].

P. tristis (P. umbrina): on 24 Ont [1341].

Pycnoporus cinnabarinus (C:204): on 20 NS
[1032]. See *Acer*.

Pyrenochaeta sp.: on 20 Ont [974].

Pythium sp.: on 16 BC [1012]; on 18 Que
[1748], NB, NS [1885], NS [F68].

P. debaryanum (C:204): on 18 NB [F60]; on
21 Sask [1848].

P. intermedium: on 4a Alta [745].

P. sylvaticum: on 18 Ont, Que, NB [1846].

P. ultimum: on 4a Alta [746]; on 21 Sask
[1848].

Ramaricium albo-ochraceum (*Trechispora
a.)*: on 18 Ont [572, 970].
Lignicolous.

Ramularia sp.: on 4 NWT-Alta [1063].

Resinicium bicolor (*Odontia b.)* (C:203):
white stringy rot, carie blanche
filandreuse: on 4 BC [1012].

Retinocyclus abietis (C:204): on 2 Alta
[53], Ont [88, 89, 91, 1827]; on 2, 4
NWT-Alta [1063]; on 4 BC [1012], Alta
[182], Que [F67]; on 4a Alta [1485].

R. olivaceus: on 2 Que [F71].

Rhinocladiella atrovirens: on 2 Ont [88];
on 4 NWT-Alta [1063].

Rhizina undulata: on 2 Sask [F70], Ont
[F72]. Causes a root-rot, typically
associated with burnt sites. See
Picea.

Rhizoctonia sp.: on 18 Que [1748], NS
[F68].

R. callae: on 21 Sask [1511].

R. endophytica var. *endophytica*
(holomorph *Ceratobasidium* sp.): on 2
Sask [1510].

R. endophytica var. *filicata*: on 2 Sask
[1510].

R. globularis (holomorph *Sebacina* sp.):
on 2 Sask [1510].

R. hiemalis (holomorph is a Discomycete):
on 2 Sask [1510].

R. repens: on 2 Man [1511]; on 21 Sask
[1511].

R. rubiginosa: on 21 Sask [1511].

R. solani (R. praticola) (holomorph
Thanatephorus cucumeris) (C:204):
damping-off, fonte des semis: on 2, 21
Sask [1511]; on 10 NB [F60]; on 18 NB
[F73], NS [1885, F72]; on 21 Sask [1848].

Rhizosphaera sp.: on 9 BC [1012].

Rosellinia sp.: on 18 Ont [1340].

R. thelena: on 4 NWT-Alta [1063], BC
[1012]; on 20 Ont [1340].

Sarcotrochila macrospora: on 4 BC [1012,
2009].

Schizophyllum commune: on 2, 4 NWT-Alta
[1063]; on 2, 18 Ont [1341]; on 4 Alta
[565]. See *Abies*.

Scirrhia pini (anamorph *Dothistroma
septospora*): on 4 BC [523, 953], Alta
[F67]; on 4, 9, 16, 24 BC [F68]; on 4,
9, 11, 12, 15, 16, 17, 23 BC [1012]; on
4a BC [F76]; on 5, 9, 11, 12, 15, 17, 24
BC [F66]; on 11, 15, 17 BC [339]; on 12,
15 BC [F67]; on 17 BC [F69].

Sclerophoma sp.: on 4, 9, 12, 15, 16 BC
[1012].

S. pithya: on 2, 18, 20 Ont [1827]; on 2,
20 Ont [1828]; on 20 Ont [1340, F68]; on
24 Ont [F67].

S. pithyophila (*Phoma acicola, P.
pinicola)* (holomorph *Sydowia
polyspora*) (C:204): on 1, 4a BC [F75];
on 2 Sask [F67]; on 2, 12 Sask, Man
[F68], on 12 BC [F64]; on 18 NS [F71];
on 18, 21 Man [F67]; on 20 Ont [1340,
F69]; on 24 BC [1012, F67], Ont [F68].

*Scoleconectria cucurbitula (Ophionectria
cylindrospora)*: on 2, 4a, 18, 20, 21
Que [1685]; on 2, 4a, 18 Ont [F71]; on
2, 18 Ont [1828, F69], Que [F67]; on 2,
18, 20 Ont [1827, F68]; on 2, 18, 20, 21
Ont [F70]; on 2, 20, 21 Ont [1340]; on
18 Ont [F66, F75]; on 20 Ont [175, F62,
F72, F73], NB, NS [F67], NB, NS, PEI
[1032], NS [F76]; on 24 Ont [F67].

S. scolecospora: on 4 NWT-Alta [1063].

Scytalidium sp.: on 18 Que [354].

S. lignicola: on 2 Ont [88, 830].

Scytinostroma galactina (*Corticium g.)*
(C:201): on 2 Ont [88, F72]; on 4, 9 BC
[1012]; on 18, 20 Ont [1341, 1964]; on
20 Ont [1963, F74], NS [1032]. See
Abies.

Septobasidium pinicola (C:204): on 20 Ont [1341, F60], Que [F63, F64, F67], NB, NS [1032]. No doubt on scale insects on pine.

Septonema chaetospira var. *pini*: on 4 Alta [183]. The species has been transferred to *Heteroconium*, but this variety has not been.

Septoria sp.: on 20 Ont [974] by inoculation.

Serpula himantioides (Merulius h., Serpula lacrimans var. *h.)* (C:203): brown cubical rot, carie brune cubique: on 2 Ont [88]; on 4, 16 BC [1012]; on 4a BC [F74]; on 20 Ont [1963]; on 24 NS [1032].

S. lacrimans: on 16 BC [1012]. See *Abies*.

Sillia sp.: on 18 Ont [1340].

Sirococcus conigenus (S. strobilinus): on 1, 4, 16 BC [1012]; on 2, 18 BC [F76], Ont [F74]; on 4 BC [835]; on 4a BC [F72]; on 7 NS [F75]; on 18 Ont [F73], Ont, Que [F75], Ont, NS [F76], NB, NS, PEI [F74], NS [1034].

Sirodothis sp. *(Pleurophomopsis* sp., *Pleurophomella* sp.*)* (holomorph *Tympanis* sp.): on 4 NWT-Alta [1063].

Sistotrema brinkmannii (Trechispora b.) (C:205): white stringy rot, carie blanche filandreuse: on 4 NWT-Alta [1063]; on 4a Alta [999]; on 18 NB [1032].

S. hirschii (Trechispora h.): on 4 NWT-Alta [1063]. Lignicolous.

Sistotremastrum suecicum: on 18 NS [1032]. See *Abies*.

Skeletocutis amorpha (Polyporus a.) (C:204): on 4 NWT-Alta [1063]; on 4, 9 BC [1012]; on 9 BC [1341]. See *Abies*.

S. alutacea (Poria a.): on 20 Ont [1341]. See *Betula*.

S. subincarnata (Poria s.): on 9 BC [1012]. See *Abies*.

Spadicoides atra: on 2 Man [1760]; on 4 Alta [791].

S. bina: on 18 Ont [443].

Sphaerellopsis filum (Darluca f.) (holomorph *Eudarluca caricis*): on 2 NB [1036].

Sphaeropsis sapinea (S. ellisii, Diplodia pinea, Macrophoma sapinea) (holomorph *Botryosphaeria laricis*): on 2, 4a, 18, 20, 21 Que [1687]; on 2, 16, 18 Ont [F64]; on 2, 18 Ont [F66]; on 2, 24 Ont [1340]; on 4 Que [F67, F69]; on 12 Nfld [F75]; on 16 BC [1012, F76]; on 18 Ont [1828, F73]; on 18, 21 Ont [1827, F74]; on 21 Que [F70].

Sporidesmium larvatum: on 20 Que [824].

Stemonitis fusca (C:204): slime mold: on 2, 4 NWT-Alta [1063]. Presumably on well-rotted wood.

Stereum sp.: on 4 NWT-Alta [1063].

S. hirsutum: on 4 NWT-Alta [1063]. See *Abies*.

Stilbospora sp.: on 4 BC [1012].

Stomiopeltis pinastri: on 9 BC [1012].

Strasseria geniculata: on 18 Que [1278]; on 21 Nfld [1278].

Strobilurus albipilatus: on senescent or buried cones of 24 BC, Ont, Que, NS [1431].

Sydowia polyspora (anamorph *Sclerophoma pithyophila*): on 1, 4, 10, 11, 12, 18, 21 BC [1012]; on 2, 4, 18, 20, 21 Que [1686, F68]; on 11 BC [F71]; on 18, 21 BC [F70]; on 20 Que [1685]; on 21 Nfld [F73].

Taeniolella sp.: on 4 NWT-Alta [1063], Alta [182]; on 16 BC [1012].

Talaromyces emersonii: on 24 BC [1723]. See note under *Picea*.

Tapesia strobilicola: on 4 Yukon [460].

Thelephora sp.: on 2 Alta [1828]; on 2, 20 Ont [1828].

T. terrestris (C:205): on 2 Que [F76]; on 2, 20 Ont [1341]; on 4 BC [1012]; on 4a BC [F71]; on 18 NS [1032]. Causes "smothering disease" of seedlings and forms ectotrophic mycorrhizae with pines.

Thermoascus aurantiacus: on 24 BC [1723].

Thermomyces lanuginosus (Humicola l.): on 24 BC [1723].

Therrya fuckelii (C:205): on 18 Ont [303, 305, 1342a, 1441, 1827, F68], Que [F61]; on 18, 20, 21 Ont [1340]; on 18, 21 Ont, Que [893]; on 21 Ont [1340].

Thyriopsis halepensis: on 2 Que [1207, F65]; on 4 Alta [F67, F69].

Thyronectria balsamea: on 2, 18 Que [F67].

Tomentella sp.: on 4 NWT-Alta [1063].

T. botryoides: on 20 Ont [1341].

T. ferruginea (T. fusca): on 4 NWT-Alta [1063]; on 24 Ont [1341, 1828].

T. molybdaea: on 2 Ont [1827].

T. ramosissima (T. fuliginea): on 2 Alta [1341].

T. ruttneri: on 24 Ont [928, 1341].

T. sublilacina: on 20, 24 Ont [1341]; on 24 Ont [1827].

Tomentellina fibrosa (Kneiffiella f.): on 4 Alta [1828].

Torula herbarum: on 2 Sask [1760].

Trechispora farinacea (Grandinia f.): on 4 NWT-Alta [1063]. See *Abies*.

T. filia (Corticium molliforme): on 4 NWT-Alta [1063]. Lignicolous.

T. pallidoaurantiaca: on 2 Mack [559]. Associated with a white rot.

T. vaga (Corticium sulphureum, Cristella s.) (C:201): on 2 Mack [970], NWT-Alta [1063], Alta [53]; on 9 BC [1012]; on 18, 24 Ont [1341]. See *Abies*.

Tremella sp.: on 4 NWT-Alta [1063].

T. encephala: on 4 NWT-Alta [1063], BC [1012]. See *Abies*.

T. lutescens (C:205): on 4 NWT-Alta [1063], BC [1012]. Lignicolous.

T. mesenterica (C:205): on 4 NWT-Alta [1063], BC [1012]. See *Acer*.

Trichaptum abietinus (Hirschioporus a., Polyporus a.) (C:204): white sap rot, carie blanche de l'aubier: on 2 Mack, Alta [53]; on 2, 4, 21, 24 NWT-Alta [1063]; on 2, 18, 20 Ont [1341]; on 2, 20 Que [1061]; on 2, 20, 21, 24 NB [1032]; on 4 BC [953, 954, 1028], Alta [951]; on 4, 4a Alta [1485]; on 4, 9, 16, 21 BC [1012]; on 4a Alta [999]; on

16 BC [1341]; on 18 Ont [1028]; on 18,
20 PEI [1032]; on 18, 20, 24 NS [1032];
on 19 Que [F65].
T. fuscoviolaceus (Hirschioporus f.): on
2 Que [1028]. See *Abies*.
T. laricinus (Hirschioporus l.): on 20
Ont [1028]. See *Abies*.
Trichia botrytis: on 2 NWT-Alta [1063],
Mack [53].
Trichoderma sp.: on 4 NWT-Alta [1063]; on
4 Que [1748].
T. lignorum: on 4 NWT-Alta [1063].
Probably a specimen of *T. viride*.
T. viride (C:205): on 2 Sask [1760], Ont
[91]; on 2, 18, 20, 21 Ont [1964,
1966]. See *Abies*.
Truncatella hartigii (Pestalotia h.): on
18 Ont [1827].
Tryblidiopsis pinastri: on 20 Ont [1340].
Tubulicrinis angustus: on 4 Que [F67].
Lignicolous.
T. calothrix (Peniophora c.): on 24
NWT-Alta [1063]. Associated with a
white rot.
*T. glebulosus (T. gracillimus, Peniophora
g.)* (C:203): on 4 NWT-Alta [1063]; on
16 BC [1012]. See *Abies*.
T. inornatus (Peniophora i.): on 24 Ont
[1935]. Lignicolous.
T. subulatus (Peniophora s.): on 2 Alta
[53]; on 2, 4 NWT-Alta [1063]. See
Abies.
Tulasnella fuscoviolacea: on 4 NWT-Alta
[1063]; on 20 PEI [1032]. Lignicolous.
T. violea: on 2 Alta [53]; on 2, 4
NWT-Alta [1063]. Lignicolous.
Tylospora fibrillosa: on 2 Ont [1341].
See *Picea*.
Tympanis sp.: on 1, 4 BC [1012]; on 2, 4
NWT-Alta [1063]; on 4 Alta [182], Que
[F67]; on 18 Ont [F60, F75]; on 18, 20
Ont [1340]; on 20 NS [1032].
T. abietina: on 2, 4, 18, 20, 21 Que
[1688]; on 4, 18, 20 by inoculation.
T. confusa (C:205): canker, chancre
typanien: on 2, 4, 18, 20, 21 Que
[1688]; on 4, 18 Que [F68]; on 20 Ont
[1340]; on 24 Que [1690].
T. hansbroughiana: canker, chancre: on
2, 4, 18, 20, 21 Que [1688].
T. hypopodia (T. piceina) (C:205):
canker, chancre: on 2 Ont [88, 89, 91,
1340, F69]; on 2, 4 NWT-Alta [1063]; on
2, 4, 18, 20, 21 Que [1688]; on 2, 21
Man [F66]; on 4 Alta [182]; on 4a Alta
[1485]; on 9 BC [1012, F63]; on 20 Ont
[1827], Que [1685], NB, NS [1032]; on 24
Sask, Que, NB, NS [1221].
T. laricina (anamorph *Sirodothis* sp.):
canker, chancre: on 2, 4, 18, 20, 21
Que [1688] by inoculation; on 24 Alta,
Ont, Que [1221].
T. neopithya: on 20 Ont [1221]; on 24
Ont, Que [1221].
T. pithya (C:205): canker, chancre: on
2, 4, 18, 20, 21 Que [1688]; on 20 Ont
[973], Que [1685, F68], Nfld [1659,
1660, 1661].
T. truncatula (T. pinastri): on 2 Ont
[F69]; on 24 Que [1221].

T. tsugae: on 24 Que [1221].
Tyromyces caesius (Polyporus c.): on 4
NWT-Alta [1063]; on 18 PEI [1032]. See
Abies.
T. fragilis (Polyporus f.): on 2 Alta
[53]; on 2, 4 NWT-Alta [1063]. See
Abies.
T. lapponicus (Polyporus l.): on 4
NWT-Alta [1063], Alta [952]. Associated
with a brown rot.
T. leucospongia (Polyporus l.) (C:204):
on 1 BC [1012, F61]; on 4 NWT-Alta
[1063]. See *Abies*.
T. mollis (Polyporus m.) (C:204): on 4
NWT-Alta [1063]; on 9 BC [1012]. See
Abies.
T. placenta (Poria monticola) (C:204): on
2 Ont [88]. Associated with a brown rot.
T. sericeomollis (Polyporus s., Poria s.)
(C:204): brown cubical rot, carie brune
cubique: on 2 Ont [88, 1341]; on 4 BC
[1012]; on 4a Alta [1485]; on 24
NWT-Alta [1063].
T. subvermispora (Poria s.) (C:204): on 4
NWT-Alta [1063], Alta [579]. See
Abies.
T. undosus (Polyporus u.) (C:204): on 9
BC [1012]; on 18 Que [F71]. See *Abies*.
Ulocladium consortiale (Stemphylium c.):
on 2 Ont [88]; on wood of 9 in water BC
[782].
Valsa friesii (anamorph *Cytospora
dubyi*): on 2 Que [F70]; on 2, 4, 18,
20, 21 Que [1689].
V. pini (C:205): canker, chancre
cytosporéen: on 2 Ont [F69, F73]; on 2,
4, 18, 20, 21 Que [1689]; on 2, 18, 20
Ont [F70]; on 2, 18, 20, 21 Ont [1340];
on 2, 21, 24 Ont [F68]; on 9 BC [1012];
on 18 Ont [1828, F71, F73]; on 18, 21
Ont [F74]; on 20 Que [1340], NB, NS
[1032].
Valsaria rubricosa: on 20 Ont [1340].
Vararia granulosa: (C:205): on 9 BC
[557]; on 16 BC [1012]. See *Abies*.
V. racemosa: on 4 Alta [557]. See
Abies.
Verticicladiella procera (C:205): on 2
Que [874]; on 20 Ont [874].
V. wagenerii: on 4 BC [1012]; on 4a BC
[F76].
Verticillium sp.: on 18 Ont [1827].
Wolfiporia extensa (Poria cocos) (C:204):
on 9, 16 BC [1012]. See *Betula*.
Xenasma filicinum (X. pseudotsugae): on
20, 24 Ont [967]; on 24 Ont [1341].
X. lloydii: on 2, 24 Ont [967].
X. tulasnelloideum: on 24 Que [969].
Lignicolous.
Xeromphalina campanella (Omphalia c.)
(C:205): white stringy rot, carie
blanche filandreuse: on 2 Ont [88]; on
4 NWT-Alta [1063]; on 24 Ont [1341], NB
[1032].
*Zalerion maritimum (Z. eistha, Z. raptor,
Z. xylestris)*: on 9 BC [782]; on 13
NS, Nfld [1126]. Both records from wood
in water.
Zythia sp.: on 2 Ont [91, 1828].

Z. resinae (Pycnidiella r.) (C:205): on 4
Alta [53, 182]; on 4a Alta [1485].

PISUM L.   LEGUMINOSAE

1. *P. sativum* L., garden pea, pois;
   imported from Eurasia.
2. Host species not named, i.e., reported
   as *Pisum* sp.

Alternaria alternata (A. tenuis auct.*)*
   (C:205): blossom and pod blight,
   brûlure alternarienne: on 1 Ont [1910].
Ascochyta pinodes (holomorph
   Mycosphaerella p.): on 1 BC, Alta,
   Ont, Que, NB, NS [107], Sask [1139,
   1095, 1097], PEI [107, 1593].
A. pisi (C:205): leaf and pod spot,
   ascochytose: on 1 BC [1816], BC, Alta,
   Que [107], Alta [338], Alta, Sask, NS
   [1594], Sask, Man, NB [336], Man, PEI
   [333], Ont [1900, 1903, 1904], Ont, NB,
   NS, PEI [107, 334], NB, NS [335].
Botrytis sp.: on 1 Sask [1139].
B. cinerea (holomorph *Botryotinia
   fuckeliana*) (C:205): gray mold,
   moisissure grise: on 1 BC [1816], BC,
   Alta, Ont, Que, NB, NS, PEI [107], Alta
   [339], Sask [1097], Ont [1910], NB, NS
   [333], NS [334, 335, 336, 337, 1594].
Cladosporium cladosporioides f. sp.
   pisicola (C. pisicola) (C:205): leaf
   spot, tache cladosporienne: on 1 BC
   [1816], Que [107].
C. myriosporum: on live pods of 2 BC
   [440].
Colletotrichum pisi (C:205): anthracnose,
   anthracnose: on 1 Alta [107], Ont [334,
   1900].
Coniella pulchella: on seeds of 1 Alta
   [1622].
Erysiphe communis (E. polygoni) (C:205):
   powdery mildew, blanc: on 1 BC [1816],
   BC, Sask [333, 339, 1257], Alta, Sask
   [334, 1594], Alta, Ont [107, 337], Alta,
   NB [336, 338, 339], Sask [1139, 1095,
   1097], Sask, Ont, NS [336, 338], Man
   [884], Man, NB [334, 1257], Ont [1903],
   Que [1257], Que, NB, NS [335], Que, PEI
   [107], NB [337], PEI [333, 336].
Fusarium sp. (C:206): fusarium wilt,
   flétrissure fusarienne: on 1 BC [335,
   338, 1816, 1817, 1818], Alta [333, 334,
   335, 337, 338, 339], Sask [339, 1139,
   1095, 1097, 1594], Ont [1460, 1461,
   1594], NB [334, 337, 339], NS [333,
   334], PEI [338].
F. oxysporum (C:206): fusarium wilt,
   flétrissure fusarienne: on 1 Alta
   [1519], PEI [858]; on 2 Ont [470].
F. oxysporum f. sp. *pisi* (C:206): near
   wilt, flétrissure fusarienne: on 1 BC,
   Alta, Ont, Que, NB, NS [107], Alta
   [338], Alta, NS [1594], Ont [167, 172,
   173, 1456], NS [337, 339], PEI [860].
F. poae (C:206): on 1 Ont [169, 1595].
F. solani (holomorph *Nectria
   haematococca*) (C:206): fusarium root
   rot, pourridié fusarien: on 1 Que [670].

F. solani f. sp. *pisi* (C:206): fusarium
   root rot, pourridié fusarien: on 1 BC,
   Ont, Que, NS, PEI [106, 107], Alta, NB
   [107], Ont [101, 103, 167, 172, 1456],
   Que [613], PEI [860]. Root rot of
   processing peas in s. Alta reduces
   annual yield by about 17%. Each year
   several irrigated fields are so heavily
   infected that pea production must be
   abandoned.
Melanospora papillata (C:206): on seeds
   of 1 Ont [1943].
Microascus longirostris (M. variabilis)
   (C:206): on seeds of 1 Ont, Que, PEI
   [308]; on 2 Que [86].
Mycosphaerella pinodes (anamorph
   Ascochyta p.) (C:206): mycosphaerella
   blight, brûlure ascochytique: on 1 Man
   [335, 1595, 1903], Ont [334, 335, 336,
   337, 338, 1900, 1902, 1903, 1904], NB
   [337], NS [335, 336].
Paecilomyces varioti (C:206): on seeds
   of 1 Ont [150].
Peronospora viciae (P. pisi) (C:206):
   downy mildew, mildiou: on 1 BC [1816,
   1817, 1818], BC, Alta, Ont, NB, NS, PEI
   [107], BC, NS [333], Alta, NS [334,
   335], Sask [1139], NS [339], PEI [336].
Phoma medicaginis var. *pinodella
   (Ascochyta p.)* (C:205): foot rot,
   pourridié ascochytique: on 1 BC, Alta,
   Ont, Que, NB, NS, PEI [107], Alta [337],
   Man, NB, PEI [1594], Ont [334, 335, 336,
   1902, 1904], Ont, Que [333], NS, PEI
   [339], PEI [1593].
Pleospora herbarum (anamorph *Stemphylium
   botryosum*) (C:206): on 2 Ont [1340].
Pythium sp. (C:206): damping-off and root
   rot, fonte des semis et pourridié
   pythien: on 1 BC [338], Alta [333, 334,
   335, 337, 338, 339, 704, 1594], Sask
   [339, 1594], Ont [1594], NB [337, 339],
   PEI [338].
P. hypogynum: on 1 Ont [1456].
P. irregulare: on 1 Ont [1456].
P. oligandrum: on 1 Ont [1456].
P. sylvaticum: on 1 Ont [1456].
P. ultimum (C:206): damping-off and root
   rot, fonte des semis et pourridié
   pythien: on 1 Man [1069], Ont [1456];
   on 2 Ont [1477].
P. vexans: on 1 Ont [1456].
Rhizoctonia sp.: on 1 Alta [337, 704],
   Sask [1097, 1139], NB [337].
R. solani (holomorph *Thanatephorus
   cucumeris*) (C:206): root rot,
   rhizoctone commun: on 1 BC [338, 1816],
   Alta [107, 333, 334, 338, 339, 1594],
   Sask [334, 339], Ont [1460, 1594], NB,
   NS [339], PEI [338].
Sclerotinia sp.: on 1 Sask [420, 1138],
   Man [1138].
S. sclerotiorum (C:206): sclerotinia rot,
   pourriture sclérotique: on 1 Sask
   [1095, 1097, 1139], Ont [334, 1900], NS
   [335].
Septoria pisi (C:206): leaf blotch, tache
   septorienne: on 1 Alta [107], Sask
   [336], Ont [107, 333, 336, 338], Que
   [107, 333, 335].

Stemphylium botryosum (holomorph
  Pleospora herbarum) (C:206): leaf
  spot, tache stemphylienne: on seeds of
  1 BC [1816], Ont [667].
*Ulocladium consortiale (Alternaria c.,
  Stemphylium c.)* (C:205): on seeds of 1
  Ont [667].
Uromyces viciae-fabae (U. fabae) (C:206):
  rust, rouille: 0 I II III on 1 BC
  [1816], BC, Ont, Que, NB, NS, PEI [107],
  Ont [337, 338, 339, 1236, 1900, 1902,
  1903], Ont, NS [334, 335, 336], NB [339,
  1594], NS [333], PEI [336, 1594].

PITTOSPORUM Banks   PITTOSPORACEAE

  Evergreen trees and shrubs native to
warm temperate, subtropical and tropical
regions of the old world. Useful
ornamentals, some species grown under glass.

1. *P. tobira* (Thunb.) Ait., Japanese
   pittosporum, pittospore de Chine;
   imported from China and Japan.

Phyllosticta sp.: on 1 BC [335, 1816].

PLANTAGO L.   PLANTAGINACEAE

1. *P. major* L., common plantain, queue de
   rat; imported from Eurasia, natzd.,
   Yukon-Mack, BC-Nfld.
2. *P. maritima* L. *(P. juncoides
   Lam.)*, goose tongue, passe-pierre;
   native, Mack, BC-Alta, Man-Nfld., Labr.
3. *P. rugelii* Dcne., Rugel's plantain,
   plantain de Rugel; native, Ont-PEI.

Coleroa plantaginis: on 3 Que [307]; on
  living leaves of *P.* sp. in c. and e.
  Canada [73].
Erysiphe sp.: powdery mildew, blanc: on
  1 BC [1012].
E. cichoracearum (C:207): powdery mildew,
  blanc: on 1 Man, Que [1257]; on 1, 2 BC
  [1816], Que [282]; on 1, 3 Ont [1257].
Mycosphaerella tassiana var.
  arthopyrenioides (C:207): on 1 Que
  [70].
Olpidium brassicae: on 1 BC [65], Ont
  [67].
Peronospora alta (C:207): downy mildew,
  mildiou: on 1 BC [1816].
Pleospora herbarum (anamorph *Stemphylium
  botryosum*): on 2 Que [282].
Rhizophydium graminis: on 1 Ont [59].
Verticillium dahliae: on 1 BC [1816,
  1978].

PLATANUS L.   PLATANACEAE

1. *P. x hybrida*: *P. occidentalis* L.
   x *P. orientalis* L. *(P. acerifolia*
   (Ait.) Willd.), London plane tree,
   platane anglais.
2. *P. occidentalis* L., sycamore, platane
   d'Occident; native, s. Ont.
3. Host species not named, i.e., reported
   as *Platanus* sp.

Apiognomonia errabunda (anamorph *Discula
  umbrinella*): on 2 Ont [1340]. See
  Fraxinus.
A. veneta (Gnomonia v.) (C:207):
  anthracnose, anthracnose: on 1 BC
  [1012]; on 3 BC [333, 337], Ont [333].
Cytospora sp.: on 3 BC [1012].
Diatrypella prominens: on 2 Ont [1340,
  F63].
*Discula umbrinella (Gloeosporium
  nervisequum)* (holomorph *Apiognomonia
  errabunda*): on 1 BC [1816]; on 2 BC
  [338]; on 3 Ont [1340, F75, F76].
*Macrodiplodiopsis desmazierii (Hendersonia
  d.)* (C:207): on 2 Ont [1340]; on 3 Ont
  [1340, 1763 on dead branches].
Valsella sp.: on 3 Ont [1828].

PLATYCLADUS Spach   CUPRESSACEAE

1. *P. orientalis* (L.) Franko cv.
   rosedale, oriental arborvitae; native,
   China and Korea, imported and planted as
   an ornamental.

Lepteutypa cupressi: on 1 (as
  Chamaecyparis rosedale) BC [1012].

PLEUROPOGON R. Br.   GRAMINEAE

1. *P. sabinei* R. Br.; native, Mack-Labr.

Hysteropezizella diminuens: on 1 Frank
  [664].
Physoderma graminis (C:208): on 1 Frank,
  Keew [1586]. Listed as "affin.
  graminis".
Ustilentyloma pleuropogonis (C:208):
  smut, charbon: on 1 Frank [1586].

PLUMERIA L.   APOCYNACEAE

1. *P. rubra* L., frangipani; imported from
   tropical America. Grown as an
   ornamental especially for the fragrance
   of the flowers.

Coleosporium plumeriae: rust, rouille:
  on 1 Alta [1824]; on *P.* sp. Alta
  [1822]. On imported plants.

POA L.   GRAMINEAE

1. *P. abbreviata* R. Br.; native,
   Yukon-Labr.
2. *P. ampla* Merr., big bluegrass; native,
   Yukon, BC-Sask.
3. *P. annua* L., annual bluegrass, pâturin
   annuel; imported from Eurasia, natzd.
   Yukon, BC-Nfld, Labr.
4. *P. arctica* R. Br.; native, Yukon-Labr,
   BC-Alta, Man-Que, Nfld.
5. *P. bulbosa* L., bulbous bluegrass;
   imported from Eurasia, introduced, BC.
6. *P. canbyi* (Scribn.) Piper; native,
   Yukon, BC-Sask, Que.
7. *P. compressa* L., Canada bluegrass,
   pâturin du Canada; imported from
   Eurasia, natzd. Yukon-Mack, BC-Nfld.

8. *P. eminens* Presl; native, sw. BC, James Bay and Hudson Bay, Que, and Nfld.
9. *P. glauca* Vahl; native, Yukon-Labr, BC-NS, Nfld.
10. *P. hartzii* Gand.; native, Mack-Frank.
11. *P. nemoralis* L., wood meadow-grass, pâturin des bois; native, Yukon-Keew, BC-NS, Nfld, Labr.
12. *P. palustris* L., fowl meadow-grass; native, Yukon-Mack, BC-Nfld, Labr.
13. *P. paucispicula* Scribn. & Merr.; native, Yukon-Mack, BC-Alta.
14. *P. pratensis* L., Kentucky bluegrass, pâturin des prés; native and introduced, Yukon-Labr, BC-Nfld. 14a. *P. p.* var. *alpigena* (Fries) Hiitonen (*P. alpigena* (Fries) Lindm. f.); native, Yukon-Labr, Man-Nfld.
15. *P. stenantha* Trin.; native, BC-Alta.
16. *P. trivialis* L., rough-stalked meadow grass; Eurasian, natzd. Yukon, BC-Alta, Man, PEI, Nfld.
17. Host species not named, i.e., reported as *Poa* sp.

Acremonium boreale: on 3 BC, Alta, Sask, Man, Ont [1707]; on 14 BC, Alta, Sask [1707].
Allophylaria pusiola (C:208): on 1, 4 Frank [664]; on 4 Frank [1543], Keew [664].
Bipolaris sorokiniana (holomorph *Cochliobolus sativus*): on 3 Ont [166]; on 14 Alta [56]; on 16 Alta [120] by inoculation; on 17 PEI [1974].
Cercosporidium graminis (*Passalora g.*, *Scolecotrichum g.*) (C:209): brown stripe, strie brune: on 6, 15 Yukon [1731]; on 7, 14 BC [1816]; on 12 NB [333, 1195].
Cladosporium herbarum (C:208): on 4 Frank [1244].
Dendrophoma poarum: on culms and inflorescences of 3 Ont [440].
Drechslera poae (*D. vagans*) (C:209): leaf blotch, tache drechsleréenne: on 9 Yukon [1731]; on 14 BC [1816], Alta [338, 339], Sask [333, 338, 1611], NB [333, 1195], Nfld [338]; on 17 Alta, Sask, Man [335].
Entyloma dactylidis (C:209): leaf smut, charbon des feuilles: on 1 BC [534].
Epicoccum purpurascens (*E. nigrum*) (C:209): on 14 NB [333, 1195].
Erysiphe graminis (C:209): powdery mildew, blanc: on 3, 12 Que [282]; on 4 Yukon [1255, 1257], Frank [1244]; on 4, 9 Frank [1586]; on 4, 9, 14a Frank, Keew [1257]; on 9 Ont [1257], Que [1255, 1257]; on 11, 12, 14 Ont [1255, 1257]; on 12 Que [1257]; on 14 Yukon [1731], Mack, Sask, Man [1257], BC [1728, 1816, 1849], Alta, Ont, Que [336], Sask [338, 1185], Sask, PEI [335], Ont, Nfld [334], NB [333, 1195]; on 17 Ont [335], Que [339], PEI [1974].
Fusarium sp.: on 14 BC [1849].
F. nivale (C:209): snow mold, moisissure des neiges: on 3, 14 Sask [1705], Que [1594]; on 14 BC [1849], Alta [956, 1594]. See *Agrotis*.

F. poae (C:209): on 14 BC [333, 1816]; on 17 Ont [169], NB [333].
Helminthosporium sp.: on 14 BC [1849], Sask [336, 1185], Man [1367].
Laetisaria fuciforme (*Corticium f.*): on 14 BC [1849].
Ligniera pilorum: on 3 Ont [60].
Low temperature basidiomycete: on 3 Alta [333]. Probably *Coprinus psychromorbidus* [1439] or *Typhula* sp.
Marasmius sp.: on 14 BC [1849].
M. oreades: on 17 Alta, Sask [335].
Olpidium brassicae: on roots of 14 Ont [68].
Pleospora sp.: on 3, 14 Sask [338].
Pseudoseptoria donacis (*Selenophoma d.*) (C:210): on 2 BC [1728]; on 12 Que [282, 1729]; on 14, 15 Yukon [1731].
P. stomaticola (*Selenophoma donacis* var. *s.*) (C:210): on 6 Yukon [1731]; on 12, 14 BC [1728].
P. everhartii (C:210): on 7 Yukon [1731].
Puccinia sp.: rust, rouille: on 14 BC [1849].
P. graminis (C:210): stem rust, rouille de la tige: II III on 7, 14 Ont [1236]; on 14 BC, Alta, Ont [333], Alta [335], Ont [335, 337], Que [336].
P. poae-nemoralis (*P. brachypodii* var. *p.*, *P. poae-sudeticae*) (C:210): leaf rust, rouille des feuilles: II III on 3, 7, 14, 17 Ont [1236]; on 4 Que [1236]; on 7, 12, 14 BC [1816]; on 9, 12 Yukon [1731]; on 12, 13, 14 BC [1012]; on 12, 14 NB [333, 1195]; on 13 BC [1728]; on 14 Sask [336, 1185]; on 17 Man [335].
P. poae-nemoralis ssp. *hyparctica* (C:210): rust, rouille: on 10, 10 X 9 Frank [1586].
P. poae-nemoralis ssp. *p.* (C:210): rust, rouille: on 4 Frank [1244]; on 4, 9 Frank [1586]; on 5 BC [1578].
P. recondita s. l. (C:210): rust, rouille: on 17 BC [1012].
P. striiformis: rust, rouille: on 14 BC [1012]. In the strict sense *P. striiformis* is a rust of *Triticeae*. It remains to be demonstrated whether it is biologically distinct from the rust on *Poa* (see Fungi Canadenses 250).
Pyrenopeziza karstenii (C:210): on 8 Que [664].
Pythium arrhenomanes: on 14 Sask [1869].
Rhizoctonia solani (holomorph *Thanatephorus cucumeris*) (C:210): on 7 BC [1729]; on 14 BC [1849], NB [336].
Sclerotinia borealis (C:210): snow mold, moisissure des neiges: on 2, 7 BC [1816].
Septoria sp.: on 14 Sask [1185].
S. macropoda (C:210): on 3 Que [1729]; on 14 Man [1367].
S. macropoda var. *septulata*: on 14 Ont [335].
S. oudemansii (C:210): on 9 Yukon [1704]; on 14 Que [1729].
Sphaerellopsis filum (*Darluca f.*) (holomorph *Eudarluca caricis*) (C:208): on 6 Yukon [1731]; on 14 Sask

[1185]. Mycoparasitic on *Puccinia poae-nemoralis*.
Typhula sp. (C:210): on 3 Sask [1705].
T. ishikariensis var. *canadensis*: on 3, 14 Sask [19].
Ustilago salvei (U. striiformis auct.*)* (C:210): stripe smut, charbon strié: on 14 NB [333]; on 17 NB [1195].

PODOPHYLLUM L.   BERBERIDACEAE

1. *P. peltatum* L., may apple, pomme de mai; native, Ont-Que, NS.

Puccinia podophylli (C:210): rust, rouille: 0 I III on 1 Ont [1236, 1341, 1827, 1828]; on *?Podophyllum* sp. Ont [1595].
Vermicularia podophylli: on fruit of 1 Ont [439].

POLYGALA L.   POLYGALACEAE

1. *P. paucifolia* Willd., bird-on-the-wing, polygala paucifolié; native, Sask-NB.

Puccinia pyrolae (C:211): rust, rouille: III on 1 Ont [1236, 1341].

POLYGONATUM Mill.   LILIACEAE

1. *P. biflorum* (Walt.) Ell., two-flowered Solomon's-seal, sceau-de-Salomon à deux fleurs; native, Sask-Que.
2. *P. pubescens* (Willd.) Pursh, downy Solomon's seal, sceau-de-Salomon pubescent; native, Ont-NS.

Macrophoma smilacina var. *smilacina*: on 1, 2 Ont [148].
M. smilacina var. *smilacis*: on 1, 2 Ont [148].

POLYGONUM L.   POLYGONACEAE

1. *P. amphibium* L., amphibious bistort, renouée amphibie; native, Yukon-Mack, BC-Nfld.
2. *P. arenastrum* Jord.; imported from Eurasia, introduced, Yukon-Mack, Alta-Nfld, Labr.
3. *P. aviculare* L., common knotweed, renouée des oiseaux; imported from Eurasia. Many Canadian reports refer to *P. arenastrum*.
4. *P. coccineum* Muhl., scarlet knotweed, renouée écarlate; native, Mack, BC-Que, NS-PEI.
5. *P. convolvulus* L., black bindweed, renouée liseron; imported from Eurasia, natzd., Yukon-Mack, BC-Nfld.
6. *P. erectum* L. (*P. achoreum* Blake), erect knotweed, renouée dressée; native, Yukon-Mack, BC-NB, Nfld.
7. *P. fowleri* Robins.; native, BC, Ont-NS, Nfld, Labr.
8. *P. hydropiperoides* Michx., false water pepper, renouée faux-poivre-d'eau; native, BC, Ont-NS.

9. *P. lapathifolium* L., dock-leaved knotweed, renouée à feuilles de Patience; native, BC-Nfld.
10. *P. pensylvanicum* (L.) Sm., pinkweed, renouée de Pennsylvanie; native, Ont-NS.
11. *P. persicaria* L., heart's-ease, fer à cheval; imported from Eurasia, introduced, BC-Nfld.
12. *P. punctatum* Ell., water smartweed, renouée ponctuée; native, BC, Sask-PEI.
13. *P. ramosissimum* Michx., bushy knotweed, renouée très rameuse; native, BC-PEI.
14. *P. scandens* L. (*P. cristatum* sensu Rouleau), climbing false buckwheat, renouée grimpante; native, Sask-Nfld.
15. *P. spergulariaeforme* Meisn.; native, s. BC.
16. *P. virginianum* L. (*Antenoron v.* (L.) Roberty & Vautier); native, Ont-Que.
17. *P. viviparum* L., alpine bistort, renouée vivipare; native, Yukon-Labr, BC-Que, Nfld.
18. Host species not named, i.e., reported as *Polygonum* sp.

Bostrichonema alpestre (B. polygoni) (C:212): on 17 Frank [1543, 1586], Alta [644], Que [282].
Cercospora avicularis (C:212): on 3 Man, Ont [1353]; on living leaves of 6 Ont [1353].
Erysiphe communis (E. polygoni) (C:212): powdery mildew, blanc: on 3 Mack, NS [1257], BC [1012, 1816]; on 3, 6 Sask, Ont [1257]; on 3, 6, 13 BC, Man, Que [1257]; on 6 Alta [1257].
Peronospora polygoni (C:212): downy mildew, mildiou: on 5 BC [1816].
Pseudorhytisma bistortae (Rhytisma b.) (C:212): on 17 Frank [1244].
Puccinia bistortae (C:212): rust, rouille: II III on 15 Frank [1239]; on 17 Yukon [460], Frank [1244, 1543, 1586], BC [1012, 1816], Alta [644], Que [282]; on 18 Man, Que [1236].
P. phragmitis (C:212): rust, rouille: 0 I on 8, 10 Ont [1236]; on 9, 10 Ont [1238].
P. polygoni-amphibii (C:212): rust, rouille: II III on 1, 4 BC [1012]; on , 4, 5, 8, 9, 14, 18 Ont [1236]; on 1, 12 BC [1816].
P. septentrionalis (C:212): on 17 Que [282].
Ramularia rufomaculans (C:212): on 1 (as *P. phytolaccaefolium*) Yukon [460].
Rhizophydium graminis: on 3 Ont [59].
Septoria polygonorum (C:212): on 11 BC [1416, 1816]; on 18 BC [1416], NB [1032].
Uromyces polygoni-avicularis (C:212): rust, rouille: 0 I II III on 3 BC [1816]; on 3, 6, 8, 13, 18 Ont [1236]; on 5 Ont [1341]; on 7, 15 BC [1012, 1816].
Ustilago bistortarum (C:213): smut, charbon: on 17 Yukon, Frank, Keew [1586], Frank [1244, 1543], Alta [644]. See next species.

U. hydropiperis (Sphacelotheca h.)
  (C:212): smut, charbon: on 17 Frank
  [1239]. Recognition of *Sphacelotheca*
  as a distinct genus does not seem
  warranted. *U. hydropiperis* occurs on
  Polygonum sect. *Persicaria*, whereas
  U. bistortarum attacks species in
  sect. *Bistorta* [971].
U. reticulata (C:213): smut, charbon: on
  14 BC [1416].
Venturia polygoni-vivipari (C:213): on
  leaves, petioles and stalks of 17 NWT,
  Que [73].
Verticillium dahliae: on 11 Ont [1087].

POLYPODIUM L.   POLYPODIACEAE

1. *P. glycyrrhiza* Eat. (*P. vulgare* var.
    occidentale (Hook.) Hulten), licorice
    fern; native, Yukon, BC.
2. *P. virginianum* L., rock polypody,
    tripe-de-roche; native, Mack, BC-Nfld.

Milesia laeviuscula (Milesina l.)
  (C:213): rust, rouille: II III on 1 BC
  [1012, 1816, 2008].
M. pycnograndis (M. polypodophila)
  (C:213): rust, rouille: II III on 2
  Ont [1236], Que [1377].

POLYSTICHUM Roth   POLYPODIACEAE

1. *P. munitum* (Kaulf.) Presl; sword fern,
    fougère épée; native, BC.

Milesia polystichi (Milesina winelandii)
  (C:213): rust, rouille: II on 1 BC
  [1012, 1816, 2008].
M. vogesiaca (Milesina v.) (C:213): rust,
  rouille: II III on 1 BC [1012, 1816,
  2008].
Taphrina faulliana (C:213): on 1 BC
  [1012, 1816].
Valdensia heterodoxa (Asterobolus
  gaultheriae) (holomorph *Valdensinia*
  h.): on 1 BC [1438].
Xenasma filicinum: on 1 BC [967].

PONTEDERIA L.   PONTEDERIACEAE

1. *P. cordata* L., pickerel weed, glaieul
    bleu; native, Ont-PEI.

Cercospora pontederiae (C:213): on leaves
  of 1 Ont [439]. Apparently a species of
  Cercosporella.

POPULUS L.   SALICACEAE

1. *P. alba* L., silver poplar, peuplier
    argenté; imported from Eurasia,
    spreading from cultivation, BC-Alta,
    Man-Nfld. 1a. *P. a.* cv. nivea.
2. *P. angustifolia* James, narrowleaf
    cottonwood, liard amer; native,
    Alta-Sask.
3. *P. balsamifera* L. (*P. candicans*
    Ait., *P. tacamahaca* Mill.), balsam
    poplar, peuplier baumier; native,
    Yukon-Keew, BC-Nfld, Labr.

4. *P.* x *berolineusis* Dippel *(P.*
    laurifolia x *P. nigra)*.
5. *P.* x *canadensis* Moench *(* P.
    deltoids x *P. nigra*, P. x *eugenei*
    Simon-Louis, *P.* x *canadensis* var.
    eugenei (Simon-Louis) Schelle).
6. *P. deltoides* Bartr., cottonwood,
    liard; native, Alta-Que. 6a. *P. d.*
    ssp. *d.* (*P. angulata* Ait.). 6b. *P.*
    d. ssp. *monilifera* (Ait.)
    Eckenwalder (*P. d.* var. *occidentalis*
    Rydb., *P. sargentii* Dode) western
    cottonwood, liard de l'Ouest; native,
    Alta-Man.
7. *P. grandidentata* Michx., largetooth
    aspen, grand tremble; native, Man-PEI.
8. *P. jackii* Sarg. (*P. balsamifera* x
    P. deltoides).
9. *P. laurifolia* Ledeb.; imported from
    Siberia.
10. *P. maximowiczii* A. Henry; imported
    from Asia.
11. *P. nigra* L., European black poplar,
    peuplier noir; imported from Eurasia.
    11a. *P. n.* var. *italica* DuRoi,
    Lombardy poplar, peuplier d'Italie.
12. *P.* x *petrowskyana* (Regel) C.K.
    Schneid. *(P. deltoidess* x *P.*
    laurifolia).
13. *P. tremuloides* Michx., trembling
    aspen, tremble; native, Yukon-Keew,
    BC-Nfld, Labr.
14. *P. trichocarpa* Torr. & Gray, western
    balsam poplar, peuplier baumier de
    l'Ouest; native, Yukon, BC-Alta.
15. *P. tristis* Fisch., brown-twig poplar,
    peuplier sombre; imported from c. Asia.
16. Hybrid Poplars: Because of extensive
    breeding programs, there are large
    numbers of hybrid poplar which are not
    common. This group includes: *P.* x
    generosa, *P.* 'I-488', *P.*
    'Brooks'10', *P.* 'northwest', etc.
17. Host species not named, i.e., reported
    as *Populus* sp.

Acanthostigma clintonii (C:214): on 17
  Ont [1340].
Acremonium sp.: on 13 Alta [1970].
A. boreale: on 13 BC, Alta, Sask [1707].
A. charticola: on 13 Alta [1970].
Acrodictys globulosa: on 13 Man [1760].
Aegerita sp.: on 14 BC [1012].
Aleurodiscus cerussatus (C:214): on 13
  NWT-Alta [1063], Alta [53]; on 17 Man
  [964]. Lignicolous.
A. lapponicus (C:214): on 17 Man [964].
  Associated with a white rot.
Alternaria sp.: on 1 Ont [1340]; on 17
  Ont [584], Que [919].
A. alternata (A. tenuis auct.) (C:214):
  on 13 Alta [1970]; on 14 BC [1012].
Alysidium resinae var. *microsporum*: on
  13 Sask [1760].
A. resinae var. *resinae*: on 13 Sask,
  Man [1760].
Amphinema byssoides: on 13 Alta [1828].
  See *Abies*.
Antrodia lenis (Poria l.): on 17 NS
  [1032]. See *Abies*.

A. serialis (Trametes s.): on 13 Ont
[1341]. See *Abies*.

A. xantha (Poria x.) (C:220): on 13 BC
[1012]; on 17 Ont [1341], NS [1032].
See *Abies*.

Apioplagiostoma populi (Plagiostoma p.)
(C:219): on 17 Ont [1340].

Aporpium carya (C:214): on 13 NWT-Alta
[1063], Alta [53]; on 13, 14 BC [1012];
on 17 Ont [1341, 1660], NS [1032]. See
Betula.

Arachnopeziza sp.: on 14 BC [1012].

Arachnophora excentrica (Acrodictys e.):
on 13 Sask, Man, [811, 1756, 1760], Man
[808].

Armillaria mellea (C:214): armillaria
root rot, pourridié-agaric: on 3, 13
Alta [1801]; on 5 BC [F60]; on 5, 11,
13, 14 BC [1012]; on 7, 13, 17 Ont
[1341]; on 13 NWT-Alta [1063], Alta
[951, F67, F72], Que [F74]; on 17 Alta
[F68], Que [919]. See *Abies*.

Arthrinium sp.: on 13 Alta [1970].
Reported as the *Arthrinium* state of
Apiospora montagnei.

Ascoryne sarcoides: on 17 Que [919].

Aspergillus sp.: on 14 BC [138]; on 17
Que [919].

Asteromella sp.: on 13 Ont [1827]. See
Phyllosticta brunnea below.

Athelia pellicularis: on 13 BC [1012].
Lignicolous. See *Abies*.

Aureobasidium sp. *(Pullularia* sp.): on
17 Ont [584].

A. pullulans: on 13 Alta [1970]; on 17
Que [919].

Bactrodesmium spilomeum: on 13 Man [1760].

Basidiodendron deminuta: on 17 Ont
[1016]. Lignicolous.

B. eyrei: on 14 BC [1016]. See *Acer*.

B. fulvum (Bourdotia grandinioides): on
14 BC [1012]. Lignicolous.

*Basidioradulum radula (Radulum
orbiculare)*: on 17 NS [1032]. See
Abies.

Beauveria bassiana (C:214): on 13 Alta
[1970]; on 14 BC [1012]. Known as a
pathogen of insects.

Biatorella sp.: on 13 NWT-Alta [1063].

Bispora sp.: on 17 Que [919].

B. betulina: on 13, 17 Man [1755, 1760].

Bisporella citrina (Helotium c.) (C:216):
on 7 Ont [1827]; on 17 BC [1012], Ont
[1340].

B. pallescens (Helotium p.): on 13
NWT-Alta [1063].

Bjerkandera adusta (Polyporus a.)
(C:219): white mottled rot, carie
blanche madrée: on 3, 13 BC [1012],
Alta [53, 1801]; on 3, 7, 13, 17 Ont
[1341]; on 3, 13, 17 NWT-Alta [1063]; on
7 Ont [830], NS [1032]; on 13 Alta
[452]; on 13, 17 NB, NS [1032], Nfld
[1659, 1660, 1661]; on 14 BC [1341]; on
17 Ont [1827], Que [919, 1377], PEI
[1032].

B. fumosa (Polyporus f.): on 13 NWT-Alta
[1063]; on 17 Ont [1341]. See *Acer*.

Boreostereum radiatum (Stereum r.): on 13
NWT-Alta [1063]. Unusual record the

fungus typically occurs on gymnosperm
wood.

Botryohypochnus isabellinus: on 14 BC
[1012]. See *Abies*.

Botrytis cinerea (holomorph *Botryotinia
fuckeliana*): on 13 Alta [1970], Sask
[1760].

chysporium britannicum: on 13 Man
[1760].

B. nigrum: on 13 Sask [827, 1760]; on 13,
17 Man [1760], Ont [827].

B. obovatum: on 17 Ont [828].

Buellia sp.: on 13 Ont [1827]. Possibly
B. stygia which occurs on *Populus*.

Cacumisporium capitulatum: on 13 Man
[1760].

Cadophora sp.: on 14 BC [1012]. Probably
a specimen of *Phialophora*.

Calathella erucaeformis: on 17 BC [1424].

Calcarisporium sp.: on 3 NWT-Alta [1063],
Mack [53].

Caliciopsis calicioides (C:214): on 3
Sask [F67], Sask, Man [F65]; on 14 BC
[1012].

Calocera cornea (C:214): on 13, 14 BC
[1012]. See *Acer*.

Candida sp.: on 17 Ont [584].

Catenularia sp.: on 17 Que [919].

Cenangium sp.: on 13 NWT-Alta [1063].

Cephalosporium sp.: on 17 Ont [584], Que
[919].

Ceraceomyces serpens: on 13 BC [1012]; on
13, 14 BC [568]; on 17 Man, Ont [568].
See *Abies*.

Ceratocystis sp.: on 3 Sask [F66]; on 13
BC [1012], Ont [1340], Que [F62].

C. capitata: on 3, 7, 13 Ont [647, 1340];
on 13 Man, Ont [1190].

C. crassivaginata: on 7, 13 Ont [647]; on
13 Man, Ont [1190], Ont [1340].

C. fimbriata: on 13 Yukon [727], BC
[1012], BC, Man, Ont [1190], Sask, Man
[1991, 1990], Sask, Man, Que [F64], Ont
[647], Ont, Que [1340], Que [F65, F76];
on 17 Que [F67].

C. moniliformis: on 13 NWT-Alta [1063].

C. pallidobrunnea: on 3, 13 Man [1190].

C. pilifera: on 3, 13 Ont [1340]; on 3,
13, 14 Ont [647]; on 13 Man [1190]; on
14 BC [1340].

C. populicola: on 13 Man [1190].

C. tremulo-aurea: on 13 Man [1190].

C. torticiliata: on 13 Man [1190].

Cercospora populina: on 3 Sask [F65].

Ceriporia purpurea (Poria p.) (C:220): on
13 BC [1012]. See *Alnus*.

C. reticulata (Poria r.) (C:220): on 13
NWT-Alta [1063], Alta [53]. See *Acer*.

C. rhodella (Poria r.) (C:220): on 17
Mack [579]. See *Pinus*.

Cerrena unicolor (Daedalea u.) (C:215):
white spongy rot, carie blanche
spongieuse: on 13 NWT-Alta [1063], Mack
[1110], Alta [53], Que [1391]; on 14 BC
[1012]; on 17 Yukon [460], BC [1855],
Ont [1341], Que [1377].

Ceuthospora sp.: on 13 Alta [1970].

Chaetomella acutiseta var. *acutiseta*:
on 13 Ont [402].

Chaetomium sp.: on 17 Ont [584].

Cheimonophyllum candidissimus (Pleurotus c.): on 17 PEI [1032]. Lignicolous.
Chikaneea holleroniae: on 13 Sask and on 17 Man [1760].
Chloridium sp.: on 3 Sask [F66].
C. chlamydosporum: on 3 Man [1760].
Chlorociboria aeruginascens (Chlorosplenium aeruginosum) (C:214): on 13 NWT-Alta [1063]; on 17 BC [1012], NB, NS [1032].
C. aeruginosa: on 13 NWT-Alta [1063]; on 13, 17 Ont [1340].
Chlorosplenium sp.: on 17 Ont [1340].
Chondrostereum purpureum (Stereum p.) (C:220): silver leaf, plomb: on 3 Alta [1801], Labr [867]; on 3, 13 NWT-Alta [1063]; on 7, 13, 17 Ont [1341]; on 11a, 13, 14 BC [1012]; on 13 Yukon [460] a sight record, Alta [452], NB [1032], NB, NS, PEI [F62], Nfld [1659]; on 13, 17 Nfld [1660, 1661]; on 17 NS [1032]. See *Abies*.
Chrysosporium sp.: on 13 Alta [1970].
Ciboria acerina: on 17 Ont [662].
C. caucus (C:214): on 17 Ont [1590], Ont, Que [662].
Ciborinia sp.: on 3 Ont [1828]; on 3, 7 Ont [1340]; on 13 BC, Man [F62], Alta [F63], Man [F64]; on 17 BC [F63], Alta [F62].
C. bifrons: on 17 Nfld [F61]. This specimen may be *C. whetzelii*.
C. pseudobifrons (C:214): on 13 NWT-Alta [1063]; on 17 Alta [55, F65].
C. whetzelii (Sclerotinia bifrons Whetz.) (C:214): ink spot, tache d'encre: on 3, 5, 11a, 13 Ont [F62]; on 3, 7, 11, 13 Que [1377]; on 3, 13 NWT-Alta [1063], Alta [53], NB, NS, PEI [F70]; on 5 NB, PEI [F63]; on 5, 7, 11a, 13, 17 NB [1032]; on 5, 13, 17 Ont [F66]; on 7, 13 Ont [F73], Ont, Que [F67], NB [F65]; on 7, 13, 17 Ont [1340]; on 11a Que [333]; on 11a, 13 Ont [F63], Que [F61], NB [F73], NS [F64]; on 11a, 13, 16 NS [1032]; on 11a, 17 Que [F62, F64]; on 13 BC [1012], BC, NS [F60], Alta [950], Alta, Sask, Man [F70, F71, F72, F73, F74], Sask, Man [F63, F66, F69], Sask, Man, Que, NS, PEI, Nfld [F65], Ont [F70], Ont, Que [F68], Ont, Que, NB [F71, F72, F74, F75], Que [282, F76], NB [F61, F62], NB, NS [F69], NB, NS, PEI, Nfld [F67, F68, F76], NB, PEI, Nfld [F63, F64], NS, Nfld [F66], NS, Labr [F74], PEI [1032], Nfld [1659, 1660, 1661, F70, F72, F73, F74]; on 13, 16 Ont [F76], Que [F73]; on 16, 17 Alta [F65]; on 17 BC [F65], BC, Alta [F64, F69], Alta [55, F66], Alta, Sask, Man [F67, F68], Sask, Man, Ont, NB [F60], Ont [F65, F72], Ont, Que [F69, F70], Ont, Que, NB, NS [F61], Que [339, 1375, F63, F66].
Cladosporium sp.: on 2 Ont [1828]; on 5 NB [1032]; on 13 Yukon [460], NS [1032]; on 14 BC [1340]; on 17 Que [919].
C. cladosporioides: on 13 Alta [1970].
C. herbarum: on 7 NB [F66]; on 13 Alta [1970], Sask, Man [1760], NS [1032]; on 14 BC [1012].

C. macrocarpum: on 3 Man [1760]; on 13 Alta [1970].
Clavaria sp.: on 17 Ont [1341]. Presumably on rotten wood.
Clavicorona pyxidata: on 17 Ont [408], Que [1385]. See *Betula*.
Climacodon septentrionalis (Steccherinum s.) (C:220): on 13 NWT-Alta [1063]; on 17 Alta [F62]. See *Acer*.
Clitocybe cyathiformis (C:214): on 14 BC [1012].
C. truncicola: on 14 BC [1012].
Coccomyces sp.: on 14 BC [1012].
C. coronatus (C:214): on 13 Ont [1340]; on 17 NS [1032].
C. tumidus: on 13 NS [1609].
Coniochaeta sp.: on 17 Ont [584].
Coniophora olivacea (C:215): on 13 BC [1012]; on 17 Ont [1341]. See *Alnus*.
Coniothyrium sp.: on 17 Ont [584].
Connersia rilstonii: on 17 Ont [1038].
Conoplea fusca (C:215): on 13 Sask, Man [1760]; on 17 Sask [784].
C. geniculata: on 13 Sask [1760]; on 17 Ont [784].
Coprinus sp.: on 14 BC [1012].
C. extinctorius: on 13 Yukon [1921].
C. micaceus (C:215): on decay in live trees of 3 Alta [1801].
Coriolopsis gallica (Trametes hispida) (C:221): on 3, 13, 17 NWT-Alta [1063]; on 13 Yukon [1110], Alta [53, 950]; on 13, 14 BC [1012]; on 14 BC [1341]; on 17 NS [1032]. See *Acer*.
Coriolus hirsutus (Polyporus h.) (C:219): white spongy rot, carie blanche spongieuse: on 3, 13, 17 Ont [1341]; on 7, 13 NB [1032]; on 13 NWT-Alta [1063], Nfld [1661]; on 17 Alta [426].
C. pubescens (Polyporus p., P. velutinus) (C:219, 220): white spongy rot, carie blanche spongieuse: on 3, 13, 17 NWT-Alta [1063]; on 13 BC [1012], Nfld [1659, 1660, 1661]; on 13, 17 Ont [1341]; on 17 Que [1385], NB, NS [1032].
C. versicolor (Polyporus v.) (C:220): on 7, 13, 17 Ont [1341]; on 13 NWT-Alta [1063], BC [1012], NB [1032]; on 17 NB, NS [1032], Nfld [1659, 1660, 1661]. See *Abies*.
C. zonatus (Polyporus z.): on 3 Alta [951]; on 3, 13 Yukon [F63, 460]; on 13 BC [1012], Alta [53, 452, 1801], NB [1032], NB, NS, PEI [F62]; on 17 Que [919], NB, NS [1032]. See *Betula*.
Corticium sp.: on 7, 13 Ont [1341]; on 13 NS [1032]; on 13, 17 NWT-Alta [1063].
Coryne sp.: on 13 Alta [452].
Costantinella micheneri: on 17 Ont [1340].
C. terrestre: on 3, 13 Man [1760].
Crepidotus cinnabarinus (C:215): on 3 Alta [1732]; on 17 Alta, Man [1017].
C. herbarum (C:215): on 14 BC [1012].
C. mollis (C. fulvotomentosus) (C:215): on 13 NS [1032]; on 14 BC [1012]; on 17 NB, NS, PEI [1032], Nfld [1659, 1660]. Associated with a white rot.
C. pubescens: on 17 PEI [1032]. Specimen possibly *C. herbarum*.

C. sepiarius (C:215): on 17 PEI [1032].
   Specimens which were the basis of this
   report and the Manitoba record in
   Conners have been redetermined
   (S.A. Redhead) as *Simocybe
   haustellaris*.
Cristinia helvetica: on 14 BC [1012].
   See *Pinus*.
Crustoderma dryinum: on 13 BC [1012].
   See *Abies*.
Crustomyces pini-canadensis (Corticium p.)
   (C:215): on 17 PEI [1032]. Lignicolous.
Cryptodiaporthe sp.: on 11a Que [F66]; on
   13 BC [1012].
C. populea (anamorph *Dothichiza
   populea*): on 5 Ont [1827, F69]; on 5,
   11a NB, NS, PEI [F68]; on 6 Que [184];
   on 11a NS [F70].
C. salicella (anamorph *Diplodina
   microsperma*): on 5, 13, 14 BC [1012].
C. salicina: on 11a Ont [1340]; on 13 Ont
   [1827]. See comment under *Salix*.
Cryptosphaeria populina (C:215): on 3
   Alta [F69]; on 3, 13, 14 BC [727]; on 13
   Yukon [727], Ont [1340]; on 13, 14 BC
   [1012].
*Cryptosporiopsis scutellata (Myxosporium
   s.)* (holomorph *Ocellaria ocellata*):
   on 13 BC [F63]; on 13, 17 Ont [1340].
Cryptosporium sp.: on 1, 17 BC [F62]; on
   17 BC [F63].
Cucurbitaria staphula (C:215): limb gall,
   tumeur des branches: on 3 Alta [F69];
   on 3, 13 Sask [12, 16]; on 13 Yukon
   [727], Alta [F68], Sask [F63]; on 14 BC
   [1012], Sask [1989]; on 17 Sask [F67],
   Sask, Man [F61, F62].
*Cylindrobasidium evolvens (Corticium e., C.
   laeve)* (C:215): on 13 Alta [1801],
   NB, NS [1032]; on 14 BC [1012]; on 17
   Que [969], NS, PEI [1032]. See *Abies*.
Cylindrocarpon destructans: on 3, 13 Man
   [1760].
C. orthosporum: on 3, 13 Man [1760].
Cylindrosporium sp.: on 3 Ont [1340].
C. populinum: on 7 Ont [1340].
C. saximontanense: on 2 Alta [F68].
Cylindrotrichum oligospermum: on 13 Man
   [1760].
Cyphellopsis anomala (Solenia a.): on 14
   BC [1012]; on 17 BC [1424], NS [1032].
   See *Alnus*.
C. confusa: on 17 BC [1424].
Cystostereum murraii: on 7 NS [1032]; on
   13 Ont [1341]; on 17 NB [1032]. See
   Acer.
Cytidia salicina (C:215): on 13 NB
   [1032]. Lignicolous.
Cytospora sp.: on 1, 3, 11a, 13 Ont
   [1340]; on 3 Sask [F67]; on 3, 5, 12,
   13, 14, 15, 16, BC [1012]; on 7, 13 Ont
   [1827]; on 13 Alta [452], NB, NS [1032];
   on 17 Yukon [460], NWT-Alta [1063], Alta
   [337], Ont [584], Que [919], NB [1032].
C. chrysosperma (holomorph *Valsa
   sordida*) (C:215): canker, chancre
   cytosporéen: on 1, 3, 5, 13, 17 Ont
   [1340]; on 1, 3, 15, 16 Sask [F68]; on
   1, 17 Alta [F65]; on 3, 13 Ont [F72]; on
   3, 13, 17 Alta [1594]; on 5 Ont [1827];

on 5, 14 BC [154]; on 6b Alta [F64]; on
   7 Man [F69]; on 11a Que [333], Nfld
   [1660, F75]; on 13 NWT [1970], NWT,
   Mack, Ont [F64], Alta [452, 952], Sask,
   Man [F62, F63], Man [F68], Ont [F70,
   F73, 1828], NS [F63], Nfld [F76]; on 13,
   17 Yukon [727], Sask [F65, F67]; on 16
   Sask, Man [1594], Man [F73]; on 17 Alta
   [338, 339, F62, F63], Sask [1151, 1595],
   Ont [1763, F69], Que [F67].
Cytosporella sp.: on 13 Ont [1828].
Dactylospora stygia (Karschia s.): on 17
   Ont [1340].
Daedaleopsis confragosa (Daedalea c.)
   (C:215): on 13 NB [1032]; on 14 BC
   [1012]. See *Acer*.
Dasyscyphus corticalis (Lachnella c.): on
   17 NS [1032].
D. virgineus (Lachnella v.): on 17 Ont
   [1340].
Datronia mollis (Trametes m.) (C:221): on
   14 BC [1012]. See *Acer*.
D. stereoides (Polyporus planellus): on
   13 Alta, Sask [579]. See *Alnus*.
*Dendrothele griseo-cana (Aleurocorticium
   g.)*: on 3 Ont [965]. Lignicolous.
D. macrospora (Aleurocorticium m.): on 17
   NWT-Alta [1063].
Dendryphion nanum: on 17 Man [1760].
Dendryphiopsis atra (holomorph
   Microthelia incrustans): on 2, 13
   Sask, Man [1760].
Diaporthe sp.: on 5, 14 BC [1012].
Dicoccum sp.: on 17 Ont [584].
Dictyosporium oblongum: on 13 Sask, Man
   [1760].
Didymium iridis: on 13 NWT-Alta [1063],
   Alta [53].
Dinemasporium canadense: on 17 Ont [1032].
Diplococcium spicatum: on 13 Man [1760].
Diplodia sp.: on 1 Ont [1340]; on 11a Que
   [F66].
D. tumefaciens (Macrophoma t.) (C:217):
   branch gall, tumeur des branches: on 3
   Alta [F65], Sask, Man [1987, 1994, F67],
   Man [1998, F73], Ont [F74]; on 3, 6, 11,
   13 Man [1992]; on 3, 13 NWT-Alta [1063],
   Alta [1995, F64], Sask [16, 1988], Sask,
   Man [F63, F64], Man [F68]; on 13 Mack
   [53], BC [F74], Alta [16, F68], Sask
   [F65], Sask, Man [F67], Man [1988, F72],
   Ont [1340], Que [F69]; on 13, 14 BC
   [1012]; on 13, 17 Alta, Ont [1988]; on
   14 BC [16, 1988, F61]; on 16 Sask, Man
   [1998]; on 17 Sask, Man [F61, F62], Man
   [F60], Ont [16, 1371].
Discina ancilis (D. perlata): on 14 BC
   [1012, F68].
Discosporium populeum (Dothichiza populea)
   (holomorph *Cryptodiaporthe populea*)
   (C:215): canker, chancre dothichizéen:
   on 1 Ont [1828]; on 1, 5, 11a Ont [339,
   F66]; on 1a, 3, 11a Ont [1340]; on 3 Ont
   [339]; on 5 Ont [F76], NB [F64]; on 5,
   11a Ont [F68], NB, PEI [1032]; on 5, 17
   Ont [F64]; on 6, 11a Ont [F74]; on 11 NB
   [1032]; on 11, 17 Que [1377]; on 11a Ont
   [338, 1827, F73, F61], Que [336, 339,
   F61, F66], NS [333, 337, 1032], Nfld
   [F75, F76]; on 11a, 16 Ont [F75]; on
   11a, 17 Ont [F62].

Dothiora sphaerioides (C:215): on 13, 17
Ont [1340].
Dothiorella canadensis: on dead branches
of 17 Ont [441].
Drepanopeziza sp.: on 3 Alta [951]; on 13
Alta [950, 952].
D. populi-albae (anamorph *Marssonina
castagnei*): on 1 BC [1012].
D. populorum (anamorph *Marssonina
populi*): on 3 Alta [644, 950]; on 6,
7, 16 Que [185]; on 13 Alta [F75], Alta,
Sask [F71, F74], Sask [F65, F66, F70,
F73], Sask, Man [F68], Man [F67, F69];
on 13, 14 BC [1012]; on 13, 17 Sask
[F67]; on 17 Alta [839].
D. tremulae (anamorph *Marssonina
brunnea*): on 7, 13, NB, NS [F69]; on
13 NB, NS [F73, F76], NB [F75].
Eichleriella deglubens (E. spinulosa)
(C:215): on 3, 13 NWT-Alta [1063], Mack
[53]; on 13 Alta [53]; on 14 BC [1012].
Associated with a white rot.
E. leveilliana: on 17 Ont [1341].
Lignicolous.
Endophragmia glanduliformis: on 13 Sask
[1760].
Endophragmiella bisbyi: on 13 Man [801,
1760].
E. biseptata (Endophragmia nannfeldtii):
on 13 Sask [1760], Man [802]; on 13, 17
Sask [802].
E. fuliginosa (Acrodictys f.): on 3, 13
Sask, Man [1756, 1760]; on 13 Sask
[808]; on 17 Sask, Man [1760], Man
[1756].
E. pallescens: on 13 Man [799].
Epicoccum sp.: on 14 BC [138]; on 17 Ont
[584].
E. purpurascens: on 3 Sask [1760]; on 13
Alta [1970].
Eutypa acharii (C:216): dieback,
dépérissement eutypéen: on 3 Sask, Man
[F67]; on 7 Ont [F64]; on 11a, 17 Ont
[1340]; on 13 Ont [1828]; on 14 BC
[1012].
Exidia glandulosa (E. spiculosa) (C:216):
on 13 Ont [1341], NB [1032], Nfld [1659,
1660]; on 14 BC [1012]. Lignicolous.
E. nucleata: on 17 NS [1032].
Lignicolous.
Exidiopsis grisea: on 17 BC [1012].
Lignicolous.
Favolus alveolaris (C:216): on 17 Ont
[1341]. See *Acer*.
Flagelloscypha citrispora: on 17 BC
[1424]. Lignicolous.
Flammula sp.: on 14 BC [1012].
Flammulina velutipes (Collybia v.)
(C:214): white spongy rot, carie
blanche spongieuse; on 13 NWT-Alta
[1063], BC [1012], Alta [1801] decay of
live trees; on stumps and dead trunks,
sometimes on dead parts of live trees of
17 Que [1385].
Flaviporus semisupina (Polyporus s.): on
17 NB [1036]. See *Betula*.
Fomes fomentarius (C:216): white mottled
rot, carie blanche madrée: on 3 BC
[F60], Ont [1377]; on 3, 13 NWT-Alta
[1063], Alta [53]; on 3, 14 BC [1012];

on 7, 17 NS [1032]; decay in live trees
of 13 Alta [1801]; on 13, 17 Ont [1341].
*Fomitopsis cajanderi (Fomes c., F.
subroseus)*: on 13 BC [1012, F67]. See
Abies.
F. pinicola (Fomes p.) (C:216): brown
cubical rot, carie brune cubique: on 3,
13, 17 NWT-Alta [1063]; on 5, 13, 17 Ont
[1341]; on 13 Alta [950, 1801 decay in
live tree]; on 13, 14 BC [1012]; on 13,
17 Alta [53], NS [1032].
F. rosea (Fomes r.): on 17 NB [1032].
See *Abies*.
Fumago sp.: on 13 NWT-Alta [1063].
Fusarium sp.: on 17 Que [919].
F. avenaceum: on 6 Que [184].
F. lateritium (C:216): canker, chancre
fusarien: on 5, 13 BC [1012].
F. solani (holomorph *Nectria
haematococca*) (C:216): on 6 Que [184].
Fusidium sp.: on 17 Que [919].
Ganoderma applanatum (Fomes a.) (C:216):
white mottled rot, carie blanche
madrée: on 3 Yukon [460, 1110], Mack,
Alta [53], Nfld [1659, 1661]; on 3, 5,
7, 11, 13 Que [1377]; on 3, 13 Sask, Man
[F67], Nfld [1660]; on 3, 13, 14
NWT-Alta [1063]; on 7, 13, 17 Ont
[1341]; on 13 NB, PEI [1032]; on 13, 14
BC [1012]; on 17 NB, NS [1032]. Poplar
shavings fermented by this fungus were
nutritionally equivalent to low quality
timothy hay when fed to livestock at
levels of 30% or less of the diet.
Geotrichum candidum: on 3, 13 Man [1760].
*Gloeocystidiellum karstenii (Gloeocystidium
k.)* (C:216): on 14 BC [1012]; on 17
Que [919], NB, PEI [1032]. See *Acer*.
G. leucoxanthum: on 13 NWT-Alta [1063],
Alta [53]. See *Acer*.
G. porosum: on 14 BC [1012]. See *Acer*.
Gloeodontia discolor (Odontia eriozona):
on 13 NWT-Alta [1063], Alta [53].
Lignicolous.
Gloeophyllum odoratum (Trametes o.): on
17 Ont [1341]. See *Abies*.
G. sepiarium (Lenzites s.) (C:217): brown
cubical rot, carie brune cubique: on 3,
13 NWT-Alta [1063]; on 13 BC [1012],
Alta [53], Que [1391], NS [1032]; on 17
NB [1032].
Gloeoporus dichrous: on 13, 14 BC
[1012]. See *Acer*.
G. pannocinctus (Poria p.): on 13 Mack
[1110]. See *Alnus*.
Gloeosporium sp.: on 11a NS [1032]; on 17
Man [F60], Que [F61].
Gloiodon strigosus: on 7 Que [F73]. See
Fraxinus.
Glyphium corrugatum: on 13 BC, Alta, Sask
[589].
G. leptothecium: on 13 Sask and on 13, 17
Alta [1759].
Godronia cassandrae f. *betulicola*: on
13 Que [684, F68]. See comment under
Alnus.
G. fuliginosa: on 13 Que [1680, 1684,
F68].
Graphium sp.: on 13 BC [1012]; on 17 Que
[919].

Guignardia populi (anamorph *Septogloeum rhopaloideum*): on 13 Yukon [F63, 477], BC [1012], Que [F75], NB, NS [1032].

Gymnopilus spectabilis (*Pholiota s.*) (C:219): on 3, 13 NWT-Alta [1063], BC [1012], Alta [1801] decay in live trees; on 13 Ont [831]; on 17 NB [1032]. See *Acer*.

Haematostereum rugosum (*Stereum r.*): on 17 NB, NS [1032]. See *Abies*.

H. sanguinolentum (*Stereum s.*): on 3 NWT-Alta [1063]. See *Abies*.

Hapalopilus nidulans (*Polyporus n.*) (C:219): on 13 BC [1012]. See *Abies*.

Haplographium sp.: on 17 Ont [584].

H. delicatum (holomorph *Hyaloscypha dematiicola*): on 13 Sask, Man [1760].

Haploporus odorus (*Trametes o.*): on 13 NWT-Alta [1063]. See *Alnus*.

Helicomyces scandens: on 13 Sask, Man [1760].

Hericium sp.: on 17 Ont [584].

H. americanum (*H. coralloides* N. Am. auct.): on 3 NWT-Alta [1063]. See *Acer*.

H. coralloides (*H. laciniatum, H. ramosum*) (C:216): white spongy rot, carie blanche spongieuse: on 3 NWT-Alta [1063], BC [F74]; on 3, 13, 14 BC [1012]; on 17 Alta [712, F62 on spruce (sic)], Que [1385].

Heterochaetella dubia (C:216): on 17 Ont [1015]. Lignicolous.

Hormodendrum sp.: on 17 Ont [584].

Hymenochaete agglutinans: on 17 PEI [1032]. A sterile state of *H. corrugata*. See *Alnus*.

H. corrugata: on 7, 13 Ont [1341]; on 17 PEI [1032]. See *Abies*.

H. spreta (C:216): on 17 BC [1012]. See *Acer*.

H. tabacina (C:216): on 7 Ont [1827]; on 13 Alta [1801], Nfld [1659, 1660, 1661]; on 13, 17 Ont [1341]; on 17 NB, NS [1032]. See *Abies*.

Hyphoderma inusitata (*Peniophora i.*) (C:218): on 14 BC [1012]. Lignicolous.

H. mutatum (*Peniophora m.*) (C:218): on 13 NB [1032]; on 13, 14 BC [1012]; on 17 Ont [1341], NB, NS, PEI [1032]. See *Acer*.

H. populneum (*Peniophora p.*) (C:218): on 13 Nfld [1659, 1660, 1661]. Associated with a white rot.

H. puberum (*Phlebia p.*): on 17 NS [1032]. See *Abies*.

H. sambuci: on 14 BC [1012]. See *Acer*.

H. setigerum: on 14 BC [1012]. See *Abies*.

Hyphodontia sp.: on 17 Ont [1341].

H. alutacea: on 17 Ont [1341]. See *Abies*.

H. arguta: on 14 BC [1012]. See *Abies*.

H. setulosa (*Steccherinum s.*): on 13 Mack [1110].

H. spathulata (*Odontia s.*) (C:218): on 13 BC [1012]. See *Abies*.

Hypochnicium analogum: on 14 BC [1012]. See *Acer*.

H. bombycinum (*Corticium b.*): on 13 Mack [1110]. See *Acer*.

H. vellereum (*Corticium v.*) (C:215): decay in live trees of 3 Alta [1801]; on 14 BC [1012]. See *Acer*.

Hypoxylon sp.: on 11a, 13 NB [1032]; on 17 NB, PEI [1032].

H. deustum: on 7 Ont [1340].

H. fuscum (C:216): on 3 Sask [F67].

H. mammatum (*H. pruinatum*) (C:216): hypoxylon canker, chancre hypoxylonien: on 3 NB [1032, F74]; on 3, 13 NWT-Alta [1063]; on 3, 17 Ont [F65]; on 7 Ont [F69, F73], Ont, NB, NS [F68]; on 7, 13 Que [778], NS [F67]; on 7, 13, 17 Ont [1340]; on 13 Mack, Alta [53], BC [137, 1012], Alta [52, 839, 950], Alta, Sask, Man [F69, F70, F74], Alta, Sask, Ont, Que, NB, NS [F73], Alta, Man, Que [F72], Sask [F65, 139], Sask, Man [F66, F67], Sask, Ont, Que [F68], Ont [357, 1827, F63, F66, F69], Ont, Que [F71], Ont, Que, NB [F75, F76], Que [728, 779, 1376, 1377], NB, NS, PEI [1032, F65, F66, F68], NS [F76]; on 13, 17 Alta, Sask, Man [F71], PEI [1032]; on 17 Mack [F62], Alta [338, F63, F65, F68], Sask, Man, Ont [F64], Ont, Que [F67], Que [1385, F66, F74]. One of the most serious diseases of poplars. Older cankers kill by girdling stems and branches. The average annual loss due to tree mortality from 1977-1981 was 11.2 million cubic metres of wood.

H. rubiginosum (C:217): on 13 Sask [F67]; on 17 Ont [1340].

H. serpens (C:217): on 13 BC [1012]; on 17 Ont [1340].

Hypsizygus ulmarius (*Pleurotus u.*) (C:219): on 13, 17 Ont [1341].

Hysterium pulicare (C:217): on 11a NS [1032].

Hysterographium fraxini: on 13 Man [F66].

H. mori (C:217): on 17 NS [1032].

Illosporium sp.: on 17 Ont [584].

Inonotus cuticularis (*Polyporus c.*) (C:219): white stringy rot, carie blanche filandreuse: on 3, 13 BC [1012]; on 7, 13, 17 Ont [1341].

I. glomeratus (*Polyporus g.*) (C:219): on 3 Ont [1341]; on 3, 13 BC [1012]. See *Acer*.

I. radiatus (*Polyporus r.*): on 13 BC [1110]. See *Acer*.

I. rheades (*Polyporus vulpinus, P. dryophilus* var. *v.*) (C:219): white pocket rot, carie blanche alvéolaire: on 7 Que [F64]; on 7, 13, 17 Ont [1341], NS [1032]; on 13 Mack [1110], BC [1012], Man [579]; on 13, 17 NB [1032]; on 17 Mack [579], BC [1014].

Irpex lacteus (*Polyporus tulipiferae*) (C:220): white spongy rot, carie blanche spongieuse: on 3 NWT-Alta [1063]; on 7, 13 Ont [1341]; on 14 BC [1341]; on 17 NB, NS [1032].

Junghuhnia nitida (*Poria attenuata, P. eupora*) (C:220): on 13 BC [1012]; on 17 Ont [1004], NB [1032]. See *Acer*.

Kirschsteiniella sp.: on 17 Que [919].

Kuehneromyces mutabilis: on 13, 14 BC
[1012].   See *Alnus*.
Laeticorticium sp.:   on 7 Ont [1341]; on
13 Ont [1827].
L. lundellii (Dendrocorticium l.):   on 3
Alta [930].   Lignicolous.
L. roseum (Aleurodiscus r.) (C:214): on 7
Ont [F66]; on 7, 13, 17 Ont [1341]; on
13 Ont [930, 1660, 1827]; on 13, 14 BC
[1012].   See *Acer*.
L. violaceum (Dendrocorticium v.):   on 17
Ont [930].   Lignicolous.
Lasiosphaeria hirsuta (C:217):   on 17 Ont
[1340].
L. pezicula: on 17 Ont [1340].
L. sulphurella: on 17 Ont [1340].
Laxitextum bicolor: on 13 Ont [1341]; on
17 NS [1032].   Lignicolous.
Leciographa gallicola (anamorph
Seimatosporium etheridgei):   on 13 BC,
Alta, Sask, Man [515].
Lentinellus cochleatus (Lentinus c.):   on
13 BC [1012].   See *Acer*.
L. vulpinus (Lentinus v.) (C:217):   on 11a
NS [1032]; on 14 BC [1012].   Similar to
L. ursinus.   The latter is associated
with a white rot.
Lenzites betulina (C:217):   on 13 NWT-Alta
[1063], Que [1391]; on 17 Ont [1341], NS
[1032].   See *Acer*.
Leptographium sp.:   on 13 BC [1012]; on 17
Que [919].
Leptosphaeria sp.:   on 13 Ont [1340].
Leucogyrophana olivascens: on 17 Ont
[581].   Associated with a brown rot.
Leucostoma nivea (Valsa n.) (anamorph
Cytospora n.) (C:221):   canker,
chancre cytosporéen: on 3, 7 NS [1032];
on 6 Que [184]; on 13 Ont [1340], NB,
NS, PEI [1032].
Libertella sp. (C:217):   on 3, 13 Alta
[1801]; on 13 Alta [452]; on 17 Ont
[584], Que [919], NB [1032].
Linospora tetraspora (C:217): leaf blight,
brûlure des feuilles: on 3 NWT-Alta
[1063], Mack, Que [F67], Alta [53],
Alta, Sask, Man [F65, F66, F67, F68,
F70, F71, F72], Alta, Man [F64], Sask
[839, F74], Sask, Man [F69, F73], Ont
[1340, 1827, F65, F68, F69, F72], Ont,
Que [79, F63, F64, F74, F76], Que [1377,
F61, F62]; on 14 BC [1012]; on 17 BC,
Alta [79].
Lophiostoma caespitosum: on 17 Ont [1340].
Lophiotricha viridicoma: on 17 Ont [1340].
Margarinomyces sp.:   on 17 Que [1377].
Marssonina sp.:   on 5 Ont [1340]; on 13
Sask, Man [F62, F63], Man [F64]; on 16
Alta, Sask [857].
M. brunnea (holomorph *Drepanopeziza
tremulae*) (C:217):   leaf spot, tache
des feuilles: on 1, 3, 13, 16 BC
[1012]; on 1, 13, 17 Yukon, BC [F62]; on
3 BC, Ont, Que, NS [1346]; on 5, 13 NS
[1032]; on 6 Ont [1346]; on 7 Ont, NB,
NS [1346]; on 7, 13 NB [1032]; on 13
Yukon, BC, Alta, Sask, Ont, Que, NB

[1346], BC [F73], NB [F62], Yukon [460];
on 17 BC [F63, F64], Que [F73].   This
pathogen can be seed borne.
M. castagnei (holomorph *Drepanopeziza
populi-albae*) (C:217):   on 1 BC, Ont,
Que, NS [1347].
M. populi (holomorph *Drepanopeziza
populorum*) (C:217):   on 1a, 3, 5, 6, 7,
11a, 12, 13, 14, 16, 17 Que [185]; on 3
NWT-Alta [1063], Ont [F69], Ont, Que
[F72], Que [1348]; on 3, 6 Ont [1348];
on 5 Ont [1827, F76]; on 6 Que [339]; on
13 Que [F66, F67], NS [333]; on 13, 14
BC [1348]; on 13 17 Ont [F64], Que
[F72]; on 17 Yukon [727], Sask, Alta
[225], Ont [1340, F76].
M. rhabdospora (holomorph *Pleuroceras
populi*) (C:217):   on 7 Ont [1340, F74].
M. tremuloides (C:217):   on 13 NWT-Alta
[1063], Mack [F61], Mack, Alta [53],
Alta [951, F70]; on 17 Yukon, Mack, Alta
[F66], Yukon, Alta [F68, F69], Alta
[F60, F62, F64, F65, F67].
Massaria salilliformis (C:217):   on 7 NS
[1032].
Melampsora sp. (C:217):   rust, rouille:
on 1 BC [1125]; on 1, 13, 16 BC [F61];
on 7, 13 NB, NS, PEI [F68]; on 13 Sask
[1341], Nfld [1660, 1661, F70]; on 17 BC
[F76], Ont [1827].
M. abietis-canadensis (C:217):   leaf rust,
rouille des feuilles: II III on 6, 7,
13, 17 Ont [1236]; on 7 Ont [1827, F69],
PEI [1032]; on 7, 13 Que [1377], NB, NS
[1032]; on 7, 13, 17 Ont [1341]; on 13
Man [F67], Que [282], Nfld [1659, 1660,
1661].
M. ?larici-tremulae (C:217):   on 17 Ont
[1236].
M. medusae (M. albertensis) (C:217):   leaf
rust, rouille des feuilles: II III on 3
Alta [952, F64, F65], Man [1838]; on 3,
5, 17 Que [1377]; on 3, 6, 6b, 13, 16,
17 Ont [1236]; on 3, 13 Ont [1828];
on 3, 13, 17 NWT-Alta [1063]; on 5, 13
Ont [1341]; on 6, 8, 13, 17 Que [F73];
on 6, 13 Ont [F70]; on 7, 13 BC [F61],
Ont, Que [F72], Que [F76], NS [1032]; on
13, 14 BC [1341, F62]; on 13, 16 Que
[F67]; on 13, 17 Alta [950]; on 13 Yukon
[F61, 460], NWT, Alta [2002], Mack
[F63], Mack, Alta [53], BC [1012, 1064,
1125, 2002, F70], Alta [644, 951, 952],
Sask, Man [F67, F68, F69], Sask, Man,
Que [F71], Ont [F73], Que [282, F65,
F74, F75], Que, NB, NS [F69, F70], PEI
[1032, F69]; on 17 Yukon, Mack, BC,
Alta, Sask, Man [2008], BC, Alta [F65],
Alta, Sask [339], Alta, Sask, Man, Que
[F70], Sask [F60], Sask, Man, Que [F66],
Man [F72], Que [F62, F69].
M. occidentalis (C:217):   leaf rust,
rouille des feuilles: on 3 Alta [950,
F70]; on 3, 14, 17 NWT-Alta [1063]; on
5, 14, 16, 17 BC [1012]; on 13, 14 BC
[F62]; on 14 BC [138, 1125, 2002, F60,
F66, F70], Alta [F61]; on 14, 17 BC
[1341, F65]; on 17 BC, Alta, Sask
[2008], Alta [338, 339, 952, F63], Sask
[F66, F67].

M. pinitorqua : on 1, 16 BC [2001] by
  inoculation; on 13 BC [F61]. This
  fungus was later shown to be *M.*
  medusae [1125].
Melanconis apocrypta (C:217): on 17 NS
  [1032].
Melanconium sp.: on 5 BC [1012]; on 13,
  17 Ont [1340]; on 17 Sask, Man [F60].
M. populinum : on 3 Ont [1340].
Melanomma sp.: on 1 BC [1012].
Menispora glauca : on 17 BC [826].
M. tortuosa : on 17 Ont [826].
Meruliopsis corium : on 14 BC [568,
  1012]. See *Acer*.
Merulius tremellosus (C:217): on NWT-Alta
  [1063], Alta [1801] decay in live trees,
  Sask, Man [568], NB [1032]; on 17 Ont
  [568, 1341, 1660]. See *Abies*.
Metulodontia nivea : on 13 BC [1012]. See
  Abies.
Monodictys levis : on 3, 13, 17 Man and on
  13 Sask [1760].
M. nitens : on 13 Sask, Man [1760].
Mucor sp.: on 17 Ont [584], Que [919].
Multiclavula mucida : on rotten wood of 17
  Que [1385].
Mycoacia fuscoatra (Odontia f.) (C:218):
  on 17 Ont [1341], NS [1032]. See *Acer*.
Mycosphaerella sp.: on 7, 17 Ont [1340];
  on 13 Nfld [1661, F68].
M. populicola (anamorph *Septoria p.*)
  (C:218): on 3 NWT-Alta [1063], Mack
  [53, F66], Alta [F67], Alta, Sask [F72],
  Sask, Man [F73], Que [1377, F61], NB
  [1032, F76]; on 3, 17 Man [1996]; on 14
  BC [1012, F71]; on 17 Man [F72], Que
  [F73].
M. populorum (anamorph *Septoria musiva*)
  (C:218): leaf spot and canker, tache
  des feuilles et chancre septorien: on 3
  Alta, Sask [F71, F72, F74], Sask, Man
  [F70], Ont [F63], Que [1377]; on 3, 11a
  Que [F73]; on 3, 13 Que [F74]; on 3, 17
  NWT-Alta [1063], Man [1996], Ont [1340];
  on 13 NB [1032]; on 14 BC [1012]; on 17
  Man [F72], Que [F62].
Nectria sp.: on 13 Ont [1340]; on 17 Ont
  [F61].
N. cinnabarina (anamorph *Tubercularia*
  vulgaris): on 13 Ont [1340], Que
  [282]; on 17 NWT-Alta [1063].
N. coryli : on 13 Que [F67].
N. dittisma : on 13 Ont [1827, F69].
N. galligena (C:218): canker, chancre
  nectrien: on 13 Man [1993], Ont [1340,
  1660]; on 13, 17 NS [1032].
N. inventa (anamorph *Verticillium*
  tenerum): on 14 BC [1012].
Nidularia sp.: on 13 NWT-Alta [1063].
Ocellaria ocellata (Pezicula o.) (anamorph
  Cryptosporiopsis scutellata) (C:218):
  on 1, 7, 13, 17 Ont [1340]; on 13
  NWT-Alta [1063], Sask, Man [F66], Ont
  [1828, F63], Que [F62], NS [1032]; on 17
  NB [1032].
Oidiodendron sp.: on 13 Alta [1970].
Oidium curtisii : on 17 BC [1012].
Oxyporus corticola (Poria c.) (C:220): on
  13 BC [1012]; on 17 NS [1032]. See
  Abies.

O. latemarginata (Poria ambigua) (C:220):
  on 17 NS [1032]. See *Acer*.
O. similis (Poria s.) : on 3 NS [1036].
  Associated with a white rot.
Paecilomyces farinosus : on 13 Alta [1970].
Panellus rigens (Panus salicinus)
  (C:218): on 17 BC [1012].
P. stipticus (Panus s.) (C:218): on 13
  Nfld [1659, 1660, 1661]; on 17 NB
  [1032]. See *Alnus*.
Panus rudis (C:218): on 13 Ont [1341]; on
  13, 14 BC [1012]. See *Acer*.
Papularia sp.: on 17 Que [919].
Parkerella populi : on 3, 13 BC [1012]; on
  13 BC [511].
Parmastomyces transmutans (Polyporus
  subcartilagineus): on 13 Mack
  [1110]. See *Abies*.
Pellidiscus pallidus : on 17 BC [1424].
Penicillium sp.: on 13 Alta [1970]; on 17
  Ont [584, 1340], Que [919].
Peniophora sp.: on 5 BC [1012]; on 13
  NWT-Alta [1063], Nfld [1659, 1660, 1661].
P. albobadia (Stereum a.) : on 13 NWT-Alta
  [1063]. Presumably a misidentification
  as the species is known only from the
  eastern United States.
P. aurantiaca (C:218): on 14 BC [1012].
  Lignicolous.
P. cinerea : on 7, 13 Ont [973]; on 17 NS
  [1032]. See *Abies*.
P. nuda : on 13 NB and on 17 PEI [1032].
  Lignicolous.
P. polygonia (Corticium p.) (C:218):
  white spongy rot, carie blanche
  spongieuse: on 7, 13 Ont [1341]; on 13
  Yukon [460], NWT-Alta [1063], BC [953,
  954], Alta [53, 452, 950, 951, 952, 1801
  decay in live trees], Man [1884], Ont
  [831], NS [1032], Labr [F73]; on 13, 14
  BC [1012]; on 17 Alta [1063, F62], Ont
  [969, 1827], Que [919], Nfld [1659,
  1660, 1661].
P. rufa (Cryptochaete r.) (C:218): on 3,
  13 Mack [53]; on 3, 13, 14 BC [1012]; 3,
  13, 17 NWT-Alta [1063], Ont [1341]; on
  7, 17 Ont [1341]; on 13 Yukon [460],
  Alta [53, 950, 951], Que [282], Nfld
  [1661]; on 13, 17 NB, NS, PEI [1032],
  Nfld [1659, 1660]; on 17 Que [969].
  Associated with a white rot of dead
  branches.
Petriella musispora : on 7 Ont [1037a].
Peyronelia sp.: on 13 Sask [1760].
Pezicula populi (Neofabraea p.) (C:218):
  on 1, 5, 13, 14 BC [1012]; on 5 Que
  [F63]; on 13 Ont [1340], Ont, Que [F62],
  Que [F68].
Peziza badioconfusa : on 17 Ont [573].
P. emileia : on 14 BC [1012].
P. repanda (C:218): on 14 BC [1012].
Phaeoisaria sparsa : on 3, 13 Man [1760].
Phaeomarasmius erinaceus : on 13 Ont
  [1434].
P. rhombosporus : on 13 Ont, NB and on 17
  Ont [1041].
Phaeoramularia maculicola (Cladosporium
  subsessile) (C:214): leaf spot, tache
  des feuilles: on 7 Ont, Que [1758], Ont
  [F63]; on 7, 13, 17 Ont [1340]; on 13

NWT-Alta [1063], BC [1012], BC, Sask,
Man, Ont, Que, Nfld [1758], Alta, Sask
[F73], Sask [F69, F71], Sask, Man,
[1760], Man [F70], Ont [F64, F68], Que
[1377], Nfld [1661]; on 14 BC [1012 as
"sessile", 1758]; on 17 Ont [1758].

Phanerochaete carnosa (Peniophora c.)
(C:218):   on 14 BC [1012]; on 17 NS
[1032].   See *Abies*.

P. laevis (Peniophora affinis):   on 17 NB
[1032].   See *Acer*.

P. sanguinea (Peniophora s.) (C:218):   on
13 BC [1012].   See *Abies*.

P. sordida (Peniophora cremea) (C:218):
on 13, 14 BC [1012].   See *Abies*.

P. tuberculata (Corticium t.) (C:215):   on
14 BC [1012].   Lignicolous.

Phellinus conchatus (Fomes c.) (C:216):
on 13 NB [1032]; on 17 Que [919].   See
Acer.

P. ferreus (Poria f.) (C:220):   on 13, 17
NB [1032].   See *Acer*.

P. ferruginosus (Poria f.) (C:220):   on 13
NWT-Alta [1063], Mack [1110], BC [1012];
on 17 NS [1032].   See *Abies*.

P. igniarius (Fomes i.) (C:216):   on 3 BC
[954, F65], NB [1032]; on 3, 5, 7, 11,
13 Que [1377]; on 3, 7, 13, 17 Ont
[1341]; on 3, 13 Alta [1801]; on 3, 13,
17 NWT-Alta [1063]; on 7 NS [1032]; on
13 Mack, Alta [53], BC [953], Alta [452,
950, 951, F72], Sask [F66], Man [1884,
F62, F72], Ont [831], Nfld [1661, F63];
on 13, 14 BC [1012]; on 13, 17 NB, NS,
PEI [1032]; on 17 Alta [F65], Man [F60],
Que [919].   See *Acer*.

P. laevigatus (Fomes igniarius var. l.,
Poria l.) (C:220):   on 3 Alta [53],
Nfld [1661]; on 13 Ont [1341]; on 13, 17
NB [1032].   See *Acer*.

P. punctatus (Poria p.) (C:220):   on 3, 17
NB [1032]; on 17 Ont [1341].   See
Abies.

P. tremulae (Fomes igniarius var.
populinus) (C:216):   decay in live
trees of 3, 13 Alta [1801]; on 13 Yukon
[728], BC [1012], BC, Alta [558], Que
[F72], Labr [F73]; on 17 Ont [584], NB
[1032].   Associated with a white rot.

P. viticola (Fomes v.):   on 14 BC [1012].
See *Abies*.

Phialocephala bactrospora (C:218):   on 14
BC [873].

P. repens: on 17 Ont [876].

Phialophora sp.:   on 13 Ont [1340]; on 17
Ont [584], Que [919].

P. fastigiata: on 17 Ont [279], Que [919].

P. lagerbergii: on 17 Que [919].

P. lignicola: on 13 Ont [1827]; on 17 Que
[919].

P. melinii: on 13 NB [279].

P. verrucosa: on 13 Sask, NB [279]; on 17
Que [919].

Phibalis fascicularis (Cenangium populneum,
Encoelia f.) (C:214, 216):   on 3 Mack
[664]; on 13 BC [1012], Que [1341], NB
[1032], Nfld [1659, 1660, 1661]; on 13,
17 Ont [1340], NS [1032]; on 17 Ont
[1827], NB [F69], PEI [1032].

P. pruinosa (Cenangium p., C. singulare)
(C:214):   sooty bark canker, chancre de
suie: on 3, 13 Yukon [727]; on 7 Ont
[1827]; on 7, 17 Ont [1340]; on 13 Yukon
[F62, 460], NWT-Alta [1063], BC [727,
1012, F74].

Phlebia sp.:   on 17 Ont [584].

P. expallens (Corticium e.) (C:215):
decay in live trees of 3 Alta [1801]; on
14 BC [1012].

P. radiata (C:218):   on 3 NWT-Alta [1063],
Mack [53]; on 13 BC [568, 1012]; on 17
Ont, Que [568].   See *Abies*.

P. rufa: on 3 Sask [568]; on 17 Alta, Que
[568].   See *Abies*.

Pholiota adiposa: on 7, 17 Ont [1341];
decay in live trees of 13 Alta [1801].
See *Acer*.

P. alnicola (Flammula a.) (C:216):   yellow
checked rot, carie jaune craquelée: on
13 BC [1012]; on 17 Que [919].   See
Abies.

P. aurivella (C:219):   on 13 NWT-Alta
[1063], BC [1012]; on 17 Que [919].   See
Abies.

P. destruens (C:219):   on 3 NWT-Alta
[1063], BC [954], decay in live trees
Alta [1801]; on 11a BC [1012]; on dead
trees of 17 Ont [1385].   Lignicolous.

P. flammans (P. kauffmaniana):   on 17 Ont
[1341].   See *Acer*.

P. limonella (P. squarrosa-adiposa):   on
13 Yukon [F64, 460].   See *Abies*.

P. squarrosa (C:219):   on 3 NWT-Alta
[1063]; on 17 Alta [463].   Associated
with a white rot.

P. squarrosoides (C:219):   on 3 NWT-Alta
[1063].   Lignicolous.

P. subsquarrosa (C:219):   decay in live
trees of 13 Alta [1801].

Phoma sp.:   on 13 BC [1012], Alta [452];
on 17 Ont [584].

Phomopsis sp.:   on 13 NB [1032].

Phragmocephala prolifera (Endophragmia
p.):   on 13 Man [1760].

Phyllosticta sp.:   on 3 Ont [F73]; on 7
Ont [1340]; on 17 NWT-Alta [1063], Que
[F67].

P. brunnea (C:219):   canker, chancre: on
3 Que [F67].   From the original
description it appears that this fungus
would now be placed in *Asteromella*.

P. intermixta (C:219):   on 3 Que [F62].
From the original description it appears
that this fungus would now be placed in
Phoma.

Phyllotopsis nidulans: on 13 Yukon [460].

Piloderma bicolor (Corticium b.) (C:215):
on 13 BC [1012].   See *Abies*.

Pleurotus sp.:   on 3, 13 NWT-Alta [1063];
on 17 NB, NS [1032].

P. ostreatus (C:219):   on 3, 13 Alta
[950]; on 3, 13, 17 NWT-Alta [1063]; on
7, 13, 17 Ont [1341]; on 11a BC
[1012]; on 13 Alta [53, F61]; on 13, 17
NB [1032]; on 17 Que [919], NS [1032].
See *Abies*.

P. salignus (C:219):   on 17 NS [1032].

P. sapidus (C:219):   on 13 NB [1032]; on
17 Ont [1341], NB, NS [1032].   See
Aesculus.

P. subareolatus (C:219): white spongy
    rot, carie blanche spongieuse: on 13
    Yukon [460], BC [1012]; on 13, 17 NB
    [1032].

Plicatura crispa (Trogia c.): on 17 Ont
    [1341].

P. nivea (Merulius n.): on 13 Sask [563,
    568]. See *Acer*.

Pollaccia elegans (holomorph *Venturia
    populina*) (C:219): on 1, 3 NB, NS
    [1032]; on 1, 3, 7, 13 NB, NS, PEI
    [F64]; on 3 Alta, Man, Ont, Que [F64],
    Sask, Man [1760], Sask, Man, Ont [F65,
    F66, F67, F68, F69], Ont [88, 1340, F61,
    F62, F63, F70, F71, F73], Ont, Que
    [F74], Ont, Nfld [F60], Nfld [1659,
    1660, 1661, F72]; on 3, 7, 11a, 13 NB
    [F63]; on 3, 7, 13 NB, NS, PEI [F65,
    F66]; on 5 Ont [F66]; on 6 Que [F74]; on
    7, 13 NS [F63]; on 13 BC [1012], Nfld
    [F65]; on 13, 17 Nfld [1660, 1661, F66,
    F70]; on 17 Alta [F62], Sask [F64], NB
    [1594], Nfld [F68].

*P. radiosa (Fusicladium r., Napicladium
    tremulae)* (holomorph *Venturia
    macularis*) (C:219): shoot blight,
    brûlure des pousses: on 1 NS [1032]; on
    1, 3, 7, 13 NB, NS, PEI [F64]; on 1, 13,
    17 Nfld [1660]; on 2 Alta [F63]; on 2,
    6b, 17 Sask [1760]; on 3, 7, 13 NB, NS,
    PEI [F65, F66]; on 3, 11a NB [F63]; on
    3, 13 Nfld [F64]; on 3, 13, 17 Nfld
    [1661]; on 7 Ont [1827], NB, NS [1032];
    on 7, 13 Ont [356, 1340, F60, F62, F63,
    F65, F67, F68, F69, F70, F72], Nfld [F61],
    NB, NS [F63], NB, NS, PEI [F67]; on 11a
    NB [1032]; on 13 BC [1340], Sask, Man
    [1760, F64, F65, F66, F67, F69], Man,
    Nfld [F62, F63], Ont [F61, F64, F66,
    F73], Ont, Nfld [F71, F74], Que [F66,
    F69], Que, Nfld [F67, F76], NB, NS, PEI
    [1032, F60], Nfld [F60, F65, F71, F72,
    F74], Labr [F74]; on 13, 16 Alta [F65];
    on 13, 17 Nfld [F66, F70, 1659]; on 17
    Yukon [727], Alta, Que [F62, F64], Sask,
    Man [F60, F68], Que [1377, F63, F65],
    NB, NS, PEI [F62], NB, PEI [1032], Nfld
    [F61, F68, F73].

Polyporus sp.: on 17 Ont [1341].

P. arcularis: on 13 Que [1389].
    Lignicolous.

P. badius (P. picipes) (C:219): on 3
    Yukon [460, 579]; on 3, 17 Mack [53]; on
    7, 17 Ont [1341]; on 13 Mack [1110], BC
    [1012]; on 13, 17 NWT-Alta [1063]. See
    Abies.

P. brumalis: on 13 Que [1389], Nfld
    [1659, 1660, 1661]. Lignicolous.

P. varius (P. elegans) (C:219): on 3
    Yukon [460]; on 13 Mack [53], PEI
    [1032]; on 17 NWT-Alta [1063], Ont
    [1341], Nfld [1659, 1660, 1661]. See
    Abies.

Poria sp.: on 3 NS [1032]; on 13 NWT-Alta
    [1063], Nfld [1659, 1661]; on 17 Ont
    [82], NB [1032], Nfld [1660].

P. aurantiaca: on 13 NWT-Alta [1063],
    Alta [53]. The name has been applied to
    several species. Specimens and reports
    should be verified.

P. subacida: on 17 Ont [1341], NS
    [1032]. See *Abies*.

Poronidulus conchifer (Polyporus c.): on
    17 NWT-Alta [1063], Alta [579].
    Associated with a white rot.

Propolis versicolor (P. faginea) (C:220):
    on 13 NS [1032].

Psathyrella sp.: on 7 Ont [1341].

P. conissans (Psilocybe c.) (C:220): on
    17 NS [1032].

P. spadicea: on stumps of 17 Que [1385].

Pseudospiropes nodosus: on 3, 13 Sask,
    Man [816, 1760].

P. simplex: on 3, 13 Man and on 13 Sask
    [817, 1760].

Pseudotomentella tristis (P. umbrina): on
    17 Alta, Ont [1828], Ont [1341].

Psilocybe sp.: on 3 NWT-Alta [1063].

*Punctularia strigosozonata (Phaeophlebia
    s., Phlebia s.)* (C:218): white spongy
    rot, carie blanche spongieuse: on 7,
    13, 17 Ont [1341]; on 13 NWT-Alta
    [1063], Alta [53, 1801], NB, NS [1032];
    on 13, 14 BC [1012]; on 17 Ont [1660],
    Que [969, 919], NB, NS, PEI [1032], Nfld
    [1659, 1661]. See *Alnus*.

*Pycnoporellus fulgens (Polyporus
    fibrillosus)*: on 3, 17 NWT-Alta
    [1063]. See *Abies*.

Pycnoporus cinnabarinus (Polyporus c.)
    (C:220): on 13 NWT-Alta [1063], Alta
    [579], NS [1032]; on 13, 17 Ont [1341];
    on 17 Ont [1181], NB [1032]. See *Acer*.

*Radulodon americanus (Radulum casearium
    auct.)* (C:220): white spongy rot,
    carie blanche spongiuse: on 3 Mack
    [53]; on 3, 13, 17 NWT-Alta [1063]; on
    13 BC [1012, 1341, 1499], Alta [53,
    1801], Ont [831], NB, NS [1032]; on 13,
    17 Ont [1341]; on 17 Alta [F62], Ont
    [1660, 1827, 1828], Que [919], NB [1032].

Radulum sp.: on 17 Ont [584]. Probably a
    specimen of *Radulodon americanus*.

*Ramaricium albo-ochraceum (Trechispora
    a.)*: on 17 Ont [572, 970].

Resupinatus silvanus (Pleurotus s.): on
    17 Yukon [460]. Now *Hohenbuehelia
    cyphelliformis*.

R. unguicularis (Pleurotus u.): on 13 NS
    [1032]. *Now Hohenbuehelia u.*

Rhytidiella baranyayi: on 13 BC [527,
    1012, F75].

R. moriformis: on 3 Yukon [460], Sask,
    Man [1994]. Causes cork-bark disease.

Sarcopodium tortuosum: on 13 Man [1760].

Schizophyllum commune (C:220): white
    spongy rot, carie blanche spongieuse:
    on 3 Sask [565]; on 3, 13, 17 NWT-Alta
    [1063]; on 13 BC [1012], Alta [53], NB
    [1032]; on 13, 17 Nfld [1659, 1660,
    1661]; on 17 Man, NB, NS [565], Ont
    [1341], NB [1032].

Schizopora paradoxa (Poria versipora):
    (C:220): on 13 NWT-Alta [1063]. See
    Abies.

Sclerophoma sp.: on 13 Ont [1827].

Sclerotium sp.: on 3 Alta [F64]; on 13
    Alta [53, F61]; on 17 Alta [F60].

Seimatosporium etheridgei (holomorph
    Leciographa gallicola): on 13 BC
    [1012].

Septogloeum sp.:   on 17 Sask [F60].
S. rhopaloideum (holomorph *Guignardia
    populi*) (C:220):   leaf blight, brûlure
    des feuilles: on 13 Yukon [F71], Sask
    [F66], Ont [1340], Ont, NB [F63], Que
    [F64, F67]; on 17 Alta [F66, F67, F68,
    F69].
Septomyxa populina: on 7, 13 Ont [79].
Septoria sp.:   on 3 Man [F63]; on 3, 13,
    17 Ont [1340]; on 6 Alta [F62]; on 13
    Que [F75]; on 16 Alta, Sask [857].
S. musiva (holomorph *Mycosphaerella
    populorum*) (C:220):   leaf spot, tache
    des feuilles: on 3 Alta, Sask [F74],
    Alta, Man [F60], Sask, Man [F65, F66,
    F67, F68, F69], Sask, Ont [F62], Man,
    Ont, Que [F64], Ont [F67, 1827], Que
    [337, F76]; on 3, 6, 7, 9, 17 Ont
    [1340]; on 3, 13, 17 Que [F72]; on 3,
    15, 16 Man [1998]; on 12 Sask [337]; on
    13 Sask [1340], Que [F67]; on 14 BC
    [F63]; on 17 Alta [225, 337, 338, 339,
    1594, 1595, F62, F63], Sask [225].
S. populi (holomorph *Mycosphaerella
    populi*):   on 3 Ont [F73].
S. populicola (holomorph *Mycosphaerella
    p.*):   (C:220):   leaf spot, tache des
    feuilles: on 3 Alta [F63], Sask [1340],
    Ont [1827, F73], Ont, Que [F76], Que
    [F64, F67]; on 3, 13, 17 Que [F72].
Septotinia populiperda (C:220):   on 13 Ont
    [F65].
Septotrullula bacilligera: on 13 Sask
    [1760].
Sirodothis populnea (*Dothiorella p.*,
    Pleurophomella spermatiospora)
    (anamorph *Tympanis spermatiospora*)
    (C:215, 219):   on 5, 13 Ont [1340, F67];
    on 13 NB [F62]; on 17 Alta [F63], Ont
    [1765, 1763] on branches.
Sistotrema sp.:   on 17 Ont [1341].
S. brinkmannii (*Trechispora b.*) (C:221):
    white stringy rot, carie blanche
    filandreuse: decay in live trees of 3,
    13 Alta [1801]; on 13 NWT-Alta [1063],
    Alta [53], Nfld [1659, 1660, 1661]; on
    14 BC [1012].   See *Abies*.
S. raduloides (*Trechispora r.*) (C:221):
    red heart rot, carie rouge du coeur: on
    13 Mack [1110]; 13, 14 BC [1012]; on 17
    Que [919].   See *Abies*.
Skeletocutis nivea (*Polyporus
    semipileatus*) (C:219):   on 13 BC
    [1012].   See *Amelanchier*.
Spadicoides atra: on 13 Man [1760].
S. bina: on 13 Sask, Man [1760]; on 14 BC
    [790].
S. canadensis: on 17 Ont [795].
Sphaeronaema sp.:   on 17 Que [919].
Spongipellis malicolus (*Polyporus spumeus
    var. malicola*): on 6 Que [F71].   See
    Fraxinus.
Spongipellis spumeus (*Polyporus s.*)
    (C:219):   on 17 NS [1032].   Associated
    with a white rot.
Sporidesmium sp.:   on 13 BC [1012].
S. achromaticum: on 3, 13 Sask [1760]; on
    13 Man [1760].
S. anglicum: on 13 Man [1760].

S. brachypus (*S. deightonii*):   on 3, 13
    Sask, Man [1760]; on 13 Man [822, F66];
    on 17 Ont [822].
S. ehrenbergii: on 3, 13 Man [1760]; on
    13 Sask [1760].
S. foliculatum: on 17 Ont [825].
S. leptosporum: on 13 Sask [1760].
S. pedunculatum: on 13 Man [1760].
Stachybotrys atra: on 3 Man [1760].
Steccherinum ciliolatum (*Hydnum c.*, *Odontia
    c.*) (C:218):   on 13 Alta [280]; on 13,
    14 BC [1012].   See *Acer*.
S. fimbriatum (*Odontia f.*) (C:218):   on 14
    BC [1012]; on 17 NS, PEI [1032].   See
    Acer.
S. ochraceum (C:220):   on 14 BC [1012].
    See *Acer*.
S. oreophilum: on 13 Mack, Alta [972].
    Associated with a white rot.
Stemonitis fusca: slime mold: on 3
    NWT-Alta [1063].   Presumably on rotted
    wood.
Stereum sp.:   on 17 Ont [584].
S. hirsutum (C:220):   on 3 NWT-Alta
    [1063].   See *Abies*.
S. ostrea (C:220):   on 17 NB [1032].   See
    Abies.
Stictis minor: on 17 Man, Ont [1608].
S. schizoxyloides: on 17 Man [1608].
Stigmina robusta: on 3, 13 Man [1760]; on
    3 Man [812]; on 17 Ont [812, 1760], NB
    [812].
Stromatoscypha fimbriata (*Porotheleum f.*)
    (C:220):   on 13, 17 Ont [1341]; on 17 NS
    [1032].   See *Betula*.
Subulicystidium longisporum (*Peniophora
    l.*) (C:218):   on 14 BC [1012]; on 17
    NWT-Alta [1063].   See *Betula*.
Taphrina sp.:   on 11a NS [1032].
T. aurea (C:221):   on 11 Que [1377]; on
    11, 11a Nfld [1659, 1660, 1661]; on 11a
    Que [1340], Nfld [1594]; on 13 Nfld
    [F73]; on 17 Nfld [1660].   Conners
    (C:221) did not think that this species
    occurred in Canada.
T. johansonii (C:221):   catkin blister,
    cloque de chatons: on 6, 13 Que [F72];
    on 7, 13 Que [1377]; on 13 Ont [1340].
T. populina (C:221):   leaf blister, cloque
    des feuilles: on 5 BC [F60]; on 5, 11a
    BC [1012]; on 11a BC, Nfld [337], Ont,
    Nfld [339], Que [F61, F71, F73], NB, NS,
    PEI [1032]; on 11a, 13 Nfld [F75]; on 13
    Nfld [338, F65]; on 13, 17 BC [336],
    Nfld [1661]; on 17 Que [336], Nfld
    [1660].
T. populi-salicis (C:221):   leaf blister,
    cloque des feuilles: on 14 BC [1012].
Tomentella asperula: on 17 Ont [929,
    1341, 1828].
T. atrorubra: on 17 Ont [1341].
T. botryoides: on 17 Ont [1341, 1827,
    1828].
T. bryophila (*Hypochnus subferrugineus*, *T.
    pallidofulva*) (C:221):   on 17 Ont [924,
    1341, 1827, 1828].
T. calcicola: on 14 BC [1012, 1341, 1828].
T. cinerascens: on 17 Ont [1341].
T. coerulea (*Hypochnus cervinus*, *T.
    papillata*):   on 13 Alta [924]; on 13,

14 BC [1341], BC, Alta [929]; on 14 BC [1012]; on 17 Ont [929, 1341, 1827].

T. crinalis (Caldesiella ferruginosa) (C:214): on 17 Ont [1341], NS [1032].

T. ellisii (T. microspora): on 17 Ont [929, 1341].

T. ferruginea (Grandinia coriaria) (C:221): on 14 BC [1012, 1341]; on 17 Man [925], Ont [1341, 1828].

T. lateritia: on 7 Ont [1828]; on 7, 17 Ont [1341].

T. neobourdotii: on 17 Ont [967, 1341, 1827, 1828].

T. nitellina: on 3, 17 Ont [1341].

T. punicea: on 17 Alta [1828].

T. ramosissima (T. fuliginea): on 17 Ont [1341, 1828].

T. ruttneri: on 17 Ont [1341, 1660, 1828].

T. sublilacina: on 17 Ont [1341, 1827].

T. subvinosa: on 17 Ont [1341].

T. terrestris (T. umbrinella): on 3, 17 Ont [1827].

T. violaceofusca (T. trachychaites): on 17 Alta [1828].

Tomentellastrum badium (T. floridanum, Tomentella atroviolacea): on 14 NWT-Alta [1012], BC [1341]; on 17 Ont [1341, 1828].

T. caesiocinereum (Tomentella c.): on 17 Ont [1341].

T. montanensis (Tomentella m.): on 13 Alta [1341].

Tomentellina fibrosa (Hypochnus canadensis, Kneiffiella f.): on 13 NWT-Alta [924]; on 17 Alta [1341, 1828].

Tomentellopsis zygodesmoides (Tomentella z.): on 13 NWT-Alta [1063].

Torula herbarum: on 13 Man [1760].

Trametes sp.: on 13 Ont [1341].

T. cervina (Polyporus biformis N. Am. auct.): on 17 Ont [1341]. See *Abies*.

T. suaveolens (C:221): on 3 Yukon [460]; on 3, 14 Yukon, BC [F63]; on 13 Que [1391, 1377]; on 13, 17 NWT-Alta [1063]; on 14 BC [1012]; on 17 Que [1385]. Associated with a white rot.

T. trogii (C:221): decay in live trees of 3 Alta [1801]; ; on 17 NB, NS [1032]. Associated with a white rot.

Trechispora alnicola: on 17 Ont [970].

T. mollusca (Poria m., P. candidissima): on 13 Mack [53], PEI [1032]; on 17 NS [1032]. See *Abies*.

T. pallidoaurantiaca: on 13 Alta [559]. See *Pinus*.

T. vaga: on 13 BC [1012]. See *Abies*.

Tremella mesenterica: on 3 Yukon [460]; on 13 NWT-Alta [1063], BC [953]; on 13, 14 BC [1012]. See *Acer*.

Trichaptum biformis (Hirschioporus pargamenus, Polyporus p.) (C:219): white spongy rot, carie blanche spongieuse: on 3 Yukon [F63, 460]; on 3, 7, 13, 17 Ont [1341]; 3, 13, 17 NWT-Alta [1063], Alta [53], on 13 Yukon [460, 1110], Mack [53, 1110], Que [1390]; on 13, 14 BC [1012]; on 13, 17 NB, NS [1032], Nfld [1659, 1660, 1661]; on 17 PEI [1032].

T. subchartaceum (Hirschioporus s., Polyporus s.) (C:220): on 13 Yukon [460, F62], BC [F62, 1012]; on 17 NWT, Alta, Sask [1823]. See *Betula*.

Trichia varia (C:221): on 13 NWT-Alta [1063], Alta [53].

Trichocladium canadense (C:221): on 17 Que [919].

Trichoderma sp.: on 14 BC [138]; on 17 Ont [584].

T. harzianum: on 13 Alta [1970]. Presumably saprophytic on bark, dead wood, etc.

T. viride: on 13 Sask, Man [1760]; on 17 Que [919]. See *Abies*..

Trichothecium roseum: on 13 Man [1760].

Trimmatostroma sp.: on 13 Ont [1827].

Tritirachium sp.: on 17 Ont [584].

Troposporella fumosa: on 3, 13 Sask, Man [1760].

Tubercularia vulgaris (holomorph *Nectria cinnabarina*): on 11a Ont [1340]; on 17 Alta [F63].

Tubeufia paludosa: on 17 Ont [1494].

Tulasnella violea: on 17 NS [1032]. Lignicolous.

Tympanis alnea (anamorph *Sirodothis inversa*): on 13 BC [1012]; on 17 BC, Ont, Que [1221].

T. heteromorpha: on 13 Que, NS [1221].

T. spermatiospora (anamorph *Sirodothis populnea*) (C:221): on 3, 13 Sask, Man [F69]; on 3, 13, 17 Ont [1340]; on 7 Man [F69]; on 13 BC [1012], Sask, Man [F67], Que [F60], NB [1032]; on 13, 17 Nfld [1659, 1660, 1661]; on 17 NWT-Alta [1063], Ont, Que, NB, NS, Nfld [1221], NS [1032].

Typhula sp.: on 14 BC [1012].

T. setipes (Pistillaria s.): on 11a BC [902, 1012].

Tyromyces aneirina (Poria a.) (C:220): on 13 NWT-Alta [1063], Mack [1110], BC [1012]; on 17 NWT, BC, Alta [1003], Ont [1827]. Associated with a white rot.

T. caesius (Polyporus c.) (C:219): on 13 BC [1012]. See *Abies*.

T. chioneus (Polyporus albellus) (C:219): on 13 Que [1390]; on 17 Ont [1341], NB [1032]. See *Abies*.

T. stipticus (Polyporus immitis): on 3 Sask [579]. See *Abies*.

T. subvermispora (Poria s.): on 17 Ont [1339]. See *Abies*.

T. tephroleucus (Polyporus t.): on 17 NS [1032]. See *Alnus*.

Ulocladium atrum: on 13 Alta [1970].

U. chartarum: on 17 Sask [1656].

Uncinula adunca (U. salicis) (C:221): powdery mildew, blanc: on 3 Mack [F67], Alta [644, 950, 951], Man, Que [1257], Ont [F71, 1660], Que [282, 333, F62, F65, F66, F76]; on 3, 6, 13 Ont [1257]; on 3, 13 Sask, Man [F66, F67, F68], Ont [1340], Que [F75], NB [1032]; on 3, 13, 17 NWT-Alta [1063]; on 3, 17 Sask [1340]; on 13 Yukon [F62, 460], Alta [53]; on 13, 17 Alta [F63]; on 14 BC [1012]; on 17 Ont [F68], Que [1377, F73].

Uredo sp.: on 1, 3, 5, 13, 16 BC [1012].
　　Probably the II state of *Melampsora*
　　spp.
Valsa sp.: on 3 Ont [1827]; on 7 Ont
　　[1828]; on 11a, 14 BC [1012]; on 13
　　Yukon [460], Ont [1340, F62], NB [1032];
　　on 17 Sask [F67].
V. ambiens (V. salicina, V. sordida)
　　(anamorph *Cytospora ambiens*) (C:221):
　　on 1, 3, 6b, 13, 17 NWT-Alta [1063]; on
　　3, 5, 7, 11, 13 Que [1377]; on 3, 13
　　Alta [F67]; on 6 Que [184, 337]; on 7,
　　11a, 13 NB, NS [1032]; on 7, 13 Ont
　　[1340]; on 11a Que [F66]; on 13 Mack,
　　Alta [53], BC [1012], PEI [1032]; on 17
　　NS [1032].
Valsaria exasperans (C:221): on 13, 17
　　Ont [1340].
V. insitiva (C:221): on 13, 17 Ont [1340].
Vararia investiens: on 17 Ont [1341].
　　See *Abies*.
Varicosporium elodeae: on 17 BC [43].
Venturia sp.: on 13 BC [F73, F74]; on 13,
　　17 BC [F72].
V. macularis (V. tremulae) (anamorph
　　Pollaccia radiosa) (C:221): on 2, 13,
　　17 NWT-Alta [1063]; on 3 NB [F73]; on 7,
　　13 Ont [1340], NB, NS, PEI [F68]; on 13
　　Yukon [460], Yukon, Alta, Man [F70,
　　F72], Mack, Alta [53, F66], BC [73, 953,
　　1012, F62, F68, F75, F76], Alta [950,
　　951, 952], Alta, Sask, Man [F71, F73],
　　Alta, Sask, NB [F74], Sask [F70], Ont,
　　NS [F76], Que [70, 282, F68, F72], Que,
　　Nfld, Labr [F75], NB [F73], NB, NS [F70,
　　F75], NB, NS, PEI [F69]; on 14 BC [F65];
　　on 17 Yukon [F69], Alta [337, F67, F69],
　　Man [839], Ont, Que, NB [73], Que [F61].
V. populina (anamorph *Pollaccia elegans*)
　　(C:221): on 3 Yukon [F61, F71, 460],
　　Yukon, Alta, Man [F69], Yukon, Man
　　[F70], Mack [53], BC, Alta [73], Alta
　　[950, 951, 952, F65, F67], Ont [88,
　　F65], Ont, Que [73, F75], Que [F61]; on
　　3, 5 Ont [1340]; on 3, 5, 17 Ont [F76];
　　on 3, 13, 14 BC [F68]; on 3, 14 BC
　　[1012]; on 3, 17 NWT-Alta [1063]; on 5
　　Ont [F66]; on 13 BC [F70]; on 13, 17 BC
　　[F69]; on 14 BC [F62]; on 17 BC [F71],
　　Alta [339, 1594], Sask [F67, F74], Man,
　　Sask [F66], Ont [F62].
Verpatinia calthicola: on 17 Que [662].
Verticillium sp.: on 13 Alta [1970]; on
　　17 Ont [584].
V. dahliae: on 13 Man [1760, F67].
Xylohypha lignicola: on 3 Man [1760]; on
　　13 Sask [1760].
X. nigrescens: on 13 Sask [1760].

PORTULACA L.   PORTULACACEAE

1. *P. grandiflora* Hook., rose-moss,
    chevalier d'onze heures; imported from
    South America, commonly cultivated.
2. *P. oleracea* L., common purslane,
    pourpier gras; imported from Eurasia,
    natzd., BC-PEI.

Fusarium acuminatum: on 1 BC [1816].
Pythium sp.: on 2 Ont [868].

P. intermedium: on 2 Ont [868].
P. irregulare: on 2 Ont [868].
Rhizophydium graminis: on 2 Ont [59].
Thielaviopsis basicola: on 2 Ont [546].
Verticillium dahliae: on 2 BC [1978].

POTENTILLA L.   ROSACEAE

1. *P. anserina* L., silverweed, argentine;
    native, Yukon-Mack, BC-Nfld.
2. *P. canadensis* L. (*P. simplex*
    Michx.), old-field-cinquefoil,
    potentille simple; native, Ont-Nfld.
3. *P. concinna* Richards.; native,
    BC-Man. 3a. *P. c.* var. *concinna*.
4. *P. diversifolia* Lehm.; native
    Yukon-Mack, BC-Alta, sw. Sask.
5. *P. flabellifolia* Hook.; native s.
    BC-se. Alta.
6. *P. fruticosa* L., shrubby cinquefoil,
    potentille frutescente; native,
    Yukon-Mack, BC-NS, Nfld.
7. *P. gracilis* Dougl.; native, Yukon,
    BC-Que. 7a. *P. g.* var.
    flabelliformis (Lehm.) Nutt.; native,
    BC-Man. 7b. *P. g.* var. *glabrata*
    (Lehm.) Hitchc. (*P. nuttallii*
    Lehm.); native, BC-Que. 7c. *P. g.*
    var. *permollis* (Rydb.) Hitchc.;
    native, BC. 7d. *P. g.* var.
    pulcherrima (Lehm.) Fern. (*P.
    pulcherrima* Lehm.) BC-Man, introduced
    Ont-Que.
8. *P. hippiana* Lehm.; native, BC-Ont.
9. *P. hookeriana* Lehm.; native,
    Yukon-Mack, BC.
10. *P. nivea* L., snowy cinquefoil,
    potentille des neiges; native,
    Yukon-Labr, BC-Alta, Man-Que, Nfld.
11. *P. norvegica* L., rough cinquefoil,
    potentille de Norvège; native,
    Yukon-Mack, BC-Nfld, Labr.
12. *P. palustris* (L.) Scop., marsh
    cinquefoil, potentille palustre; native,
    Yukon-Keew, BC-Nfld, Labr.
13. *P. pensylvanica* L., prairie
    cinquefoil, potentille de Pennsylvanie;
    native, Yukon-Mack, BC-Que, NS, Nfld,
    Labr. 11a. *P. p.* var. *pectinata*
    (Raf.) Boivin (*P. pectinata* Raf.),
    coast cinquefoil, potentille pectinée;
    native, Mack, Alta-Que, NS, Nfld, Labr.
14. *P. pulchella* R. Br.; native,
    Yukon-Labr, Man-Que, Nfld.
15. *P. recta* L., rough-fruited cinquefoil,
    potentille dressée, imported from
    Eurasia, introduced, BC-Nfld.
16. *P. tridentata (Soland.)* Ait.,
    three-toothed cinquefoil, potentille
    tridentée; native, Mack, Alta-Nfld, Labr.
17. Host species not named, i.e., reported
    as *Potentilla* sp.

Frommea obtusa (Uredo o.) (C:223): rust,
　　rouille: 0 I II III on 2 Ont [1236]; on
　　7 Sask, cited as "Moose Jaw, N.W. Terr."
　　[437]; on 17 Ont [1341].
Gnomonia comari (anamorph *Zythia
　　fragariae*): on 17 Nfld [79].
Leptosphaeria doliolum (C:223): on 11 Que
　　[70].

Marssonina potentillae (C:223): on 1 BC
[1816]; on 12 Mack [53].
Peronospora potentillae (C:223): downy
mildew, mildiou: on 7b BC [1816].
Phragmidium sp.: rust, rouille: on 13 BC
[1012, 1816]; on 17 Ont [1828].
P. andersonii (C:223): rust, rouille: 0
I II III on 6 Yukon [460], Mack [53], BC
[1012, 1816], BC, Alta [644], Alta [951,
952, 1839], Ont [1236], Que [282]; on 6,
17 NWT-Alta [1063].
P. biloculare: rust, rouille: on 5 BC
[1579].
P. boreale: rust, rouille: on 4 BC, Alta
[1579].
P. ivesiae: rust, rouille: I II III on
3a, 8 Alta [1579]; on 7 BC [1012]; on 7,
7b BC [1816]; on 7, 7c BC [1579]; on 7a,
7b, 7d BC, Alta, Sask [1579]; on 8 Sask
[1579]; on 11 Ont [1236, 1579]; on 15
Ont [1257, 1828], Ont, Que [1579]; on 17
Ont [1827].
P. potentillae (C:223): rust, rouille: 0
I II III on 3a, 13 Alta [1566]; on 7 BC
[1012, 1816]; on 8, 13 Sask BC [1566];
on 9, 13 Yukon, Mack [1566]; on 10 Que
[282]; on 10, 11a Que [1566]; on 13 Mack
[53], BC, Man [1566]; on 15 Ont [1236,
1341, 1828]; on 17 Yukon [460], NWT-Alta
[1063], Alta [53, 951].
Plagiostoma lugubre: on 12 BC [79].
Pleospora moravica (C:223): on 6 Que [70].
Pucciniastrum potentillae (C:223): rust,
rouille: on 16 Alta [2008, F67], Man
[2008], Ont [1236, 1341], Que [1236].
Scopinella solani: on 14 Frank [1039].
Sphaerotheca macularis (C:223): powdery
mildew, blanc: on 6 Ont [1254], Que
[1257]; on 6, 11, 12 Ont [1257]; on 12
NS [1254].
Venturia palustris: on living and
overwintered leaves of 17 Que [73].
V. potentillae (Coleroa p.) (C:223): on
living and overwintered leaves and
stalks of 14 NWT [73].
Verticillium dahliae: on 7b BC [1816,
1978].

PRENANTHES L.   COMPOSITAE

1. *P. alata* (Hook.) Dietr.; native,
BC-Alta.
2. *P. alba* L., rattlesnake-root, prenante
blanche; native, Sask-Que.
3. *P. altissima* L., tall
rattlesnake-root, prenante très haute;
native, Man-Que, NS.

Puccinia dioicae (C:224): rust, rouille:
on 2 Ont [1236].
P. insperata (P. variabilis var. *i.)*
(C:224): rust, rouille: on 1 BC
[1816]; on 2 BC [1012].
P. orbicula (C:224): rust, rouille: 0 I
II III on 2, 3 Ont [1236]; on *P.* sp.
Ont [1236], NB [1032].
Sphaerotheca fuliginea (C:224): powdery
mildew, blanc: on 3 Ont, Que [1257].

PRIMULA L.   PRIMULACEAE

Perennial herbs, low, mostly boreal or
alpine, confined to northern hemisphere,
only one or two in s. South America.

Ramularia primulae (C:224): leaf spot,
tache foliaire: on *P.* spp. BC [1816].

PRUNELLA L.   LABIATAE

1. *P. vulgaris* L., heal-all, brunelle;
native (partly introduced), BC-Alta,
Man-Nfld, Labr.

*Linospora brunellae (Ceuthocarpon b.,
Leptosphaeria hesperia)* (C:224): on 1
BC [72, 1816].
Ophiobolus prunellae: on 1 BC [1012,
1620].

PRUNUS L.   ROSACEAE

1. *P. americana* Marsh., wild plum,
prunier d'Amérique; native, Sask-Que.
2. *P. armeniaca* L., apricot, abricotier;
imported from China.
3. *P. avium* (L.) L., sweet cherry,
cerisier de France; imported from
Eurasia.
4. *P. besseyi* Bailey, sand cherry,
cerisier de sable; native, Sask-Man.
5. *P. cerasifera* Ehrh., myrobalan plum,
prunier myrobalan; imported from Asia.
5a. *P. c.* cv. atropurpurea (*P.
pissardii* Carrière), Pissard plum,
prunier mycobolan pourpre.
6. *P. cerasus* L., sour cherry, cerisier
commun; origin uncertain.
7. *P. domestica* L., plum, prunier
domestique; imported from Eurasia. 7a.
P. d. var. *insititia* (L.) Schneid.
(*P. insititia* L.), damson plum,
créquier.
8. *P. dulcis* (P. Mill.) D.A. Webb (*P.
amydalus* Batsch, *P. tangutica*
Batal.), almond, amandier; imported from
w. Asia.
9. *P. emarginata* (Dougl.) Walp., bitter
cherry, cerisier amer; native, BC.
10. *P. glandulosa* Thunb., dwarf flowering
almond, cerisier glanduleux; imported
from Japan and China.
11. *P. japonica* Thunb., Chinese bush
cherry, cerisier du Japon; imported from
China and Korea.
12. *P. laurocerasus* L., cherry-laurel,
laurier-cerise; imported from Eurasia.
13. *P. mahaleb* L., perfumed cherry, bois
de Sainte-Lucie; imported from Eurasia.
14. *P. nigra* Ait., Canada plum, guignier;
native, Man-NS.
15. *P. padus* L., bird cherry, cerisier à
grappes; imported from Eurasia. 15a.
P. p. var. *commutata* Dipp., May-day
tree.
16. *P. pensylvanica* L. f., pin cherry,
cerisier d'été; native, Mack, BC-Nfld,
Labr.

17. *P. persica* (L.) Batsch, peach, pêcher;
imported from China. 17a. *P. p.* var.
nicipersica (Suckow) C.K. Schneid.
(*P. persica* var. *nectarine* (Ait. f.)
Maxim.), nectarine, nectarine.
18. *P. pumila* L., sand cherry, cerisier de
sable; native, Man-NB. 18a. *P. p.*
var. *depressa* (Pursh) Gleason (*P.
depressa* Pursh), dwarf cherry, cerisier
déprimé; native, Ont-NB.
19. *P. salicina* Lindl., Japanese plum,
prunier japonais; imported from China.
20. *P. serotina* Ehrh., black cherry,
cerisier d'automne; native, Ont-NS.
21. *P. serrulata* Lindl., Oriental cherry,
cerisier à feuilles dentées en scie;
imported from Asia.
22. *P. spinosa* L., sloe, prunellier;
imported from Eurasia.
23. *P. subcordata* Benth., Sierra plum,
prunier du Pacifique; imported from sw.
USA.
24. *P. subhirtella* Miq., Higan cherry;
imported from Japan. 24a. *P. s.* cv.
pendula.
25. *P. tomentosa* Thunb., Nanking cherry;
imported from c. Asia.
26. *P. triloba* Lindl., flowering almond,
amandier rose; imported from China.
27. *P. virginiana* L., choke cherry,
cerisier de Virginie; native, Mack,
BC-Que, NS, Nfld. 27a. *P. v.* var.
demissa (Nutt.) Torr. (*P. demissa*
(Nutt.) Walp.), western choke cherry,
cerisier à grappes de la côte du
Pacifique; native, BC-Alta. 27b. *P.
v.* var. *virginiana*; native, Mack,
BC-Que, NS, Nfld.
28. Host species not named, i.e., reported
as *Prunus* sp.

Alternaria sp. (C:225): on 7 Ont [334],
NS [591]; on 28 Ont [1930].
A. alternata (*A. tenuis* auct.): on 3 BC
[1099]; on submerged leaves of 12 BC
[5]; on 13 Ont [3].
Aleurodiscus canadensis: on 16 Ont
[1341]. See *Abies*.
Antrodia lindbladii (*Poria cinerascens*)
(C:227): on 9 BC [1012]. See *Abies*.
Apiosporina morbosa (*Apiospora m.*,
Dibotryon m.) (C:226): black knot,
nodule noir: on 2, 3, 6, 7, 7a, 22, 23,
27a BC [1816]; on 3 Nfld [F62]; on 6 NB
[1594], PEI [334]; on 6, 20, 27, 28, Que
[338]; on 7 BC [334, 335, 1817, 1818],
Ont [334, 339, 591], Que, NB, NS, PEI
[335, 336] also NB, NS, PEI [334] and NS
[591], Nfld [336]; on 7, 8, 22, 23, 27
BC [1012]; on 7a Nfld [1661]; on 14 Ont
[1340], NS [334, 1032]; on 15 Alta [336,
337, 339]; on 15a Alta [1594], Man
[335]; on 15a, 16, 27, 28 NWT-Alta
[1063]; on 16 Alta [293], Sask, Man
[1594, F67], Ont [338, F64, F65], Que
[70, 282, 1377, 1376, 1385, F74], NB
[F61, F72], NS [293], PEI [338, 339,
1032, F64, F65, F66, F67, F69], Nfld
[337, 339, F62, F64, F65, F66, F68, F69,
F70, F71, F72, F75] and NB, NS [335,

336, 338, 339, 1032, F62, F63, F64, F65,
F66, F67, F69, F75]; on 16, 20, 27 Ont
[339, 1340, F66]; on 16, 28 Nfld [1659,
1660, 1661, F63, F74, F76]; on 17 NS
[591]; on 20 Que [338, F65], NB, NS
[1032, F69, F75], PEI [F69]; on 27 Alta
[53, 839, 950], Man [1594], Ont [293,
F64], Que [293, 338, 1385], NB [F72], NS
[335, F62], PEI [F64], Nfld [1661, F65]
and Sask, Man [F63, F67], NB, NS [336,
1032, F63, F64, F75]; on 28 BC [338,
339], Alta, Man [333], Ont [335, 338,
1340, 1594, 1827, F65, F66, F67, F74],
Que [333, 334, 336, 337, 338, 339,
1340], NB, NS, PEI [333, 338, 1032],
Nfld [333, 335, 336, 338] and NB [334,
335, 337, 339, 1612, 1595], NS [339,
1594, F62]. See *Phaeostoma* below.
Aplosporella sp. (*Haplosporella* sp.):
on 16 Nfld [1659, 1660].
Armillaria mellea (C:225): root rot,
pourridié-agaric: on 3, 5, 6 BC [1816];
on 5 BC [1012]; on 13 Ont [3]; on 20 Ont
[F63]. See *Abies*.
Ascocoryne inflata: on 9 BC [1012].
A. sarcoides (*Coryne s.*): on 17 Ont
[1930].
Aureobasidium sp. (*Pullularia* sp.):
on 17 Ont [1930].
A. pullulans (*Pullularia p.*) (C:225): on
3, 6 BC [1816]; on 28 BC [334].
Basidioradulum radula (*Hyphoderma r.*,
Radulum orbiculare): on 28 Ont [1828],
PEI [1032]. See *Abies*.
Bjerkandera fumosa (*Polyporus f.*): on 12
BC [1012]. See *Acer*.
Blumeriella jaapii (*Coccomyces hiemalis*,
Higginsia h.) (anamorph *Phloeosporella
padi*) (C:225): shot hole, criblure: on
3 BC, Ont, NB [333], NS [335, 336, 337,
1594]; on 6 Ont [334], NB [1594], PEI
[335] and Ont, NB, Nfld [337]; on 12 BC
[1012]; on 16 Alta, Sask, Man, NB, NS,
PEI [F66], Sask, Man [F67, F68] also
Sask [F71, F73], Man [339, F70], Que
[F61, F75, F76], NB, NS, PEI [1032, F64,
F65] also NB, NS [F63] and NB [F72]; on
16, 27 Ont [339, 1340, F64]; on 20 NB
[336, 1032], NS [1032, F63]; on 27 Alta
[1594, F67], Sask, Man [F66, F67, F68],
NB [1032], Nfld [1661, F65, F74] and
Sask [F71, F72, F73, F74], Man [339,
F70]; on 28 Ont, NB, NS, PEI [334], Nfld
[1661] and Ont [339], NB [335, 336, 338,
339, F73].
Botrytis cinerea (holomorph *Botryotinia
fuckeliana*) (C:225): gray mold,
moisissure grise: on 2, 3, 6 BC [1816];
on 27 BC [F70]; on 28 BC [334, 339], NS
[337, 1032].
Calosphaeria minima (C:225): on 28 Ont
[1340, F60].
C. princeps: on 16 Ont [1340, F60].
C. pulchella: on 16 Ont [1340].
Candida sp.: on 16 Ont [1827].
Cephalothecium roseum: on 7, 17 NS
[591]. Usually listed as a questionable
synonym of *Trichothecium roseum*.
Cerocorticium cremoricolor (*Corticium
c.*): on 28 NS [1032]. Lignicolous.

Cerrena unicolor (Daedalea u.) (C:226):
on 28 NS [1032]. See *Acer*.

Chlorociboria aeruginosa: on 20 Ont
[1340]. Lignicolous, typically on
rotted wood.

Chlorosplenium chlora: on 7 BC [1012].

Chondrostereum purpureum (Stereum p.)
(C:228): silver leaf, plomb: on 7, 9,
12 BC [1816]; on 9 BC [1012]; on 28 PEI
[1032]. See *Abies*.

Cladosporium sp.: on 9 BC [1012]; on 14
NB [1032]; on 16 Ont [1340]; on 27 Ont
[1827].

C. carpophilum (Fusicladium c.) (holomorph
Venturia c.) (C:226): scab,
tavelure: on 28 Que [337].

C. macrocarpum: on 27 Sask [1760].

Coccomyces sp.: on 16 Que [1377].

Coltricia perennis (Polyporus p.): on 16
NS [1032]. Mycorrhizal with conifers,
typically on sandy soil, this specimen
was presumably on rotten wood.

Coniothyrium sp. (C:226): on 7, 17 NS
[591].

C. fuckelii (holomorph *Leptosphaeria
coniothyrium*): on 17, 28 Ont [1930].

Conoplea olivacea (C. sphaerica) (C:226):
on 28 Ont [784].

Coriolus hirsutus (Polyporus h.) (C:227):
white spongy rot, carie blanche
spongieuse: on 2 BC [1012]; on 2, 7 BC
[1816]; on 20 Ont [1341]; on 28 NS, PEI
[1032].

C. versicolor (Polyporus v.) (C:227): on
9 BC [1012, 1341]; on 16 Nfld [1659,
1660, 1661]; on 16, 28 PEI [1032]; on
20, 28 Ont [1341]; on 27, 28 NS [1032].
See *Abies*.

Corollospora maritima: on submerged
leaves of 12 BC [5].

Cryptodiaporthe salicella (anamorph
Diplodina microsperma): on 28 BC
[1012].

Cylindrocarpon sp.: on submerged leaves
of 12 BC [5].

C. destructans: on 13 Ont [3].

Cylindrosporium sp. (C:226): on 2 BC
[339, 1816]; on 16 PEI [1032], Nfld
[1661]; on 16, 27 NB, NS [1032].

Cytospora sp. (C:226): canker, chancre
cytosporéen: on 2 Ont [1594]; on 16,
20, 28 Ont [1340]; on 27 Yukon [460].

C. chrysosperma (holomorph *Valsa
sordida*): on 16 Nfld [F76].

C. cincta (holomorph *Leucostoma
cincta*): on 2, 17 Ont [397].

C. pruni: on dead branches of 27 Ont
[439].

C. rubescens (C. leucostoma) (holomorph
Leucostoma persoonii) (C:226):
dieback, dépérissement: on 17 Ont [397].

Daedaleopsis confragosa (Daedalea c.)
(C:226): on 20 Ont [1341]; on 28 BC
[1012]. See *Acer*.

Daldinia vernicosa (C:226): on 28 Ont
[1340].

*Dendrothele maculata (Aleurocorticium
m.)*: on 27, 28 Ont [965]. Lignicolous.

Dendryphiopsis atra (holomorph
Microthelia incrustans): on 28 Man
[1760].

Dermea cerasi (anamorph *Foveostroma
drupacearum*) (C:226): on 9 BC [1012,
1816]; on 16 Ont [664], NS [1032]; on
16, 20, 28 Ont [1340]; on 27 Sask, Man
[F67]; on 28 Man [F66], NB, NS [1032].

D. padi (C:226): on 16 Ont [1340, F60].

D. prunastri (C:226): on 27, 28 Ont
[F60]; on 28 Ont [1340], NS, PEI [1032].

Diaporthe ?perniciosa: on 13 Ont [3].

D. pruni (C:226): on 28 Ont [F60].

D. prunicola (C:226): on 16, 20, 27, 28
Ont [1340]; on 27, 28 Ont [F60].

Diatrype albopruinosa (C:226): on 27 Man
[1354].

Diatrypella discoidea (C:226): on 28 NS
[1032].

Dictyosporium pelagicum: on submerged
leaves of 12 BC [5].

Diplodia sp.: on 27 Ont [1340].

Eutypella sorbi: on 27 NWT-Alta [1063].
A later homonym of *E. sorbi* (Schum.:
Fr.) Sacc.

Exidia glandulosa: on 28 Nfld [1659,
1660]. Lignicolous.

E. recisa (C:226): on 20 NS [1032].
Lignicolous.

Fomes fomentarius: on 16 Nfld [1661].
See *Acer*.

Fomitopsis cajanderi (Fomes c.): on 16
Que [F74]. See *Abies*.

F. pinicola (Fomes p.) (C:226): on 7 BC
[1816]; on 20 Ont [1150]. See *Abies*.

Foveostroma sp.: on 28 BC [1012].
Reported as *Micropera* sp. but that
name is invalid.

F. drupacearum (Micropera d.) (holomorph
Dermea cerasi) (C:227): on 27 Ont
[1340]; on 28 BC, Ont, NS [1763].

Fumago sp.: on 16 NB [1032].

Fusarium sp. (C:226): on 7, 17 NS [591].

F. oxysporum (C:226): on 13 Ont [3].

F. poae: on 13 Ont [3].

Fusicoccum sp.: on 17 Ont [1930].

F. amygdali: on 17 Ont [1930].

Ganoderma applanatum (C:226): on 3, 6, 7
BC [1816]. See *Abies*.

G. tornatum (G. applanatum var.
brownii.) (C:227): on 5 BC [1007,
1816, F61]. Lignicolous.

Gloeodes pomigena (C:227): on 28 NS
[1032].

Gloeophyllum sepiarium (Lenzites s.)
(C:227): on 9 BC [1012]; on 28 NS
[1032]. See *Abies*.

Gloeosporium sp.: on 28 NB [1032].

Godronia confertus (G. urceolus var. *c.)*
C:227): on 9 BC [1012]; on 16 Ont
[F66], Que [1684]; on 16, 28 Ont [656,
1340]; on 27 Que [F68]; on 28 Que [656],
NS [1032].

Graphostroma platystoma: on 20 Que [1343].

Haematostereum gausapatum (Stereum g.):
on 16 Ont [1341]. Lignicolous.

H. rugosum (Stereum r.): on 16 NS [1036];
on 16, 27 NS [1032]. See *Abies*.

Hapalopilus nidulans (Polyporus n.)
(C:227): on 9 BC [1012]. See *Abies*.

Helotium sp.: on 28 Ont [1340]. The
specimen is probably a species of
Calycella.

Humicola alopallonella: on submerged
leaves of 12 BC [5].
Hymenochaete sp.:   on 27 Ont [1341].
Lignicolous.
H. tabacina (C:227):   on 20 Ont [1341]; on
28 NB, NS [1032], Nfld [1659, 1660,
1661].   See *Abies*.
Hyphoderma setigerum (Peniophora aspera)
(C:227):   on 9 BC [1012]; on 28 NS
[1032], Nfld [1659, 1660, 1661].   See
Abies.
Hypholoma fasciculare (Naematoloma f.):
on 5a BC [1012].   See *Abies*.
Hypoxylon fuscum: on 16 Ont [1340].
H. howeianum: on 28 Ont [1340].
H. multiforme (C:227):   on 9 BC [1012]; on
28 NB [1032].
H. vogesiacium: on 9 BC [F70].
Inonotus dryophilus (Polyporus d.):   on 16
NB [1036].   Associated with a white rot.
I. radiatus (Polyporus r.):   on 20 Ont
[1341].   See *Acer*.
Irpex lacteus (Polyporus tulipiferae)
(C:227):   on 2 BC [1012, 1816]; on 16 NB
[1032]; on 16, 20, 28 Ont [1341]; on 28
NS [1032].   See *Acer*.
*Laeticorticium violaceum (Dendrocorticium
v.)*:   on 28 Ont [930].   Lignicolous.
Laetiporus sulphureus (Polyporus s.)
(C:227):   on 20 Ont [1341]; on 28 NS
[1032].   See *Abies*.
Leucostoma cincta (Valsa c.) (C:229):
peach canker, chancre cytosporéen: on
17 Ont [335, 844, 1183, 1924, 1931]; on
17, 28 Ont [930].
L. persoonii (Valsa leucostoma) (C:229):
dieback, dépérissement: on 7, 17, 17a
BC [1816]; on 16 Ont [1340]; on 16, 27
NS [1032]; on 17 Ont [335, 844, 1183,
1930, 1924].
Lulworthia medusa: on 12 BC [5].   Usually
on *Spartina*.
Massaria conspurcata (C:227):   on 27 Man
[F67]; on 27, 28 Ont [1340]; on 28 Man
[F66].
M. pruni (C:227):   on 28 NS [1032].
Merismodes fasciculatus (Cyphella f.):   on
28 Ont [1341].   Lignicolous.
Monilia sp.:   on 16 Ont [1827], NB [1032].
M. fructigena auct.:   on 16 NS [1032].
Monilinia demissa (C:227):   brown rot,
pourriture brune: on 2, 27a BC [336];
on 27 BC [1012, 1816]; on 27a BC [334,
335].
M. fructicola (C:227):   brown rot, and
blossom and twig blight, pourriture
brune: on 2, 3, 4, 6, 7, 17, 17a, 21 BC
[1816]; on 2 Ont [334]; on 2, 17 BC
[336, 337, 339, 1594, 1595]; on 3 Ont
[333], NS [1594]; on 3, 6, 7a, 14, 17,
18, 28 Ont [432]; on 4, 25, 27 Man
[432]; on 6 NB [335, 1594]; on 6, 14, 28
Que [432]; on 7 NB [336]; on 14 Ont
[1340, 1827]; on 16 NB, NS [1032]; on
17, 28 Ont [333, 335, 338]; on 17 Ont
[337], Que [334], NS [333, 334, 1594]
and Ont, NS [335, 336, 339]; on 26 BC
[336], NS [333]; on 27 BC [333]; on 28
BC [334, 337, 339, 432, 1594], Ont [334,
335, 337, F74], Que [333], NB [335], NS
[334, 339], PEI [337, 339].

M. laxa (C:227):   blossom and twig blight,
and brown rot, pourriture brune: on 2,
28 BC [339]; on 2, 3, 6, 7, 10, 11, 24a
BC [1816]; on 3 BC [1594]; on 24a BC
[334]; on 28 BC [333].
M. padi (C:227):   blossom and twig blight,
brûlure sclérotique: on 6 PEI [333]; on
27 BC [1012].
M. seaveri (C:227):   on 20 Ont [337, 429],
Que [899].
Monodictys pelagica: on submerged leaves
of 12 BC [5].
Nectria sp.:   on 16 Que [1376]; on 28 Que
[282].
N. cinnabarina (anamorph *Tubercularia
vulgaris*) (C:227):   nectria dieback,
dépérissement nectrien: on 12 BC
[1012]; on 15a NWT-Alta [1063]; on 16
Nfld [F76]; on 27 Sask [F67, F69], NS
[1032]; on 28 Sask [F66].
N. galligena: on 27 Que [F60].
Nipterella parksii (Belonidium p.):   on 7
BC [1012].
Nowakowskiella elegans: on submerged
leaves of 12 BC [5].
Odontia sp.:   on 9 BC [1012].
Oidium sp.:   powdery mildew, blanc: on 2
BC [1816].
Panellus serotinus (Pleurotus s.):   on 28
NB [1032].   See *Acer*.
Papulaspora halina: on submerged leaves
of 12 BC [5].
Penicillium sp.:   on 7 NS [591].
Peniophora cinerea (C:227):   on 16, 27, 28
Ont [1341]; on 27 NS [1032]; on 28 NB
[1032].   See *Abies*.
P. incarnata (C:227):   on 9 BC [1012].
Lignicolous.
Phaeostoma sphaerophila: on 28 Que
[142].   Mycoparasitic on stroma of
Apiosporina morbosa.   Could be a
biological control agent.
Phellinus ferreus (Poria f.) (C:227):   on
9 BC [1012].   See *Acer*.
P. ferruginosus (Poria f.):   on 28 NB
[1032].   See *Abies*.
P. laevigatus (Fomes igniarius var. *l.,
Poria l., Poria prunicola)* (C:228):   on
16 Que [1376], NB, NS [1032]; on 16, 20,
27, 28 Ont [1341]; on 28 Que [282,
1377], NB, NS, PEI [1032], Nfld [1659,
1660, 1661].   See *Acer*.
P. punctatus (Poria p.):   on 20 Ont
[1341]; on 27 NS [1032].   See *Abies*.
Phialophora malorum: on 13 Ont [3].
Phlebia radiata (C:227):   on 9 BC [568,
1012].   See *Abies*.
P. rufa: on 9 BC [568, 1012]; on 20 Ont
[568].   See *Abies*.
*Phloeosporella padi (Cylindrosporium
hiemalis, C. lutescens, C. padi, C.
prunophorae)* (holomorph *Blumeriella
jaapii*) (C:226):   on 3, 6, 7, 9 BC
[1816]; on 16 Ont [F74, F75]; on 27 Ont
[1340]; on 28 Sask, Ont [1763], NS
[1032].
Phoma sp.:   on 7 NS [591].
Phomopsis sp.:   on 7 NS [591]; on 16 Ont
[1340].

P. perniciosa :   on 28 NS [337]. Possibly
an anamorph of *Diaporthe eres* .

Phyllosticta sp.:   on 9 BC [1816].

P. circumscissa (C:227):   leaf spot, tache
des feuilles: on 28 Que [333].

P. virginiana (C:227):   on 1 Ont [1340].

Phytophthora cactorum (C:227):   fruit rot
or collar rot, mildiou du collet: on 2,
3, 6, 17 BC [1816]; on 17 BC [339]; on
28 BC [333, 334, 335, 1074 by
inoculation].

P. cambivora :   on 28 BC [1074] by
inoculation.

P. cryptogea :   on 28 BC [1074] by
inoculation.

Pirex concentricus (Irpex owensii) :   on 28
BC [1012]. Lignicolous. See *Alnus* .

Pleospora sp.:   on 12 BC [1012].

Plicatura crispa (Trogia c.) :   on 16 Ont
[1341]. Lignicolous.

Podosphaera sp.:   powdery mildew, blanc:
on 15a Alta [1594].

P. clandestina (P. oxyacanthae) (C:227):
powdery mildew, blanc: on 3 BC, NS
[1594]; on 3, 6, 9, 17, 27a BC [1816];
on 6 BC [338, 339], Ont [1403], NS [339,
1594] and Man, Ont [1257]; on 14 Ont,
Que, NS [1257] and NS [1032]; on 17 BC
[333]; on 18 Man, Ont [1257]; on 18a Ont
[1340]; on 20 Ont [1828]; on 27 BC, Man,
Ont, Que, NB [1257] and Alta [F66],
Sask, Man [F67], Ont [1340, 1828], NB
[1032, F65]; on 28 BC [336, 337], Ont
[333, F67], Que [336, 337, 338, 339], NB
[339, 1032], PEI [1594] and BC, Ont,
Que, NB [1257].

Polyporus brumalis :   on 16 Que [1389], NS
[1032], Nfld [1659, 1660, 1661].
Lignicolous.

Poria sp.:   on 28 NB [1032]. Lignicolous.

P. subacida (C:228):   on 9 BC [1012]. See
Abies .

Porodisculus pendulus (Polyporus pocula) :
on 16 Nfld [1659, 1660, 1661].
Lignicolous.

Pseudospiropes nodosus :   on 9 BC [816].

*Punctularia strigosozonata (Phaeophlebia
s., Phlebia s.)* :   on 16 Ont [1341]; on
28 Que [969], NB [1032]. See *Alnus* .

Pycnoporus cinnabarinus (Polyporus c.)
(C:228):   on 16 Ont [1341]; on 28 NB,
PEI [1032]. See *Acer* .

Rabenhorstia sp.:   on 16 NB [1032].

"Radulum" pallidum :   on 16 PEI [1032].
Lignicolous. See *Abies* .

Rhizoctonia solani (holomorph
Thanatephorus cucumeris):   on 13 Ont
[3].

Rhizopus sp.:   on 17 BC [1595].

R. nigricans (C:228):   fruit rot,
moisissure chevelue: on 17 BC [333,
334, 335, 336, 339, 1594], Ont [339].

R. stolonifer :   on 17 BC [1816].

Schizophyllum commune (C:228):   white
spongy rot, carie blanche spongieuse:
on 2 BC [1012]; on 16 NS [1032]; on 16,
20 Ont [1341].

Schizopora paradoxa (Poria versipora)
(C:228):   on 9 BC [1012]. See *Abies* .

Schizosaccharomyces sp.:   on 17 Ont [1930].

Sclerotinia sp.:   on 3 BC [1012].

Septoria pruni :   on 27 NS [335, 336].

Sistotrema brinkmannii (Trechispora b.)
(C:229):   on 9 BC [1012]; on 28 NB
[1032]. See *Abies* .

Sphaerotheca pannosa (C:228):   powdery
mildew, blanc: on 17 Ont [334, 1257];
on 17, 17a BC [1816].

Sporotrichum maritimum :   on submerged
leaves of 12 BC [5]. The type is lost
and the fungus was considered to have
been "insufficiently described" [890].
The name has been declared nomen dubium.

Stachybotrys atra :   on submerged leaves of
12 BC [5].

Steccherinum fimbriatum (Odontia f.) :   on
27 NS [1032, 1036]. See *Acer* .

Stereum hirsutum (C:228):   on 16 Nfld
[1659, 1660, 1661]; on 28 BC [1012].
See *Abies* .

*Stigmina carpophila (Coryneum c.,
Sciniatosporium c.)* (C:228):   blight,
brûlure ou criblure: on 2 BC [336], Ont
[333, 334]; on 2, 3, 6, 7, 8, 12, 17 BC
[1816]; on 2, 12, 17 BC [333, 335]; on 3
BC [338]; on 4, 28 Sask [1127]; on 17 BC
[337, 339, 1594], Ont [339]; on 28 BC
[333, 1127], Sask [1760], Que [333].

Stromatocrea cerebriforme (holomorph
Hypocreopsis lichenoides):   on 16 Que
[230].

Taphrina sp.:   on 16 NS [1032]; on 20 NB
[1032].

T. communis (C:228):   plum pockets,
pochette: on 1 NS [1032, 1925]; on 7 BC
[1816], Man [335], Ont [336], NS [334,
335, 590]; on 14 Ont [1340]; on 19 NS
[338, 339, 590]; on 28 Sask [337], Que
[339], NB [337, 339, 1594], NS [333,
336, 1594] and Sask, Man, Que [333, 338].

T. confusa (C:228):   pockets, cloque: on
27 BC [1012], Ont [1340], NB [1032,
F66], NS [590, 1032], Nfld [1659, 1660,
1661, F65]; on 27a BC [1816]; on 28 NB
[1032], Nfld [1661].

T. deformans (C:228):   leaf curl, cloque:
on 8, 17 BC [1594]; on 17 BC [333, 334,
336, 338, 339, 1012], Ont [337], Que
[334], NS [335, 336, 338, 1594] and Ont,
NS [333, 334, 339]; on 17, 17a BC
[1816]; on 28 NB, NS, PEI [1032].

T. flectans :   on 9 BC [1012]; on 28 BC
[F68].

T. pruni (C:228):   plum pockets,
pochette: on 1 Ont [1828]; on 14 Ont
[1340], NB [1032]; on 16 NS [1032]; on
27, 28 Sask, Man [F68]; on 28 NB, NS,
PEI [1032].

T. wiesneri (T. cerasi) (C:228):
witches'-broom, cloque-balai de
sorcière: on 3, 6, 9 BC [1816]; on 3,
28 BC [334]; on 9 BC [1012]; on 16 Que
[1377, F69], NS [335, 590], PEI [337,
1032, F64], Nfld [1659, 1660, 1661] and
NB, NS [336, 1032, F63, F64]; on 16, 28
Ont [1340]; on 20 NB [338, F65]; on 28
BC [333].

Tomentella sublilacina :   on 28 Ont [1341,
1827], Que [1341].

Tranzschelia discolor (C:229): rust, rouille: on 7 BC [1816]; on 28 BC [335]. Ziller [2008] did not accept *T. discolor* as part of the mycoflora of w. Canada and did not mention its occurrence in other parts of North America. Perhaps specimens so named should be referred to *T. pruni-spinosae* var. *discolor*.

T. pruni-spinosae (C:229): rust, rouille: II III on 1, 17, 20, 27 Ont [1236]; on 16 Man [F67], Ont [1341, F65].

T. pruni-spinosae var. *discolor*: rust, rouille: on 7 BC [1012, 2008]. Ziller [2008] has pointed out that this variety should be treated as distinct from *T. discolor*.

Trichaptum abietinum (Hirschioporus a., Polyporus a.) (C:227): on 9 BC [1012]; on 28 NS [1032]. See *Abies*.

T. biformis (Polyporus pargamenus) (C:227): on 20 Ont [1341]; on 27 NB [1032]; on 28 NB, NS, PEI [1032]. See *Acer*.

Trichoderma sp.: on 13 Ont [3].

T. hamatum: on 13 Ont [3].

T. viride: on 13 Ont [3]. See *Abies*.

Troposporella fumosa: on 28 Man [1760].

Tubercularia sp.: on 28 Ont [1340].

T. vulgaris (holomorph *Nectria cinnabarina*): on 15a Alta [F63].

Tubulicrinis glebulosus (Peniophora gracillima) (C:227): on 28 NS [1032]. See *Abies*.

Tympanis prunicola (C:229): on 16 Ont [1340, 1828]; on 27 Sask [F67], Que [F68]; on 28 Ont, Que, NB [1221], NS [1032].

T. spermatiospora (anamorph *Sirodothis populnea*): on 28 Nfld [1661].

Tyromyces chioneus (Polyporus albellus) (C:227): on 9 BC [1012]; on 16 NB [1032], Nfld [1659, 1660, 1661]; on 28 PEI [1032]. See *Abies*.

T. guttulatus (Polyporus g.) (C:227): on 9 BC [1012]. See *Abies*.

Valsa sp. (C:229): canker, chancre cytosporéen: on 16, 28 Ont [1340]; on 17 Ont [334, 336, 337, 339, 1594, 1595].

Vararia investiens: on 28 Ont [557]. See *Abies*.

Venturia pruni: on overwintered leaves of 16 Que [73].

Verticillium sp. (C:229): wilt, flétrissure verticillienne: on 2, 17, 28 Ont [97, 96].

V. albo-atrum (C:229): wilt, flétrissure verticillienne: on 2, 17, 28 Ont [98]; on 17 Ont [334]; on 28 BC [334], Ont [38].

V. dahliae (C:229): wilt, flétrissure verticillienne: on 2 BC [333]; on 2, 3, 6 BC [335, 336, 337, 339, 1816, 1978]; on 2, 6, 17 BC [1980]; on 2, 17, 28 BC [38]; on 17 Ont [333, 336, 337, 339, 1816, 1978], Ont [333, 338, 1087].

Zalerion maritimum: on submerged leaves of 12 BC [5].

PSATHYROSTACHYS Nevski   GRAMINEAE

See discussion of generic circumscriptions under *Agropyron*.

1. *P. juncea* (Fisch.) Nevski (*Elymus j.* Fisch.), Russian wild rye; imported from Eurasia.

Alternaria alternata (A. tenuis auct.) (C:106): on 1 Sask [337].

Ascochyta sp.: on 1 Alta [133].

Bipolaris sorokiniana (Helminthosporium sativum) (holomorph *Cochliobolus sativus*) (C:106): on 1 Alta [120, 133], Sask [961].

Erysiphe graminis (C:106): powdery mildew, blanc: on 1 Alta [133], Sask [1257].

Puccinia impatienti-elymi (P. recondita in part) (C:107): leaf rust, rouille des feuilles: on 1 Alta [133]. See comment under *Elymus*.

Septoria elymi (C:107): on 1 Sask [337].

PSEUDOTSUGA Carr.   PINACEAE

1. *P. menziesii* (Mirbel) Franco (*P. taxifolia* (Lamb.) Britt.), douglas fir, sapin de Douglas; native, BC-Alta.
2. Host species not named, i.e., reported as *Pseudotsuga* sp.

Aleurodiscus amorphus (C:231): on 1 NWT-Alta [1063], BC [964], Alta [950]. See *Abies*.

A. farlowii (C:231): on 1 BC [964, 1012]. Lignicolous.

A. lividocoeruleus: on 1 BC [964, 1012]. See *Abies*.

A. penicillatus (C:231): on 1 BC [964, 1012]. See *Abies*.

A. spiniger: on 1 BC [964, 1012]. Lignicolous.

A. weirii (C:231): on 1 BC [964, 1012]. Lignicolous.

Alternaria alternata (A. tenuis auct.): on 1 BC [F68].

A. maritima: on submerged panels of 1 BC [782].

Amylostereum chailletii: on 1 BC [1012, 1721]. See *Abies*.

Antrodia carbonica (Poria c.) (C:232): on 1 BC [1003, 1012, 1144, 1341, 1721]. See *Abies*.

A. heteromorpha (Trametes h.): on 1 NWT-Alta [1063], BC [1012]. See *Abies*.

A. lenis (Poria l.) (C:232): on 1 BC [1012]. See *Abies*.

A. lindbladii (Poria cinerascens) (C:232): on 1 BC [1012, 1341]. See *Abies*.

A. serialis (Trametes s.): on 1 NWT-Alta [1063], BC [1012]. See *Abies*.

A. sinuosa (Poria s.): on 1 BC [1012]. See *Picea*.

A. variiformis (Trametes v.): on 1 BC [1012, 1063]. See *Abies*.

A. xantha (Poria x.) (C:233): brown cubical rot, carie brune cubique: on 1 BC [1012, 1721].

Aplosporella sp. *(Haplosporella* sp.*)*:
on 1 BC [510, 1012].
Armillaria mellea (C:231): root rot,
pourridié-agaric: on 1 NWT-Alta [1063],
BC [852, 953, 1012, 1144, 1328, 1721,
F60, F66, F67, F76], Alta [952, F62,
F65, F66], Nfld [1659, 1660, 1661, 1663,
F68, F69]. See *Abies*.
Ascocoryne sarcoides (Coryne s.): on 1
NWT-Alta [1063], BC [1012, 1144], Alta
[952].
Asterodon ferruginosus (C:231): on 1 BC
[1012]. See *Abies*.
Athelia pellicularis: on 1 BC [1012].
Lignicolous. See *Abies*.
Atichia sp. (holomorph *Seuratia* sp.):
on 1 BC [1012].
Aureobasidium sp.: on 1 BC [525].
Aureobasidium pullulans: on 1 BC [1012,
1064].
Auriscalpium vulgare (C:231): on fallen
cones of 1 BC [1012].
Bactrodesmium obliquum: on 1 BC [1012,
1760].
Basidioradulum radula (Hyphoderma r.): on
1 BC [1012]. See *Abies*.
Biatorella resinae (anamorph *Zythia
r.*): on 1 BC [1012]. Associated with
resin exudates and cankers.
Botryobasidium subcoronatum: on 1 BC
[1012]. See *Betula*.
B. vagum: on 1 BC [1012]. See *Abies*.
Botryohypochnus isabellinus: on 1 BC
[1012]. See *Abies*.
Botryosphaeria pseudotsugae: on 1 BC
[510, 916, F75].
B. tsugae: on 1 BC [F68].
Botrytis cinerea (holomorph *Botryotinia
fuckeliana*): on 1 BC [1012, F76].
Caliciopsis pinea: on 1 BC [495 by
inoculation, 1012].
C. pseudotsugae (C:231): canker, chancre
caliciopsien: on 1 BC [493, 495, 1012,
1340].
Caloscypha fulgens (anamorph
Geniculodendron pyriforme): on stored
seeds of 1 BC [1749]. See *Picea*.
Cephaloascus fragrans: on 1 BC [1012].
Ceratocystis sp.: on 1 BC [1012].
C. allantospora: on 1 BC [1190].
C. angusticollis: on 1 BC [1190].
C. davidsonii: on 1 BC [1190].
C. fasciata: on 1 BC [1190].
C. leucocarpa: on 1 BC [1190].
C. pseudominor: on 1 BC [1190].
C. sagmatospora: on 1 BC [1190].
Ceriosporopsis halima: on submerged wood
of 1 BC [782].
Chaetomium elatum (C:231): on 1 BC [1012].
Chalara ungeri: on 1 BC [1165].
Chondropodium pseudotsugae (C:231): on 1
BC [1012].
Chondrostereum purpureum: on 1 BC
[1012]. See *Abies*.
Ciboria rufofusca: on 1 BC [663].
Restricted to cone scales.
Cirrenalia macrocephala: on submerged
wood of 1 BC [782].
Cladosporium herbarum: on 1 BC [1012].

Claussenomyces pseudotsugae (Tympanis p.)
(C:233): on 1 BC [1012, F64].
Coccomyces pseudotsugae: on 1 BC [510,
1012, 1609, F75].
Colpoma crispum: on 1 BC [1012].
Coltricia cinnamomea (Polyporus c.): on 1
BC [1012]. Typically on sandy soil.
C. perennis (Polyporus p.) (C:232): on 1
BC [1012]. See *Abies*.
*Confertobasidium olivaceo-album (Athelia
fuscostrata)*: on 1 BC [1012].
Lignicolous.
Coniochaeta sp.: on 1 BC [1012].
Coniophora fusispora: on 1 BC [1012,
F68]. See *Picea*.
C. olivacea (C:231): on 1 BC [1012]. See
Alnus.
C. puteana (C:231): on 1 BC [1012,
1721]. See *Abies*.
Connersia rilstonii (Pseudoeurotium r.):
on submerged wood of 1 BC [782].
Coprinus atramentarius: on 1 BC [1012].
Coriolus pubescens (Polyporus p.)
(C:232): on 1 BC [1012]. See *Abies*.
C. versicolor (Polyporus v.): on 1 BC
[1012, 1721, F65]. See *Abies*.
Coronophora sp.: on 1 BC [1012].
Crepidotus herbarum (C:231): on 1 BC
[1012].
Crucibulum laeve: bird's nest fungus: on
1 BC [1012]. Typically on rotting wood.
Crustoderma dryinum: on 1 BC [1012]. See
Abies.
*Crustomyces pini-canadensis (Corticium
p.)*: on 1 BC [1012]. Lignicolous.
Cryptoporus volvatus (Polyporus v.)
(C:232): on 1 [180, 1012, 1063, 1150].
See *Abies*.
Cryptosporium sp.: on 1 BC [1012].
*Cylindrocarpon destructans (C.
radicicola)*: on 1 BC [156, 1750, F70,
F71].
C. didymum: on 1 BC [156].
Cytospora sp.: on 1 NWT-Alta [1063], BC
[1012, F60], Alta [952].
C. kunzei (holomorph *Leucostoma k.*): on
1 BC [F75].
Dacrymyces capitatus (Dacryomitra nuda)
(C:231): on 1 BC [1012]. See *Alnus*.
D. palmatus: on 1 BC [1012]. See *Abies*.
Dacryobolus karstenii: on 1 BC [1012].
See *Pinus*.
Dasyscyphus sp.: on 2 BC [1340].
*Dendrothele incrustans (Aleurocorticium
i.)*: on 1 BC [965, 1012]. Lignicolous.
Dermea sp.: on 1 BC [F64].
D. pseudotsugae (anamorph *Foveostroma
boycei*): on 1 BC [499, 512, 852, 1012,
1098, F67, F71, F76].
D. tetrasperma (anamorph *Gelatinosporium
lunaspora*): on 1 BC [512, 1012].
Diaporthe lokoyae (anamorph *Phomopsis
l.*): on 1 BC [501, 852, 1012, 1098,
F68, F71, F74].
Dichomera gemmicola: on 1 BC [526, 1012,
1763, F72].
Dichomitus squalens (Polyporus anceps)
(C:232): on 1 BC [1012, 1341]. See
Abies.

Discina ancilis (D. perlata) : on 1 BC [1012, F68].

Durandiella pseudotsugae (C:231): on 1 BC [492, 1012].

Echinodontium tinctorium (C:231): Indian paint fungus: on 1 NWT-Alta [1063], BC [1012]. See *Abies*.

Epicoccum sp.: on 1 BC [1012, 1064].

Euantennaria rhododendri (Limacinia alaskensis) (C:232): sooty mold, fumagine: on 1 BC [1012].

Exidia saccharina : on 1 BC [1012]. Lignicolous.

Exidiopsis macrospora : on 1 BC [1012]. Lignicolous.

Flagelloscypha citrispora : on 1 BC [1424]. Lignicolous.

Fomitopsis cajanderi (Fomes c., F. subroseus) (C:231): brown cubical rot, carie brune cubique: on 1 NWT-Alta [1063], BC [954, 1012, 1031, 1341].

F. officinalis (Fomes o.) (C:231): brown cubical rot, carie brune cubique: on 1 NWT-Alta [1063], BC [954, 1012], Alta [951].

F. pinicola (Fomes p.) (C:231): brown cubical rot, carie brune cubique: on 1 NWT-Alta [1063], BC [954, 1012, 1721], Alta [950].

F. rosea (Fomes r.) (C:231): brown cubical rot, carie brune cubique: on 1 BC [1012].

Foveostroma boycei (Micropera b.) (holomorph *Dermea pseudotsugae*): on 2 BC [659].

Fumago sp.: on 1 BC [1012].

Fusarium sp. (C:231): damping-off, fonte des semis: on 1 BC [F75].

F. oxysporum (C:231): on 1 BC [156, 155].

F. oxysporum f. sp. *pini* (C:231): on 1 BC [1012].

F. oxysporum var. *redolens (F. redolens)* (C:231): on 1 BC [156].

Fusicoccum sp.: on 1 BC [1012].

Ganoderma applanatum (C:231): white mottled rot, carie blanche madrée: on 1 BC [1012, 1721]. See *Acer*.

G. oregonense (C:231): on 1 BC [1012]. See *Abies*.

Gelatinosporium fosteri (Micropera f.) : on 1 BC [512, 1012].

G. pinicola (Micropera p.) : on 1 BC [512, 1012].

G. sinuatum : on 1 BC [517].

G. uncinatum (Micropera u.) : on 1 BC [512, 1012].

Gloeocystidiellum furfuraceum (Corticium f.): on 1 BC [1012]. Lignicolous.

G. propinquum : on 1 BC [1012]. Lignicolous.

Gloeophyllum odoratum (Trametes o.) (C:233): brown cubical rot, carie brune cubique: on 1 BC [1012].

G. sepiarium (Lenzites s.) (C:232): brown cubical rot, carie brune cubique: on 1 NWT-Alta [1063], BC [1012, 1341, 1721].

Graphium sp.: on 1 BC [F117, 1012, 1144].

Griseoporia carbonaria (Trametes c.) (C:233): on 1 BC [1012, 1341]. See *Pinus* .

Gymnopilus aeruginosus (Pholiota a.) (C:232): on 1 BC [1012]. Lignicolous.

Gyromitra esculenta (C:231): on 1 BC [1012]. Probably on rotted wood.

Haematostereum rugosum (Stereum r.): on 1 NWT-Alta [1063]. See *Abies*.

H. sanguinolentum (Stereum s.) (C:233): red heart-rot, carie rouge du sapin: on 1 BC [1012, 1144, 1341, 1721, F66, F67]. See *Abies*.

Halonectria milfordensis: on submerged wood of 1 BC [782].

Helotium resinicola (anamorph *Stilbella* sp.): on 1 BC [54, 1012, F69]. An Ascomycete which needs to be redisposed.

Heterobasidion annosum (Fomes a.) (C:231): white pocket rot, maladie du rond: on 1 BC [852, 1012, 1143, 1144, 1721, 1916, F68]. See *Abies*.

Humicola alopallonella: on submerged wood of 1 BC [782].

Hymenochaete fuliginosa (C:231): on 1 BC [1012]. See *Picea*.

H. tabacina: on 1 BC [1012]. See *Abies*.

Hyphoderma praetermissum (H. tenue): on 1 BC [1012, 1721]. See *Abies*.

H. resinosum: on 1 BC [1012]. See *Picea*.

Hyphodontia barba-jovis: on 1 BC [1012]. See *Abies*.

H. subalutacea: on 1 BC [1012]. See *Abies*.

Hypholoma capnoides (Naematoloma c.): on 1 BC [1012, F70]. See *Abies*.

H. fasciculare (Naematoloma f.) (C:232): on 1 BC [1012]. See *Abies*.

Inonotus subiculosus (Poria s.) (C:233): on 1 BC [1012]. See *Abies*.

Ischnoderma resinosum (Polyporus r.) (C:232): on 1 BC [1012]. See *Abies*.

Lachnellula ciliata: on 1 BC [1012].

L. pseudotsugae: on 1 BC [1012].

Laeticorticium minnsiae: on 1 BC [1012]. Lignicolous.

Laetiporus sulphureus (Polyporus s.) (C:232): brown cubical rot, carie brune cubique: on 1 BC [1012, 1721].

Leocarpus fragilis: on 1 BC [1063].

Leucogyrophana mollusca: on 1 BC [1012] and on boards of old building [571]; on 2 BC [581].

L. pinastri: on 2 BC [581, 571].

L. romellii: on 1 BC [571]. See *Abies*.

Leucostoma kunzei (anamorph *Cytospora k.*): on 1 BC [1012].

Limacinia moniliformis var. *quinqueseptata* (C:232): sooty mold, fumagine: on 1 BC [1012].

Lophium mytilinum (C:232): on 1 BC [1012].

Lulworthia medusa: on submerged wood of 1 BC [782].

Mariannaea elegans: on 1 BC [153].

Melampsora sp.: rust, rouille: on 1 BC [F76].

M. medusae (M. albertensis) (C:232): needle rust, rouille des aiguilles: on 1 NWT-Alta [1063], BC [953, 1012, 1064, 1125 by inoculation, 1341, 2002 by inoculation, F61, F62, F64, F65, F66, F67, F68, F73, F76], Alta [950, 952, 1999].

M. occidentalis (C:232): needle rust,
    rouille des aiguilles: on 1 NWT-Alta
    [1063], BC [1012, 1125 by inoculation,
    2002 by inoculation, F64, F65, F74],
    Alta [952].
Melampsorella caryophyllacearum: on 1 BC
    [F67]. Usually the coniferous host is
    Abies [2008].
Meruliopsis taxicola (Poria t.): on 1 BC
    [1012, 1721]. See *Abies*.
Merulius tremellosus: on 1 BC [568, 1012,
    1721]; on 2 BC [568]. See *Abies*.
Microthelia linderi (M. maritima): on
    submerged wood of 1 BC [782].
Monodictys pelagica: on submerged wood of
    1 BC [782].
Naematoloma sp.: on 1 BC [1012].
    Specimens probably referrable to
    Hypholoma.
Nectria sp.: on 1 BC [1012].
Neournula nordmanensis: on 1 BC [1224].
Nidula niveotomentosa (N. microcarpa):
    bird's nest fungus: on 1 BC [1012].
    Typically on rotted wood.
Nitschkia molnarii: on 1 BC [518].
Odontia lactea (C:232): on 1 BC [1012].
    See comment under *Abies*.
Onnia circinata (Polyporus tomentosus var.
    c.) (C:232): red butt rot, carie
    rouge alvéolaire: on 1 BC [1012]. See
    Abies.
O. tomentosa (Polyporus t. var.
    t.): on 1 BC [1012]. See *Abies*.
Ostenia obducta (Polyporus osseus)
    (C:232): on 1 BC [1012]. See *Abies*.
Paecilomyces carneus: on 1 BC [151].
Panellus mitis (Pleurotus m.) (C:232): on
    1 BC [1012].
P. serotinus (Pleurotus s.): on 1 BC
    [1012, F66]. See *Acer*.
P. stipticus (Panus s.): on 1 BC [1012].
    See *Alnus*.
Paxillus atrotomentosus (C:232): on 1 BC
    [1012]. Associated with a brown rot.
P. panuoides (C:232): on 1 BC [1012].
    Associated with a brown rot. Now in
    Tapinella.
Penicillium sp.: on 1 BC [155, 1144].
Peniophora piceae (P. separans) (C:232):
    on 1 BC [1012]. See *Abies*.
P. pseudopini (C:232): on 1 BC [1012,
    1941]. See *Abies*.
P. septentrionalis (C:232): red heart
    rot, carie rouge du coeur: on 1 BC
    [1012]. See *Abies*.
Pezicula livida (anamorph
    Cryptosporiopsis abietina): on 1 BC
    [1012, F73].
P. populi (Neofabraea p.): on 1 BC [1012].
Pezizella chapmanii: on 1 BC [1012, 1946].
Phacidiopycnis pseudotsugae (holomorph
    Potebniamyces coniferarum): on 1
    NWT-Alta [1063].
Phacidium abietis (C:232): on 1 BC [954,
    1012, 1444].
P. infestans: on 1 NWT-Alta [1063], BC
    [954, F60]. Records are suspect. See
    note under *Abies*.
Phaeocryptopus gaeumannii (C:232): needle
    cast, rouge: on 1 BC [73, 852, 1012,
    1340]. Typically on live needles.

Phaeolus schweinitzii (Polyporus s.)
    (C:232): brown cubical rot, carie brune
    cubique: on 1 BC [1012, 1341, F76].
Phanerochaete sanguinea (Peniophora s.)
    (C:232): on 1 BC [1012]. See *Abies*.
P. carnosa (Peniophora c.) (C:232): on 1
    BC [1012]. See *Abies*.
Phellinus ferreus (Poria f.) (C:232):
    white rot, carie blanche: on 1 BC
    [1012].
P. ferrugineofuscus (Poria f.): on 1 BC
    [1012]. See *Abies*.
P. nigrolimitatus (Fomes n.) (C:231):
    white pocket rot, carie blanche
    alvéolaire: on 1 BC [1012].
P. pini (Fomes p.) (C:231): red ring rot,
    carie blanche alvéolaire: on 1 NWT-Alta
    [1063], BC [953, 1012, 1341], Alta
    [950], NB [1032]. See *Abies*.
P. punctatus (Poria p.): on 1 BC [1012].
    See *Abies*.
P. repandus (Fomes r.) (C:231): on 1 BC
    [1012, 1341, F67]. Associated with a
    white pocket rot.
P. robustus (Fomes r.) (C:231): white
    spongy rot, carie blanche spongieuse:
    on 1 BC [1012].
P. tsugina (Poria t.): on 1 BC [F65].
    See *Abies*.
P. viticola (Fomes v.): on 1 BC [1012].
    See *Abies*.
P. weirii (Poria w.) (C:233): yellow ring
    rot, carie jaune annelée: on 1 BC [87,
    157, 852, 1012, 1142, 1912, 1913, 1915,
    1917, F63, F75]. See *Abies*. Causes a
    root rot which is a major problem in the
    young stands of coastal BC.
Phlebia centrifuga (P. albida auct.*)*:
    on 1 BC [1012]. See *Abies*.
P. livida (Corticium l.): on 1 BC [1012,
    1721]. See *Abies*.
P. phlebioides (Peniophora p.) (C:232):
    on 1 BC [1012, 1721]. See *Abies*.
P. radiata: on 1 BC [568, 1012]. See
    Abies.
P. segregata (Peniophora livida): on 1 BC
    [1012]. Lignicolous.
Phlebiopsis gigantea (Peniophora g.)
    (C:232): white sap rot, carie blanche
    de l'aubier: on 1 BC [1012, 1721]. See
    Abies.
Pholiota decorata (Flammula d.): on 1 BC
    [1012].
P. destruens: on 1 BC [1012]. See
    Populus.
Phoma sp.: on 1 BC [1012].
Phomopsis sp.: on 1 BC [1340].
P. lokoyae (teleomorph *Diaporthe l.*):
    on 1 NWT-Alta [1063], BC [1012, 1340,
    F67], Alta [951].
P. porteri: on 1 BC [510, 1012].
Phragmoporthe pseudotsugae: on 1 BC [510,
    1012, F75].
Piloderma bicolor (Corticium b.) (C:231):
    on 1 BC [1012]. See *Abies*.
Pistillaria sp.: on 1 BC [1012].
    Presumably on fallen needles.
Pithya vulgaris: on 1 BC [1012].
*Pleurocybella porrigens (Pleurotus p., P.
    albolanatus)* (C:232): on 1 BC [1012].
    Lignicolous.

Plicatura crispa (Trogia c.) (C:233): on 1 BC [1012]. Lignicolous.

Polyporus hirtus (C:232): on 1 BC [1012]. Presumably associated with a white rot.

P. varius (P. elegans) (C:232): on 1 BC [1012]. See *Abies*.

Poria rivulosa (P. albipellucida): on 1 BC [1012, 1341, 1721]. See *Abies*.

P. subacida (C:233): on 1 BC [M160, 1721]. See *Abies*.

P. tenuis var. *pulchella* (C:232): on 1 BC [1012]. See *Pinus*.

P. zonata: on 1 BC [1012]. Associated with a white pocket rot.

Potebniamyces coniferarum (anamorph *Phacidiopycnis pseudotsugae*): on 1 BC [1012, F66].

Pragmopycnis pithya (holomorph *Pragmopora p.*): on 1 BC [1765, 1763, F75].

Pragmopora pithya (anamorph *Pragmopycnis p.*): on 1 BC [953, 1012]; on 2 BC [657].

Propolis sp.: on 1 BC [1012].

Pseudohydnum gelatinosum (C:233): on 1 BC [1012]. Probably on rotted wood.

Pycnoporellus fulgens (Polyporus fibrillosus) (C:232): on 1 BC [1012, 1179]. See *Abies*.

Pythium sp.: on 1 BC [1012, F70].

P. intermedium: on 1 Alta [745].

Raffaelea canadensis: on 1 BC [109].

Ramaria stricta var. *concolor*: on 1 BC [1012]. On dead wood.

Resinicium bicolor (Odontia b.) (C:232): white stringy rot, carie blanche filandreuse: on 1 BC [1012, 1144, 1721].

Retinocyclus abietis (C:233): leader dieback, dépérissement: on 1 BC [1012].

Rhabdocline pseudotsugae (C:233): needle cast, rouge: on 1 NWT-Alta [1063], BC [338, 852, 953, 1234, 1926, 1927, F60, F61, F62, F63, F64, F65, F66, F67, F68, F69, F73, F74, F75, F76], Alta [951, 952, F61, F67, F69].

R. pseudotsugae ssp. *epiphylla*: on 1 BC [1012, 1234].

R. pseudotsugae ssp. *pseudotsugae*: on 1 BC [1012], Alta [F69].

R. weirii: on 1 BC [F69] and BC, Alta [1234].

R. weirii ssp. *oblonga*: on 1 NS [F75].

R. weirii ssp. *obovata*: on 1 BC [1012, 1234].

R. weirii ssp. *weirii*: on 1 BC [1012], Alta [F69].

Rhabdogloeum sp.: on 1 BC [1340].

R. pseudotsugae (C:233): on 1 BC [1012, 1340]; on 2 BC [1164]. Typically associated with *Rhabdocline*.

Rhinocladiella mansonii: on 1 BC [1012].

Rhizina undulata: on 1 BC [562, 564, 1012, F67, F68, F70]. Caused mortality of seedlings planted on recently clear-cut then burned sites.

Rigidoporus sp.: on 1 BC [1721].

R. nigrescens (Poria n.) (C:232): on 1 BC [1012, F61]. See *Acer*.

Rosellinia herpotrichioides (C:233): needle blight, brûlure des aiguilles: on 1 BC [1012].

Schizopora paradoxa (Poria versipora) (C:233): on 1 BC [1012]. See *Abies*.

Sclerophoma sp.: on 1 BC [1012, F67].

S. pithya: on 1 BC [1827].

S. pithyophila (holomorph *Sydowia polyspora*): on 1 BC [1098, F67, F71, F75, 525].

Scytinostroma arachnoideum (C:233): on 1 BC [555, 1012]. Associated with a white rot.

S. galactinum (Corticium g.) (C:231): white stringy rot, carie blanche filandreuse: on 1 BC [1012, 1721].

Scytinostromella humifaciens (Peniophora h.) (C:232): on 1 BC [1012].

Sebacina sp.: on 1 BC [1012].

Septonema dendryphioides: on 1 BC [1012].

Serpula himantioides: on 1 BC [1012]. See *Abies*.

S. incrassata (Poria i.): on 1 BC [1012, F67]. See *Picea*.

S. lacrimans: on 1 BC [1012]. See *Abies*.

Seuratia millardetii: on 1 BC [1042].

Sirococcus conigenus (S. strobilinus): on 1 BC [1012].

Sirodothis sp. (holomorph *Tympanis* sp.): on 1 BC [1012].

Sistotrema brinkmannii (Trechispora b.): on 1 BC [1012, 1144, 1721]. See *Abies*.

Skeletocutis amorpha (Polyporus a.) (C:232): on 1 BC [1012]. See *Abies*.

S. subincarnata (Poria s.) (C:233): on 1 BC [1012]. See *Abies*.

Sorocybe resinae: on 1 BC [1012].

Sparassis radicata (C:233): root rot, pourridié sparassien: on 1 BC [1012]. Associated with a brown rot of roots and butts of live trees. The decayed wood is very similar to wood decayed by *Phaeolus schweinitzii*.

Steccherinum ochraceum (C:233): on 1 BC [1012]. See *Acer*.

Stereum ostrea (C:233): on 1 BC [1012]. See *Abies*.

Strigopodia batistae: on 1 BC [786].

Strobilurus albipilatus: on senescent and buried cones of 1 BC [1431].

S. trullisatus: on 1 BC [1431]. Reported in Lowe [1012] and Conners [C:321] as *Collybia albipilata* and *C. conigenoides* but the specimens are *S. trullisatus* [1431].

Sydowia polyspora (anamorph *Sclerophoma pithyophila*): on 1 BC [1012].

Thelephora sp.: on 1 BC [1012].

T. terrestris: on 1 BC [F68]. See *Pinus*.

Therrya pseudotsugae: on stems of 1 BC [519].

Tiarosporella pseudotsugae: on 1 BC [1012], Alta [1947].

Tomentella ellisii (T. microspora): on 1 BC [929, 1341, 1827].

T. sublilacina: on 1 BC [1012, 1341].

Tomentellina fibrosa (Kneiffiella f.): on 1 BC [1012, 1341].

Trechispora mollusca (T. onusta): on 1 BC [1012].

Tremella encephala (C:233): on 1 BC
[1012]. See *Abies*.
*Trichaptum abietinus (Hirschioporus a.,
Polyporus a.)* (C:232): white pitted
sap rot, carie blanche de l'aubier: on
1 NWT-Alta [1063], BC [1012, 1721], Alta
[951]. See *Abies*.
T. laricinus (Hirschioporus l.): on 1 BC
[1028]. See *Abies*.
Trichoderma sp.: on 1 BC [1144].
T. viride: on 1 BC [155]. See *Abies*.
Trichothecium roseum: on 1 BC [1012, F60].
Truncatella sp.: on 1 BC [1012].
Tubulicrinis glebulosus (T. gracillimus):
on 1 BC [1012]. See *Abies*.
Tympanis laricina: on 1 BC [1012].
Tyromyces balsameus (Polyporus b.)
(C:232): brown cubical rot, carie brune
cubique: on 1 BC [1012, F61].
T. caesius (Polyporus c.) (C:232): on 1
BC [1341]. See *Abies*.
T. fragilis (Polyporus f.) (C:232): on 1
BC [1012, 1341]. See *Abies*.
T. guttulatus (Polyporus g.): on 1 BC
[1012]. See *Abies*.
T. leucospongia (Polyporus l.) (C:232):
on 1 BC [1012]. See *Abies*.
T. mollis (Polyporus m.) (C:232): on 1 BC
[1012]. See *Abies*.
T. perdelicatus (Polyporus p.) (C:232):
on 1 BC [1012]. Associated with a brown
rot.
T. placenta (Poria p., P. monticola)
(C:232): on 1 BC [1012, 1721]. See
Pinus.
T. sericeomollis (Polyporus s.): on 1 BC
[1012]. See *Abies*.
T. stipticus (Polyporus s., P. immitis):
on 1 BC [1012]. See *Abies*.
T. undosus (Polyporus u.) (C:232): on 1
BC [1012]. See *Abies*.
Ulocladium consortiale: on submerged wood
of 1 BC [782].
Valsa sp.: on 1 BC [193].
V. abietis (C:233): canker, chancre
cytosporéen: on 1 BC [1012].
Vararia granulosa (C:233): on 1 BC [1012,
1721]. See *Abies*.
Verticicladiella abietina (C:233): on 1
BC [874].
V. brachiata (C:233): on 1 BC [874, 1012].
V. wagenerii: on 1 BC [1012, F76].
*Vesiculomyces citrinus (Gloeocystidiellum
c.)*: on 1 BC [1012]. See *Abies*.
Wolfiporia extensa (Poria cocos) (C:232):
on 1 BC [1012, 1721]. See *Betula*.
Wrightoporia lenta (Poria l.): on 1 BC
[1012]. See *Abies*.
Xenasma filicinum: on 1 BC [1012].
X. rallum: on 1 BC [1012, 967].
Lignicolous.
X. tulasnelloideum: on 1 BC [1012].
Lignicolous.
Xenasmatella inopinata (Corticium i.)
(C:231): on 1 BC [1012].
Xenomeris abietis: on 1 BC [525, 852,
1012, 1098, F69].
Xeromphalina campanella (C:233): white
stringy rot, carie blanche filandreuse:
on 1 BC [1012, 1721].

Zalerion maritimum: on submerged wood of
1 BC [782].

PSILOTUM Sw.   PSILOTACEAE

Epiphytic, clump-forming, rootless,
leafless plants widely distributed in the
tropics and subtropics.

1. *P. nudum* (L.) Beauv., whisk fern;
   imported, e. North America.

Gliocladium roseum: on 1 Ont [336, 837].

PSORALEA L.   LEGUMINOSAE

1. *P. argophylla* Pursh; native, Alta, Man.

Fusicladium psoraleae: on 1 Man [829].
Uromyces psoraleae (C:233): rust,
rouille: on 1 Man [1838].

PTERIDIUM Scop.   POLYPODIACEAE

1. *P. aquilinum* (L.) Kuhn, bracken,
   fougère d'aigle; native, BC-Alta,
   Man-Nfld. 1a. *P. a.* var.
   latiusculum (Desv.) Underw. (*P.
   latiusculum* (Desv.) Hieron.); native,
   BC-Alta, Man-Nfld. 1b. *P. a.* var.
   pubescens Underw.; native, BC-Alta,
   Ont-Que.
2. Host species not named, i.e., reported
   as *Pteridium* sp.

*Cryptomycella pteridis (Gloeosporium
leptospermum, G. obtegens)* (holomorph
Cryptomycina p.): on 1a Que [282]; on
2 Ont [1340].
Cryptomycina pteridis (anamorph
Cryptomycella p.) (C:234): on 1a Ont
[1340], Que [282]; on 1b BC [1012, 1816].
Flagelloscypha citrispora: on 1 BC [1424].
Mycosphaerella aquilina: on 1 Ont [1340].
M. pteridicola: on 1 Ont [1340].
Myiocopron sp.: on 2 Ont [1827].
Pellidiscus pallidus: on 1 BC [1424].
Rhopographus filicinus: on 1a Ont [1827].
Thelephora terrestris: on 1 BC [1012].
See *Pinus*.
Uredinopsis sp.: rust, rouille: on 2 NS
[1036].
U. hashiokai (C:234): rust, rouille: on
1b BC [1012, 1816, 2008].
U. pteridis (U. macrosperma) (C:234):
rust, rouille: on ?1 Que [F66]; on 1b
BC [1012, 1816, 2008]; on 2 BC [1341].
Valdensia heterodoxa (holomorph
Valdensinia h.): on 1 BC [1438].

PUCCINELLIA Parl.   GRAMINEAE

1. *P. angustata* (R. Br.) Rand. & Redf.;
   native, Frank.
2. *P. hauptiana* (Krecz.) Kitagawa;
   native, Yukon, Alta, Ont.

Botrytis cinerea (holomorph *Botryotinia
fuckeliana*) (C:234): on 1 Frank [664,
1586].

Erysiphe graminis (C:234):   powdery
   mildew, blanc: on 1 Frank [1255, 1586].
Pseudoseptoria everhartii (Selenophoma e.)
   (C:234): on 2 BC [1731]. The host
   plant is not known from BC.
Spermospora subulata (C:234): on 2 BC
   [1731]. See above.
Ustilago salvei (U. striiformis auct.):
   on 1 Frank [1586].

PYCNANTHEMUM Michx.   LABIATAE

   Erect perennial herbs of North America
with a pungent, mint-like odor when crushed.

1. *P. flexuosum* (Walt.) BSP.; native, s.
   Ont.
2. *P. virginianum* (L.) Durand & Jackson,
   Virginia mountain-mint, pycnanthème de
   Virginie; native, Ont-Que.

Puccinia menthae: rust, rouille: 0 I II
   III on 1, 2 Ont [1236].

PYRACANTHA M. Roem.   ROSACEAE

1. *P. coccinea* Roemer (*Cotoneaster
   pyracantha* (L.) Spach), scarlet fire
   thorn, buisson ardent écarlate; imported
   from Eurasia.

Nectria cinnabarina (anamorph
   Tubercularia vulgaris): on 1 NWT-Alta
   [1063].
Spilocaea pyracanthe (Fusicladium p.)
   (C:235): scab, tavelure: on 1 BC
   [1816].

PYROLA L.   PYROLACEAE

1. *P. asarifolia* Michx., pink
   wintergreen, pyrole à feuilles d'Asaret;
   native, Yukon-Mack, BC-Nfld. 1a. *P.
   a.* var. *a.* (*P. bracteata* Hook.);
   native, transcontinental but less
   northern than *P. a.* var. *purpurea*.
   1b. *P. a.* var. *purpurea* (Bunge)
   Fern., bog wintergreen, pyrole des
   marais; native, transcontinental.
2. *P. dentata* Sm.; native, BC.
3. *P. elliptica* Nutt., shinleaf, pyrole
   elliptique; native, BC-Nfld.
4. *P. minor* L., lesser wintergreen,
   pyrole mineure; native, Yukon-Keew,
   BC-Nfld, Labr.
5. *P. picta* Sm. (*P. aphylla* Sm.),
   white-veined shinleaf; native, BC.
6. *P. rotundifolia* L., roundleaf pyrola,
   pyrole à feuilles rondes; native,
   Ont-Nfld. 6a. *P. r.* var. *americana*
   (Sw.) Fern. (*P. americana* Sweet),
   American wintergreen, pyrole d'Amérique;
   native, Ont-Nfld.
7. *P. secunda* L., one-sided pyrola,
   pyrole unilatérale; native, Yukon-Keew,
   BC-Nfld, Labr.
8. *P. virens* Schweigger, greenish
   wintergreen, pyrole à fleurs verdâtre;
   native, Yukon-Mack, BC-Nfld, Labr.

9. Host species not named, i.e., reported
   as *Pyrola* sp.

Chrysomyxa pirolata (C. pyrolae) (C:235):
   rust, rouille: on 1 Yukon, Alta [1402];
   on 1, 1a, 5, 7, 8 BC [1816]; on 1a, 1b,
   2, 5, 7, 8 BC [1012]; on 1, 3, 4, 6a, 7,
   8, 9 Ont [1236]; on 1, 4 Alta [644]; on
   1, 7, 9 Ont [1341]; on 1, 7, 8, 9
   NWT-Alta [1063]; on 1, 8, 9 Alta [53,
   951, 952]; on 3, 6a, 7 Que [1377]; on 4
   Yukon [460]; on 7 Yukon [7], Mack [F67];
   on 7, 9 BC [954]; on 7, 8, 9 Mack [53];
   on 8 Mack [168, F65]; on 9 Yukon, NWT,
   BC, Alta, Sask, Man [2008], Alta [950,
   F61], Ont [1828]. Alternate stages on
   Picea spp.
Lophodermium pyrolae (C:235): on 1a BC
   [1816]; on 1b BC [1012].
Melampsora epitea: on 9 Nfld [1661].
Mycosphaerella minor (C:235): on 7 Que
   [70].
M. tassiana var. *arthopyrenioides*
   (C:235): on 7 Que [70].
Pucciniastrum pyrolae (C:235): rust,
   rouille: II III on 1 Mack [53, F63],
   Alta [951]; on 1, 1a, 7, 8 BC [1816]; on
   1a, 7, 8 BC [1012]; on 1, 4, 7, 8, 9
   NWT-Alta [1063]; on 3, 7 Ont [1236]; on
   4 Que [1236]; on 4, 9 Alta [F64]; on 7
   Yukon [7, 460], Que [282]; on 7, 8 Alta
   [952]; on 7, 8, 9 Mack, Alta [53]; on 9
   Yukon, NWT, BC, Alta, Sask, Man [2008],
   BC [644], Alta [F61].

PYRUS L.   ROSACEAE

1. *P. arbutifolia* (L.) L. (*Aronia a.*
   (L.) Elliott), red chokeberry; Man-NS,
   PEI, Nfld.
2. *P. communis* L., common pear, poirier;
   imported from Eurasia.
3. *P. longipes* Coss. & Dur.; imported
   from Algeria.

Alternaria alternata: on 2 NS [990].
Apiosporina collinsii: on *P.* sp. Nfld
   [1661]. Known in Canada only on
   Amelanchier [292], perhaps the host
   was misdetermined.
Botryosphaeria obtusa (Physalospora o.):
   on 2 NS [335, 336, 1594].
Botrytis cinerea (holomorph *Botryotinia
   fuckeliana*) (C:235): gray mold,
   moisissure grise: on 2 BC [1816].
Coniothyrium pirinum (C:236): frog-eye
   spot, tache ocellée: on 2 Que [339].
*Cryptosporiopsis malicorticis (Gloeosporium
   perennans)* (holomorph *Pezicula m.*):
   on 2 BC [338].
Cylindrocarpon heteronema (C. mali): on 2
   BC [1816].
Cytospora sp. (C:236): canker, chancre
   cytosporéen: on 2 BC [1816].
Fusicladium pyrorum (holomorph *Venturia
   pirina*): on 2 BC [1816].
Gloeodes pomigena (C:236): sooty blotch,
   tache de suie: on 2 Ont [334], NS [336].
Gymnosporangium clavariiforme (C:236):
   rust, rouille: on 2 BC [1816], NS

[1262]. The host for [1262] was
redetermined (J.A. Parmelee, 1983) as
Cydonia vulgaris and the other record
probably was on *C. vulgaris*.
G. clavipes (C:236): rust, rouille: 0 I
on 1 NB [1240]; on *P.* sp. NS [1032].
G. fuscum (C:236): trellis rust, rouille
grillagée du poirier: 0 I on 2 BC [335,
337, 338, 1012, 1198, 1240, 1253, 1816,
2000, 2008]; on *P.* sp. BC [181, 1247,
F60]; and on *Pirophorum* BC [833].
G. globosum: rust, rouille: on 2 Ont
[1268]. The rust and the host have been
redetermined (J.A. Parmelee, 1983) as
G. juniperi-virginianae and *Malus*
sp., respectively.
Monilinia fructicola (C:236): brown rot,
pourriture brune: on 2 BC [336].
M. laxa (C:236): blossom and twig blight,
pourriture brune: on 2 BC [338, 1816].
Mucor piriformis: on 2 BC [1006].
Nectria cinnabarina (anamorph
Tubercularia vulgaris): on 2 NS
[1594].
N. galligena (C:236): European canker,
chancre européen: on 2 BC [1416].
Phialophora sp.: on 2 BC [1816].
P. ?malorum (C:236): side rot, pourriture
phialophoréene: on 2 BC [334].
Phomopsis ambigua: on 2 NS [335].
Phytophthora cactorum (C:236): collar
rot, mildiou du collet: on 2 BC [333,
339, 1073, 1074, 1816], Ont [336], Ont,
NS [335], NS [1594].
P. cambivora: on 2 BC [1074, 1816].
P. cryptogea: on 2 BC [1074].
Podosphaera leucotricha (C:236): powdery
mildew, blanc: on 2 BC [333, 336,
1816], NS [335].
Rhizopus nigricans (C:236): on 2 BC
[336], Ont [335].
Sclerotinia sclerotiorum: on 2 BC [339].
Taphrina bullata (C:236): leaf blister,
cloque des feuilles: on 2, 3 BC [1816].
Valsa sp.: on 2 BC [1012].
Venturia inaequalis: on 2 BC [1012].
Perhaps a specimen of *V. pirina*.
V. pirina (anamorph *Fusicladium pyrorum*)
(C:236): scab, tavelure: on 2 BC
[1012], BC, Ont, Que, NB, NS [290, 333],
BC, Ont, NB [334, 335], BC, NB [338,
1594], NB [337, 339], NB, NS [336], PEI
[290], cosmopolitan on leaves, fruit and
twigs [73].

QUERCUS L.   FAGACEAE

1. *Q. alba* L., white oak, chêne blanc;
   native, s. Ont-s. Que, NS.
2. *Q. bicolor* Willd., swamp white oak,
   chêne bleu; native, s. Ont-s. Que
   (Montreal area and s.).
3. *Q. coccinea* Muench., scarlet oak,
   chêne écarlate; native in e. USA,
   formerly thought to occur in Ont.
4. *Q. garryana* Dougl., Garry oak, chêne
   de Garry; native, e. coast Vancouver
   Island and lower Fraser River, BC.
5. *Q. macrocarpa* Michx., bur oak, chêne à
   gros glands; native, e. Sask-NB.
6. *Q. muhlenbergii* Engelm., chinquapin
   oak, chêne jaune; native, s. Ont.
7. *Q. nigra* L., water oak, chêne
   aquatique; imported from e. USA.
8. *Q. palustris* Muench., pin oak, chêne
   des marais; native, s. Ont.
9. *Q. prinus* L., chestnut oak, chêne
   châtaignier; native, s. Ont.
10. *Q. robur* L., English oak, chêne
    rouvre; imported from Eurasia.
11. *Q. rubra* L., red oak, chêne rouge;
    native, Ont-PEI. 11a. *Q. r.* var.
    borealis (Michx. f.) Farw., northern
    red oak, chêne rouge du Nord; native,
    Ont-PEI.
12. *Q. velutina* Lam., black oak, chêne
    noir; native, s. Ont.
13. Host species not named, i.e., reported
    as *Quercus* sp.

Acremonium boreale: on 5 BC, Alta, Sask
[1707].
Aleurodiscus canadensis: on 11a Ont
[1341]. See *Abies*.
Amphiporthe raveneliana (*Cryptodiaporthe
densissima, Diaporthe leiphaemia* var.
r.) (C:237): on 1, 11 Ont [F60, F61];
on 1, 11a Ont [1828]; on 1, 11a, 13 Ont
[1340]; on 11a Ont [1827]; on 13 Ont
[1440, F63].
Antrodia oleracea (*Poria o.*): on 11 Que
[F71]. Associated with a brown rot.
A. xantha (*Poria x.*): on 11 Que [F71]; on
13 Ont [1341]. See *Abies*.
Apiognomonia errabunda (anamorph *Discula
umbrinella*): on 1, 2, 5, 6, 11a, 13
Ont [1340]; on 5 Man [F65]; on 5, 13 NB
[1032]; on 11a NB, NS [1032]. See
Fraxinus.
A. quercina (*Gnomonia q.*) (C:237):
anthracnose, anthracnose: on 11a Que
[336]; on 13 Ont [79].
A. veneta (*Gnomonia v.*): on 1 Que [1377].
Armillaria mellea (C:237): root rot,
pourridié-agaric: on 4 BC [1012]; on 5
Sask, Man [F67]; on 11 Ont [F63]; on 13
Que [1377]. See *Abies*.
Articulospora tetracladia: on 11 BC [43].
Botryobasidium sp.: on 13 Ont [1828].
Botryosphaeria melanops (C:237): on 11,
13 Ont [1614].
B. quercuum (C:237): on 11 Ont [F68]; on
11, 13 [1614]; on 11a Ont [1340, 1827],
NS [1614].
Byssocorticium atrovirens: on 13 Ont
[1827].
Cacumisporium capitulatum: on 13 Ont
[586].
Catenuloxyphium semiovatum: on 13 Ont
[798].
Ceratocystis acericola: on 11a Ont [1340].
C. stenospora: on 11a Ont [647, 1340].
C. tenella: on 11a Ont [647, 1340].
Cerrena unicolor (*Daedalea u.*): on 13 Ont
[1341]. See *Acer*.
Chalara brevispora: on 13 Ont [1165].
Cladosporium sp.: on 5, 11a Ont [1340];
on 11 Que [F64].
Coccomyces coronatus: on 11a Ont [1340];
on 13 Ont [1609].

C. triangularis: on 1 Ont [1609].
C. tumidus: on 11 Ont [1609].
Colpoma quercinum: on 1, 2, 13 Ont [1340].
Coniophora sp.: on 13 Ont [1341].
C. puteana: on 13 Ont [1828], NS [1032].
   See *Abies*.
Conoplea globosa: on 11a Ont [1340]; on
   13 Ont [784].
Coriolus hirsutus (Polyporus h.) (C:238):
   white spongy rot, carie blanche
   spongieuse: on 4 BC [1012]; on 11a Ont
   [1341].
C. pubescens (Polyporus p.) (C:238): on
   13 Ont [1341]. See *Abies*.
C. versicolor (Polyporus v.) (C:238): on
   1, 11a Ont [1341]; on 4 BC [1012]; on
   11a NS [1032]. See *Abies*.
Corticium sp.: on 13 Ont [1827].
Coryneum japonicum: on 1, 13 Ont [1762];
   on bark of 13 Ont [1763].
C. megaspermum: on bark of 13 Man [1763].
C. megaspermum var. *cylindricum*: on 13
   Ont [1762, 1763].
C. megaspermum var. *megaspermum*: on 5
   Man [1762]; on 13 Ont [1762].
C. umbonatum (C. canadense, C. kunzei)
   (holomorph *Pseudovalsa longipes*)
   (C:237): on 1 Ont [195, 1762]; on 1, 5,
   13 Ont [1340]; on 1, 11 Ont [F71]; on 1,
   11a Ont [1828]; on 11 Ont, Que [F72]; on
   13 Ont [1763].
Cristella sp.: on 13 Ont [1341].
   Probably *Sistotrema* or *Trechispora*.
Cronartium quercuum (C:237): rust,
   rouille: II III on 11 Ont [F69]; on 11,
   12 Ont [735, 1236]; on 11a Ont [1341,
   1827]; on 13 Ont [1236, 1377].
Cryptocline cinerescens (Gloeosporium c.)
   (C:237): on 11 NS [F60]; on 11a NS
   [1032].
Cytospora intermedia: on 5 Ont [1340]; on
   11 Ont [F70]; on 11a Ont [1828].
Daedaleopsis confragosa (Daedalea c.): on
   11a Ont [1341]. See *Acer*.
Daedalea quercina (C:237): on 11a NB
   [1032]; on 11a, 13 Ont [1341]. See
   Fagus.
*Dendrothele alliacea (Aleurocorticium
   a.)*: on 1 Ont and on 5 Man [965].
   Lignicolous.
D. candida (Aleurocorticium c.): on 4 BC
   [965, 1012]. Lignicolous.
D. griseo-cana (Aleurocorticium g.): on 5
   Man and on 13 Ont [965]. Lignicolous.
D. microspora (Aleurocorticium m.): on
   13 Ont [965].
Diaporthe sp.: on 11a Ont [1340].
D. leiphaemia (C:237): on 10 Ont [339];
   on 11a NB [1032]. Probably a
   Amphiporthe and a European species
   [79].
Diatrype albopruinosa: on 13 Ont [1354].
Diplodia quercus (C:237): on 11 Ont
   [F60]; on 11a Ont [1340].
Diplodina sp.: on 11a Ont [1340, 1828].
*Discula umbrinella (D. quercina,
   Gloeosporium q.)* (holomorph
   Apiognomonia errubunda) (C:237): leaf
   spot, tache des feuilles: on 1 Ont
   [339, F69, F76]; on 1, 5 Ont [1827]; on

5 NB [F63]; on 11 Ont [F74, F75], NB, NS
   [F65], NB, NS, PEI [F66], NS [F72]; on
   11a NB, NS [338].
Dothiorella sp.: on 1, 11a Ont [1828].
D. advena (C:237): on 11 Ont [F67, F68],
   Que [F72]; on 11a Ont [1340].
D. quercina: on 1, 7, 11a, 13 Ont [1340];
   on 4 BC [1012, 1340]; on 11 Ont [F68,
   F70, F71, F74, F76]; on 13 Ont [F63].
Exidia glandulosa (E. spiculosa): on 11a
   Ont [1341]. Lignicolous.
E. recisa: on 13 Que [570]. Lignicolous.
Exidiopsis grisea: on 4 BC [1012].
   Lignicolous.
Favolus alveolaris: on 11a Ont [1341].
   See *Acer*.
Fumago vagans: on 13 Ont [798]. See
   comment under *Corylus*.
Fusicoccum sp.: on 11a Ont [1828].
F. ellisianum (C:237): on 4 BC [1012].
Ganoderma applanatum (Fomes a.) (C:237):
   on 1, 11a Ont [1341]; on 4 BC [1012]; on
   11a Que [1377]; on 11a, 13 NS [1032].
   See *Acer*.
G. lucidum: on 1 Ont [1341]. See *Abies*.
Gloeocystidiellum leucoxanthum: on 4 BC
   [1012]. See *Acer*.
Gloeosporium sp.: on 1 Ont [1827]; on 5
   Man [F60]; on 11a Ont [1340, 1828].
Gnomonia setacea: on 1, 5 Ont [1340].
Graphostroma platystoma: on 13 Ont, Que,
   NS [1343].
Grifolia frondosa (Polyporus f.) (C:238):
   on 13 Ont [F66], NB, NS [1032].
   Associated with a white rot.
Haematostereum gausapatum (Stereum g.)
   (C:238): on 4 BC [1012]; on 11a Ont
   [1341]. See *Alnus*.
Helminthosporium velutinum: on 5 Man
   [810].
Hercospora taleola (Diaporthe t.)
   (C:237): on 1 Ont [1440]; on 1, 5, 11a
   Ont [1340]; on 11 Ont [F60].
Hericium americanum (H. coralloides N.
   Amer. auct.*)*: on 4 BC [1012]; on 11a
   Ont [1341]; on live or dead trunks of 13
   Que [1385].
H. erinaceus: on 11a Ont [1341]. See
   comment under *Acer*.
Hydnochaete olivaceum (Irpex cinnamomea)
   (C:238): on 1, 11a Ont [1341]; on 11a
   NB, NS [1032]; on 13 NS [1032]. See
   Acer.
Hymenochaete agglutinans: on 11a NS
   [1032]. A sterile state of *H.
   corrugata*. See *Alnus*.
H. corrugata: on 11a Ont [1341]. See
   Abies.
H. tabacina: on 11a Ont [1341]. See
   Abies.
Hypocrea citrina: on 13 PEI [1036].
Hyphoderma amoenum: on 4 BC [1012].
   Lignicolous.
Hyphodontia arguta: on 13 Ont [1341].
   See *Abies*.
H. crustosa: on 4 BC [1012]. See *Abies*.
Hypoxylon deustum: on 11a, 13 NS [1032].
H. mammatum: on 1, 11a Ont [1340].
H. mediterraneum (C:238): on 11a Ont
   [1340].

H. rubiginosum: on 13 Ont [1340].
Inonotus cuticularis (Polyporus c.)
(C:238): on 1 Ont [1341]; on 4 BC
[1012]. See *Acer*.
Irpex lacteus (Polyporus tulipiferae)
(C:238): on 11a Ont [1341]. See *Acer*.
Kavinia himantia (Mycoacia h.) (C:238):
on 4 BC [1012]. See *Acer*.
*Laeticorticium roseocarneum
(Dendrocorticium r.)*: on 13 NS [930].
Lignicolous.
L. roseum: on 11a Ont [1341]. See *Acer*.
Laetiporus sulphureus (Polyporus s.)
(C:238): brown cubical rot, carie brune
cubique: on 4 BC [1012]; on 11a Que
[1377]; on 11a, 12, 13 Ont [1341]; on
11a, 13 NB, NS [1032]; on 13 Que [1385].
Lentaria micheneri: on dead trees of 13
Que [1385].
Lentinellus cochleatus: on 13 NS [1032].
See *Acer*.
Lenzites betulina (C:238): on 13 NS
[1032]. See *Acer*.
Leptosphaeria sp.: on 11a Ont [1340].
Marasmius capillaris: on dead leaves and
twigs of 13 Ont [1385].
Massariovalsa sudans (Melanconis s.): on
1 Ont [1340, F63].
Marssonina martini (Marssonia m.)
(C:238): on 1 Ont [195], Que [1377]; on
5 Man [F65]; on 9 Ont [1340].
M. quercina: on 11a Ont [1340].
Menispora ciliata: on 13 Que [826].
M. tortuosa: on 13 Que [826].
Meruliopsis corium: on 4 BC [568]. See
Acer.
Microsphaera penicillata (M. alni)
(C:238): powdery mildew, blanc: on 4
BC [1816]; on 5 Sask, Man [F67], Man,
Ont [1257]; on 10 BC [335], Ont [1257];
on 11 Ont, Que, NS [1257]; on 11a NB, NS
[1032]; on 13 Ont, Que, PEI [1257], Que
[1377], PEI [1032].
Mollisia cinerea: on 13 Ont [1340].
Mycena capillaris: on dead leaves of 13
Ont [1385].
M. stylobatus: on leaves (?fallen) of 13
Que [1385].
M. subcoerulea: on rotten wood and bark
of 13 Ont [1385].
Mycosphaerella punctiformis: on 5 Ont
[1340].
M. spleniata: on 11a Ont [1340].
Nectria sp.: on 3 BC [1012]; on 13 Que
[1377].
N. episphaeria: on 13 Man [175].
Nodulisporium sp.: on 11 Ont [973].
Nummularia repanda: on 13 Man, Ont [862].
Odontia sp.: on 6 Ont [1341].
Oidium sp.: powdery mildew, blanc: on 10
BC [1816].
Omphalotus olearius: on stumps and buried
roots of 13 Que [1385].
Panellus stipticus (Panus s.) (C:238): on
11a Ont [1341]. See *Alnus*.
Panus torulosus: on 4 BC [1012, F66].
Lignicolous.
Peniophora sp.: on 11a NS [1032].
P. cinerea (C:238): on 1 Ont [1341]; on
11 Ont [973]; on 11a NB [1032]. See
Abies.

P. incarnata (C:238): on 4 BC [1012].
Lignicolous.
Pestalozzina unicolor (Pestalotia u.): on
5 Man [F65], Ont [1340].
Phaeobulgaria inquinans: on 11a Ont
[1828].
Phellinus conchatus (Fomes c.): on 1,
11a, 13 Ont [1341]. See *Acer*.
P. everhartii (Fomes e.) (C:237): on 1
Ont [F63]; on 11 Ont [F65, F67]; on 11a,
13 Ont [1341]; on 13 Ont [F66]. See
Betula.
P. ferreus (Poria f.) (C:238): on 4 BC
[1012]; on 11a, 13 NS [1032]. See
Acer.
P. ferruginosus (Poria f.): on 11a NS
[1032]. See *Abies*.
P. igniarius (Fomes i.) (C:237): white
trunk rot, carie blanche du tronc: on
1, 11a Ont [1341].
Phlebia martiana (P. atkinsoniana): on 11
Que [F65]. Lignicolous.
P. radiata (C:238): on 4 BC [1012]; on 13
Ont [568]. See *Abies*.
P. rufa: on 4 BC [1012]. See *Abies*.
Pholiota squarrosa (C:238): on 13 NS
[1032]. See *Populus*.
Phomopsis sp.: on 11a NS [1032].
P. quercina (Myxosporium lanceola): on 10
Ont [336].
Phyllactinia guttata: powdery mildew,
blanc: on 8 Ont [1257].
Phyllosticta livida: on 5 Man [F67].
P. phomiformis (C:238): on 5 Ont [1340].
Physalospora glandicola: on 11 Ont [F62].
Polyporus varius (P. elegans): on 11a Ont
[1341]. See *Abies*.
Poria sp.: on 11a Ont [1341]; on 13 NS
[1032].
P. compacta (Polyporus c.) (C:238): on 11
Que [F73]; on 13 Ont [1153] NS [1032].
P. medulla-panis: on 13 NS [1036].
Associated with a white rot.
P. subacida: on 13 NS [1032]. See
Abies.
Propolis versicolor (P. faginea) (C:238):
on 13 NS [1032].
Pseudospiropes simplex: on 13 Man [817].
Pseudotomentella griseopergamacea: on 13
Ont [1828].
Pseudovalsa longipes (anamorph *Coryneum
umbonatum*) (C:238): on 1 Ont [F60]; on
1, 12, 13 Ont [1340]; on 11a NB, NS
[1032]; on 13 Ont [79].
Pycnoporus cinnabarinus (Polyporus c.):
on 11a Ont [1341]. See *Acer*.
Pyrenochaeta sp.: on 11 Ont [973].
Pythium vexans: from leaf of 10 BC [44].
Ramaricium flavomarginatum: on 4 BC
[572]. Associated with a white rot.
Schizophyllum commune: on 1, 11a, 13 Ont
[1341]. See *Abies*.
Schizopora paradoxa (Poria versipora)
(C:238): on 4 BC [1012]. See *Abies*.
Scopuloides hydnoides (Peniophora h.)
(C:238): on 4 BC [1012]. See *Fagus*.
Spadicoides grovei: on 13 Ont [793].
Spongipellis unicolor (Polyporus obtusus)
(C:238): white spongy rot, carie
blanche spongieuse: on 1, 11a, 13 Ont
[1341]; on 5 Ont [F63]; on 13 Que [1377].

Sporidesmium foliculatum: on 13 Que [825].
S. hormiscioides: on 13 Que [823].
S. leptosporum: on 5 Man [1760].
Steccherinum fimbriatum (Odontia f.): on
    13 Ont [1341]. See *Acer*.
S. ochraceum: on 4 BC [1012]. See *Acer*.
Stereum sp.: on 11a NS [1032].
S. hirsutum (C:238): on 4 BC [1012]; on
    11a Ont [1341]; on 13 NS [1032]. See
    Abies.
S. ostrea: on 4 BC [1012]. See *Abies*.
*Stromatocypha fimbriata (Porotheleum
    f.)*: on 1 Ont [1341]. See *Betula*.
Subulicystidium longisporum (Peniophora l.)
    (C:238): on 4 BC [1012]. See *Betula*.
Taeniolella alta: on 13 Que [814].
Taphrina sp.: on 11 NB, NS [F70]; on 13
    NB [1032].
T. caerulescens (C:238): leaf blister,
    cloque des feuilles: on 1 Ont [F74]; on
    4 BC [1012, 1816, F60]; on 5 NB [1032];
    on 11 Alta [F60], Que [335, F62, F65,
    F71, F73, F76], Que, NB [F61], NB, NS
    [F65, F66]; on 11a Ont [1340, 1827,
    1828], Que [339, 1377], NB, NS [1032];
    on 13 NWT-Alta [1063], Que [333, 336],
    NB, NS, PEI [1032], NS [1925], PEI [334].
Tomentella botryoides: on 11a Ont [1341].
*T. bryophila (Hypochnus subferrugineus, T.
    pallidofulva)*: on 11a Ont [924]; on
    11a, 13 Ont [1341].
T. coerulea (T. papillata): on 13 Ont
    [929, 1341].
T. neobourdotii: on 13 Ont [927, 1341].
T. ramosissima (T. fuliginea): on 11a, 13
    Ont [1341].
Trechispora vaga (Corticium sulphureum):
    on 13 NS [1032]. See *Abies*.
Tremella foliacea: on dead wood of 13 Que
    [1385].
Trichaptum biformis (Polyporus pargamenus)
    (C:238): on 11a Ont [1341]. See *Acer*.
Tricladium splendens: on 13 BC [43].
Tubakia dryina (Actinopelte d.) (C:237):
    leaf spot, tache des feuilles: on 11a
    NB [1032].
Tyromyces chioneus (Polyporus albellus):
    on 13 NS [1032]. See *Abies*.
T. spraguei (Polyporus s.) (C:238): on 11
    Que [F71]; on 11a Que [373]. Associated
    with a brown rot.
Valsa sp.: on 11a Ont [1828].
Xeromphalina kauffmanii: on stumps of 13
    Que [1385].
Xylobolus frustulatus (Stereum f.)
    (C:238): on 11a NS [1032]. Lignicolous.
X. subpileatus (Stereum s.): on 11a, 13
    NS [1032]. Lignicolous.

RANUNCULUS L.   RANUNCULACEAE

1. *R. abortivus* L., kidneyleaf buttercup,
    renoncule abortive; native, Yukon-Mack,
    BC-Nfld, Labr.
2. *R. acris* L., tall buttercup, bouton
    d'or; imported from Eurasia, natzd.,
    Mack, BC-Nfld.
3. *R. cymbalaria* Pursh, seaside
    crowsfoot, renoncule cymbalaire; native,
    Yukon-Mack, BC-Nfld.

4. *R. glaberrimus* Hook., sagebrush
    buttercup; native, BC-Sask.
5. *R. hyperboreus* Rottb.; native,
    Yukon-Labr, BC-Alta, Man, Que, Nfld.
6. *R. nivalis* L.; native, Yukon-Labr, BC,
    Que.
7. *R. pedatifidus* Sm.; native,
    Yukon-Labr, BC-Sask, Ont-Que, Nfld. 7a.
    R. p. var. *leiocarpus* (Trautv.)
    Fern.; native, Yukon-Labr, BC.
8. *R. repens* L., creeping buttercup,
    bassin d'or; imported from Eurasia,
    natzd., BC-Alta, Ont-NS, Nfld, Labr.
9. *R. sceleratus* L., cursed crowsfoot,
    herbe de feu; native, Yukon-Mack, BC-PEI.
10. *R. sulphureus* Soland.; native,
    Mack-Labr.
11. *R. uncinatus* D. Don; native, Mack,
    BC-Alta.

Cylindrosporium sp. (C:239): on 5 Frank
    [1586].
Entyloma ficariae (C:239): smut,
    charbon: on 11 BC [1816]; on *R.* sp.
    BC [1012].
E. microsporum (C:239): smut, charbon:
    on 8 BC [1816].
Erysiphe communis (E. polygoni) (C:239):
    powdery mildew, blanc: on 1, 2, 9 Ont
    [1257]; on 2 Que, NB, NS [1257].
Leptotrochila ranunculi (Fabraea r.): on
    2 Nfld [1340].
Mycosphaerella ranunculi (C:239): leaf
    spot, tache foliaire: on 6, 10 Frank
    [1586].
M. tassiana var. *t.* (C:239): leaf spot,
    tache foliaire: on 2 Que [70].
Nodulosphaeria modesta: on 2 Que [70].
Peronospora ficariae (C:239): downy
    mildew, mildiou: on *R.* sp. BC [1816].
Pseudopeziza sp.: on *R.* sp. BC [1816].
    Possibly *P. singularis* (see Fungi
    Canadenses No. 229).
Puccinia blyttiana (C:240): rust,
    rouille: III on 7a Keew [1236].
P. eatoniae var. *ranunculae* (C:240):
    rust, rouille: 0 I on 1 Ont [1236].
P. recondita s.l. (C:240): rust,
    rouille: 0 I on 2 Ont [1236]; on 3, 4
    BC [1012, 1816].
Ramularia aequivoca (C:240): leaf spot,
    tache ramularienne: on 11 BC [1816].

RAPHANUS L.   CRUCIFERAE

1. *R. sativus* L., radish, radis; a
    cultigen, a garden-escape, BC, Man,
    Ont-Nfld.

Alternaria brassicae (C:240): black spot,
    tache noire: on 1 Alta [1594].
A. brassicicola: on 1 Alta [1594].
A. raphani (C:240): leaf and pod spot,
    pourriture noire: on 1 BC [1816], Alta
    [1594], Sask [339, 1320].
Aphanomyces raphani (C:240): black root,
    racine noire: on 1 BC [334, 1816], NS
    [335].
Peronospora parasitica (P. brassicae)
    (C:240): downy mildew, mildiou: on 1

BC [1816], Ont [1452], Que [335, 337, 338, 339, 1641, 1647, 1652].

Plasmodiophora brassicae (C:240): club root, hernie: on 1 BC [1816].

Rhizoctonia solani (holomorph *Thanatephorus cucumeris*) (C:240): damping-off or root rot, fonte des semis ou rhizoctone commun: on 1 BC [1816], NS [338].

Stemphylium botryosum (holomorph *Pleospora herbarum*) (C:240): on seeds of 1 Ont [667].

Streptomyces scabies (C:240): scab, gale commune: on 1 Sask [334].

Ulocladium consortiale (Stemphylium c.): on seeds of 1 Ont [667].

RAPHIOLEPIS Lindl.   ROSACEAE

1. *R. indica* (L.) Lindl., Indian hawthorn; evergreen shrub, imported from s. China.

Diplocarpon mespili (D. maculatum) (anamorph *Entomosporium mespili*): on 1 BC [1416].

RHAMNUS L.   RHAMNACEAE

1. *R. alnifolia* L'Hér., alder-leaved buckthorn, nerprun à feuilles d'aulne; native, BC-Nfld.
2. *R. cathartica* L., common buckthorn, nerprun commun; imported from Eurasia, introduced, Sask-PEI.
3. *R. frangula* L., alder buckthorn, nerprun bourdaine; imported from Eurasia, introduced, Man-PEI.
4. *R. infectoria* L., Persian berry buckthorn, nerprun des teinturiers; imported from s. Europe and Asia Minor.
5. *R. japonica* Maxim., Japanese buckthorn, nerprun du Japon; imported from Japan.
6. *R. pallasii* Fisch. & C. A. Mey.; imported from w. Asia.
7. *R. purshiana* DC., cascara, cascara; native, BC.
8. *R. saxatilis* Jacq. (*R. willdenourana* Roem. & Schult.), rock buckthorn, nerprun des rochers; imported from Europe. 8a. *R. s.* ssp. *tinctoria* (Waldst. & Kit.) Nyman (*R. chlorophora* Decne.).
9. *R. spathulifolia* Fisch. & C.A. Mey.; imported from w. Asia.
10. *R. utilis* Decne., Chinese buckthorn, nerprun de Chine; imported from China.
11. Host species not named, i.e., reported as *Rhamnus* sp.

Melasmia sp.: on 11 NB [1032].

Phyllosticta rhamni (C:241): leaf spot, tache foliaire: on 7 BC [1816].

Puccinia coronata (C:241): crown rust, rouille des feuilles: on 1 Ont [1827], NS [F68], Nfld [1659, 1660, 1661]; on 1, 2 NB [333]; on 1, 2, 4, 5, 6, 8, 8a, 9, 10, 11 Ont [1236]; on 1, 2, 7 BC [1012]; on 1, 2, 11 Ont [1341]; on 1, 3 NB [1032]; on 1, 7 BC [1816]; on 2 Man [688], Ont [641], NB [335, 1196]; on 2, 11 NB, NS, PEI [1032]; on 3 Sask [F60], Ont [333, 335]; on 7, 11 BC, Alta, Sask, Man [2008]; on 11 Man [839], Ont [338, 339, 472, 1828, F73].

P. coronata var. *coronata (P. c.* f. sp. *agrostidis, P. c.* f. sp. *calamagrostidis)* (C:241): rust, rouille: on 1 NB [1195]; on 3 Ont [337], Ont, NB, NS [334].

P. coronata var. *avenae (P. c.* f. sp. *avenae)* (C:241): on 2 Man, Ont [473], Ont [475, 686, 689].

Schizophyllum commune (C:241): on 7 BC [1012].

Tubercularia vulgaris (holomorph *Nectria cinnabarina*): on 7 BC [1012].

RHEUM L.   POLYGONACEAE

1. *R. rhaponticum* L., rhubarb, rhubarbe; imported from Asia, occasionally a garden-escape, BC, Sask-NS.

Ascochyta rhei (C:241): leaf spot, tache ascochytique: on 1 Alta, Ont [1594], Ont [337, 339], Ont, NS [338], NS [335, 336], PEI [333].

Botrytis cinerea (holomorph *Botryotinia fuckeliana*) (C:241): gray mold, moisissure grise: on 1 Que [334, 337].

Colletotrichum dematium (C. erumpens) (C:241): anthracnose, anthracnose: on 1 BC [1816], Alta, Que [338].

Erysiphe communis (E. polygoni) (C:241): powdery mildew, blanc: on 1 BC, Sask [336].

Fusarium sp. (C:241): on 1 NB [1594].

Phytophthora sp.: on 1 Sask [1186].

Pythium sp.: on 1 Alta [339].

Rhizoctonia solani (holomorph *Thanatephorus cucumeris*) (C:241): crown rot, rhizoctone: on 1 Man [333].

Sclerotinia sclerotiorum: on *R.* sp. Sask [892].

RHINANTHUS L.   SCROPHULARIACEAE

1. *R. borealis* (Sterneck) Druce, northern rattle, rhinanthe boréal; native, Yukon-Keew, BC-Alta, Man-Que, NS, Nfld, Labr.
2. *R. crista-galli* L., common rattle, rhinanthe crête-de-coq; imported from Eurasia, introduced, Mack, BC-Nfld.

Cronartium sp.: rust, rouille: on 1 BC [1816].

C. coleosporioides (aecial state *Peridermium stalactiforme*) (C:242): rust, rouille: on 2 NWT-Alta [1063], BC, Que [735, 1760], Alta [735, F64], Que [1377].

C. coleosporioides f. *coleosporioides*: rust, rouille: on 2 BC [1012].

Peridermium harknessii: on 2 Que [F60]. This rust does not occur on *Rhinanthus*. The specimens were, no doubt, a *Cronartium* sp.

RHODODENDRON L.   ERICACEAE

1. *R. albiflorum* Hook., white
   rhododendron, azalée blanche; native,
   BC-Alta.
2. *R. augustinii* Hemsl.; imported from
   China.
3. *R. calostrotum* Balf. f. & Farr. (*R.
   riparium* F.K. Ward); imported from
   Burma.
4. *R. campylogynum* Franch.; imported from
   China.
5. *R. canadense* (L.) Torr., rhodora,
   rhodora; native, Que-Nfld.
6. *R. catawbiense* Michx., catawba
   rhododendron; imported from s. USA.
7. *R. ciliatum* Hook. f., fringed
   rhododendron; imported from the
   Himalayas.
8. *R. dichroanthum* Diels. 8a. *R. d.*
   ssp. *scyphocalyx* (Balf. f. & Farr.)
   Cowan (*R. scyphocalyx* Balf. f. &
   Farr.); imported from China and Burma.
9. *R. fastigiatum* Franch.; imported from
   China.
10. *R. ferrugineum* L., rock rhododendron,
    rosage des Alpes; imported from Europe.
11. *R. hirsutum* L., garland rhododendron,
    rhododendron hirsute; imported from
    Europe.
12. *R. impeditum* Balf. f. & W.W. Sm.;
    imported from China.
13. *R. indicum* (L.) Sweet, Indian azalea,
    azalée de l'Inde; imported from Japan.
14. *R. japonicum* Suringar; imported from
    Japan.
15. *R. kiusianum* Mak. (*R. obtusum* f.
    japonicum (Maxim.) E.H. Wils.),
    Kyushu azalea; imported from Japan.
16. *R. lapponicum* (L.) Wahlenb., lapland
    rosebay, rhododendron de Laponie;
    native, Yukon-Labr, BC, Man-Que, Nfld.
17. *R. macrophyllum* D. Don, west coast
    rhododendron; native, BC.
18. *R. obtusum* (Lindl.) Planch., Hiryu
    azalea; imported from Japan. 18a. *R.
    o.* var. *hinodegiri*; ornamental
    species, grown in nurseries in BC.
19. *R. pemakoense* F.K. Ward; imported from
    Tibet.
20. *R. ponticum* L.; imported from the
    Mediterranean region.
21. *R. racemosum* Franch.; imported from
    China.
22. *R. radicans* Balf. f. & Farr.; imported
    from Tibet.
23. *R. smirnowii* Trautv., Smirnow
    rhododendron; imported from Eurasia.
24. *Rhododendron* hybrids.
25. Host species not named, i.e., reported
    as *Rhododendron* sp.

Botryosphaeria rhodorae: on 6 Ont [1349];
   on 25 NS [1349].
Botrytis cinerea (holomorph *Botryotinia
   fuckeliana*) (C:242): gray mold,
   moisissure grise: on 25 NS [337, 338].
Cercospora handelii: on 6 Ont [1350]; on
   13 NS [337].

Chrysomyxa ledi (C:242): rust, rouille:
   on 25 NB, NS [1032].
C. ledi var. *rhododendri* (C:242): rust,
   rouille: on 3, 4, 8a, 10, 11, 16, 17,
   19, 22, 24 BC [1012]; on 16 Frank
   [1244], BC [F62]; on 24 BC [1816]; on 25
   NWT, BC, Man [2008].
C. piperiana (C:242): rust, rouille: on
   17 BC [1012, 1816, 2008]; on 24 BC
   [1012, 2008].
*Colletotrichum gloeosporioides
   (Gloeosporium rhododendri)* (holomorph
   Glomerella cingulata) (C:242): leaf
   spot or blight, anthracnose: on 14, 24
   BC [1816]; on 25 BC [335].
Diplodina eurhododendri (C:242): leaf
   spot, tache diplodinéenne: on 23 NS
   [337]; on 24 BC [1816]; on 25 NS [338].
Exobasidium sp.: on 15 BC [1012].
E. burtii (C:242): on 1 BC [1012, 1816],
   Alta [952, F66].
E. canadense (C:242): on 5 Que [1385],
   Nfld [1341].
E. vaccinii (C:242): red leaf, feuille
   rouge: on 5 NS [1032]; on 5, 25 NB
   [1032]; on 13, 14, 17, 18, 24 BC [1816];
   on 17 BC [1012]; on 25 BC [333, 335,
   336, 337], BC, Ont [334].
Godronia cassandrae f. *cassandrae*: on 5
   Que [1684] by inoculation.
Microsphaera penicillata: powdery mildew,
   blanc: on 25 Ont, NS [1257].
Mycosphaerella rhododendri: leaf spot,
   tache foliaire: on 25 BC [1594].
*Pestalotiopsis sydowiana (Pestalotia
   rhododendri)* (C:242): leaf spot, tache
   pestalotienna: on 24 BC [1816]; on 25
   BC [1012].
Phyllosticta sp.: on 24 BC [1816].
Phytophthora sp.: on 18a, 20 BC [23] by
   inoculation.
P. cinnamomi (C:242): on 20 BC [25, 23];
   on 25 BC [1012].
*Protoventuria rhododendri (Antennularia
   r.)*: on 25 BC, Que [73].
Pucciniastrum vaccinii (P. myrtilli)
   (C:242): rust, rouille: on 24 NS [334].
Seimatosporium arbuti: on 25 BC [1359].
S. rhododendri (Coryneum r.) (C:242):
   leaf spot, tache foliaire: on 24 BC
   [1816].
Septoria azaleae (C:242): angular leaf
   spot, tache angulaire: on 25 BC [1595].

RHUS L.   ANACARDIACEAE

1. *R. aromatica* Ait., fragrant sumac,
   sumac aromatique; native, Alta-Sask,
   Ont-Que.
2. *R. glabra* L., smooth sumac, vinaigrier
   glabre; native, BC, Sask-Que.
3. *R. radicans* L., poison ivy, herbe à la
   puce; native, BC-PEI.
4. *R. toxicodendron* L., poison oak;
   native, e. USA. Not native to Canada,
   records probably refer to *R. radicans*.
5. *R. typhina* L., staghorn sumac,
   vinaigrier; native, Ont-PEI.
6. Host species not named, i.e., reported
   as *Rhus* sp.

Amphiporthe aculeans (Cryptodiaporthe a.,
Diaporthe spiculosa) (C:242): on 5 Que
[F76]; on 5, 6 Ont [1340]; on 6 Ont
[F60].
Aplosporella sumachi (Aplosporella s.):
on 5 Ont [1340].
Botryodiplodia sp.: on 5 Ont [1828].
Ciliciopodium sp.: on 5 Ont [1828].
Cladosporium aromaticum: on 5 Que [339].
Coryneum sp.: on 2 BC [1012, 1816].
Godronia rhois: on 5 Ont, Que [656], Que
[F68].
Hypocreopsis lichenoides (anamorph
Stromatocrea cerebriforme): on 5 Ont
[230].
Irpex lacteus (Polyporus tulipiferae): on
6 Ont [1341].
Phaeocalicium curtisii: on 5 Ont [1812].
Phloeospora irregularis (Cylindrosporium
i.) (C:242): on 3 Ont [1340, F62]; on
4 Ont [195].
Pileolaria brevipes (C:243): rust,
rouille: II III on 1, 3 Ont [1236]; on
3 BC, Man, Ont, Que, NS [1284].
Schizophyllum commune: on 5 Ont [1341].
Sphaerotheca macularis (C:243): powdery
mildew, blanc: on 5 Ont [1254], Ont,
Que [1257].
Steccherinum fimbriatum (Odontia f.): on
5, 6 NB [1032]. See *Acer.*
Stigmina negundinis: on 2 Man [1066].
S. pallida (Sciniatosporium p., Coryneum
rhoinum) (C:242): on 2 BC [1127], Man
[1066], Ont [F67]; on 2, 6 Ont [1340];
on 5 Ont [F62]; on 6 BC [1762].
Tubercularia vulgaris (holomorph *Nectria*
cinnabarina) (C:243): dieback,
dépérissement: on 6 Que [334].
Valsa ceratosperma (V. ceratophora)
(C:243): on 5 Ont [F62].
Verticillium sp.: on 5 Ont [96].
V. albo-atrum: on 5 PEI [335].

RHYNCHOSPORA Vahl   CYPERACEAE

1. *R. alba* (L.) Vahl, beaked rush,
rhynchospore blanc; native, BC, Sask,
Ont-Nfld.

Uromyces rhyncosporae (C:243): rust,
rouille: II III on 1 Ont [1236].

RIBES L.   SAXIFRAGACEAE

1. *R. alpinum* L., alpine currant,
grosellier des Alpes; imported from
Europe.
2. *R. americanum* P. Mill., American black
currant, gadellier américain; native,
Alta-NS.
3. *R. aureum* Pursh, golden currant,
gadellier doré; native, BC-Sask.
4. *R. bracteosum* Dougl., stink currant;
native, BC.
5. *R. cereum* Dougl., squaw currant;
native, BC-Alta.
6. *R. cynosbati* L., dogberry, grosellier
des chiens; native, Ont-Que.
7. *R. diacanthum* Pallas, Siberian
currant; imported from n. Asia.

8. *R. divaricatum* Dougl., straggly
gooseberry; native, BC-Alta. 8a. *R.*
d. var. *inerine* (Rydb.) McMim.;
native, BC.
9. *R. glandulosum* Grauer, skunk currant,
gadellier glanduleux; native,
Yukon-Mack, BC-Nfld.
10. *R. grossularia* L., European
gooseberry, grosellier cultivé; imported
from Europe.
11. *R. howellii* Greene; native, BC.
12. *R. hudsonianum* Richards. (*R.*
petiolare Dougl.), wild black currant,
gadellier de l'Hudson; native,
Yukon-Mack, BC-Que.
13. *R. irriguum* Dougl., Idaho gooseberry;
native, BC.
14. *R. lacustre* (Pers.) Poir, bristly
black currant, gadellier lacustre;
native, Yukon-Mack, BC-Nfld, Labr.
15. *R. laxiflorum* Pursh, trailing black
currant; native, Yukon, BC-Alta.
16. *R. lobbii* Gray, gummy gooseberry;
native, BC.
17. *R. missouriense* Nutt., Missouri
gooseberry; native in c. USA.
18. *R. nevadense* Kellogg, Sierra currant;
native in sw. USA.
19. *R. nigrum* L., black currant, gadellier
noir; imported from Eurasia.
20. *R. niveum* Lindl.; introduced.
21. *R. odoratum* H. Wendl., buffalo
currant, gadelier odorant; imported from
USA, natzd., e. and s. Ont, and s. BC.
22. *R. orientale* Desf., oriental currant,
groseillier d'Orient; imported from
Eurasia.
23. *R. oxyacanthoides* L., northern
gooseberry, groseillier sauvage; native,
Yukon-Mack, BC-Nfld. 23a. *R. o.* var.
calcicola Fern. (*R. hirtellum* var.
calcicola Fern.); Ont-Nfld. 23b. *R.*
o. var. *hirtellum* (Michx.) Scoggan
(*R. hirtellum* Michx.); native,
Sask-Nfld. 23c. *R. o.* var. *saxosum*
(Hook.) Cov. (*R. hirtellum* var.
saxosum (Hook.) Fern.); native,
Alta-Man, Que-Nfld.
24. *R. rubrum* L., red currant, gadellier
rouge; imported from Eurasia, cultivated
in Europe but rarely in North America.
25. *R. sanguineum* Pursh, Winter currant,
gadellier sanguin; native, BC.
26. *R. sativum* (Reichenb.) Syme (*R.*
vulgare Lam.), red garden currant,
gadellier cultivé; imported from Europe.
27. *R. setosum* Lindl., Missouri
gooseberry; native, Alta-Que.
28. *R. triste* Pallas, swamp red currant,
gadellier rouge sauvage; native,
Yukon-Mack, BC-Nfld. 28a. *R. t.* var.
albinervum (Michx.) Fern.
29. *R. viscossissimum* Pursh, sticky
currant; native, BC-Alta.
30. *R. watsonianum* Koehne; native, BC.
31. Host species not named, i.e., reported
as *Ribes* sp.

Apiognomonia alniella var. *ribis*: on
23c Que [79].

Armillaria sp.: on 1 Que [F61]. Probably
   A. mellea.
A. mellea: on 1 NB, NS [F75]. See
   Abies.
Botryosphaeria obtusa (C:243): on 19 Ont
   [1614].
Botrytis cinerea (holomorph *Botryotinia
   fuckeliana*) (C:244): gray mold,
   moisissure grise: on 10, 26 BC [1816].
Cronartium ribicola (C:244): white pine
   blister rust, rouille vésiculeuse du pin
   blanc: II III on 1 Que [735]; on 2, 6,
   9 Ont, Que [735, 1236, 1377]; on 3, 10,
   26 BC, Ont, Que [735]; on 3, 10, 28b Que
   [1377]; on 4, 8, 14, 15, 16, 19, 23, 25,
   26, 29 BC [1012, 1816]; on 5, 8, 11, 13,
   15, 16, 18, 20, 25, 27, 29, 30 BC [735];
   on 6, 7, 21 Man [735]; on 8a, 12 BC
   [735, 1012]; on 9 Que [337], Nfld [F76];
   on 9, 19, 23 Man, NB, NS, PEI [735]; on
   9, 19, 31 Ont [1341]; on 9, 23b Nfld
   [1659, 1660, 1661]; on 10 BC [1816]; on
   10, 12, 19 Nfld [735]; on 10, 23b NB, NS
   [735]; on 14 Alta [F72], Nfld [F73,
   F74]; on 14, 19, 23, 28 BC, Alta, Ont,
   Que [735]; on 14, 19, 23b, 26, 28 Ont,
   Que [1236, 1377]; on 17, 20, 21 Ont
   [735, 1236]; on 19 Alta [F69], Sask
   [336], Nfld [337, 338, 1660]; on 19, 24
   Que [338, 339]; on 19, 24, 26 BC [1341];
   on 21 Ont, Que, NS [735]; on 21, 31 Man
   [F62]; on 23, 31 NWT-Alta [1063], Ont
   [1236], Que [282 host given as "*R.
   ?palustris*"]; on 23b, 29 Alta [735]; on
   24 NB [333]; NS [335]; on 26 NS, PEI
   [735]; on 31 BC [953, 954, F70], BC,
   Alta, Man [2008], Alta [951, F75], Alta,
   NB, NS, PEI [F66], Sask, Man [F63], Que
   [335], NB [334, 336, 338, 1594], NB, NS
   [1032, F67], NS [333], Nfld [1661].
Cylindrosporium ribis: on 31 NWT-Alta
   [1063].
Discostroma massarina (Clathridium m.)
   (anamorph *Seimatosporium
   ribis-alpini*)(C:244): on 19 Man
   [1629]; on 31 Ont [1629].
Dothidea ribesia (C:244): on 9 Que
   [1685]; on 19 Ont [1340].
Drepanopeziza ribis (Pseudopeziza r.)
   (anamorph *Gloeosporidiella r.*)
   (C:244): anthracnose, anthracnose: on
   10, 26 BC [1816]; on 10, 31 Que [339];
   on 31 Que, Nfld [333].
D. variabilis (anamorph *Gloeosporidiella
   v.*) (C:244): anthracnose,
   anthracnose: on 1 Ont [334, 336], Ont,
   Que [338]; on 31 Ont, Que, NS [337], Que
   [336, 339].
Ganoderma applanatum: on 25 BC [1012].
Gloeosporidiella bartholomaei: on 29 BC
   [1012, 1816].
G. ribis (Gloeosporium r.) (holomorph
   Drepanopeziza r.): causing leaf
   lesions on 31 Sask, Ont, Que [1763], Que
   [F66, F69], NB, NS [1032].
G. variabilis (holomorph *Drepanopeziza
   v.*): on 1 Ont [335], Ont, Que, NS
   [333]; on 31 Que [335].
Gnomonia ribicola: on 31 BC, Que [79].

Godronia cassandrae f. *ribicola*
   (C:244): on 9, 31 Ont [656].
G. davidsonii (C:244): on 9 Que [1525];
   on 31 Nfld [656].
G. ribis (C:244): on 31 Ont [1340].
G. uberiformis (anamorph *Topospora u.*)
   (C:244): on 31 Alta [644], Ont [656].
*Melampsora ribesii-purpureae (M. epitea
   race r.)* (C:244): rust, rouille: 0 I
   on 8, 14 BC [1012]; on 31 Yukon, BC,
   Alta [2008], Alta [F63], Ont [1236].
Mycosphaerella ribis (C:244): leaf spot,
   tache septorienne: on 10 Nfld [337]; on
   31 NWT-Alta [1063], NS [333].
Nectria cinnabarina (anamorph
   Tubercularia vulgaris) (C:244):
   dieback, dépérissement nectrien: on 1
   NB [1032]; on 19 Ont [1340]; on 31 Sask
   [F69].
Otthia spiraeae: on 31 Alta [644].
Phyllosticta grossulariae (C:244): leaf
   spot, tache foliaire: on 26 BC [1816].
   Host given as "var. *albescens*".
Plasmopara ribicola (C:244): downy
   mildew, mildiou: on 31 Ont [333].
*Protoventuria grossulariae (Antennularia
   g., Venturia g.)* (C:245): on 31 Que
   [70, 73 on overwintered leaves].
Puccinia sp.: rust, rouille: on 14 Que
   [282]; on 31 NWT-Alta [1063].
P. caricina (P. caricis, P. caricis var.
   grossulariata) (C:245): rust,
   rouille: 0 I on 2, 6, 9, 12, 14, 19,
   23, 23a 31 Ont [1236]; on 4, 8, 14, 28
   BC [1012, 1816]; on 6, 31 Ont [1341]; on
   9 BC [1012]; on 9, 31 Ont [1828]; on 10
   BC [1816], Que [335, 339]; on 19 Que
   [338]; on 31 NWT-Alta [1063], Alta [53,
   339, 952], Ont [F60], Que, NB, NS, Nfld
   [333].
P. parkerae (C:245): rust, rouille: III
   on 8, 14 BC [1012, 1816]; on 14 Alta
   [644]; on 31 NWT-Alta [1063], BC [953,
   954].
P. ribis (C:245): rust, rouille: III on
   28 BC [1012, 1816], Ont [1236], Que
   [282].
Ramularia coalescens: on 4 BC [1351].
Septoria ribis: on 1 Que [333]; on 10, 26
   BC [1816]; on 31 Ont [1340].
S. sanguinea (C:245): leaf spot, tache
   septorienne: on 25 BC [1816].
Sphaerotheca mors-uvae (C:245): powdery
   mildew, blanc: on 1 Ont [336]; on 1, 2,
   10, 19, 31 Ont [1257]; on 1, 31 Alta
   [334]; on 9, 13 Mack [1257]; on 10 BC,
   Ont, Que, Nfld [334, 335], Alta, Que
   [339], Que, Nfld [337]; on 10, 19, 26 BC
   [1816]; on 10, 31 Que [336, 1257]; on 14
   BC [1257]; on 19 BC [338], BC, Alta,
   Nfld [339]; on 31 Alta [337, 338, 1594],
   Sask, Que [333], Ont [334].
Stictis pustulata: on 31 Ont [1608].
Thyronectria berolinensis (C:245):
   dieback, dépérissement: on 23 Ont
   [1340]; on 31 Man, Ont [175].
Topospora uberiformis (holomorph *Godronia
   u.*) (C:245): on 31 Alta [644], Ont
   [1763] on twigs.

RICINUS L.   EUPHORBIACEAE

1. *R. communis* L., castor bean, ricin
   commun; imported from the tropics.

Verticillium dahliae: on 1 BC [1978].

ROBINIA L.   LEGUMINOSAE

1. *R. pseudo-acacia* L., black locust,
   robinier: imported from e. USA,
   occasionally a garden escape, Ont-PEI.
2. Host species not named, i.e., reported
   as *Robinia* sp.

Aglaospora profusa (A. anomia): on 1 Ont
   [80, 1624], NB [1036].
Alternaria tenuissima: on 1 NB [1036].
Arthrobotrys oligospora: on 1 NB [1036].
Camarosporium robiniae (C:245): on 1 BC
   [1012], Ont [1828, F63]; on 2 Ont [1340].
Cucurbitaria elongata: on 1 Ont [1340],
   NB, NS [1036].
*Cucurbidothis pithyophila (Cucurbitaria
   p.)*: on 1 NB [1036].
Diaporthe oncostoma: on 1 NB [1036].
Fusarium sp.: on 1 NB [1036].
Irpex lacteus (Polyporus tulipiferae): on
   1 NB [1036]. See *Acer*.
Lopharia cinerascens: on 1 NB [1036].
Massaria sp.: on 2 Ont [1340]. Probably
   Aglaospora profusa which is very
   common on *Robinia*.
Myxosporium sp.: on 1 Ont [1828].
Nectria cinnabarina (anamorph
   Tubercularia vulgaris): on 1 NB
   [1036, 1860, F73, F74].
Phomopsis sp.: on 2 Ont [1340].
Stigmina trimera (Coryneum t.): on 1 NB
   [1032].

ROMNEYA Harv.   PAPAVERACEAE

1. *R. coulteri* Harvey, California tree
   poppy; imported from California.

Sclerotinia sclerotiorum (C:245): wilt,
   fletrissure sclerotique: on 1 BC [1816].

RORIPPA Scop.   CRUCIFERAE

1. *R. islandica* (Oeder) Borbas,
   watercress, cresson des marais; native,
   Yukon-Mack, BC-Nfld, Labr.

Albugo candida (A. cruciferarum) (C:245):
   on *R.* sp. Sask [1319].
Myrothecium roridum: on seeds of 1 Alta
   [1797].

ROSA L.   ROSACEAE

1. *R. acicularis* Lindl., prickly rose,
   églantier; native, Yukon-Mack, Alta-NB.
   1a. *R. a.* ssp. *sayi* (Schwein.) W.H.
   Lewis (*R. a.* var. *bourgeauiana*
   Crép.).
2. *R. arkansana* Porter; native, BC-Man.
   2a. *R. a.* var. *suffulta* (Greene)
   Cockerell; native, Alta-Man.

3. *R. blanda* Ait., meadow rose, rosier
   sauvage; native, Mack, Sask-NS.
4. *R.* x *borboniana* Desp. (*R.
   chinensis* x *R. damascena*), borbon
   rose or hybrid perpetual rose.
5. *R. carolina* L., pasture rose; native,
   Ont-PEI.
6. *R. centifolia* L., cabbage rose, rose à
   cent feuilles; imported from Eurasia.
   6a. *R. c.* var. *muscosa* (Ait.) Ser.,
   moss cabbage rose, rosier mousseux.
7. *R. chiensis* Jacq., China rose, rosier
   de l'Inde; imported from China.
8. *R. damascens* Mill., Damask rose,
   rosier de Belgique; imported from Asia
   Minor.
9. *R.* x *dilecta* Rehder (*R. odorata* x
   R. borboniana), hybrid tea roses; a
   large group which forms many of the
   roses in florists' shops and gardens.
10. *R. gallica* L., French rose, rosier de
   Provins; imported from Eurasia.
11. *R. gymnocarpa* Nutt., dwarf rose,
   rosier nain; native, BC.
12. *R. helenae* Rehd. & E.H. Wils.;
   imported from China.
13. *R. moschata* J. Herrm., musk rose,
   rosier musqué; imported from the
   Mediterranean region.
14. *R. multiflora* Thunb., Japanese rose,
   rosier multiflore; imported from Japan
   and Korea.
15. *R. nutkana* Presl, Nootka rose, rosier
   de Nutka; native, BC. 15a. *R. n.* var.
   hispida Fern. (*R. spaldingii*
   Crépin); native, BC. 15b. *R. n.* var.
   nutkana; native, BC.
16. *R. palustris* Marsh., swamp rose,
   rosier palustre; native, Ont-NS.
17. *R. pisocarpa* Gray, cluster rose;
   native, BC.
18. *R. richardii* Rehd.; imported from
   Ethiopia.
19. *R. rubrifolia* Jill., red-leaf rose,
   rosier à feuilles rouges; imported from
   Europe.
20. *R. rugosa* Thunb., rough rose, rosier
   rugueux; imported from China and Japan.
21. *R. setigera* Michx., prairie rose;
   native, Ont-Que. 21a. *R. s.* var.
   tomentosa Torr. & Gray; native, Ont.
22. *R. wichuraiana* Crép., memorial rose,
   rosier de Wichura; imported from Asia.
23. *R. woodsii* Lindl.; native, Yukon-Mack,
   BC-Que. 23a. *R. w.* var.
   ultramontanum (S. Wats.) Jepson. 23b.
   R. w. var. *woodsii*; native, BC-Man.
24. *Rosa* hybrids.
25. Host species not named, i.e., reported
   as *Rosa* sp.

Botryosphaeria stevensii (anamorph
   Diplodia mutila): on 25 Ont [675].
Botrytis cinerea (holomorph *Botryotinia
   fuckeliana*) (C:246): gray mold,
   moisissure grise: on 24 BC [1816]; on
   25 BC [1012], Que [335, 337], NS [333].
Cercospora rosicola: leaf spot, tache
   foliaire: on 6a BC [1816]; on 25 NB
   [1194].

Cladosporium macrocarpum :   on 25 Man
    [1760].
Coleosporium miniatum :   on 3 Ont [437].
    The specimen is either *Phragmidium
    rosae-pimpinellifoliae* or *P.
    mucronatum* .
Coniothyrium fuckelii (holomorph
    Leptosphaeria coniothyrium):   on 14
    Ont [338]; on 24 BC [1816].
Conoplea fusca :   25 Sask [1760].
Cryptosporium minimum (C:246):   brown
    canker, chancre brun: on 24 BC [1816].
Cytospora sp.:   on 15 BC [1012].
C. ambiens (holomorph *Valsa a.*)
    (C:246):   dieback, dépérissement: on 25
    Que [338].
Cytosporella sp.:   on 15 BC [1012].
Diplocarpon rosae (C:246):   black spot,
    tache noire: on 20 Ont [171, 1778]; on
    24 BC [1816]; on 25 BC, Ont, PEI [337],
    Sask, Que [333, 337, 339], Sask, NB
    [1594], Man, NB [335], Ont, NB, Nfld
    [334], Ont, NS [336], NB [339, 1194],
    NB, NS, PEI [333, 338].
Discostroma corticola (Clathridium c.)
    (anamorph *Seimatosporium
    lichenicola*)(C:246):   on 25 Man [1629].
Dothiorella sp.:   on 25 Ont [675].
Leptosphaeria coniothyrium (C:246):   stem
    canker, chancre: on 25 NS [335, 336].
Mucilago crustacea (M. spongiosa var.
    solida) :   on 25 Ont [337].   Listed as
    "affin. var. *solida* ".
Mycosphaerella rosicola (C:246):   leaf
    spot, tache foliaire: on 25 NB [333].
Nectria cinnabarina (anamorph
    Tubercularia vulgaris) (C:246):   on 25
    Ont [335].
Ophiobolus rubellus :   on 16 Ont [1620].
Peronospora sparsa (C:246):   downy mildew,
    mildiou: on 24 BC [1816].
Phoma sp.:   on 25 Alta [336].
Phragmidium sp. (C:246):   rust, rouille:
    on 1, 19, 25 NWT-Alta [1063]; on 25 Mack
    [53], Alta [338], Man [334], Man, Ont,
    NS, PEI [333], Ont [337, 1341], NB, NS,
    PEI [1032].
P. americanum (C:246):   rust, rouille: 0
    I II III on 1, 3, 5, 21a, 25 Ont [1236];
    on 1, 25 Ont [1341, 1828]; on 25 Alta
    [644], Que [335], NB, NS [1032].
P. fusiforme (C:246):   rust, rouille: 0 I
    II III on 1 Mack [53]; on 1, 25 NWT-Alta
    [1063]; on 1a, 2a, 25 Ont [1236]; on 15,
    15a BC [1012, 1816]; on 25 Yukon [460].
P. fusiforme var. *novi-boreale* :   rust,
    rouille: on 1a Yukon, Mack, Sask, Man,
    Ont [1569]; on 1a, 23b Alta [1569]; on
    11, 15a, 15b, 17, 23a BC [1569].
P. montivagum (C:246):   rust, rouille: on
    17 BC [1012, 1816].
P. mucronatum (C:246):   rust, rouille: 0
    I II III on 3, 25 Ont [1236]; on 25 BC
    [644], Alta, Nfld [1594].
P. rosae-arkansanae (C:247):   rust,
    rouille: on 17 BC [1012, 1816]; on 25
    Yukon [460].
P. rosae-californicea (C:247):   rust,
    rouille: on 1 BC [644]; on 11, 15 BC
    [1012, 1816]; on 15 BC [1341].

*P. rosae-pimpinellifoliae (P.
    subcorticinum)* (C:247):   rust,
    rouille: 0 I II III on 3, 25 Ont
    [1236]; on 24 BC [1816]; on 25 Que [335].
P. speciosum (C:247):   rust, rouille: 0 I
    III on 1, 25 Ont [1236]; on 23 Man
    [1838]; on 25 Ont [1341], NB, NS [1032].
P. tuberculatum (C:247):   rust, rouille:
    on 4, 6, 9, 12, 13, 18, 20, 24 BC [1012].
Pilobolus crystallinus (C:247):   on 24 BC
    [1816].
Pseudomassaria sepincolaeformis :   on 16,
    25 Ont [1616].
Seimatosporium caudatum (C:247):   canker,
    chancre: on 25 BC [1621].
S. discosioides (C:247):   leaf spot, tache
    des feuilles: on 3 Ont [1613]; on 25
    Ont, Que [1613].
*S. lichenicola (Coryneum microsticta,
    Coryneopsis m.)* (holomorph *Discostroma
    corticola*):   on 15 BC [1012]; on 15, 24
    BC [1816].
S. rosae (C:247):   on 25 Ont [1613].
Sphaceloma rosarum (C:247):   anthracnose,
    anthracnose: on 24 BC [1816].
Sphaerotheca macularis (C:247):   powdery
    mildew, blanc: on 24 BC [1816].
S. pannosa (C:247):   powdery mildew,
    blanc: on 5, 14 Ont [1257]; on 20, 25
    Ont, Que [1257]; on 22 Nfld [335]; on 25
    BC [338, 339], BC, Alta, NS [336], BC,
    Sask, NB [1594], Alta, Que, NS [335],
    Ont [469], Ont, Que, NB, NS, PEI [334],
    Ont, NB, NS [333], Que [337], NB [1194].
Stereum sp.:   on 25 BC [1012].
Verticillium sp. (C:247):   wilt,
    flétrissure verticillienne: on 25 Ont
    [97, 96].
V. albo-atrum (C:247):   wilt, flétrissure
    verticillienne: on 25 Ont [38].

RUBUS L.   ROSACEAE

1. *R. acaulis* Michx., dewberry, ronce
    acaule; native, Yukon-Mack, BC-Que, Labr.
2. *R. alleghaniensis* Porter (*R.
    glandicaulis* Blanch.), high blackberry,
    ronce alléghanienne; native, Ont-PEI.
3. *R. arcticus* L., arctic bramble, mûres
    rouges; native, Yukon, BC-Alta.
4. *R. canadensis* L., Canada blackberry,
    ronce du Canada; native, Ont-Nfld.
5. *R. chamaemorus* L., bake-apple, mûres
    blanchee; native, Yukon-Labr, BC-Nfld.
6. *R. flagellaris* Willd., flagellate
    dewberry, ronce à flagelles; native,
    Ont-NB.
7. *R. idaeus* L. (*R. i.* ssp.
    sacchalinensis Lévl.), red raspberry,
    framboisier; native, Yukon-Keew,
    BC-Nfld, Labr. 7a. *R. i.* var.
    aculeatissimus Regel & Tiling (*R.
    melanolasius* Dick); native, Yukon-Mack,
    BC, Sask. 7b. *R. i.* var. *peramoenus*
    (Greene) Fern.; native, Yukon, BC. 7c.
    R. i. var. *strigosus* (Michx.) Maxim.
    (*R. strigosus* Michx.); native,
    transcontinental.
8. *R. laciniatus* Willd., evergreen
    blackberry; imported from Europe, a
    garden escape, BC.

9. *R. x neglectus (R. idaeus x R.
   occidentalis)*, purple raspberry.
10. *R. occidentalis* L., black raspberry,
    mûrier; native, BC, Ont-NB, PEI-Nfld.
    10a. *R. o.* var. *leucodermis* (Dougl.)
    Focke (*R. leucodermis* Dougl.);
    native, BC.
11. *R. odoratus* L., purple-flowering
    raspberry, calottes; native, Ont-NS.
12. *R. parviflorus* Nutt., thimbleberry,
    ronce parviflore; native, BC-Alta, Ont.
13. *R. pedatus* Sm., trailing raspberry,
    framboisier rampant; native, Yukon,
    BC-Alta.
14. *R. pensylvanicus* Poir. (*R. frondosus*
    Bigel.); native, Ont-NS, Nfld.
15. *R. procerus* P.J. Muell., Himalayan
    blackberry; imported from Eurasia, a
    garden escape, BC.
16. *R. pubescens* Raf. *(R. triflorus*
    Richards.)*, dwarf raspberry,
    catharinettes; native, BC-Nfld.
17. *R. spectabilis* Pursh, salmonberry,
    ronce remarquable; native, BC.
18. *R. ursinus* Cham. & Schlecht. (*R.
    macropetalus* Dougl.), pacific dewberry;
    native, BC. 18a. *R. u.* var.
    loganobaccus (Bailey) Bailey (*R.
    loganobaccus* Bailey), loganberry, ronce
    framboise.
19. *R. vitifolius* Cham. & Schlecht.;
    imported from California.
20. Host species not named, i.e., reported
    as *Rubus* sp.

Alternaria sp.: on 7 BC [372].
Armillaria mellea (C:248):   root rot,
   pourridié-agaric: on 12, 20 BC [1816];
   on 15 BC [1012].
Botrytis cinerea (holomorph *Botryotinia
   fuckeliana*) (C:248):   gray mold,
   moisissure grise: on 7 BC [366, 370,
   372, 487, 488, 1595, 1817, 1818], BC,
   Que, NS [335], Alta, Que [338, 339],
   Que, NB [337], Que, NS [334], NB, NS,
   PEI [1594], NS [333, 986]; on 20 BC
   [1816].
Chalara sp.: on 20 Man [1760].
Chlamydozyma zygota (holomorph
   Metschnikowia reukaufii): on 7c Que
   [1968].
Cladosporium sp.: on 7 BC [372].
C. herbarum: on fruit of 7 in storage NS
   [986].
Clypeosphaeria hendersoniae: on 7a Ont
   [1340].
Coleroa chaetomium (C:248): on 12, 13 BC
   [288]; on 13 BC [1012], Alta [288]; on
   living leaves 20 BC [73].
C. rubicola (*Stigmatea r.*) (C:250): on 7
   Ont [1340]; on 7a Sask, Ont [294]; on 7c
   Man, Ont, NS [294]; on 20 BC [294], Man,
   Ont [73].
Coniothyrium fuckelii (holomorph
   Leptosphaeria coniothyrium) (C:248):
   on 7 Ont [1827]; on 20 BC [1816].
Crucibulum laeve: bird's nest fungus: on
   20 NB [1036]. See *Pseudotsuga*.
Cryptodiaporthe vepris (Apioporthe v.)
   (C:248): on 7 NS [599].

Cylindrocladium scoparium (C:248): on 20
   BC [1816].
Dasyscyphus bicolor (Lachnella b.): on 12
   BC [1012, 1816]; on 20 Ont [1340], Que
   [282].
D. bicolor var. *rubi*: on 20 Ont [1340].
D. sulphureus: on 12 BC [1012].
D. virgineus (Lachnella v.): on 7a Ont
   [1340].
Didymella applanata (anamorph *Phoma* sp.)
   (C:248): spur blight, brûlure des
   dards: on 7 BC [291], BC, Alta, Sask,
   NB, NS [339], BC, Sask, Ont [291], Alta,
   Sask, Que, NS [337], Alta, Ont, NS
   [334], Alta, NB [338], Sask, Ont, NS
   [333], Que, NB, NS [336], Que, NS [335];
   on 7, 18a BC [1594]; on 20 BC [1816],
   Ont, Que [168].
Discostroma corticola (anamorph
   Seimatosporium lichenicola) (C:248):
   on 20 Man [1629].
Elsinoe veneta (C:248):   anthracnose,
   anthracnose: on 7 BC, Ont [333], Ont
   [334], Que [335, 336], Que, NB [334,
   338], NB, NS [333, 335, 336, 337, 339,
   1594], NS [710]; on 20 BC [1816].
Fabraea cincta (C:248): on 13 BC [1357].
Fusarium sp. (C:248): on 2 Nfld [1594].
F. avenaceum (C:248): on 20 BC [1816].
Gnomonia comari (anamorph *Zythia
   fragariae*): on 5 Que [79].
G. rubi: on 20 Ont, NS [79].
Gymnoconia peckiana (C:249):   orange rust,
   rouille orangée: 0 I III on 1, 2, 7a,
   7c, 10, 14, 16, 20 Ont [1236]; on 4, 20
   Nfld [1659, 1660, 1661]; on 7 Que [339];
   on 7, 20 Ont [1341]; on 16 BC [7]; on 20
   NWT-Alta [1063], Alta [53], NB, NS, PEI
   [1032].
Hapalosphaeria deformans (C:249):   dry
   berry, anther and stigma blight, brûlure
   des drupéoles: on 18a, 20 BC [1816].
Helminthosporium velutinum: on 20 Ont
   [810].
Hymenochaete tabacina: on 20 NS [1032].
Hymenoscyphus separabilis (Helotium s.):
   on 20 Que [1340].
Hyphoderma pallidum: on 17 BC [1012].
Hypholoma fasciculare (Naematoloma f.)
   (C:249): on 20 BC [1816].
Karstenia rubicola (Coccomyces r.): on 7c
   Ont [440, 1609].
Kuehneola uredinis (C:249):   rust,
   rouille: 0 I II III on 6, 20 Ont
   [1236]; on 8 BC [335, 338, 1012, 1816].
Kunkelia nitens (Caeoma luminatum): on 16
   Ont [437].
Leptosphaeria coniothyrium (anamorph
   Coniothyrium fuckelii) (C:249):   cane
   blight, brûlure de la tige: on 7 Alta,
   NS [339], Que [334], NS [333].
Lophodermium rubiicola: on 15 BC [1012].
Monilia sp.: on 2 Ont [1827].
Mycosphaerella rubi (C:249): on 7 NS
   [337, 339]; on 12, 17, 18 BC [1012]; on
   18a BC [337]. There is some question
   whether *Septoria rubi* is the anamorph
   or not, see Conners (p. 250).
Nectria cinnabarina (anamorph
   Tubercularia vulgaris) (C:249):

dieback or coral spot, dépérissement nectrien: on 20 NWT-Alta [1063].
Nidula candida (C:249): bird's nest fungus: on 17 BC [1012].
Oidium sp.: powdery mildew, blanc: on 12 BC [1816].
Omphalina ericetorum: on 17 BC [1012]. A lichenized Basidiomycete, coincidentally on *Rubus*.
Penicillium sp.: on 7 BC [372], NS [986] on fruit in storage.
Peniophora cinerea: on 20 NS [1032].
P. incarnata: on 17 BC [1012].
Peronospora rubi (C:249): downy mildew, mildiou: on 12, 17, 19 BC [1816].
Pezicula rubi (C:249): on 10 Ont [1340].
Phellinus ferreus (Poria f.): on 17 BC [1012].
Phragmidium sp.: rust, rouille: on 2, 20 Ont [1341]; on 20 Ont [1828], NS [1032].
P. arcticum (C:249): rust, rouille: 0 I II III on 1 Mack, Alta, Que [1575], Que [1236].
P. occidentale (C:249): rust, rouille: 0 I II III on 11 Ont, Que [1576]; on 12 NWT-Alta [1063], BC [954, 1012, 1816], BC, Alta, Ont [1576]; on 16 Mack [53].
P. rubi-idaei (C:249): yellow rust, rouille jaune: 0 I II III on 7 Yukon [460], BC, Que [333, 334, 339], Que [335, 338], NB [339, 1594], NS [337]; on 7, 7a, 7c Ont [1236]; on 7, 7b BC [1012]; on 7a, 10, 20 BC [1816]; on 7a, 20 NWT-Alta [1063]; on 12 Alta [950]; on 16 Mack [53]; on 20 BC [954].
P. rubi-odorati (C:249): rust, rouille: I II III on 11 Ont, Que [1236]; on 20 Ont [1341].
Phytophthora sp. (C:249): root rot, pourridié des racines: on 7 BC [333]; on 20 BC [1816].
P. fragariae: on 18a BC [1816].
Polyporus badius (P. picipes): on 17 BC [1012].
Pucciniastrum americanum (C:249): late yellow rust, rouille jaune tardive: II III on 7 BC [1012], Que [1377, F60], Que, NS, PEI [335], NB, NS [336], NS [333], NS, PEI [334]; on 7a, 7c, 20 Ont [1236]; on 9 Que [1236]; on 20 BC, Man [2008], Alta [644].
P. arcticum (C:249): rust, rouille: II III on 3 Yukon [460], BC [1816, F62]; on 3, 12, 16 BC [1012]; on 3, 16 Yukon, Mack, BC, Alta, Sask, Man [2008]; on 16 Mack [F66], Mack, Alta [53], Ont [1236], Que [1377]; on 20 BC [954], Alta [951].
Pyrenopeziza rubi (C:249): on 20 Ont [1340].
Rhizoctonia solani (holomorph *Thanatephorus cucumeris*) (C:250): root rot, rhizoctone commun: on 7 Que [334].
Rhizopus sp.: on 7 BC [370, 372], NS [986].
Seimatosporium lichenicola (Hendersonia rubi) (holomorph *Discostroma corticola*): on 20 BC [1816].
Septoria rubi (C:250): leaf spot, tache septorienne: on 7 BC, NS [335], Que

[334]; on 7, 10, 10a, 12, 17, 18, 18a, 19, 20 BC [1816]; on 18a BC [339, 1594]. See comment under *Mycosphaerella rubi*.
Sphaerotheca macularis (S. humuli) (C:250): powdery mildew, blanc: on 2, 4, 7, 11, 16, 20 Ont [1257]; on 5 Que [282, 1254]; on 5, 7, 11 Que [1257]; on 7 BC [333, 488], Alta, Que [338], Sask [1254], Que [339], Que, NS [335], NS [337, 1594]; on 7, 17, 19 BC [1816]; on 16 Ont [1254]; on 17 BC [1012].
Stomiopeltis borealis: on 10 BC [72].
Sydowiella depressula: on 20 BC [79].
Typhula erythropus: on 12 BC [901].
Valdensia heterodoxa (Asterobolus gaultheriae) (holomorph *Valdensinia h.*): on 12, 18 BC [1438]; on 13 BC [1426].
Valsa sp.: on 17, 20 BC [193].
V. ceratosperma (V. ceratophora) (C:250): on 7c Que [70]; on 17 BC [193]; on 20 Ont [1340].
Venturia chamaemori (Sphaerella c.) (C:250): on overwintered leaves of 5 Labr [73].
Verticillium sp. (C:250): wilt or blue stem, flétrissure verticillienne: on 7 Ont [96].
V. albo-atrum (C:250): wilt or blue stem, flétrissure verticillienne: on 7 BC [1816], Ont [98], Que [339], Que, NS [333, 334], NB [1594], NS [335, 336]; on 20 Ont [38].
V. dahliae (C:250): wilt or blue stem, flétrissure verticillienne: on 7 Ont [1087]; on 10 Ont [38].

RUDBECKIA L.   COMPOSITAE

1. *R. hirta* L., black-eyed susan, marguerite jaune; imported from midwestern USA, introduced BC-Nfld. Most plants are referable to *R. hirta* var. *pulcherrima* Farw. (*R. serotina* Nutt.).
2. *R. laciniata* L., coneflower, rudbeckie laciniée; native, Man-Que, NS.

Erysiphe cichoracearum (C:251): powdery mildew, blanc: on 1, 2 Ont [1257]; on 2 BC [1816].
Uromyces rudbeckiae (C:251): rust, rouille: III on 2 Ont [1236].

RUMEX L.   POLYGONACEAE

1. *R. acetosella* L., sheep sorrel, oseille; imported from Eurasia, natzd. Yukon, BC-Nfld, Labr.
2. *R. crispus* L., curly dock, requette; imported from Eurasia, introduced, Yukon, BC-Nfld.
3. *R. obtusifolia* L., bitter dock, patience; imported from Eurasia, introduced BC, Ont-Nfld.
4. *R. orbiculatus* Gray, water dock, patience orbiculée; native, BC-Nfld. 4a. *R. o.* var. *borealis* Rech. f. (*R. britannica* aucts. non L.).

5. *R. verticillatus* L., swamp dock,
   patience verticillée; native, Ont-Que.

Erysiphe communis (E. polygoni): powdery
   mildew, blanc: on 2 Ont [1257].
Puccinia acetosae (C:251): rust,
   rouille: II III on 1 Ont [1236].
P. ornata (C:251): rust, rouille: III on
   4, 5, *R*. sp. Ont [1236].
Ramularia rubella (C:251): on 2 Que [282].
Venturia canadensis: on 1 BC, Que [301]
   and overwintered leaves and stalks BC,
   Que, Nfld [73].
V. rumicis (C:251): on 1 Que [70]; on 3
   BC, Que, Nfld [300]; on spots in living
   and dying leaves of *R*. sp. BC, Nfld
   [73].

SAGITTARIA L.   ALISMATACEAE

   Aquatic, mostly perennial herbs of
tropical and temperate regions especially in
the w. hemisphere.

1. *S. latifolia* Willd., wapat, wapaton;
   native, BC-NS.

Doassansia intermedia (C:252): on *S*.
   sp. Ont [1341].
Gloeosporium confluens (C:252): on leaves
   of 1 Ont [440]. The correct generic
   placement for this species is uncertain
   [20].

SAINTPAULIA H. Wendl.   GESNERIACEAE

1. *S. ionantha* H. Wendl., common African
   violet, violette africaine; imported
   from Africa.

Botrytis cinerea (holomorph *Botryotinia
   fuckeliana*): gray mold, moisissure
   grise: on 1 Que [334].
Erysiphe cichoracearum (C:252): powdery
   mildew, blanc: on 1 Alta [1594], Ont
   [1257].
Phytophthora nicotianae var. *n*. : on 1
   Alta [204], Ont [205].
Pythium sp.: on 1 BC [1816].

SALIX L.   SALICACEAE

1. *S. alaxensis* (Anderss.) Coville;
   native, Yukon-Frank, BC-Man.
2. *S. alba* L., white willow, saule blanc;
   imported from Eurasia, natzd.,
   Man-Nfld. 2a. *S. a.* var. *vitellina*
   (L.) Stokes (*S. vitellina* L.), golden
   willow, osier jaune. 2b. *S. a.* var.
   v. cv. Pendula (*S. a.* var. *tristis*
   (Ser.) Gaudin).
3. *S. amygdaloides* Anderss., peachleaf
   willow, saule à feuilles de pêcher;
   native, BC-Que.
4. *S. arbusculoides* Anderss. (*S.
   acutifolia* Hook.); Yukon-Mack, Keew,
   BC-Man.
5. *S. arctica* Pallas (*S. torulosa*
   Trautv.), arctic willow, saule arctique;
   native, Yukon-Labr, BC-Alta, Que, Nfld.

6. *S. arctophila* Cockerell; native,
   Yukon-Labr, Sask-Que, Nfld.
7. *S. babylonica* L., weeping willow,
   saule pleureur; imported from China.
8. *S. bebbiana* Sarg. (*S. rostrata*
   Richards. non Thuill.), beaked willow,
   chaton; native, Yukon-Mack, BC-Nfld,
   Labr.
9. *S.* x *blanda* Anderss., Niobe willow,
   probably *S. babylonica* x *S. fragilis*.
10. *S. candida* Fluegge, hoary willow,
   saule tomenteux; native, Yukon-Mack,
   BC-Nfld, Labr. 10a. *S. c.* f.
   denudata (And.) Rouleau; native, range
   of species.
11. *S. cinerea* L., gray willow, saule
   cendré; imported from Eurasia.
12. *S. cordata* Michx., heart-shaped
   willow, saule à feuilles en coeur;
   native, Man-Nfld. 12a. *S. c.* var.
   rigida (Muhl.) Carey (*S. rigida*
   Muhl.), erect willow, saule rigide;
   native, Man-Nfld.
13. *S. discolor* Muhl., large pussy willow,
   petit minou; native, BC-Nfld, Labr.
14. *S. exigua* Nutt. (*S. rubra* Richards.
   non Huds.), sandbar willow; native,
   Yukon-Mack, BC-NB.
15. *S. fragilis* L., crack willow, saule
   cassant; imported from Eurasia, spread
   from cultivation, Alta, Man-Nfld.
16. *S. glauca* L., gray-leaved willow,
   saule vert-de-mer; native, BC-Que, Nfld.
17. *S. herbacea* L., herbaceous willow,
   saule herbacé; native, Mack-Labr, Man,
   Que, Nfld.
18. *S. humilis* Marsh., small pussy willow,
   saule humble; native, Sask-Nfld.
19. *S. lasiandra* Benth., Pacific willow,
   saule laurier de l'Ouest; native, Yukon,
   BC-Sask.
20. *S. laurentiana* Fern.; native, Ont-NB,
   Nfld. 20a. *S. l.* f. *glaucophylla*
   (Bebb.) Boivin (*S. glaucophylla*
   Bebb.); native, Ont-NB, Nfld.
21. *S. lucida* Muhl., shining willow, saule
   laurier; native, Sask-Nfld, Labr.
22. *S. maccaliana* Rowlee; native,
   Yukon-Mack, BC-Que.
23. *S. myrtillifolia* Anderss.; native,
   Yukon-Mack, BC-NB, Nfld.
24. *S. nigra* Marsh., black willow, saule
   noir; native, Ont-NB.
25. *S. pedicillaris* Pursh, bog willow,
   saule pédicellé; native, Yukon, BC-Nfld,
   Labr.
26. *S. pellita* Anderss., silky willow,
   saule saliné; native, Sask-NS, Nfld,
   Labr.
27. *S. pentandra* L. (*S. laurifolia*
   Wesm.), bayleaf willow, saule laurier
   européen; imported from Europe.
28. *S. petiolaris* Sm., slender willow,
   saule pétiolé; native, BC-PEI.
29. *S. planifolia* Pursh ssp. *planifolia*;
   native, Yukon-Labr, e. BC-Alta.
30. *S. purpurea* L., basket willow, saule
   pourpre; imported from Eurasia,
   introduced, Ont-Nfld. 30a. *S. p.* cv.
   nana.

31. *S. reticulata* L., netted willow, saule
réticulé; native, Yukon-Labr, BC-Man.
32. *S. scouleriana* Barratt, Scouler
willow, saule de Scouler; native,
Yukon-Mack, BC-Man. 32a. *S. s.* var.
poikila Schn.; native, Yukon, BC.
32b. *S. s.* var. *thompsonii* Ball;
native BC.
33. *S. sitchensis* Sanson, Sitka willow,
saule de Sitka; native, Yukon, BC-Alta.
34. *S. x smithiana* Willd. (*S. x
fruticosa* Doell., *S. caprea* L. x
S. viminalis L.).
35. *S. triandra* L. (*S. amygdalina* L.),
almond-leaved willow, osier amandier;
imported from Europe.
36. Host species not named, i.e., reported
as *Salix* sp.

Acremonium boreale: on 2a, 4 BC, Alta,
Sask [1707].
Acrospermum cuneolum: on 36 BC [1012].
Acrotheca sp.: on 36 BC [1012].
Aleurodiscus cerussatus (C:253): on 36
Man [964]. Lignicolous.
A. oakesii (C:253): on 36 NS [964, 1032,
1036]. Lignicolous.
Allantoporthe tessella (*Diaporthe t.*)
(C:254): on 36 Sask [F66], Ont [79,
1340, F60], NS [1032].
Anthostoma melanotes (C:253): on 36 NS
[1032].
Armillaria mellea (C:253): on 36 BC
[1012], Ont [1341]. See *Abies*.
Articulospora tetracladia: on 36 BC [43].
Basidiodendron cinerea: on ?36 BC
[1016]. Lignicolous.
Basidioradulum radula (*Hyphoderma r.*): on
36 BC [1012]. See *Abies*.
Bispora betulina: on 36 Sask [1760].
B. effusa: on 36 Alta [1755, 1760].
Bisporella citrina: on 36 BC [1012].
Bjerkandera adusta (*Polyporus a.*)
(C:256): on 36 NB, NS [1032]. See
Abies.
Camarosporium salicinum: on 36 Sask [F66].
Cenangium betulinum: on 36 Yukon [664].
Cephalosporium sp.: on 7 BC [1012].
Ceriporia purpurea (*Poria p.*): on 36 Sask
[579]. See *Alnus*.
C. rhodella (*Poria r.*): on 36 BC [1012].
See *Pinus*.
Cercospora salicina (C:253): on 13 Nfld
[1659, F76]; on 13, 36 Nfld [1660,
1661]; on 36 Nfld [F68, F73, F74].
Cerrena unicolor (*Daedalea u.*) (C:254):
on 36 Yukon [460, F62], Que [1855]. See
Acer.
Chondrostereum purpureum (*Stereum p.*)
(C:257): on 7 Ont [1341]; on 7, 36 NB
[1036]; on 36 BC [1012], NB, NS [1032].
See *Abies*.
Ciboria caucus (C:253): on 36 Ont [662,
1590].
Ciborinia sp.: on 28, 36 Sask, Man [1766].
C. foliicola (C:253): on 36 BC [953,
1012, F73], Alta [951, F68, F73], Sask,
Man [1766, F66, F67, F68, F71], Man
[F69], Que [1766, F60, F61, F67, F70].
Cladosporium sp.: on 36 NB [1032].

C. herbarum (C:253): on 36 Yukon [460],
NB [1032], NS [1036].
C. macrocarpum: on 36 Sask, Man [1760].
Clavicorona pyxidata: on rotten wood of
36 Que [1385]. See *Betula*.
Colletotrichum sp.: on 19 BC [1012].
Coltricia perennis (*Polyporus p.*)
(C:256): on 36 BC [1012].
Columnophora rhytismatis (C:253): on 5
Frank [1586]. Possibly a *Stigmina*.
Coniophora puteana: on 36 Nfld [1659,
1660, 1661]. See *Abies*.
Conoplea geniculata: on 36 NWT-Alta
[1063], Alta [53].
Coprinus micaceus: on 36 BC [1012]. See
Populus.
Coriolopsis gallica (*Trametes hispida*)
(C:257): on 7 BC [1012]. See *Acer*.
Coriolus hirsutus (*Polyporus h.*) (C:256):
on 36 BC [1012], NS [1032]. See *Abies*.
C. versicolor (*Polyporus v.*) (C:256): on
36 BC [1012]. See *Abies*.
C. zonatus (*Polyporus z.*): on 36 BC
[1012]. See *Betula*.
Crepidotus mollis (*C. fulvotomentosus*)
(C:253): on 36 BC [1012]. See
Populus.
Crustomyces subabruptus (*Cystostereum
s.*): on 36 BC [1012]. Lignicolous.
Cryptodiaporthe apiculata: on 36 BC
[1012, F70].
C. salicella (anamorph *Diplodina
microsperma*) (C:253): on 7 Ont [339];
on 36 BC [1012, F70], Que [79, F66].
C. salicina (C:253): on 7 Ont [1827], NB,
NS [1036, F75]; on 36 Ont [1340, F60,
F66], NB, NS [F68], NB, NS, PEI [1032,
F70], NS [F73]. Barr [79] referred
Wehmeyer's collections to the preceding
species but Currey's fungus, *C.
salicina*, is distinct with wider spores
[386].
Cryptomyces maximus (C:253): on 36 BC
[1012], Alta [F68], Sask, Man [F67].
Cryptosporiopsis scutellata (holomorph
Ocellaria ocellata): on 36 Sask [F65].
Cucurbidothis pithyophila: on 36 Yukon
[460], BC [1012, F64, F67].
Cylindrobasidium evolvens (*Corticium
laeve*) (C:253): on 32, 36 BC [1012];
on 36 NS [1032]. See *Abies*.
Cylindrocarpon orthosporum: on 36 Man
[1760].
Cylindrosporium salicinum (C:253): on 36
Que [F61].
Cyphella sp.: on 36 Ont [1341].
Cyphellopsis anomala: on 36 BC [1012].
Cytidia flocculenta: on 16 Yukon [1110].
This specimen may be *Auriculariopsis
ampla*.
C. salicina (C:253): on 1 Yukon [1110];
on 36 Yukon [460], BC [1012], Ont
[1341], Que [282], NB, NS [1032], Nfld
[1659, 1660]. Lignicolous.
Cytospora sp. (C:253): on 7 BC [726], Ont
[1827], Que [F70], NB [1032, 1036], Nfld
[F76]; on 7, 19, 36 BC [1012]; on 13 NS
[1032]; on 27 Que [F65]; on 36 Yukon
[460], NWT-Alta [1063], Mack [53], BC
[F60], BC, Ont [1340], Alta [337], Ont
[1828], NB, PEI [1032].

C. chrysosperma (holomorph *Valsa
sordida*) (C:253): on 2b, 7 Ont [339];
on 5 Que [339]; on 7 BC [1340, 1816,
F60], Ont [F73], Nfld [1660]; on 9 BC
[335]; on 27 Alta, Que [338]; on 36 Alta
[339, 1594], Ont [336, 1340, 1828, F64].

C. pulcherrima (C:254): on 36 BC [1340].

C. salicis (C:254): on 7 NB [F64]; on 13
Alta [950], Nfld [1660, F74]; on 36 Nfld
[1661, F76].

Daedaleopsis confragosa (*Daedalea c.*)
(C:254): on 36 Yukon [460, F62],
NWT-Alta [1063], Mack [53], BC [1012],
Ont [1341], NB, NS [1032]. See *Acer*.

Datronia stereoides (*Polyporus
planellus*): on 36 Yukon [1110]. See
Alnus.

Dendrothele griseo-cana (*Aleurocorticium
g.*): on 36 Ont [965]. Lignicolous.

Dendryphiopsis atra (holomorph
Microthelia incrustans): on 36 Sask
[1760].

Dentocorticium sulphurellum (*Laeticorticium
s.*): on 36 Ont [555, 930].
Lignicolous.

Diaporthe sp.: on 36 BC [193].

Diatrype bullata (C:254): dieback,
dépérissement: on 36 BC [1012].

Diplodina salicicola: on 7 Que [F70].

D. microsperma (*D. salicis*, *Septomyxa s.*)
(holomorph *Cryptodiaporthe salicella*)
(C:254, 257): on 36 BC [1012, F62], Man
[F66], Que [338, 1377], NB, NS [1032].

Discella carbonacea (C:254): on 7 Ont
[1827]; on 36 Sask, Man [F67], Ont
[1340, F74], Que [F66]. Specimens are
probably *Diplodina microsperma* [1763].

Discostroma corticola (*Griphosphaeria c.*)
(anamorph *Seimatosporium
lichenicola*): on 36 Man [F68].

Dothiorella pyrenophora var. *salicis*
(C:254): on 36 Sask [F65, F67].

Drepanopeziza salicis (C:254): leaf spot,
tache des feuilles: on 36 Man [F65].

D. sphaerioides (C:254): on 7, 36 NB
[1036]; on 36 Sask [F66].

Eichleriella deglubens (*E. spinulosa*): on
36 Ont [1341]. See *Populus*.

Endophragmiella biseptata (*Endophragmia
nannfeldtii*): on 36 Sask [1760].

Epipolaeum longisetosa: on 36 Que, Labr
[73].

Exidia glandulosa f. *populi*: on 36
Yukon [460], BC [1012]. Lignicolous.

E. recisa (C:254): on 36 Yukon [460],
NWT-Alta [1063], Alta [53]. Lignicolous.

Exidiopsis grisea: on 36 BC [1012].
Lignicolous.

Favolus alveolaris (C:254): on 36
NWT-Alta [1063], Ont [1341]. See *Acer*.

Flammulina velutipes: on stumps and dead
trunks, sometimes on dead parts of live
trees of 36 Que [1385]. See *Populus*.

Fomes sp.: on 36 NWT-Alta [1063].

Fusarium sp.: on 2a BC [1012]; on 7 NB
[1032].

F. lateritium: on 36 BC [1012].

Fusicladium sp.: on 36 BC [1012], Que
[282].

Fusicoccum sp. (C:254): on 5 Frank [1586].

Ganoderma applanatum (*Fomes a.*): on 36 BC
[1012], Man [F67], Que [1377]. See
Abies.

Gloeocystidiellum porosum: on 36 BC
[1012]. See *Acer*.

Gloeoporus dichrous (*Polyporus d.*)
(C:256): on 36 Ont [1341]. See *Acer*.

Gloeosporium sp.: on 2a, 7 BC [F61]; on
36 BC [1012], Sask [F60], Ont [1340],
NB, NS, PEI [1032].

Glyphium elatum: on 36 BC [589].

Godronia cassandrae (anamorph *Fusicoccum
putrefaciens*): on 36 Que [1678].

G. cassandrae f. *betulicola*: on 36 Que
[1684, F68]. See comment under *Alnus*.

G. fuliginosa (C:254): on 36 BC [1012,
F65], Sask, Ont [656], Que [1680, 1684,
F68].

Graphium sp.: on 36 Ont [1340].

Gymnopilus spectabilis (*Pholiota s.*)
(C:256): on 36 NS [1032]. See *Acer*.

Hapalopilus nidulans (*Polyporus n.*)
(C:256): on 19 BC [1012]; on 36 BC, Ont
[1341], Nfld [1659, 1660, 1661]. See
Abies.

Haploporus odorus (*Trametes o.*) (C:257):
on 36 NWT-Alta [1063], Mack, Alta [579],
BC [1012, F60], BC, Sask [1014], Alta
[1341], Sask [F67]. See *Alnus*.

Helminthosporium sp.: on 36 BC [1012].

Hendersonia sp.: on 36 Ont [1340]. See
comments under *Abies*.

Heteroconium tetracoilum (*Septonema t.*):
on 36 Man [1760].

Hymenochaete tabacina (C:254): on 36 BC
[1012], Ont [1341], NB, NS [1032], Nfld
[1659, 1660, 1661]. See *Abies*.

Hyphodontia pallidula: on 36 BC [1012].
See *Abies*.

H. quercina: on 36 BC [1012].
Lignicolous.

Hypochnicium bombycinum (*Corticium b.*)
(C:253): on 36 NS [1032]. See *Acer*.

Hypocreopsis lichenoides (anamorph
Stromatocrea cerebriforme): on dead
branches of 36 Que [1385].

Hypoxylon sp.: on 36 NS [1032].

H. fuscum: on 36 Sask [F67].

H. mammatum (C:254): on 36 BC [1012],
Sask, Man [F66], Ont [1340], NS [1032],
Nfld [1659, 1660, 1661].

H. rubiginosum: on 36 Sask [F67].

Ingoldia craginiformis: on 36 BC [43].

Intextomyces contiguus (*Corticium c.*)
(C:253): on 36 NWT-Alta [1063], Mack
[53], BC [1012]. Lignicolous.

Irpex lacteus (*Polyporus tulipiferae*)
(C:256): on 36 BC [1012], Que [1390],
NB [1032]. See *Acer*.

Junghuhnia nitida (*Poria attenuata*): on
36 NS [1032]. See *Acer*.

Kabatiella boreale (*Gloeosporium b.*)
(C:254): on 36 Que [1377].

Lachnella alboviolascens (*Cyphella a.*):
on 36 BC [1012, 1424]. Lignicolous.

Laeticorticium lundellii (*Dendrocorticium
l.*): on 36 Yukon [930]. Lignicolous.

Lentinus suavissimus: on 36 Que [1385].
Lignicolous.

Lenzites betulina (C:254): on 36 NS
    [1032]. See *Acer*.
Leptosphaeria sp.: on 36 Ont [1340].
Leucostoma nivea (Valsa n.) (anamorph
    Cytospora n.): on 36 NWT-Alta [1063],
    NB [1032].
L. persoonii (Valsa leucostoma) (anamorph
    Cytospora rubescens): on 27 Que [337].
Linospora capreae: on overwintered leaves
    of 36 BC [79].
L. insularis: on 36 NWT [79].
Lophodermium versicolor (C:255): on 36
    Ont [1828].
Lycoperdon pusillum (C:255): puffball:
    on 36 NS [1032]. Probably on rotted
    wood.
Marssonina kriegeriana (holomorph
    Drepanopeziza triandrae) (C:255):
    anthracnose, anthracnose: on 7 BC [334,
    336, 1012, 1816]; on 36 NB [334, 1032].
M. lindii (C:255): on 5 Frank [1586].
M. populi (holomorph *Drepanopeziza
    populorum*) (C:255): on 36 NS, PEI
    [1032].
M. salicicola (C:255): on 7 BC [726,
    1012].
M. santonensis: on 36 Sask [F66, F67].
Melampsora sp.: rust, rouille: on 36 BC
    [F76], Alta [1594], Ont [1341, 1827,
    1828, F73], NB, NS, PEI [F68], NS, PEI
    [F67], Nfld [1659, 1660, 1661].
M. abietis-canadensis: rust, rouille: on
    36 Ont [1828]. Doubtful host record,
    probably on *Populus*.
M. abieti-capraearum (C:255): rust,
    rouille: II III on 8, 12, 14, 18, 21,
    24, 36 Ont [1236]; on 8, 36 Nfld [1659,
    1660, 1661]; on 13 Nfld [F76]; on 32a,
    32b BC [1012]; on 36 Mack, BC, Alta,
    Sask, Man [2008], Sask, Man [F67, F69],
    Man [F68], Ont [1341, F73], Que [1377],
    Nfld [F71], Nfld, Labr [F74].
M. arctica (C:255): rust, rouille: II
    III on 6, 12 Que [1236]; on 33 BC
    [1012]; on 36 Frank, BC, Alta [2008].
M. epitea (C:255): rust, rouille: II III
    on 5 Frank [1249, 1544, 1586]; on 5, 6,
    17, 31, 36 Frank [1244]; on 10, 10a, 13
    Que [282]; on 13, 18 NS [1036]; on 13,
    24 NB [1036]; on 13, 26 Nfld [1659,
    1660, 1661]; on 15, 19, 31 BC [1012]; on
    22 Alta [950]; on 36 Yukon [460],
    NWT-Alta [1063], Mack, Alta [53], Mack,
    Ont, Que [F67, F72], BC [954, F70], Alta
    [951, 952, 1402], Ont [1341, F65, F68],
    Que [F69, F71, F73], Que, Nfld, Labr
    [F74], NB, NS, PEI [1032].
M. epitea f. sp. *tsugae* (C:255): rust,
    rouille: II III on 32a, 33 BC [1012,
    2008].
M. medusae: rust, rouille: II III on 36
    Alta [952], NB, NS [F70].
M. paradoxa (M. bigelowii) (C:255): rust,
    rouille: II III on 3, 8, 12, 13, 21,
    24, 36 Ont [1236]; on 13 NS [1036]; on
    19 BC [1012, 2007]; on 36 Yukon, Mack,
    BC [2008], NWT-Alta [1063], Alta [952],
    Alta, Sask, Man [2008, F69, F70], Sask,
    Man [F63, F67, F68, F71, F73], Sask
    [F62, F74], Man [F64, F66], Ont [F74],

Que [1377], NB, NS [1032], NB, PEI
    [F69], PEI [1036], Nfld [1660, F72].
M. ribesii-purpurea (C:255): on 19 BC
    [1341]; on 32 BC [1012]; on 36 Yukon,
    BC, Alta [2008].
M. salicina: on 36 Ont [437]. Probably a
    specimen of *M. epitea*.
Melanomma pulvis-pyrius (C:255): on 36 BC
    [1012].
Melzericium udicola: on 36 BC [1012].
Merismodes sp.: on 36 BC [1012].
Meruliopsis ambiguus: on 36 Alta [568].
    See *Alnus*.
M. corium: on 31 BC [1012]; on 36 Sask
    [568]. See *Acer*.
Monostichella salicis (Gloeosporium s.)
    (C:254): anthracnose, anthracnose: on
    36 Ont [1340, 1763], Que [1377], NS
    [1032].
Mycoacia uda: on 36 BC [1012]. See
    Acer.
Mycosphaerella maculiformis: on 36 Que
    [70].
Nectria sp.: on 33 BC [1012].
N. cinnabarina (anamorph *Tubercularia
    vulgaris*) (C:255): on 2b Que [335]; on
    36 Sask [F67].
N. coccinea (C:255): on 36 BC [1012].
N. coryli (C:255): on 36 NS [1032].
N. galligena (C:255): on 36 NS [1032].
*Nematogonium ferrugineum (Gonatorrhodiella
    highlei)*: on 36 BC [1012].
Neta patuxentica: on 36 Man [1760].
Nidula niveotomentosa: bird's nest
    fungus: on 36 BC [1012]. See
    Pseudotsuga.
Ocellaria ocellata (Pezicula o.) (anamorph
    Cryptosporiopsis scutellata) (C:256):
    on 36 BC [1012], Sask, Man [F66, F67],
    Ont [1340, F67], NS [1032].
Orbilia sp.: on 36 NB [1032].
Peniophora cinerea (C:256): on 36 Ont
    [973, 1341]. See *Abies*.
P. incarnata: on 36 NS [1032]. See
    Abies.
P. rufa (Cryptochaete r.): on 36 Ont
    [1341]. See *Populus*.
Phaeoisaria sparsa: on 36 Yukon [460],
    Man [1760].
Phaeomarasmius erinaceus: on 36 Yukon,
    BC, Ont [1434], Labr [867]. See
    Populus.
P. erinacellus (Pholiota e.) (C:256): on
    36 NS [1032].
P. rhombosporus: on 36 Que [1041]. See
    Populus.
*Phaeostalagmus tenuissimus (Verticillium
    t.)*: on 36 Sask [1760, F67].
Phellinus conchatus (Fomes c.) (C:254):
    on 36 NS [1032]. See *Acer*.
P. ferreus (Poria f.) (C:256): on 19 BC
    [1012]. See *Acer*.
P. ferruginosus (Poria f.) (C:256): on 36
    NS [1032]. See *Abies*.
P. igniarius (Fomes i.) (C:254): white
    trunk rot, carie blanche du tronc: on
    19 BC [1341]; on 29 Labr [867]; on 36
    NWT-Alta [1063], Mack [53], BC [1012],
    NB [1032], Nfld [1659, 1660, 1661]. See
    Acer.

P. laevigatus (Fomes igniarius var.
l.): on 36 NB [1032]. See *Acer.*
P. punctatus (Poria p.) (C:256): on 36
Ont [1827], NB, NS [1032]. See *Abies.*
P. robustus (Fomes r.): on 8 Mack, Sask
[579]; on 36 Alta, Sask [579]. See
Abies.
Phibalis furfuracea (Encoelia f.): on 36
NWT-Alta [1063], Alta [952].
Phlebia deflectens (Corticium d.)
(C:253): on 36 BC [1012], NS [1032].
See *Acer.*
P. radiata: on 36 BC [1012]. See *Abies.*
P. rufa: on 36 BC [568, 1012]. See
Abies.
Pholiota aurivella (C:256): on 36 BC
[1012]. See *Abies.*
P. aurivelloides: on 32 BC [1012].
P. squarrosoides (C:256): on 36 NS
[1032]. See *Acer.*
Phomopsis sp.: on 36 BC [1012].
Phyllactinia guttata: powdery mildew,
blanc: on 13 Nfld [1659, 1660, 1661].
Probably *Uncinula adunca.*
Phyllosticta apicalis (C:256): on 36 Que
[F61].
Physalospora miyabeana (C:256): black
canker or blight, chancre noire: on 2a
NS [336]; on 2a, 2b, 9, 15, 24 NS [711];
on 2a, 7 BC [334, 1012]; on 7 BC [1816],
Que [339]; on 12, 13 Que [1377]; on 36
BC, Ont [1340], BC, Que [335, 336, F62],
Que [F63], NB, NS [339, F60, F66, F69,
F70, F71], NB, NS, PEI [336, 1032, F61,
F62, F63, F64, F65, F67, F72], NS [335,
1151, 1594, F68]. Almost always
associated with *Venturia saliciperda*
(anamorph *Pollaccia saliciperda*).
Pleospora sp.: on 36 BC [1012].
Pleuroceras helveticum: on dead leaves of
36 Que [79].
P. pleurostyla: on overwintered leaves of
36 Labr [79].
Pleurophomopsis salicicola: on 7 BC
[726]. The generic name is a synonym of
Sirodothis [1763], and this species is
probably a *Sirodothis.*
Pleurotus ostreatus (C:256): on 36 BC
[1012]. See *Abies.*
Plicatura nivea: on 36 Mack, Sask [563,
568]. See *Acer.*
Pluteus salicinus: on stumps of 36 Que
[1385].
Podosphaera clandestina: powdery mildew,
blanc: on 36 NS [1032]. Probably
Uncinula adunca.
Pollaccia saliciperda (Fusicladium s.)
(holomorph *Venturia s.*) (C:256):
scab, brûlure du saule: on 2a, 2b, 9,
15, 24 NS [711]; on 2a, 7 BC [F61]; on
2a, 34, 35 BC [F60]; on 2a, 36 NS [337];
on 2b, 30a Que [335]; on 7 BC [333,
1816], Que [339, F66]; on 7, 27 Que
[333]; on 12, 13 Que [1377]; on 27, 36
Que [337]; on 36 BC, Que [F62], BC, Que,
NS [335], Man [1760, F68], Ont [1340,
1827, F68, F70, F72, F73], Ont, Nfld
[F64], Que [F63, F74, F76], Que, NS, PEI
[333], NB, NS [339, F60, F66], NB, NS,
PEI [338, 1032, F61, F62, F63, F64, F65,

F67], NS [1594], Nfld [1659, 1660,
1661]. Usually associated with
Physalospora miyabeana.
Polyporus badius (P. picipes) (C:256): on
36 NWT-Alta [1063], BC [1012], NB
[1032]. See *Abies.*
P. brumalis: on dead branch of 36 Yukon
[579]. See *Abies.*
P. melanopus: on 36 Yukon [1110].
Reported to be mycorrhizal with *Pinus.*
P. squamosus: on 36 NWT-Alta [1063]. See
Acer.
P. varius (P. elegans) (C:256): on 36
Yukon [1110], NWT-Alta [1063], Mack
[579, 1110], BC [1012], Alta [952], Ont
[1341], NS [1032], Nfld [1659, 1660,
1661]. See *Abies.*
Poria sp.: on 36 NWT-Alta [1063], Mack
[53].
P. subacida: (C:256): on 33 BC [1012].
See *Abies.*
Pseudospiropes simplex: on 36 Sask [817],
Sask, Man [1760].
Pseudotomentella atrofusca: on 36 BC
[1828].
*Punctularia strigosozonata (Phaeophlebia
s.)*: on 36 NB [1032]. See *Alnus.*
Ramularia rosea (C:256): leaf spot, tache
des feuilles: on 36 Sask [F65], Sask,
Man [F69], Man [F67, F68], Que [F61].
Ramulaspera salicina: on 36 BC [1012,
F64].
Rhabdospora salicella: on 36 Ont [1340,
F67].
Rhytisma sp.: on 5 Frank [1239].
R. salicinum (C:256): tar spot, tache
goudronneuse: on 5 Frank [1586]; on 6,
17 Frank [1244]; on 13 Que [282], NS
[1032], Nfld [F76]; on 13, 36 Nfld
[1659, 1660, 1661]; on 32 BC [1012]; on
36 Yukon [460], NWT-Alta [1063], Mack,
Alta [53], BC [953, 954], BC, Ont
[1340], Alta [644, 951, 952], Alta, Sask
[F71], Sask, Man [F60, F63, F69], Sask,
Man, Ont [F65, F66, F67, F68], Man [F62,
F64, F70], Ont, Que [F72], Ont, NS
[F64], Que [1341, F67, F75, F76], NB,
NS, PEI [1032], Nfld [F63, F64, F65].
Schizophyllum commune: on 36 Ont [1341],
NB [1032]. See *Abies.*
Schizopora paradoxa (Poria versipora)
(C:256): on 36 BC [1012]. See *Abies.*
Schizoxylon compositum: on 36 Ont [1608].
Sclerophoma sp.: on 27 BC [1012].
Septoria sp.: on 36 BC [1012].
S. didyma: on 36 Sask, Man [F67, F68],
Man [F66].
S. salicicola (C:257): on 36 Man [F65].
*Skeletocutis nivea (Polyporus
semipileatus)* (C:256): on 36 BC
[1012]. See *Amelanchier.*
Spadicoides bina: on 36 Sask, Man [1760].
Sphaceloma murrayae: on 36 Man [F66].
*Spongipellis malicolus (Polyporus spumeus
var. malicola)*: on 24 Que [F71]. See
Fraxinus.
Sporidesmium achromaticum: on 36 Sask,
Man [1760].
S. leptosporum: on 36 Man [1760].
Stictis radiata (C:257); on 36 BC [1012].

*Taeniolella stilbospora (Septonema
    atrum)*: on 36 NS [1032].
Tapesia fusca (C:257): on 36 Ont [1340].
*Tectella patellaris (Panellus p., Panus
    operculatus)*: on 36 Yukon [1108], NB
    [1032].
Tetracladium setigerum: on 36 BC [43].
Tomentella botryoides (C:257): on 36 NS
    [1032].
T. umbrinospora: on 36 Ont [1341, 1828].
Tomentellina fibrosa (Kneiffiella f.): on
    36 Alta [1828].
Trametes suaveolens (C:257): on 36 Yukon,
    BC [F63], Yukon [460], NWT-Alta [1063],
    Ont [1341], Que [1377, 1385], NB, NS
    [1032]. See *Populus*.
T. trogii: on 36 NS [1032]. See
    Populus.
*Trichaptum biformis (Polyporus
    pargamenus)*: on 36 Mack [53], Ont
    [1341], NB [1032]. See *Acer*.
Trimmatostroma sp.: on 36 Ont [1827].
T. salicis: on 36 Sask, Man [1760], Man
    [F65].
Tubercularia sp.: on 36 Ont [1340].
T. vulgaris (holomorph *Nectria
    cinnabarina*) (C:257): on 36 BC [1012].
Tulasnella violea: on 36 BC [1012]. See
    Alnus.
Tympanis alnea (anamorph *Sirodothis
    inversa*): on 36 Alta [1221].
T. heteromorpha: on 36 Ont, Que, NB
    [1221].
T. salicina (C:257): on 36 Man [F67], Que
    [1221].
T. saligna (anamorph *Sirodothis s.*)
    (C:257): on 36 NWT-Alta [1063], Alta
    [952].
Tyromyces chioneus (Polyporus albellus)
    (C:256): on 36 NS [1032]. See *Abies*.
Uncinula adunca (U. salicis) (C:257):
    powdery mildew, blanc: on 2, 3, 7, 15,
    25, 36 Que [1377]; on 8, 12, 12a, 13,
    18, 20a, 21, 36 Ont [1257]; on 8, 12a,
    13, 21, 36 Que [1257]; on 13 Nfld [1659,
    1660, 1661]; on 16, 25 BC [1257]; on 23
    Yukon [1257]; on 32 BC [1012]; on 36
    Yukon [460], NWT-Alta [1063], Mack [53],
    Sask, Man [F63, F66, F67, F68], Man
    [1257], Ont [1340, F68], Que [F72, F73],
    NB, NS, PEI [1032]. See under
    Phyllactinia guttata and *Podosphaera
    clandestina*, above.
Valsa sp.: on 32 BC [1012]; on 36 Sask,
    Man [F65]; Ont [1340].
V. ambiens (V. salicina, V. sordida)
    (anamorph *Cytospora a.*) (C:257):
    canker, chancre cytosporéen: on 7 BC
    [1012], NB [1036]; on 24, 36 Ont [1340];
    on 36 NWT-Alta [1063], Ont [1828, F65],
    Que [1377, F68], NB, NS [1032].
V. ceratosperma: on 36 NWT-Alta [1063].
V. translucens (C:257): on 36 NWT-Alta
    [1063].
Valsella salicis (V. fertilis, Valsa s.)
    (C:257): on 36 Ont [1340, 1440, 1828,
    F60], Que [F68].
Vararia effuscata: on 36 Ont [1341]. See
    Acer.
Varicosporium elodeae: on 36 BC [43].

Venturia chlorospora (C:257): on
    overwintered leaves and seed capsules
    of 36 BC, Que, Labr [73], Que [70].
V. minuta: on overwintered leaves of 36
    BC, Que [73].
V. saliciperda (anamorph *Pollaccia
    saliciperda*): on 2a, 7 BC [334]; 2a,
    7, 34, 36 BC [1012]; on 2a, 36 NS [336];
    on 7 BC [F76]; on 7, 36 NB, NS [1036];
    on 36 BC, Que, NB, PEI [336], Ont, Que,
    NB [F75], Que, NB, NS [F69], NB, NS
    [F70, F71, F73], NB, NS, PEI [F72, F76],
    NS [F68], PEI [1036], Canada [73].
V. subcutanea: on overwintered leaves and
    catkins of 30 NWT [73].
Xylaria hypoxylon: on 36 BC [1012].

SALSOLA L.   CHENOPODIACEAE

1. *S. kali* L. var. *tenuifolia* Tausch,
    Russian thistle, chardon de russie;
    imported from Eurasia, introduced,
    BC-PEI.

Verticillium dahliae: on 1 BC [1816,
    1978].

SALVIA L.   LABIATAE

1. *S. officinalis* L., common sage, sauge;
    imported from Eurasia.

Sclerotinia sclerotiorum: on 1 NB [333,
    1194].

SAMBUCUS L.   CAPRIFOLIACEAE

1. *S. canadensis* L., common elder, sureau
    blanc; native, Man-PEI.
2. *S. cerulea* Raf. (*S. glauca* Nutt.),
    blue elder, sureau bleu; native, BC.
3. *S. nigra* L., European elder, sureau
    noir; imported from Eurasia,
    occasionally spreading from cultivation,
    s. Ont.
4. *S. racemosa* L., European red elder,
    sureau rouge à grappes; native,
    BC-Nfld. 4a. *S. r.* var. *arborescens*
    (Torr. & Gray) Gray; native, BC. 4b.
    S. r. var. *pubens* (Michx.) Koehne
    (*S. pubens* Michx.), catberry, sureau
    rouge; native, Man-NS.
5. Host species not named, i.e., reported
    as *Sambucus* sp.

Cerocorticium confluens (Corticium c.):
    on 5 NS [1032]. See *Abies*.
Dendropleella hirta (C:258): on 4b Que
    [70].
Dendryphion comosum: on 4b BC [818].
Diaporthe sp.: on 4b BC [193].
Dothidea sambuci: on 4b, 5 Ont [1340]; on
    5 Ont [F66].
Endophragmiella socia: on 5 BC [808].
Exidia nucleata: on 5 NS [1032]. See
    Acer.
Fenestella vestita (C:258): on 5 Ont
    [1340, F62].
Fusarium sp. (C:258): crown rot,
    pourridié fusarien: on 5 Que [F66].

Helminthosporium velutinum: on 4b BC
[810].
Hymenochaete tabacina (C:258): on 5 BC
[1012]. See *Abies*.
Hyphoderma sambuci (Peniophora s.)
(C:258): on 4a BC [1012]; on 5 Que
[F64], NS [1032]. See *Acer*.
Hypochnicium bombycinum (Corticium b.):
on 5 NS [1032]. See *Acer*.
Microsphaera grossulariae (C:258):
powdery mildew, blanc: on 1, 5 Que
[1257]; on 1, 4a, 5 Ont [1257]; on 4b, 5
NS [1257].
M. penicillata (C:258): powdery mildew,
blanc: on 5 NB [339]. Probably a
specimen of *M. grossulariae*.
Nectria cinnabarina (anamorph
Tubercularia vulgaris) (C:258): on 4b
Sask [F69], Nfld [1659, 1660, 1661]; on
5 NWT-Alta [1063], Mack [53], Sask [336].
Pellidiscus pallidus: on 5 BC [1424].
See *Alnus*.
Peniophora incarnata (C:258): on 5 BC
[1012], NS [1032]. See *Abies*.
Phragmocephala elliptica: on 5 BC [807].
Phytophthora citricola: on 3, 5 Alta
[1594]; on 5 Alta [339].
Pseudospiropes nodosus: on 5 BC [816].
Puccinia bolleyana (C:258): rust,
rouille: 0 I on 1 Ont [1828], NS
[1032]; on 1, 3, 5 Ont [1236]; on 4b
Sask [F74]; on 4b, 5 Ont [1341]; on 5
Ont [F60, F62], Ont, Que [F67].
Pythium sp.: on 5 Alta [338].
Ramularia sambucina (C:258): on 4 Que
[F61].
Schizopora paradoxa (Poria versipora)
(C:258): on 4a BC [1012]. See *Abies*.
Septoria sambucina (C:258): leaf spot,
tache septorienne: on 2 BC [1012]; on
2, 4 BC [1816]; on 5 NWT-Alta [1063],
Alta [338].
Stigmina sp.: on 5 Ont [1340].
S. pedunculata (Sciniatosporium p.)
(C:258): on 2 BC [1127].
Tubercularia sp.: on 5 NWT-Alta [1063].
T. ulmea: on 4b Sask [F67].
Vermicularia sambucina: on dead young
stems of 5 Ont [440].
Verticillium albo-atrum: on 5 Que [F65].

SANGUINARIA L.   PAPAVERACEAE

1. *S. canadensis* L., bloodroot,
    sang-dragon; native, Man-NS. Woodland
    perennial herb native to e. North
    America.

Stromatinia sanguinariae: on 1 Que [661].

SANGUISORBA L.   ROSACEAE

1. *S. canadensis* L., Canadian burnet,
    herbe à pisser; native, Yukon, BC,
    Que-NS, Labr. la. *S. c.* ssp.
    latifolia (Hook.) Calder & Taylor (*S.
    sitchensis* Mey.); native, Yukon, BC.
2. *S. officinalis* L. (*S. microcephala*
    Presl), great burnet; native,
    Yukon-Mack, BC.

Isariopsis bulbigera (C:259): on 1a BC
[1012].
Sphaerotheca macularis (C:259): powdery
mildew, blanc: on 1 Que, Nfld [1254];
on 1a BC [1254]; on 2 BC [1012, 1816].

SANICULA L.   UMBELLIFERAE

1. *S. gregaria* Bickn., yellow sanicle,
    sanicle grégaire; native, Ont-NS.
2. *S. marilandica* L., black snakeroot,
    sanicle de Maryland; native, BC-Nfld.

Phyllosticta vexans: on 1 Ont [195].
Puccinia marylandica (C:259): rust,
rouille: 0 I II III on 2 BC [1551],
Ont, Que [1236, 1551].

SARRACENIA L.   SARRACENIACEAE

1. *S. purpurea* L., pitcher plant, petits
    cochons; native, Mack, BC-Nfld.

Glomerella cingulata (anamorph
Colletotrichum gloeosporioides)
(C:259): on 1 Que [70].
Thozetella canadensis: on *S*. sp.
growing in greenhouse Ont [1162].

SASSAFRAS Nees   LAURACEAE

Aromatic trees or shrubs, native, s. Ont.

Cladosporium sp.: on *S*. sp. Ont [1340].

SATUREJA L.   LABIATAE

1. *S. douglasii* (Benth.) Briq.
    (*Micromeria d.* Benth.), yerba buena;
    native, BC.
2. *S. glabella* (Michx.) Briq. (*S.
    arkansana* (Nutt.) Briq.); native, Ont.
3. *S. hortensis* L., summer savory,
    sarriette des jardins; imported from
    Eurasia.
4. *S. vulgaris* (L.) Fritsch (*Clinopodium
    v.* L.); wild basil, grand basilic
    sauvage; imported from Eurasia.

Alternaria sp.: on 3 NS [602].
Cephalosporium sp.: on 3 NS [602].
Colletotrichum coccodes: on 3 NS [602].
Gibberella cyanogena (anamorph *Fusarium
sulphureum*): on 3 NS [602].
Mortierella ramanniana var. *ramanniana*:
on 3 NS [602].
Puccinia menthae (C:259): rust, rouille:
0 I II III on 1 BC [1012, 1816]; on 2, 4
Ont [1236]; on 4 Ont [1828].
Pyrenochaeta sp.: on 3 NS [602].
Pythium oligandrum: on 3 NS [602].
Ulocladium atrum: on 3 NS [602].
Veticillium dahliae: on 3 NS [602].

SAXIFRAGA L.   SAXIFRAGACEAE

1. *S. aizoides* L., yellow mountain
    saxifrage; native, Yukon-Labr, BC-Alta,
    Man-Que, NS, Nfld.

2. *S. aizoon* Jacq., livelong saxifrage,
saxifrage aizoon; native, Mack,
Frank-Labr, Ont, NS, Nfld. 2a. *S. a.*
var. *neogaea* Butters; native, n. Que,
Labr, Nfld.
3. *S. bronchialis* L.; native, Yukon-Mack,
BC-Alta.
4. *S. caespitosa* L., California
saxifrage; native, Mack-Labr, BC-Alta.
5. *S. callosa* Sm. *(S. lingulata*
Bellardi, *S. lantoscana* Boiss. &
Reut.)*; imported from s. Europe.
6. *S. cernua* L., nodding saxifrage;
native, Yukon-Labr, BC-Alta, Ont-Que.
7. *S. cotyledon* L.; imported from Europe.
8. *S. ferruginea* Graham; native,
Yukon-Mack, BC-Alta.
9. *S. flagellaris* Sternb. & Willd.;
native, Yukon-Frank, BC-Alta. 9a. *S.*
f. ssp. *platysepala* (Trautv.)
Porsild; native, Frank.
10. *S. hieracifolia* Waldst. & Kit.;
native, Yukon-Frank.
11. *S. hirculus* L., yellow marsh
saxifrage; native, Yukon-n. Que,
Man-Ont. 11a. *S. h.* ssp. *propinqua*
(R. Br.) Love & Love.
12. *S. lyallii* Engl.; native, Yukon-Mack,
BC-Alta. 12a. *S. l.* ssp. *hultenii*
(Calder & Savile) Calder & Taylor. 12b.
S. l. ssp. *l.*
13. *S. mertensiana* Bong.; native, BC-Alta.
14. *S. nelsoniana* D. Don; native,
Yukon-Keew, BC-Alta. 14a. *S. n.* ssp.
carlottae (Calder & Savile) Hulten
(S. punctata ssp. *c.* Calder &
Savile); native, BC. 14b. *S. n.* ssp.
cascadensis (Calder & Savile) Hulten
(S. punctata ssp. *c.* Calder &
Savile); native, BC. 14c. *S. n.* ssp.
porsildiana (Calder & Savile) Hulten
(S. punctata ssp. *porsildiana* Calder
& Savile); native, Yukon-Keew, BC-Alta.
15. *S. nivalis* L., alpine saxifrage;
native, Yukon-Labr. 15a. *S. n.* var.
tenuis Wahlenb. *(S. tenuis*
(Wahlenb.) Sm.); native, range of
species but more southern.
16. *S. occidentalis* S. Wats.; native,
BC-Sask. 16a. *S. o.* var. *rufidula*
(Small) Hitchc.; native, BC.
17. *S. odontoloma* Piper *(S. arguta*
aucts. non Don); native, BC.
18. *S. oppositifolia* L., purple mountain
saxifrage, saxifrage à feuilles
opposées; native, Yukon-Labr, BC-Alta,
Man, Que, Nfld.
19. *S. reflexa* Hook.; native, Yukon-Mack,
BC.
20. *S. rivularis* L., alpine brook
saxifrage; native, Yukon-Labr, BC-Alta,
Man, Que, Nfld.
21. *S. serpyllifolia* Pursh; native,
Yukon-Mack.
22. *S. stellaris* L. *(S. foliolosa* R.
Br.), star saxifrage, saxifrage étoilée;
native, Yukon-Labr, Nfld.
23. *S. tolmei* Torr. & Gray; native, BC.
24. *S. tricuspidata* Rottb.; native,
Yukon-Labr, BC-Ont.

25. *S. virginiensis* Michx., early
saxifrage, passe-pierre; native, Man-NB.

Arcticomyces warmingii (C:260): on 1, 18
Frank [1544]; on 18 Frank [1543, 1586],
Que [1544].
Botrytis cinerea (holomorph *Botryotinia*
fuckeliana) (C:260): on 20 Frank [664].
Fabraea sp. (C:260): on 11a Frank [664].
Specimen probably a *Leptotrochila*.
Melampsora arctica (C:260): rust,
rouille: 0 I on 4, 18 Frank [2008].
M. epitea (C:260): rust, rouille: on 18
Frank [1239, 1586].
M. hirculi: rust, rouille: on 11 Keew
[1584].
Mycosphaerella allicina (M. saxifragae)
(C:260): on 4, 6, 11 Keew, on 11a, 15,
22 Frank and on 20 NWT [1586]; on 22
Frank [1244].
Pseudomassaria inconspicua (C:261): on 2a
Nfld, on 3 BC, on 18, 24 Frank, and on
24 Que [71].
Puccinia austroberingiana: rust,
rouille: on 14a BC [1560]. The
specimens were intermediate between ssp.
austroberingiana and ssp.
saxifragarum.
P. austroberingiana ssp.
austroberingiana: rust, rouille: on
6, 14a, 14b BC and on 14c Yukon, Mack,
Keew [1560]. Host (14a) was
intermediate between ssp. *carlottae*
and ssp. *porsildiana*.
P. austroberingiana ssp. *saxifragarum*:
rust, rouille: on 12a BC [1560].
P. fischeri (C:261): rust, rouille: on
18 Frank [1543, 1544, 1560, 1586], Keew
[1560].
P. heucherae (C:261): rust, rouille: on
17 BC [1012]; on 17, *S.* sp. BC [1816].
P. heucherae var. *cordillerana*: rust,
rouille: on 12 BC [1560]; on 12a Yukon,
BC, Alta [1560], BC [1012]. Host (12)
determined as intermediate between ssp.
hultenii and ssp. *lyallii*.
P. heucherae var. *diffusistriata*: rust,
rouille: on 23 BC [1560].
P. heucherae var. *heucherae* (C:261):
rust, rouille: on 12 BC [1816]; on 12a,
14b BC [1560]. Host (12a) determined as
S. lyallii ssp. *hultenii* x *S.*
odontoloma.
P. pazschkei (C:261): rust, rouille: III
on 5, 7 Ont [1236]; on 24 NWT [1341].
P. pazschkei var. *ferrugineae*: rust,
rouille: on 8 BC [1560].
P. pazschkei var. *heterisiae* (C:261):
rust, rouille: on 13 BC [1560].
P. pazschkei var. *jueliana* (C:261):
rust, rouille: on 1 Frank, Keew [1560].
P. pazschkei var. *oppositifoliae*
(C:261): rust, rouille: on 18 Que
[1560].
P. pazschkei var. *pazschkei*: rust,
rouille: on 5, 7 Ont [1560]. Plants
were from a greenhouse.
P. pazschkei var. *tricuspidatae*
(C:261): rust, rouille: on 3, 24 BC
[1560]; on 10 Frank [1543]; on 24 Yukon,

Mack, Frank, Keew [1560], Frank [1244, 1586].

P. saxifragae: rust, rouille: on 6 Frank, Mack, on 15 Frank, on 20 Labr [1560]. The specimens were intermediate between var. *curtipes* and var. *longior*.

P. saxifragae var. *curtipes*: rust, rouille: on 10 Frank, Keew, on 12b Alta, on 15 Mack, Frank, Keew, on 15, 16, 16a BC, on 15a Frank, BC, Alta, on 16, 16a BC, on 19, 20, 21 Yukon, on 20 BC, Que, on 25 Ont, Que [1560]. Host 16 was determined as intermediate between ssp. *occidentalis* and ssp. *rufidula*.

P. saxifragae var. *longior*: rust, rouille: on 6 Mack, Frank, Keew, BC, Que, on 20 Frank, BC, Alta, Labr [1560].

P. saxifragae var. *saxifragae (P. heucherae* var. *s.)* (C:261): rust, rouille: on 6, 15a Frank [1586]; on 10 Frank [1543]; on 15 Frank [1239, 1544]; on 15a Frank, BC [1544]; on 20, 25 Que [1236].

Pyrenopeziza svalbardensis (Pseudopeziza s.) (C:261): on 11a Frank [664, 1586].

Sphaerotheca fuliginea: powdery mildew, blanc: on 24 Man, on 25 Ont [1257].

Synchytrium rubrocinctum (C:261): on 1 Keew, on 6, 9a, 18 Frank [1586].

Wentiomyces fimbriatus (Venturia f.) (C:261): on 4 NWT, BC [73].

SCHIZACHYRIUM (Nees) Trin.   GRAMINEAE

1. *S. scoparium* (Michx.) Nash (*Andropogon s.* Michx.), broom-beardgrass, schizachyrium à balais; native, se. BC-NS.

Puccinia andropogonis (C:33): rust, rouille: II III on 1 Ont [1236].

P. ellisiana (C:33): rust, rouille: II III on 1 Ont [1236].

SCIRPUS L.   CYPERACEAE

1. *S. americanus* Pers., three-square, scirpe d'Amérique; native, BC-Nfld.
2. *S. atrocinctus* Fern., black-girded wool-grass, scirpe à ceinture noir; native, BC-Nfld.
3. *S. atrovirens* Willd., blackish bulrush, scirpe noirâtre; native, Alta-Nfld.
4. *S. caespitosus* L., tufted club-rush, scirpe gazonnaute; native, Mack-Labr, BC-Nfld.
5. *S. cyperinus* (L.) Kunth, common wool grass, scirpe souchet; native, Man-Nfld.
6. *S. fluviatilis* (Torr.) Gray, river bulrush, scirpe des rivières; native, Alta-NB.
7. *S. lacustris* L.; native, Yukon-Keew, BC-Nfld. 7a. *S. l.* ssp. *glaucus* (Sm.) Hartm. (*S. acutus* Muhl.); native, Yukon-Keew, BC-Nfld. 7b. *S. l.* ssp. *validus* (Vahl) Koyama (*S. validus* Vahl); native, BC, NB-PEI.
8. *S. maritimus* var. *m.* L., saline marshes and shores; natzd., BC-NS, PEI.

8a. *S. m.* var. *paludosus* (Nels.) Kukk.; native, Mack, BC-NS.
9. *S. microcarpus* Presl (*S. rubrotinctus* Fern.), red-tinged bulrush, scirpe à noeuds rouges; native, BC-Nfld, Labr.
10. *S. pedicillatus* Fern., pedicillate wool grass, scirpe pédicelle; native, Ont-NS.

Arthrinium sporophleum: on 4 BC [1012].

Melanotus caricicola: on 9 Que [1437].

Puccinia angustata (C:262): rust, rouille: II III on 2 NB, Nfld, on 2, 3, 5 Ont, on 2, 5, 9, 10 Que, on 5 NS, on 9 Alta [1559]; on 2, 3, 5, 9, 10, *S.* sp. Ont [1236].

P. eriophori-alpini: rust, rouille: on 4 BC [1559].

P. mcclatchieana (C:262): rust, rouille: on 3, 9 Ont [1559]; on 9 BC [1012, 1816], BC, Alta, Sask, Que, NS [1559].

P. obtecta (C:262): rust, rouille: II III on 1, 5, 7a, 7b, *S.* sp. Ont [1236]; on 7a, 7b Ont, on 7b Sask [1559].

P. osoyoosensis: rust, rouille: II on 1 BC, Alta [1559].

Uromyces americanus (C:262): rust, rouille: II III on 7b Ont [1236], Ont, NS [1559].

U. lineolatus (C:262): rust, rouille: II III on 6 Ont [1236].

U. lineolatus ssp. *nearcticus*: rust, rouille: on 6 Ont, Que, on 8 PEI, on 8a BC Sask, Que [1559].

SCUTELLARIA L.   LABIATAE

1. *S. galericuta* L. (*S. epilobiifolia* Hamilton), common skullcap, toque; native, Yukon, BC-Nfld.
2. *S. lateriflora* L., mad-dog skullcap, scutellaire latériflore; native, BC-Nfld.
3. *S. parvula* Michx., small skullcap, scutellaire petite; native, Ont-Que.

Erysiphe cichoracearum (C:262): powdery mildew, blanc: on 1, 2, 3 Ont, Que [1257].

E. communis (E. polygoni) (C:262): powdery mildew, blanc: on 2 Ont [1257].

SECALE L.   GRAMINEAE

1. *S. cereale* L., rye, seigle; a cultigen.
2. *S. montanum* Guss.; introduced from sw. Asia and escaped from cultivation.

Acremonium boreale: on 1 BC, Alta, Sask [1707].

Alternaria sp.: on 1 Maritimes [272].

Bipolaris sorokiniana (holomorph *Cochliobolus sativus*) (C:262): common root rot, piétin commun: on 1 Sask [333, 334], Ont [338]; on 2 Alta [120] by inoculation.

Cladosporium sp.: on 1 Maritimes [272].

Claviceps sp.: on 1 Sask [1781].

C. purpurea (C:262): ergot, ergot: on 1 BC [1816], Alta [707, 708], Alta, Sask

[333, 336, 338], Sask [339, 1726], Sask, Man, Ont, NB [335], Sask, Ont, Que [1594], Man [1364], Que [337], Que, NS, PEI [334], NB [1196], NB, NS, PEI [854], NB, PEI [333], Nfld [338, 339], Maritimes [272]; on flowers of *S*. sp. Que [1385].

Cochliobolus sativus (anamorph *Bipolaris sorokiniana*): on 1 NB, NS, PEI [854].
Colletotrichum dematium: on 1 Alta [1595].
C. graminicola (C:263): anthracnose, anthracnose: on 1 Alta [693].
Drechslera tritici-repentis (holomorph *Pyrenophora t.*). (C:263): leaf blotch, tache des feuilles: on 1 Alta, Ont [1611].
Erysiphe graminis (C:263): powdery mildew, blanc: on 1 BC [334, 1816], BC, Alta, Sask [1594], BC, Sask, Man, Ont, NS [1257], BC, Man [335], BC, NB, NS [333], Alta [1333], NB [1196], NS [1255].
Fusarium sp. (C:263): on 1 Alta [703], Sask [334], Ont [338], NS, PEI [854].
Myriosclerotinia borealis: on 1 Sask [892].
Polymyxa graminis: on 1 Ont [60].
Puccinia graminis (C:263): stem rust, rouille de la tige: II III on 1 BC [335, 1816], BC, Alta, Que [339, 621], BC, Man, Ont, Que [619, 623, 626, 627, 629, 630, 634, 636], BC, Ont [336, 622, 628, 633, 567], Alta [619, 626, 636], Alta, Sask [338], Sask [336, 630], Man [628, 633], Man, Ont, Que [635], Ont [337, 1236], Que, NS [622], NB, NS [626, 652], NB, PEI [333], NS [619, 623, 627, 628, 634, 635], PEI [334, 627].
P. graminis ssp. *graminis* var. *stakmanii* (*P. graminis* f. sp. *secalis*) (C:263): stem rust, rouille de la tige: on 1 BC, Alta, Ont, Que [640], BC, Man, Ont, Que [639, 620, 625, 631, 1533], BC, Ont [615], Alta, Sask [620], Sask [631], Ont [641], NB [639, 1196], NS [1533], NS, PEI [639, 640].
P. recondita (*P. secalina*) (C:263): leaf rust, rouille des feuilles: II III on 1 BC [1012, 1816], BC, Alta, Man, Ont, Que [639, 1533], BC, Alta, Ont [1594], BC, Ont, Que [640], BC, PEI [334], BC, Nfld [333, 335], Alta [1331, 1334], Sask [339], Ont [1236], NB [1196], NB, NS, PEI [333, 639, 640], NS [1533], Nfld [336, 337, 639]. The name *P. recondita* has been used for what is now known to be a species complex. The fungus on *Secale* is *P. recondita* in the strict sense. Taxa segregated from the complex include *P. impatienti-elymi* and *P. triticina*.
Pythium sp. (C:263): browning root rot, piétin brun: on 1 Alta [703, 1595].
Rhizopus sp.: on 1 Alta [703].
Rhynchosporium secalis (C:263): scald, tache pâle: on 1 BC [1816], Alta [338, 1331, 1333], NB [333, 1196], Nfld [335, 336].
Septoria secalis (C:263): speckled leaf spot, tache septorienne: on 1 Alta [339], Sask [338].

Stemphylium botryosum (holomorph *Pleospora herbarum*) (C:263): on seeds of 1 Ont [667].
Urocystis occulata (C:263): stem smut, charbon de la tige: on 1 Alta [338]. Found in 34% of the fields surveyed in s. Alta in 1978 whereas it had been rare before 1971. Shifting to more resistant varieties and treating seed with a fungicide should keep losses to acceptable levels. Stem smut caused losses of nearly $1.25 million in s. Alta from 1977-1981 [377].

SEDUM L.   CRASSULACEAE

1. *S. spathulifolium* Hook.; native, BC.

Puccinia umbilici (C:264): rust, rouille: on 1 BC [1012, 1816].

SEMPERVIVIUM L.   CRASSULACEAE

1. *S. tectorum* L., hens-and-chickens, joubarbe des toits; imported from Europe.

Endophyllum sempervivi (C:264): rust, rouille: 0 I III on 1 BC [1816]; on 1, *S*. sp. Ont [1236].

SENECIO L.   COMPOSITAE

1. *S. aureus* L., golden ragwort, senéçou dorée; native, Man-Nfld.
2. *S. cruentus* DC., common cineraria, cinéraire hybride; imported from Canary Islands.
3. *S. triangularis* Hook.; native, Yukon-Mack, BC-Alta.
4. *S. vulgaris* L., common groundsel, grand mouron; imported from Eurasia, natzd., Yukon-Mack, BC-Nfld.

Erysiphe communis (*E. polygoni*): powdery mildew, blanc: on 1 Ont [1257].
Oidium sp.: powdery mildew, blanc: on 2 BC [1816].
Puccinia angustata (C:264): rust, rouille: 0 I on 1 Ont [1236]; on 3 BC [1012, 1816].
P. eriophori: rust, rouille: on 3 BC [1559].
P. recedens (C:264): rust, rouille: III on 1 BC [1012, 1816], Ont [1236]; on *S*. sp. NWT, Que [720].
P. senecionis: rust, rouille: on *S*. sp. Alta [720].
Ramularia senecionis (C:264): on 3 Alta [950].
Rhizophydium graminis: on 4 Ont [59].
Sclerotinia sclerotiorum: on 4 Sask [1140].
Verticillium dahliae: on 4 BC [1816, 1978].

SEQUOIADENDRON Buchholz   TAXODIACEAE

1. *S. giganteum* (Lindl.) Buchh., giant redwood, séquoia géant; native to California. Planted in Canada in mild climatic zones as ornamental trees.

Botrytis sp.: on 1 BC [1012].

SETARIA Beauv.   GRAMINEAE

1. *S. glauca* (L.) Beauv. (*S. lutescens*
   (Weigel) Stuntz), pigeon-grass, foin
   sauvage; imported from Eurasia, natzd.,
   BC-PEI.
2. *S. italica* (L.) Beauv., german millet,
   millet des oiseaux; imported from
   Eurasia, natzd., BC, Ont-Que, NS.
3. *S. viridis* (L.) Beauv., bottle-grass,
   mil sauvage; imported from Eurasia,
   natzd., BC-Nfld.

Bipolaris sorokiniana (*Helminthosporium*
   sativum) (holomorph *Cochliobolus*
   sativus) (C:264): on 3 Sask [961].
Olpidium brassicae: on 1 Ont [67].
Sclerophthora macrospora: downy mildew,
   mildiou: on *S*. sp. Man [338].
Stemphylium botryosum (holomorph
   Pleospora herbarum) (C:265): on seeds
   of 2 Ont [667].
Ustilago neglecta (C:265): smut,
   charbon: on 1 BC [333, 1816]; on 3 BC
   [336]; on *S*. sp. BC [334].

SHEPHERDIA Nutt.   ELAEAGNACEAE

1. *S. argentea* (Pursh) Nutt., buffalo
   berry, graines de boeuf; native,
   Alta-Man.
2. *S. canadensis* (L.) Nutt., soapberry,
   sheperdie du Canada; native, Yukon-Mack,
   BC-Nfld.

Oidium sp.: powdery mildew, blanc: on 2
   NWT-Alta [1063].
Poria fraxinophila (*Fomes f.*): on 1
   NWT-Alta [1063], Alta [F65].
Puccinia caricis-shepherdiae (C:265):
   rust, rouille des carex: on 1, 2 BC
   [1816]; on 2 NWT-Alta [1063], Mack [53],
   BC [953, 1012], Alta [951, 1402], Ont
   [1236, 1828].
P. coronata (C:265): crown rust, rouille
   couronnée des graminées: on 2 Yukon
   [7, 460], BC [1012, 1816], Alta [644],
   Sask [1839].
P. coronata var. *bromi*: rust, rouille:
   I on 2 Sask [1839].
Septoria shepherdiae (C:265): leaf spot,
   tache septorienne: on 2 BC [1012, 1816].
Valsa sp.: on 2 Alta [644].
V. ambiens (anamorph *Cytospora a.*): on
   2 Ont [644].

SIBBALDIA L.   ROSACEAE

1. *S. procumbens* L.; native, Yukon-Labr,
   BC, Que, Nfld.

Diplocarpon sp.: on 1 Alta [644]. The
   anamorph was stated to be *Marssonina*
   sp.

SIDALCEA Gray   MALVACEAE

1. *S. hendersonii* S. Wats.,
   alkali-mallow; native, BC.

Ramularia sidalceae (C:265): on 1 BC
   [1816].

SILENE L.   CARYOPHYLLACEAE

1. *S. acaulis* (L.) Jacq., moss-campion,
   silène acaule; native, Yukon-Labr,
   BC-Alta, Que, NS, Nfld. 1a. *S. a.*
   var. *exscapa* (All.) DC.; native,
   Yukon-Labr, Que, Nfld.
2. *S. alba* (Mill.) E.H.L. Krause
   (*Lychnis a.* Mill.), white campion,
   compagnon blanc; imported from Eurasia.
3. *S. armeria* L., none-so-pretty, silène
   arméria; imported from Europe, a garden
   escape, BC, Ont-NS, Nfld.
4. *S. douglasii* Hook.; native, BC. 4a.
   S. d. var. *douglasii*.
5. *S. menziesii* Hook.; native,
   Yukon-Mack, BC-Man.
6. *S. noctiflora* L., night-flowering
   catchfly, fleurde nuit; from Europe;
   occurs as weed, BC-NS, PEI, Nfld.
7. *S. sorensensis* (Boivin) Bocquet
   (*Lychnis triflora* R. Br.); native,
   Yukon-Frank, BC.
8. *S. uralensis* (Rupr.) Bocquet. 8a. *S.*
   u. ssp. *apetala* (L.) Bocquet
   (*Lychnis a.* L., including *L. a.* var.
   arctica (Fries) Cody); native,
   Yukon-Labr, BC-Alta, Man-Que.

Acremonium boreale: on 6 BC, Alta, Sask
   [1707].
Diplocarpon agrostemmatis: on 1a Keew
   [664].
Mycosphaerella densa: on 8a Frank [1586].
Nectria pedicularis (C:265): on 1 Que
   [70].
Phyllosticta lychnidis: on 2 BC [1816].
Septoria sp. (C:266): on 3 BC [1816].
S. silenes (C:266): on 5 BC [1012, 1816].
Uromyces inaequialtus: rust, rouille: on
   4a BC [1578].
Ustilago violacea (C:266): smut,
   charbon: on 1a, 8a Frank [1239].
U. violacea var. *violacea*: smut,
   charbon: on 1a Frank [1244]; on 1a, 7,
   8a Frank [1586]; on 8a Frank [1543].
Wentiomyces fimbriatus (*Venturia f.*)
   (C:266): on 1 Que [70, 73].

SILPHIUM L.   COMPOSITAE

1. *S. perfoliatum* L., cup plant; native,
   s. Ont.

Puccinia silphii (C:266): rust, rouille:
   III on 1 Ont [1236, 1241, 1260].
Uromyces silphii: rust, rouille: 0 I on
   S. sp. Ont [1236].

SINNINGIA Nees   GESNERIACEAE

1. *S. speciosa* (Lodd.) Hiern., gloxinia,
   gloxinie; imported from Brazil.

Phytophthora nicotianae var. *n.*: on 1
   Ont [205].

SISYMBRIUM L.   CRUCIFERAE

1. *S. altissimum* L., tumble mustard,
    herbe roulante; imported from Eurasia,
    introduced, Yukon-Mack, BC-Nfld.
2. *S. loesellii* L., tall hedge mustard,
    sisymbre élevé; imported from Eurasia,
    introduced, BC-Que.
3. *S. officinale* (L.) Scop., hedge
    mustard, herbe au chantre; introduced,
    BC-Alta, Man-Nfld.

Albugo candida (A. cruciferarum) (C:266):
    on 1 BC [1012], Sask [1321, 1594]; on 1,
    3 BC [1816].
Alternaria brassicae: on 1, 2 Sask [1323].
Leptosphaeria maculans (anamorph
    Plenodomus lingam): on seeds of 1
    Sask [1325].
Mycosphaerella tassiana var. *tassiana*:
    on 2 Sask [1327].
Selenophoma sp.: on 1, 2 Sask [1323].
Verticillium dahliae: on 1 BC [38, 1816,
    1978].

SISYRINCHIUM L.   IRIDACEAE

1. *S. douglasii* Dietr.; native, s.
    Vancouver Island, BC.
2. *S. montanum* Greene, blue-eyed grass,
    bermudienne montagnarde; native, s.
    Yukon-Labr, s. into USA.

Mycosphaerella tassiana var. *arctica*
    (C:266): on 2 Que [70].
Uromyces probus: rust, rouille: on 1 BC
    [1012, 1816].

SIUM L.   UMBELLIFERAE

1. *S. suave* Walt. (*S. cicutaefolium*
    Schrank), water parsnip, berle; native,
    Mack, BC-Nfld.

Uromyces americanus: rust, rouille: on 1
    Ont, Que [1559].
U. lineolatus (C:266): rust, rouille: 0
    I on 1 Ont [1403].

SKIMMIA Thunb.   RUTACEAE

1. *S. japonica* Thunb., Japanese skimmia,
    skimmie du Japon; imported from Japan.
    An evergreen shrub.

Nectria sp.: on 1 BC [1595].

SMELOWSKIA C.A. Mey.   CRUCIFERAE

1. *S. borealis* (Greene) Drury & Rollins;
    native, Yukon-Mack. 1a. *S. b.* var.
    b.; native, Yukon.
2. *S. calycina* (Steph.) Mey.; native,
    Mack, BC-Alta. 2a. *S. c.* var.
    americana (Regel & Harder) Drury &
    Rollins; native, BC-Alta.

Puccinia aberrans: rust, rouille: III on
    2 Alta [644]; on 2a BC, Alta [1563].

P. codyi: rust, rouille: on 1a Mack
    [1567].
P. monoica: rust, rouille: 0 I on 2 Alta
    [644].

SMILACINA Desf.   LILIACEAE

1. *S. racemosa* (L.) Desf., false
    Solomon's seal, faux sceau-de-Salomon;
    native, BC-PEI.
2. *S. stellata* (L.) Desf., star-flowered
    false Solomon's seal, smilacine étoilée;
    native, Yukon-Mack, BC-Nfld, Labr.

Cylindrosporium smilacinae (C:266): on 1
    BC [1816], BC, Que [1545].
Phyllosticta convallariae (C:266): on 1
    Ont [1340]. Probably a specimen of *P.*
    pallidior.
P. pallidior: on 1 Ont, Que [149]; on 2
    Man, Ont [149].
Puccinia sessilis (C:267): rust,
    rouille: 0 I on 2 Ont [1236].
P. sporoboli (C:267): rust, rouille: 0 I
    on 2 Ont [1236].
Ramularia smilacinae (C:267): on 1 BC
    [1545, 1816].

SMILAX L.   LILIACEAE

1. *S. herbacea* L., carrion-flower, raisin
    de couleuvre; native, Sask-NB. 1a. *S.*
    h. var. *h. (S. peduncularis* Muhl.*)*;
    native, Sask-NB. 1b. *S. h.* var.
    lasioneura (Hook.) DC.; native,
    Sask-Ont.
2. *S. rotundifolia* L. (*S. aspera* DC.),
    common greenbrier; native, s. Ont.
3. *S. tamnoides* L. var. *hispida* (Muhl.)
    Fern. (*S. hispida* Muhl.), china-root;
    native, s. Ont.

Ascochyta fuscopapillata: on 1 Ont [195].
A. londonensis: on 1 Ont [195].
A. smilacigena (*Macrophoma pellucida*,
    Stagonospora p., S. smilacigena): on 1
    Ont [195].
Didymella pellucida (*Sphaerella p.)*: on 1
    Ont [195].
Macrophoma smilacina var. *smilacina* (*M.*
    smilacis): on 1 Ont [195], Ont, Que,
    NB [148].
M. smilacina var. *smilacis*: on 1 Man,
    Ont, NB [148]; on 1a Ont [148]; on 1b
    Sask, Man [148].
Metasphaeria dearnessii (C:267): on 2 Ont
    [195].
Phaeoseptoria canadensis: on 1 Ont [195].
Phyllosticta hispidae (C:267): on leaves
    of 3 Ont [440].
P. londonensis: on 1 Ont [195]. Probably
    a *Phoma* sp.
P. pellucida: on 1 Ont [195]. Possibly
    Phoma sp.
P. smilacigena: on 1 Ont [195]. A
    species of *Phoma*.
Pleosphaerulina canadensis: on 1 Ont
    [195].
Septoria pellucida: on 1 Ont [195].

SOLANUM L.   SOLANACEAE

1. *S. dulcamara* L., climbing nightshade,
   morelle douce-amère; imported from
   Eurasia, introduced BC-Alta, Man-PEI.
2. *S. melongena* L.; imported from Africa
   and Asia. 2a. *S. m.* var.
   esculentum Nees, eggplant, aubergine.
3. *S. nigrum* L., black nightshade,
   tue-chien; imported from Eurasia,
   introduced BC-Nfld.
4. *S. sarachoides* Sendtner; imported from
   South America, introduced, BC-Alta, Man,
   Que.
5. *S. tuberosum* L., potato, pomme de
   terre; imported from South America.

Alternaria solani (C:267):   early blight,
   brûlure alternarienne: on 5 BC [1595,
   1816], BC, Alta, Sask, Ont, Que, NB, NS
   [333, 334, 335, 336, 337, 338], BC,
   Alta, Que, NB [339, 1594], Sask [1594],
   Man [333, 335, 336, 337, 338], Que [348,
   1641, 1647, 1651, 1652, 1654], NS [339],
   PEI, Nfld [333, 336, 338], Nfld [334,
   335].
Aplosporella dulcamara: on 1 Ont [1340].
Armillaria mellea (C:267):   root rot,
   pourridié-agaric: on 5 BC [1816]. See
   Abies.
Botryotinia fuckeliana (anamorph *Botrytis
   cinerea*): on *S.* sp. Ont [666a].
Botrytis cinerea (holomorph *Botryotinia
   fuckeliana*) (C:267):   gray mold,
   moisissure grise: on 5 BC [1816], Ont
   [334], Que [348], Que, NB, PEI [333],
   Que [336, 337, 338].
Cephalosporium sp.: on 5 Man [758, 1595].
C. acremonium: on 5 Que [552]. See
   comment under *Avena*.
Colletotrichum sp.: on 5 [334], from two
   carloads of potatoes three days after
   arrival in Ont from NB.
C. coccodes (C:267):   black dot or
   anthracnose, dartrose: on 5 Alta, Que
   [338, 1594], Sask [446], Sask, Que [335,
   894], Man [758, 1595], Ont, Que, NS
   [334], Que [336, 337, 723, 1537, 1538,
   1539], Que, NS, PEI [333], NS [338, 339].
C. gloeosporioides (holomorph *Glomerella
   cingulata*): on 2a Que [339].
Cylindrocarpon sp.: on 5 Man [758, 1595].
*Doratomyces stemonitis (Stysanus
   tubericola)*: on pieces of 5 in a moist
   chamber Ont [440].
Fusarium sp. (C:269):   on 5 BC [213, 1817,
   1818], BC, Alta [338], BC, Alta, Sask
   [333, 334, 336, 337, 339], Alta, Sask,
   Ont, Que [335, 1594], Man [333, 334,
   335, 338], Man, Que [339], Ont [1460],
   Ont, Que [333, 334, 337, 338], Ont, Que,
   NB, NS [336], NB, NS, PEI [333, 334,
   335], NS, PEI [337, 338, 339], Nfld
   [334, 337]. In the case of [334] see
   under *Colletotrichum* sp.
F. avenaceum (C:269): on 5 Que [552].
F. coeruleum (C:269): on 5 BC [334,
   1816], Que [1595], NS [336, 339].
F. oxysporum (C:270): on 5 Que [552,
   723], Nfld [684].

F. oxysporum var. *redolens (F.
   redolens)*: (C:270): on 5 BC [1816].
F. oxysporum f. sp. *tuberosi* (C:270):
   on 5 BC [1816], Que [348], PEI [333].
F. sambucinum (C:269): on 5 PEI [41].
F. solani (holomorph *Nectria
   haematococca*) (C:270): on 5 BC [1816].
F. solani var. *eumartii* (C:270):
   stem-end rot, nécrose fusarienne du
   talon: on 5 BC [1594].
F. sulphureum (F. sambucinum f. 6
   Wr.) (holomorph *Gibberella
   cyanogena*) (C:269): on 5 BC [1816],
   Sask [334], NS [339], PEI [334, 336].
Helicobasidium purpureum (anamorph
   Rhizoctonia crocorum): on 5 PEI [339].
Helminthosporium solani (H. atrovirens)
   (C:270): silver scurf, tache argentée:
   on 5 BC [1816], Ont, Que [333, 334, 335,
   336, 337], Que [339, 1537, 1538, 1539,
   1540, 1541, 1594], NS [333], Nfld [336].
Oidium sp.: powdery mildew, blanc: on 3,
   4 BC [1816].
Ophiobolus erythrosporus: on 5 Ont [1620].
O. rubellus: on 1 Ont [1620].
Phacidiopycnis tuberivora (Phomopsis t.)
   (C:270): stem-end hard rot, pourriture
   ferme du talon: on 5 BC [1594, 1816].
Phoma sp. (C:270): on 5 NB [334, 336].
P. exigua (P. tuberosa) (C:270): on 5 BC
   [1816].
P. lycopersici (Ascochyta l.) (holomorph
   Didymella l.) (C:267): leaf spot,
   ascochytose: on 5 BC [1816]. Recent
   studies [1763] indicate that the species
   occurs only on *Lycopersicon*. *Phoma
   exigua* is very similar.
Phomopsis vexans (C:270): phomopsis
   blight, brûlure phomopsienne: on 2a
   Ont [337].
Phytophthora erythroseptica (C:270): pink
   rot, pourriture rose: on 5 BC [339,
   1594, 1816, 1818], Que [333], Que, NS
   [335].
P. infestans (C:271): late blight,
   mildiou: on 3, 5 BC [1816]; on 5 BC
   [46, 1818], BC, Ont, Que, NB, NS, Nfld
   [333, 334, 335, 336, 338], BC, Ont, Que,
   NS [337, 339], Alta, Nfld [339], Man, NB
   [337], Que [348, 551, 1299, 1640, 1641,
   1642, 1643, 1644, 1645, 1646, 1647,
   1648, 1650, 1651, 1652, 1653, 1654],
   Que, NB, NS, PEI, Nfld [1594], Que, NS
   [1024], PEI [179, 210, 211, 212, 213,
   214, 215, 216, 217, 218, 219, 220, 333,
   334, 1595].
Pythium sp. (C:272): on 2a Ont [868] by
   inoculation.
P. irregulare: on 2a Ont [868] by
   inoculation.
P. sulcatum: on 2a Ont [868] by
   inoculation.
P. sylvaticum: on 2a Ont [868] by
   inoculation.
P. torulosum: on 2a Ont [868] by
   inoculation.
P. ultimum (C:272): leak or seed-piece
   decay, pourriture aqueuse ou pourriture
   pythienne du planton: on 5 BC [334,
   1816], BC, Alta, Que [336, 337], BC,

Sask [335], BC, Sask, Que [333, 1594], Ont [335], Que [339], Que, PEI [334, 338], NS [336], NS, PEI [333, 339], Nfld [337].

Rhizoctonia crocorum (holomorph *Helicobasidium purpureum*) (C:272): violet root rot, rhizoctone violet: on 5 BC [1816].

R. solani (holomorph *Thanatephorus cucumeris*): on 5 BC [1984], BC, Sask, Man, Que, PEI [1595], Sask [243], Man [758], Ont [1458], Que [48, 49, 348, 552, 723, 1641, 1652, 1654, 1594, 1785].

Sclerotinia minor: on 5 BC [1816], Que [892].

S. sclerotiorum (C:272): sclerotinia stalk rot, pourriture sclerotique: on 5 BC [1816], Ont [334], Que [333], Nfld [334, 335].

Spongospora subterranea (C:272): powdery scab, gale poudreuse: on 5 BC [1816], Que [337, 339, 1594], Que, NS [333, 334, 338], Que, Nfld [335, 336], Nfld [333].

Stilbella flavescens: on 5 Que [2].

Streptomyces scabies (C:272): common scab, gale commun: on 5 BC [1816, 1817, 1818], BC, Alta, Sask [333, 334, 335, 336, 339, 1594], BC, Alta, PEI [338], Alta, Sask [337], Man, Ont, Que [333, 334], Ont [200, 1460], Ont, Que [336, 337, 1594], Que [339], NB, NS [333, 334, 335, 338, 339], NB, Nfld [1594], NS, Nfld [336, 337], PEI [336, 1595], PEI, Nfld [334, 339], Nfld [333, 335].

Synchytrium endobioticum (C:273): wart, tumeur verruqueuse: on 5 Nfld [333, 334, 336, 337, 338, 339, 682, 683, 1193, 1594], Nfld, Labr [335, 1192].

Thanatephorus cucumeris (*Pellicularia filamentosa*) (anamorph *Rhizoctonia solani*) (C:270): rhizoctonia, rhizoctonie: on 5 Mack [334], BC [1816], BC, Alta [338], BC, Alta, Sask [333, 334, 335, 336, 337, 339, 1594], Man [333, 335, 338], Ont, Que [336, 339, 1594], Ont, Que, NB, NS [333, 334, 335, 337, 338], Que [1651, 1653], NS [336, 339], PEI, Nfld [337, 339], Nfld [333, 334, 335, 336, 338].

Ulocladium consortiale (*Stemphylium c.*): on 5 BC [1816].

Verticillium sp. (C:273): verticillium wilt, flétrissure verticillienne: on 2 Ont [96]; on 2a Ont [1080], Que [1594]; on 2a, 5 Ont [97]; on 5 BC, Alta, Sask, Ont, Que, NS [337], BC, Sask, Que, NS, PEI [339], Sask [1594], Man [1295], Ont [203].

V. albo-atrum (C:273): verticillium wilt, flétrissure verticillienne: on 2 Man, Ont [38], Ont [1081]; on 2, 2a, 5 Ont [333]; on 2a Ont [98]; on 2a, 5 Que [336]; on 5 BC [1816], BC, Alta, Sask [333, 334, 336], BC, Alta [338], Alta, Sask [335], Sask [446], Man [333, 758, 1595], Man, Ont [334, 335, 338, 339], Ont [201, 336, 337, 676, 1024, 1079, 1087], Que [392, 393, 1594, 1639, 1647, 1652, 1654, 1784], Que, NB, PEI [38,

333, 334, 335], Que, NS, PEI [338], NB, NS [336], NS [333, 335], NS, Nfld [334], PEI [98, 1480].

V. dahliae (C:273): verticillium wilt, flétrissure verticillienne: on 1, 2, 5 BC [38]; on 2 BC [333], Ont [38, 1083, 1154]; on 2, 3, 4, 5 BC [1816, 1978]; on 2, 5 Ont [676, 1087], Que [38, 393]; on 2a BC [336, 339], BC, NS [337], Ont [334, 335, 338, 1082, 1421]; on 2a, 5 Ont [337, 339, 1594], Que [392]; on 5 Ont [1461], Que [676, 1832].

V. nigrescens: on 5 Ont [201, 337, 676, 749, 1082, 1087].

V. tenerum (*V. lateritium*) (holomorph *Nectria inventa*): on 2, 5 Que [39] by inoculation.

V. tricorpus: on 5 Man [1295].

Volutella sp.: on 5 Man [758, 1595].

SOLIDAGO L.   COMPOSITAE

1. *S. bicolor* L., silverrod, verge d'or bicolore; native, Man-NS.
2. *S. caesia* L., blue-stemmed goldenrod, verge d'or bleuâtre; native, Ont-NB.
3. *S. canadensis* L., Canada goldenrod, verge d'or du Canada; native, Yukon-Mack, BC-Nfld, Labr. 3a. *S. c.* var. *c.* (*S. lepida* sensu Fern.); native, transcontinental. 3b. *S. c.* var. *gilvocanescens* Rydb. (*S. dumetorum* Lunnell); native, transcontinental. 3c. *S. c.* var. *salebrosa* (Piper) Jones (*S. lepida* var. *elongata* (Nutt.) Fern., and var. *fallax* Fern.); native, transcontinental. 3d. *S. c.* var. *scabra* (Muhl.) Torr. & Gray (*S. altissima* L.); native, Ont-NS. 3e. *S. c.* var. *neomexicana* (Gray) Cronq.; native, Yukon, BC-Man.
4. *S. flexicaulis* L., zigzag goldenrod, verge d'or à tige flexible; native, Ont-PEI.
5. *S. gigantea* Ait., giant goldenrod, verge d'or géante; native, Mack, BC-PEI.
6. *S. glomerata* Michx.; imported from e. USA.
7. *S. graminifolia* (L.) Salisb. (*S. lanceolata* L.), grassleaf goldenrod, verge d'or à feuilles de graminées; native, Mack BC-Nfld.
8. *S. hispida* Muhl., hairy goldenrod, verge d'or hispide; native, Sask-Nfld.
9. *S. juncea* Ait., early goldenrod, verge d'or jonciforme; native, Man-PEI.
10. *S. macrophylla* Pursh, largeleaf goldenrod, verge d'or à grandes feuilles; native, Ont-Nfld, Labr.
11. *S. missouriensis* Nutt., Missouri goldenrod, verge d'or de Missouri; native, BC-Ont.
12. *S. mollis* Bartl.; native, Alta-Man.
13. *S. multiradiata* Ait.; native, Yukon-Labr, BC, Sask-NS, Nfld.
14. *S. nemoralis* Ait., gray goldenrod, verge d'or des bois; native, BC-PEI.
15. *S. patula* Muhl.; native, s. Ont.
16. *S. riddellii* Frank; native, Man-Ont.

17. *S. rigida* L., rigid goldenrod, verge
d'or rigide; native, Alta-Ont.
18. *S. rugosa* Ait., rough goldenrod, verge
d'or rugueuse; native, Ont-Nfld.
19. *S. sempervirens* L., seaside goldenrod,
verge d'or tonjours verte; native,
Que-Nfld.
20. *S. shortii* Torr. & Gray; imported from
Kentucky.
21. *S. spathulata* DC.; native, Yukon-Mack,
BC-NS. 21a. *S. s.* ssp. *randii*
(Porter) Cronq. var *racemosa* (Greene)
Gleason (*S. racemosa* Greene); native,
Ont-NB. 21b. *S. s.* ssp. *s.* var.
nana (Gray) Cronq. (*S. decumbens*
Greene); native, Yukon, BC-Alta.
22. *S. speciosa* Nutt.; native, s. Ont.
23. *S. squarrosa* Muhl., bracted goldenrod,
verge d'or squarieuse; native, Ont-NB.
24. *S. uliginosa* Nutt., bog goldenrod,
verge d'or des marais; native, Ont-Nfld,
Labr.
25. Host species not named, i.e., reported
as *Solidago* sp.

Botrytis cinerea (holomorph *Botryotinia
fuckeliana*): on 25 Sask [1760].
Cercospora sp.: on 25 BC [1816].
C. stomatica: on 25 Man [380].
C. tertia: on 3b Man [380].
Cercosporella dearnessii: on 3 Ont [195].
C. virgaureae: on 3, 7, 8 Ont [1352]; on
3, 7 Ont [380]; on 3b, 8 Man [380]; on
17 Sask [1352].
Coleosporium sp.: on 3c BC [1012].
C. asterum (C. solidaginis) (C:278):
rust, rouille: II III on 2, 3, 3b, 3c,
3d, 4, 5, 6, 8, 11, 12, 13, 14, 15, 16,
17, 18, 19, 20, 21a, 22, 23, 24, 25 Ont
[1236]; on 3 BC [2007]; on 3, 4, 8, 25
Ont [1828]; on 3, 7, 25 Ont [1341]; on
3a BC [1816]; on 3a, 13, 21 BC [1012];
on 3e Yukon [460]; on 13 Yukon [460] as
DAOM 25896; on 18 NS [1036]; on 21b Mack
[53, 1402]; on 21b, 25 Alta [53]; on 25
Mack [F65], NWT-Alta [1063], BC [953],
Alta [951, 952, F61], Sask, Man, Ont
[F67], Sask, NB, NS [F63], Que [282,
1377], NB [F64], NB, NS [1032], Nfld
[1659, 1660, 1661].
C. delicatulum (C:278): rust, rouille:
II III on 7 Ont [1236], Que [F61]; on 25
Que [F67].
C. tussilaginis s. 1.: rust, rouille: on
10, 18 Que [282]. These collections
would today be referred to *C. asterum*.
Diaporthe linearis (C:278): on 25 Ont
[1340].
Erysiphe cichoracearum (C:278): powdery
mildew, blanc: on 3, 3d Ont [1257].
Hymenoscyphus herbarum (Helotium h.): on
25 Ont [1340].
H. scutulus (Helotium s.): on 25 BC
[1012], Ont [1340].
Leptosphaeria ogilviensis (C:278): on 25
Que [70].
L. planiuscula (C:278): on 25 Ont [1340].
Leptothyrium tumidulum: on live leaf
petioles of 17 Ont [1502].

Mazzantia biennis: on overwintered leaves
of 9 Ont [79].
Mollisia atrata: on 25 Ont [1340].
Mycovellosiella minax: on 3 Ont [381].
Ophiobolus acuminatus: on 25 Ont [1620].
O. anguillidus: on 25 Ont [1340, 1620].
O. erythrosporus: on 25 Ont [1620].
O. fulgidus (C:278): on 25 Man [1620].
O. niesslii: on 25 Ont [1620].
Ovularia occulta: on leaves and stems of
17 Ont [1502].
Pleospora helvetica: on 25 Que [70].
Puccinia dioicae (C:278): rust, rouille:
on 1, 18 NS [1036]; on 25 BC [1012], NB,
NS [1032].
P. stipae (C:278): rust, rouille: on 3a
NWT-Alta [1063], BC [1816].
P. virgae-aureae (C:278): rust, rouille:
III on 7 Ont [1236]; on 18 NS [1036].
Pyrenopeziza artemisiae var. *solidaginis*
(C:278): on 25 Ont [1340].
P. solidaginis: on 25 Ont [1340].
Rhinotrichum herbicolum: on dead stems of
3 Ont [440].
Rhytisma solidaginis: on 7, 25 Ont
[1340]. Not a true *Rhytisma*.
Contrary to Conners (p. 278) there is a
fungus involved [221]. Possible names
for it are *Botryodiplodia gallicola*
and *Sclerotium asteris*.
Sclerotinia sp.: on 3c Sask [420], Sask,
Man [1138].
Uromyces sommerfeltii (U. solidaginis)
(C:279): rust, rouille: on 3a BC
[1012, 1816]; on 25 BC [7].

SONCHUS L.   COMPOSITAE

1. *S. arvensis* L., field sow-thistle,
laiteron des champs; imported from
Eurasia, introduced, Mack, BC-Nfld.
2. *S. asper* (L.) Hill, spiny-leaved
sow-thistle, chaudronnet; imported from
Eurasia, introduced Yukon, BC-NS, PEI,
Nfld, Labr.
3. *S. oleraceus* L., common sow-thistle,
laiteron potager; imported from Eurasia,
introduced, Mack, BC-Nfld.

Alternaria sonchi: on 3 BC [1816].
Bremia lactucae (C:279): downy mildew,
mildiou: on 3 BC [1816].
Mycosphaerella tassiana var. *tassiana*:
on *S.* sp. Sask [1327].
Sclerotinia sp.: on 1 Sask [420]; on 1, 2
Sask, Man [1138].
S. sclerotiorum (C:279): on 3 Man [750];
on *S.* sp. Sask [1319].
Selenophoma sp.: on 1 Sask [1323].
Sphaerotheca fuliginea: powdery mildew,
blanc: on 3 Ont [1257].

SORBOPYRUS Schneid.   ROSACEAE

1. *S. auricularia* (Knopp) Schneid.
Hybrid from *Pyrus communis* x *Sorbus
aria*. Originated before 1620.

Taphrina bullata (C:279): leaf blister,
cloque des feuilles: on 1 BC [1816].

SORBUS L.   ROSACEAE

1. *S. americana* Marsh. (*Pyrus a.*
   (Marsh.) DC.), American mountain ash,
   maskonabina; native, Ont-Nfld, Labr.
2. *S. aria* (L.) Crantz, whitebeam
   mountain ash, alisier blanc; imported
   from Europe.
3. *S. aucuparia* L. (*Pyrus a.* (L.)
   Gaertn.), European mountain ash, sorbier
   des oiseleurs; imported from Eurasia.
4. *S. decora* (Sarg.) Schneid. (*Pyrus d.*
   (Sarg.) Hyland), showy mountain ash,
   sorbier plaisant; native, Sask-Nfld,
   Labr.
5. *S. occidentalis* (S. Wats.) Greene;
   native, BC.
6. *S. scopulina* Greene, sorbier des
   Rocheuses; native, Yukon-Mack, BC-Sask.
7. *S. sitchensis* M.J. Roem., Sitka
   mountain ash, sorbier de Sitka; native,
   Yukon, BC-Alta. 7a. *S. s.* var.
   grayii (Wenzig) Hitchc. (*Pyrus
   occidentalis* Wats.); native, BC. 7b.
   S. s. var. *sitchensis*; native,
   Yukon, BC-Alta.
8. Other hosts: *S. alnifolia* (Sieb. &
   Zucc.) K. Koch; imported from e. Asia.
9. Host species not named, i.e., reported
   as *Sorbus* sp.

Armillaria mellea (C:279): root rot,
   pourridié-agaric: on 1 Que [1376]; on 3
   BC [1816]; on 9 BC [6]. See *Abies*.
*Basidioradulum radula (Radulum
   orbiculare)*: on 4 Nfld [1659, 1660].
   See *Abies*.
Botryosphaeria obtusa (C:279): canker,
   chancre: on 1 NB [1032], NS [1036]; on
   1, 2, 9 Ont [1614].
Cercospora sp.: on 7 BC [1012, 1816].
Cerrena unicolor (Daedalea u.) (C:279):
   on 1 BC [1012, 1341]. See *Acer*.
Chondrostereum purpureum (Stereum p.): on
   1 Ont [1341]. See *Abies*.
Coccomyces coronatus: on 9 Ont [1340].
Coniothyrium fuckelii (holomorph
   Leptosphaeria coniothyrium): on 9 Man
   [F66].
Coriolopsis gallica (Trametes hispida):
   on 4 NWT-Alta [1063]. See *Acer*.
Coriolus hirsutus (Polyporus h.): on 3, 9
   Ont [1341]. See *Abies*.
C. pubescens (Polyporus p.) (C:280): on 3
   BC [1816]. See *Abies*.
Crepidotus mollis (C. fulvotomentosus):
   on 1, 9 NS [1032]. See *Populus*.
Cucurbitaria sorbi (C:279): on 1 Ont
   [1340].
Cytospora sp. (C:279): canker, chancre:
   on 1 Alta [1594], Alta, Man [339]; on 3
   NB [1032]; on 9 BC [1012], Alta [338],
   Ont [1340], Ont, Que [336].
C. chrysosperma (holomorph *Valsa
   sordida*) (C:279): canker, chancre
   cytosporéen: on 9 Ont [1340].
C. cincta (holomorph *Leucostoma c.*): on
   3 Ont [1828].
C. rubescens (C. leucostoma) (holomorph
   Leucostoma persoonii) (C:279):

canker, chancre cytosporéen: on 1 Man,
   Ont [339]; on 3 BC [1816]; on 9 BC
   [333], Alta [336, F64].
Daedaleopsis confragosa (Daedalea c.): on
   4 Nfld [1659, 1660, 1661]. See *Acer*.
Dermea ariae (Dermatea a.) (C:279): on 1
   Ont, Que, NS [654], NB, NS [1032]; on 1,
   9 Ont [1340]; on 9 Sask [F68].
Diaporthe impulsa (C:279): on 1 Ont
   [1340], NS [1032].
Diplocarpon mespili (Fabraea maculata)
   (anamorph *Entomosporium mespili*)
   (C:280): leaf spot, tache des
   feuilles: on 1 NS [1036]; on 5 BC
   [1816]; on 9 Alta [F68].
Dothiora sorbi (C:279): on 1, 9 Ont
   [1340].
Durandiella lenticellicola (C:280): on 1
   NB [F65], NS [1032].
Entomosporium mespili (E. maculatum)
   (holomorph *Diplocarpon mespili*): on
   4, 9 Ont [1827]; on 9 Ont [F68].
Eutypella sorbi (C:280): on 1 Ont [1340];
   on 9 NWT-Alta [1063].
Exidia glandulosa (E. spiculosa): on 1
   Ont [1341]. See *Abies*.
E. saccharina: on 9 Ont [1341]. See
   Pinus.
Flammulina velutipes (Collybia v.): on 9
   Ont [1150]. See *Fagus*.
Fusarium sp.: on 1 NB [F72].
F. lateritium f. sp. *mori* (C:280): on 9
   Ont [1340, F63, F67, F69].
Gloeosporium sp.: on 1 NS [1032]; on 9
   Man [F60].
Guignardia sp.: on 7 BC [1012].
Gymnosporangium sp.: rust, rouille: on
   1, 3, 4 Nfld [1659, 1660, 1661]; on 9
   Alta [951], Sask, Man [F63], Man [F64],
   Que [F75], Nfld [337].
G. aurantiacum: rust, rouille: on 9 Que
   [1377], Nfld [F64]. See comment under
   Juniperus.
G. clavipes (C:280): rust, rouille: 0 I
   on 1 Ont [1341], NS [1263]; on 1, 3, 4
   Ont [1236]; on 3, 9 NS [1032]; on 5, 7
   BC [1816]; on 7a, 7b BC [1012]; on 9
   Alta, Sask [F71].
G. cornutum (C:280): rust, rouille: 0 I
   on 1 Man, Nfld [335], Que [282], NB, NS
   [1032], NS [1264], Nfld [339, F69, F70,
   F71, F73, F76]; on 1, 3, 4 Ont [1264];
   on 1, 4 Ont [1236], Que, Nfld [1264]; on
   1, 4, 9 Nfld [1659, 1660, 1661]; on 1, 9
   Ont [1341]; on 4 Ont [333, 334]; on 5, 7
   BC [1816]; on 6 Alta [952], Sask [1264];
   on 6, 7 BC, Alta [1264]; on 6, 9 Alta,
   Ont by inoculation [1247]; on 7 Alta
   [644]; on 7a, 7b BC [1012]; on 9 BC
   [1247], Alta [951], Man [1240, F67], Ont
   [F71, F73], Ont, Nfld [1240, F67, F68,
   F74], Que, NB, NS [1240], Nfld [F65,
   F66].
G. globosum: rust, rouille: on 9 Que
   [1377]. Probably a specimen of *G.
   cornutum*.
G. nootkatense (C:280): rust, rouille: 0
   I on 5, 7 BC [1816]; on 7a BC [1247]; on
   7a, 7b [1012, 2008].

G. tremelloides (C:280): rust, rouille:
0 I on 5, 7 BC [1816]; on 6 Alta [952],
Ont by inoculation [1247]; on 6, 7 BC,
Alta [1266], Alta [1247]; on 6, 9
NWT-Alta [1063], BC [1381], Alta [950];
on 7a, 7b BC [1012]; on 9 BC [644, 954],
Alta [2008].
Hyphoderma setigerum (Peniophora aspera):
on 9 NB [1032]. See *Abies*.
Hypoxylon fuscum: on 1 Ont [1340].
Irpex lacteus (Polyporus tulipiferae)
(C:280): on 9 NB [1032]. See *Acer*.
Isariopsis sp.: on 9 Que [F60].
Leucostoma cincta (Valsa c.) (anamorph
Cytospora c.): on 1 Ont [1828]; on 1,
9 Ont [1340]; on 9 Ont [F62, F71], Que
[F61].
L. massariana: on 1, 3 Que [338]; on 9
Que [F65].
L. persoonii (anamorph *Cytospora
rubescens*): on 1 NS [1032]; on 3 NB
[1032]; on 9 NWT-Alta [1063], Alta [F63].
Lophodermium tumidum (C:280): on 1 NB, NS
[1032]. Probably specimens of *L.
aucupariae*, see [1609 page 96].
Melanomma pulvis-pyrius (M. subsparsum)
(C:280): on 1 Que [70], NS [1032].
Micropera sorbi (C:280): on 1, 9 Ont
[1340]. The generic name is invalid and
this species should be transferred, see
[401].
Mycosphaerella sp.: on 1 Nfld [1661]; on
3 Nfld [1660, F73].
M. maculiformis: on 1 Que [70].
Nectria sp.: on 8 Que [1377].
N. cinnabarina (anamorph *Tubercularia
vulgaris*) (C:280): on 1 Alta [338]; on
1, 9 NB [1032]; on 3, 9 NS [1032]; on 4
Nfld [1659, 1660, 1661]; on 9 NWT-Alta
[1063], BC [1012], Sask [F69], Sask, Man
[F67].
N. galligena (C:280): on 9 Ont, Que
[336], NS [1032].
Nummularia discincola: on 9 Ont [1340].
Oidium sp.: powdery mildew, blanc: on 3
BC [1594].
Paxillus sp.: on 1 Nfld [1660].
Phaeoisariopsis sorbi: on 1 Ont [1217];
on 1, 4 Que [1217].
Phellinus ferreus (Poria f.): on 1 Nfld
[1659, 1660, 1661]. See *Acer*.
P. igniarius (Fomes i.) (C:280): on 9 Ont
[1341]. See *Acer*.
Phomopsis sp.: on 9 Ont [1827].
P. sorbicola: on 9 Man [F67].
Phyllactinia guttata (C:280): powdery
mildew, blanc: on 7 BC [1012, 1816].
Phyllosticta minima: on 1 Nfld [1661].
Only known from *Aceraceae*. This
specimen may be *P. sorbi*.
P. sorbi: on 3 NS [1032].
Pleurotus subareolatus: on 7 BC [1012].
See *Acer*.
Pseudospiropes simplex: on 4 Man [817,
1760].
Pycnoporus cinnabarinus: on 9 Nfld [1659,
1660, 1661]. See *Acer*.
Septoria musiva (holomorph *Mycosphaerella
populorum*): on 3 Ont [1594].
Typically on *Populus*.

S. sorbi (holomorph *Mycosphaerella
aucupariae*): on 3 Ont [1340, F67].
Sphaeropsis malorum: on 1 NB [1032]. See
comment under *Malus*.
Tubercularia vulgaris (holomorph *Nectria
cinnabarina*) (C:280): on 9 BC [1012],
Alta [F63].
Tympanis conspersa: on 9 Man [F67], Man,
Que, NB [1221].
Valsa sp.: on 9 Ont [1340].
V. ambiens (V. sordida) (anamorph
Cytospora a.): on 9 NWT-Alta [1063],
Que [1377], NB [1032].
V. amphibola (C:280): on 1 NS [1032].

SORGHUM Moench   GRAMINEAE

1. *S. bicolor* (L.) Moench (*S. vulgare*
Pers.), sorghum, sorgho; imported from
Africa.

Ustilago sorghi (C:280): covered kernel
smut, charbon couvert: on 1 BC [1816].

SPARGANIUM L.   SPARGANIACEAE

1. *S. eurycarpum* Engelm., giant bur-reed,
rubanier à gros fruits; native, Mack,
BC-Nfld.

Uromyces sparganii (C:280): rust,
rouille: II III on 1 Ont [1236], Ont,
Que, NS [1287].

SPARTINA Schreb.   GRAMINEAE

1. *S. alterniflora* Loisel., salt marsh
grass, herbe salée; native, Que-PEI,
Nfld.
2. *S. gracilis* Trin., alkali cordgrass;
native, Mack, BC-Man.
3. *S. pectinata* Link (*S. michauxiana*
Hitchc.), fresh-water cordgrass, chaume;
native, BC-Alta, Man-Nfld.
4. Host species not named, i.e., reported
as *Spartina* sp.

Buergenerula spartinae: on 3, 4 NB [891];
on 4 NB [553].
Cladosporium sp.: on 4 NS [553].
Drechslera tritici-repentis (holomorph
Pyrenophora t.) (C:281): on 3 Ont
[1611].
Leptosphaeria albopunctata: on 4 NB [553].
L. marina: on 4 NB, NS [553].
L. neomaritima: on 4 NS [553].
Lulworthia sp.: on 4 NB, NS [553].
Mycosphaerella sp.: on 4 NB [553].
Phaeosphaeria typharum: on 1 NS [553]; on
4 Que, NB, NS [553].
Phoma sp.: on 1 NS [553]; on 4 NB, NS
[553].
Phyllachora graminis: on 4 Ont [1828].
Puccinia seymouriana (C:281): rust,
rouille: II III on 3 Ont [1236].
P. sparganioides (P. peridermiospora)
(C:281): rust, rouille: on 3 Que
[1377], NB [1036]; on 3, 4 Ont [1236];
on 4 NB, NS [1032].
Septoria spartinae: on 1 BC [1728].

Stagonospora sp.: on 1 NS [553]; on 4 NB
[553].
Uromyces acuminatus var. *spartinae*:
rust, rouille: on 4 NS [1036].

SPERGULA L.   CARYOPHYLLACEAE

1. *S. arvensis* L., corn spurrey,
    spargoute des champs; imported from
    Eurasia, natzd., Yukon-Mack, BC-Alta,
    Ont-Nfld.

Puccinia arenariae (C:281): rust,
rouille: III on 1 Ont [1236].

SPHAERALCEA St-Hil.   MALVACEAE

1. *S. coccinea* (Nutt.) Rydb., scarlet
    globe-mallow; native, BC-Man.

Puccinia schedonnardi: rust, rouille: on
1 Alta [1274].
P. sherardiana (C:281): rust, rouille:
on 1 Alta, Sask, Man [1275].

SPHENOPHOLIS Scribner   GRAMINEAE

1. *S. intermedia* Rydb., slender wedge
    grass, sphenopholis pâle; native, Mack,
    BC-Nfld.

Puccinia eatoniae var. *ranunculae*:
rust, rouille: II III on 1 Ont [1236].

SPINACEA L.   CHENOPODIACEAE

1. *S. oleracea* L., spinach, épinard;
    imported from sw. Asia.

Albugo occidentalis: on 1 BC [337, 1816].
Botrytis cinerea (holomorph *Botryotinia
    fuckeliana*) (C:281): on 1 Ont [1460,
    1594].
Cladosporium variabile (Heterosporium v.)
    (C:281): leaf spot, tache
    hétérosporienne: on 1 BC [1816].
Fusarium oxysporum f. sp. *spinaciae*: on
    1 Ont [1454].
Ligniera pilorum: on 1 Ont [60].
Olpidium brassicae: on 1 Ont [67].
Peronospora effusa (C:281): downy mildew,
    mildiou: on 1 Que [348].
P. farinosa (C:281): downy mildew,
    mildiou: on 1 BC [338, 339, 1816, 1817,
    1818].
Phytophthora megasperma (C:282): black
    root rot, pourriture des racines: on 1
    BC [1816].
Polymyxa betae f. sp. *betae*: on 1 Ont
    [60].
Puccinia aristidae (C:282): rust,
    rouille: on 1 BC [1816].
Pythium ultimum (C:282): damping off,
    fonte des semis: on 1 BC [1816].
*Stagonospora atriplicis (Ascochyta
    chenopodii)* (C:281): seed spot,
    ascochytose: on 1 BC [1816].
Stemphylium botryosum (holomorph
    Pleospora herbarum) (C:282): leaf
    spot, tache stemphylienne: on 1 BC
    [1816], Ont [667] from seeds.

Ulocladium consortiale (Stemphylium c.):
    on seeds of 1 Ont [667].

SPIRAEA L.   ROSACEAE

1. *S. alba* DuRoi (*S. salicifolia* aucts.
    non L.), meadowsweet, thé du Canada;
    native, Alta-Nfld. 1a. *S. a.* var.
    latifolia (Ait.) Ahles (*S. l.* (Ait.)
    Borkh.); native, Man-Nfld.
2. *S.* x *arguta* Zabel (*S.* x
    muliflora Zab. x *S. thunbergii*
    Siebold).
3. *S. betulifolia* Pallas (including var.
    lucida (Dougl.) Hitchc. *(S. lucida*
    Dougl.*)*), shiny-leaf meadowsweet,
    spirée à feuilles de Bouleau; native,
    BC-Sask.
4. *S. douglasii* Hook., Douglas spiraea,
    spirée de Douglas; native, BC.
5. *S. tomentosa* L., steeplebush, reine
    des prés; native, Man-PEI.
6. *S.* x *vanhouttei* (Briot) Zab. (*S.
    cantonensis* Lour. x *S. trilobata*
    L.), bridal wreath.
7. Host species not named, i.e., reported
    as *Spiraea* sp.

Cylindrosporium filipendulae (C:282): on
    3 BC [644, 953, 1012, 1816].
C. spiraeicola (C:282): leaf spot,
    cylindrosporienne: on 1 BC [1816].
Diaporthe viburni var. *spiraeicola*
    (C:282): on 7 Ont [1340].
Fusarium equiseti (C:282): on 6 Que [334].
Godronia spiraeae (C:282): on 1a Que
    [1525]; on 7 BC, Ont, Que [656].
Helminthosporium ?spiraeae: on 2 Que
    [337].
H. velutinum: on 4 BC [810].
Mollisia cinerea: on 7 Ont [1340].
Nectria cinnabarina (anamorph
    Tubercularia vulgaris) (C:282):
    dieback or coral spot, dépérissement
    nectrien: on 7 Que [333, 338], NB
    [1036].
Phloeospora dearnessii: on 1 Ont [195].
Podosphaera clandestina (C:282): powdery
    mildew, blanc: on 1 Ont [1257]; on 1a,
    5 Ont, Que, NS [1257].
Tomentella ellisii (T. ochracea): on 7
    Ont [929, 1341, 1828].

STACHYS L.   LABIATAE

1. *S. cooleyae* Heller; native, BC.
2. *S. mexicana* Benth. (*S. ciliata*
    Dougl.); native w. USA. Possibly
    misidentifications of *S. cooleyae*.
3. *S. palustris* L., marsh hedge-nettle,
    crapaudine; native, Yukon-Mack,
    BC-Nfld. 3a. *S. p.* var. *pilosa*
    (Nutt.) Fern. (*S. scopulorum* Greene),
    woundwort, ortie morte; native,
    transcontinental.

Erysiphe sp.: powdery mildew, blanc: on
    1 BC [1012].
E. cichoracearum (C:283): powdery mildew,
    blanc: on 2 BC [1816]; on 3 Ont [1257].

E. galeopsidis (C:283): powdery mildew,
blanc: on 3, 3a BC, Alta [1257].
Phyllosticta palustris: on live leaves of
3 Ont [441].

STAPHYLEA L.   STAPHYLEACEAE

1. *S. trifolia* L., American bladdernut,
staphylier; native, s. Ont, s. Que, a
shrub or small tree.

Aplosporella staphylina: on dead plants
of 1 Que [441].
Nectria atrofusca: on 1 Ont, Que [143].

STELLARIA L.   CARYOPHYLLACEAE

1. *S. calycantha* (Ledeb.) Bong., northern
starwort, stellaire calycanthe; native,
Yukon-Labr, BC-Nfld. 1a. *S. c.* var.
c. (*S. borealis* Bigelow); native,
Yukon-Mack, BC-Nfld, Labr.
2. *S. crispa* Cham. & Schlecht.; native,
Yukon, BC-Alta.
3. *S. graminea* L., stitchwort, mouron des
champs; imported from Eurasia,
introduced, BC, Man-Nfld.
4. *S. longipes* complex: 4a. *S.
crassipes* Hulten; native, Keew-Labr,
Nfld. 4b. *S. edwardsii* R. Br. (*S.
ciliatosepala* Trautv.); native,
transcontinental. 4c. *S. laeta*
Richards.; native, transcontinental.
4d. *S. longipes* Goldie, long-stalked
chickweed, stellaire à longs pedicelles;
native, Yukon-Labr, BC-NB. 4e. *S.
monantha* Hulten; native,
transcontinental.
5. *S. media* (L.) Cyrillo, chickweed,
mouron des oiseaux; imported from
Eurasia, natzd., Yukon-Mack, BC-Nfld,
Labr.
6. Host species not named, i.e., reported
as *Stellaria* sp.

Laetinaevia stellariae (C:283): on 4a, 4b
Frank [664].
Leptosphaeria stellariae (C:283): on 6
NWT-Alta [1063].
*Melampsorella caryophyllacearum (M.
cerastii)* (C:283): rust, rouille: II
III on 1 NWT-Alta [1063], Alta [F63]; on
1, 2, 4, 5 BC [1012]; on 1a, 2, 4e, 5 BC
[1816]; on 3, 5 Ont [1236]; on 4, 4e Que
[1236]; on 5 BC [2007]; on 5, 6 Que
[1377]; on 6 BC, Alta, Sask, Man [2008].
*Mycosphaerella tassiana (Sphaerella
stellarianearum)* (C:283): on dead
stems and leaves of 4 n. Que [437].
Olpidium brassicae: isolated from
infected rootlets of 5 BC [1977].
Puccinia arenariae (C:283): rust,
rouille: on 1, 4e BC [1578]; on 4c Alta
[1578]; on 4e Frank [1543]; on 6
NWT-Alta [1063], Man, Que [1236].
Pythium sp.: isolated from infected
rootlets of 5 BC [1977].
Ramularia stellariae (C:283): on 1a BC
[1816].

Septoria stellariae (C:283): on 5 BC
[1816].
Ustilago violacea var. *stellariae*
(C:283): smut, charbon: on 4b Frank
[1244]; on 4b, 4c, 4e Frank [1586].
U. violacea var. *violacea*: smut,
charbon: on 4b Frank [1244].

STEPHANOMERIA Nutt.   COMPOSITAE

1. *S. tenuifolia* (Torr.) Hall; native, BC.

Puccinia harknessii var. *stephanomeriae
(P. harknessii)*: rust, rouille: on 1
BC [1012].

STIPA L.   GRAMINEAE

1. *S. comata* Trin. & Rupr., spear grass
or needle-and-thread; native, Yukon,
BC-Man, introduced, Ont.
2. *S. occidentalis* var. *minor* (Vasey)
Hitchc. (*S. columbiana* Macoun); Yukon,
BC-Sask.

Puccinia stipae (C:284): rust, rouille:
II III on 1 BC [644].
Septoria secalis var. *stipae* (C:284):
on 2 Yukon [1731].

STREPTOPUS Rich.   LILIACEAE

1. *S. roseus* Michx., rose twisted-stalk,
rognons de coq; native, BC, Ont-Nfld,
Labr. 1a. *S. r.* var. *curvipes*
(Vail) Fassett; native, BC.

Septoria streptopodis: on 1 Ont [1545].
Tuburcinia clintoniae (C:284): on 1a BC
[1545, 1816]. The generic name is
considered to be synonymous with
Urocystis but this species has not
been transferred to *Urocystis*.

STROPHOSTYLES Elliott   LEGUMINOSAE

1. *S. helvola* (L.) Ell., wild bean,
haricot sauvage; native, Ont-Que.

Uromyces appendiculatus (U. phaseoli):
rust, rouille: 0 I II III on 1 Ont
[1236].

SYMPHORICARPOS Duham.   CAPRIFOLIACEAE

1. *S. albus* (L.) Blake, snowberry, graine
d'hiver; native, Mack, BC-Que.
2. *S. mollis* Nutt.; native, BC. The BC
plant is referable to *S. m.* var.
hesperius (Jones) Cronq. (*S. h.* G.N.
Jones).
3. *S. occidentalis* Hook., wolfberry,
graine de loup; native, BC-Ont.
4. *S. orbiculatus* Moench, Indian currant;
imported from the USA.

Cercospora symphoricarpi (C:284): leaf
spot, tache cercosporéenne: on 1 BC
[1012, 1816].

Griphosphaerioma kansensi (anamorph
   Labridella cornu-cervae):   on 3 Man
   [1612].
Labridella cornu-cervae (Pestalotia c.)
   (holomorph *Griphosphaerioma
   kansensis*):   on twigs of 3 Man [1757,
   1763].
Microsphaera diffusa (C:285):   powdery
   mildew, blanc: on 1 BC [333, 334, 339,
   1816], Que [338]; on 1, 4 Ont [1257].
Phomopsis sp.:   on *S.* sp. Que [F65].
Puccinia crandallii (C:285):   rust,
   rouille: on 1 BC [1012, 1816]; on 3
   NWT-Alta [1063], Alta [644].
P. symphoricarpi (C:285):   rust, rouille:
   on 1, 2 BC [1012]; on 2 BC [1816].
Rhizogene impressa: on *S.* sp. Sask [75].
Septoria symphoricarpi (C:285):   leaf
   spot, tache septorienne: on 1 BC [1816].
Sphaceloma symphoricarpi (C:285):
   anthracnose, anthracnose: on 1 NS [336].
Sphaerotheca sp.:   powdery mildew, blanc:
   on 1 BC [1012].

SYMPHYTUM L.   BORAGINACEAE

1. *S. officinale* L., common comfrey,
    herbe à la coupure; eurasian,
    introduced, BC-Alta, Ont-NS, Nfld.

Calyptella capula: on fallen stems of 1
   BC [1424].

SYRINGA L.   OLEACEAE

1. *S. reticulata* (Blume) Hara, Japanese
    tree lilac; imported from Japan. 1a.
    S. r. var. *mandshurica* (Maxim.) Hara
    (*S. amurensis* Rupr.), Amur lilac,
    lilas de l'Amour; imported from
    Manchuria.
2. *S. vulgaris* L., lilac, lilas; imported
    from Europe, well established as an
    escape from cultivation, Sask, Ont-PEI,
    Nfld.
3. Host species not named, i.e., reported
    as *Syringa* sp.

Alternaria sp.:   on 3 NB [1032].
Armillaria mellea: on 3 BC [1012].   See
   Abies.
Ascochyta syringae (C:285):   leaf spot,
   tache ascochytique: on 2 BC [1816]; on
   3 PEI [1032].
Botrytis cinerea (holomorph *Botryotinia
   fuckeliana*) (C:285):   gray mold,
   moisissure grise: on 2 Que [335]; on 3
   NB, PEI [1032].
Cladosporium herbarum: on 2 BC [1012]; on
   3 BC [337].
Coriolus versicolor (Polyporus v.):   on 2
   BC [1012], Que [1390]. See *Abies*.
Crucibulum laeve: bird's nest fungus: on
   2 NS [1036].
*Dendrothele griseo-cana (Aleurocorticium
   g.)*:   on 2 Ont [965]. See *Acer*.
Microsphaera penicillata (M. alni)
   (C:285):   powdery mildew, blanc: on 1a,
   2 Ont [334]; on 2 Ont [1257], Ont, Que
   [338], Ont, NB [333], Que, NS [334], NB
   [1194], NS [336]; on 3 Ont, NB [337].

Pestalotiopsis funerea (Pestalotia f.):
   on 2 Ont [336]. A common parasite on
   Pinaceae, this specimen may have been
   misidentified.
Phomopsis depressa: on 3 Man [F65].
Phytophthora citricola: on 2 Alta [338].
Sclerophoma sp.:   on 3 BC [1012].
Verticillium albo-atrum: on 2 Que [335];
   on 3 Nfld [1594].

TAENIDIA (Torr. & Gray) Drude UMBELLIFERAE

1. *T. integerrima* (L.) Drude, yellow
    pimpernel, ténidia à feuilles entières;
    native, Ont-Que.

*Cercosporidium punctiforme (Didymaria
   platyspora)*:   on 1 Ont [379].
Puccinia angelicae (C:285):   rust,
   rouille: 0 I II III on 1 Ont [1236,
   1551].

TAGETES L.   COMPOSITAE

1. *T. erecta* L., Aztec marigold, rose
    d'Inde; imported from Mexico and Central
    America.
2. *T. patula* L., French marigold, petit
    oeillet d'Inde; imported from Mexico and
    Guatemala.
3. *T. tenuifolia* Cav. (*T. signata*
    Bartl.), signet marigold; imported from
    Mexico and Central America. 3a. *T. t.*
    cv. pumila; a group name applying to
    dwarf cultivars.

Botrytis cinerea (holomorph *Botryotinia
   fuckeliana*) (C:285):   gray mold,
   moisissure grise: on 2 Ont [163]; on 3a
   BC [1816].
Fusarium sp. (C:285):   foot rot, pourridié
   fusarien: on 1 NS [339].
Phytophthora sp. (C:285):   stem rot,
   pourridié phytophthoréen: on 3a BC
   [1816].
P. cryptogea (C:285):   stem rot, pourridié
   phytophthoréen: on *T.* sp. Ont [333].
Sclerotinia sclerotiorum (C:286):
   sclerotinia rot, pourriture
   sclérotique: on 1 Sask [1140].

TARAXACUM Zinn   COMPOSITAE

1. *T. ceratophorum* (Ledeb.) DC., rough
    dandelion, pissenlit tuberculé; native,
    Yukon-Mack, BC-Que, Nfld.
2. *T. hyparticum* Dahlst.; native, Mack,
    Frank.
3. *T. kok-saghyz* Rodin, Russian
    dandelion; imported from Turkestan.
4. *T. lacerum* Greene (*T. arctogenum*
    Dahlst.), rough dandelion, pissenlit
    tuberculé; native, Yukon-Labr, BC-Que,
    Nfld.
5. *T. laevigatum* (Willd.) DC. (*T.
    erythrospermum* Andrz.), red-seeded
    dandelion, pissenlit lisse; imported
    from Europe, introduced Mack, BC-NS.
6. *T. latilobum* DC.; native, Que-NB, Nfld.

7. *T. officinale* Weber, common dandelion, pissenlit; imported from Eurasia, natzd. Yukon-Mack, BC-Nfld.
8. *T. phymatocarpum* Vahl *(T. pumilum* Dahlst.); native, Yukon-Frank.

Botrytis cinerea (holomorph *Botryotinia fuckeliana*): on 3 BC [1816].
Cercosporella angustana (holomorph *Mycosphaerella taraxaci*): on 4 Frank [1586].
Mycosphaerella taraxaci (anamorph *Cercosporella angustana*) (C:286): on 2, 4, 8 Frank [1586].
Olpidium brassicae: on 7 Ont [67].
Puccinia hieracii (C:287): rust, rouille: 0 I II III on 1, 4, 6, 7 Ont [1236]; on 7 BC [1012, 1816], Man [1838], Ont [1341]; on *T.* sp. Yukon [7].
Ramularia taraxaci (C:287): on 7 BC [1816].
Sphaerotheca fuliginea (S. humuli var. *f.)* (C:287): powdery mildew, blanc: on 2, 4, 8 Frank [1586]; on 7 Man [1838], Ont [1257].
S. macularis (C:287): powdery mildew, blanc: on 7 BC [1816].
Synchytrium taraxaci: on 7 BC [1816].
Ulocladium consortiale (Stemphylium c.): from seeds of 3 Ont [667].
Verticillium dahliae: on *T.* sp. BC [1816, 1978].

TAXUS L.   TAXACEAE

1. *T. baccata* L., English yew, if d'Europe; imported from Europe, n. Africa and Asia. la. *T. b.* cv. fastigiata, Irish yew.
2. *T. brevifolia* Nutt., western yew, if de l'ouest; native, BC-Alta.
3. *T. canadensis* Marsh., Canada yew, buis de sapin; native, Man-Nfld.
4. *T. cuspidata* Siebold & Zucc., Japanese yew, if japonais; imported from Japan, Korea and Manchuria.

Acanthostigmella pellucida: on 3 Que [78].
Armillaria mellea: on 2 BC [1012]. See *Abies*.
Asteridiella pitya: on 2 BC [1012, 1358].
Dendrothele incrustans (Aleurocorticium i.): on 2 BC [965, 1012]. See *Abies*.
Dothiora taxicola (Sphaerulina t.) (C:287): on 2 BC [954, 1012, F67]; on 3 NB [1032], Nfld [1659, 1660, 1661].
Hymenochaete badioferruginea (C:287): on 2 BC [1012]. See *Abies*.
H. fuliginosa (C:287): on 2 BC [1012]. See *Picea*.
H. tabacina (C:287): on 3 NS [1032]. See *Abies*.
Lepidoderma carestianum (C:287): on *T.* sp. Que [F60].
Limacinia sp.: on 2 BC [1012].
Nectria coccinea: on 2 Nfld [F75].
Phacidium taxicola (C:287): on 2 BC [1012, F64], Que [F68]; on 3 Ont [1340], Que [1681]; on *T.* sp. Ont [1444].

Phomopsis sp. (C:287): on la, 2 BC [1012].
Physalospora gregaria: on 2 BC [1012].
Phytophthora sp. (C:287): on *T.* sp. BC [333].
P. cinnamomi (C:287): on 1 BC [1012]; on 1, 4 BC [23, 25, 337].
Spiropes helleri: on 2 BC [820].
Valsa taxi: on 3 Que [1689].

TELLIMA R. Br.   SAXIFRAGACEAE

1. *T. grandiflora* (Pursh) Dougl., fringe-cup; native, BC.

Puccinia austroberingiana ssp. *austroberingiana (P. heucherae* var. *a.)* (C:287): rust, rouille: on 1 BC [1012, 1560, 1816].
P. austroberingiana ssp. *saxifragarum*: rust, rouille: on 1 BC [1560].
P. heucherae (C:287): rust, rouille: on 1 BC [1012].

THALICTRUM L.   RANUNCULACEAE

1. *T. alpinum* L., alpine meadow rue, pigamon alpin; native, Yukon-Mack, n. Que-Labr, BC.
2. *T. dasycarpum* Fisch. & Lall., purple meadow rue; native, BC-Que.
3. *T. dioicum* L., early meadow rue, pigamon dioique; native, Ont-Que.
4. *T. occidentale* Gray, western meadow rue; native, Yukon-Mack, BC-Sask.
5. *T. pubescens* Pursh, tall meadow rue, grand pigamon; native, Ont-Nfld, Labr.
5a. *T. p.* var. *p.* *(T. polygamum* Muhl.); native, Ont-Nfld.
6. *T. sparsiflorum* Turcz.; native, Yukon-Mack, BC-Ont.
7. Host species not named, i.e., reported as *Thalictrum* sp.

Didymosphaeria thalictri (C:288): on dead stems of 5a Ont [440].
Diplodia thalictri: on dead stems of 5a Ont [440].
Erysiphe communis (E. polygoni) (C:288): powdery mildew, blanc: on 5a Ont [1257].
Mollisia atrata: on 7 Ont [1340].
Ophiobolus niesslii: on 7 Ont [1620].
O. rubellus: on 7 Ont [1340, 1620].
Puccinia sp.: rust, rouille: on 4 NWT-Alta [1063].
P. gigantispora: rust, rouille: on 1 BC [1012, 1816].
P. recondita s.1. (C:288): rust, rouille: 0 I on 2, 5a Ont [1236]; on 3 Ont [1341]; on 4 BC [1012, 1816]; on 4, 6, 7 NWT-Alta [1063]; on 5a Nfld [1659, 1660, 1661].
Pyrenopeziza thalictri: on 5a Ont [1340].
Streptotinia caulophylli: on 5a Que [430].
Tranzschelia anemones (C:288): rust, rouille: 0 III on 3 Ont [1236, 1341]; on 5a Ont [1236].
T. thalictri (C:288): on 5a NB [1032].

THELYPTERIS Schmid.   POLYPODIACEAE

1. *T. palustris* (Salisb.) Scholt
   (*Dryopteris thelypteris* (L.) Gray),
   marsh-fern; found in low woods and
   swampy ground, Man-PEI, Nfld.

Uredinopsis atkinsonii: rust, rouille:
   II II III on 1 Ont [1236].

THLASPI L.   CRUCIFERAE

1. *T. arvense* L., pennycress, tabouret
   des champs; imported from Eurasia,
   introduced, Yukon-Mack, BC-Nfld.

Acremonium boreale: on 1 BC, Alta, Sask
   [1707].
Albugo candida (A. cruciferarum): on 1
   Sask [1319].
Alternaria brassicae (C:289): on 1 Sask
   [339, 1323].
A. raphani: on 1 Sask [1323].
Colletotrichum dematium: on 1 Sask [1323].
Fusarium sp.: on 1 Sask [1323]. Reported
   as "*Fusarium* complex".
Leptosphaeria maculans (anamorph
   Plenodomus lingam): on 1 Sask [1321,
   1322, 1325].
Mycosphaerella tassiana var. *tassiana*:
   on 1 Sask [1327].
Plenodomus lingam (Phoma l.) (holomorph
   Leptosphaeria maculans) (C:288): on 1
   Alta, Sask [339], Sask [1319, 1871].
Sclerotinia sclerotiorum: on 1 Alta
   [1320], Sask [1319].
Verticillium dahliae: on 1 BC [38, 1978].

THUJA L.   CUPRESSACEAE

1. *T. occidentalis* L., eastern white
   cedar, cèdre blanc; native, Man-PEI.
   1a. *T. o.* cv. fastigiata.
2. *T. orientalis* L., oriental cedar,
   thuja d'Orient; imported from China and
   Korea. 2a. *T. o.* cv. stricta.
3. *T. plicata* Don, western red cedar,
   cèdre de l'ouest; native, BC-Alta. 3a.
   T. p. cv. atrovirens.
4. Host species not named, i.e., reported
   as *Thuja* sp.

Aleurodiscus botryosus (C:288): on 1 Ont,
   Que [964]. See *Acer*.
A. canadensis (C:288): on 1 Ont [1341],
   Que [964]. See *Abies*.
A. lividocoeruleus: on 1 Ont [964]; on 3
   BC [964, 1012]. See *Abies*.
A. penicillatus (C:288): on 3 BC [1012].
   See *Abies*.
A. tsugae (C:288): on 1 Ont, Que [964];
   on 3 BC [964, 1012]. Lignicolous.
Alternaria maritima: on submerged panels
   of 3 BC [782].
Amphinema byssoides: on 1 Que [969]; on 3
   BC [1012]. See *Abies*.
Amyloathelia amylacea (Corticium a.)
   (C:288): on 1 Ont [743], Ont, Que
   [964], Que [969]; on 3 BC [743, 964,
   1012].

Amylocorticium cebennense: on 3 BC [1012,
   1939]. See *Abies*.
Amylostereum chailletii: on 3 BC [1012].
   See *Abies*.
Antrodia lenis (Poria l.) (C:290): on 3
   BC [1012]. See *Abies*.
A. lindbladii (Poria cinerascens)
   (C:290): on 3 BC [1012]. See *Abies*.
A. serialis (Trametes s.): on 1 NWT-Alta
   [1063]. See *Abies*.
A. sinuosa (Poria s.): on 3 BC [1012].
   See *Picea*.
A. xantha (Poria x.) (C:290): on 1 Ont
   [1341]; on 3 BC [1012]. See *Abies*.
Armillaria mellea (C:288): root rot,
   pourridié-agaric: on 1 Ont [1341, F66],
   Que [F62]; on 3 BC [1012, F74]. See
   Abies.
Asterostroma andinum (C:288): on 3 BC
   [1012].
Athelia decipiens: on 1 NS [1032]. See
   Pinus.
A. pellicularis: on 3 BC [1012]. See
   Abies.
Aureobasidium pullulans: on 1 Que [F61].
Auricularia auricula (A. auricularis)
   (C:288): on 1 Ont [1341], NB [423,
   1032]; on 3 BC [423, 1012]. See *Abies*.
Basidiodendron cinerea: on 1 BC [1016].
   See *Abies*.
B. nodosa: on 1 Ont [1016]. See *Abies*.
B. rimosa: on 1 Ont [1016].
Botryobasidium subcoronatum: on 3 BC
   [1012]. See *Betula*.
B. vagum: on 3 BC [1012]. See *Abies*.
Brevicellicium exile: on 1 Ont [742].
   See *Picea*.
Ceratocystis allantospora: on 1 BC [1190].
C. angusticollis: on 1 BC [1190].
C. pilifera: on 1 Man [1190].
C. sagmatospora: on 1 BC [1190].
Cerocorticium notabile (Corticium n.)
   (C:289): on 3 BC [1012]. See *Picea*.
*Chloroscypha limonicolor (Kriegeria
   enterochroma)* (C:289): on 1 Ont [1340].
C. seaveri (Kriegeria s.): on 3 BC [1012,
   F65].
Chondrostereum purpureum: on 3 BC
   [1012]. See *Abies*.
Cladosporium sp.: on 1 Ont [1828].
C. herbarum: on 2a BC [1012].
C. macrocarpum: on 4 Sask [1760].
Collybia acervata (C:288): on 3 BC
   [1012]. See *Picea*.
Coltricia perennis (Polyporus p.)
   (C:289): on 3 BC [1012]. See *Abies*.
*Confertobasidium olivaceo-album (Athelia
   fuscostrata)*: on 3 BC [1012]. See
   Abies.
Coniophora arida (C. betulae) (C:288): on
   3 BC [1012]. See *Acer*.
C. arida var. *suffocata (C. suffocata)*
   (C:288): on 3 BC [1012]. See *Abies*.
C. puteana (C:288): on 3 BC [1012]. See
   Abies.
Coniothyrium sp.: on 1a BC [1012].
Conoplea fusca: on 1 Man [1760].
C. juniperi: on 1 Man [1760].
Coriolus hirsutus (Polyporus h.) (C:289):
   on 3 BC [1012]. See *Abies*.

Coriolus versicolor (Polyporus v.)
(C:289): on 3 BC [1012]. See *Abies*.
Coryneum thujinum: on 3 BC [F67].
Crepidotus herbarum (C:289): on 3 BC
[1012]. See *Populus*.
Crucibulum laeve (C. vulgare): bird's
nest fungus: on 1 NB [1032]. See
Pseudotsuga.
Cylindrobasidium corrugum (Corticium c.)
(C:289): on 3 BC [1012]. See *Abies*.
Cylindrocephalum sp.: on 3 BC [1854].
Cytospora sp.: on 2a, 3 BC [1012].
Cytosporella sp.: on 1 Ont [1828].
Dacryobolus karstenii: on 3 BC [1012].
See *Pinus*.
Datronia stereoides (Polyporus planellus)
(C:289): on 3 BC [1012]; on 4 BC
[1014]. See *Alnus*.
Dendrothele griseo-cana (Aleurocorticium
g.): on 1 Ont [965]. See *Acer*.
D. incrustans (Aleurocorticium i.): on 3
BC [965, 1012]. See *Acer*.
D. pachysterigmata (Aleurocorticium p.):
on 1 Ont [965].
Delphinella tsugae (Mycosphaerella
peckii): on 3 BC [1012].
Dermea balsamea (anamorph *Foveostroma*
abietinum): on 3 BC [1012].
Diaporthe lokoyae (anamorph *Phomopsis*
l.): on 3 BC [508, 1012, F70].
Didymascella thujina (Keithia t.)
(C:289): needle blight, brûlure des
aiguilles: on 1 Ont [303, 305], Ont,
Que [1340, F64, F66], NB, PEI [1032,
F60]; on 3 NWT-Alta [1063], BC [954,
1012, 1340, 1416, F74], Ont [F66]; on 3a
BC [1012, F70]; on 4 NB [1032].
Discina ancilis (D. perlata) (C:289): on
3 BC [1012, F68].
Dothiorella sp.: on 1a BC [1012].
Echinodontium tinctorum: on 3 BC [1012].
See *Abies*.
Epicoccum purpurascens (E. nigrum): on 1
Ont [1827]; on 4 Sask [1760].
Euantennaria sp.: on 3 BC [798, 1012].
E. rhododendri (Limacinia alaskensis)
(C:289): on 3 BC [1012].
Fibricium rude (F. greschikii): on 3 BC
[1012]. See *Abies*.
Fomitopsis pinicola (Fomes p.) (C:289):
brown cubical rot, carie brune cubique:
on 3 BC [1012]. See *Abies*.
F. rosea (Fomes r.): on 1 Ont [1341].
See *Abies*.
Fusarium sp.: on 1 Ont [1828].
Gloeocystidiellum furfuraceum (Corticium
f.) (C:289): on 3 BC [1012]. See
Abies.
G. propinquum: on 3 BC [1012]. See
Pseudotsuga.
Gloeophyllum odoratum (Trametes o.)
(C:290): on 3 BC [1012]. See *Abies*.
G. sepiarium (Lenzites s.) (C:289): brown
cubical rot, carie brune cubique: on 3
BC [1012]. See *Abies*.
Gloeoporus dichrous: on 3 BC [1012]. See
Acer.
Graphium sp.: on 1 Ont [1340].
Griseoporia carbonaria (Trametes c.)
(C:290): on 3 BC [1012]. See *Pinus*.

Gymnopilus liquiritiae (Flammula l.)
(C:289): on 3 BC [1012].
Haematostereum sanguinolentum: on 3 BC
[1012]. See *Abies*.
Heterobasidion annosum (Fomes a.) (C:289):
root rot, maladie du rond: on 3 BC
[1012]. See *Abies*.
Humicola alopallonella: on submerged
panels of 3 BC [782].
Hymenochaete cinnamomea (C:289): on 3 BC
[1012]. See *Alnus*.
H. fuliginosa (C:289): on 3 BC [1012].
See *Picea*.
H. tabacina (C:289): on 1 NB [1032]; on 3
BC [1012, 1341]. See *Abies*.
Hyphoderma praetermissum (H. tenue): on 3
BC [1012]. See *Abies*.
H. sambuci: on 3 BC [1012]. See *Acer*.
Hyphodontia alutacea: on 3 BC [1012].
See *Abies*.
H. aspera: on 3 BC [1012, 1341]. See
Pinus.
H. pallidula: on 3 BC [1012]. See
Abies.
H. spathulata (Odontia s.): on 1, 4 NB
[1032]. See *Abies*.
Hypholoma fasciculare (Naematoloma f.)
(C:289): on 3 BC [1012]. See *Abies*.
Inonotus subiculosus (Poria s.) (C:290):
on 1 NB [1032]. See *Abies*.
Ischnoderma resinosum (Polyporus r.): on
1 Ont [1341]. See *Abies*.
Junghuhnia collabens (Poria rixosa): on 4
Ont [1004]. See *Abies*.
Kabatina thujae: on 1 Que [1377, F60,
F61]; on 3a BC [508].
Kirschsteiniella thujina: on 3 BC [1859].
Laetiporus sulphureus (Polyporus s.)
(C:289): on 3 BC [1012]. See *Abies*.
Lentinus lepideus (C:289): on 3 BC
[1012]. Now in *Neolentinus*. See
Pinus.
Lepteutypa cupressi: on 2, 3 BC [1012];
on 3 BC [F73, F74].
Leucogyrophana mollusca: on 3 BC [571,
1012]. See *Abies*.
L. pulverulenta: on 3 BC [571]. See
Pinus.
Lophiostoma thujae (C:289): on 1 Que [70].
Lophodermium thujae (C:289): on 1 NB
[F60]; on 1, 4 NB [1032].
Lophophacidium hyperboreum: on 1 Que
[1692].
Lulworthia medusa: on submerged panels of
3 BC [782].
Macrophoma magnifructa: on 1 Ont [1340].
M. thujana: on 3 BC [1012, F73].
Meruliopsis albostramineus (Ceraceomerulius
a.): on 3 BC [568, 1012]. See *Abies*.
Microspora sp.: on 1 BC [508].
Mitrula borealis: on dead foliage of 1
Ont [1428].
M. elegans: on 3 BC [1012].
M. lunulatospora: on dead foliage of 1
Ont [1428].
Mycena thujina: on trunks of 1 Ont [1385].
Myxosporium sp.: on 1 Que [F66]. May be
referable to *Cryptosporiopsis*
abietina, anamorph of *Pezicula livida*.
Neournula nordmanensis: on 3 BC [1224].

Nidula niveotomentosa: bird's nest
   fungus: on 3 BC [1012]. See
   Pseudotsuga.
Odontia lactea (C:289): on 3 BC [1012,
   1341]. See comment under *Abies*.
Omphalina ericetorum: on 3 BC [1012].
Oxyporus cuneatus (*Polyporus c.*) (C:289):
   on 3 BC [1012, 1341]; on 4 BC [1014].
Panus rudis (C:289): on 1 Ont [1341]; on
   3 BC [1012]. See *Acer*.
Patellaria atrata (C:289): on 3 BC [1012].
Paxillus panuoides (C:289): on 3 BC
   [1012]. See *Pseudotsuga*.
Pestalotia funerea var. *macrochaeta*: on
   3 BC [1012]. The variety is referable
   to *Pestalotiopsis* but the combination
   has not been validly published.
P. thujae: on 3 BC [1012]. See comment
   under *Chamaecyparis*.
Pestalotiopsis funerea (*Pestalotia f.*)
   (C:289): on 1 Ont [F60, F67], Que [F66,
   F74]; on 1, 4 Ont [1340], NB [1032].
Pezicula livida (anamorph
   Cryptosporiopsis abietina): on 1 Man
   [F66].
Phacidium sp.: on 1 Que [F62].
Phaeolus schweinitzii (*Polyporus s.*)
   (C:289): brown cubical rot, carie brune
   cubique: on 3 BC [1012]. See *Abies*.
Phanerochaete carnosa (*Peniophora c.*)
   (C:289): on 1 Que [969]; on 3 BC
   [1012]. See *Abies*.
P. sanguinea (*Peniophora s.*) (C:289): on
   1 BC [1012]. See *Abies*.
P. sordida (*Peniophora cremea*) (C:289):
   on 1 Que [969]. See *Abies*.
Phellinus ferreus (*Poria f.*) (C:290): on
   3 BC [1012]. See *Acer*.
P. ferrugineofuscus (*Poria f.*) (C:290):
   on 1 Man [579]; on 3 BC [1012]. See
   Abies.
P. nigrolimitatus (*Fomes n.*) (C:289):
   white pocket rot, carie blanche
   alvéolaire: on 3 BC [1012]. See
   Abies.
P. pini (*Fomes p.*) (C:289): red ring rot,
   carie blanche alvéolaire: on 3 BC
   [1012]. See *Abies*.
P. punctatus (*Poria p.*): on 3 BC [1012].
   See *Abies*.
P. repandus (*Fomes r.*): on 3 BC [1012,
   F67]. See *Pseudotsuga*.
P. viticola (*Fomes v.*): on 3 BC [1012,
   1341]. See *Abies*.
P. weirii (*Poria w.*) (C:290): yellow ring
   rot, carie jaune annelée: on 3 BC
   [1012, 1142, 1341]. See *Abies*.
Phialophora botulispora: on 4 Ont [279].
Phlebia centrifuga (*P. albida* auct.)
   (C:289): on 3 BC [1012]. See *Abies*.
P. flavoferruginea (*Peniophora f.*)
   (C:289): on 3 BC [1012, 1341]. See
   Pinus.
Phlogiotis helvelloides: on 4 Ont
   [1341]. Typically on well-rotted wood.
Pholiota decorata (*Flammula d.*) (C:289):
   on 3 BC [1012]. See *Pseudotsuga*.
Phoma sp.: on 3 BC [1012].
Phomopsis sp.: on 3 BC [1012].

Phomopsis juniperovora (C:289): blight,
   brûlure phomopsienne: on 1 BC [508,
   F73]; on 1a BC [1012].
Piloderma bicolor (*Corticium b.*) (C:288):
   on 3 BC [1012]. See *Abies*.
Pithya cupressina (C:289): on 3 BC [1012].
Pleospora herbarum (anamorph *Stemphylium
   botryosum*): on 2a BC [1012].
Pleurocybella porrigens (*Pleurotus p.*):
   on 1 Ont [1341]. See *Abies*.
Pleurotus sp.: on 3 BC [1012].
Polyporus hirtus: on 3 BC [1012, F61].
   See *Abies*.
P. varius (*P. elegans*) (C:289): on 3 BC
   [1012]. See *Abies*.
Poria sp.: on 1 Ont [1341]; on 4 NB
   [1032].
P. rivulosa (*P. albipellucida*) (C:290):
   on 3 BC [1012, 1854]. See *Abies*.
P. sequoiae (C:290): on 3 BC [1012].
P. subacida (C:290): on 1 Ont [1341], Que
   [1377], NB [1032]; on 3 BC [1012]. See
   Abies.
Psathyrella sp.: on 3 BC [1012].
Pseudotomentella humicola: on 1 Ont [926,
   1341].
P. mucidula (*Tomentella m.*): on 1 Ont
   [1881]; on 3 BC [1012, 1341].
P. tristis: on 1 Ont [1828].
P. vepallidospora: on 3 BC [1012, 1341].
Ramaricium alboflavescens: on 1, 4 Ont
   [572]. See *Abies*.
Resinicium bicolor (*Odontia b.*) (C:289):
   white stringy rot, carie blanche
   filandreuse: on 3 BC [1012]. See
   Abies.
Riessia semiophora: from old wharf of 3
   BC [783].
Schizophyllum commune (C:290): on 3 BC
   [1012]. See *Abies*.
Scirrhia conigena (*Mycosphaerella c.*): on
   1 Ont [1340].
Sclerophoma pithyophila (holomorph
   Sydowia polyspora): on 4 Ont [F68].
Scytinostroma galactinum: on 1 Ont
   [1341]; on 3 BC [1012]. See *Abies*.
S. ochroleucum (C:290): on 3 BC [1012].
Scytinostromella humifaciens (*Peniophora
   h.*) (C:289): on 3 BC [1012]. See
   Pseudotsuga.
Seiridium sp.: on 1 BC [508]; on 1a BC
   [1012].
S. cardinale: on 3 BC [508].
Serpula himantioides: on 3 BC [1012].
   See *Abies*.
S. lacrimans: on 3 BC [1012]. See
   Abies.
Sistotrema brinkmannii: on 3 BC [1012].
   See *Abies*.
Skeletocutis nivea (*Polyporus
   semipileatus*) (C:289): on 3 BC
   [1012]. See *Amelanchier*.
S. subincarnata (*Poria s.*) (C:290): on 1
   NB [1032]; on 3 BC [1012]. See *Abies*.
Sphaeropsis sapinea (*Macrophoma s.*): on 1
   Que [1687].
Sporidesmium achromaticum: on 1 Man
   [1760].
S. coronatum: on 1 Man [1760].
S. larvatum: on 1 Ont [824].

Stigmina thujina (Sciniatosporium t.): on
   3 BC [1012, 1127].
Subulicium lautum: on 1 Ont [744].
Sydowia polyspora (anamorph *Sclerophoma
   pithyophila*): on 1 Que [1686, F68]; on
   3 BC [1012].
Taeniolella rudis: on 1 Ont, Que [815].
Thelephora sp.: on 3 BC [1828].
T. terrestris: on 3 BC [1012, 1341]. See
   Pinus.
Tomentella sp.: on 1 Ont [1341].
T. avellanea: on 3 BC [1012, 1341].
T. brevispina: on 1 Ont [1341].
T. bryophila (T. pallidofulva): on 1, 4
   Ont [1828].
T. coerulea (T. papillata): on 1 Ont
   [929, 1341, 1828].
T. ellisii (T. microspora): on 1 Ont
   [929, 1341]; on 3 BC [929, 1341]; on 4
   Ont [1827].
T. ferruginea: on 3 BC [1012].
T. nitellina: on 1 Ont [1341]; on 3 BC
   [1341].
T. ramosissima (T. fuliginea): on 3 BC
   [1012, 1341].
T. subclavigera: on 1 Ont [1341].
T. sublilacina: on 1 Ont [1341, 1827]; on
   3 BC [1341].
*T. terrestris (T. badiofusca, T.
   umbrinella)*: on 1 Ont [1341, 1827]; on
   3 BC [1827].
*Tomentellastrum badium (Tomentella
   fimbriata)*: on 1 Ont [1341].
*Tomentellina fibrosa (Hypochnus canadensis,
   Kneifiella f.)*: on 1 Ont [1341]; on 3
   BC [924].
*Trechispora mollusca (Poria candidissima,
   T. onusta)* (C:290): on 3 BC [1012]; on
   4 NB [1032]. See *Abies*.
T. vaga: on 3 BC [1012]. See *Abies*.
*Trichaptum abietinus (Hirschioporus a.,
   Polyporus a.)* (C:289): on 1 NB [1032];
   on 3 BC [1012, 1341]; on 4 NB, NS
   [1032]. See *Abies*.
Trichoderma sp.: on 3 BC [1854].
Triposporium sp.: on 4 NB [1032].
*Truncatella angustata (Pestalotia
   truncata)*: on 3 BC [1012].
Tubercularia sp.: on 3 BC [1012].
Tubulicrinis chaetophorus: on 3 BC
   [1012]. See *Picea*.
T. glebulosus (T. gracillimus): on 3 BC
   [1012]. See *Abies*.
Tympanis laricina (anamorph *Sirodothis*
   sp.): on 1 Que [1221].
T. hypopodia (T. piceina): on 1 Que
   [1688].
T. pithya: on 1 Que [1688].
Tyromyces balsameus (Polyporus b.)
   (C:289): on 1 Que [1377]; on 3 BC
   [1012]. See *Abies*.
T. caesius (Polyporus c.) (C:289): on 3
   BC [1012]. See *Abies*.
T. placenta (Poria p., P. monticola)
   (C:290): on 3 BC [1012]. See *Pinus*.
T. sericeomollis (Polyporus s., Poria s.)
   (C:290): on 3 BC [1003, 1012, 1341].
   See *Abies*.
T. stipticus (Polyporus immitis) (C:289):
   on 3 BC [1012]. See *Abies*.

T. undosus (Polyporus u.) (C:289): on 3
   BC [1012]. See *Abies*.
Valsa sp.: on 3 BC [193].
V. abietis: on 3 BC [1012].
V. thujae: on 1 Que [1689].
Vararia pallescens (C:290): on 3 BC
   [1012]. See *Acer*.
V. racemosa (C:290): on 3 BC [1012]. See
   Abies.
Velutarina rufo-olivacea: on 3 BC [508,
   1012].
*Vesiculomyces citrinus (Gloeocystidiellum
   c.)*: on 3 BC [1012]. See *Abies*.
Vestigium felicis: on 3 BC [1361].
Xeromphalina campanella (C:290): on 3 BC
   [1012]. See *Abies*.
Zalerion maritimum: on submerged panels
   of 3 BC [782].

TIARELLA L.   SAXIFRAGACEAE

1. *T. cordifolia* L., foamflower, tiarella
    à feuilles cordées; native, Ont-NS.
2. *T. trifoliata* L., three-leaf
    foamflower, tiarelle trifoliée; native,
    BC-Alta. 2a. *T. t.* ssp. *t.* var.
    laciniata (Hook.) Wheelock (*T. l.*
    Hook.), cut leaf foamflower, tiarelle
    laciniée; native, BC. 2b. *T. t.* ssp.
    unifoliata (Hook.) Kern. (*T. u.*
    Hook.), leafy foamflower, tiarelle
    monofoliée; native, BC-Alta.

Puccinia heucherae (C:290): rust,
   rouille: on 2b BC [1012].
P. heucherae var. *heucherae* (C:290):
   rust, rouille: III on 1 Ont [1236,
   1560]; on 2 X 2a hybrid, 2, 2b BC
   [1560]; on 2, 2b BC [1816]; on 2b Alta
   [1560].

TILIA L.   TILIACEAE

1. *T. americana* L., (*T. glabra* Veut.),
    basswood, bois blanc; native, Sask-NB.
2. *T. cordata* Mill., littleleaf linden,
    tilleul à petites feuilles; imported
    from Europe.
3. *T. europaea* L., European linden,
    tilleul de Hollande; imported from
    Europe.
4. *T. platyphyllos* Scop., bigleaf linden,
    tilleul à grandes feuilles d'Europe;
    imported from Europe.
5. Host species not named, i.e., reported
    as *Tilia* sp.

Albatrellus peckianus (Leucoporus p.): on
   1 Que [1385]. See *Fagus*.
Apiognomonia tiliae (Gnomonia t.)
   (anamorph *Gloeosporium t.*) (C:291):
   on 1 Que [333], NB [F63]; on 1, 2, 5 NS,
   PEI [1032]; on 1, 5 NB [1032]; on 2 Nfld
   [333]; on 3 PEI [333, 334].
Ascotremella faginea: on dead trunks of
   1 Que [1385].
Asteroma canadense: on 1 Ont [195].
Bjerkandera adusta (Polyporus a.)
   (C:291): on 1 Ont [1341]. See *Abies*.

Catenuloxyphium semiovatum: on live
leaves, branches and trunks of 1 Ont
[798].
Cenococcum graniforme: on 1 Ont [1227,
1228, 1229].
Ceratocystis spinulosa: on 1 Ont [647];
on 1, 5 Ont [1340].
Cercospora microsora (C:290): leaf spot,
tache des feuilles: on 1 Que [339,
1377], NB, NS, PEI [1032], NS [333], PEI
[F71]; on 5 BC [1816], NB, PEI [1036].
Ceriporia spissa (Poria s.): on 1 Ont
[1341]. See *Acer*.
Cerrena unicolor (Daedalea u.): on 1 Ont
[1341]; on 5 Ont?, Que [1855]. See
Acer.
Cladosporium sp. (C:290): on 3 BC [1012,
1816].
Coriolus hirsutus (Polyporus h.) (C:291):
on 1 Ont [1341]; on 3 NB [1032]. See
Abies.
C. versicolor (Polyporus v.): on 1, 5 NB
[1032]; on 5 Ont [1341]. See *Abies*.
Coronicium alboglaucum: on 5 Ont [967].
See *Acer*.
Cristinia helvetica (Corticium h.): on 5
Ont [1341]. See *Pinus*.
Corynespora olivacea: on 1 Man [442,
1760], Ont [F67, F68]; on 1, 5 Ont [442,
1340]; on 5 Ont [1762]. Apparently the
same fungus is also named
Helminthosporium oligosporum.
Cryptodiaporthe sp.: on 1 Que [F63].
Cryptospora sp.: on 1, 5 Ont [1340].
C. tiliae: on 1, 5 Ont [1340].
Cytospora sp.: on 5 Ont [1340].
Daldinia concentrica: on 1 Ont [1340].
Dendrophoma tiliae: on 1 Ont [F63]; on 5
Ont [1340].
*Dendrothele griseo-cana (Aleurocorticium
g.)*: on 1 Ont [965]. See *Acer*.
D. microspora (Aleurocorticium m.): on 1
Ont, Que [965]. See *Betula*.
Diaporthe sp.: on 1 Que [F69].
Discula umbrinella (D. quercina)
(holomorph *Apiognomonia errabunda*):
on 1 NB, NS, PEI [1036]; on 5 PEI [1036].
Doratomyces stemonitis: on submerged wood
of 5 NB [788].
Elsinoe tiliae (C:291): anthracnose,
anthracnose: on 1, 5 NS [1032]; on 4 NS
[333].
Endophragmiella subolivacea: on 1 Que
[804, 808].
Flammulina velutipes (Collybia v.)
(C:290): white spongy rot, carie
blanche spongieuse: on 1 Ont [1150].
See *Fagus*.
Fumago vagans: on 5 Ont [798]. See
comment under *Corylus*.
Ganoderma applanatum (Fomes a.) (C:291):
white mottled rot, carie blanche
spongieuse: on 1 Ont [1341], Que
[1377]. See *Abies*.
Gloeosporium tiliae (holomorph
Apiognomonia t.) (C:291):
anthracnose, anthracnose: on 1 PEI
[F64]; on 4 BC [1012]; on 5 BC [1816],
Ont [336].
Graphostroma platystoma: on 1 Que [1343].

Helminthosporium oligosporum: on 1 Que
[303, 305]. See under *Corynespora
olivacea*.
Hercospora tiliae: on 1 Que [1685, F71];
on 5 Ont [1340].
Holwaya gigantea: on 1 Ont [1340].
H. mucida (Chlorosplenium canadense): on
5 Ont [405, 898].
Hygrophorus chrysodon: on 1 Ont [1230].
Hyphoderma setigerum (Peniophora aspera):
on 1 NS [1032]. See *Abies*.
Irpex lacteus (Polyporus tulipiferae): on
1 Ont [1341]. See *Acer*.
Ischnoderma resinosum (Polyporus r.): on
1 Ont [1341]. See *Abies*.
Kavinia himantia: on 1 Ont [1341]. See
Acer.
Lachnella tiliae (C:291): on 1 Ont
[F62]; on 5 Ont [1340].
Macrophoma sp.: on 1 Ont [1828].
M. tiliacea: on 1 Ont [1340].
Melanconis sp.: on 5 Ont [1340].
Menispora glauca: on 5 Ont [1828].
Meruliopsis corium: on 5 Ont [568]. See
Acer.
Mycena meliigena (M. corticola Amer.
auct.): on 1 Que [F66].
Nectria cinnabarina (anamorph
Tubercularia vulgaris) (C:291): on 1
Ont [1340], PEI [F72]; on 1, 5 PEI
[1032]; on 5 Que [335].
Oedemium didymum (holomorph
Chaetosphaerella fusca): on 1 Ont
[821].
Olpitrichum patulum: on dead twigs of 1
Ont [765].
Ophiognomonia melanostyla: on 5 Ont [79].
Patellaria atrata: on 5 Ont [1340].
Penicillium sp.: on 5 Ont [1340].
Peniophora cinerea: on 1 Ont [1341]. See
Abies.
Peziza varia: on rotting branches of 1
Ont [574].
Phellinus gilvus (Polyporus g.): on 1 Ont
[1341]. See *Abies*.
P. igniarius (Fomes i.) (C:291): white
trunk rot, carie blanche du tronc: on
1, 5 Ont [1341]. See *Acer*.
Pholiota sp.: on 1 Que [1377].
Phragmodiaporthe tiliacea (Diaporthe t.)
(C:291): on 1 Ont [F63]; on 1, 5 Ont
[1340].
Phyllosticta tiliae (C:291): on 1 Ont
[1340].
Pleurotus ostreatus: on 1 Que [1377].
See *Abies*.
Plicatura crispa (Trogia c.): on 1 Ont
[1341]. See *Acer*.
P. nivea: on 1 Ont [563, 568, 1341]. See
Acer.
Sphaeronaema robiniae: on 1 Ont [1340].
Tomentella sp.: on 1 Ont [1341].
T. cinerascens: on 1 Ont [1341].
T. griseoviolacea: on 1 Ont [929, 1341,
1828].
T. neobourdotii: on 1 Ont [1341].
T. pilosa: on 5 Ont [1881].
T. ramosissima (T. fuliginea): on 1 Ont
[1341].

T. rubiginosa (T. subrubiginosa): on 1
Ont [1881].
T. ruttneri: on 1 Ont [1341]; on 5 Ont
[928].
Tubercularia sp.: on 5 Ont [1340].
T. vulgaris (holomorph *Nectria
cinnabarina*): on 1 NB, PEI [F64]; on 5
Que [F62].
Uncinula clintonii (C:291): powdery
mildew, blanc: on 1 Ont, Que [1257],
Que [333, 1377].
Xenasma pulverulentum: on 1 Ont [967].
Lignicolous.

TOFIELDIA Huds.   LILIACEAE

1. *T. pusilla* (Michx.) Pers. (*T.
palustris* Am. auct., *T. borealis*
Wahlenb.); native, Yukon-Labr., se. BC,
Man, Ont-Que.

Chalara fusidioides: on old leaves of 1
Frank [1165].

TOLMIEA Torr. & Gray   SAXIFRAGACEAE

1. *T. menziesii* (Pursh) Torr & Gray,
pickaback plant; native, BC.

Puccinia heucherae var. *heucherae*
(C:291): rust, rouille: on 1 BC [1012,
1560, 1816].

TORREYOCHLOA Church   GRAMINEAE

Weak-stemmed perennial grass of wet
habitats, native to North America and e.
Asia.
1. *T. pallida* (Torr.) Church (*Glyceria
pallida* (Torr.) Trin.); native,
Ont-Que, NS.
2. *T. pauciflora* (Presl) Church,
(*Glyceria p.* Presl); native, Yukon,
BC-Alta.

Puccinia coronata: rust, rouille: on 2
BC [1012].
Stagonospora glycericola: on 1 Que [1729].

TOVARA Adans.   POLYGONACEAE

1. *T. virginiana* (L.) Raf., jumpseed,
native, s. Ont (n. to Waterloo Co.)-Que
(n. to Beauport, near Quebec City),
reported from PEI and NS.

Puccinia polygoni-amphibii: rust,
rouille: II III on 1 Ont [1236].

TRAGOPOGON L.   COMPOSITAE

1. *T. porrifolius* L., salsify or oyster
plant, salsifis cultivé; imported from
Europe, a garden escape; BC-Alta, Man-NS.
2. *T. pratensis* L., goat's beard,
salsifis des prés; imported from
Eurasia, introduced, BC-PEI.

Albugo tragopogonis (C:291): white rust,
albugine: on 1 Que [339]; on 2 BC
[1816], Que [714].
Erysiphe cichoracearum: powdery mildew,
blanc: on 2 Ont [1257].
Mycosphaerella tassiana var. *tassiana*:
on 2 Sask [1327].
Puccinia hysterium: rust, rouille: on 2
Ont, Que [1286].
Verticillium dahliae: on 2 BC [1816,
1978].

TRAUTVETTERIA Fisch. & Mey.   RANUNCULACEAE

1. *T. caroliniensis* (Walt.) Vail,
tassel-rue; native, BC. The BC plant
may be referred to as *T. c.* var.
occidentalis (Gray) Hitchc. (*T.
grandis* Nutt.), false bugbane, traut
vettérie.

Puccinia pulsatillae (C:292): rust,
rouille: on 1 BC [1816].
Ramularia trautvetteriae (C:292): on 1 BC
[1012, 1816].

TRIDENS Roem. & Schult.   GRAMINEAE

1. *T. flavus* (L.) Hitchc. (*Triodia f.*
(L.) Smyth.); imported from e. USA.

Puccinia windsoriae (C:295): rust,
rouille: II III on 1 Ont [1236].

TRIENTALIS L.   PRIMULACEAE

1. *T. borealis* Raf., American
starflower, trientale boréale; native,
Yukon, BC-Nfld, Labr. 1a. *T. b.* ssp.
b. (*T. americana* (Pers.) Pursh);
native, Man-Nfld, Labr. 1b. *T. b.*
ssp. *latifolia* (Hook.) Hult. (*T.
latifolia* Hook.); native, Yukon, BC.
2. *T. europaea* L., European starflower;
native, Mack, BC-Alta. Most of the
Canadian plants are referable to 2a. *T.
e.* ssp. *arctica* (Fisch.) Hult. (*T.
arctica* Fisch.).

Puccinia karelica (C:292): rust,
rouille: 0 I on 1a Ont [1236]; on 2 BC
[1550, 1816]; on *T.* sp. NS [1550].
P. karelica ssp. *karelica*: rust,
rouille: on 2 BC [1550].
P. karelica ssp. *laurentina*: rust,
rouille: on 1 Ont, NS, Nfld [1550].
P. limosae (P. caricina var. *l.)*
(C:292): rust, rouille: on 2a BC
[1012].
Septoria trientalis (S. increscens)
(C:292): on 1 Que [282]; on 1b BC
[1012, 1816].
Urocystis trientalis (Tuburcinia t.)
(C:292): smut, charbon: on 1b BC
[1012, 1816].

TRIFOLIUM L.   LEGUMINOSAE

1. *T. dubium* Sibth., suckling clover,
trèfle douteux; imported from Europe,
introduced, BC, Ont, NB-PEI.

2. *T. hybridum* L., alsike clover, trèfle
    hybride; imported from Eurasia,
    introduced, Yukon-Mack, BC-Nfld, Labr.
3. *T. incarnatum* L., crimson clover,
    trèfle incarnat; imported from Europe,
    escaped from cultivation, BC, Ont.
4. *T. microcephalum* Pursh; native, BC.
5. *T. pratense* L., red clover, trèfle
    rouge; imported from Eurasia, natzd.,
    Yukon-Mack, BC-Nfld, Labr.
6. *T. procumbens* L., low hop clover,
    trèfle couché; imported from Europe,
    introduced, BC, Sask-Nfld.
7. *T. repens* L., white clover, trèfle
    blanc; imported from Eurasia,
    introduced, Yukon-Mack, BC-Nfld, Labr.
8. *T. subterraneum* L., subterranean
    clover, trèfle souterrain; imported from
    Europe.
9. *T. wormskjoldii* Lehm. (*T. fimbriatum*
    Lindl.); native, BC.
10. Host species not named, i.e., reported
    as *Trifolium* sp.

Absidia sp.: on 5 PEI [1973].
Acremonium boreale: on 2, 5, 7 BC, Alta,
    Sask and on 2, 7 Man, Ont [1707].
Alternaria sp.: on 5 Que [34], PEI [1973].
Ascochyta sp.: on 5 Que [34].
A. viciae (A. meliloti) (C:292): black
    stem, tige noire: on 2, 5 Alta [119,
    121, 123, 124].
Aspergillus sp.: on 5 Que [34].
Aureobasidium caulivorum (Kabatiella c.)
    (C:293): northern anthracnose,
    anthracnose septentrionale: on 2, 5 BC
    [1816], BC, Alta [334]; on 5 BC, Alta
    [337, 338], Alta [117, 119, 121, 123,
    124, 339], Alta, PEI [336, 1594], Sask
    [694], PEI [335, 1974]; on 10 Alta,
    Sask, NB [333], NB [1195].
Botrytis sp.: on 5 Que [34], PEI [1973].
B. cinerea (holomorph *Botryotinia*
    fuckeliana) (C:293): on 5 PEI [1974].
Camposporium sp.: on 5 PEI [1973].
Cercospora zebrina (C:292): leaf spot,
    rayure nervale: on 5 PEI [335]; on 7
    Que [32]; on 10 Que [335].
Chaetomella sp.: on 5 Que [34].
Chaetomium sp.: on 5 PEI [1973, 1975].
C. cochliodes: on 5 Que [34].
C. crispatum: on 5 Que [34].
C. dolichotrichum: on 5 Que [34].
C. globosum: on 5 Que [34].
Cladosporium sp.: on 5 Que [34], PEI
    [1973].
Colletotrichum sp.: on 5 Que [34].
C. trifolii: on 5 PEI [1974].
Coniothyrium sp.: on 5 Que [34].
Cylindrocarpon sp.: on 5 PEI [1973].
C. ehrenbergii (C:292): root rot, chancre
    des racines: on 5 Sask [334], Que
    [34]. This name is considered to be a
    nomen nudum [247].
C. destructans (C. radicicola): on 5 Que
    [34].
Diplodia sp.: on 5 Que [34].
Drechslera phlei: on 5 Ont [674].
Erysiphe communis (E. polygoni) (C:293):
    powdery mildew, blanc: on 2 BC, Alta,
    PEI [338], Alta [1594]; on 2, 5 BC, Alta

[334, 336, 337], BC, Ont [1257], Alta
    [119, 121, 123, 124, 339], Que [337,
    339], PEI [335, 336, 1974]; on 2, 5, 7
    Que [32]; on 5 BC, Alta, NB, PEI [333],
    BC, Sask [335], Sask [694], Sask, Man,
    Que, NB, NS [1257], NB [339, 1195], NS
    [337], PEI [334, 1594]; on 6 Ont [1257];
    on 7 BC [337], Que [339], PEI [336,
    338]; on 10 Sask, NB, NS [333].
Fusarium sp.: on 2 Que [32]; on 5 PEI
    [1594, 1973, 1975].
F. avenaceum (C:293): on 2 Que [31, 338,
    339].
F. culmorum (C:293): on 2 Que [31, 338,
    339].
F. oxysporum (C:293): on 2 Que [31, 338,
    339]; on 5 Que [34], PEI [1810] by
    inoculation.
F. roseum: on 5 Que [34], PEI [1810] by
    inoculation.
F. solani (holomorph *Nectria*
    haematococca) (C:293): on 5 Que [34],
    PEI [1810] by inoculation.
Gliocladium sp.: on 5 PEI [1973, 1975].
Leptosphaeria pratensis (anamorph
    Stagonospora meliloti) (C:293): on 2
    Alta [124]; on 2, 7 Alta [119, 121, 123].
Leptosphaerulina trifolii (Pseudoplea t.)
    (C:293): pepper spot, tacheture noire:
    on 2, 5, 7 Alta [121], PEI [1974]; on 2,
    7 Alta [119, 123, 124]; on 5, 7 BC [339,
    1594], PEI [335].
Low-temperature Basidiomycete (C:292): snow
    mold, moississure des neige; winter
    crown rot, pourridié hibernal: on 2
    Sask [335]; on 2, 5 Alta [336]. Sask
    [334]; on 5 BC [1816], BC, Alta [334];
    on 10 Alta [333]. See comment under
    Medicago.
Macrosporium sp.: on 5 PEI [1973].
Microsphaera penicillata: powdery mildew,
    blanc: on 6 Que, NS [1257].
Monilia sitophila (holomorph *Neurospora*
    s.)): on 5 Que [34].
Mucor sp.: on 5 Que [34], PEI [1973].
Mycosphaerella killianii (Cymadothia
    trifolii) (anamorph *Polythrincium t.*)
    (C:292): sooty blotch, tache de suie:
    on 2 Alta [119], Alta, PEI [338]; on 2,
    5 Alta [1594], Que [337], PEI [336]; on
    2, 5, 7 Alta, Que [339], Que [32], PEI
    [335, 1974]; on 2, 7 Alta [121, 123,
    124]; on 2, 5, 7, 9 BC [1816]; on 5, 7
    NB [1195]; on 7 Alta [335], NB [333]; on
    7, 9 BC [1012]; on 10 NS [333].
Nigrospora sp.: on 5 PEI [1973].
Oidium sp.: powdery mildew, blanc: on 2
    BC [1816].
Papularia sp.: on 5 Que [34], PEI [1973].
Penicillium sp.: on 5 Que [34], PEI
    [1973, 1975].
Peronospora trifoliorum (C:293): downy
    mildew, mildiou: on 1 BC [1816].
Peziza ostracoderma: on 5 Que [34].
Phoma sp.: on 2 Alta [334]; on 2, 5 PEI
    [336]; on 5 Que [34], PEI [1973, 1975].
P. medicaginis (Ascochyta imperfecta): on
    2, 5 Alta [338]; on 5 Alta [1594], Sask
    [694]. The name appeared in [1594] as
    A. "medicaginis Pk.".

P. sclerotioides (Plenodomus meliloti)
(C:293): brown root rot,
pourridié-plénodome: on 10 Yukon [337],
Alta [336].
P. medicaginis var. *pinodella (P.
trifolii)* (C:293): black stem, tige
noire: on 2 Alta [1594]; on 2, 5 PEI
[1974]; on 5 Que [32, 339].
P. terrestris (Pyrenochaeta t.): on 5 Que
[34].
Phytophthora megasperma: on 5 Ont [240]
by inoculation.
Polythrincium trifolii (holomorph
Mycosphaerella killianii): on 10 Man
[1760].
Pseudopeziza trifolii (C:293): common
leaf spot, taché commune: on 2, 5 NB
[1195]; on 5 BC [1012].
P. trifolii f. sp. *trifolii-pratensis*
(C:293): on 2, 5 Que [337], PEI [336];
on 2, 5, 7 PEI [1974]; on 5 BC [1816],
Que [32, 339], PEI [338]; on 10 NB [333].
P. trifolii f. sp. *trifolii-repentis*
(C:293): on 7 BC [1816].
Rhizoctonia sp.: on 5 Que [34], PEI
[1973, 1975].
R. leguminicola: on 5 Alta [125].
Black-patch, not previously known in
Canada, was found in 1977 naturally
infecting fields in Alta.
R. solani (C:293): stunt, nanisme: on 5
Ont [243].
Rhizopus sp.: on 5 Que [34], PEI [1973].
Sclerotinia sp.: on 5 Sask, Man [1138],
Que [34].
S. sclerotiorum (C:293): sclerotinia stem
rot, pourriture sclérotique: on 2, 5
Alta [892]; on 5 Sask [1140]; on 10
received from BC [334].
S. trifoliorum (S. trifolii) (C:293):
sclerotinia wilt, flétrissure
sclérotique: on 1, 5, 8 BC [1816]; on
2, 5 PEI [1974]; on 5 Que [32, 63 by
inoculation, 339, 391, 450], PEI [338,
1594, 1976].
*Stagonospora meliloti (Ascochyta
medicaginis)* (holomorph *Leptosphaeria
pratensis*): on 2, 5 Alta [339]; on 10
Alta [338].
S. recedens (C:293): leaf spot, tache
stagonosporéenne: on 5 Alta [119, 121,
123, 124]; on 5, 7 BC [1816].
Stemphylium sp.: on 5 Que [34].
S. botryosum (holomorph *Pleospora
herbarum*) (C:293): leaf spot, tache
stemphylienne: on seeds of 5 Ont [667].
S. sarcinaeforme (C:293): target spot,
tache zonée: on 2 BC, Alta [334]; on 5
BC [1816], Ont [419, 667 on seeds], Que
[32, 339], PEI [334, 335, 336, 1974],
PEI, Nfld [333]; on 10 Sask [333].
Stilbella flavescens: on roots of 10 Que
[451].
Stysanus medius: on 5 Que [34].
Trichoderma sp.: on 5 PEI [1973].
T. album: on 5 Que [34].
T. viride: on 5 Que [34]. See *Abies*.
Uromyces minor (C:293): rust, rouille:
on 1, 4 BC [1012, 1816].
U. nerviphilus (C:293): vein rust,
rouille des nervures: 0 I III on 7 BC

[1012, 1816], BC, Ont, Que [336], Ont
[1236], PEI [1974].
U. striatus (C:294): rust, rouille: II
III on 6, 10 Ont [1237].
U. trifolii-repentis: rust, rouille: on
2 Que [282]; on 5 BC [1012].
U. trifolii-repentis var. *fallens (U.
fallens)* (C:293): rust, rouille: on 5
Que [32].
U. trifolii-repentis var. *t.-r. (U.
trifolii)* (C:294): rust, rouille: on
2 BC [336, 337], Alta [119, 338, 339,
1594], PEI [335]; on 2, 3, 5, 7, 8 BC
[1816]; on 2, 5 Ont [1237], Que [337];
on 2, 5, 7 Alta [118], Ont [1236], PEI
[336, 338, 1974]; on 2, 7 Alta [121,
123, 124, 337]; on 5 Que [336, 339], NB
[1195], PEI [334, 335, 1594]; on 7 Alta
[336]; on 10 Ont [1341], NB, NS, PEI
[333].
Verticillium sp. (C:294): wilt,
flétrissure verticillienne: on 5 Que
[1296]; on 10 Que [1295].
V. albo-atrum: on 2, 5, 7 Que [36]; on 7
Que [35]; on 10 Ont [37].
V. dahliae: on 2, 5, 7 Que [36]; on 5 Ont
[35, 36, 38]; on 7 BC [1978], Que [38].
V. nigrescens: on 5 Que [39, 38].
Wardomyces anomala: on 5 Que [34].
Zygorhynchus sp.: on 5 PEI [1973].

TRIGLOCHIN L.   ALISMATACEAE

Perennials of cosmopolitan range with
rush-like fleshy basal leaves. Two species
in Canada.

Pleospora triglochinicola (P. maritima)
(anamorph *Stemphylium t.*) (C:294):
Conners listed *P. maritima* as a
synonym of *P. herbarum*. Now it is
considered distinct with a quite
different *Stemphylium* state.
Venturia juncaginearum (Mycosphaerella j.)
(anamorph *Asteroma j.*) (C:294): on
T. sp. NWT, Nfld [73].

TRILLIUM L.   LILIACEAE

1. *T. erectum* L., purple trillium, trille
rouge; native, Man-NS.
2. *T. grandiflorum* (Michx.) Salisb.,
white trillium, trille blanc; native,
Ont-Que.
3. *T. ovatum* Pursh, western wakerobin;
native, BC-Alta.

Urocystis trillii (C:294): smut,
charbon: on 3 BC [1816].
Uromyces halstedii: rust, rouille: 0 I
on 1, 2 Que [1236].

TRISETUM L.   GRAMINEAE

1. *T. spicatum* (L.) Richter, spike
trisetum, trisète à épi; native,
Yukon-Labr, BC-Nfld.

Puccinia poae-nemoralis (C:295): rust,
rouille: II III on 1 Que [1236].

P. recondita s.l. (C:295): rust,
rouille: on 1 Yukon [1731].

TRITICOSECALE Wittmack   GRAMINEAE

A cross between *Triticum* and *Secale*,
usually referred to as *Triticale* Muntzing.

Claviceps purpurea: ergot, ergot: on
   T. sp. Alta [1594].
Puccinia graminis: rust, rouille: on
   T. sp. Sask [634], Man [631, 632, 633].
Ustilago nuda: smut, charbon: on *T.*
   sp. Alta [1594].

TRITICUM L.   GRAMINEAE

North American wheats were originally
introduced from cultivation in Europe and
came with the early settlers. More recently
wheats have been introduced from centres of
origin in w. Asia via breeding. Most
commercial varieties (99%) bred and released
in Canada or USA have been for North
American culture and marketing.

1. *T. aestivum* L., common wheat, blé
   commun; the major western crop, includes
   soft or hard, red or white grains. In
   Canada in 1982 common wheat was planted
   on 27 million acres and the yield was
   843 million bushels.
2. *T. compactum* Host, club wheat;
   presently no commercial acreage in
   Canada, although grown in w. USA.
3. *T. monococcum* L., einkorn; no
   commercial acreage in Canada.
4. *T. turgidum* L. emend Bowden (*T.*
   dicoccum Schrank, *T. durum* Desf., *T.*
   carthlicum Nevski), durum wheat,
   Persian wheat, wild emmer, blé durum,
   blé de Perse, emmer sauvage; durum wheat
   is grown on a large scale in w. Canada.
   This type furnishes the bulk of wheat
   for the manufacture of semolina, which
   is made into spaghetti, macaroni and
   similar products.
5. *T.* x *Agropyron*, hybrid.
6. Host species not named, i.e., reported
   as *Triticum* sp.

Absidia sp.: on seeds of 1 in storage Man
   [1665].
A. corymbifera: on seeds of 6 in storage
   Man [1896].
A. glauca: on seeds of 1 in storage Sask,
   Man [1895], Man [1665].
A. lichtheimii: on 1 Man [1665].
A. orchidis: on seeds of 1 in storage
   Sask, Man [1895], Man [1665].
A. ramosa: on seeds of 6 in storage Man
   [1896].
Acremonium boreale: on 1 BC, Alta, Sask
   [1707].
Alternaria sp.: on 1 Sask [1594], Man
   [335, 336, 338], NB, NS, PEI [854]; on
   1, 4 Man [334]; on 4 Sask [335].
A. alternata (A. tenuis auct.) (C:295):
   on 1 Sask, Man [1895], Sask, NB [333],
   Man [1114, 1665 on seeds in storage], NB
   [1196]; on 6 Man [1896].

Ascochyta sorghi (C:296): leaf spot,
   tache ascochytique: on 1 Alta [1594],
   Alta, Sask [333, 336, 337], Sask [334,
   335, 338, 339]; on 4 Sask [337].
Aspergillus sp.: on 1 Alta [222], Man
   [337, 1894 on seeds].
A. amstelodami: on seeds of 6 in storage
   Man [1896].
A. candidus: on seeds of 6 in storage Man
   [1896].
A. flavus: on seeds of 1 Sask, Man
   [1895]; on seeds of 6 in storage Man
   [1665, 1896].
A. fumigatus: on seeds of 1 Sask, Man
   [1895]; on 6 Man [1896] in storage.
A. repens: on seeds of 6 in storage Man
   [1896].
A. ruber (A. sejunctus): on seeds of 6 in
   storage Man [1896].
A. versicolor: on seeds of 1 in storage
   Sask, Man [1895]; on seeds of 6 in
   storage Man [1665, 1896].
Aureobasidium sp. (*Pullularia* sp.): on
   6 Ont [1308]. In greenhouse.
Bipolaris sorokiniana (Helminthosporium
   sativum, H. sorokinianum) (holomorph
   Cochliobolus sativus) (C:296): on 1
   BC [334, 1816], Alta [223, 1333, 1602
   inocluated on seedlings], Alta, Sask
   [333, 334, 338], Alta, Sask, Man [335,
   336, 337, 339], Sask [47 inoculated onto
   seeds, 248, 701, 961, 962, 1655], Sask,
   Man [1071, 1790, 1895 on seeds in
   storage], Man [672, 1070, 1114], Ont
   [334, 337, 338], PEI [855]; on 2, 4, 6
   Sask [697]; on 4 Sask [335, 1790]; on 6
   Sask [698, 700], Man [1896] on seeds in
   storage. See comment under *Drechslera*
   tritici-repentis.
B. victoriae (holomorph *Cochliobolus*
   v.): on 1 BC [1816].
Botryotrichum piluliferum: on seeds of 6
   in storage Man [1896].
Cephalosporium sp.: on 1 Ont [1496].
C. acremonium: on 1 Man [1114]. See
   comment under *Avena*.
Cercosporella herpotrichioides (C:297):
   eye spot, tache ocellée: on 1 BC
   [1816], Que [236].
Chaetomium sp.: on 1 Ont [1496].
C. dolichotrichum: on seeds of 6 in
   storage Man [1896].
C. funicolum: on seeds of 1 in storage
   Man [1665].
Cladosporium sp.: on 1 Sask, Man [1895],
   Ont [1496]; on 4 Sask [335].
C. cladosporioides (Hormodendrum c.): on
   1 Man [1665]; on 6 Man [1896] on seeds
   in storage.
C. herbarum (C:297): on 1 Man [1665] on
   seeds in storage, NB [333, 1196].
Claviceps purpurea (C:297): ergot,
   ergot: on 1 BC [1816], Alta [223, 334,
   335, 337, 1331, 1333], Alta, Sask [338,
   339], Sask, Man [337, 1594], Sask, NS
   [333], Ont [334], Que [335]; on 1, 4
   Alta [1598], Sask [336], Sask, Man [335,
   1514], Man [1364]; on flowers of 6 Que
   [1385].

Cochliobolus sativus (anamorph *Bipolaris
    sorokiniana*) (C:297): on 1 Alta, Sask
    [959, 963, 1512, 1594, 1689], Sask [199,
    247, 960, 1116, 1595, 1876, 1877], Man
    [963, 1116, 1689], Ont [260], NB, NS,
    PEI [854]. The average loss caused by
    Common Root Rot in Sask was 7.4% for
    1969-1978. The epidemiology of Common
    Root Rot in Manitou wheat in Man was
    described by Verma et al.[1879]
Colletotrichum dematium: on 1 Alta [1595].
C. graminicola (C:298): anthracnose,
    anthracnose: on 1 Alta [338, 339, 693,
    1174], Que [335].
Coprinus sp.: on 1 Man [1820]. Possibly
    C. psychromorbidus.
Cylindrocarpon sp.: on 1 Ont [1496]; on 6
    Ont [1308].
*Drechslera avenacea (Helminthosporium
    avenae)* (holomorph *Pyrenophora
    chaetomioides*): on seeds of 1 in
    storage Sask, Man [1895].
D. teres (Helminthosporium t.) (holomorph
    Pyrenophora t.) (C:298): on seeds of
    1 in storage Sask, Man [1895].
D. tritici-repentis (holomorph
    Pyrenophora t.-r.) (C:298): leaf
    blotch, tache foliaire: on 1 Sask [336,
    1869], Sask, Man [1071], Man [339, 672,
    1070, 1611], Ont [334]; on 1, 4 Alta
    [1611], Sask [1870]; on 1, 5 Ont [1611];
    on 4 Sask [339]; on 6 Alta, Sask, Man
    [1790]. Leaf spots, caused by *D.
    tritici-repentis*, *Bipolaris
    sorokiniana* and *Septoria avenae* f.
    sp. *triticea*, caused losses of 2.6
    million bushels or 3.5% of the potential
    production in Man in 1971 [672].
*Embellisia chlamydospora
    (Pseudostemphylium c.)*: on 6 NWT
    [1657], Mack [1618].
Epicoccum purpurascens (E. nigrum)
    (C:298): on 1 Alta [338], Man [1114],
    Ont [333, 1504]; on 6 Man [1760].
Erysiphe graminis (C:298): powdery
    mildew, blanc: on 1 BC, Alta [333, 334,
    338, 339], BC, Ont, Que [639, 640,
    1533], Alta [639, 640, 1331, 1333,
    1595], Alta, Ont [335, 336, 337, 1594],
    Sask [335], Ont [260, 334, 338, 339,
    843, 1255], Ont, NS [1257], NB [854],
    NS, PEI [854, 857], PEI [1594],
    Maritimes [272]; on 6 Man [1838].
E. graminis f. sp. *tritici* (C:298):
    powdery mildew, blanc: on 1 NB, NS, PEI
    [273, 854], PEI [855, 856, 861].
Fusarium sp. (C:298): on 1 BC [334], Alta
    [223, 335], Alta, Sask [333, 334, 336,
    337, 338, 339, 959, 1512, 1513, 1594],
    Sask [701, 960, 1595], Man [335, 339,
    1114, 1513], Ont [335, 337, 338, 1496],
    Ont, NS, PEI [1594], NB, NS, PEI [854],
    Maritimes [272]; on 1, 4 Sask [335]; on
    6 Sask [698], Ont [1308]. Snow mold
    caused by *Fusarium* and *Typhula*
    species caused severe damage (i.e.,
    total loss in some fields) to
    soft-white-winter wheat in e. and c. Ont
    in the spring of 1978.

F. acuminatum (C:298): on 1 Alta, Sask
    [1512], Man [1114].
F. avenaceum (C:298): on 1 Man [1114].
F. culmorum (C:298): on 1 Alta [1602],
    Alta, Sask [959, 1512, 1513], Sask
    [1595, 1814], Man [1513], Maritimes
    [272].
F. equiseti (C:298): on 1 Alta, Sask
    [1512].
F. nivale: on 1 Alta [956, 1594]. See
    Agrostis.
F. oxysporum (C:298): on 1 Alta, Sask
    [1512, 1513], Man [1114, 1513].
F. poae (C:298): on 1 Man [335, 1114].
F. sacchari: on 1 Man [1114].
F. sambucinum (C:298): on 1 Man [1114].
F. sporotrichioides (C:298): on 1 Man
    [1114].
F. sulphureum: on 1 Man [1114].
F. tricinctum: on 1 Man [1114].
Gaeumannomyces graminis (Ophiobolus g.)
    (C:299): take-all, piétin-échaudage:
    on 1 BC [1816], Alta [333, 334, 1331,
    1602], Alta, Sask [337, 338, 339], Sask
    [1594, 1595], Sask, Que [336], Ont
    [337], NB, NS [859].
G. graminis var. *tritici*: on 1 BC,
    Sask, Ont, Que [1619], Maritimes [273].
Gibberella zeae (anamorph *Fusarium
    graminearum*): on 1 Alta, Man [337].
Gliocladium sp.: on 1 Ont [1496] in
    greenhouse.
Gonatobotrys simplex: on overwintered
    plants of 1 Man [1114]; on 6 in storage
    Man [1896].
Helminthosporium sp.: on 6 Ont [1308] in
    greenhouse.
Hormodendrum sp.: on seeds of 1 in
    storage Sask, Man [1895].
*Hymenula cerealis (Cephalosporium
    gramineum)* (C:297): cephalosporium
    stripe, strie céphalosporienne: on 1
    Alta, Ont [334], Ont [333, 843].
Lagena radicicola (C:299): on 1 Ont
    [1670]; on 6 Alta, Ont [69].
Ligniera pilorum: on 1 Ont [60].
Mucor sp.: on 1 Ont [1496] in greenhouse.
Myriosclerotinia borealis (Sclerotinia b.)
    (C:302): snow mold, moisissure nivale:
    on 6 Sask [892].
Nigrospora sp.: on 1 Sask [701].
N. oryzae (holomorph *Khuskia o.*): on 6
    Man [1760, 1896 in storage].
N. sphaerica: on 1 Man [334].
Olpidium brassicae (C:299): root
    necrosis, nécrose des radicelles: on 1
    Ont [68, 1670]; on 6 Alta, Ont [69].
Penicillium sp.: on 1 Alta [222], Sask
    [701], Sask, Man [1895] on seeds in
    storage, Man [337, 1114, 1894 on seeds],
    Ont [1496] in greenhouse; on 6 Ont
    [1308] in greenhouse.
P. aurantiogriseum (P. cyclopium): on
    seeds of 6 in storage Man [1665, 1896].
P. chrysogenum: on 1 Man [1665]; on seeds
    of 6 in storage Man [1896].
P. funiculosum: on seeds of 6 in storage
    Man [1665, 1896].
P. melinii: on 1 Man [1665]; on seeds of
    6 in storage Man [1896].

P. patulum: on seeds of 6 in storage Man [1896].

P. piceum: on seeds of 6 in storage Man [1896].

P. viridicatum: on seeds of 6 in storage Man [1896].

Periconia sp.: on 1 Ont [1496]; on 6 Ont [1308] in greenhouse.

Phoma sp.: on 1 Ont [1496]; on 6 Ont [1308] in greenhouse.

Plenodomus sp.: on 1 Alta [337].
Probably *Phoma sclerotioides (Plenodomus meliloti)* which was recorded for the first time on wheat and other cereals in Canada in 1979 in Alta and Sask.

Polymyxa graminis (C:299): root-hair necrosis, nécrose des poils absorbants: on 1 Ont [60, 68, 1670]; on 6 Ont [69].

Pseudoseptoria stomaticola (Selenophoma donacis var. *s.)*: on 1 PEI [1534].

Puccinia graminis (C:300): rust, rouille: on 1 BC [1816], Alta [1331], Alta, Sask, Ont [335, 336, 338], Alta, Sask, NB [333], Sask, Ont, Que [334, 337], Man [635], Man, Que [336], Ont [1236]; on 4 Sask [338].

P. graminis ssp. *graminis* var. *graminis (P. g.* f. sp. *tritici)* (C:300): stem rust, rouille de la tige: on 1 BC [619, 625, 627, 628, 630, 631, 633, 639, 1533], BC, Alta, Sask, Man, Ont [621, 632, 640], Alta [339, 615, 620, 626], Sask [623, 634, 1071, 1594], Sask, Man, Ont [339, 615, 619, 620, 633, 1533], Man [635, 672, 1070, 1071, 1595], Man, Ont, Que [639, 1594], Ont [641, 843], Que [615, 619, 620, 623, 626, 629, 630, 633, 1533], Que, NS [621, 640], NB [619, 1196], NB, NS, PEI [854], NS [628, 1533]; on 1, 3, 4 Man, Ont, Que [625]; on 1, 4 BC [626], Alta [628], Sask, Man, Ont [626, 627, 628, 630, 631], Man, Ont [623, 629], Man, Ont, Que [622, 634], Que [626, 630]; on 4 Man [633].

P. graminis ssp. *graminis (P. g.* var. *tritici)* (C:300): stem rust, rouille de la tige: on 6 Sask [1451].

P. triticina (P. recondita N. Amer. auct., *P. r.* f. sp. *tritici)* (C:302): leaf rust, rouille des feuilles: on 1 BC [1526, 1816], BC, Alta, Sask, Man, Ont, Que [639, 640, 1515, 1516, 1517, 1518, 1520, 1533], BC, Alta, Sask [334, 339], BC, Ont, Que [335, 1519], Alta [223, 337, 1331, 1333, 1522], Alta, Sask, Ont [333, 1594], Alta, Sask, Que, NS [1521], Sask [335, 1523, 1527], Sask, Man [1071, 1528, 1531], Sask, Man, Ont, Que [1529, 1530], Man [339, 672, 1070, 1519, 1594, 1595], Ont [334, 337, 641, 1236, 1531], Que, NB [1524], NB [1196], NB, NS, PEI [333, 639, 640, 854, 1516, 1533], NB, NS, Nfld [1515], NS [1519, 1520, 1522], PEI [1525], Nfld [333, 339, 1533]; on 1, 4 BC [1522], BC, Man, Ont [1521, 1523, 1524, 1525, 1527], Alta, Sask [1524, 1525], Alta, Que, NS [1527], Sask [336, 337, 338], Sask, Man, Ont,

Que [1522, 1526], Que [1523], Que, NB [1525], NS [1524]; on 6 Man [484]. This fungus, often labelled *P. recondita* when on *Triticum*, differs from *P. recondita* which occurs on *Secale*, in having an unrelated aecial host and a distinct, although related telial host, and there are slight to marked differences between the spore types of the two fungi (Savile, in litt.). The name *P. triticina* has been widely accepted in Europe and Conners (loc. cit.) seems to have been the first to recognize it in North America. The second most important disease, after bacterial black chaff, leaf rust in Man in 1971 caused a loss of 1.1 million bushels or 1.6% of the potential production [672].

P. striiformis (C:301): stripe rust, rouille jaune striée: on 1 BC [1816], Alta [335, 336, 337, 338, 339, 1594].

Pyrenophora semeniperda (anamorph *Podosporiella verticillata*): on 1 Alta, Sask [1617].

P. tritici-repentis (anamorph *Drechslera t.)* (C:302): on 1 Sask [1369]; on 1, 5 Ont [1611].

Pythium sp. (C:302): browning root rot, piétin brun: on 1 Alta, Sask, Ont [339], Sask [1869, 1870], Ont [1670]; on 6 Alta, Ont [69], Ont [1308].

P. aphanidermatum: on 6 Ont [69].

P. aristosporum (C:302): on 6 Ont [69].

P. arrhenomanes (C:302): browning root rot, piétin brun: on 1 Alta [334], Sask [335, 1869]; on 6 Alta, Ont [69].

P. graminicola (C:302): browning root rot, piétin brun: on 1 Alta, Sask [338], Sask [337].

P. paroecandrum: on 6 Ont [69].

P. tardicrescens (C:302): browning root rot, piétin brun: on 1 Alta [1595]; on 6 Alta, Ont [69].

P. torulosum: on 6 Ont [69].

P. ultimum (C:302): browning root rot, piétin brun: on 6 Ont [69].

P. vanterpoolii: on 6 Ont [69].

P. vexans: on 6 Ont [69].

P. volutum (C:302): browning root rot, piétin brun: on 6 Alta, Ont [69].

Rhizoctonia sp.: on 1 in greenhouse Ont [1496].

R. solani (holomorph *Thanatephorus cucumeris*) (C:302): sharp eye spot, rhizoctone ocellé: on 1 BC [1816].

Rhizophydium graminis (C:302): on 1 Ont [68, 1670]; on 4 Ont [59]; on 6 Alta, Ont [69].

Rhizopus sp.: on 1 Sask, Man [1895] on seeds in storage, Ont [1496] in greenhouse; on 6 Man [1896] in storage, Ont [1308] in greenhouse.

R. arrhizus: on 1 Man on overwintered plants [1114] and on stored seeds [1665].

Sclerophthora macrospora (C:302): downy mildew, mildiou: on 1 NB [333, 1196].

Selenophoma donacis: on 4 Sask [339].
Does not occur on *Triticum*. Pehaps a specimen of *Pseudoseptoria stromaticola*.

Septoria sp.: on 1 BC, Alta, Ont, Que
[639, 640, 1533], Alta, Sask [333], Sask
[337], Sask, Man, NB, Nfld [639], Man
[1594], Ont [1504], NS [333], NS, Nfld
[640, 1533], PEI [1533]; on 1, 4 Sask
[339].

S. tritici (C:303): speckled leaf blotch,
tache septorienne: on 1 BC [1816], Alta
[333, 223, 336], Alta, Sask, Man, Ont
[334], Ont [843, 1594]; on 1, 4 Sask
[335, 336].

Stagonospora avenae f. sp. *triticea*
(*Septoria a.* f. sp. *triticea)*
(holomorph *Phaeosphaeria avenaria* f.
sp. *triticea*): on 1 Alta [338, 1331],
Sask [1595], Sask, Man [1071], Sask, Ont
[338, 1594], Man [335, 672, 1070], Ont
[336, 843], NB, NS, PEI [854]; on 6
Alta, Sask, Man [1790], Sask [339]. See
comment under *Drechslera*
tritici-repentis.

S. nodorum (Septoria n.) (C:302): glume
blotch, tache des glumes: on 1 BC [339,
1816], Alta [223], Alta, Sask [334, 337,
339], Alta, Sask, Ont [338, 1594], Alta,
Man, NB [333], Sask [335, 336], Ont
[843], NB [1196], NB, NS, PEI [273], NS,
PEI [854], PEI [861], Maritimes [272].

Stemphylium botryosum (holomorph
Pleospora herbarum): on seeds of 1
Ont [667].

Streptomyces sp.: on 1 Sask, Man [1895],
Man [1665] on seeds in storage.

S. griseus: on 1 Man [1665].

Thanatephorus cucumeris (Pellicularia
filamentosa, P. praticola) (anamorph
Rhizoctonia solani): on base of stems
of 1 Sask [1874].

Tilletia caries (T. foetens, T. foetida, T.
tritici) (C:303): common bunt, carie:
on 1 BC [1816], Alta [333, 334, 339],
Man [1021 by inoculation onto seeds,
1020, 1022], Ont [531], Que [337]; on 6
Sask [1838].

T. controversa (C:303): dwarf bunt, carie
naine: on 1 BC [1252, 1816], Alta [333,
335, 339], Ont [333, 1252, 1908].

Trichoderma sp.: on 1 in greenhouse Ont
[1496].

Trichothecium roseum (C:303): on seeds of
1 in storage Man [1665].

Typhula sp. (C:303): blight, brûlure: on
1 BC [1816].

Ustilago tritici (C:303): loose smut,
charbon nu: on 1 BC [1816], Alta [223,
334], Alta, Man, Ont [338], Sask [333,
334], Sask, Man [1176], Que [337], NB
[854], NS, PEI [259, 854], PEI [855]; on
1, 4 Sask [335, 336, 338], Sask, Man
[337, 339, 1594], Man [334, 1070, 1071];
on 4 Man [336]; on 6 Sask [1838].

Volutella sp.: on 1 in greenhouse Ont
[1496].

Wojnowicia hirta (W. graminis) (C:303):
basal rot, piétin tardif: on 1 Sask
[1761].

TROPAEOLUM L.   TROPAELACEAE

1. *T. majus* L., garden nasturtium,
capucine; imported from South America.

Verticillium dahliae: on 1 BC [1978].

TSUGA Carr.   PINACEAE

1. *T. canadensis* (L.) Carr., eastern
hemlock, pruche; native, Ont-PEI.
2. *T. heterophylla* (Raf.) Sarg., western
hemlock, pruche occidentale; native, BC.
3. *T. mertensiana* (Bong.) Carr., mountain
hemlock, pruche de Mertens; native, BC.
4. Host species not named, i.e., reported
as *Tsuga* sp.

Acremonium tsugae: on 2 BC [1012].

Aleurodiscus amorphus (C:304): on 3 BC
[964]. Probably a specimen of *A.*
grantii, see [576]. See *Abies*.

A. aurantius: on 4 Ont [964].

A. farlowii (C:304): on 1 Ont, Que
[964]. See *Pseudotsuga*.

A. penicillatus (C:304): on 1 Ont [964];
on 2 BC [964, 1341]; on 2, 3 BC [1012].
See *Abies*.

A. spiniger (C:304): on 2 BC [964,
1012]. See *Larix*.

A. weirii (C:304): on 2 BC [964, 1012].
See *Pseudotsuga*.

Alternaria alternata (A. tenuis auct.):
on 1 Que [457].

Amylostereum chailletii: on 2 BC [1012,
1914]; on 4 NS [1032]. See *Abies*.

Anomoporia myceliosa (Poria m.) (C:307):
on 2 BC [1012]. See *Picea*.

Antennatula lumbricoidea (Hyphosoma
abietis): on 2 BC [1012].

Antrodia heteromorpha (Coriolellus h.,
Trametes h.) (C:305): on 1, 4 NS
[1032]; on 2 BC [1012, 1341]. See
Abies.

A. lenis (Poria l.) (C:306): on 1 Ont
[1341]; on 2 BC [1012]; on 4 NS [1032].
See *Abies*.

A. lindbladii (Poria cinerascens): on 2
BC [1012]. See *Abies*.

A. sinuosa (Poria s.): on 2 BC [1012].
See *Picea*.

Armillaria mellea (C:304): on 1 Ont
[F63], NB [1032, F65]; on 1, 4 Ont
[1341]; on 2 BC [1012, 1144, 1341]; on 4
NS [1032]. See *Abies*.

Ascobolus epimyces: on 2 BC [1012].
Presumably on rotted wood.

Ascoconidium tsugae (holomorph *Sageria*
t.): on 2 BC [498, 1165, F66]; on 2, 3
BC [1012]; on 3 BC [509].

Ascocoryne sarcoides: on 2 BC [455, 1012,
1144].

Asterodon ferruginosus (C:304): on 2 BC
[1012, 1341]. See *Abies*.

Atropellis pinicola: on 2 BC [1446].

A. piniphila: on 2 Alta [1446].

Aureobasidium pullulans (Pullularia p.):
on 1 Que [457].

Bactrodesmium obliquum: on 2 BC [1012,
1760].

Basidiodendron pini: on 1, 4 Ont [1016].
See *Pinus*.

Basidioradulum radula (Hyphoderma r.,
Radulum orbiculare) (C:307): on 1 Ont
[1341], NS [1032]; on 2 BC [1012]. See
Abies.

Biatorella resinae (anamorph *Zythia r.*): on 2 infected with dwarf mistletoe BC [51, 1012].

Bothrodiscus obscurus: on 1 Nfld [1163].

Botryosphaeria tsugae: on 2 BC [494, 501, 507, 1012, F64, F68].

Botrytis sp.: on 2 BC [F70].

B. cinerea (holomorph *Botryotinia fuckeliana*): on 2, 3 BC [F76].

Caliciopsis sp. (C:305): on 2 BC [1063].

C. orientalis (C:305): on 1 Ont [493, 1340].

C. pseudotsugae (C:305): on 2 BC [51 host infected with dwarf mistletoe, 493, 1012, F62].

Capnodium spongiosum: on 2 infected with dwarf mistletoe BC [51].

Cephaloascus fragrans (*Ascocybe grovesii*): on 2 BC [1012, F65].

Cephalosporium sp. (C:305): on 2 BC [1012].

Ceratocystis nigra: on 1 Ont [647, 1340].

C. piceae: on 1 Que [457].

Ceuthospora sp.: on 2 BC [1012].

Chaetoderma luna: on 2 BC [1012]. See *Abies*.

Chlorociboria aeruginascens ssp. *aeruginascens*: on 2 BC [1012].

Chondropodium sp.: on 2 BC [1012].

Chondrostereum purpureum: on 2 BC [1012]. See *Abies*.

Clitocybula familia: on dead trunks of 1 Que [1385]. See *Larix*.

Coccomyces heterophyllae: on 2 BC [500, 1012, 1609].

Colpoma crispum: on 2 BC [1012, 1340].

Columnocystis abietina (*Stereum a.*) (C:307): on 2 BC [1012, 1341]. See *Abies*.

Coniophora sp.: on 2 BC [1012].

C. arida: on 4 NB [1032]. See *Acer*.

C. puteana (C:305): on 1 NB [1032]; on 2 BC [1012]. See *Abies*.

Coniosporium punctoideum: on 2 BC [1012].

Conoplea juniperi var. *robusta* (C:305): on 2 BC [1012, F63].

Coriolus versicolor (*Polyporus v.*) (C:306): on 1 Ont [1341]; on 2 BC [1012]; on 4 NS [1032]. See *Abies*.

Cristella sp.: on 1 Ont [1341]. Probably referrable to *Sistotrema* or *Trechispora*.

Crustoderma dryinum: on 2 BC [1012]. See *Abies*.

Cryptoporus volvatus (*Polyporus v.*) (C:306): on 2 BC [1012]. See *Abies*.

Cryptosporiopsis abietina (*Myxosporium a.*) (holomorph *Pezicula livida*) (C:306): on 2 BC [1012].

Cucurbidothis pithyophila: on 2 BC [1012, F67].

Cylindrobasidium evolvens (*Corticium laeve*): on 2 BC [1012]. See *Abies*.

Dacrymyces capitatus (*Dacryomitra nuda*) (C:305): on 2 BC [1012].

D. palmatus (C:305): on 1 Ont [1341]; on 1, 4 NS [1032]. See *Abies*.

Dacryobolus sudans: on 2 BC [1012]. See *Picea*.

Delphinella strobiligena (C:305): on 1 Ont [1340]; on 4 Ont [F62].

D. tsugae (*Mycosphaerella t.*): on 1 Ont [1340].

Dermea sp.: on 2 BC [455 from dead branches, 499].

D. balsamea (anamorph *Foveostroma abietinum*) (C:305): on 1 Ont [1340], Que [1690], NB, NS [1032]; on 2 BC [1012]; on 4 Ont [F62].

Diaporthe lokoyae (anamorph *Phomopsis l.*): on 2 BC [501, 1012, F68].

Dichomitus squalens (*Polyporus anceps*) (C:306): on 2 BC [1012]. See *Abies*.

Dimerium balsamicola (*Dimeriella b.*): on 2 BC [464].

Diplodia sp.: on 2 BC [F61].

Diplodina sp.: on 2 BC [F66].

Discocainia treleasei (*Atropellis t.*) (C:305): on 2 BC [1012, 1340, 1446, F66].

Dothiora taxicola (*Sphaerulina t.*): on 4 NB [1032].

Durandiella tsugae: on 2 BC [51 host infected with dwarf mistletoe, 1012].

Echinodontium tinctorium (C:305): on 2 BC [455, 458, 459, 1063, 1160]; on 2, 3 BC [1012]. See *Abies*.

Epicoccum purpurascens (*E. nigrum*): on 1 Que [457]; on 2 infected with dwarf mistletoe BC [51].

Epipolaeum tsugae (*Dimeriella t.*): on 2 BC [464, 1012, 1340, 1615].

Fabrella tsugae (C:305): on 1 Ont [1340, F63], NS [1032, F64].

F. tsugae ssp. *grandispora* (C:305): on 2 BC [897, 1012].

F. tsugae ssp. *tsugae* (C:305): on 4 Que [F64].

Femsjonia peziziformis (*F. luteoalba*): on 2 BC [1012]. See *Abies*.

Fomitopsis cajanderi (*Fomes c.*): on 1 NS [1032]. See *Abies*.

F. pinicola (*Fomes p.*) (C:305): on 1 Ont [1341], NB, NS [1032]; on 2 BC [1150]; on 2, 3 BC [1012]; on ? 4 Ont [1150]. See *Abies*.

F. rosea (*Fomes r.*) (C:305): on 1 Ont [1341]; on 2 BC [1012]. See *Abies*.

Foveostroma abietinum (*Gelatinosporium a.*) (holomorph *Dermea balsamea*) (C:305): on 1 Ont [1340].

Fusidium sp.: on 1 Que [457].

Ganoderma applanatum (C:305): on 1 Ont [1341]; on 2 BC [1012, 1341]. See *Abies*.

G. lucidum (C:305): on 1, 4 NB, NS [1032]. See *Abies*.

G. oregonense (C:305): on 2 BC [1012, 1341, 1660]. See *Abies*.

G. tsugae (C:305): on 1 Ont [1341]; on 2 BC [1012]. See *Tsuga*.

Gelatinosporium griseo-lanatum: on 2 BC [517].

G. pinicola (*Micropera p.*): on 2 BC [1012].

G. stillwellii: on 2 BC [517].

Gloeocystidiellum furfuraceum (*Corticium f.*) (C:305): on 2 BC [1012]. See *Abies*.

G. ochraceum: on 2 BC [1012]. See
Abies.
Gloeophyllum sepiarium (Lenzites s.)
(C:305): on 1, 4 NS [1032]; on 2 BC
[1012]. See *Abies*.
G. trabeum (Lenzites t.) (C:306): brown
cubical rot, carie brune cubique: on 1
Ont [1341, F63]. See *Malus*.
Gloeoporus dichrous: on 2 BC [1012]; on 4
log BC [568]. See *Acer*.
Gloiodon sp.: on 2 BC [1012].
Graphium sp.: on 1 Ont [647]; on 2 BC
[1144].
Guepiniopsis alpina: on 2 BC [1012]. See
Alnus.
G. chrysocomus (C:305): on 2 BC [1012].
See *Abies*.
Gymnopilus sapineus (Flammula s.): on 4
Ont [1306].
Gyromitra esculenta: on 2 BC [1012].
Probably on rotted wood.
Haematostereum sanguinolentum (Stereum s.)
(C:307): white stringy rot, carie
blanche filandreuse: on 1 Ont [1341],
Que [1377], NB, NS [1032]; on 2 BC
[1144, 1341, 1914]; on 2, 3 BC [1012];
on 3 BC [F65]. See *Abies*.
Helotium sp.: on 2 infected with dwarf
mistletoe BC [51].
H. resinicola (anamorph *Stilbella* sp.):
on 2 BC [54, 1012, F69]. An Ascomycete
which needs to be redisposed.
Henningsomyces pubera: on ? 4 bark BC
[1424].
Hericium abietis (C:305): yellow pitted
rot, carie jaune alvéolaire: on 2, 3 BC
[1012]. See *Abies*.
H. americanum (H. coralloides N. Amer.
auct.*)*: on 2 BC [1012]. See *Acer*.
Herpotrichia juniperi: on 2, 3 BC [1012].
Heterobasidion annosum (Fomes a.)
(C:305): on 2 BC [1012, 1143, 1144,
1914, 1916, F67]; on 4 BC [F68]. See
Abies.
Hohenbuehelia angustata: on 1 Que
[1385]. See *Betula*.
H. petaloides (Pleurotus p.) (C:306): on
2 BC [1012]. See *Alnus*.
Hymenochaete badioferruginea (C:305): on
2 BC [1012]. See *Abies*.
H. fuliginosa (C:305): on 2 BC [1012].
See *Picea*.
H. tabacina (C:305): on 2 BC [1012]. See
Abies.
Hyphoderma sambuci (Peniophora s.)
(C:306): on 1, 4 NS [1032]. See *Acer*.
H. setigerum: on 2 BC [1012]. See
Abies.
Hyphodontia arguta: on 2 BC [1012]. See
Abies.
Hypholoma capnoides (Naematoloma c.): on
2 BC [1012]. See *Abies*.
Hypocrea sp.: on 2 BC [1012].
H. citrina: on 1 NS [1032].
Inonotus dryadeus (Polyporus d.) (C:306):
on 2 BC [1012]. See *Abies*.
Ischnoderma resinosum (Polyporus r.)
(C:306): on 1 Ont [1341], NS [1032]; on
2 BC [1012]. See *Abies*.

Junghuhnia collabens (Poria rixosa): on 2
BC [1004, 1012]. See *Abies*.
Kuehneromyces lignicola (Naucoria l.)
(C:305): on 2 BC [1012]. See *Betula*.
Lachnellula agassizii (Dasyscyphus a.)
(C:305): on bark and dead wood of 1 Que
[1385]; on 2 BC [51 host infected with
dwarf mistletoe, 1012].
Laeticorticium minnsiae (Aleurodiscus m.)
(C:304): on 1 Ont [1341], Ont, Que
[930]; on 2 BC [930]. See *Abies*.
Laetiporus sulphureus (Polyporus s.)
(C:306): on 2 BC [1012]. See *Abies*.
Lasiosphaeria spermoides: on 4 Ont [1340].
Laurilia sulcata: on 2 BC [1012]. See
Abies.
Lentinellus ursinus: on 1 Que [1385].
See *Abies*.
L. vulpinus (Lentinus v.): on 2 BC
[1012]. See *Populus*.
Leucogyrophana mollusca: on 2 BC [1012].
See *Abies*.
L. romellii: on 1 Ont [571]. See *Abies*.
Leucostoma kunzei (anamorph *Cytospora
k.*): on 1 Que [457].
Lophium mytilinum: on 2 infected with
dwarf mistletoe BC [51].
Lophodermium sp. (C:306): on 2 BC [1012].
Lycoperdon perlatum: puff-ball: on 2 BC
[1012]. Probably on well-rotted wood.
L. pyriforme: puff-ball: on 2 BC
[1012]. Probably on well-rotted wood.
Macrophoma sp. (holomorph *Botryosphaeria
tsugae*): on 2 BC [494].
Melampsora abietis-canadensis (C:306):
rust, rouille: on 1 Ont [1827, 1828,
F63, F66, F68, F70, F71, F72, F73, F74],
Que [1377], NB, NS [F69]; on 1, 4 Ont
[1341], NS [1032]; on 4 Ont [F64, F65,
F67].
M. epitea f. sp. *tsugae* (C:306): needle
rust, rouille des aiguilles: 0 I on 2
BC [F73]; on 2, 3 BC [1012, 2008].
M. farlowii (C:306): rust, rouille: on 1
NS [1032].
M. medusae: rust, rouille: on 3 BC
[1012, 2002 by inoculation]; on 4 BC
[F65].
Meruliopsis taxicola: on 2 BC [1012].
See *Abies*.
Merulius tremellosus (C:306): on 2 BC
[1012]; on 2, 4 BC [568]. See *Abies*.
Microlychnus epicorticis: on 2 BC [506,
1012].
Micropera sp.: on 1 Que [F66]. Probably
a specimen of *Foveostroma*.
Mycena sp.: on 2 BC [1012].
M. epipterigea (C:306): on 4 NS [1032].
M. epipterigea var. *lignicola* (C:306):
on 4 NS [1032].
M. lilacifolia: on rotten wood of 1 Que
[1385].
Mytilidion tortile: on 2 infected with
dwarf mistletoe BC [51].
Nectria neomacrospora (N. fuckeliana var.
macrospora) (anamorph *Cylindrocarpon
cylindroides*): on 2 BC [528, 1012].
Commonly associated with open resinous
cankers on dwarf mistletoe swellings.

Neournula nordmanensis: on 2 BC [1224].
Nidula candida: bird's nest fungus: on 2 BC [1012]. Probably on rotted wood.
Nidularia sp.: bird's nest fungus: on 2 BC [1012]. Probably on rotted wood.
Nipterella tsugae: on 2 BC [514].
Nitschkia molnarii: on 2 BC [518].
Odontia lactea (C:306): on 2 BC [1012]. See comment under *Abies*.
Onnia circinata (Polyporus c., P. tomentosus var. *c.)* (C:306): on 1 Que [1377]; on 2 BC [1012]. See *Abies*.
O. tomentosa (Polyporus t. var. *t.)*: on 2 BC [1012]. See *Abies*.
Panus sp.: on 2 BC [1012].
Panellus serotinus (Pleurotus s.) (C:306): on 1 NS [1032]; on 2, 3 BC [1012]; on 4 NB [1032]. See *Acer*.
P. stipticus (Panus s.) (C:306): on 3 BC [1012]. See *Alnus*.
Paxillus atrotomentosus (C:306): on 4 Ont [1306]. Typically on rotted wood. See *Pseudotsuga*.
P. panuoides (C:306): on 2 BC [1012, F62]. See *Pseudotsuga*.
Penicillium sp.: on 1 Que [457]; on 2 BC [1144].
Peniophora cinerea (C:306): on 2 BC [1012]. See *Abies*.
P. perexigua (C:306): on 1 Ont [1341].
P. piceae (P. separans) (C:306): on 2 BC [1012]. See *Abies*.
Pestalopeziza tsugae: on 2 BC [514].
Phaeocryptopus sp.: on 2 BC [1012].
Phaeolus schweinitzii (Polyporus s.) (C:306): on 1 Ont [1341], Que [1377]; on 2 BC [1012]. See *Abies*.
Phanerochaete sanguinea (Peniophora s.) (C:306): on 2 BC [1012]. See *Abies*.
Phellinus ferreus (Poria f.) (C:306): on 2 BC [1012]. See *Acer*.
P. ferrugineofuscus (Poria f.) (C:306): on 2 BC [1012]. See *Abies*.
P. nigrolimitatus (Fomes n.) (C:305): on 2 BC [1012]. See *Abies*.
P. pini (Fomes p.) (C:305): on 1 Ont [1341], Que [1377]; on 2 BC [1341, F76]; on 2, 3 BC [1012]. See *Abies*.
P. punctatus (Poria p.): on 1 Ont and on 2 BC [1341]; on 2, 3 BC [1012]. See *Abies*.
P. repandus (Fomes r.): on 2 BC [1012, F67]. See *Pseudotsuga*.
P. robustus (Fomes r.) (C:305): white spongy rot, carie blanche spongieuse: on 2 BC [1012]. See *Abies*.
P. tsugina (Poria t.) (C:307): on 1 Que [1377]. See *Abies*.
P. viticola (Fomes v.): on 2 BC [1012]. See *Abies*.
P. weirii (Poria w.) (C:307): yellow ring rot, carie jaune annelée: on 2 BC [1012, 1912, 1915]. See *Abies*.
Phialocephala fusca (C:306): on 2 BC [877, 1012].
Phialophora melinii: on 1 Ont and on 2 BC [279], reported from NB but the DAOM specimen is from BC.
Phlebia centrifuga (P. albida auct.*)* (C:306): on 2 BC [1012]. See *Abies*.

P. georgica (Rogersella eburnea): on 4 NS [740].
P. livida: on 2 BC [1012]. See *Abies*.
P. phlebioides (Peniophora p.): on 2 BC [1012]. See *Abies*.
P. radiata (C:306): on 2 BC [568, 1012]. See *Abies*.
P. segregata (Peniophora livida): on 2 BC [1012]. See *Picea*.
Phlebiopsis gigantea (Peniophora g.): on 2 BC [1012]. See *Abies*.
Pholiota alnicola (Flammula a.) (C:305): on 2 BC [387, 1012, F60]. See *Abies*.
P. aurivella (C:306): on 2 BC [1012]. See *Abies*.
Phoma sp.: on 2 BC [1012].
Phomopsis sp.: on 2 BC [1012].
Phymatotrichum fungicola: on 2 BC [507].
Piloderma bicolor (Corticium b.) (C:305): on 2 BC [1012]; on 4 Ont [1341]. See *Abies*.
Pluteus admirabilis: on 4 Ont [1306]. See *Acer*.
Polyporus badius (P. picipes) (C:306): on 1 Ont [1341]; on 2 BC [1012]. See *Abies*.
P. varius (P. elegans) (C:306): on 4 NS [1032]. See *Abies*.
Poria rivulosa: on 2 BC [1012]. See *Abies*.
P. subacida (C:307): on 1 Ont [1341], Que [1377], NS [1032]; on 2 BC [1012, 1144]. See *Abies*.
P. zonata (C:307): on 2 BC [554]. See *Pseudotsuga*.
Pseudohydnum gelatinosum (C:307): on 1 Ont [1341]; on 2 BC [1012]. See *Picea*.
Pseudomerulius aureus (Merulius a.): on 2 BC [568, 1012]. See *Pinus*.
Pseudoplectania nigrella: on 4 Ont [1340].
Pterula sp.: on 2 BC [1012].
Pucciniastrum vaccinii (C:307): needle rust, rouille des aiguilles: 0 I on 1 Ont [1236, 1341, F63, F68], NB [F64, F68], NB, NS [1032, F69], NS [F63, F70]; on 1, 2, 3 BC [1012]; on 2 BC [F76]; on 4 Que [F60], NS [1032].
Pycnoporellus alboluteus (Polyporus a.) (C:306): on 2 BC [1012]. See *Abies*.
P. fulgens (Polyporus fibrillosus) (C:306): on 2 BC [1012, 1341]. See *Abies*.
Ramaricium albo-ochraceum: on 1 Ont [572]. See *Abies*.
Resinicium bicolor (Odontia b.) (C:306): on 2 BC [1012, 1144]; on 4 NS [1032]. See *Abies*.
Retinocyclus abietis: on 1 Que [457]; on 2 infected with dwarf mistletoe BC [51].
R. olivaceus (C:307): on 2 BC [1012, F61].
Rhizina undulata (C:307) : on 2 BC [1012]. See *Picea*.
Rhizothyrium abietis (holomorph *Rhizocalyx a.)* (C:307): on 1 Ont [363].
Rigidoporus nigrescens (Poria n.) (C:307): white spongy rot, carie blanche spongieuse: on 2 BC [1012]. See *Acer*.

Sageria tsugae (anamorph *Ascoconidium t.*):   on 2 BC [509, 1012, F75].

Schizophyllum commune: on 1 Ont [1341]. See *Abies*.

Sclerophoma sp.:   on 2 BC [1012].

Scytinostroma galactinum: on 2 BC [1012]; on 4 NB [1032]. See *Abies*.

S. ochroleucum (C:307):   on 2 BC [1012]. See *Salix*.

Serpula himantioides (*S. lacrimans* var. *h.*):   on 1 NS [1032]; on 2 BC [1012]. See *Abies*.

Sirococcus conigenus (*S. strobilinus*):   on 2 BC [505, 1012, F73, F74, F75, F76].

Sirodothis sp.:   on 2 BC [1012].

Sistotrema brinkmannii: on 2 BC [1012, 1144]. See *Abies*.

S. raduloides: on 2 BC [1012].   See *Abies*.

Skeletocutis amorpha (*Polyporus a.*) (C:306):   on 2 BC [1012]. See *Abies*.

S. subincarnata (*Poria s.*) (C:307):   on 2 BC [1012]. See *Abies*.

Sorocybe resinae: on 2 BC [1012].

Sphaeropsis sapinea (*Macrophoma s.*):   on 1 Que [1687] by inoculation.

Steccherinum sp.:   on 2 BC [1012].

Stereum ostrea (C:307):   on 2 BC [1012]. See *Abies*.

Stomiopeltis sp.:   on 2 BC [1012].

Subulicium lautum: on 1 Ont [744].

Sydowia polyspora (anamorph *Sclerophoma pithyophila*):   on 1 Que [1686]; on 2 BC [1012, F70].

Thelephora terrestris (C:307):   on 2 BC [1012]. See *Pinus*.

Therrya tsugae: on 2 BC [519].

Thysanophora longispora: on 1 Ont, Que [872].

T. penicillioides: on 1 Ont [872].

Tomentella cinerascens: on 1 Ont [1341].

T. griseoviolacea: on 1 Ont [929, 1341, 1828].

T. olivascens: on 1 Ont [1341].

T. sublilacina: on 2 BC [1012, 1341]; on 4 Ont [1828].

T. terrestris (*T. badiofusca*, *T. umbrinella*):   on 1 Ont [1341, 1827, 1828].

Trechispora farinacea (*Cristella f.*):   on 4 Ont [1341]. See *Abies*.

T. mollusca (*T. onusta*):   on 1 Que [970]; on 2 BC [1012]. See *Abies*.

Trichaptum abietinus (*Hirschioporus a.*, *Polyporus a.*) (C:306):   white sap rot, carie blanche de l'aubier: on 1 Ont [1341], Que [1028]; on 1, 4 NB, NS [1032]; on 2 BC [1012]. See *Abies*.

T. fuscoviolaceus (*Hirschioporus f.*):   on 1 Que [1028]. See *Abies*.

Trichoderma sp.:   on 2 BC [1144].

Tricholomopsis decora: on 1 Que [1385]. See *Picea*.

Trichosporium sp.:   on 2 BC [1012]. The generic name has been declared illegitimate, the specimen should be reexamined.

Truncatella angustata (*T. truncata*):   on 2 infected with dwarf mistletoe BC [51].

Truncocolumella rubra (C:307):   on 2 BC [1012].

Tubulicrinis glebulosus (*T. gracillimus*):   on 2 BC [1012]. See *Abies*.

Tympanis sp.:   on 1 Que [457].

T. abietina: on 1 Que [1688] by inoculation.

T. confusa: on 1 Que [1688] by inoculation.

T. hansbroughiana: on 1 Que [1688] by inoculation.

T. hypopodia (*T. piceina*):   on 1 Que [1221], 1688 by inoculation].

T. laricina (anamorph *Sirodothis* sp.):   on 1 Que [1688] by inoculation.

T. prunicola: on 1 Ont [1221].

T. truncatula: on 1 Que [1688] by inoculation.

T. tsugae (C:307):   on 1 Ont, Que [1221], Que [1688, F68].

Typhula sp.:   on 2 BC [501].

T. abietina: on 2 BC [1012].

Tyromyces balsameus (*Polyporus b.*) (C:306):   on 1 Que [1377]; on 2 BC [1012]. See *Abies*.

T. borealis (*Polyporus b.*) (C:306):   on 1 NS [1032]. See *Abies*.

T. caesius (*Polyporus c.*) (C:306):   on 2 BC [1012]; on 4 Ont [1827]. See *Abies*.

T. fragilis (*Polyporus f.*) (C:306):   on 2 BC [1012]. See *Abies*.

T. guttulatus (*Polyporus g.*) (C:306):   on 2 BC [1012]. See *Abies*.

T. mollis (*Polyporus m.*):   on 2 BC [1012]. See *Abies*.

T. placenta (*Poria p.*, *P. monticola*) (C:307):   on 2 BC [1012]. See *Pinus*.

T. sericeomollis (*Polyporus s.*) (C:307):   on 2 BC [1012]. See *Abies*.

T. tephroleucus (*Polyporus t.*) (C:306):   on 2 BC [1012]. See *Alnus*.

T. undosus (*Polyporus u.*) (C:306):   on 2 BC [1012]. See *Abies*.

Uraecium holwayi (holomorph *Pucciniastrum vaccinni*) (C:307):   needle rust, rouille des aiguilles: 0 I on 1 BC [F64]; on 1, 2, 3 BC [2008]; on 2 BC [F63]; on 2, 3 BC [728]. Causes discoloration and premature shedding of hemlock needles.

Valsa sp.:   on 2 BC [193].

V. abietis (C:307):   canker, chancre cytosporéen: on 1 Que [1689]; on 2 BC [193, 1012].

V. friesii (anamorph *Cytospora dubyi*):   on 1 Que [1689] by inocluation.

V. pini: on 1 Que [1689] by inoculation.

Vararia granulosa (*Grandinia g.*) (C:305, 307):   on 2 BC [1012]. See *Abies*.

V. pallescens (C:307):   on 2 BC [1012]. See *Acer*.

Verticillium sp. (C:307):   on 2 BC [1012].

Vesiculomyces citrinum (*Gloeocystidiellum c.*):   on 4 NS [1032]. See *Abies*.

Wolfiporia extensa (*Poria cocos*) (C:306):   on 2 BC [1012]. See *Betula*.

Xenasmatella inopinata: on 1 Ont [744]. See *Pseudotsuga*.

Xenomeris abietis: on 2 BC [525, 1012, F69].

Xeromphalina campanella (C:307): on 2 BC
    [1012]. See *Abies*.

TULIPA L.   LILIACEAE

1. *T. gesnerana* L., tulip, tulipe;
    imported from e. Europe and Asia Minor.

Armillaria mellea (C:307): dry rot,
    pourridié-agaric: on 1 BC [1816].
Botrytis sp.: on 1 BC [1816].
B. cinerea (holomorph *Botryotinia*
    fuckeliana) (C:307): gray mold,
    moisissure grise: on 1 BC [334, 1816],
    NS [334].
B. tulipae (C:307): fire, feu: on 1 BC
    [337, 1301, 1816], BC, NS [333, 336,
    338], Alta [335, 339], Ont [336], Que,
    NS [335, 337], NB [333, 1194], Nfld
    [338]. Extended periods of moist
    weather favor disease development.
Penicillium sp. (C:308): bulb rot,
    pourriture du bulbe: on 1 BC [1816].
Phytophthora sp.: on 1 BC [1816].
Pythium ultimum (C:308): root and bulb
    rot, pourridié pythien: on 1 BC [1816].
Rhizoctonia solani (holomorph
    Thanatephorus cucumeris) (C:308):
    rhizoctonia, rhizoctonie: on 1 BC
    [1816].
Sclerotinia sclerotiorum (C:308): stem
    rot, pourridié sclérotique: on 1 BC
    [1816].
Sclerotium tuliparum (C:308): gray bulb
    rot, pourriture grise du bulbe: on 1 BC
    [1816] as *Sclerotinia*, Ont [333].
Typhula sp. (C:308): snow mold,
    moisissure nivéale: on 1 BC [1816].

TYPHA L.   TYPHACEAE

1. *T. angustifolia* L., narrow-leaved
    cattail, massette; native, Man-PEI.
2. *T. latifolia* L., common cattail,
    quenouille; native, Yukon-Mack, BC-Nfld.
3. Host species not named, i.e., reported
    as *Typha* sp.

Arthrinium phaeospermum: on 2 Sask [1760].
Cladosporium macrocarpum: on 3 Man [1760].
Galerina subbadipes: on rotted debris of
    3 Que [1385].
Haligena spartinae: on drift stems and
    leaves of 3 BC [782].
Lophodermium typhinum: on 2 Ont [1340].
Monodictys pelagica: on stems of 3 BC
    [782].
Paraphaeosphaeria michotii: on old leaves
    2 Ont [1621].
Phaeosphaeria punctillum: on 2 Ont [760].
Psathyrella typhae: on 1 Ont [1429].
Remispora hamata: on 3 BC [782].
Stagonospora vitensis: on 3 BC [782].
Typhula latissima: on stems of 2 Que
    [1385].

ULMUS L.   ULMACEAE

1. *U. americana* L., white elm, orme
    blanc; native, Sask-PEI.

2. *U. glabra* Huds. (*U. montana* With.),
    Wych elm, orme de montagne; imported
    from Eurasia.
3. *U. x hollandica* Mill.: *U.*
    carpinifolia Ruppius ex Suckow x *U.*
    glabra x *U. plotii* Druce; Dutch elm.
4. *U. parvifolia* Jacq., Chinese elm, orme
    de Chine; imported from China and Japan.
5. *U. procera* Salisb., English elm, orme
    champêtre; imported from Europe.
6. *U. pumila* L., dwarf elm, orme nain;
    imported from e. Asia.
7. *U. rubra* Muhl. (*U. fulva* Michx.),
    slippery elm, orme gras; native, Ont-NB.
8. *U. thomasii* Sarg. (*U. racemosa*
    Thomas), rock elm, orme liège; native,
    Ont-Que.
9. Host species not named, i.e., reported
    as *Ulmus* sp.

Acarosporina microspora (*Schizoxylon m.*):
    on 9 Que [1249].
Acremoniella atra: on 1 Man [1760].
Aleurodiscus cerussatus (C:308): on 9 Man
    [964]. See *Acer*.
Anthostoma gastrinum (C:308): on 1 Ont
    [1340]; on 9 Ont [F63].
Anthostomella suberumpens?: on 1 Ont
    [1340].
Apiognomonia errabunda (anamorph *Discula*
    umbrinella): on 6 Man [F66]. See
    Fraxinus.
Apioporthe apiospora (C:308): on 1 Ont
    [1340, F60, F65]. Probably a species of
    Anisogramma.
Armillaria mellea (C:308): on 1 NS
    [1036]; on 9 Que [1377]. See *Abies*.
Arthrobotrys dolioformis: on 1 Man [1760].
Bjerkandera adusta (*Polyporus a.*)
    (C:309): on 1 Ont [1341]. See *Abies*.
Botryodiplodia sp.: on 1 BC [1012], Ont
    [1828].
Calcarisporium arbuscula: on 1 Man
    [1760].
Camarosporium ulmi: on dead branches of 9
    Ont [440].
Catenuloxyphium semiovatum: on leaves of
    living 9 Ont [798].
Catinella nigro-olivacea: on 9 Ont [1340].
Cephalosporium sp. (C:308): wilt,
    flétrissure: on 1 Sask, Man [F60]; on 9
    NB [F68].
Ceratocystis albida: on 1 Man [1190].
C. ulmi (C:308): Dutch elm disease,
    maladie de l'orme: on 1 Ont [645, 647,
    896, 1340, 1782, 1783, 1827, 1828, F67],
    Que [584, 1105, 1204, 1213, 1214, 1379,
    1380, 1381, 1382, 1383, 1384, 1388,
    1393, F63, F68], NB [1032], NS [1036];
    on 1, 5 Ont [F66]; on 5 Ont [339]; on 9
    Man [895, F75, F76], Ont [334, F69],
    Ont, Que, NB [335, 336, 1219, F62], Ont,
    Que, NB, NS [895, F70, F71, F72, F73,
    F74, F75, F76], Ont, NB [337, 338, 339,
    1594, 1595, F60, F61, F63, F64, F65,
    F68], Que [535, 943], NB [535, 1737,
    F66, F67], NB, NS [1032, F69]. The
    disease caused by this fungus has,
    despite efforts to control it, decimated
    the population of native elms in s. and

e. Ont and s. Que, and is now progressing through the Maritime Provinces. In s. Ont about 70% of the elm has died, a loss of about 213 million cubic metres of wood.

Cerrena unicolor (Daedalea u.) (C:309): on 1 Ont [1341], NB [1032]; on 9 BC [1012]. See *Acer*.

Chalara thielavioides (Chalaropsis t.) (C:308): black mold, moisissure noire: on 6 Que [1377].

Ciboria acericola: on 9 Ont, Que [662].

C. acerina: on 9 Ont [302, 434].

Circinotrichum maculiforme: on 1 Man [1760].

Climacodon septentrionalis (Hydnum s., Steccherinum s.): on 9 Que [1377, 1385], NB [1032]. See *Acer*.

Coriolus hirsutus (Polyporus h.) (C:309): on 1 Ont [1341], NB [1032]. See *Abies*.

C. versicolor (Polyporus v.): on 1 NB [1032]; on 9 Ont [426]. See *Abies*.

Cryptosporella hypodermia (C:309): on 1 Man [F60], Man, Ont [1340]; on 6 Ont [1828].

Cylindrobasidium evolvens (Corticium laeve): on 1 NB [1032]. See *Abies*.

Cyphellopsis anomala (Solenia a.) (C:309): on 1 Ont [1341]; on 9 NS [1032]. See *Betula*.

Cytospora sp.: on 4 Ont [338]; on 9 NB, NS [1032].

C. ambiens (holomorph *Valsa a.*) (C:309): on 1 NS [F63].

C. carbonacea: on 4 Ont [1340].

Daldinia concentrica (C:309): on 9 NS [1032].

Datronia mollis (Trametes m.) (C:310): on 9 NS [1032]. See *Acer*.

Dendrothele acerina (Aleurocorticium a.): on 9 Ont [965].

D. alliacea (Aleurocorticium a.): on 9 Ont, Que [965]. See *Acer*.

D. griseo-cana (Aleurocorticium g.): on 9 Ont [965]. See *Acer*.

D. microspora (Aleurocorticium m.): on 9 Ont [965]. See *Betula*.

Deuterophoma ulmi (Dothiorella u.) (C:309): wilt, flétrissure céphalosporienne: on 1 Sask, Man [F60], Que [F65, F66], NS [1032]; on 9 Man [F74], Que [1377].

Diatrype alboproinosa: on 1 Ont [1354]; on 9 Man, Ont, Que [1354]. On dead branches and twigs.

Diplodia melaena (C:309): on 1 Sask [F66].

Eutypella scoparia: on 9 Ont [1340].

E. stellulata: on 9 Ont [1340].

Exidia glandulosa (E. spiculosa): on 1 Ont [1341]. See *Abies*.

Flammulina velutipes (Collybia v.) (C:308): on 1 Que [1385] on stumps and dead trunks, sometimes on dead parts of live trees, NB, NS [1032]. See *Populus*.

Fomes fomentarius: on 1 NB [1032]. See *Acer*.

Fusarium sp. (C:309): on 1 Man [F72].

Ganoderma applanatum (Fomes a.) (C:309): on 1 NB [1032]; on 1, 8, 9 Ont [1341];

on 9 Que [1377], PEI [1032]. See *Abies*.

G. resinaceum (G. sessile): on 1, 9 Ont [1341]. See *Acer*.

Gloeocystidiellum sp.: on 9 Ont [1341].

Gloeosporium sp.: on 2 NB [1032].

Gonatobotrys simplex: on 1 Man [1760].

Graphium penicilloides: on 1 Sask, Man [1760].

Helminthosporium velutinum: on 7 Ont [810].

Hericium coralloides (H. ramosum): on dead trunks of 9 Que [1385]. See *Acer*.

H. erinaceus: on 1 NB [1032]. See comment under *Acer*.

Heterobasidion annosum (Fomes a.): on 1 Ont [1341, 1419]. See *Abies*.

Heterochaetella bispora (C:309): on 9 Ont [1015].

H. brachyspora (C:309): on 9 Ont [1015].

H. dubia (C:309): on 9 Ont [1015]. See *Acer*.

Hyalinia breviasca: on 9 Ont [1340].

Hymenochaete sp.: on 1 Ont [1341].

H. tabacina (C:309): on 1 Ont [1827]; on 9 NS [1032]. See *Abies*.

Hyphoderma heterocystidium (Peniophora h.): on 9 Ont [1341]. See *Acer*.

Hypochnicium vellereum (Corticium v.) (C:309): on 9 NS [1032]. See *Acer*.

Hypocreopsis lichenoides (anamorph *Stromatocrea cerebriforme*): on 1 Que [230].

Hypoxylon deustum: on 1 Ont [1340].

H. papillatum: on 9 Ont [1340].

Hypsizygus ulmarius (H. tessulatus, Pleurotus t., P. ulmarius) (C:309): on 1 Ont [1341]; on 9 Que [1377, 1385], NB [1032, 1036]. See *Acer*.

Ischnoderma resinosum (Polyporus r.) (C:309): on 1 Ont [1341]. See *Abies*.

Laetiporus sulphureus (Polyporus s.): on living trunk of 1 Que [1389]. See *Abies*.

Lasiosphaeria ovina: on 9 Ont [1340].

Lenzites betulina: on 1 NB [1032]. See *Acer*.

Leucogyrophana mollusca: on 1 Que [571]. See *Abies*.

Lopharia cinerascens (Stereum c.): on 1 Ont [1341], Que [F70].

Massaria zanthoxyli: on 9 Ont [80].

Meruliopsis corium: on 1 Ont [568]. See *Acer*.

Mycena algeriensis: on trunks of 9 Ont [1385]. See *Alnus*.

M. roseipallens: on debris of 9 Ont [1385]. See *Alnus*.

Mycoacia aurea (M. stenodon): on 9 Ont [1341]. See *Alnus*.

Mycosphaerella ulmi (anamorph *Phloeospora u.*) (C:309): leaf spot, tache des feuilles: on 1 NS [F61]; on 1, 9 NS [1032].

Nectria canadensis (Calonectria c.) (anamorph *Cilicipodium grayanum*): on 9 Ont [1495].

N. cinnabarina (anamorph *Tubercularia vulgaris*) (C:309): coral spot, dépérissement nectrien: on 1, 4, 9

NWT-Alta [1063]; on 1, 6 NB, NS [1032];
on 2, 9 PEI [1032]; on 4 Nfld [335,
1659, 1660, 1661, F61]; on 4, 9 NB
[1032]; on 6 Ont [1340, F61], Ont, NB
[333], Que [335, 339, 1377], Nfld [336,
338, F62, F63, F65]; on 9 Ont [1828],
Nfld [F64].
N. galligena (C:309): nectria canker,
chancre nectrien: on 1 Sask [F63], Que
[F60].
Orbilia luteorubella: on 9 Ont [1340].
Panellus serotinus (Pleurotus s.): on 9
NB [1032]. See *Acer*.
Panus conchatus (P. torulosus) (C:309):
on 9 NS [1032]. See *Alnus*.
Peniophora sp.: on 1 NB [1032].
P. aurantiaca: on 9 Ont [1341]. See
Alnus.
P. cinerea (C:309): on 1 Ont [1341], NB
[1032]; on 9 BC [1012]. See *Abies*.
Peziza varia: on rotting branches of 1
Que [574].
*Phaeostalagmus cyclosporus (Verticillium
c.)*: on 1 Man [F68].
Phellinus conchatus (Fomes c.): on 1 Ont
[1341]. See *Acer*.
P. ferreus (Poria f.) (C:309): on 1 NS
[1032]. See *Acer*.
P. igniarius (Fomes i.) (C:309): on 1, 9
Ont [1341]; on 9 Que [1377], NB [1032].
See *Acer*.
P. laevigatus (Poria l.): on 1 Ont
[1341]. See *Acer*.
Phialophora sp.: on 1 Ont [1827].
Phlebiopsis roumeguerii (Phlebia r.): on
1 Que [F67].
Phloeospora ulmi (holomorph
Mycosphaerella u.): on 1, 9 NS
[1032]; on 2 Nfld [1661]; on 9 Ont
[1767].
Phoma sp.: on 1 Ont [973, 1828].
Phyllosticta melaleuca: on 1 Que [335].
Pithomyces chartarum: on 1 Man [1760].
Pleurotus ostreatus: on 1 NB [1036]; on
1, 9 Ont [1341]; on 9 Que [1377]. See
Abies.
P. sapidus: on 1 NB [1032]. See
Aesculus.
Podosphaera kunzei: powdery mildew,
blanc: on leaves of seedlings of 9 Ont
[437]. Probably a specimen of *Uncinula
macrospora* which is represented in DAOM
by specimens from Hull, Que (142864) and
Belleville, Ont (142865).
Polyporus sp.: on 1 Ont [1827].
P. squamosus (C:309): on 1 NB [1032]; on
1, 9 Ont [1341]; on 9 Que [1377], NS
[1032]. During the 1970s and early
1980s this fungus has been rather common
on large, dead elms killed 3-5 years
earlier by Dutch Elm Disease. See
Acer.
Poronidulus conchifer (Polyporus c.)
(C:309): on 1, 9 Ont [1341], NB [1032];
on 9 NS [1032]. See *Populus*.
Pseudotomentella tristis (P. umbrina): on
9 Ont [1341].
*Punctularia strigosozonata (Phaeophlebia
s.)*: on 9 NS [1032]. See *Alnus*.
Pyrenochaeta sp.: on 3 BC [1012].

Rosellinia compressa: on decorticated
trunk of 9 Ont [440].
R. pulveracea: on 9 Ont [1340].
Sphaeropsis ulmicola (C:309): twig
canker, chancre des rameaux: on 9 NB
[1032].
Spongipellis unicolor (Polyporus obtusus)
(C:309): on 1 Ont [1341]. See
Quercus.
Sporidesmium brachypus: on 9 Ont [822].
S. folliculatum: on 9 Ont [825].
Stegophora ulmea (Gnomonia u.) (C:309):
leaf spot, tache des feuilles: on 1
Sask, Man [F68], Man [F63, F67, F70],
Ont [F61], Que [F72, F73], NB [F63,
336], NB, NS [F64], NB, NS, PEI [F65,
F66]; on 1, 6 Que [1377]; on 1, 6, 8, 9
Ont [1340]; on 1, 7, 8 Que [1374]; on 1,
9 NB, NS [1032]; on 4 Ont [335, 338,
339], Que [336]; on 6 Sask [337, F73],
Man [F66], Ont [336], NB [1032]; on 9
Alta [336], Man [1595, F69], Ont [79],
Ont, Que [334], PEI [1032], Nfld [1661].
Stereum hirsutum: on 9 Ont [1341]. See
Abies.
*Stigmina compacta (Sciniatosporium c.,
Thyrostroma c.)* (C:309): twig blight,
brûlure des rameaux: on 9 Que [1127,
1377].
S. pulvinata: on 4 Ont [1340]; on 6 Ont
[1827].
Stromatocrea cerebriforme (holomorph
Hypocreopsis lichenoides): on 1 Que
[230].
Taphrina ulmi (C:309): leaf blister,
cloque des feuilles: on 9 Que [1377].
Thielaviopsis basicola: on 1 Man [1760].
Trechispora farinacea (Cristella f.): on
9 Ont [1341]. See *Abies*.
Trichoderma viride: on 1 Man [1760]. See
Abies.
Trichothecium roseum: on 1 Sask, Man
[1760]; on 4, 6 NB [1032]; on 6 NS
[1032].
Tubercularia sp.: on 4 NB [1032]; on 9 BC
[1012].
T. nigricans: on 4 Ont [1340, 1594]; on 6
Ont [F67].
T. ulmea (C:310): coral spot, brûlure des
rameaux: on 6 Alta, Que [337], Sask
[F67], Que [334, F66]; on 9 Que [338],
Nfld [337].
T. vulgaris (holomorph *Nectria
cinnabarina*) (C:310): on 1 Alta [F61];
on 6 Ont [F70]; on 9 Alta [F64], Ont
[1340].
Uncinula macrospora: powdery mildew,
blanc: on 9 Ont, Que [1257].
Valsa ambiens (anamorph *Cytospora a.*)
(C:310): on 9 NS [1032].
Verticillium sp. (C:310): wilt,
flétrissure verticillienne: on 1 Man
[F72]; on 9 Que [1377], NB [1032].
V. albo-atrum: on 1 Sask, Man [F68], Man
[1760]; on 9 Man [1594, 1595], Ont [F62].
V. dahliae: on 1 Sask, Man [F68], Man
[1760, F67]; on 9 Man [1594, 1595].
V. lecanii: on 1 NB [1036].
Volvariella bombycina: on 9 Que [1385] on
wounds. See *Acer*.

Xylaria polymorpha (C:310): on 1 Ont
    [1340]. Typically on buried wood.
Xylohypha lignicola: on 1 Man [1760].

URTICA L.   URTICACEAE

1. *U. dioica* L., stinging nettle, ortie
    piquante; native, Yukon-Mack, BC-PEI,
    Labr. 1a. *U. d.* ssp. *gracilis*
    (Ait.) Seland.; native,
    transcontinental. 1b. *U. d.* ssp. *g.*
    var. *gracilis* (*U. gracilis* Ait.);
    native, transcontinental. 1c. *U. d.*
    ssp. *g.* var. *lyallii* (Wats.) Hitchc.
    (*U. lyallii* Wats.); native, BC-Alta.
    1d. *U. d.* ssp. *g.* var. *procera*
    (Muhl.) Weddell (*U. procera* Muhl.);
    native, Sask-NS.
2. *U. urens* L., burning nettle, ortie
    brûlante; native, BC-Alta, Man-Nfld.
3. Host species not named, i.e., reported
    as *Urtica* sp.

Coprinus sp.: on 1 Man [1820]. Possibly
    C. psychromorbidus.
Dendryphion comosum: on dead stems of 3
    BC [818].
Lachnella villosa: on fallen stems of 3
    BC [1424].
Leptosphaeria acuta (C:310): on 1c BC
    [1012, 1816].
Ophiobolus erythrosporus: on 1d Ont
    [1620].
Puccinia caricina (*P. caricis-urticata*)
    (C:310): rust, rouille: 0 I on 1b, 1c,
    1d, 3 Ont [1236]; on 1c BC [1012, 1816];
    on 2 NWT-Alta [1063].
Ramularia urticae (C:310): on 1c BC
    [1012, 1816].
Sclerotinia sclerotiorum (C:310): on 1b
    Sask [1140].
Typhula erythropus: on 1c BC [901].

UVULARIA L.   LILIACEAE

1. *U. grandiflora* Sm., large-flowered
    bellwort, uvulaire à grandes feuilles;
    native, Ont-Que.
2. *U. perfoliata* L., perfoliate bellwort,
    uvulaire perfoliée; native, Ont.

Botrytis cinerea (holomorph *Botryotinia*
    fuckeliana) (C:310): gray mold,
    moisissure grise: on 2 Que [1545].
Macrophoma smilacina var. *smilacina*: on
    1 Que [148].

VACCINIUM L.   ERICACEAE

1. *V. alaskaense* Howell; native, BC.
2. *V. angustifolium* Ait., lowbush
    blueberry, bleuet nain; native,
    Man-Nfld, Labr. 2a. *V. a.* var.
    laevifolium House (*V. pennsylvanicum*
    Lam.); native, Man-NS, Nfld. 2b. *V.*
    a. var. *nigrum* (Wood) Dole; native,
    Man-NS, Nfld.
3. *V. caespitosum* Michx., dwarf bilberry,
    bleuet maganés; native, Yukon-Mack,
    BC-NS, Nfld.

4. *V. corymbosum* L., highbush blueberry,
    bleuet en corymbes; native, Ont-NS.
5. *V. macrocarpon* Ait., large cranberry,
    grande canneberge; native, Ont-Nfld.
6. *V. membranaceum* Dougl., mountain
    huckleberry, airelle à feuilles
    membraneuses; native, Mack, BC-Alta, Ont.
7. *V. myrtilloides* Michx. (*V. canadense*
    Kalm), sour-top blueberry, airelle
    fausse-myrtille; native, Mack, BC-Nfld,
    Labr.
8. *V. myrtillus* L., low bilberry,
    myrtille; native, BC-Alta.
9. *V. nubigenum* Fern., born of the
    clouds; e. Que, n. Nfld.
10. *V. ovalifolium* Sm., mathers, airelle
    bleue; native, Yukon, BC, Que, NS, Nfld,
    Labr.
11. *V. ovatum* Pursh, box blueberry, bleuet
    ovale; native, BC.
12. *V. oxycoccos* L., small cranberry,
    grisettes; native, Yukon-Keew, BC-Nfld,
    Labr. 12a. *V. o.* var. *intermedium*
    Gray (*Oxycoccos quadripetalus* Gilib.);
    native, Mack, BC, Man-Nfld, Labr. 12b.
    V. o. var. *o.* (*Oxycoccus*
    microcarpus Turcz.); native,
    Yukon-Keew, BC-Que, Labr.
13. *V. parvifolium* Sm., red bilberry,
    airelle rouge; native, BC.
14. *V. scoparium* Leib., grouseberry,
    airelle à tige mince; native, BC.
15. *V. uliginosum* L., bog bilberry, bleuet
    traînard; native, Yukon-Labr, BC-Nfld.
16. *V. vitis-idaea* L., bog cranberry,
    airelle vigne d'Ida; native, Yukon-Labr,
    BC-Nfld. (including *V. v.-i.* var.
    minis Lodd.)
17. Host species not named, i.e., reported
    as *Vaccinium* sp.

Acremonium strictum: on 2 Que [188].
Aleurodiscus aurantius (C:310): on 17 BC
    [964].
Alternaria alternata: on 2 Que [188]; on
    5 NS [601].
A. tenuissima: on 2 Que [188].
Apiospora montagnei: on leaves of 5 NS
    [601].
Arthrinium phaeospermum: from rotted
    fruit of 5 NS [601].
Aspergillus niger: on 2 Que [188].
Aureobasidium pullulans (C:311): on 2 Que
    [188]; on 5 NS [996, 601 from leaves and
    rotted fruit].
Bifusepta tehonii (C:311): on 2 Que and
    on 2, 7 Ont [361].
Botrytis cinerea (holomorph *Botryotinia*
    fuckeliana) (C:311): gray mold,
    moisissure grise: on 2 Que [188]; on 2,
    4 NS [335]; on 4 BC [1595], NS [333,
    1594]; on 5 NS [601, 607]; on 17 NB
    [336, 1594].
Cephaloascus albidus: isolated from
    cranberry pumace of 12 NS [907].
Ceuthospora lunata (C:311): from rotted
    fruit of 5 NS [601, 607].
Chaetomium brevipilium: from rotted fruit
    on 5 NS [601].
C. fusisporale: on 2 Que [188].

C. globosum :   on 2 Que [188].

C. spirale :   from rotted fruit of 5 NS [601].

Chrysomyxa sp.:   rust, rouille: on 16 Ont [1341].

C. ledi var. *vaccinii* (C:311):   rust, rouille: II III on 13 BC [1012, 1816, 2008].

Cladosporium sp.:   on 13 BC [1012].

C. herbarum :   on 2 Que [188].

Cochliobolus sativus (anamorph *Bipolaris sorokiniana*):   on leaves of 5 NS [601].

Coniothyrium sp.:   on 2 Que [188].

Curvularia clavata :   from rotted fruit of 5 NS [601].

Cylindrocarpon destructans :   on 2 Que [188].

Dasyscyphus virginellus :   on 13 BC [1012, 1816].

Diaporthe vaccinii (anamorph *Phomopsis v.*) (C:311):   canker, chancre phomopsien: on 2 Que [188]; on 4 NS [334]; on 5 NS [550, 607, 996, 1595].

Discohainesia oenotherae (Pezizella o.) :   from rotted fruit of 5 NS [601].

Dwayalomella vaccinii :   on 2 Que [189].

Eupenicillium brefeldianum :   on 2 Que [188].

E. lapidosum :   on 2 Que [188].

Exobasidium sp.:   on 6 NWT-Alta [1063].

E. cordilleranum :   on 10 BC [1816].

E. cordilleranum var. *cordilleranum* (C:311):   on 10 BC [1012].

E. cordilleranum var. *minor* (C:311):   on 6 BC [1012, 1816].

E. parvifolii (C:311):   on 10 BC [1816]; on 10, 13 BC [1012].

E. vaccinii (C:311):   red leaf, rouge: on 2 Que [337], NB [1594], NB, NS, Nfld [339], NS [338]; on 5 BC [339]; on 5, 10, 14 BC [1816]; on 17 BC, Nfld [333], Ont [1341], Que [1385], NB, Nfld [334], Nfld [336, 337, 338]. Nannfeldt [1170] has reassessed the European species of *Exobasidium* and refers to some Canadian collections, but redisposition of the above literature records must be made following study of the specimens.

E. vaccinii var. *vaccinii* (C:311):   on 13, 14, 16 BC [1012]; on 15 Yukon [460]; on 16 Yukon [460] as DAVFP 14696 at DAOM, BC [1816].

E. vaccinii-uliginosi (C:311):   on 3 BC [1012, 1816]; on 15 Yukon [460]; on 17 Que [1385].   See comment under *E. vaccinii* .

E. vaccinii-uliginosi var. *vaccinii-uliginosi* (C:311):   on 15 Frank [1244].

Fusarium oxysporum :   on 2 Que [188].

F. solani (holomorph *Nectria haematococca*):   on 2 Que [188].

Fusicladium sp.:   on 2 Que [188].

Fusicoccum putrefaciens (holomorph *Godronia cassandrae*) (C:311):   on 2 Que [188]; on 4 Que [333], NS [214, 335, 337, 983, 987, 991]; on 5 NB [337]; on 17 BC, NS [334], NS [336], NS, Nfld [333].

Gelasinospora reticulospora :   on 2 Que [188].

G. tetrasperma :   on 2 Que [188].

Gibbera elegantula :   on 10 BC [1012]; on overwintered twigs and leaves of 17 BC [73].

G. myrtilli (C:311):   on 10 Que [70]; on 17 BC, Que [73].

Gliocladium roseum :   on 2 Que [188].

Gloeosporium minus (C:311):   from rotted fruit of 5 NS [601].

Glomerella cingulata (anamorph *Colletotrichum gloeosporioides*) (C:311):   anthracnose, anthracnose: on 5 NS [607].

G. cingulata-vaccinii :   from decayed fruit of 5 NS [996].

Godronia cassandrae (anamorph *Fusicoccum putrefaciens*) (C:311):   canker, chancre: on 2, 4, 7 Que [1678]; on 4 BC [1816]; on 17 BC [1817, 1818].

G. cassandrae f. *vaccinii* (C:311):   canker, chancre: on 2 NS [988]; on 2, 4, 7 Que [1678]; on 2, 4, 7, 9, 15 Que [1684]; on 4 BC [339, 656], BC, NS [338], NS [983]; on 5 NS [607, 601 isolated from leaves, stored and rotted fruit, 996, 1594]; on 17 BC [1594], BC, Ont, Que, NS [656].

G. cassandrae var. *vaccinii* :   on 2 Que [188].

G. urceoliformis (C:311):   on 2, 4, 7, 15 Que [1684] by inoculation; on 15 Labr [656].

Guignardia vaccinii (C:311):   on 5 NS [601, 607, 996, 1594, 1595]; on 17 NS [336].

Helminthosporium sp.:   on 13 BC [1012, 1816].

Hymenochaete tabacina (C:311):   on 17 BC [1012].

Lophodermium maculare (C:311):   on 15 Frank [1543]; on 17 NWT-Alta [1063].

L. melaleucum (C:311):   on 11 BC [1012].

L. melaleucum var. *melaleucum* :   on 11 BC [1816].

Meliola sp.:   on 13 BC [1012].

Meliola nidulans (C:311):   on 16 Que [70].

Microsphaera penicillata :   powdery mildew, blanc: on 17 BC [333].   Probably a specimen of *M. vaccinii* .

M. vaccinii (*M. penicillata* var. *v.*) (C:311):   powdery mildew, blanc: on 2 Que [337], NS [336, 338]; on 2, 4, 7 Ont [1257]; on 4 NS [337, 1594]; on 5, 13 BC [1816]; on 7 Que [1257]; on 13 BC [1012]; on 17 Nfld [333].

Microsphaeriopsis olivacea (Coniothyrium o.) :   from rotted fruit of 5 NS [601].

Monilia sp.:   on 13 BC [1012, 1816].

Monilinia oxycocci (C:311):   hard rot, pourriture sclérotique: on 5 BC [1816]; on 17 BC [333, 334].

M. vaccinii-corymbosi (Sclerotinia v.) (C:311):   twig and blossom blight, pourriture sclérotique: on 2 Que [188], Que, NS [337], NB [335]; on 2a, 2b, 7 NB, NS [976]; on 4 BC [337, 338, 1302, 1303, 1304, 1595, 1816], BC, NS [339, 1594]; on 4, 17 NS [334]; on 5 NS [996]; on 17 BC [1817, 1818], NB, NS [333].

Mortierella nana: on 2 Que [188].
Mucor sp.: on leaves of 5 NS [601].
M. hiemalis: on 2 Que [188].
M. ?luteus: on 2 Que [188].
Mycosphaerella minor (C:311): on 10 Que [70].
M. vacciniicola (M. nigro-maculans): on 5 BC [339].
M. vaccinii (C:311): on 15 Frank [1543].
Myrothecium sp.: from rotted fruit of 5 NS [601].
Nipterella parksii (Belonidium p.) (C:311): on 17 BC [1012].
Oidiodendron ?griseum: on 2 Que [188].
Paecilomyces marquandii: on 2 Que [188].
P. varioti: on 2 Que [188].
Papulaspora anomala: on leaves of 5 NS [601].
Penicillium sp. (C:311): on 5 NS [601, 607, 996]; on 17 NS [336].
P. chrysogenum (P. notatum): on 2 Que [188].
P. thomii: from stored and rotted fruit of 5 NS [601].
P. variabile: from stored fruit of 5 NS [601].
Peniophora incarnata (C:311): on 17 BC [1012].
Pestalotia vaccinii (C:311): leaf spot and fruit rot, pestalotiose: on 5 NS [607].
Pestalotiopsis sydowiana (Pestalotia s.): on 2 Que [188]; on stored fruit of 5 NS [601].
Phoma glomerata: on 2 Que [188].
Phomatospora clarae-bonae (Plectosphaera c.) (C:312): on 16 Que [70, 74 on dead leaves].
Phomopsis vaccinii (holomorph *Diaporthe v.*) (C:312): dieback brûlure phomopsienne: on 2 Que [188]; on 4 NS [339].
Phyllosticta putrefaciens: from rotted fruit of 5 NS [601].
P. vaccinii: on 4 BC [339].
Physalospora vaccinii (Acanthorhynchus v.) (C:311): blotch rot, pourriture tachetée: on 2 Que [188]; on 5 NS [601, 607, 996, 1594]; on 17 NS [336].
P. vitis-idaeae (C:312): on 16 Que [70, 74 on dead leaves].
Phytophthora cinnamomi: on 17 Ont [1497] by inoculation.
Pithoascus intermedius: on 2 NS [609].
Podosphaera myrtillina (C:312): powdery mildew, blanc: on 6 Alta [1257]; on 15 NWT, Man, Que [1257]. Conners has this name as a synonym of *P. clandestina* but see Parmelee [1257].
Pseudomassaria leucothoes var. *borealis*: on 16 Que, Nfld [71].
Pseudophacidium ledi: on 2 Que [1685].
Pucciniastrum goeppertianum (C:312): witches'-broom rust, rouille-balai des sorcière: on 2 NB [339], NS [1036], Nfld [1594]; on 2, 7 NB [1859, 1861, 1862], NB, NS, PEI [1857]; on 2a, 7 Que [1377]; on 2a, 16, 17 Ont [1236]; on 3 Nfld [1659, 1660, 1661]; on 3, 14, 16, 17 NWT-Alta [1063]; on 3, 17 Alta [F65];

on 4 NS [333, 335]; on 6, 10, 11, 13, 14, 16 BC [1012, 1816]; on 7 NB [1036]; on 8 BC [953], Alta [737]; on 8, 14 BC [954]; on 8, 14, 17 Alta [951, 952]; on 10, 13 BC [1341]; on 12, 16 Que [282]; on 16 Yukon [460, F62], Mack [F66], Mack, Alta [53], Que [1341]; on 17 Yukon, Mack, BC, Alta, Sask, Man [2008], BC [644], Alta [53, 950, F66], Que [1392], NB [333, 334, 335, 336, F73], NB, NS [1032, 1594], PEI [1032], Nfld [333, 334, 337, 338, 339].
P. vaccinii (P. myrtilli) (C:312): leaf rust, rouille de la pruche: on 2 NS [335]; on 2a, 4, 17 Ont [1236]; on 3 Alta [F65]; on 3, 6, 7, 13, 14, 16 BC [1012, 1816]; on 3, 6, 7, 16, 17 NWT-Alta [1063]; on 4 NS [337]; on 10, 12a BC [1012]; on 16 Yukon [460], Mack [53], Que [1341]; on 17 Yukon, BC [728], Mack [F64], Alta [F63], Ont [1341], Que [F60], NS [334].
Ramularia effusa (C:312): leaf spot, tache ramularienne: on 17 Nfld [1659, 1660, 1661].
R. vaccinii: on 17 Ont [1340].
Rhizoctonia solani (holomorph *Thanatephorus cucumeris*): on 2 Que [188].
Rhizopus nigricans: on 2 Que [188].
R. sexualis: on 2 Que [188].
Rhytisma vaccinii (C:312): on 17 NB [1032].
Sclerotinia oxycocci auct.: on 5 BC [1595, 1817, 1818]. *S. oxycocci* Wor. is an obligate synonym of *Monilinia oxycocci*.
Seimatosporium lichenicola (Coryneum microsticta): on 2 Que [188].
Septoria sp.: on 17 NS [339].
Sordaria fimicola: on 2 Que [188].
Sphaeronaema pomorum: on 5 NS [1594].
Sporomega degenerans (Colpoma d.): on 17 Ont [1340].
Sporonema oxycocci (C:312): on 5 NS [601, 607, 996]; on 17 NS [336].
Stigmatea conferta (Gibbera c., G. compacta, Venturia c.) (C:311): leaf spot, tache des feuilles: on 5 BC [1816], NS [601, 984, 1594, 1595]; on 7 NWT-Alta [1063]; on 12a, 12b BC [1012]; on 17 NWT, Que, NS [73], Que [70].
Terriera cladophila (Lophodermium c.) (C:311): on 13 BC [1012, 1816].
Thelephora terrestris (C:312): on 17 BC [1012]. See *Pinus*.
Thysanophora penicillioides: from rotted fruit of 5 NS [601].
Trichoderma koningii: on 2 Que [188].
T. viride: on 2 Que [188]. See *Abies*.
Trichurus spiralis: on seeds of 17 Que [1623].
Truncatella angustata (Pestalotia truncata, Truncatella t.): on 2 Que [188]; on 5 NS [601].
Uniseta flagellifera: on 2 Que [188].
Valdensia heterodoxa (holomorph *Valdensinia h.*): on 1, 6 BC [1426]; on 2, 7 Que [1430]; on 10, 13 BC [1438].
Verticillium dahliae: on 2 Que [188].

VERATRUM L.   LILIACEAE

1. *V. viride* Ait. (*V. eschschotzii*
    Gray), white hellebore, tabac du diable;
    native, Yukon-Mack, BC-Alta, Ont-NB,
    Labr.

Cercosporella veratri (C:312): on 1 BC
    [1545, 1816].
Phyllosticta melanoplaca (C:312): on 1 BC
    [1545, 1816].
Puccinia veratri (C:312): rust, rouille:
    on 1 BC [1012, 1545, 1816].

VERBASCUM L.   SCROPHULARIACEAE

1. *V. thapsus* L., common mullein,
    bouillon blanc; imported from Eurasia,
    introduced, BC-Alta, Man-Nfld.

Ophiobolus erythrosporus: on 1 Ont [1620].
O. mathieui: on *V.* sp. Ont [1620].
O. niesslii: on *V.* sp. Ont [1620].
Ramularia variabilis: on 1 Ont, Que
    [1554].

VERBENA L.   VERBENACEAE

1. *V. hastata* L., blue vervain, verveine
    bleue; native, BC, Sask-NS.
2. *V. urticifolia* L., white vervain,
    verveine blanche; native, Sask, Ont-NB.

Erysiphe cichoracearum: powdery mildew,
    blanc: on 1 Que [1257]; on 1, 2 Ont
    [1257].
Ophiobolus anguillidus: on decaying stems
    of 1 Ont [1620].

VERNONIA Schreb.   COMPOSITAE

1. *V. altissima* Nutt.; native in s. Ont.

Coleosporium vernoniae: rust, rouille:
    on 1 Ont [1236].
Erysiphe cichoracearum: powdery mildew,
    blanc: on 1 Ont [1257].
Puccinia longipes: rust, rouille: on 1
    Ont [1259].

VERONICA L.   SCROPHULARIACEAE

1. *V. alpina* L., alpine speedwell;
    native, Yukon-Labr, BC-Alta, Que, Nfld.
    1a. *V. a.* var. *unalaschcensis* C. &
    S. (*V. wormskjoldii* Roemer &
    Schultes); native, transcontinental.
2. *V. americana* (Raf.) Schwein., American
    brooklime, véronique américaine; native,
    Yukon-Mack, BC-Nfld.
3. *V. arvensis* L., corn speedwell,
    véronique des champs; imported from
    Eurasia, introduced, BC, Ont-NS, Nfld,
    Labr.
4. *V. cusickii* Gray; native, BC.
5. *V. incana* L., woolly speedwell;
    imported from n. Asia.
6. *V. latifolia* L., Hungarian speedwell,
    véronique à larges feuilles; imported
    from Asia.

7. *V. longifolia* L., long-leaved
    speedwell, véronique à longues feuilles;
    imported from Eurasia, introduced,
    Alta-Sask, Ont-Nfld.
8. *V. officinalis* L., common speedwell,
    véronique officinale; imported from
    Eurasia, natzd., BC, Ont-Nfld. 8a. *V.
    o.* var. *tournefortii* (Vill.)
    Reichenb. (*V. tournefortii* Vill.);
    imported from Eurasia, natzd. NS-Nfld.
9. *V. serpyllifolia* L., thyme-leaved
    speedwell, veronique à feuilles de
    Serpolet. 9a. *V. s.* var. *humifusa*
    (Dickson) Vahl; native, BC-Sask, Ont-NS,
    Nfld, Labr. 9b. *V. s.* var. *s.*;
    imported from Eurasia, introduced Yukon,
    BC, Man-Nfld.

Cylindrosporella sp.: on 1a, 4 BC [1554].
Discogloeum veronicae (Gloeosporium v.)
    (C:313): leaf spot, anthracnose: on 8a
    BC [1816].
Entyloma linariae (C:313): on 2 Alta, Que
    [1554].
Isariopsis veronicae: on 1a BC [1554].
Peronospora grisea (C:313): downy mildew,
    mildiou: on 8 NB [333, 1194]; on 9a BC,
    Ont [1554].
Puccinia albulensis (C:313): rust,
    rouille: on 1a BC [1012].
P. albulensis ssp. *albulensis*: rust,
    rouille: on 1a BC [1552, 1555], Alta
    [1555].
P. albulensis ssp. *cascadensis*: rust,
    rouille: on 1a BC [1552]; on 4 BC
    [1555].
Ramularia veronicae (C:313): leaf spot,
    tache ramularienne: on 6 Ont [1554].
Schroeteria delastrina: on 3 BC, Ont
    [1554].
Sphaerotheca fuliginea (C:313): powdery
    mildew, blanc: on 5 Man [1257]; on 7
    Ont, Que [1257]; on 9 [1554].
S. macularis (S. humuli) (C:313): powdery
    mildew, blanc: on 8 NB [333, 1242].

VIBURNUM L.   CAPRIFOLIACEAE

1. *V. acerifolium* L., maple-leaved
    viburnum, viorne à feuilles d'érable;
    native, Ont-Que.
2. *V. alnifolium* Marsh., hobblebush, bois
    d'orignal; native, Ont-PEI.
3. *V. cassinoides* L., wild raisin,
    bleuets sains; native, Ont-Nfld.
4. *V. edule* (Michx.) Raf., squashberry,
    pimbina; native, Yukon-Mack, BC-NS,
    Nfld, Labr.
5. *V. lentago* L., nannyberry, alisier;
    native, Sask-Que.
6. *V. opulus* L., guelder rose, obier;
    imported from Eurasia.
7. *V. rafinesquianum* Schultes, downy
    arrow-wood, viorne de Rafinesque;
    native, Man-Que.
8. *V. tinus* L., laurestinus viburnum,
    laurier-tin; imported from the
    Mediterranean region.
9. *V. trilobum* Marsh., highbush
    cranberry, viorne trilobée; native,
    BC-Nfld.

10. Host species not named, i.e., reported
as *Viburnum* sp.

Aplosporella viburni (Sphaeropsis v.):
on leaves of 5 Ont [439].
Ascochyta viburni (C:313): leaf spot,
tache foliaire: on 4 BC [1816].
Botrytis cinerea (holomorph *Botryotinia
fuckeliana*) (C:313): gray mold,
moisissure grise: on 6 NS [1032].
Cenangium ellisii: on 3 Ont [1340].
Coleosporium viburni (C:314): rust,
rouille: II III on 3 NB, NS [1036,
F74]; on 3, 5, 7 Ont [1236]; on 5 Man,
Ont, Que [2008]; on 10 Que [F60, F67].
Conoplea globosa: on 5 Man [1760].
Dermea viburni (anamorph *Corniculariella
hystricina*) (C:314): on 3 Que [654];
on 3, 5 Ont [654]; on 3, 10 Ont [1340].
Godronia fuckeliana: on 6 Ont [656]; on
10 Ont [F66].
G. viburni: on 4 Que [1684]; on 10 Ont
[1340].
Massaria lantanae (M. plumigera var.
tetraspora) (C:314): on 1 Ont [1340];
on 1, 5, 6 Ont [1627]; on 10 Ont [80].
Microsphaera penicillata (M. alni)
(C:314): powdery mildew, blanc: on 1,
7 Ont [1257]; on 5 Man [1257], Ont
[1660]; on 5, 9 Ont, Que [1257]; on 5,
10 Ont [1340]; on 8 BC [337].
Mycosphaerella punctiformis: on 1 Ont
[1340].
Oidium sp.: powdery mildew, blanc: on 8
BC [1816].
Pezicula minuta (C:314): on 1, 10 Ont
[1340].
Phyllosticta punctata (C:314): on live
leaves of 6 Ont [439].
P. viburni: on leaves of 5 Ont [439].
Puccinia linkii (C:314): rust, rouille:
on 4 NWT-Alta [1063], Mack [53], BC
[1012, 1816], Ont [1236], Que [282,
F61]; on 4, 10 Ont [1341]; on 9 NS
[1032].
Stictis fusca (C:314): on twigs of 10
?Man [1608].
Tympanis fasciculata (C:314): on 5 Ont
[1340], NS [1925]; on 10 Ont, Que [1221].
T. grovesii: on 3 Ont, Que, NS [1221].
T. rhabdospora: on 10 Ont, Que, NS [1221].
Typhula viburni: on dead leaves of 3 Que
[1385].
Verticillium albo-atrum (C:314): wilt,
flétrissure verticillienne: on 10 Ont
[333].

VICIA L.   LEGUMINOSAE

1. *V. americana* Muhl., wild vetch, vesce
d'Amérique; native, Mack, BC-NB. 1a.
V. a . var. *oregana* Nutt.
2. *V. cracca* L., cow vetch, jargeau;
imported from Eurasia, natzd., Yukon,
BC-Nfld.
3. *V. faba* L., broad bean, fève des
marais; imported from Africa and Asia.
4. *V. gigantea* Hook., giant vetch;
native, BC.

5. *V. hirsuta* (L.) S.F. Gray, hoary tare,
vesce hérissée; imported from Eurasia,
introduced, BC-Alta, Ont-PEI.
6. *V. sativa* L., spring vetch, vesce
cultivée; imported from Europe,
introduced, BC, Man-Nfld. 6a. *V. s.*
var. *angustifolia* (L.) Wahlenb. (*V.
angustifolia* L.), narrow-leaved vetch,
vesce à feuilles étroites; imported from
Eurasia, introduced, BC, Man-Nfld.
7. *V. villosa* Roth, winter vetch, vesce
velue; imported from Europe, escaped
from cultivation, BC, Man-Que, NS.
8. Host species not named, i.e., reported
as *Vicia* sp.

Absidia sp.: on seeds of 3 Man [1368].
Alternaria sp. (C:314): on 3 Sask [1095,
1096 from seeds and leaves] and on seeds
Man [1368].
A. alternata: on seeds of 3 Man [1897].
Arthrinium sp.: on seeds of 3 Sask [1096].
Ascochyta sp.: on 6, 6a BC [1816].
A. fabae (C:314): leaf spot,
ascochytose: on 3 BC [1816], Sask [1096
on seeds], Man [886, 1368 on seeds], NS
[606].
A. pisi (C:314): leaf and pod spot,
ascochytose du pois: on 1, 5 BC [1816].
Aspergillus sp.: on seeds of 3 Sask
[1096], Man [1368].
A. candidus: on 3 Man [1897].
A. flavus: on seeds of 3 Man [1897].
A. glaucus group: on seeds of 3 Man
[1897].
A. nidulans: on seeds on 3 Man [1897].
A. versicolor: on seeds on 3 Man [1897].
Aureobasidium sp. (*Pullularia* sp.): on
seeds of 3 Sask [1096].
Botrytis cinerea (holomorph *Botryotinia
fuckeliana*) (C:314): gray mold or
chocolate spot, moisissure grise: on 3
BC [1816], Alta [339]; on seeds of 3
Sask [1096], Man [1368].
B. fabae: on 3 Sask [1096], Man [1368]
from seeds, NS [606].
Cephalosporium sp.: on 3 Sask [1096].
Chaetomium sp.: on seeds of 3 Sask [1096].
Cladosporium sp.: on leaves and seeds of
3 Sask [1096], and on seeds of 3 Man
[1368].
C. cladosporioides (C:315): on seeds of 3
Man [1897].
Colletotrichum sp.: on 3 Sask [1096].
Coniella pulchella: on 3 Sask [1096], Ont
[1622].
Epicoccum sp.: on seeds of 3 Sask [1096],
Man [1368].
Erysiphe communis (E. polygoni) (C:315):
powdery mildew, blanc: on 3 Sask
[1095]. Inoculations [884] of *V. faba*
and *V. americana* using *E. communis*
showed these hosts to be
"nonsusceptible" and the authors
questioned the validity of earlier
reports of *Erysiphe* on *Vicia*.
Flagelloscypha citrispora: on dead ?4 BC
[1424].
Fusarium sp. (C:315): foot rot, pourridié
fusarien: on seeds of 3 Sask [1096], Man
[1368].

F. oxysporum f. sp. *fabae* (C:315):
    wilt, flétrissure fusarienne: on 3 Que
    [311 inoculated soil, 334, 335, 337].
Gliocladium sp.: on seeds of 3 Sask
    [1096].
Microsphaera sp.: powdery mildew, blanc:
    on 1, 3 Man [884] by inoculation.
M. penicillata (C:315): powdery mildew,
    blanc: on 2 Que, NS [1257].
M. penicillata var. *ludens*: powdery
    mildew, blanc: on 1, 3 Man [884] by
    inoculation; on 3 Sask [1096, 1137], Man
    [883, 885].
Mucor sp.: on seeds of 3 Man [1368, 1897].
Nigrospora sp.: on seeds of 3 Sask [1096].
Oidium sp.: powdery mildew, blanc: on 3
    BC [1816].
Ovularia sp.: on 6a BC [1816].
Paecilomyces sp.: on seeds of 3 Sask
    [1096].
Pellidiscus pallidus: on dead 4 BC [1424].
Penicillium sp.: on seeds of 3 Sask
    [1096], Man [1368, 1897].
P. aurantiogriseum (P. cyclopium): on
    seeds of 3 Sask [1096].
Peronospora narbonensis (C:315): downy
    mildew, mildiou: on 1a BC [1816].
P. viciae (C:315): downy mildew,
    mildiou: on 6 BC [1816].
Pestalotia sp.: on 3 Sask [1096].
    Probably referrable to *Pestalotiopsis*
    or *Truncatella*.
Pythium sp.: on 3 Sask [1096]; on 7 BC
    [1816].
Rhizoctonia sp.: on 3 Sask [1095, 1096].
Rhizopus sp.: on seeds of 3 Man [1368,
    1897].
Sclerotinia sclerotiorum (C:315):
    sclerotinia rot, pourriture
    sclérotique: on 3 Sask [1096], Man
    [1368] from seeds.
Scopulariopsis brevicaulis: on seeds of 3
    Man [1897].
Scytalidium sp.: on 3 Sask [1096].
Stemphylium sp.: on seeds of 3 Sask
    [1096].
S. botryosum (holomorph *Pleospora
    herbarum*) (C:315): on 3 BC [1816], Ont
    [667].
Streptomyces sp.: on seeds of 3 Man
    [1897].
Thamnidium sp.: on seeds of 3 Sask [1096].
Trichoderma sp.: on seeds of 3 Sask
    [1096].
Ulocladium consortiale (Stemphylium c.):
    on seeds of 3 Ont [667].
Uromyces sp.: rust, rouille: on 3 Man
    [1368]; on 8 Alta [1594].
U. coloradensis (C:315): rust, rouille:
    0 I III on 1 Ont [1236].
U. coloradensis var. *maritimus*: rust,
    rouille: on 1a BC [1012, 1816].
U. viciae-fabae (U. fabae, Aecidium album)
    (C:315): rust, rouille: 0 I II III on
    1 NWT-Alta [1063], Ont [437]; on 2 Ont
    [1237]; on 2, 8 Ont [1236]; on 3 Sask
    [1096], on 6, 6a BC [1816]; on 6a BC
    [1012].
Verticillium sp.: on 3 Sask [1096].

VINCA L.   APOCYNACEAE

1. *V. major* L., large periwinkle, grande
    pervenche; imported from Europe, a
    garden escape, BC.

Puccinia vincae (C:315): rust, rouille:
    0 II III on 1 Ont [1236].

VIOLA L.   VIOLACEAE

1. *V. adunca* Sm., early blue violet,
    violette des chiens; native, Yukon-Mack,
    BC-Nfld, Labr.
2. *V. blanda* Willd., sweet white violet,
    violette agréable; native, Man-Que.
3. *V. canadensis* L., Canada violet,
    violette du Canada; native, Mack,
    BC-NS. 3a. *V. c.* var. *rugulosa*
    (Greene) Hitchc. (*V. r.* Greene);
    native, Mack, BC-Que.
4. *V. conspersa* Reichenb., American dog
    violet, violette décombante; native,
    Man-NS.
5. *V. cucullata* Ait., blue marsh violet,
    violette cucullée; native, Ont-Nfld.
6. *V. epipsila* Ledeb.; native,
    Yukon-Mack, BC, Man. The North American
    plant is referable to *V. e.* ssp.
    repens (Turcz.) Becker.
7. *V. eriocarpa* Schwein. (*V.
    pensylvanica* auct. non Michx.), smooth
    yellow violet, violette de Pennsylvanie;
    native, Man-PEI.
8. *V. glabella* Nutt., stream violet;
    native, BC-Alta.
9. *V. howellii* Gray; native, BC.
10. *V. incognita* Brainerd, large-leaved
    white violet, violette méconnue; native,
    Ont-Nfld.
11. *V. langsdorfii* (Regel) Fisch., Alaska
    violet; native, Yukon-BC.
12. *V. mackloskeyi* Lloyd; native,
    Mack-Keew, BC-Alta, Man-Nfld, Labr.
    12a. *V. m.* var. *pallens* (Banks)
    Hitchc. (*V. pallens* (Banks) Brainerd),
    white snowdrops, violette pâle; native,
    Man-Nfld.
13. *V. nephrophylla* Greene, bog violet,
    violette néphrophylle; native, BC-Nfld.
14. *V. orbiculata* Geyer, round-leaved
    violet; native, BC-Alta.
15. *V. palustris* L., alpine marsh violet,
    violette des marais; native, BC-Que,
    Nfld, Labr.
16. *V. pedatifida* G. Don., larkspur
    violet; native, Alta-Ont.
17. *V. pubescens* Ait., downy yellow
    violet, violette pubescente; native,
    Man-Que.
18. *V. renifolia* Gray, kidneyleaf violet,
    violette réniforme; native, Yukon,
    BC-Nfld, Labr.
19. *V. sagittata* Ait., arrow-leaved
    violet, violette sagittée; native,
    Ont-PEI. 19a. *V. s.* var. *ovata*
    (Nutt.) Torr. & Gray (*V. fimbriatula*
    Sm.); native, Ont-PEI.
20. *V. selkirkii* Pursh, Selkirk's violet,
    violette de Selkirk; native, Yukon,
    BC-Nfld, Labr.

21. *V. septentrionalis* Greene, northern
    blue violet, vilette septentrionale;
    native, BC, Sask, Ont-Nfld.
22. *V. soraria* Willd., woolly blue violet,
    violette parente; native, Man-Que.
23. *V. tricolor* L., pansy, pensée;
    imported from Eurasia, a garden escape,
    BC-NS (including cultigens such as *V.
    x wittrockiana* Gams).
24. Host species not named, i.e., reported
    as *Viola* sp.

Ascochyta violae (C:316): leaf spot,
    tache ascochytique: on 23 BC [1816].
Botrytis cinerea (holomorph *Botryotinia
    fuckeliana*) (C:316): gray mold,
    moisissure grise: on 23 BC [1816], NB
    [333, 1194, 1594]; on 24 Ont [163].
Cercospora granuliformis (C:316): on 23
    NS [334, 338].
C. violae (C:316): leaf spot, tache
    cercosporéenne: on 23 BC [1816], NS
    [333].
Erysiphe communis (E. polygoni): powdery
    mildew, blanc: on 24 Que [1257].
Fusarium culmorum (C:316): on 23 BC
    [1816].
Mycocentrospora acerina (Centrospora a.)
    (C:316): on 23 NS [333, 334, 335, 338].
Myrothecium roridum (C:316): crown and
    stem rot, pourriture de la tige: on 23
    BC [1816].
Oidium sp.: powdery mildew, blanc: on 23
    BC [1816].
Puccinia sp.: rust, rouille: on 24 Ont
    [1341].
P. canadensis (C:316): rust, rouille: on
    14 BC [1816], BC, Alta [1570].
P. ellisiana (C:316): rust, rouille: 0 I
    on 24 Ont [1236].
P. fergussonii (C:316): rust, rouille:
    on 6 Yukon [1571]; on 11, 15 BC [1571];
    on 15 BC [1012, 1816].
P. glacieri: rust, rouille: on 8 BC
    [1574].
P. ornatula (C:316): rust, rouille: on 8
    BC [1012, 1573, 1578, 1816].
P. violae (P. aegra, Aecidium v.)
    (C:316): rust, rouille: 0 I II III on
    1 NWT-Alta [1063]; on 1, 2, 3, 4, 7,
    12a, ? 13, 17, 20, 21, 22, 24 Ont
    [1236]; on 3, 8, 23 BC [1012]; on 3a, 8,
    9, 23 BC [1816]; on 7 Ont [1828]; on 18
    Ont [437]; on 24 Yukon [460], BC [644,
    953], Ont [1827], Que [282], NB [1032].
P. violae ssp. *americana*: rust,
    rouille: on 1 Mack; on 1, 3, 6, 8, 9,
    11, 15, 18 BC, on 1, 3, 13 Alta, Sask,
    on 1, 5, 20, 21 Que, on 3, 7, 12a, 17,
    22 Ont, Que, on 4 Ont, on 10 NB, on 10,
    12a, 19a NS, on 16, 18 Alta, on 10, 12a,
    18 PEI. All from [1572].
P. violae ssp. *violae*: rust, rouille:
    on 23 BC [1012].
Ramularia lactea (C:316): leaf spot,
    tache ramularienne: on 23 BC [1816].
Sphaerotheca sp.: powdery mildew, blanc:
    on 23 Alta [1594].
S. macularis (C:316): powdery mildew,
    blanc: on 23 BC [333], Alta, Sask [339].

Synchytrium sp. (C:316): on 8 BC [1012,
    1816].

VITIS L.   VITACEAE

1. *V. labrusca* L., fox grape, vigne
   Isabelle; imported from e. USA.
2. *V. piasezkii* Maxim.; imported from
   China.
3. *V. riparia* Michx., wild grape, vigne
   sauvage; native, Man-NS.
4. *V. vulpina* L., winter grape, vigne
   foxée; imported from e. USA.
5. *V.* spp., cultivated grapes, vignes
   cultivée: There are three major groups
   of cultivated grapes in Canada: 5a.
   American cultivars, derived principally
   from North American species. An example
   is the Concord grapes. 5b. French
   hybrids, cultivars derived from both *V.
   vinifera* and the American cultivars.
   5c. *V. vinifera* L., wine grape, la
   vigne; the grape commonly cultivated in
   Europe.
6. Host species not named, i.e., reported
   as *Vitis* sp.

Alternaria sp.: on 6 Ont [231].
Botryosphaeria obtusa (Physalospora o.)
    (C:317): on 5 Ont [1614]; on 6 Ont
    [231].
B. stevensii (anamorph *Diplodia mutila*)
    (C:317): on 6 Ont [1614].
Botrytis cinerea (holomorph *Botryotinia
    fuckeliana*): gray mold, moisissure
    grise: on 5 BC [335]; on 5c BC [1816].
    In 1979 in 3 cultivars 20% of the
    Ontario crop was rotten by late
    September.
Cryptosporella viticola: in Ont on 5b
    (Seibel 10878) [336], 5c [337], 6
    [231]. In association with the
    Sphaeropsis state of *Physalospora
    obtusa* and *Mycosporium* sp., *C.
    viticola* is a secondary saprophyte
    [336] (see C:317 under *Fusicoccum
    viticola*, and [1413]).
Discostroma corticola (Clathridium c.)
    (anamorph *Seimatosporium lichenicola*)
    (C:317): on 2 Ont [1613].
Elsinoe ampelina (anamorphs *Gloeosporium
    ampelophagum, Sphaceloma ampelinum*)
    (C:317): anthracnose, anthracnose: on
    5c NS [338].
Eutypa armeniacae: on 5 Ont [1819]; on 6
    Ont [1121].
Fusarium sp.: on 6 Ont [231].
Gliocladium sp.: on 6 Ont [231].
Guignardia bidwellii (anamorph
    Phyllosticta ampelicida) (C:317):
    black rot, pourriture noire: on 5 (var.
    Agawam) Ont [334]; on 5c BC [1816]; NS
    [337]; on 6 Que [336].
Lophiostoma winteri: on 5c Ont [234].
Pestalotia sp.: on 6 Ont [231].
*Pestalotiopsis quadriciliata (Pestalotia
    q.)*: on 4 Ont [195].
Phoma sp.: on 6 Ont [231].
Phomopsis viticola (Fusicoccum v.)
    (C:317): dead arm, branche moribonde:

in Ont on 5 [1819], 5a, 5b [335], 5b
[334], 6 [333].
Plasmopara viticola (C:317): downy
mildew, mildiou: on 5 Ont [339]; on 5a,
5b Ont [334, 335]; on 5a Ont, Que [333].
Roesleria hypogeae: on 5c Ont [338].
Sphaeropsis malorum: on 6 Ont [231]. See
comment under *Malus*.
Uncinula necator (C:317): powdery mildew,
blanc: on 1, 3, 6 Ont [1257]; on 3 Que
[1257]; on 5 BC [335], Ont [333]; on 5a,
5b Ont [334, 335]; on 5b BC [334]; on 5c
BC [339, 1594, 1816], Ont [337, 339]; on
6 BC [1257], NS [334].
Verticillium dahliae: on 1 BC [1816,
1978].

WALDSTEINIA Willd.   ROSACEAE

1. *W. fragarioides* (Michx.) Tratt.,
   barren strawberry, waldsteinie
   faux-fraisier; native, Ont-NB.

Puccinia waldsteiniae (C:317): rust,
rouille: III on 1 Ont [1236, 1341].

WOODWARDIA Sm.   POLYPODIACEAE

1. *W. virginica* (L.) Sm., Virginia chain
   fern, woodwardie de Virginie; native,
   Ont-PEI.

Uredinopsis arthurii (C:318): rust,
rouille: on 1 Que [1377]. See comment
under *Abies*.

WYETHIA Nutt.   COMPOSITAE

Coarse, tap-rooted perennial herbs
native to the s. USA.

Puccinia balsamorhizae: rust, rouille:
on *W*. sp. BC [1249].

XANTHIUM L.   COMPOSITAE

1. *X. chinense* Mill., cocklebur,
   lampourde de Chine; native, BC-Que.
2. *X. echinatum* Murr., sea burdock,
   glouteron; native, Sask-Que.
3. *X. italicum* Moretti, cocklebur,
   glouteron; native, BC-NB, PEI.
4. *X. strumarium* L., cocklebur,
   glouteron; native, BC-Que. 4a. *X. s.*
   var. *canadense* (Mill.) Torr. & Gray
   (*X. canadense* Mill.).
5. Host species not named, i.e., reported
   as *Xanthium* sp.

Puccinia canaliculata: rust, rouille: on
3 BC [1816].
P. osoyoosensis: rust, rouille: on 4a BC
[1559].
P. xanthii (C:318): rust, rouille: III
on 1 Sask, Ont [1261]; on 1, 2, 3, 4 Ont
[1236]; on 3 Sask, Ont [1245]; on 3, 5
Ont [1261]; on 4 Man [1245, 1261].
Verticillium dahliae: on 1 Ont [1087]; on
4a BC [1978].

YUCCA L.   AGAVACEAE

1. *Y. filamentosa* L., Adam's-needle,
   yucca filamenteux; imported from se. USA.
2. *Y. recurvifolia* Salisb.; imported from
   s. USA, planted, s. Ont.

*Microsphaeropsis concentrica (Coniothyrium
c.)* (C:318): leaf spot, tache zonale:
on 1 BC [1816]; on 2 Ont [1130].

ZANTHOXYLUM L.   RUTACEAE

Prickly, deciduous or evergreen shrubs
or trees with aromatic bark.

1. *Z. americanum* Mill., northern prickly
   ash, frêne épineux; native, Ont-Que.

Massaria zanthoxyli: on 1 Ont [1628]; on
Z. sp. Ont [80].
Phyllactinia guttata: powdery mildew,
blanc: on 1 Ont, Que [1257].

ZEA L.   GRAMINEAE

1. *Z. mays* L., maize, blé d'Inde;
   imported from South and Central
   America. 1a. *Z. m.* var. *rugosa*
   Bonsf., sweet corn.

Absidia sp.: on overwintered seeds 1 Ont
[462].
Acremonium strictum: on overwintered
seeds 1 Ont [462].
Alternaria sp.: on 1 Ont [1772, 1948].
A. alternata (A. tenuis auct.*)* (C:318):
on 1 Man [1114], Ont [349, 462 on seeds].
Arthrinium phaeospermum: on 1 Man [1760].
Aspergillus sp.: on overwintered seeds of
1 Ont [462]; on 1a Que [1300].
A. candidus: on overwintered plants of 1
Man [1114], on overwintered seeds of 1
Ont [462].
A. flavus: on overwintered plants of 1
Man [1114]; on overwintered seeds of 1
Ont [462].
A. fumigatus: on overwintered seeds of 1
Ont [462].
A. glaucus group: on overwintered plants
of 1 Man [1114]; on overwintered seeds
of 1 Ont [462].
A. terreus: on overwintered seeds of 1
Ont [462].
Aureobasidium zeae (Kabatiella z.): on 1
Ont [536, 1595], Que [237, 239].
Bipolaris maydis (Helminthosporium m.)
(holomorph *Cochliobolus
heterostrophus*) (C:319): on 1 Ont
[170, 540, 542, 1508], Que [237], PEI
[859].
B. sorokiniana (holomorph *Cochliobolus
sativus*) (C:318): on overwintered
plants of 1 Man [1052].
Cephalosporium acremonium (C:318): on
overwintered plants of 1 Man [1114].
See comment under *Avena*.
Chaetomium sp.: on overwintered seeds of
1 Ont [462].

C. globosum (C:318): on 1 Ont [462 on
    overwintered seeds, 1340].
Cladosporium sp.: on overwintered seeds
    [462] and on injured and uninjured
    kernels [1772] of 1 Ont.
C. cladosporioides (Hormodendrum c.)
    (C:318): on overwintered plants of 1
    Man [1114]; on seeds of 1 Ont [349, 337,
    1883].
Dendryphion nanum: on 1 Man [819, 1760].
Drechslera phlei: on 1 Ont [1611].
    Usually on *Phleum*, but occurs as a
    saprophyte on other genera.
Epicoccum purpurascens: on overwintered
    plants of 1 Man [1114]; on overwintered
    seeds of 1 Ont [462].
Exserohilum turcicum (Bipolaris t.,
    Helminthosporium t.) (holomorph
    Setosphaeria t.) (C:318): leaf spot,
    tache foliaire: on 1 Man, Ont [337],
    Ont [116, 1882], Que [237]; on 1, la Ont
    [334, 335, 336].
Fusarium sp. (C:319): root, stalk and ear
    rot, pourriture fusarienne: on 1 Man
    [1114] on overwintered plants, Ont
    [1460, 1772 on injured and uninjured
    kernels, 1948], Ont, NB [1594], Que
    [337].
F. acuminatum (C:318): on overwintered
    plants of 1 Man [1114].
F. avenaceum: on overwintered plants of 1
    Man [1114].
F. culmorum (C:319): on 1 BC [1816].
F. equiseti (C:319): root, stem and ear
    rot, pourriture fusarienne: on 1 Ont
    [1948].
F. graminearum (holomorph *Gibberella*
    zeae) (C:319): on 1 Ont [349, 337 on
    seeds, 338, 339, 462 on overwintered
    seeds, 1145, 1166, 1167, 1594, 1771,
    1772 on uninjured kernels, 1882, 1883].
F. moniliforme (C:319): on 1 Ont [349,
    337, 338, 339, 1771, 1772, 1948, 1882,
    1949]; on la Ont [335].
F. moniliforme var. *subglutinans* (C:319):
    on overwintered seeds of 1 Ont [462].
F. oxysporum (C:318): on overwintered
    plants of 1 Man [1114]; on 1 Ont [349 on
    seeds, 1948].
F. poae (C:319): on overwintered seeds of
    1 Man [1114], Ont [462].
F. roseum: on 1 Ont [1008].
F. roseum f. sp. *cerealis*: on 1 Ont
    [1147]. This may be *F. graminearum*.
F. sacchari: on overwintered plants of 1
    Man [1114].
F. sambucinum: on overwintered plants of
    1 Man [1114].
F. solani (holomorph *Nectria*
    haematococca) (C:319): on 1 Ont
    [1772].
F. sporotrichioides: on overwintered
    plants of 1 Man [1114].
F. sulphureum: on overwintered plants of
    1 Man [1114].
F. tricinctum: on 1 Man [1114] on
    overwintered plants, Ont [1772, 1883].
    In Wall [1883] the authorities for the
    species name are given as Snyder and

Hansen. The Snyder and Hansen species
    concept was broad and included, for
    example, *F. poae*.
Gibberella zeae (anamorph *Fusarium*
    graminearum) (C:319): on 1 Ont [335,
    336, 447, 448, 538, 539, 541, 1146,
    1949], Que [237, 238].
Gonatobotrys simplex: on overwintered
    plants of 1 Man [1114].
Kabatiella sp.: on 1 Ont [542].
Microascus longirostris: on seeds of 1
    Ont [308, 462].
Mucor sp.: on overwintered plants of 1
    Man [1114] and from injured kernels Ont
    [1772].
Nigrospora oryzae (holomorph *Khuskia o.*)
    (C:319): on 1 Sask [339], Man [1114] on
    overwintered plants.
Paecilomyces varioti (C:318): on seeds of
    1 Ont [150].
Penicillium sp.: on 1 Man [1114] on
    overwintered plants, Ont [462 on
    overwintered seeds, 1772].
Phoma sp.: on overwintered seeds of 1 Ont
    [462].
P. terrestris (Pyrenochaeta t.): on 1 Ont
    [1948].
Phyllosticta sp.: on 1 Ont [536, 542,
    1595].
P. maydis: on 1 Ont [1700].
Puccinia sorghi (C:319): rust, rouille:
    II III on 1 Ont [337, 1236, 1341, 1882],
    Que [194, 339, 1103, 1104].
Pythium arrhenomanes (C:319): on 1 Ont
    [1948].
P. irregulare: on 1 Ont [868] by
    inoculation.
P. torulosum: on 1 Ont [868] by
    inoculation.
Rhizophydium graminis: on 1, la Ont [59].
Rhizopus sp.: on injured kernels of 1 Ont
    [1772].
R. arrhizus: on overwintered plants of 1
    Man [1114].
Sclerophthora macrospora: on 1 Ont [536,
    1595].
Setosphaeria turcica (Trichometasphaeria
    t.) (anamorph *Bipolaris t.*): on la
    Ont [115].
Sphacelotheca reiliana: on 1 Ont [1019].
    First reported in Canada in the fall of
    1979 from s. Ont. Subsequently the
    fungus has been found in nearly 70
    fields of field and sweet corn in BC,
    Ont and Que. Typically the percentage
    of infection was traced to light but in
    Ont several fields had 30% infection.
Stenocarpella maydis (Diplodia m.)
    (C:318): dry rot, pourriture sèche: on
    1 BC [339], Ont [1752].
Torula herbarum: on 1 Man [1760].
Trichoderma sp.: on 1 Ont [1772 from
    injured and uninjured kernels, 1948].
T. hamatum: on 1 Ont [1072].
T. harzianum: on 1 Ont [1072].
T. koningii: on 1 Ont [1072, 1769].
Ustilago maydis (U. zeae) (C:318): smut,
    charbon: on 1 BC [1816], BC, Alta, Sask
    [339], Sask, Ont [337, 1594], Sask, Que,
    PEI [338], Man, Ont, Que [335], Ont

[1460, 1882], Que [237, 1385], Que, NB,
NS [1594], NB [337]; on la Alta, Sask
[334], Sask [333], Ont, NS [336].

Volutella ciliata: on 1 Ont [1340].
Reported as *V. "setula"*, no doubt a
lapsus calami.

ZINNIA L.   COMPOSITAE

1. *Z. elegans* Jacq., common zinnia,
   zinnia; imported from Mexico.
2. Host species not named, i.e., reported
   as *Zinnia* sp.

Alternaria zinniae ("*A. zinniae* Pape")
   (C:319): alternaria blight, brûlure
   alternarienne: on 1 BC [1816], Ont
   [1944]; on 2 Ont [333, 334], Que [335],
   NB [333, 1194], NS [333, 336]. Pape's
   name is not validly published because
   the description lacked both a Latin
   description and designation of a type
   specimen.
Botrytis cinerea (holomorph *Botryotinia
   fuckeliana*) (C:319): gray mold,
   moisissure grise: on 1 BC [1816], Ont
   [163].
Erysiphe cichoracearum (C:319): powdery
   mildew, blanc: on 1 Ont [338], Ont, Que
   [337, 1257]; on 2 NS [333].
E. communis: powdery mildew, blanc: on 2
   Ont, Que [334]. Probably specimens of
   E. cichoracearum.
Fusarium sp. (C:319): wilt, flétrissure
   fusarienne: on 1 BC [1816].
Phytophthora cryptogea (C:319): foot rot,
   mildiou du pied: on 1 BC [1816]; on 2
   Ont [333].
Rhizoctonia solani (holomorph
   Thanatephorus cucumeris): on 1 Ont
   [338].
Sclerotinia sclerotiorum (C:319): stem
   rot, pourriture sclérotique: on 2 NS
   [333], PEI [334].

ZIZANIA L.   GRAMINEAE

1. *Z. aquatica* L., wild rice, riz
   sauvage; native, Sask-PEI.

Claviceps sp.: on 1 Alta [1781].
Drechslera catenaria (C:320): on 1 Ont
   [1611].

ZIZIA Koch   UMBELLIFERAE

1. *Z. aurea* (L.) Koch, golden alexanders,
   zizie dorée; native, Sask-NS.

Puccinia angelicae (C:320): rust,
   rouille: on 1 Man, Que [1236], Que
   [1551].

ZOSTERA L.   ZOSTERACEAE

1. *Z. marina* L., eel grass, mousse de
   mer; native, Keew, BC, Man-Nfld, Labr.

Lulworthia halima (C:320): on 1 Que, NB,
   NS [1620].

ZYGADENUS Michx.   LILIACEAE

1. *Z. elegans* Pursh (*Z. glaucus* Nutt.),
   white camas, zigadène élégant; native,
   Yukon-Mack, BC-NB.
2. *Z. venenosus* Wats., death camas;
   native, BC-Sask.

Puccinia atropuncta (C:320): rust,
   rouille: II III on 1 Ont [1236], Que
   [1545].
P. grumosa (C:320): rust, rouille: on 1
   Yukon [7, 460], Yukon, BC, Alta [1545],
   BC [1816].
Uromyces zygadeni (C:320): rust,
   rouille: on 2 BC [1545, 1816].

ZYGOCACTUS K. Schum.   CACTACEAE

Flat-stemmed short jointed cacti often
cultivated for bright showy flowers;
imported from Brazil.

Fusarium sp.: on Z. sp. Que [338].

BIBLIOGRAPHY

C.   CONNERS, I.L. 1967. An Annotated
      Index of Plant Diseases in
      Canada. Can. Dept. Agr. Publ.
      1251. Ottawa. 381 p.
F60. Forest Insect and Disease Survey, Ann.
      Report for 1960. Published in
      1961. Can. Forestry Service,
      Ottawa.
F61. F.I.D.S., Ann. Report 1961. Publ.
      1962.
F62. F.I.D.S., Ann. Report 1962. Publ.
      1963.
F63. F.I.D.S., Ann. Report 1963. Publ.
      1964.
F64. F.I.D.S., Ann. Report 1964. Publ.
      1965.
F65. F.I.D.S., Ann. Report 1965. Publ.
      1966.
F66. F.I.D.S., Ann. Report 1966. Publ.
      1967.
F67. F.I.D.S., Ann. Report 1967. Publ.
      1968.
F68. F.I.D.S., Ann. Report 1968. Publ.
      1969.
F69. F.I.D.S., Ann. Report 1969. Publ.
      1970.
F70. F.I.D.S., Ann. Report 1970. Publ.
      1971.
F71. F.I.D.S., Ann. Report 1971. Publ.
      1972.
F72. F.I.D.S., Ann. Report 1972. Publ.
      1973.
F73. F.I.D.S., Ann. Report 1973. Publ.
      1974.
F74. F.I.D.S., Ann. Report 1974. Publ.
      1975.
F75. F.I.D.S., Ann. Report 1975. Publ.
      1977.
F76. F.I.D.S., Ann. Report 1976. Publ.
      1979.
1.   AGNIHOTRI, V.P. & O. VAARTAJA. 1967.
      Can. J. Bot. 45: 1031-1040.
2.   ALKHOURY, I. & R.H. ESTEY. 1979.
      Phytoprotection 60: 171.
3.   ALLEN, W.R. 1974. Can. J. Plant
      Sci. 54: 431-432.
4.   AMMIRATI, J., J. TRAQUAIR, S. MARTIN,
      W. GILLON & J. GINNS. 1979.
      Mycologia 71: 310-321.
5.   ANASTASIOU, C.J. & L.M. CHURCHLAND.
      1969. Can. J. Bot. 47: 251-257.
6.   ANDERSON, J.B. & R.C. ULLRICH. 1979.
      Mycologia 71: 402-414.
7.   ANDERSON, J.P. 1952. Iowa State
      Coll. J. Sci. 26: 507-526.
8.   ANDERSON, T.R. 1980. Can. Plant Dis.
      Survey 60: 33-34.
9.   ARNOLD, R.H. 1967. Can. J. Bot.
      45: 783-801.
10.   ARNOLD, R.H. 1970. Can. J. Bot.
      48: 1525-1540.
11.   ARNOLD, R.H. 1971. Can. J. Bot.
      49: 2187-2196.
12.   ARNOLD, R.H. 1974. Fungi Canadenses
      No. 17, National Mycological
      Herbarium, Agriculture Canada,
      Ottawa.

13.   ARNOLD, R.H. 1974. Fungi Canadenses
      No. 30, National Mycological
      Herbarium, Agriculture Canada,
      Ottawa.
14.   ARNOLD, R.H. 1975. Fungi Canadenses
      No. 70, National Mycological
      Herbarium, Agriculture Canada,
      Ottawa.
15.   ARNOLD, R.H. & J.C. CARTER. 1974.
      Mycologia 66: 191-197.
16.   ARNOLD, R.H. & R.C. RUSSELL. 1960.
      Mycologia 52: 499-512.
17.   ARNOLD, R.H. & A.E. STRABY. 1973.
      Can. Plant Dis. Survey 53:
      183-186.
18.   ARNOTT, J.T. 1974. Can. J. Forest
      Res. 4: 69-75.
19.   ÅRSVOLL, K. & J. D. SMITH. 1979.
      Can. J. Bot. 56: 348-364.
20.   ARX, J. 1970. Biblio. Mycol. 24:
      1-203.
21.   ATKINSON, G.F. 1905. J. Mycol. 11:
      248-267.
22.   ATKINSON, R.G. 1961. Can. J. Bot.
      39: 1531-1536.
23.   ATKINSON, R.G. 1965. Can. J. Bot.
      43: 1471-1475.
24.   ATKINSON, R.G. 1965. Can. J. Plant
      Sci. 45: 609-611.
25.   ATKINSON, R.G. 1965. Can. Phytopath.
      Soc. Proc. 32: 10.
26.   ATKINSON, R.G. 1970. Can. J. Plant
      Sci. 50: 565-568.
27.   ATKINSON, R.G. 1980. Can. J. Plant
      Sci. 60: 747-749.
28.   ATKINSON, R.G. & R.M. ADAMSON. 1977.
      Can. J. Plant Sci. 57: 675-680.
29.   ATKINSON, R.G. & J.G. TRELAWNY.
      1962. Can. Plant Dis. Survey
      42: 151-154.
30.   AUBE, C. 1965. Phytoprotection 46:
      123-124.
31.   AUBE, C. 1966. Can. Plant Dis.
      Survey 46: 11-13.
32.   AUBE, C. 1967. Can. Plant Dis.
      Survey 47: 25-27.
33.   AUBE, C. 1969. Can. J. Plant Sci.
      49: 737-742.
34.   AUBE, C. & J. DESCHENES. 1967. Plant
      Disease Rep. 51: 573-577.
35.   AUBE, C. & W.E. SACKSTON. 1963. Can.
      Phytopath. Soc. Proc. 30: 10.
36.   AUBE, C. & W.E. SACKSTON. 1964. Can.
      J. Plant Sci. 44: 427-432.
37.   AUBE, C. & W.E. SACKSTON. 1964.
      Phytoprotection 45: 44.
38.   AUBE, C. & W.E. SACKSTON. 1965. Can.
      J. Bot. 43: 1335-1342.
39.   AUBE, C. & W.E. SACKSTON. 1966.
      Phytoprotection 47: 91.
40.   AYERS, G.W. 1972. Can. Plant Dis.
      Survey 52: 77-81.
41.   AYERS, G.W. & G.C. RAMSAY. 1961.
      Can. Plant Dis. Survey 41:
      199-202.
42.   BANDONI, R.J. 1963. Can. J. Bot.
      41: 467-474.

43. BANDONI, R.J. 1972. Can. J. Bot. 50: 2283-2285.

44. BANDONI, R.J. & D.J.S. BARR. 1976. Trans. Mycol. Soc. Japan 17: 220-225.

45. BANDONI, R.J., J.R. MAZE & J.P. DELANGE. 1979. Syesis 12: 105-106.

46. BANHAM, F.L. 1960. Can. J. Plant Sci. 40: 165-171.

47. BANTING, J.D., Y.S. WU & L.H. SHEBESKI. 1961. Can. J. Plant Sci. 41: 137-152.

48. BANVILLE, G.-J. 1970. Phytoprotection 51: 152.

49. BANVILLE, G.-J. 1971. Can. Phytopath. Soc. Proc. 38: 13.

50. BARANYAY, J.A. 1963. Can. Phytopath. Soc. Proc. 30: 10.

51. BARANYAY, J.A. 1966. Can. J. Bot. 44: 597-604.

52. BARANYAY, J.A. 1967. Forestry Chronicle 43: 372-380.

53. BARANYAY, J.A. 1968. Can. Dept. Forestry & Rural Devel. Publ. 1238. Ottawa.

54. BARANYAY, J.A. & A. FUNK. 1969. Can. J. Bot. 47: 1011-1014.

55. BARANYAY, J.A. & Y. HIRATSUKA. 1967. Can. J. Bot. 45: 189-191.

56. BARANYAY, J.A. & R.B. SMITH. 1974. Can. J. Forest Res. 4: 361-365.

57. BARANYAY, J.A. & G.R. STEVENSON. 1964. Forestry Chronicle 40: 350-361.

58. BARR, D.J.S. 1968. Can. J. Bot. 46: 1087-1091.

59. BARR, D.J.S. 1973. Can. Plant Dis. Survey 53: 191-193.

60. BARR, D.J.S. 1979. Can. J. Plant Path. 1: 85-94.

61. BARR, D.J.S. 1980. Can. J. Plant Path. 2: 116-118.

62. BARR, D.J.S. 1980. Fungi Canadenses No. 176, National Mycological Herbarium, Agriculture Canada, Ottawa.

63. BARR, D.J.S. & E.O. CALLEN. 1963. Phytoprotection 44: 18-24.

64. BARR, D.J.S. & V.E. HADLAND-HARTMANN. 1977. Can. J. Bot. 55: 3063-3074.

65. BARR, D.J.S. & V.E. HARTMANN. 1977. Can. J. Bot. 55: 1221-1235.

66. BARR, D.J.S. & C.J. HICKMAN. 1967. Can. J. Bot. 45: 423-430.

67. BARR, D.J.S. & W.G. KEMP. 1975. Can. Plant Dis. Survey 55: 77-82.

68. BARR, D.J.S. & J.T. SLYKHUIS. 1969. Can. Plant Dis. Survey 49: 112-113.

69. BARR, D.J.S. & J.T. SLYKHUIS. 1976. Can. Plant Dis. Survey 56: 77-81.

70. BARR, M.E. 1961. Can. J. Bot. 39: 307-325.

71. BARR, M.E. 1964. Mycologia 56: 841-862.

72. BARR, M.E. 1967. Can. J. Bot. 45: 1041-1046.

73. BARR, M.E. 1968. Can. J. Bot. 46: 799-864.

74. BARR, M.E. 1970. Mycologia 62: 377-394.

75. BARR, M.E. 1972. Contrib. Univ. Mich. Herb. 9: 523-638.

76. BARR, M.E. 1976. Mycologia 68: 611-621.

77. BARR, M.E. 1977. Mycologia 69: 952-966.

78. BARR, M.E. 1977. Mycotaxon 6: 17-23.

79. BARR, M.E. 1978. The Diaporthales in North America. Mycol. Soc. Amer. Memoir No. 7. Cramer, Lehre, Germany. 232 p.

80. BARR, M.E. 1979. Mycotaxon 9: 17-37.

81. BARRON, G.L. 1961. Can. J. Bot. 39: 1573-1578.

82. BARRON, G.L. 1962. Can. J. Bot. 40: 589-607.

83. BARRON, G.L. 1962. Can. J. Bot. 40: 1603-1613.

84. BARRON, G.L. & C. BOOTH. 1966. Can. J. Bot. 44: 1057-1061.

85. BARRON, G.L., R.F. CAIN & J.C. GILMAN. 1961. Can. J. Bot. 39: 837-845.

86. BARRON, G.L., R.F. CAIN & J.C. GILMAN. 1961. Can. J. Bot. 39: 1629-1632.

87. BARTON, G.M. 1967. Can. J. Bot. 45: 1545-1552.

88. BASHAM, J.T. 1966. Can. J. Bot. 44: 275-295.

89. BASHAM, J.T. 1966. Can. J. Bot. 44: 849-860.

90. BASHAM, J.T. 1973. Can. J. Bot. 51: 1379-1392.

91. BASHAM, J.T. 1975. Can. J. Forest Res. 5: 706-721.

92. BASHAM, J.T. & H.W. ANDERSON. 1977. Can. J. Bot. 55: 934-976.

93. BASHAM, J.T. & L.D. TAYLOR. 1965. Plant Disease Rep. 49: 771-774.

94. BASHAM, J.T., H.M. GOOD & L.D. TAYLOR. 1969. Can. J. Bot. 47: 1629-1634.

95. BASHAM, J.T., J. HUDAK, D. LACHANCE, L.P. MAGASI & M.A. STILLWELL. 1976. Can. J. Forest Res. 6: 406-414.

96. BASU, P.K. 1961. Can. J. Bot. 39: 165-196.

97. BASU, P.K. 1962. Can. Phytopath. Soc. Proc. 29: 12.

98. BASU, P.K. 1962. Can. Plant Dis. Survey 42: 246-252.

99. BASU, P.K. 1974. Can. Plant Dis. Survey 54: 24-25.

100. BASU, P.K. 1974. Can. Plant Dis. Survey 54: 45-51.

101. BASU, P.K. 1975. Can. Phytopath. Soc. Proc. 42: 19.

102. BASU, P.K. 1976. Plant Disease Rep. 60: 1037-1040.

103. BASU, P.K. 1978. Can. Plant Dis. Survey 58: 5-8.

104. BASU, P.K. 1980. Can. Plant Dis. Survey 60: 23-24.

105. BASU, P.K., C.S. LIN. & M.R. BINNS. 1977. Can. J. Plant Sci. 57: 1091-1097.

106. BASU, P.K., N.J. BROWN, R. CRETE, C.O. GOURLEY, H.W. JOHNSTON, H.S. PEPIN & W.L. SEAMAN. 1976. Can. Plant Dis. Survey 56: 25-32.

107. BASU, P.K., R. CRETE, A.G. DONALDSON, C.O. GOURLEY, J.H. HAAS, F.R. HARPER, C.H. LAWRENCE, W.L. SEAMAN, H.N.W. TOMS, S.I. WONG & R.C. ZIMMER. 1973. Can. Plant Dis. Survey 53: 49-57.

108. BATRA, L.R. 1963. Mycologia 55: 508-520.

109. BATRA, L.R. 1967. Mycologia 59: 976-1017.

110. BEAUCHAMP, P. & G.J. PELLETIER. 1979. Phytoprotection 60: 167.

111. BEDARD, R. & R.O. LACHANCE. 1970. Phytoprotection 51: 72-77.

112. BENEDICT, W.G. 1960. Can. Phytopath. Soc. Proc. 27: 12.

113. BENEDICT, W.G. 1966. Can. J. Plant Sci. 46: 553-560.

114. BENEDICT, W.G. 1971. Can. J. Bot. 49: 1721-1726.

115. BENEDICT, W.G. 1976. Can. J. Bot. 54: 552-555.

116. BENEDICT, W.G. 1979. Can. J. Bot. 57: 1809-1814.

117. BERKENKAMP, B. 1969. Can. J. Bot. 47: 453-456.

118. BERKENKAMP, B. 1969. Can. Plant Dis. Survey 49: 65.

119. BERKENKAMP, B. 1971. Can. Plant Dis. Survey 51: 96-100.

120. BERKENKAMP, B. 1971. Phytoprotection 52: 52-57.

121. BERKENKAMP, B. 1972. Can. Plant Dis. Survey 52: 51-55.

122. BERKENKAMP, B. 1972. Can. Plant Dis. Survey 52: 62-63.

123. BERKENKAMP, B. 1973. Can. Plant Dis. Survey 53: 11-15.

124. BERKENKAMP, B. 1974. Can. Plant Dis. Survey 54: 111-115.

125. BERKENKAMP, B. 1977. Can. Plant Dis. Survey 57: 65-67.

126. BERKENKAMP, B. 1980. Can. J. Plant Sci. 60: 1039-1040.

127. BERKENKAMP, B. & H. BAENZIGER. 1962. Can. Plant Dis. Survey 42: 265.

128. BERKENKAMP, B. & H. BAENZIGER. 1969. Can. J. Plant Sci. 49: 181-183.

129. BERKENKAMP, B. & K. DEGENHARDT. 1974. Can. Plant Dis. Survey 54: 35-36.

130. BERKENKAMP, B. & H. VAARTNOW. 1972. Can. J. Plant Sci. 52: 973-976.

131. BERKENKAMP, B., L.P. FOLKINS & J. MEERES. 1969. Plant Disease Rep. 53: 348-349.

132. BERKENKAMP, B., L.P. FOLKINS & J. MEERES. 1972. Can. Plant Dis. Survey 52: 1-3.

133. BERKENKAMP, B., L.P. FOLKINS & J. MEERES. 1973. Can. Plant Dis. Survey 53: 36-38.

134. BERKENKAMP, B., L.P. FOLKINS & J. MEERES. 1978. Can. J. Plant Sci. 58: 893-894.

135. BERNIER, C.C. 1972. Can. Plant Dis. Survey 52: 108.

136. BHATT, G.C. & G.L. BARRON. 1965. Can. Phytopath. Soc. Proc. 32: 11.

137. BIER, J.E. 1965. Can. J. Bot. 43: 877-883.

138. BIER, J.E. 1965. Forestry Chronicle 41: 306-313.

139. BIER, J.E. & M.H. ROWAT. 1963. Can. J. Bot. 41: 1585-1596.

140. BIER, J.E., P. SALISBURY & R. WALDIE. 1948. Canada Dept. Agr., Ottawa. Tech. Bull. 66 (Publ. 804).

141. BIGELOW, H.E. & M.E. BARR. 1960. Rhodora 62: 187-198.

142. BIGELOW, H.E. & M.E. BARR. 1962. Rhodora 64: 127-137.

143. BIGELOW, H.E. & M.E. BARR. 1963. Rhodora 65: 289-309.

144. BIGELOW, H.E. & M.E. BARR. 1966. Rhodora 68: 175-191.

145. BIGELOW, H.E. & M.E. BARR. 1969. Rhodora 71: 177-203.

146. BIRD, C.J. & D.W. GRUND. 1979. Proc. Nova Scotian Institute Sci. 29(1): 1-131.

147. BISSETT, J. 1977. Fungi Canadenses No. 102, National Mycological Herbarium, Agriculture Canada, Ottawa.

148. BISSETT, J. 1979. Can. J. Bot. 57: 2071-2081.

149. BISSETT, J. 1979. Can. J. Bot. 57: 2082-2095.

150. BISSETT, J. 1979. Fungi Canadenses No. 151, National Mycological Herbarium, Agriculture Canada, Ottawa.

151. BISSETT, J. 1979. Fungi Canadenses No. 152, National Mycological Herbarium, Agriculture Canada, Ottawa.

152. BISSETT, J. 1979. Fungi Canadenses No. 153, National Mycological Herbarium, Agriculture Canada, Ottawa.

153. BISSETT, J. 1979. Fungi Canadenses No. 160, National Mycological Herbarium, Agriculture Canada, Ottawa.

154. BLOOMBERG, W.J. 1962. Can. J. Bot. 40: 1271-1280.

155. BLOOMBERG, W.J. 1965. Forestry Chronicle 41: 182-187.

156. BLOOMBERG, W.J. 1966. Can. J. Bot. 44: 413-420.

157. BLOOMBERG, W.J. & G.W. WALLIS. 1979. Can. J. Forest Res. 9: 76-81.

158. BOHAYCHUK, W.P. & R.D. WHITNEY. 1973. Can. J. Bot. 51: 801-815.

159. BOLTON, A.T. 1963. Can. J. Bot. 41: 237-241.

160. BOLTON, A.T. 1963. Can. J. Bot. 41: 503-543.

161. BOLTON, A.T. 1964. Can. Plant Dis. Survey 44: 267.
162. BOLTON, A.T. 1966. Phytoprotection 47: 84-88.
163. BOLTON, A.T. 1976. Can. J. Plant Sci. 56: 861-864.
164. BOLTON, A.T. 1977. Can. J. Plant Sci. 57: 87-92.
165. BOLTON, A.T. 1978. Can. Plant Dis. Survey 58: 83-86.
166. BOLTON, A.T. & W.E. CORDUKES. 1979. Can. J. Plant Sci. 59: 1113-1116.
167. BOLTON, A.T. & A.G. DONALDSON. 1972. Can. J. Plant Sci. 52: 189-196.
168. BOLTON, A.T. & J.B. JULIEN. 1961. Can. Plant Dis. Survey 41: 261-264.
169. BOLTON, A.T. & V.W. NUTTALL. 1968. Can. J. Plant Sci. 48: 161-166.
170. BOLTON, A.T. & W.L. SEAMAN. 1972. Can. Plant Dis. Survey 52: 70-71.
171. BOLTON, A.T. & F.J. SVEJDA. 1979. Can. Plant Dis. Survey 59: 38-42.
172. BOLTON, A.T., A.G. DONALDSON & V.W. NUTTALL. 1970. Can. Plant Dis. Survey 50: 107-108.
173. BOLTON, A.T., V.W. NUTTALL & L.H. LYALL. 1965. Can. J. Plant Sci. 45: 343-348.
174. BONAR, L. 1962. Mycologia 54: 395-399.
175. BOOTH, C. 1959. Commonw. Mycol. Inst., England, Mycol. Paper 73.
176. BOOTH, C. 1966. Commonw. Mycol. Inst., England, Mycol. Paper 104.
177. BOOTH, T. 1969. Syesis 2: 141-161.
178. BOOTH, T. 1971. Syesis 4: 197-208.
179. BOOTSMA, A. 1979. Can. Plant Dis. Survey 59: 63-66.
180. BORDEN, J.H. & M. MCCLAREN. 1970. Syesis 3: 145-154.
181. BORNO, C. & B.J. VAN DER KAMP. 1975. Can. J. Bot. 53: 1266-1269.
182. BOURCHIER, R.J. 1961. Can. J. Bot. 39: 1373-1385.
183. BOURCHIER, R.J. 1961. Can. J. Bot. 39: 1781-1784.
184. BOYER, M.G. 1961. Can. J. Bot. 39: 1195-1204.
185. BOYER, M.G. 1961. Can. J. Bot. 39: 1409-1427.
186. BOYER, M.G. 1967. Can. J. Bot. 45: 501-513.
187. BRADBURY, J.F. & R.W. FISHER. 1964. Plant Disease Rep. 48: 104-105.
188. BRISSON, J.D., J.F. PAUZE & V. LAVOIE. 1974. Aspects mycologiques et histopathologiques du dépérissement des tiges de Vaccinium angustifolium Aiton. Cahier no. 13. Faculté sci. agr. et alimen., Univ. Laval, Ste-Foy, Qué.
189. BRISSON, J.D., K.A. PIROZYNSKI & J.F. PAUZE. 1975. Can. J. Bot. 53: 2866-2871.
190. BRODIE, H.J. 1968. Can. Field-Nat. 75: 41-42.
191. BRODIE, H.J. 1970. Can. J. Bot. 48: 847-849.
192. BROUGH, S. & R.J. BANDONI. 1975. Syesis 8: 301-303.
193. BROUGH, S.G. 1974. Can. J. Bot. 52: 1853-1859.
194. BROWN, R.I. 1966. Can. Plant Dis. Survey 46: 143.
195. BUBAK, Fr. 1916. Hedwigia 58: 15-34.
196. BUCHANNON, K.W. & W.C. MCDONALD. 1965. Can. J. Plant Sci. 45: 189-193.
197. BUCHANNON, K.W. & H.A.H. WALLACE. 1962. Can. J. Plant Sci. 42: 534-536.
198. BURDSALL, H.H. 1969. Mycologia 61: 915-923.
199. BURRAGE, R.H. & R.D. TINLINE. 1960. Can. J. Plant Sci. 40: 672-679.
200. BUSCH, L.V. 1961. Can. Plant Dis. Survey 41: 167-168.
201. BUSCH, L.V. 1967. Can. Plant Dis. Survey 47: 76-77.
202. BUSCH, L.V. & G.L. BARRON. 1963. Can. J. Plant Sci. 43: 166-173.
203. BUSCH, L.V. & L.E. PHILPOTTS. 1972. Can. Phytopath. Soc. Proc. 39: 26.
204. BUSCH, L.V. & E.A. SMITH. 1976. Can. Phytopath. Soc. Proc. 43: 33.
205. BUSCH, L.V. & E.A. SMITH. 1978. Can. Plant Dis. Survey 58: 73-74.
206. BUZZELL, R.I., J.H. HAAS, L.G. CRAWFORD & O. VAARTAJA. 1977. Can. Plant Dis. Survey 57: 68-70.
207. CAIN, R.F. 1961. Can. J. Bot. 39: 1633-1666.
208. CAIN, R.F. 1961. Can. J. Bot. 39: 1667-1671.
209. CAIN, R.F. & L.K. WERESUB. 1957. Can. J. Bot. 35: 119-131.
210. CALLBECK, L.C. 1960. Can. Plant Dis. Survey 40: 56-58.
211. CALLBECK, L.C. 1960. Can. Plant Dis. Survey 40: 87-91.
212. CALLBECK, L.C. 1962. Can. Plant Dis. Survey 42: 185-188.
213. CALLBECK, L.C. 1963. Can. Plant Dis. Survey 43: 201-205.
214. CALLBECK, L.C. 1964. Can. Plant Dis. Survey 44: 244-247.
215. CALLBECK, L.C. 1967. Can. Plant Dis. Survey 47: 99-100.
216. CALLBECK, L.C. 1969. Can. Plant Dis. Survey 49: 14-15.
217. CALLBECK, L.C. 1969. Can. Plant Dis. Survey 49: 75-77.
218. CALLBECK, L.C. 1972. Can. Plant Dis. Survey 52: 30-31.
219. CALLBECK, L.C. 1972. Can. Plant Dis. Survey 52: 151-152.
220. CALLBECK, L.C. 1974. Can. Plant Dis. Survey 54: 21-23.
221. CAMP, R.R. 1981. Can. J. Bot. 59: 2466-2477.
222. CAMPBELL, W.P. 1962. Can. Plant Dis. Survey 42: 195-201.

223. CAMPBELL, W.P., L.J. PIENING & D.W.
     CREELMAN. 1961. Can. Plant Dis.
     Survey 41: 369-370.
224. CARLSON, L.W. 1969. Plant Disease
     Rep. 53: 100.
225. CARLSON, L.W. 1972. Can. Plant Dis.
     Survey 52: 99-100.
226. CARMICHAEL, J.W. 1962. Can. J. Bot.
     40: 1137-1173.
227. DELETED.
228. CARPENTER, S.E. & K.P. DUMONT. 1978.
     Mycologia 70: 1223-1238.
229. CASAGRANDE, F. & G.B. OUELLETTE.
     1971. Can. J. Bot. 49: 155-159.
230. CAUCHON, R. & G.B. OUELLETTE. 1964.
     Mycologia 56: 453-455.
231. CHAMBERLAIN, G.C., R.S. WILLISON, J.L.
     TOWNSHEND & J.H. de RONDE.
     1964. Can. J. Bot. 42: 351-355.
232. CHAREST, P.M., G.B. OUELLETTE & F.J.
     PAUZE. 1979. Phytoprotection
     60: 164.
233. CHAREST, P.M., G.B. OUELLETTE & F.J.
     PAUZE. 1980. Phytoprotection
     61: 119.
234. CHESTERS, C.G.C. & A. BELL. 1970.
     Commonw. Mycol. Inst., England,
     Mycol. Paper 120.
235. CHEW, P.S. & R. HALL. 1978. Can.
     Phytopath. Soc. Proc. 45: 34.
236. CHEZ, D. 1974. Phytoprotection 55:
     38-42.
237. CHEZ, D. & M. HUDON. 1975.
     Phytoprotection 56: 90-95.
238. CHEZ, D., M. HUDON & M.S. CHIANG.
     1977. Phytoprotection 58: 5-17.
239. CHEZ, D., M. HUDON, R. MARTIN & G.
     PELLETIER. 1979.
     Phytoprotection 60: 162.
240. CHI, C.C. 1966. Plant Disease Rep.
     50: 451-453.
241. CHI, C.C. 1967. Can. Phytopath. Soc.
     Proc. 34: 18.
242. CHI, C.C. 1970. Can. Plant Dis.
     Survey 50: 109-110.
243. CHI, C.C. & W.R. CHILDERS. 1965.
     Plant Disease Rep. 49: 512-515.
244. CHI, C.C. & W.R. CHILDERS. 1966.
     Plant Disease Rep. 50: 695-698.
245. CHIANG, M.S. & R. CRETE. 1972. Can.
     Plant Dis. Survey 52: 45-50.
246. CHINN, D.A. & R.H. ESTEY. 1966.
     Phytoprotection 47: 66-72.
247. CHINN, S.H.F. 1976. Can. J. Plant
     Sci. 56: 199-201.
248. CHINN, S.H.F., B.J. SALLONS & R.J.
     LEDINGHAM. 1962. Can. J. Plant
     Sci. 42: 720-727.
249. CHURCHLAND, L.M. & M. McCLAREN.
     1972. Can. J. Bot. 50:
     1269-1273.
250. CLARK, G.H. 1962. Can. Plant Dis.
     Survey 42: 260.
251. CLARK, G.H. & R.N. WENSLEY. 1961.
     Can. Plant Dis. Survey 42: 363.
252. CLARK, R.V. 1962. Can. Phytopath.
     Soc. Proc. 29: 12.
253. CLARK, R.V. 1963. Can. Plant Dis.
     Survey 43: 27-32.
254. CLARK, R.V. 1964. Can. J. Plant
     Sci. 44: 488-490.
255. CLARK, R.V. 1966. Can. Plant Dis.
     Survey 46: 105-109.
256. CLARK, R.V. 1966. Can. J. Plant
     Sci. 46: 603-609.
257. CLARK, R.V. 1968. Can. Plant Dis.
     Survey 48: 134-135.
258. CLARK, R.V. 1971. Can. Plant Dis.
     Survey 51: 71-75.
259. CLARK, R.V. 1971. Can. Plant Dis.
     Survey 51: 131-133.
260. CLARK, R.V. 1977. Can. Plant Dis.
     Survey 57: 45-48.
261. CLARK, R.V. 1978. Can. Plant Dis.
     Survey 58: 33-38.
262. CLARK, R.V. 1979. Can. J. Plant
     Path. 1: 113-117.
263. CLARK, R.V. 1979. Can. J. Plant
     Path. 2: 37-38.
264. CLARK, R.V. 1980. Can. J. Plant
     Path. 2: 213-216.
265. CLARK, R.V. & F.L. DRAYTON. 1960.
     Can. J. Bot. 38: 103-110.
266. CLARK, R.V. & H.W. JOHNSTON. 1973.
     Can. J. Plant Sci. 53: 471-475.
267. CLARK, R.V. & V.R. WALLEN. 1969.
     Can. Plant Dis. Survey 49: 60-64.
268. CLARK, R.V. & F.J. ZILLINSKY. 1960.
     Can. J. Bot. 38: 93-102.
269. CLARK, R.V. & F.J. ZILLINSKY. 1960.
     Can. Plant Dis. Survey 40: 1-9.
270. CLARK, R.V. & F.J. ZILLINSKY. 1962.
     Can. J. Plant Sci. 42: 620-627.
271. CLARK, R.V., C.O. GOURLEY, H.W.
     JOHNSTON, L.J. PIENING, G.
     PELLETIER, J. SANTERRE & H.
     GENEREUX. 1975. Can. Plant Dis.
     Survey 55: 36-43.
272. CLOUGH, K.S. & H.W. JOHNSON. 1978.
     Can. Plant Dis. Survey 58: 95-96.
273. CLOUGH, K.S. & H.W. JOHNSTON. 1978.
     Can. Plant Dis. Survey 58:
     97-98.
274. CLOUGH, K.S. & Z.A. PATRICK. 1972.
     Can. J. Bot. 50: 2251-2253.
275. CLOUGH, K.S. & K.R. SANDERSON. 1979.
     Can. Plant Dis. Survey 59: 19-21.
276. CLOUTIER, A. 1971. Phytoprotection
     52: 91.
277. COHEN, Y. & W.E. SACKSTON. 1972.
     Phytoprotection 53: 50.
278. COHEN, Y. & W.E. SACKSTON. 1973.
     Can. J. Bot. 51: 15-22.
279. COLE, G.T. & B. KENDRICK. 1973.
     Mycologia 65: 661-688.
280. COLOTELO, N. & H. NETOLITZKY. 1964.
     Can. J. Bot. 42: 1467-1469.
281. COMEAU, A. & G.J. PELLETIER. 1976.
     Can. J. Plant Sci. 56: 13-19.
282. COOKE, W.B. & R. POMERLEAU. 1964.
     Mycologia 56: 607-618.
283. CORLETT, M. 1966. Can. J. Bot. 44:
     1141-1149.
284. CORLETT, M. 1967. Can. J. Bot. 45:
     221-227.
285. CORLETT, M. 1968. Can. J. Bot. 46:
     1303-1307.
286. CORLETT, M. 1970. Canada Agric.,
     Ottawa 15(2): 24-25.

287. CORLETT, M. 1971. Can. J. Bot. 49: 39-40.

288. CORLETT, M. 1974. Fungi Canadenses No. 20, National Mycological Herbarium, Agriculture Canada, Ottawa.

289. CORLETT, M. 1974. Fungi Canadenses No. 35, National Mycological Herbarium, Agriculture Canada, Ottawa.

290. CORLETT, M. 1974. Fungi Canadenses No. 36, National Mycological Herbarium, Agriculture Canada, Ottawa.

291. CORLETT, M. 1974. Fungi Canadenses No. 49, National Mycological Herbarium, Agriculture Canada, Ottawa.

292. CORLETT, M. 1975. Fungi Canadenses No. 76, National Mycological Herbarium, Agriculture Canada, Ottawa.

293. CORLETT, M. 1976. Fungi Canadenses No. 84, National Mycological Herbarium, Agriculture Canada, Ottawa.

294. CORLETT, M. 1977. Fungi Canadenses No. 92, National Mycological Herbarium, Agriculture Canada, Ottawa.

295. CORLETT, M. 1978. Fungi Canadenses No. 121, National Mycological Herbarium, Agriculture Canada, Ottawa.

296. CORLETT, M. 1979. Fungi Canadenses No. 131, National Mycological Herbarium, Agriculture Canada, Ottawa.

297. CORLETT, M. 1979. Fungi Canadenses No. 149, National Mycological Herbarium, Agriculture Canada, Ottawa.

298. CORLETT, M. & A. CARTER. 1976. Fungi Canadenses No. 88, National Mycological Herbarium, Agriculture Canada, Ottawa.

299. CORLETT, M. & J. CHONG. 1977. Can. J. Bot. 55: 5-7.

300. CORLETT, M. & K.N. EGGER. 1980. Fungi Canadenses No. 181, National Mycological Herbarium, Agriculture Canada, Ottawa.

301. CORLETT, M. & K.N. EGGER. 1980. Fungi Canadenses No. 182, National Mycological Herbarium, Agriculture Canada, Ottawa.

302. CORLETT, M. & M.E. ELLIOTT. 1974. Can. J. Bot. 52: 1459-1463.

303. CORLETT, M. & E.G. KOKKO. 1975. Can. J. Bot. 53: 1338-1341.

304. CORLETT, M. & E.G. KOKKO. 1977. Fungi Canadenses No. 93, National Mycological Herbarium, Agriculture Canada, Ottawa.

305. CORLETT, M. & E.G. KOKKO. 1977. Microscopica Acta 79: 39-42.

306. CORLETT, M. & E.G. KOKKO. 1978. Fungi Canadenses No. 114, National Mycological Herbarium, Agriculture Canada, Ottawa.

307. CORLETT, M. & E.G. KOKKO. 1978. Fungi Canadenses No. 120, National Mycological Herbarium, Agriculture Canada, Ottawa.

308. CORLETT, M. & G.A. NEISH. 1980. Fungi Canadenses No. 180, National Mycological Herbarium, Agriculture Canada, Ottawa.

309. CORLETT, M. & R.G. ROSS. 1979. Can. J. Plant Path. 1: 79-84.

310. CORLETT, M., J. CHONG & E.G. KOKKO. 1976. Can. J. Microbiol. 22: 1144-1152.

311. COULOMBE, L.J. 1961. Can. Plant Dis. Survey 41: 191-193.

312. COULOMBE, L.J. 1966. Phytoprotection 47: 146.

313. COULOMBE, L.J. 1968. Phytoprotection 49: 135.

314. COULOMBE, L.J. 1969. Phytoprotection 50: 7-15.

315. COULOMBE, L.J. 1969. Phytoprotection 50: 23-31.

316. COULOMBE, L.J. 1971. Phytoprotection 52: 90.

317. COULOMBE, L.J. 1975. Phytoprotection 56: 55-66.

318. COULOMBE, L.J. 1976. Phytoprotection 57: 23-32.

319. COULOMBE, L.J. 1976. Phytoprotection 57: 33-35.

320. COULOMBE, L.J. 1976. Phytoprotection 57: 41-46.

321. COULOMBE, L.J. 1977. Phytoprotection 58: 39-45.

322. COULOMBE, L.J. 1977. Phytoprotection 58: 127.

323. COULOMBE, L.J. 1978. Phytoprotection 59: 55-64.

324. COULOMBE, L.J. 1979. Phytoprotection 60: 79-92.

325. COULOMBE, L.J. & A. JACOB. 1980. Phytoprotection 61: 48-54.

326. COUTURE, L. 1980. Can. Plant Dis. Survey 60: 8-10.

327. COUTURE, L. & G.J. PELLETIER. 1975. Phytoprotection 56: 31-41.

328. COUTURE, L. & J.C. SUTTON. 1978. Can. J. Bot. 56: 2162-2170.

329. COUTURE, L. & J.C. SUTTON. 1978. Phytoprotection 59: 65-75.

330. COUTURE, L. & J.C. SUTTON. 1978. Phytoprotection 59: 189.

331. COUTURE, L. & J.C. SUTTON. 1980. Can. Plant Dis. Survey 60: 59-61.

332. CRANE, J.L. & J.D. SCHOKNECHT. 1977. Can. J. Bot. 55: 3013-3019.

333. CREELMAN, D.W. 1961. Can. Plant Dis. Survey 41: 31-121.

334. CREELMAN, D.W. 1962. Can. Plant Dis. Survey 42: 23-102.

335. CREELMAN, D.W. 1963. Can. Plant Dis. Survey 43: 61-130.

336. CREELMAN, D.W. 1964. Can. Plant Dis. Survey 44: 1-82.

337. CREELMAN, D.W. 1965. Can. Plant Dis. Survey 45: 42-78.

338. CREELMAN, D.W. 1966. Can. Plant Dis. Survey 46: 37-76.

339. CREELMAN, D.W. 1967. Can. Plant Dis. Survey 47: 31-71.
340. CRETE, R. & E.G. BEAUCHAMP. 1968. Phytoprotection 49: 41-48.
341. CRETE, R. & M.S. CHIANG. 1967. Plant Disease Rep. 51: 991-992.
342. CRETE, R. & E.J. HOGUE. 1974. Phytoprotection 55: 96.
343. CRETE, R. & L. TARTIER. 1971. Can. Plant Dis. Survey 51: 88-90.
344. CRETE, R. & L. TARTIER. 1973. Phytoprotection 54: 32-42.
345. CRETE, R. & L. TARTIER. 1971. Phytoprotection 52: 91.
346. CRETE, R. & L. TARTIER. 1971. Phytoprotection 52: 92.
347. CRETE, R., J. LALIBERTE & J.J. JASMIN. 1963. Can. J. Plant Sci. 43: 349-354.
348. CRETE, R., L. TARTIER & T. SIMARD. 1971. Can. Plant Dis. Survey 51: 66-70.
349. CUDDY, T.F. & V.R. WALLEN. 1965. Can. Plant Dis. Survey 45: 33-34.
350. CUMMINS, G.B. & H.C. GREENE. 1966. Mycologia 58: 702-721.
351. CUMMINS, G.B., M.P. BRITTON & J.W. BAXTER. 1969. Mycologia 61: 924-944.
352. CURREN, T. 1969. Can. J. Bot. 47: 2108-2109.
353. CUSSON, Y. & D. LACHANCE. 1974. Phytoprotection 55: 17-28.
354. CUSSON, Y. & D. LACHANCE. 1974. Phytoprotection 55: 96.
355. DANCE, B.W. 1961. Can. J. Bot. 39: 875-890.
356. DANCE, B.W. 1961. Can. J. Bot. 39: 1429-1435.
357. DANCE, B.W. 1968. Plant Disease Rep. 52: 659-661.
358. DANCE, B.W. & J.D. CAFLEY. 1974. Plant Disease Rep. 58: 677-679.
359. DARKER, G.D. 1963. Can. J. Bot. 41: 1383-1388.
360. DARKER, G.D. 1963. Can. J. Bot. 41: 1389-1394.
361. DARKER, G.D. 1963. Mycologia 55: 812-818.
362. DARKER, G.D. 1964. Can. J. Bot. 42: 1005-1009.
363. DARKER, G.D. 1965. Can. J. Bot. 43: 11-14.
364. DARKER, G.D. 1967. Can. J. Bot. 45: 1445-1449.
365. DAUBENY, H.A. 1961. Can. J. Plant Sci. 41: 239-243.
366. DAUBENY, H.A. 1978. Can. J. Plant Sci. 58: 283-285.
367. DAUBENY, H.A. & H.S. PEPIN. 1962. Can. Plant Dis. Survey 42: 212-213.
368. DAUBNEY, H.A. & H.S. PEPIN. 1964. Can. Plant Dis. Survey 44: 257-258.
369. DAUBENY, H.A. & H.S. PEPIN. 1965. Can. J. Plant Sci. 45: 365-368.
370. DAUBENY, H.A. & H.S. PEPIN. 1974. Can. J. Plant Sci. 54: 511-516.
371. DAUBENY, H.A. & H.S. PEPIN. 1974. Plant Disease Rep. 58: 1024-1027.
372. DAUBENY, H.A., J.A. FREEMAN & H.S. PEPIN. 1974. Plant Disease Rep. 58: 391-395.
373. DAVID, A. 1980. Bull. Soc. Linnéenne Lyon 49: 1-56.
374. DAVIDSON, A.G. & D.E. ETHERIDGE. 1963. Can. J. Bot. 41: 759-765.
375. DAYAL, R. & G.L. BARRON. 1969. Can. Phytopath. Soc. Proc. 35: 20.
376. DE HOOG, G.S. & E.J. HERMANDIES-NIJHOF. 1977. Centraalbureau voor Schimmelcultures, Baarn, Netherlands, Studies in Mycol. 15.
377. DEGENHARDT, K.J., F.R. HARPER & T.G. ATKINSON. 1982. Can. J. Plant Path. 4: 375-380.
378. DEGENHARDT, K.J., W.P. SKOROPAD & Z.P. KONDRA. 1974. Can. J. Plant Sci. 54: 795-799.
379. DEIGHTON, F.C. 1967. Commonw. Mycol. Inst., England, Mycol. Paper 112.
380. DEIGHTON, F.C. 1973. Commonw. Mycol. Inst., England, Mycol. Paper 133.
381. DEIGHTON, F.C. 1974. Commonw. Mycol. Inst., England, Mycol. Paper 137.
382. DEIGHTON, F.C. 1979. Commonw. Mycol. Inst., England, Mycol. Paper 144.
383. DEJARDIN, R.A. & E.W.B. WARD. 1971. Can. J. Bot. 49: 339-347.
384. DENBY, L.G. & G.E. WOOLLIAMS. 1962. Can. J. Plant Sci. 42: 681-685.
385. DENISON, W.C. 1972. Mycologia 64: 609-623.
386. DENNIS, R.W.G. 1978. British Ascomycetes. Cramer, Vaduz.
387. DENYER, W.B.G. 1960. Can. J. Bot. 38: 909-920.
388. DESMARTEAU, R., L. TARTIER & A.L. DEVAUX. 1972. Phytoprotection 53: 117-119.
389. DEVAUX, A.L. 1970. Phytoprotection 51: 150.
390. DEVAUX, A.L. 1978. Phytoprotection 59: 19-27.
391. DEVAUX, A.L. & R.H. ESTEY. 1963. Phytoprotection 44: 50-51.
392. DEVAUX, A.L. & W.E. SACKSTON. 1964. Phytoprotection 45: 132.
393. DEVAUX, A.L. & W.E. SACKSTON. 1966. Can. J. Bot. 44: 803-811.
394. DEVAUX, A.L. & L.M. TARTIER. 1978. Phytoprotection 59: 185.
394a. DEVRIES, B. 1966. Can. J. Bot. 44: 1228-1230.
395. DHANVANTARI, B.N. 1967. Can. J. Bot. 45: 1525-1543.
396. DHANVANTARI, B.N. 1967. Can. Plant Dis. Survey 47: 21.
397. DHANVANTARI, B.N. 1968. Can. J. Plant Sci. 48: 401-404.
398. DHARNE, C.G. 1965. Phytopath. Zeitschr. 53: 101-144.
399. DICKINSON, C.H. 1968. Commonw. Mycol. Inst., England, Mycol. Paper 115.

272

400. DICKOUT, R.S. & D.J. ORMROD. 1972.
Can. Plant Dis. Survey 52: 109.
401. DICOSMO, F. 1978. Can. J. Bot. 56:
1665-1690.
402. DICOSMO, F. & G.T. COLE. 1980. Can.
J. Bot. 58: 1129-1137.
403. DIEHL, W.W. 1950. *Balansia* and the
Balansiae in America. USDA
Monogr. 4.
404. DIRKS, V.A., T.R. ANDERSON & E.F.
BOLTON. 1980. Can. J. Plant
Path. 2: 179-183.
405. DIXON, J.R. 1974. Mycotaxon 1:
65-104.
406. DOBBELER, P. 1978. Mitteilungen Bot.
Staatssammlung Munchen 14: 1-360.
407. DOBBELER, P. 1979. Mitteilungen Bot.
Staatssammlung Munchen 15:
193-221.
408. DODD, J.L. 1972. Mycologia 64:
737-773.
409. DORWORTH, C.E. 1972. Can. J. Bot.
50: 751-765.
410. DORWORTH, C.E. 1973. Can. J. Forest
Res. 3: 161-164.
411. DORWORTH, C.E. 1974. Can. J. Bot.
52: 919-922.
412. DORWORTH, C.E. 1975. Plant Disease
Rep. 59: 272-273.
413. DORWORTH, C.E. 1979. Can. J. Forest
Res. 9: 316-322.
414. DORWORTH, C.E. & J. KRYWIENCZYK.
1975. Can. J. Bot. 53:
2506-2525.
415. DORWORTH, C.E., J. KRYWIENCZYK & D.D.
SKILLING. 1977. Plant Disease
Rep. 61: 887-890.
416. DOSTALER, D. & G.J. PELLETIER. 1978.
Phytoprotection 59: 189.
417. DOSTALER, D., G.J. PELLETIER & L.
COUTURE. 1980. Phytoprotection
61: 19-25.
418. DUCZEK, L.J. & J.A. BUCHAN. 1979.
Can. Phytopath. Soc. Proc. 46:
56.
419. DUCZEK, L.J. & V.J. HIGGINS. 1976.
Can. J. Bot. 54: 2609-2619.
420. DUCZEK, L.J. & R.A.A. MORRALL. 1971.
Can. Plant Dis. Survey 51:
116-121.
421. DUMONT, K.P. 1973. Mycologia 65:
175-191.
422. DUMONT, K.P. 1976. Mycologia 68:
233-267.
423. DUNCAN, E.G. & J.A. MACDONALD. 1967.
Mycologia 59: 803-818.
424. DURAN, R. & G.W. FISCHER. 1961. The
genus *Tilletia*. Wash. State
Univ., Pullman, Wash.
425. DURAVETZ, J.S. & J.F. MORGAN-JONES.
1971. Can. J. Bot. 49:
1267-1272.
426. EDWARDS, R. & L.L. KENNEDY. 1973.
Can. J. Bot. 51: 2385-2393.
427. ELLIOTT, M.E. 1962. Can. J. Bot.
40: 1197-1201.
428. ELLIOTT, M.E. 1964. Can. J. Bot.
42: 1065-1070.
429. ELLIOTT, M.E. 1965.
Greenhouse-Garden-Grass (Can.
Agr., Ottawa) 5(2): 7-10.
430. ELLIOTT, M.E. 1969. Can. J. Bot.
47: 1895-1898.
431. ELLIOTT, M.E. 1973. Fungi Canadenses
No. 1, National Mycological
Herbarium, Agriculture Canada,
Ottawa.
432. ELLIOTT, M.E. 1974. Fungi Canadenses
No. 38, National Mycological
Herbarium, Agriculture Canada,
Ottawa.
433. ELLIOTT, M.E. & R. CAUCHON. 1976.
Fungi Canadenses No. 90, National
Mycological Herbarium,
Agriculture Canada, Ottawa.
434. ELLIOTT, M.E. & M. CORLETT. 1972.
Can. J. Bot. 50: 2153-2156.
435. ELLIOTT, M.E. & M. KAUFERT. 1974.
Can. J. Bot. 52: 467-472.
436. ELLIOTT, M.E. & C. RICHARD. 1971.
Mycologia 63: 648-652.
437. ELLIS, J.B. & B. EVERHART. 1885. J.
Mycol. 1: 85-87.
438. ELLIS, J.B. & B. EVERHART. 1892. The
North American Pyrenomycetes.
Publ. by the authors, Newfield,
New Jersey.
439. ELLIS, J.B. & J. DEARNESS. 1893.
Can. Record of Sci. 5: 266-272.
440. ELLIS, J.B. & J. DEARNESS. 1897.
Proc. Can. Inst. NS 1: 89-93.
441. ELLIS, J.B. & J. DEARNESS. 1899.
Trans. Can. Inst. 6: 637-639.
442. ELLIS, M.B. 1960. Commonw. Mycol.
Inst., England, Mycol. Paper 76.
443. ELLIS, M.B. 1963. Commonw. Mycol.
Inst., England, Mycol. Paper 93.
444. ELLIS, M.B. 1965. Commonw. Mycol.
Inst., England, Mycol. Paper 103.
445. ELLIS, M.B. 1966. Commonw. Mycol.
Inst., England, Mycol. Paper 106.
446. EMMOND, G.S. & R.J. LEDINGHAM. 1972.
Can. J. Plant Sci. 52: 605-611.
447. ENERSON, P.M. & R.B. HUNTER. 1980.
Can. J. Plant Sci. 60:
1123-1128.
448. ENERSON, P.M. & R.B. HUNTER. 1980.
Can. J. Plant Sci. 60:
1459-1465.
449. ERIKSSON, J. & L. RYVARDEN. 1975.
The Corticiaceae of North Europe,
Vol. 3. Fungiflora, Oslo.
450. ESTEY, R.H. 1959. Soc. Qué.
Protection Plantes 41: 83-86.
451. ESTEY, R.H. & V. PHOKTHAVI. 1977.
Phytoprotection 58: 129.
452. ETHERIDGE, D.E. 1961. Can. J. Bot.
39: 799-816.
453. ETHERIDGE, D.E. 1965. Can.
Phytopath. Soc. Proc. 32: 12.
454. ETHERIDGE, D.E. 1969. Can. J. Bot.
47: 457-479.
455. ETHERIDGE, D.E. & H.M. CRAIG. 1976.
Can. J. Forest Res. 6: 299-318.
456. ETHERIDGE, D.E. & L.A. MORIN. 1963.
Can. J. Bot. 41: 1532-1534.
457. ETHERIDGE, D.E. & L.A. MORIN. 1967.
Can. J. Bot. 45: 1003-1010.
458. ETHERIDGE, D.E., H.M. CRAIG & S.H.
FARRIS. 1979. Can. Phytopath.
Soc. Proc. 38: 14.

459. ETHERIDGE, D.E., H.M. CRAIG & L.D.
     TAYLOR. 1969. Can. Phytopath.
     Soc. Proc. 36: 15.
460. EVANS, D., D.P. LOWE & R.S. HUNT.
     1978. Can. Forestry Service,
     Victoria. Report BC-X-169.
461. EVANS, G. & C.D. MCKEEN. 1975. Can.
     J. Plant Sci. 55: 857-859.
462. FARNWORTH, E.R. & G.A. NEISH. 1980.
     Can. J. Plant Sci. 60: 727-731.
463. FARR, E.R., O.K. MILLER & D.F. FARR.
     1977. Can. J. Bot. 55:
     1167-1180.
464. FARR, M.L. 1963. Mycologia 55:
     226-246.
465. FINLAYSON, D.G. & C.J. CAMPBELL.
     1971. Can. Plant Dis. Survey
     51: 122-126.
466. FISCHER, G.W. 1953. Manual of the
     North American smut fungi.
     Ronald Press, New York.
467. FISCHER, L.A. 1967. Phytoprotection
     48: 23-43.
468. FISHER, R.W. & E.A. KERR. 1964. Can.
     J. Plant Sci. 44: 133-138.
469. FISHER, R.W., G.C. CHAMBERLAIN & W.G.
     KEMP. 1960. Plant Disease Rep.
     44: 273-275.
470. FLEISCHMANN, G. 1963. Can. J. Bot.
     41: 1569-1584.
471. FLEISCHMANN, G. 1963. Can. J. Bot.
     41: 1613-1615.
472. FLEISCHMANN, G. 1963. Can. Plant
     Dis. Survey 43: 168-172.
473. FLEISCHMANN, G. 1964. Can. J. Bot.
     42: 1151-1157.
474. FLEISCHMANN, G. 1965. Can. Plant
     Dis. Survey 45: 15-18.
475. FLEISCHMANN, G. 1965. Plant Disease
     Rep. 49: 132-133.
476. FLEISCHMANN, G. 1966. Can. Plant
     Dis. Survey 46: 22-23.
477. FLEISCHMANN, G. 1967. Can. Plant
     Dis. Survey 47: 11-13.
478. FLEISCHMANN, G. 1968. Can. Plant
     Dis. Survey 48: 14-16.
479. FLEISCHMANN, G. 1968. Can. Plant
     Dis. Survey 48: 99-101.
480. FLEISCHMANN, G. 1969. Can. Plant
     Dis. Survey 49: 91-94.
481. FLEISCHMANN, G. 1971. Can. Plant
     Dis. Survey 51: 14-16.
482. FLEISCHMANN, G. 1972. Can. Plant
     Dis. Survey 52: 15-16.
483. FLEISCHMANN, G., D.J. SAMBORSKI & B.
     PETURSON. 1963. Can. J. Bot.
     41: 481-487.
484. FORSYTH, F.R. & B. PETURSON. 1960.
     Plant Disease Rep. 44: 208-211.
485. FORTIN, J.A., Y. PICHE & M. LALONDE.
     1980. Can. J. Bot. 58: 361-365.
486. FREEMAN, J.A. 1964. Can. Plant Dis.
     Survey 44: 96-104.
487. FREEMAN, J.A. 1965. Can. Plant Dis.
     Survey 45: 107-110.
488. FREEMAN, J.A. & H.S. PEPIN. 1967.
     Can. Plant Dis. Survey 47:
     104-107.
489. FREEMAN, J.A. & H.S. PEPIN. 1969.
     Can. Plant Dis. Survey 49: 139.
490. FREVE, A. & C. RICHARD. 1978.
     Phytoprotection 59: 188.
491. FULLERTON, R.A. & J. NIELSEN. 1974.
     Can. J. Plant Sci. 54: 253-257.
492. FUNK, A. 1962. Can. J. Bot. 40:
     331-335.
493. FUNK, A. 1963. Can. J. Bot. 41:
     503-543.
494. FUNK, A. 1964. Can. J. Bot. 42:
     769-775.
495. FUNK, A. 1964. Plant Disease Rep.
     48: 677.
496. FUNK, A. 1965. Can. J. Bot. 43:
     45-48.
497. FUNK, A. 1965. Can. J. Bot. 43:
     929-932.
498. FUNK, A. 1966. Can. J. Bot. 44:
     219-222.
499. FUNK, A. 1967. Can. J. Bot. 45:
     1803-1809.
500. FUNK, A. 1967. Can. J. Bot. 45:
     2263-2266.
501. FUNK, A. 1968. Can. J. Bot. 46:
     601-603.
502. FUNK, A. 1969. Can. J. Bot. 47:
     751-753.
503. FUNK, A. 1969. Can. J. Bot. 47:
     1509-1511.
504. FUNK, A. 1970. Can. J. Bot. 48:
     1023-1025.
505. FUNK, A. 1972. Plant Disease Rep.
     56: 645-647.
506. FUNK, A. 1973. Can. J. Bot. 51:
     1249-1250.
507. FUNK, A. 1973. Can. J. Bot. 51:
     1643-1645.
508. FUNK, A. 1974. Can. Plant Dis.
     Survey 54: 166-168.
509. FUNK, A. 1975. Can. J. Bot. 53:
     1196-1199.
510. FUNK, A. 1975. Can. J. Bot. 53:
     2297-2302.
511. FUNK, A. 1976. Can. J. Bot. 54:
     868-871.
512. FUNK, A. 1976. Can. J. Bot. 54:
     2852-2856.
513. FUNK, A. 1978. Can. J. Bot. 56:
     245-247.
514. FUNK, A. 1978. Can. J. Bot. 56:
     1575-1578.
515. FUNK, A. 1979. Can. J. Bot. 57:
     4-6.
516. FUNK, A. 1979. Can. J. Bot. 57:
     7-10.
517. FUNK, A. 1979. Can. J. Bot. 57:
     765-767.
518. FUNK, A. 1979. Can. J. Bot. 57:
     2113-2115.
519. FUNK, A. 1980. Can. J. Bot. 58:
     1291-1294.
520. FUNK, A. 1980. Can. J. Bot. 58:
     2447-2449.
521. FUNK, A. & J.A. BARANYAY. 1973. Can.
     Plant Dis. Survey 53: 182.
522. FUNK, A. & J. KUIJT. 1970. Can. J.
     Bot. 48: 1481-1483.
523. FUNK, A. & A.K. PARKER. 1966. Can.
     J. Bot. 44: 1171-1176.
524. FUNK, A. & A.K. PARKER. 1972. Can.
     J. Bot. 50: 1623-1625.

274

525. FUNK, A. & R.A. SHOEMAKER. 1971. Mycologia 63: 567-574.
526. FUNK, A. & B.C. SUTTON. 1972. Can. J. Bot. 50: 1513-1518.
527. FUNK, A. & H. ZALASKY. 1975. Can. J. Bot. 53: 752-755.
528. FUNK, A., R.B. SMITH & J.A. BARANYAY. 1973. Can. J. Forest Res. 3: 71-74.
529. FURLAN, V. & J.A. FORTIN. 1973. Naturaliste Can. 100: 467-477.
530. FUSHTEY, S.G. 1960. Can. Plant Dis. Survey 40: 65.
531. FUSHTEY, S.G. 1961. Can. J. Plant Sci. 41: 568-577.
532. FUSHTEY, S.G. 1975. Can. Plant Dis. Survey 55: 87-90.
533. FUSHTEY, S.G. 1980. Can. Plant Dis. Survey 60: 25-31.
534. FUSHTEY, S.G. & D.K. TAYLOR. 1977. Can. Plant Dis. Survey 57: 29-30.
535. GAGNON, C. 1961. Can. J. Bot. 39: 1087-1093.
536. GALES, L.F. & C.G. MORTIMORE. 1969. Can. Plant Dis. Survey 49: 128-131.
537. GAMS, W. 1971. Cephalosporium-artige Schimmelpilze. G. Fischer, Stuttgart.
538. GATES, L.F. 1970. Can. J. Plant Sci. 50: 679-684.
539. GATES, L.F. 1970. Can. Phytopath. Soc. Proc. 37: 22.
540. GATES, L.F. & B. BOLWYN. 1972. Can. Plant Dis. Survey 52: 64-69.
541. GATES, L.F. & C.G. MORTIMORE. 1972. Can. J. Plant Sci. 52: 929-935.
542. GATES, L.F., C.D. MACKEEN, C.G. MORTIMORE, J.C. SUTTON & A.T. BOLTON. 1971. Can. Plant Dis. Survey 51: 32-37.
543. GAYED, S.K. 1969. Can. Plant Dis. Survey 49: 70-74.
544. GAYED, S.K. 1971. Can. Plant Dis. Survey 51: 142-144.
545. GAYED, S.K. 1972. Can. J. Plant Sci. 52: 103-106.
546. GAYED, S.K. 1972. Can. J. Plant Sci. 52: 869-873.
547. GAYED, S.K. 1976. Phytoprotection 57: 109-115.
548. GAYED, S.K. 1978. Can. Plant Dis. Survey 58: 104-106.
549. GAYED, S.K. & M.C. WATSON. 1975. Can. Plant Dis. Survey 55: 31-35.
550. GAYED, S.K., D.J.S. BARR & L.K. WERESUB. 1978. Can. Plant Dis. Survey 58: 15-19.
551. GENEREUX, H. 1961. Can. Plant Dis. Survey 41: 361-362.
552. GENEREUX, H. & C. AUBE. 1966. Can. Plant Dis. Survey 46: 9-10.
553. GESSNER, R.V. & J. KOHLMEYER. 1976. Can. J. Bot. 54: 2023-2037.
554. GILBERTSON, R.L. 1960. Can. J. Bot. 38: 87-91.
555. GILBERTSON, R.L. 1962. Mycologia 54: 658-677.
556. GILBERTSON, R.L. 1965. Mycologia 57: 845-871.
557. GILBERTSON, R.L. 1965. Papers Michigan Acad. Sci., Arts, Letters 50: 161-184.
558. GILBERTSON, R.L. 1979. Mycotaxon 9: 51-89.
559. GILBERTSON, R.L. & A.B. BUDINGTON. 1970. Mycologia 62: 673-678.
560. GILBERTSON, R.L., G.B. CUMMINS & E.D. DARNELL. 1979. Mycotaxon 10: 49-92.
561. GILLESPIE, T.J. & J.C. SUTTON. 1979. Can. J. Plant Path. 1: 95-99.
562. GINNS, J. 1968. Plant Disease Rep. 52: 579-580.
563. GINNS, J. 1970. Can. J. Bot. 48: 1039-1043.
564. GINNS, J. 1974. Can. J. Forest Res. 4: 143-146.
565. GINNS, J. 1974. Fungi Canadenses No. 42, National Mycological Herbarium, Agriculture Canada, Ottawa.
566. GINNS, J. 1974. Fungi Canadenses No. 52, National Mycological Herbarium, Agriculture Canada, Ottawa.
567. GINNS, J. 1975. Fungi Canadenses No. 68, National Mycological Herbarium, Agriculture Canada, Ottawa.
568. GINNS, J. 1976. Can. J. Bot. 54: 100-167.
569. GINNS, J. 1976. Fungi Canadenses No. 85, National Mycological Herbarium, Agriculture Canada, Ottawa.
570. GINNS, J. 1976. Fungi Canadenses No. 86, National Mycological Herbarium, Agriculture Canada, Ottawa.
571. GINNS, J. 1978. Can. J. Bot. 56: 1953-1973.
572. GINNS, J. 1979. Bot. Notiser 132: 93-102.
573. GINNS, J. 1980. Fungi Canadenses No. 168, National Mycological Herbarium, Agriculture Canada, Ottawa.
574. GINNS, J. 1980. Fungi Canadenses No. 169, National Mycological Herbarium, Agriculture Canada, Ottawa.
575. GINNS, J. 1980. Fungi Canadenses No. 174, National Mycological Herbarium, Agriculture Canada, Ottawa.
576. GINNS, J. 1982. Can. Field-Nat. 96: 131-138.
577. GINNS, J. 1982. Opera Bot. 61: 1-61.
578. GINNS, J. 1983. Mycotaxon 18: 439-442.
579. GINNS, J. & R. MACRAE. 1971. Can. J. Bot. 49: 899-902.
580. GINNS, J. & S. SUNHEDE. 1978. Bot. Notiser 131: 167-173.
581. GINNS, J. & L.K. WERESUB. 1976. Memoirs New York Bot. Garden 28: 86-97.

582. GOCHNAUER, T.A. & S.J. HUGHES. 1976. Can. Ent. 108: 985-988.
583. GOCHNAUER, T.A., S.J. HUGHES & J. CORNER. 1972. Can. Agric., Ottawa 17(2): 36-37.
584. GOOD, H.M. & J.I. NELSON. 1962. Can. J. Bot. 40: 615-627.
585. GOOD, H.M., J.T. BASHAM & S.D. KADZIELAWA. 1968. Can. J. Bot. 46: 27-36.
586. GOOS, R.D. 1969. Mycologia 61: 52-56.
587. GOOS, R.D. 1969. Mycologia 61: 1048-1053.
588. GOOSSEN, P. G. & W.E. SACKSTON. 1964. Phytoprotection 45: 133.
589. GOREE, H. 1974. Can. J. Bot. 52: 1265-1269.
590. GOURLEY, C.O. 1961. Can. Plant Dis. Survey 41: 174.
591. GOURLEY, C.O. 1962. Can. J. Plant Sci. 42: 122-129.
592. GOURLEY, C.O. 1963. Can. J. Plant Sci. 43: 462-468.
593. GOURLEY, C.O. 1966. Can. J. Plant Sci. 46: 531-536.
594. GOURLEY, C.O. 1968. Can. J. Plant Sci. 48: 267-272.
595. GOURLEY, C.O. 1968. Can. Plant Dis. Survey 48: 32-36.
596. GOURLEY, C.O. 1969. Can. J. Bot. 47: 945-950.
597. GOURLEY, C.O. 1971. Can. Plant Dis. Survey 51: 135-137.
598. GOURLEY, C.O. 1972. Can. J. Bot. 50: 49-51.
599. GOURLEY, C.O. 1972. Can. Plant Dis. Survey 52: 85-87.
600. GOURLEY, C.O. 1975. Can. J. Plant Sci. 55: 439-442.
601. GOURLEY, C.O. 1979. Can. Plant Dis. Survey 59: 15-17.
602. GOURLEY, C.O. 1979. Can. Plant Dis. Survey 59: 18.
603. GOURLEY, C.O. 1979. Can. Plant Dis. Survey 59: 80.
603a. GOURLEY, C.O. 1983. Proc. Nova Scotian Inst. Sci. 32: 75-295.
604. GOURLEY, C.O. & D.L. CRAIG. 1968. Can. Plant Dis. Survey 48: 93-94.
605. GOURLEY, C.O. & R.W. DELBRIDGE. 1972. Can. Plant Dis. Survey 52: 97-98.
606. GOURLEY, C.O. & R.W. DELBRIDGE. 1973. Can. Plant Dis. Survey 53: 79-82.
607. GOURLEY, C.O. & K.A. HARRISON. 1969. Can. Plant Dis. Survey 49: 22-26.
608. GOURLEY, C.O. & A.A. MACNAB. 1964. Can. J. Plant Sci. 44: 544-549.
609. GOURLEY, C.O. & N.L. NICKERSON. 1979. Can. J. Bot. 57: 1218-1219.
610. GRAHAM, J.H. & E.S. LUTTRELL. 1961. Phytopathology 51: 680-693.
611. GRAHAM, K.M., M.P. HARRISON & C.E. SMITH. 1962. Can. Plant Dis. Survey 42: 205-207.
612. GRAHAM, K.M., R.A. SHOEMAKER & S.R. COLPITTS. 1964. Can. Plant Dis. Survey 44: 113-117.
613. GRAY, B. & W.E. SACKSTON. 1979. Phytoprotection 60: 163.
614. GREEN, G.J. 1963. Can. Plant Dis. Survey 43: 173-176.
615. GREEN, G.J. 1963. Can. Plant Dis. Survey 43: 177-182.
616. GREEN, G.J. 1963. Can. Plant Dis. Survey 43: 196.
617. GREEN, G.J. 1964. Can. J. Bot. 42: 1653-1664.
618. GREEN, G.J. 1965. Can. Plant Dis. Survey 45: 19-22.
619. GREEN, G.J. 1965. Can. Plant Dis. Survey 45: 23-29.
620. GREEN, G.J. 1966. Can. Plant Dis. Survey 46: 27-32.
621. GREEN, G.J. 1967. Can. Plant Dis. Survey 47: 5-8.
622. GREEN, G.J. 1968. Can. Plant Dis. Survey 48: 9-13.
623. GREEN, G.J. 1968. Can. Plant Dis. Survey 48: 104-106.
624. GREEN, G.J. 1969. Can. Plant Dis. Survey 49: 78-79.
625. GREEN, G.J. 1969. Can. Plant Dis. Survey 49: 83-87.
626. GREEN, G.J. 1971. Can. Plant Dis. Survey 51: 20-23.
627. GREEN, G.J. 1972. Can. Plant Dis. Survey 52: 11-14.
628. GREEN, G.J. 1972. Can. Plant Dis. Survey 52: 162-167.
629. GREEN, G.J. 1974. Can. Plant Dis. Survey 54: 11-15.
630. GREEN, G.J. 1975. Can. Plant Dis. Survey 55: 51-57.
631. GREEN, G.J. 1976. Can. Plant Dis. Survey 56: 15-18.
632. GREEN, G.J. 1976. Can. Plant Dis. Survey 56: 119-122.
633. GREEN, G.J. 1978. Can. Plant Dis. Survey 58: 44-48.
634. GREEN, G.J. 1979. Can. Plant Dis. Survey 59: 43-47.
635. GREEN, G.J. 1980. Can. J. Plant Path. 2: 241-245.
636. GREEN, G.J. & V.M. BENDELOW. 1961. Can. J. Plant Sci. 41: 431-435.
637. GREEN, G.J. & A.B. CAMPBELL. 1979. Can. J. Plant Path. 1: 3-11.
638. GREEN, G.J. & R.I.H. MCKENZIE. 1964. Can. J. Plant Sci. 44: 418-426.
639. GREEN, G.J. & D.J. SAMBORSKI. 1961. Can. Plant Dis. Survey 41: 1-21.
640. GREEN, G.J. & D.J. SAMBORSKI. 1962. Can. Plant Dis. Survey 42: 1-18.
641. GREEN, G.J., F.J. ZILLINSKY, R.V. CLARK & D.J. SAMBORSKI. 1962. Can. Plant Dis. Survey 42: 129-134.
642. GREEN, G.T., T. JOHNSON & J.N. WELSH. 1961. Can. J. Plant Sci. 41: 153-165.
643. GREENE, H.C. & G.B. CUMMINS. 1967. Mycologia 59: 47-57.
644. GREMMEN, J. & J.A. PARMELEE. 1973. Can. Forestry Service, Edmonton. Report NOR-X-44.
645. GRIFFIN, H.D. 1964. Plant Disease Rep. 48: 615.

646. GRIFFIN, H.D. 1965. Forestry
     Chronicle 42: 295-300.
647. GRIFFIN, H.D. 1968. Can. J. Bot.
     46: 689-718.
648. GRIFFIN, H.D. 1969. Can. J. Bot.
     47: 761-771.
649. GROSS, H.L. 1969. Forestry
     Chronicle 45: 326.
650. GROSS, H.L. 1978. Can. J. Forest
     Res. 8: 47-53.
651. GROSS, H.L., A.R. EK & R.F. PATTON.
     1980. Can. J. Forest Res. 10:
     190-198.
652. GROSS, H.L., R.F. PATTON & A.R. EK.
     1980. Can. J. Forest Res. 10:
     199-208.
653. GROTH, J.V. 1975. Can. J. Bot. 53:
     2233-2239.
654. GROVES, J.W. 1940. Mycologia 32:
     736-751.
655. GROVES, J.W. 1954. Can. J. Bot.
     32: 116-144.
656. GROVES, J.W. 1965. Can. J. Bot.
     43: 1195-1276.
657. GROVES, J.W. 1967. Can. J. Bot.
     45: 169-181.
658. GROVES, J.W. 1968. Can. J. Bot.
     46: 1273-1278.
659. GROVES, J.W. 1968. Mycologia 60:
     718-719.
660. GROVES, J.W. 1969. Can. J. Bot.
     47: 1319-1331.
661. GROVES, J.W. & M.E. ELLIOTT. 1961.
     Bull. Res. Council Israel 10D:
     150-156.
662. GROVES, J.W. & M.E. ELLIOTT. 1961.
     Can. J. Bot. 39: 215-230.
663. GROVES, J.W. & M.E. ELLIOTT. 1969.
     Friesia 9: 29-36.
664. GROVES, J.W. & M.E. ELLIOTT. 1971.
     Rep. Kevo Subarctic Res. Sta.
     8: 22-30.
665. GROVES, J.W. & S.C. HOARE. 1953.
     Can. Field-Nat. 67: 95-102.
666. GROVES, J.W. & S.C. HOARE. 1954.
     Can. Field-Nat. 68: 1-8.
666a. GROVES, J.W. & C.A. LOVELAND. 1953.
     Mycologia 45: 415-425.
667. GROVES, J.W. & A.J. SKOLKO. 1944.
     Can. J. Res., C, 22: 190-199.
668. GROVES, J.W. & S.C. THOMSON. 1955.
     Can. Field-Nat. 69: 44-51.
669. GROVES, J.W., S.C, THOMSON & M.
     PANTIDOU. 1958. Can.
     Field-Nat. 72: 133-138.
670. GUTTMAN, S. & W.E. SACKSTON. 1979.
     Phytoprotection 60: 163.
671. HAAS, J.H. & B. BOLWYN. 1972. Can.
     J. Plant Sci. 52: 525-533.
672. HAGBORG, W.A.F., A.W. CHIKO, G.
     FLEISCHMANN, C.C. GILL, G.J.
     GREEN, J.W. MARTENS, J.J. NIELSON
     & D.J. SAMBORSKI. 1972. Can.
     Plant Dis. Survey 52: 113-118.
673. HAGEN, P.O. & A.G. ROSE. 1961. Can.
     J. Microbiol. 7: 287-294.
674. HAHN, N.J., E.A. PETERSON & R.A.
     SHOEMAKER. 1966. Can. Plant
     Dis. Survey 46: 99.

675. HALL, R. 1970. Can. Plant Dis.
     Survey 50: 124-125.
676. HALL, R. 1975. Can. J. Bot. 53:
     452-455.
677. HALL, R. 1976. Can. Phytopath. Soc.
     Proc. 43: 29.
678. HALL, R. 1979. Can. Phytopath. Soc.
     Proc. 46: 57.
679. HAMET-AHTI, L. 1972. Syesis 5:
     83-85.
680. HAMILTON, D.G., R.V. CLARK, A.E.
     HANNAH & R. LOISELLE. 1960.
     Can. J. Plant Sci. 40: 713-720.
681. HAMMOND, G.H. 1961. Can. Field-Nat.
     75: 41-42.
682. HAMPSON, M.C. 1977. Can. Plant Dis.
     Survey 57: 75-78.
683. HAMPSON, M.C. & N.F. HAARD. 1980.
     Can. J. Plant Path. 2: 143-147.
684. HAMPSON, M.C., K.G. PROUDFOOT & C.R.
     KELLY. 1976. Can. Plant Dis.
     Survey 56: 73.
685. HARDER, D.E. 1975. Can. Plant Dis.
     Survey 55: 63-65.
686. HARDER, D.E. 1976. Can. Plant Dis.
     Survey 56: 19-22.
687. HARDER, D.E. 1976. Can. Plant Dis.
     Survey 56: 129-131.
688. HARDER, D.E. 1978. Can. J. Bot.
     56: 214-224.
689. HARDER, D.E. 1978. Can. Plant Dis.
     Survey 58: 39-43.
690. HARDER, D.E. 1979. Can. Plant Dis.
     Survey 59: 35-37.
691. HARDER, D.E. 1980. Can. J. Plant
     Path. 2: 249-252.
692. HARDER, D.E. & R.I.H. MCKENZIE.
     1974. Can. Plant Dis. Survey
     54: 16-18.
693. HARDER, D.E. & W.P. SKOROPAD. 1968.
     Can. Plant Dis. Survey 48:
     39-42.
694. HARDING, H. 1969. Can. Plant Dis.
     Survey 49: 126-127.
695. HARDING, H. 1970. Can. Plant Dis.
     Survey 50: 126-129.
696. HARDING, H. 1970. Can. Plant Dis.
     Survey 50: 142-144.
697. HARDING, H. 1971. Can. J. Bot. 49:
     281-287.
698. HARDING, H. 1972. Can. J. Bot. 50:
     1805-1810.
699. HARDING, H. 1972. Can. Plant Dis.
     Survey 52: 149-150.
700. HARDING, H. 1973. Can. J. Bot. 51:
     9-13.
701. HARDING, H. 1973. Can. J. Bot. 51:
     2514-2516.
702. HARDING, H. & R.A.A. MORRALL. 1973.
     Can. Plant Dis. Survey 53: 60.
703. HARPER, A.M., A.D. SMITH & F.R.
     HARPER. 1969. Can. J. Plant
     Sci. 49: 531-533.
704. HARPER, F.R. 1968. Plant Disease
     Rep. 52: 565-568.
705. HARPER, F.R. & P. BERGEN. 1976. Can.
     Plant Dis. Survey 56: 48-52.
706. HARPER, F.R. & L.J. PIENING. 1974.
     Can. Plant Dis. Survey 54: 1-5.

707. HARPER, F.R. & W.L. SEAMAN. 1980.
Can. J. Plant Path. 2: 222-226.
708. HARPER, F.R. & W.L. SEAMAN. 1980.
Can. J. Plant Path. 2: 227-231.
709. HARRISON, K.A. 1961. Can. Plant Dis.
Survey 41: 175-178.
710. HARRISON, K.A. 1962. Can. Plant Dis.
Survey 42: 122-123.
711. HARRISON, K.A. 1965. Can. Plant Dis.
Survey 45: 94-95.
712. HARRISON, K.A. 1973. Michigan Bot.
12: 177-194.
713. HARRISON, K.A. & C.L. LOCKHART.
1963. Can. Plant Dis. Survey
43: 33-38.
714. HARTMANN, H. & A.K. WATSON. 1980.
Can. J. Plant Path. 2: 173-175.
715. HAWKSWORTH, D.L. 1971. Commonw.
Mycol. Inst., England, Mycol.
Paper 126.
716. HAWN, E.J. 1961. Can. Plant Dis.
Survey 42: 371.
717. HENNEBERT, G.L. 1962. Can. J. Bot.
40: 1203-1216.
718. HENNEBERT, G.L. & J.W. GROVES. 1963.
Can. J. Bot. 41: 341-370.
719. HENNEN, J.F. 1967. Mycologia 59:
264-273.
720. HENNEN, J.F. & G.B. CUMMINS. 1969.
Mycologia 61: 340-356.
721. HENRY, A.W. & C. ELLIS. 1971. Can.
Plant Dis. Survey 51: 76-79.
722. HENRY, A.W. & D. STELFOX. 1971. Can.
Phytopath. Soc. Proc. 38: 16.
723. HENRY, J.V. & G.J. PELLETIER. 1979.
Phytoprotection 60: 172.
724. HESLER, L.R. & A.H. SMITH. 1965.
North American species of
Crepidotus. Hafner, New York.
725. HILDEBRAND, A.A. & L.W. KOCH. 1952.
Can. Phytopath. Soc. Proc. 20:
17.
726. HILL, J.T., D.J. ORMROD & R.J.
COPEMAN. 1977. Can. Plant Dis.
Survey 57: 71-74.
727. HINDS, T.E. & T.H. LAURENT. 1978.
Plant Disease Rep. 62: 972-975.
728. HIRATSUKA, Y. 1965. Can. J. Bot.
43: 475-478.
729. HIRATSUKA, Y. 1970. Can. J. Bot.
48: 433-435.
730. HIRATSUKA, Y. 1973. Mycologia 65:
137-144.
731. HIRATSUKA, Y. & A. FUNK. 1976. Plant
Disease Rep. 60: 631.
732. HIRATSUKA, Y. & E.J. GAUTREAU. 1966.
Plant Disease Rep. 50: 419.
733. HIRATSUKA, Y. & P.J. MARUYAMA. 1968.
Plant Disease Rep. 52: 650-651.
734. HIRATSUKA, Y. & P.J. MARUYAMA. 1976.
Plant Disease Rep. 60: 241.
735. HIRATSUKA, Y. & J.M. POWELL. 1976.
Can. Forestry Service, Ottawa.
Tech. Report 4.
736. HIRATSUKA, Y. & A. TSUNEDA. 1978.
Can. Phytopath. Soc. Proc. 45:
36.
737. HIRATSUKA, Y., L.E. McARTHUR & F.J.
EMOND. 1967. Can. J. Bot. 45:
1913-1915.
738. HIRATSUKA, Y., W. MORF & J.M. POWELL.
1966. Can. J. Bot. 44:
1639-1643.
739. HIRATSUKA, Y., A. TSUNEDA & L.
SIGLER. 1979. Plant Disease
Rep. 63: 512-513.
740. HJORTSTAM, K. & E. HOGHOLEN. 1980.
Mycotaxon 10: 265-268.
741. HJORTSTAM, K. & K.-H. LARSSON. 1977.
Mycotaxon 5: 475-480.
742. HJORTSTAM, K. & K.-H. LARSSON. 1978.
Mycotaxon 7: 117-124.
743. HJORTSTAM, K. & L. RYVARDEN. 1979.
Mycotaxon 10: 201-209.
744. HJORTSTAM, K. & L. RYVARDEN. 1979.
Mycotaxon 9: 505-519.
745. HOCKING, D. 1970. Can. Plant Dis.
Survey 50: 121-123.
746. HOCKING, D. 1971. Can. J. Forest
Res. 1: 208-215.
747. HODGES, C.S. 1962. Mycologia 54:
62-69.
748. HOES, J.A. 1969. Can. Plant Dis.
Survey 49: 27.
749. HOES, J.A. 1971. Can. J. Bot. 49:
1863-1866.
750. HOES, J.A. & H.C. HUANG. 1976. Can.
Plant Dis. Survey 56: 75-76.
751. HOES, J.A. & E.G. KENASCHUK. 1968.
Can. Plant Dis. Survey 48: 153.
752. HOES, J.A. & E.D. PUTT. 1962. Can.
Plant Dis. Survey 42: 256-257.
753. HOES, J.A. & E.D. PUTT. 1963. Can.
Plant Dis. Survey 43: 210-211.
754. HOES, J.A. & E.D. PUTT. 1964. Can.
Plant Dis. Survey 44: 236-237.
755. HOES, J.A. & I.H. TYSON. 1963. Plant
Disease Rep. 47: 836.
756. HOES, J.A. & I.H. TYSON. 1966. Plant
Disease Rep. 50: 62-63.
757. HOES, J.A. & D.E. ZIMMER. 1976.
Plant Disease Rep. 60:
1010-1013.
758. HOES, J.A. & R.C. ZIMMER. 1968. Can.
Plant Dis. Survey 48: 152-153.
759. HOES, J.A., G.W. BRUEHL & C.G. SHAW.
1965. Mycologia 57: 904-912.
760. HOLM, L. 1957. Symb. Bot. Upsal.
14: 1-188.
761. HOLM, L. 1961. Svensk Bot. Tidsk.
55: 73.
762. HOLM, L. 1963. Svensk Bot. Tidskr.
57: 129-144.
763. HOLM, L. 1968. Svensk Bot. Tidskr.
62: 217-242.
764. HOLM, L. 1975. Svensk Bot. Tidskr.
69: 143-160.
765. HOLUBOVA-JECHOVA, V. 1974. Folia
Geobot. Phytotax., Praha 9:
425-432.
766. HOLUBOVA-JECHOVA, V. 1982. Folia
Geobot. Phytotax., Praha 17:
295-327.
767. HOPKINS, J.C. 1961. Can. J. Bot.
39: 1521-1529.
768. HOPKINS, J.C. 1963. Can. J. Bot.
41: 1535-1545.
769. HORNER, R.M. 1952. Can. Phytopath.
Soc. Proc. 20: 17.

770. HORNER, R.M. 1953. Mycological Investigations of Birch in Ontario. *In*: Report of the Symposium on Birch Dieback. Can. Dept. Agr., Sci. Service, Forest Biology Div., Ottawa. Part 2: 144-146.

771. HORNER, R.M. 1955. Can. Phytopath. Soc. Proc. 23: 16-17.

772. HOWARD, R.J. 1979. Can. Phytopath. Soc. Proc. 46: 60.

773. HOWARD, R.J. & R.A.A. MORRALL. 1971. Can. Phytopath. Soc. Proc. 38: 17.

774. HUANG, H.C. 1978. Can. J. Bot. 56: 2243-2246.

775. HUANG, H.C. 1980. Can. J. Plant Path. 2: 26-32.

776. HUANG, H.C. & J. DUECK. 1980. Can. J. Plant Path. 2: 47-52.

777. HUANG, H.C. & J.A. HOES. 1980. Plant Disease 64: 81-84.

778. HUBBES, M. 1966. Can. J. Bot. 44: 365-386.

779. HUBBES, M. & R. d'ASTOUS. 1967. Can. J. Bot. 45: 1145-1153.

780. HUDAK, J. & P. SINGH. 1970. Can. Plant Dis. Survey 50: 99-101.

781. HUDAK, J. & R.E. WELLS. 1974. Forestry Chronicle 50: 74-76.

782. HUGHES, G.C. 1969. Syesis 2: 121-140.

783. HUGHES, G.C. 1973. Syesis 6: 233-238.

784. HUGHES, S.J. 1960. Can. J. Bot. 38: 659-696.

785. HUGHES, S.J. 1965. N. Zealand J. Bot. 3: 136-150.

786. HUGHES, S.J. 1968. Can. J. Bot. 46: 1099-1107.

787. HUGHES, S.J. 1968. Can. J. Bot. 46: 939-943.

788. HUGHES, S.J. 1971. Annellophores. *In*: B. Kendrick ed., Taxonomy of Fungi Imperfecti. pp. 132-140.

789. HUGHES, S.J. 1971. Phycomycetes, Basidiomycetes, and Ascomycetes as Fungi Imperfecti. *In*: B. Kendrick ed., Taxonomy of Fungi Imperfecti. pp. 7-36.

790. HUGHES, S.J. 1973. Fungi Canadenses No. 4, National Mycological Herbarium, Agriculture Canada, Ottawa.

791. HUGHES, S.J. 1973. Fungi Canadenses No. 5, National Mycological Herbarium, Agriculture Canada, Ottawa.

792. HUGHES, S.J. 1973. Fungi Canadenses No. 6, National Mycological Herbarium, Agriculture Canada, Ottawa.

793. HUGHES, S.J. 1973. Fungi Canadenses No. 7, National Mycological Herbarium, Agriculture Canada, Ottawa.

794. HUGHES, S.J. 1973. Fungi Canadenses No. 8, National Mycological Herbarium, Agriculture Canada, Ottawa.

795. HUGHES, S.J. 1973. Fungi Canadenses No. 9, National Mycological Herbarium, Agriculture Canada, Ottawa.

796. HUGHES, S.J. 1974. Fungi Canadenses No. 34, National Mycological Herbarium, Agriculture Canada, Ottawa.

797. HUGHES, S.J. 1974. N. Zealand J. Bot. 12: 299-356.

798. HUGHES, S.J. 1976. Mycologia 68: 693-820.

799. HUGHES, S.J. 1978. Fungi Canadenses No. 122, National Mycological Herbarium, Agriculture Canada, Ottawa.

800. HUGHES, S.J. 1978. Fungi Canadenses No. 123, National Mycological Herbarium, Agriculture Canada, Ottawa.

801. HUGHES, S.J. 1978. Fungi Canadenses No. 124, National Mycological Herbarium, Agriculture Canada, Ottawa.

802. HUGHES, S.J. 1978. Fungi Canadenses No. 125, National Mycological Herbarium, Agriculture Canada, Ottawa.

803. HUGHES, S.J. 1978. Fungi Canadenses No. 126, National Mycological Herbarium, Agriculture Canada, Ottawa.

804. HUGHES, S.J. 1978. Fungi Canadenses No. 129, National Mycological Herbarium, Agriculture Canada, Ottawa.

805. HUGHES, S.J. 1978. N. Zealand J. Bot. 16: 311-370.

806. HUGHES, S.J. 1979. Fungi Canadenses No. 132, National Mycological Herbarium, Agriculture Canada, Ottawa.

807. HUGHES, S.J. 1979. Fungi Canadenses No. 150, National Mycological Herbarium, Agriculture Canada, Ottawa.

808. HUGHES, S.J. 1979. N. Zealand J. Bot. 17: 139-188.

809. HUGHES, S.J. 1980. Fungi Canadenses No. 162, National Mycological Herbarium, Agriculture Canada, Ottawa.

810. HUGHES, S.J. 1980. Fungi Canadenses No. 163, National Mycological Herbarium, Agriculture Canada, Ottawa.

811. HUGHES, S.J. 1980. Fungi Canadenses No. 164, National Mycological Herbarium, Agriculture Canada, Ottawa.

812. HUGHES, S.J. 1980. Fungi Canadenses No. 167, National Mycological Herbarium, Agriculture Canada, Ottawa.

813. HUGHES, S.J. 1980. Fungi Canadenses No. 183, National Mycological Herbarium, Agriculture Canada, Ottawa.

814. HUGHES, S.J. 1980. Fungi Canadenses No. 184, National Mycological Herbarium, Agriculture Canada, Ottawa.

815. HUGHES, S.J. 1980. Fungi Canadenses
No. 185, National Mycological
Herbarium, Agriculture Canada,
Ottawa.
816. HUGHES, S.J. & J.C. COOKE. 1979.
Fungi Canadenses No. 144,
National Mycological Herbarium,
Agriculture Canada, Ottawa.
817. HUGHES, S.J. & J.C. COOKE. 1979.
Fungi Canadenses No. 145,
National Mycological Herbarium,
Agriculture Canada, Ottawa.
818. HUGHES, S.J. & J.C. COOKE. 1979.
Fungi Canadenses No. 146,
National Mycological Herbarium,
Agriculture Canada, Ottawa.
819. HUGHES, S.J. & J.C. COOKE. 1979.
Fungi Canadenses No. 147,
National Mycological Herbarium,
Agriculture Canada, Ottawa.
820. HUGHES, S.J. & J.C. COOKE. 1980.
Fungi Canadenses No. 161,
National Mycological Herbarium,
Agriculture Canada, Ottawa.
821. HUGHES, S.J. & G.L. HENNEBERT. 1963.
Can. J. Bot. 41: 773-809.
822. HUGHES, S.J. & W.I. ILLMAN. 1974.
Fungi Canadenses No. 57, National
Mycological Herbarium,
Agriculture Canada, Ottawa.
823. HUGHES, S.J. & W.I. ILLMAN. 1974.
Fungi Canadenses No. 58, National
Mycological Herbarium,
Agriculture Canada, Ottawa.
824. HUGHES, S.J. & W.I. ILLMAN. 1974.
Fungi Canadenses No. 59, National
Mycological Herbarium,
Agriculture Canada, Ottawa.
825. HUGHES, S.J. & W.I. ILLMAN. 1974.
Fungi Canadenses No. 60, National
Mycological Herbarium,
Agriculture Canada, Ottawa.
826. HUGHES, S.J. & W.B. KENDRICK. 1963.
Can. J. Bot. 41: 693-718.
827. HUGHES, S.J. & E.G. KOKKO. 1975.
Fungi Canadenses No. 69, National
Mycological Herbarium,
Agriculture Canada, Ottawa.
828. HUGHES, S.J. & E.G. KOKKO. 1977.
Fungi Canadenses No. 108,
National Mycological Herbarium,
Agriculture Canada, Ottawa.
829. HUGHES, S.J. & K.A. PIROZYNSKI.
1972. Can. J. Bot. 52:
2521-2534.
830. HULME, M.A. & J.K. SHIELDS. 1972.
Can. J. Bot. 50: 1421-1427.
831. HUNT, K., J.T. BASHAM & J.A.
KEMPERMAN. 1978. Can. J. Forest
Res. 8: 181-187.
832. HUNT, R.S. 1980. Mycotaxon 11:
233-240.
833. HUNT, R.S. & H.J. O'REILLY. 1978.
Plant Disease Rep. 62: 659-660.
834. HUNTLEY, J.H., J.D. CAFLEY & E.
JORGENSON. 1961. Forestry
Chronicle 37: 228-236.
835. ILLINGWORTH, K. 1973. Can. J. Forest
Res. 3: 585-589.

836. ILLMAN, W.I. 1960. Can. Plant Dis.
Survey 40: 45.
837. ILLMAN, W.I. 1964. Plant Disease
Rep. 48: 68.
838. ILLMAN, W.I. 1972. Can. Plant Dis.
Survey 52: 110.
839. IVES, W.G.H. 1972. Can. Forestry
Service, Edmonton. Report
NOR-X-12.
840. JABBAR MIAH, M.A. & W.E. SACKSTON.
1970. Phytoprotection 51: 1-16.
841. JABBAR MIAH, M.A. & W.E. SACKSTON.
1970. Phytoprotection 51: 17-35.
842. JACKSON, R.S. 1972. Can. J. Bot.
50: 869-875.
843. JAMES, W.C. 1971. Can. Plant Dis.
Survey 51: 24-31.
844. JAMES, W.C. & T.R. DAVIDSON. 1971.
Can. Plant Dis. Survey 51:
148-153.
845. JARVIS, W.R. 1977. Canada Dept.
Agric., Ottawa. Monograph No.
15.
846. JARVIS, W.R. & K. SLINGSBY. 1977.
Plant Disease Rep. 61: 728-730.
847. JARVIS, W.R. & H.J. THORPE. 1976.
Plant Disease Rep. 60:
1027-1031.
848. JARVIS, W.R. & H.J. THORPE. 1980.
Plant Disease 64: 309-310.
849. JARVIS, W.R., H.J. THORPE & B.H.
MACNEILL. 1975. Can. Plant
Dis. Survey 55: 25-26.
850. JENG, R.S. & J.C. KRUG. 1977. Can.
J. Bot. 55: 83-95.
851. JHOOTY, J.S. & W.E. MCKEEN. 1962.
Plant Disease Rep. 46: 218-219.
852. JOHNSON, A.L.S., G.W. WALLIS & R.E.
FOSTER. 1972. Forestry
Chronicle 48: 316-319.
853. JOHNSON, D.W. & J.E. KUNTZ. 1978.
Can. J. Bot. 56: 1518-1525.
854. JOHNSON, H.W. 1969. Can. Plant Dis.
Survey 49: 122-125.
855. JOHNSTON, H.W. 1970. Plant Disease
Rep. 54: 91-93.
856. JOHNSTON, H.W. 1972. Can. Plant Dis.
Survey 52: 82-84.
857. JOHNSTON, H.W. 1974. Can. Plant Dis.
Survey 54: 71-73.
858. JOHNSTON, H.W. & J.A. CUTCLIFFE.
1969. Can. Plant Dis. Survey
49: 140.
859. JOHNSTON, H.W. & L.S. THOMPSON.
1972. Can. Plant Dis. Survey
52: 19.
860. JOHNSTON, H.W., J.A. IVANY & J.A.
CUTCLIFFE. 1980. Plant Disease
64: 942-943.
861. JOHNSTON, H.W., J.A. MACLEOD & H.G.
NASS. 1979. Can. Phytopath.
Soc. Proc. 46: 61.
862. JONG, S.C. & C.R. BANJAMIN. 1971.
Mycologia 63: 862-876.
863. JORGENSEN, E. & J.D. CAFLEY. 1961.
Forestry Chronicle 37: 394-400.
864. JÜLICH, W. 1972. Willdenowia, Beih.
7: 1-283.

865. JULIEN, J.B. 1961. Can. Plant Dis.
      Survey 41: 329.
866. JULIEN, J.B. 1966. Phytoprotection
      47: 73-77.
867. KALLIO, P. 1980. Centre for Northern
      Studies and Research, McGill
      University, Montreal. Subarctic
      Res. Paper 30.
868. KALU, N.N., J.C. SUTTON & O.
      VAARTAJA. 1976. Can. J. Plant
      Sci. 56: 555-561.
869. KANKAINEN, E. 1969. Karstenia 9:
      23-34.
870. KEANE, E.M. & W.E. SACKSTON. 1970.
      Can. J. Plant Sci. 50: 415-422.
871. KEMP, W.G. & D.J.S. BARR. 1978.
      Phytopath. Zeitschr. 91:
      203-217.
872. KENDRICK, W.B. 1961. Can. J. Bot.
      39: 817-832.
873. KENDRICK, W.B. 1961. Can. J. Bot.
      39: 1079-1085.
874. KENDRICK, W.B. 1962. Can. J. Bot.
      40: 771-797.
875. KENDRICK, W.B. 1962. Can. J.
      Microbiol. 8: 639-647.
876. KENDRICK, W.B. 1963. Can. J. Bot.
      41: 573-577.
877. KENDRICK, W.B. 1963. Can. J. Bot.
      41: 1015-1023.
878. KENDRICK, W.B. & A.C. MOLNAR. 1965.
      Can. J. Bot. 43: 39-44.
879. KENDRICK, W.B., G.T. COLE & G.C.
      BHATT. 1968. Can. J. Bot. 46:
      591-596.
880. KENNEDY, L.L. & A.W. STEWART. 1967.
      Can. J. Bot. 45: 1597-1604.
881. KENNETH, R. & P.K. ISAAC. 1964. Can.
      J. Plant Sci. 44: 182-187.
882. KHALIL, M.A.K. 1976. Forestry
      Chronicle 53: 150-154.
883. KHARBANDA, P.D. & C.C. BERNIER.
      1975. Can. Phytopath. Soc.
      Proc. 42: 20.
884. KHARBANDA, P.D. & C.C. BERNIER.
      1977. Can. J. Plant Sci. 57:
      745-749.
885. KHARBANDA, P.D. & C.C. BERNIER.
      1977. Plant Disease Rep. 61:
      1051-1053.
886. KHARBANDA, P.D. & C.C. BERNIER.
      1979. Can. J. Plant Sci. 59:
      661-666.
887. KHARBANDA, P.D. & C.C. BERNIER.
      1979. Plant Disease Rep. 63:
      662-663.
888. KLINE, D.M. & R.R. NELSON. 1963.
      Plant Disease Rep. 47: 890-894.
889. KOHLMEYER, J. & E. 1974. Mycologia
      66: 77-86.
890. KOHLMEYER, J. & E. 1979. Marine
      mycology. Academic Press, N.Y.
891. KOHLMEYER, J. & R.V. GESSNER. 1976.
      Can. J. Bot. 54: 1759-1766.
892. KOHN, L.M. 1979. Mycotaxon 9:
      365-444.
893. KOKKO, E.G. & R.A. SHOEMAKER. 1974.
      Fungi Canadenses No. 48, National
      Mycological Herbarium,
      Agriculture Canada, Ottawa.
894. KOKKO, E.G. & R.A. SHOEMAKER. 1974.
      Fungi Canadenses No. 51, National
      Mycological Herbarium,
      Agriculture Canada, Ottawa.
895. KONDO, E.S. 1979. Can. Phytopath.
      Soc. Proc. 46: 62.
896. KONDO, E.S., D.N. ROY & E. JORGENSEN.
      1973. Can. J. Forest Res. 3:
      548-555.
897. KORF, R.P. 1962. Mycologia 54:
      12-33.
898. KORF, R.P. & G.S. ABAWI. 1971. Can.
      J. Bot. 49: 1879-1883.
899. KORF, R.P. & S.C. GRUFF. 1978.
      Mycotaxon 7: 185-203.
900. KOSKE, R.E. 1972. Can. J. Bot. 50:
      2565-2567.
901. KOSKE, R.E. 1974. Mycologia 66:
      298-318.
902. KOSKE, R.E. & P.W. PERRIN. 1971.
      Can. J. Bot. 49: 695-697.
903. KOWALSKI, D.T. & A.A. HINCHEE. 1972.
      Syesis 5: 95-97.
904. KOZAR, F. & H.J. NETOLITZKY. 1975.
      Can. J. Bot. 53: 972-977.
905. KUIJT, J. 1969. Can. J. Bot. 47:
      1359-1365.
906. KUKKONEN, I. 1961. Can. J. Bot.
      39: 155-164.
907. KURTZMAN, C.P. 1977. Mycologia 69:
      547-555.
907a. LACHANCE, D. 1970. Can. J. Bot.
      48: 447-452.
908. LACHANCE, D. 1972. Phytoprotection
      53: 49.
909. LACHANCE, D. 1974. Phytoprotection
      55: 95.
910. LACHANCE, D. 1975. Can. J. Forest
      Res. 5: 130-138.
911. LACHANCE, D. 1977. Phytoprotection
      58: 77-83.
912. LACHANCE, D. 1977. Phytoprotection
      58: 129.
913. LACHANCE, D. 1977. Phytoprotection
      58: 130.
914. LACHANCE, D. 1978. Forestry
      Chronicle 54: 20-23.
915. LACHANCE, D. 1979. Can. J. Forest
      Res. 9: 287-289.
916. LACHANCE, D. 1979. Phytoprotection
      60: 168.
917. LACHANCE, D. & L. MORIN. 1972.
      Phytoprotection 53: 34-41.
918. LACHANCE, E. 1967. Phytoprotection
      48: 134.
919. LAFLAMME, G. & M. LORTIE. 1973. Can.
      J. Forest Res. 3: 155-160.
920. LAKSHMANAN, M. & T.C. VANTERPOOL.
      1967. Can. J. Bot. 45: 847-853.
921. LANGENBERG, W.J., T.J. GILLESPIE &
      J.C. SUTTON. 1976. Can.
      Phytopath. Soc. Proc. 43: 30.
922. LANIER, L. 1968. Bull. Soc. Mycol.
      France 83: 959-979.
923. LARSEN, M.J. 1964. Can. J. Bot.
      42: 1167-1172.
924. LARSEN, M.J. 1965. Can. J. Bot.
      43: 1485-1510.
925. LARSEN, M.J. 1966. Mycologia 58:
      597-613.

926. LARSEN, M.J. 1968. Mycologia 60: 547-552.

927. LARSEN, M.J. 1968. Mycologia 60: 1178-1184.

928. LARSEN, M.J. 1969. Mycologia 61: 670-679.

929. LARSEN, M.J. 1970. Mycologia 62: 256-271.

930. LARSEN, M.J. & R.L. GILBERTSON. 1977. Norw. J. Bot. 24: 99-121.

931. LARSEN, M.J. & B. ZAK. 1978. Can. J. Bot. 56: 1122-1129.

932. LARSEN, M.J., F.F. LOMBARD & P.E. AHO. 1979. Can. J. Forest Res. 9: 31-38.

933. LARTER, E.N. & H. ENNS. 1962. Can. J. Plant Sci. 42: 69-77.

934. LAUT, J.G., B.C. SUTTON & J.J. LAWRENCE. 1966. Plant Disease Rep. 50: 208.

935. LAVALLEE, A. 1964. Can. J. Bot. 42: 1495-1502.

936. LAVALLEE, A. 1964. Phytoprotection 45: 46.

937. LAVALLEE, A. 1965. Phytoprotection 46: 163-168.

938. LAVALLEE, A. 1966. Phytoprotection 47: 92.

939. LAVALLEE, A. 1967. Phytoprotection 48: 1-3.

940. LAVALLEE, A. 1968. Forestry Chronicle 44(4): 5-10.

941. LAVALLEE, A. 1969. Phytoprotection 50: 16-22.

942. LAVALLEE, A. 1971. Phytoprotection 52: 91-92.

943. LAVALLEE, A. 1971. Phytoprotection 52: 125-130.

944. LAVALLEE, A. 1974. Forestry Chronicle 50: 228-232.

945. LAVALLEE, A. & A. BARD. 1971. Can. J. Forest Res. 1: 113-120.

946. LAVALLEE, A. & G. BARD. 1973. Can. J. Forest Res. 3: 251-255.

947. LAVALLEE, A. & G. BARD. 1978. Phytoprotection 59: 132-136.

948. LAVALLEE, A. & M. LORTIE. 1968. Forestry Chronicle 44(2): 5-10.

949. LAVALLEE, A. & M. LORTIE. 1971. Phytoprotection 52: 112-118.

950. LAWRENCE, J.J. & Y. HIRATSUKA. 1972. Can. Forestry Service, Edmonton. Report NOR-X-20.

951. LAWRENCE, J.J. & Y. HIRATSUKA. 1972. Can. Forestry Service, Edmonton. Report NOR-X-21.

952. LAWRENCE, J.J. & Y. HIRATSUKA. 1972. Can. Forestry Service, Edmonton. Report NOR-X-22.

953. LAWRENCE, J.J. & Y. HIRATSUKA. 1972. Can. Forestry Service, Edmonton. Report NOR-X-28.

954. LAWRENCE, J.J. & Y. HIRATSUKA. 1972. Can. Forestry Service, Edmonton. Report NOR-X-27.

955. LAZAROVITS, G., C.H. UNWIN & E.W.B. WARD. 1980. Plant Disease 64: 163-165.

956. LEBEAU, J.B. 1968. Can. Plant Dis. Survey 48: 130-131.

957. LEBEAU, J.B. & E.J. HAWN. 1961. Can. Plant Dis. Survey 41: 317-320.

958. LEBREAU, J.B., M.W. CORMACK & E.W.B. WARD. 1961. Can. J. Plant Sci. 41: 744-750.

959. LEDINGHAM, R.J. 1961. Can. J. Plant Sci. 41: 479-486.

960. LEDINGHAM, R.J. 1970. Can. J. Plant Sci. 50: 175-179.

961. LEDINGHAM, R.J. & S.H.F. CHINN. 1964. Can. J. Plant Sci. 44: 47-52.

962. LEDINGHAM, R.J., B.J. SALLANS & A. WENHARDT. 1960. Can. J. Plant Sci. 40: 310-316.

963. LEDINGHAM, R.J., T.G. ATKINSON, J.S. HORRICKS, J.T. MILLS, L.J. PIENING & R.D. TINLINE. 1973. Can. Plant Dis. Survey 53: 113-122.

964. LEMKE, P.A. 1964. Can. J. Bot. 42: 213-282.

965. LEMKE, P.A. 1964. Can. J. Bot. 42: 723-768.

966. LEMKE, P.A. 1969. Mycologia 61: 57-76.

967. LIBERTA, A.E. 1960. Mycologia 52: 884-913.

968. LIBERTA, A.E. 1965. Mycologia 57: 459-464.

969. LIBERTA, A.E. 1966. Mycologia 58: 927-933.

970. LIBERTA, A.E. 1973. Can. J. Bot. 51: 1871-1892.

971. LINDEBERG, B. 1959. Symb. Bot. Upsal. 16(2): 1-175.

972. LINDSEY, J.P. & R.L. GILBERTSON. 1977. Mycologia 69: 193-197.

973. LINZON, S.N. 1962. Forestry Chronicle 38: 497-504.

974. LINZON, S.N. 1968. Can. J. Bot. 46: 1565-1574.

975. LOCHHEAD, A.G. & F.D. COOK. 1961. Can. J. Bot. 39: 7-19.

976. LOCKHART, C.L. 1961. Can. J. Plant Sci. 41: 336-341.

977. LOCKHART, C.L. 1961. Can. Plant Dis. Survey 41: 366.

978. LOCKHART, C.L. 1962. Can. Plant Dis. Survey 42: 254.

979. LOCKHART, C.L. 1963. Can. Plant Dis. Survey 43: 23-26.

980. LOCKHART, C.L. 1965. Can. Plant Dis. Survey 45: 11-12.

981. LOCKHART, C.L. 1968. Can. Plant Dis. Survey 48: 128-129.

982. LOCKHART, C.L. 1970. Can. Phytopath. Soc. Proc. 37: 24.

983. LOCKHART, C.L. 1970. Can. Plant Dis. Survey 50: 93-94.

984. LOCKHART, C.L. 1970. Can. Plant Dis. Survey 50: 108.

985. LOCKHART, C.L. & E.W. CHIPMAN. 1963. Can. J. Plant Sci. 43: 503-507.

986. LOCKHART, C.L. & D.L. CRAIG. 1964. Can. Plant Dis. Survey 44: 105-108.

987. LOCKHART, C.L. & D.L. CRAIG. 1967. Can. Plant Dis. Survey 47: 17-20.

988. LOCKHART, C.L. & R.W. DELBRIDGE. 1972. Can. Plant Dis. Survey 52: 119-121.

989. LOCKHART, C.L. & R.W. DELBRIDGE. 1972. Can. Plant Dis. Survey 52: 140-142.

990. LOCKHART, C.L. & F.R. FORSYTH. 1974. Can. Plant Dis. Survey 54: 101-102.

991. LOCKHART, C.L. & F.R. FORSYTH. 1976. Can. Plant Dis. Survey 56: 35.

992. LOCKHART, C.L. & K.A. HARRISON. 1962. Can. Plant Dis. Survey 42: 107-110.

993. LOCKHART, C.L. & A.A. MACNAB. 1966. Can. Plant Dis. Survey 46: 88-89.

994. LOCKHART, C.L. & R.G. ROSS. 1961. Can. Plant Dis. Survey 41: 280-285.

995. LOCKHART, C.L., C.A. EAVES & F.R. FORSYTH. 1970. Can. Plant Dis. Survey 50: 90-92.

996. LOCKHART, C.L., I.V. HALL & R.A. MURRAY. 1973. Can. Plant Dis. Survey 53: 99-100.

997. LOCKHART, C.L., A.A. MACNAB & B. BOLWYN. 1969. Can. Plant Dis. Survey 49: 46-48.

998. LOISELLE, R. 1964. Can. J. Plant Sci. 44: 259-262.

999. LOMAN, A.A. 1962. Can. J. Bot. 40: 1545-1559.

1000. LOMAN, A.A. 1963. Can. Phytopath. Soc. Proc. 30: 13.

1001. LOMAN, A.A. 1970. Can. J. Bot. 48: 1303-1308.

1002. LOMAN, A.A. & G.D. PAUL. 1963. Forestry Chronicle 39: 422-435.

1003. LOMBARD, F.F. & R.L. GILBERTSON. 1965. Mycologia 57: 43-76.

1004. LOMBARD, F.F. & R.L. GILBERTSON. 1966. Mycologia 58: 827-845.

1005. LOPATECKI, L.E. & H. BURDON. 1966. Can. J. Plant Sci. 46: 633-638.

1006. LOPATECKI, L.E. & W. PETERS. 1972. Can. J. Plant Sci. 52: 875-879.

1007. LOPATECKI, L.E. & S.W. PORRITT. 1969. Can. J. Plant Sci. 49: 719-711.

1008. LOPEZ, M.E. & C.L. FERGUS. 1965. Mycologia 57: 897-903.

1009. LORTIE, M. 1964. Naturaliste Can. 91: 241-248.

1010. LORTIE, M. 1965. Forestry Chronicle 41: 112-115.

1011. LORTIE, M. & J.E. KUNTZ. 1963. Can. J. Bot. 41: 1203-1210.

1012. LOWE, D.P. 1977. Can. Forestry Service, Victoria. Report BC-X-32.

1013. LOWE, J.L. 1975. Mycotaxon 2: 1-83.

1014. LOWE, J.L. & R.L. GILBERTSON. 1961. Mycologia 53: 474-511.

1015. LUCK-ALLEN, E.R. 1960. Can. J. Bot. 38: 559-569.

1016. LUCK-ALLEN, E.R. 1963. Can. J. Bot. 41: 1025-1052.

1017. LUTHER, B.S. & S.A. REDHEAD. 1981. Mycotaxon 12: 417-430.

1018. LUTTRELL, E.S. 1963. Mycologia 55: 643-674.

1019. LYNCH, K.V., L.V. EDGINGTON & L.V. BUSCH. 1980. Can. J. Plant Path. 2: 176-178.

1020. MACHACEK, J.E. & H.A.H. WALLACE. 1961. Can. Plant Dis. Survey 41: 301-305.

1021. MACHACEK, J.E. & H.A.H. WALLACE. 1962. Can. Plant Dis. Survey 42: 189-194.

1022. MACHACEK, J.E. & H.A.H. WALLACE. 1963. Can. Plant Dis. Survey 43: 188-193.

1023. MACHACEK, J.E. & H.A.H. WALLACE. 1964. Can. Plant Dis. Survey 44: 248-253.

1024. MACLACHLAN, D.S. 1964. Can. Plant Dis. Survey 44: 254-256.

1025. MACLAREN, R.B. & J.D.E. STERLING. 1964. Can. J. Plant Sci. 44: 157-160.

1026. MACLEOD, D.M. 1960. Can. Phytopath. Soc. Proc. 27: 14.

1027. MACNAB, A.A. & C.O. GOURLEY. 1962. Can. Plant Dis. Survey 42: 238-245.

1028. MACRAE, R. 1967. Can. J. Bot. 45: 1371-1398.

1029. MADHOSINGH, C. 1961. Can. J. Microbiol. 7: 553-567.

1030. MADHOSINGH, C. 1962. Can. Plant Dis. Survey 42: 155-158.

1031. MADHOSINGH, C. 1964. Can. J. Bot. 42: 1677-1683.

1032. MAGASI, L.P. 1966. Can. Forestry Service, Fredericton. Report M-X-7.

1033. MAGASI, L.P. 1972. Plant Disease Rep. 56: 245-246.

1034. MAGASI, L.P. 1975. Plant Disease Rep. 59: 623-624.

1035. MAGASI, L.P. & J.M. MANLEY. 1974. Plant Disease Rep. 58: 892-894.

1036. MAGASI, L.P. & J.M. MANLEY. 1975. Can. Forestry Service, Fredricton. Report M-X-52.

1037. MAGASI, L.P., S.A. MANLEY & R.E. WALL. 1975. Plant Disease Rep. 59: 664.

1037a. MALLOCH, D. 1970. Mycologia 62: 727-740.

1038. MALLOCH, D. 1974. Fungi Canadenses No. 32, National Mycological Herbarium, Agriculture Canada, Ottawa.

1039. MALLOCH, D. 1976. Fungi Canadenses No. 82, National Mycological Herbarium, Agriculture Canada, Ottawa.

1040. MALLOCH, D. & R.F. CAIN. 1970. Can. J. Bot. 48: 1815-1825.

1041. MALLOCH, D. & S.A. REDHEAD. 1979. Fungi Candenses No. 142, National Mycological Herbarium, Agriculture Canada, Ottawa.

1042. MALLOCH, D. & C.T. ROGERSON. 1977. Can. J. Bot. 55: 1505-1509.

1043. MANOCHA, M.S. 1965. Can. J. Bot. 43: 1329-1333.

1044. MARTENS, J.W. 1967. Can. Plant Dis. Survey 47: 9-10.
1045. MARTENS, J.W. 1968. Can. Plant Dis. Survey 48: 17-19.
1046. MARTENS, J.W. 1969. Can. Plant Dis. Survey 49: 88-90.
1047. MARTENS, J.W. 1971. Can. Plant Dis. Survey 51: 11-13.
1048. MARTENS, J.W. 1972. Can. Plant Dis. Survey 52: 171-172.
1049. MARTENS, J.W. 1974. Can. Plant Dis. Survey 54: 19-20.
1050. MARTENS, J.W. 1975. Can. Plant Dis. Survey 55: 61-62.
1051. MARTENS, J.W. 1978. Can. Plant Dis. Survey 58: 51-52.
1052. MARTENS, J.W. 1979. Can. Plant Dis. Survey 59: 70-72.
1053. MARTENS, J.W. 1980. Can. J. Plant Path. 2: 253-255.
1054. MARTENS, J.W. & P.K. ANEMA. 1972. Can. Plant Dis. Survey 52: 17-18.
1055. MARTENS, J.W. & G.J. GREEN. 1966. Can. Plant Dis. Survey 46: 24-26.
1056. MARTENS, J.W. & G.J. GREEN. 1968. Can. Plant Dis. Survey 48: 102-103.
1057. MARTENS, J.W. & R.I.H. MCKENZIE. 1976. Can. Plant Dis. Survey 56: 23-24.
1058. MARTENS, J.W. & R.I.H. MCKENZIE. 1976. Can. Plant Dis. Survey 56: 126-128.
1059. MARTENS, J.W., G. FLEISCHMANN & R.I.H. MCKENZIE. 1972. Can. Plant Dis. Survey 52: 122-125.
1060. MARTIN, K.J. & R.L. GILBERTSON. 1977. Mycotaxon 6: 43-77.
1061. MARTIN, K.J. & R.L. GILBERTSON. 1978. Mycotaxon 7: 337-356.
1062. MARTIN, K.J. & R.L. GILBERTSON. 1980. Mycotaxon 10: 479-501.
1063. MCARTHUR, L.E. 1966. Can. Forestry Service, Edmonton. Report A-X-4.
1064. MCBRIDE, R.P. 1969. Can. J. Bot. 47: 711-715.
1065. McDONALD, W.C. 1961. Can. Plant Dis. Survey 41: 194-198.
1066. MCDONALD, W.C. 1966. Can. J. Bot. 44: 43-46.
1067. MCDONALD, W.C. 1970. Can. Plant Dis. Survey 50: 113-117.
1068. MCDONALD, W.C. & K.W. BUCHANNON. 1964. Can. Plant Dis. Survey 44: 118-119.
1069. MCDONALD, W.C. & H.H. MARSHALL. 1961. Can. Plant Dis. Survey 41: 275-279.
1070. MCDONALD, W.C., J.W. MARTENS, G.J. GREEN, D.J. SAMBORSKI, G. FLEISCHMANN & C.C. GILL. 1969. Can. Plant Dis. Survey 49: 114-121.
1071. MCDONALD, W.C., J.W. MARTENS, J. NIELSON, G.J. GREEN, D.J. SAMBORSKI, G. FLEISCHMANN, C.C. GILL, A.W. CHIKO & R.J. BAKER. 1971. Can. Plant Dis. Survey 51: 105-110.

1072. MCFADDEN, A.G. & J.C. SUTTON. 1975. Can. J. Plant Sci. 55: 579-586.
1073. MCINTOSH, D.L. 1960. Plant Disease Rep. 44: 262-264.
1074. MCINTOSH, D.L. 1964. Can. J. Bot. 42: 1411-1415.
1075. MCINTOSH, D.L. 1969. Plant Disease Rep. 53: 816-817.
1076. MCINTOSH, D.L. 1975. Can. Plant Dis. Survey 55: 109-116.
1077. MCINTOSH, D.L. & K. LAPINS. 1966. Can. J. Plant Sci. 46: 619-623.
1078. MCKAY, M.C.R. & B.H. MACNEILL. 1979. Can. J. Plant Path. 1: 76-78.
1079. MCKEEN, C.D. 1970. Can. Phytopath. Soc. Proc. 37: 25.
1080. MCKEEN, C.D. 1972. Can. Phytopath. Soc. Proc. 39: 35.
1081. MCKEEN, C.D. & W.B. MOUNTAIN. 1960. Can. J. Bot. 38: 789-794.
1082. MCKEEN, C.D. & W.B. MOUNTAIN. 1966. Can. Phytopath. Soc. Proc. 33: 14.
1083. MCKEEN, C.D. & W.B. MOUNTAIN. 1967. Can. J. Plant Sci. 47: 1-10.
1084. MCKEEN, C.D. & R.M. SAYRE. 1964. Can. J. Plant Sci. 44: 466-470.
1085. MCKEEN, C.D. & H.J. THORPE. 1967. Can. Phytopath. Soc. Proc. 34: 22.
1086. MCKEEN, C.D. & H.J. THORPE. 1968. Can. J. Bot. 46: 1165-1171.
1087. MCKEEN, C.D. & H.J. THORPE. 1973. Can. J. Plant Sci. 53: 615-622.
1088. MCKEEN, C.D. & R.N. WENSLEY. 1962. Can. J. Microbiol. 8: 769-784.
1089. MCKEEN, W.E. 1974. Can. Phytopath. Soc. Proc. 41: 26.
1090. MCKEEN, W.E. 1975. Can. Plant Dis. Survey 55: 12-14.
1091. MCKEEN, W.E. 1977. Can. J. Bot. 55: 12-16.
1092. MCKEEN, W.E. 1977. Can. J. Bot. 55: 44-47.
1093. MCKEEN, W.E. & J.A. TRAQUAIR. 1980. Can. J. Plant Path. 2: 42-44.
1094. MCKEEN, W.E. & R.C. ZIMMER. 1964. Can. J. Bot. 42: 665-668.
1095. MCKENZIE, D.L. & R.A.A. MORRALL. 1973. Can. Plant Dis. Survey 53: 187-190.
1096. MCKENZIE, D.L. & R.A.A. MORRALL. 1975. Can. Plant Dis. Survey 55: 1-7.
1097. MCKENZIE, D.L. & R.A.A. MORRALL. 1975. Can. Plant Dis. Survey 55: 97-100.
1098. MCMINN, R.G. & A. FUNK. 1970. Can. J. Bot. 48: 2123-2127.
1099. MCPHEE, W.J. 1980. Plant Disease 64: 847-849.
1100. MEAD, H.W. 1962. Can. J. Bot. 40: 263-271.
1101. MEAD, H.W. 1963. Can. J. Bot. 41: 312-314.
1102. MEAD, H.W. & M.W. CORMACK. 1961. Can. J. Bot. 39: 793-797.
1103. MEDERICK, F.M. & W.E. SACKSTON. 1970. Phytoprotection 51: 36-45.

1104. MEDERICK, F.M. & W.E. SACKSTON. 1972. Can. J. Plant Sci. 52: 551-557.

1105. MEHRAN, A.R. & R. POMERLEAU. 1966. Naturaliste Can. 93: 351-353.

1106. METCALFE, D.R. 1962. Can. J. Plant Sci. 42: 176-189.

1107. METCALFE, D.R., V.M. BENDELOW & W.H. JOHNSTON. 1963. Can. J. Plant Sci. 43: 222-225.

1108. MILLER, O.K. 1968. Mycologia 60: 1190-1203.

1109. MILLER, O.K. 1970. Michigan Bot. 9: 17-30.

1110. MILLER, O.K. & R.L. GILBERTSON. 1969. Mycologia 61: 840-844.

1111. MILLS, J.T. 1970. Can. J. Bot. 48: 541-546.

1112. MILLS, J.T. 1980. Can. J. Plant Sci. 60: 831-839.

1113. MILLS, J.T. & G.J. BOLLEN. 1976. Can. J. Bot. 54: 2893-2902.

1114. MILLS, J.T. & Ch. FRYDMAN. 1980. Can. Plant Dis. Survey 60: 1-7.

1115. MILLS, J.T. & A. TEKARIZ. 1974. Can. Plant Dis. Survey 54: 65-70.

1116. MILLS, J.T. & K. SCHREIBER. 1970. Can. Plant Dis. Survey 50: 80-83.

1117. MILLS, J.T. & H.A.H. WALLACE. 1968. Can. J. Plant Sci. 48: 587-594.

1118. MILLS, J.T. & H.A.H. WALLACE. 1969. Can. J. Plant Sci. 49: 543-548.

1119. MILLS, J.T. & H.A.H. WALLACE. 1971. Can. Plant Dis. Survey 51: 154-158.

1120. MINTER, D.W. 1980. Can. J. Bot. 58: 906-917.

1121. MOLLER, W.J., A.J. BRAUN, J.K. UYEMOTO & A.N. KASIMATIS. 1977. Plant Disease Rep. 61: 422-423.

1122. MOLNAR, A.C. 1961. Plant Disease Rep. 45: 854-855.

1123. MOLNAR, A.C. 1965. Can. J. Bot. 43: 563-570.

1124. MOLNAR, A.C. & R.G. MCMINN. 1960. Forestry Chronicle 36: 50.

1125. MOLNAR, A.C. & B. SIVAK. 1964. Can. J. Bot. 42: 145-158.

1126. MOORE, R.T. & S.P. MEYERS. 1962. Can. J. Microbiol. 8: 407-416.

1127. MORGAN-JONES, G. 1971. Can. J. Bot. 49: 993-1009.

1128. MORGAN-JONES, G. 1971. Can. J. Bot. 49: 1011-1013.

1129. MORGAN-JONES, G. 1971. Can. J. Bot. 49: 1363-1365.

1130. MORGAN-JONES, G. 1974. Can. J. Bot. 52: 2575-2579.

1131. MORGAN-JONES, J.F. & R.L. HULTON. 1977. Can. J. Bot. 55: 2605-2612.

1132. MORGAN-JONES, J.F. & R.L. HULTON. 1979. Mycologia 71: 1043-1052.

1133. MORRALL, R.A.A. 1977. Can. J. Bot. 55: 8-11.

1134. MORRALL, R.A.A. & J. DUECK. 1979. Can. Phytopath. Soc. Proc. 46: 65.

1135. MORRALL, R.A.A. & D.L. MCKENZIE. 1974. Plant Disease Rep. 58: 342-345.

1136. MORRALL, R.A.A. & D.L. MCKENZIE. 1975. Can. Plant Dis. Survey 55: 69-72.

1137. MORRALL, R.A.A. & D.L. MCKENZIE. 1977. Can. J. Plant Sci. 57: 281-283.

1138. MORRALL, R.A.A., L.J. DUCZEK & J.W. SHEARD. 1972. Can. J. Bot. 50: 767-786.

1139. MORRALL, R.A.A., D.L. MCKENZIE, L.J. DUCZEK & P.R. VERMA. 1972. Can. Plant Dis. Survey 52: 143-148.

1140. MORRALL, R.A.A., J. DUECK, D.L. MCKENZIE & D.C. MCGEE. 1976. Can. Plant Dis. Survey 56: 56-62.

1141. MORRALL, R.A.A., J.W. SHEPPARD & J.D. BISSETT. 1979. Can. Phytopath. Soc. Proc. 46: 65.

1142. MORRISON, D.J. 1969. Can. J. Bot. 47: 1605-1610.

1143. MORRISON, D.J. & A.L.S. JOHNSON. 1970. Forestry Chronicle 46: 200-202.

1144. MORRISON, D.J. & A.L.S. JOHNSON. 1978. Can. J. Forest Res. 8: 177-180.

1145. MORTIMORE, C.G. & L.F. GATES. 1969. Can. J. Plant Sci. 49: 723-729.

1146. MORTIMORE, C.G. & L.F. GATES. 1972. Can. Plant Dis. Survey 52: 93-96.

1147. MORTIMORE, C.G. & R.E. WALL. 1965. Can. J. Plant Sci. 45: 487-492.

1148. MORTON, F.J. & G. SMITH. 1963. Commonw. Mycol. Inst., England, Mycol. Paper 86.

1149. MOUNCE, I. 1927. In: H.T. Gussow ed., Report of the Dominion Botanist for the year 1926. Can. Dept. Agr., Ottawa. pp. 20-24.

1150. MOUNCE, I. 1929. In: H.T. Gussow ed., Report of the Dominion Botanist for the year 1927. Can. Dept. Agr., Ottawa. pp. 41-44.

1151. MOUNCE, I. 1930. In: H.T. Gussow ed., Report of the Dominion Botanist for the year 1928. Can. Dept. Agr., Ottawa. pp. 40-44.

1152. MOUNCE, I. 1931. In: H.T. Gussow ed., Report of the Dominion Botanist for the year 1929. Can. Dept. Agr., Ottawa. pp. 32-34.

1153. MOUNCE, I., R. MACRAE & M.K. NOBLES. 1938. In: H.T. Gussow ed., Progress Report of the Dominion Botanist for the years 1935 to 1937 inclusive. Can. Dept. Agr., Ottawa. p. 71.

1154. MOUNTAIN, W.B. & C.D. MCKEEN. 1965. Can. J. Bot. 43: 619-624.

1155. MUIR, J.A. 1967. Plant Disease Rep. 51: 798-799.

1156. MUIR, J.A. 1972. Can. J. Forest Res. 2: 413-416.

1157. MUIR, J.A. 1977. Can. J. Forest Res. 7: 579-583.

1158. MUIR, J.A. 1977. Can. J. Forest
        Res. 7: 589-594.
1159. MUELLER, E. & R.A. SHOEMAKER. 1965.
        Can. J. Bot. 43: 1343-1345.
1160. MUNRO, D.D. 1971. Can. J. Forest
        Res. 1: 32-34.
1161. NAG RAJ, T.R. 1973. Can. J. Bot.
        51: 2463-2472.
1162. NAG RAJ, T.R. 1976. Can. J. Bot.
        54: 1370-1376.
1163. NAG RAJ, T.R. 1979. Can. J. Bot.
        57: 2489-2496.
1164. NAG RAJ, T.R. & G. MORGAN-JONES.
        1973. Can. J. Bot. 51: 565-569.
1165. NAG RAJ, T.R. & W.B. KENDRICK. 1975.
        A monograph of *Chalara* and
        allied genera. Wilfrid Laurier
        University Press, Ont.
1166. NAIK, D.M. & L.V. BUSCH. 1978. Can.
        J. Bot. 56: 1113-1117.
1167. NAIK, D.M., L.V. BUSCH & G.L. BARRON.
        1978. Can. J. Plant Sci. 58:
        1095-1097.
1168. NANNFELDT, J.A. 1977. Bot. Notiser
        130: 351-375.
1169. NANNFELDT, J.A. 1979. Symb. Bot.
        Upsal. 22(3): 1-41.
1170. NANNFELDT, J.A. 1981. Symb. Bot.
        Upsal. 23(2): 1-72.
1171. NANNFELDT, J.A. & B. LINDEBERG.
        1965. Svensk Bot. Tidskr. 59:
        189-210.
1172. NETOLITZKY, H. & N. COLOTELO. 1965.
        Can. J. Bot. 43: 615-616.
1173. NETOLITZKY, H.J. & N. COLOTELO.
        1965. Mycologia 57: 977-979.
1174. NETOLITZKY, H.J. & W.P. SKOROPAD.
        1971. Can. J. Plant Sci. 51:
        49-51.
1175. NEUHAUSER, K.S. & R.L. GILBERTSON.
        1971. Mycologia 63: 722-735.
1176. NIELSEN, J. 1969. Plant Disease
        Rep. 53: 393-395.
1177. NIELSEN, J. 1972. Can. Plant Dis.
        Survey 52: 56-57.
1178. NIELSEN, J. 1977. Can. J. Plant
        Sci. 57: 611-612.
1179. NIEMELÄ, T. 1980. Karstenia 20:
        1-15.
1180. NOBLES, M.K. 1953. Can. J. Bot.
        31: 748.
1181. NOBLES, M.K. & B.P. FREW. 1962. Can.
        J. Bot. 40: 987-1016.
1182. NORTHOVER, J. 1975. Plant Disease
        Rep. 59: 357-360.
1183. NORTHOVER, J. 1978. Plant Disease
        Rep. 62: 706-709.
1184. NOVIELLO, C. 1963. Can. Plant Dis.
        Survey 43: 163-165.
1185. NOVIELLO, C. 1963. Can. Plant Dis.
        Survey 43: 215-216.
1186. NOVIELLO, C. & R.J. LEDINGHAM. 1965.
        Can. J. Bot. 43: 537-544.
1187. OBERWINKLER, F. & R. BANDONI. 1981.
        Can. J. Bot. 59: 1034-1040.
1188. OHENOJA, E. 1972. The Musk-ox. No.
        10. p. 67.
1189. OLCHOWECKI, A. & H.J. BRODIE. 1968.
        Can. J. Bot. 46: 1423-1429.
1190. OLCHOWECKI. A. & J. REID. 1974. Can.
        J. Bot. 52: 1675-1711.
1191. OLIVER, D.M. & D.K. HOSFORD. 1979.
        Mycotaxon 9: 277-284.
1192. OLSEN, O.A. 1961. Can. Plant Dis.
        Survey 41: 148-155.
1193. OLSEN, O.A. 1964. Can. Phytopath.
        Soc. Proc. 31: 14.
1194. ORLOB, G.B. 1960. Can. Plant Dis.
        Survey 40: 66-71.
1195. ORLOB, G.B. 1960. Can. Plant Dis.
        Survey 40: 78-86.
1196. ORLOB, G.B. & R.H.E. BRADLEY. 1960.
        Can. Plant Dis. Survey 40: 92-97.
1197. ORMROD, D.J. 1976. Can. Plant Dis.
        Survey 56: 69-72.
1198. ORMROD, D.J. 1977. Can. Phytopath.
        Soc. Proc. 44: 43.
1199. ORMROD, D.J. & J.F. CONROY. 1970.
        Can. Plant Dis. Survey 50:
        110-111.
1200. ORMROD, D.J. & W.D. CHRISTIE. 1972.
        Plant Disease Rep. 56: 53-55.
1201. ORMROD, D.J., E.C. HUGHES & R.A.
        SHOEMAKER. 1970. Can. Plant
        Dis. Survey 50: 111-112.
1202. ORMROD, D.J., T.A. SWANSON & G.N.
        SMITH. 1977. Can. Plant Dis.
        Survey 57: 49-51.
1203. ORR, G.F. 1977. Mycotaxon 6: 33-42.
1204. OUELLET, C.E. & R. POMERLEAU. 1965.
        Can. J. Bot. 43: 85-96.
1205. OUELLETTE, G.B. 1965. Plant Disease
        Rep. 49: 909.
1206. OUELLETTE, G.B. 1966. Can. J. Bot.
        44: 1117-1120.
1207. OUELLETTE, G.B. 1966. Mycologia
        58: 322-325.
1208. OUELLETTE, G.B. 1966.
        Phytoprotection 47: 147-148.
1209. OUELLETTE, G.B. 1967.
        Phytoprotection 48: 82-85.
1210. OUELLETTE, G.B. 1967.
        Phytoprotection 48: 86-91.
1211. OUELLETTE, G.B. 1967.
        Phytoprotection 48: 101-106.
1212. OUELLETTE, G.B. 1970.
        Phytoprotection 51: 151-152.
1213. OUELLETTE, G.B. 1972.
        Phytoprotection 53: 48.
1214. OUELLETTE, G.B. 1980.
        Phytoprotection 61: 118.
1215. OUELLETTE, G.B. & G. BARD. 1966.
        Phytoprotection 47: 147.
1216. OUELLETTE, G.B. & G. BARD. 1966.
        Plant Disease Rep. 50: 722-724.
1217. OUELLETTE, G.B. & R. CAUCHON. 1972.
        Mycologia 64: 649-654.
1218. OUELLETTE, G.B. & F. COTE. 1968.
        Phytoprotection 49: 135-136.
1219. OUELLETTE, G.B. & E. GAGNON. 1960.
        Can. J. Bot. 38: 235-241.
1220. OUELLETTE, G.B. & L.P. MAGASI. 1966.
        Mycologia 58: 275-280.
1221. OUELLETTE, G.B. & K.A. PIROZYNSKI.
        1974. Can. J. Bot. 52:
        1889-1911.
1222. OUELLETTE, G.B., G. BARD & R.
        CAUCHON. 1971. Phytoprotection
        52: 119-124.

1223. OUELLETTE, G.B., J.M. CONWAY & G.
     BARD. 1965. Forestry Chronicle
     41: 444-453.
1224. PADEN, J.W. & E.E. TYLUKI. 1968.
     Mycologia 60: 1160-1168.
1225. PADEN, J.W., J.R. SUTHERLAND & T.A.D.
     WOODS. 1978. Can. J. Bot. 56:
     2375-1379.
1226. PARADIS, R.O. 1979. Phytoprotection
     60: 69-78.
1227. PARK, J.Y. 1970. Can. J. Bot. 48:
     1339-1341.
1228. PARK, J.Y. 1971. Can. J. Bot. 49:
     95-97.
1229. PARK, J.Y. 1971. Can. J. Bot. 49:
     1291-1292.
1230. PARK, J.Y. & B. MOORE. 1971.
     Mycologia 63: 403-405.
1231. PARKER, A.K. 1970. Can. J. Bot.
     48: 837-838.
1232. PARKER, A.K. & D.G. COLLIS. 1966.
     Forestry Chronicle 42: 160-161.
1233. PARKER, A.K. & A.L.S. JOHNSON. 1960.
     Forestry Chronicle 36: 30-45.
1234. PARKER, A.K. & J. REID. 1969. Can.
     J. Bot. 47: 1533-1545.
1235. PARMASTO, E., F. KOTLABA & Z. POUZAR.
     1980. Ceska Mykol. 34: 208-213.
1236. PARMELEE, J.A. 1960. Can. Dept.
     Agr., Ottawa. Publ. 1080.
1237. PARMELEE, J.A. 1962. Can. J. Bot.
     40: 491-510.
1238. PARMELEE, J.A. 1963. Mycologia 55:
     133-141.
1239. PARMELEE, J.A. 1963. Mycological
     studies in 1961. In: F.
     Muller et al., Axel Heiberg
     Island Research Reports,
     Preliminary Report 1961-1962,
     McGill University, Montréal. pp.
     173-181.
1240. PARMELEE, J.A. 1965. Can. J. Bot.
     43: 239-267.
1241. PARMELEE, J.A. 1967. Can. J. Bot.
     45: 2267-2327.
1242. PARMELEE, J.A. 1968. Can. Plant Dis.
     Survey 48: 150-151.
1243. PARMELEE, J.A. 1968. Fungi in the
     Canadian Arctic. In: Can.
     Dept. Indian Affairs No.
     Develop. "North" magazine,
     March-April issue.
1244. PARMELEE, J.A. 1969. Can.
     Field-Nat. 83: 48-53.
1245. PARMELEE, J.A. 1969. Can. J. Bot.
     47: 1391-1402.
1246. PARMELEE, J.A. 1969. Mycologia 61:
     401-404.
1247. PARMELEE, J.A. 1971. Can. J. Bot.
     49: 903-926.
1248. PARMELEE, J.A. 1971. Can. Plant Dis.
     Survey 51: 91-92.
1249. PARMELEE, J.A. 1972. Can. J. Bot.
     50: 1457-1459.
1250. PARMELEE, J.A. 1973. Can. Plant Dis.
     Survey 53: 147-148.
1251. PARMELEE, J.A. 1974. Fungi
     Canadenses No. 28, National
     Mycological Herbarium,
     Agriculture Canada, Ottawa.
1252. PARMELEE, J.A. 1974. Fungi
     Canadenses No. 33, National
     Mycological Herbarium,
     Agriculture Canada, Ottawa.
1253. PARMELEE, J.A. 1974. Fungi
     Canadenses No. 43, National
     Mycological Herbarium,
     Agriculture Canada, Ottawa.
1254. PARMELEE, J.A. 1975. Fungi
     Canadenses No. 63, National
     Mycological Herbarium,
     Agriculture Canada, Ottawa.
1255. PARMELEE, J.A. 1975. Fungi
     Canadenses No. 71 National
     Mycological Herbarium,
     Agriculture Canada, Ottawa.
1256. PARMELEE, J.A. 1976. Fungi
     Canadenses No. 89, National
     Mycological Herbarium,
     Agriculture Canada, Ottawa.
1257. PARMELEE, J.A. 1977. Can. J. Bot.
     55: 1940-1983.
1258. PARMELEE, J.A. 1977. Fungi
     Canadenses No. 95, National
     Mycological Herbarium,
     Agriculture Canada, Ottawa.
1259. PARMELEE, J.A. 1977. Fungi
     Canadenses No. 97, National
     Mycological Herbarium,
     Agriculture Canada, Ottawa.
1260. PARMELEE, J.A. 1977. Fungi
     Canadenses No. 98, National
     Mycological Herbarium,
     Agriculture Canada, Ottawa.
1261. PARMELEE, J.A. 1977. Fungi
     Canadenses No. 99, National
     Mycological Herbarium,
     Agriculture Canada, Ottawa.
1262. PARMELEE, J.A. 1978. Fungi
     Canadenses No. 115, National
     Mycological Herbarium,
     Agriculture Canada, Ottawa.
1263. PARMELEE, J.A. 1978. Fungi
     Canadenses No. 116, National
     Mycological Herbarium,
     Agriculture Canada, Ottawa.
1264. PARMELEE, J.A. 1978. Fungi
     Canadenses No. 117, National
     Mycological Herbarium,
     Agriculture Canada, Ottawa.
1265. PARMELEE, J.A. 1978. Fungi
     Canadenses No. 118, National
     Mycological Herbarium,
     Agriculture Canada, Ottawa.
1266. PARMELEE, J.A. 1978. Fungi
     Canadenses No. 119, National
     Mycological Herbarium,
     Agriculture Canada, Ottawa.
1267. PARMELEE, J.A. 1979. Fungi
     Canadenses No. 134, National
     Mycological Herbarium,
     Agriculture Canada, Ottawa.
1268. PARMELEE, J.A. 1979. Fungi
     Canadenses No. 135, National
     Mycological Herbarium,
     Agriculture Canada, Ottawa.
1269. PARMELEE, J.A. 1979. Fungi
     Canadenses No. 136, National
     Mycological Herbarium,
     Agriculture Canada, Ottawa.

1270. PARMELEE, J.A. 1979. Fungi
      Canadenses No. 137, National
      Mycological Herbarium,
      Agriculture Canada, Ottawa.
1271. PARMELEE, J.A. 1979. Fungi
      Canadenses No. 138, National
      Mycological Herbarium,
      Agriculture Canada, Ottawa.
1272. PARMELEE, J.A. 1979. Fungi
      Canadenses No. 139, National
      Mycological Herbarium,
      Agriculture Canada, Ottawa.
1273. PARMELEE, J.A. 1979. Fungi
      Canadenses No. 140, National
      Mycological Herbarium,
      Agriculture Canada, Ottawa.
1274. PARMELEE, J.A. 1980. Fungi
      Canadenses No. 172, National
      Mycological Herbarium,
      Agriculture Canada, Ottawa.
1275. PARMELEE, J.A. 1980. Fungi
      Canadenses No. 173, National
      Mycological Herbarium,
      Agriculture Canada, Ottawa.
1276. PARMELEE, J.A. 1980. Fungi
      Canadenses No. 182, National
      Mycological Herbarium,
      Agriculture Canada, Ottawa.
1277. PARMELEE, J.A. 1980. Fungi
      Canadenses No. 188, National
      Mycological Herbarium,
      Agriculture Canada, Ottawa.
1278. PARMELEE, J.A. & R. CAUCHON. 1979.
      Can. J. Bot. 57: 1660-1662.
1279. PARMELEE, J.A. & M. CORLETT. 1973.
      Rept. Tottori Mycol. Inst.
      (Japan) 10: 189-201.
1280. PARMELEE, J.A. & P. deCARTERET.
      1977. Fungi Canadenses No. 96,
      National Mycological Herbarium,
      Agriculture Canada, Ottawa.
1281. PARMELEE, J.A. & P.M. deCARTERET.
      1980. Fungi Canadenses No. 171
      National Mycological Herbarium,
      Agriculture Canada, Ottawa.
1282. PARMELEE, J.A. & P.M. deCARTERET.
      1980. Fungi Canadenses No. 186,
      National Mycological Herbarium,
      Agriculture Canada, Ottawa.
1283. PARMELEE, J.A. & P.M. deCARTERET.
      1980. Fungi Canadenses No. 187,
      National Mycological Herbarium,
      Agriculture Canada, Ottawa.
1284. PARMELEE, J.A. & M.E. ELLIOTT. 1974.
      Fungi Canadenses No. 50, National
      Mycological Herbarium,
      Agriculture Canada, Ottawa.
1285. PARMELEE, J.A. & Y. HIRATSUKA. 1970.
      Can. J. Bot. 48: 1002-1004.
1286. PARMELEE, J.A. & D. MALLOCH. 1972.
      Mycologia 64: 922-924.
1287. PARMELEE, J.A. & D.B.O. SAVILE.
      1954. Mycologia 46: 823-836.
1288. PARNIS, E.M. & W.E. SACKSTON. 1972.
      Phytoprotection 53: 50.
1289. PARSONS, T.R. 1962. Can. J. Bot.
      40: 523-524.
1290. PATRICK, Z.A. & L.W. KOCH. 1961.
      Can. Plant Dis. Survey 41:
      374-375.
1291. PATRICK, Z.A. & L.W. KOCH. 1963.
      Can. J. Bot. 41: 747-758.
1292. PATRICK, Z.A., E.A. KERR & D.L.
      BAILEY. 1971. Can. J. Bot.
      49: 189-193.
1293. PECK, C.H. 1896. Bull. Torrey Bot.
      Club 23: 411-420.
1294. PELLETIER, G. 1972. Phytoprotection
      53: 48-49.
1295. PELLETIER, G. & R. HALL. 1971. Can.
      J. Bot. 49: 1293-1297.
1296. PELLETIER, G. & R. HALL. 1971.
      Phytoprotection 52: 131-142.
1297. PELLETIER, G.J. 1974. Can.
      Phytopath. Soc. Proc. 41: 27.
1298. PELLETIER, G.J., A. COMEAU & L.
      COUTURE. 1974. Phytoprotection
      55: 9-12.
1299. PELLETIER, R.L. & GWO-CHEN LI. 1966.
      Phytoprotection 47: 147.
1300. PELLETIER, R.-L. & J.K. TWUMASI.
      1969. Phytoprotection 50: 124.
1301. PEPIN, H.S. & J.E. BOSHER. 1962.
      Can. Plant Dis. Survey 42:
      159-161.
1302. PEPIN, H.S. & D.J. ORMROD. 1968.
      Can. Plant Dis. Survey 48:
      132-133.
1303. PEPIN, H.S. & D.J. ORMROD. 1974.
      Plant Disease Rep. 58: 840-843.
1304. PEPIN, H.S. & H.N.W. TOMS. 1969.
      Can. Plant Dis. Survey 49:
      105-107.
1305. PERRON, J.P. 1965. Phytoprotection
      46: 61-64.
1306. PETERSEN, R.H. 1963. Bull. Torrey
      Bot. Club 90: 260-264.
1307. PETERSEN, R.H. 1976. Mycotaxon 3:
      358-362.
1308. PETERSON, E.A. 1961. Can. J.
      Microbiol. 7: 1-6.
1309. PETRIE, G.A. 1973. Can. Plant Dis.
      Survey 53: 19-25.
1310. PETRIE, G.A. 1973. Can. Plant Dis.
      Survey 53: 26-28.
1311. PETRIE, G.A. 1973. Can. Plant Dis.
      Survey 53: 83-87.
1312. PETRIE, G.A. 1973. Can. Plant Dis.
      Survey 53: 88-92.
1313. PETRIE, G.A. 1974. Can. Plant Dis.
      Survey 54: 31-34.
1314. PETRIE, G.A. 1974. Can. Plant Dis.
      Survey 54: 155-165.
1315. PETRIE, G.A. 1975. Can. Plant Dis.
      Survey 55: 19-24.
1316. PETRIE, G.A. 1978. Can. Plant Dis.
      Survey 58: 15-19.
1317. PETRIE, G.A. 1978. Can. Plant Dis.
      Survey 58: 99-103.
1318. PETRIE, G.A. 1979. Can. J. Plant
      Sci. 59: 899-901.
1319. PETRIE, G.A. & T.C. VANTERPOOL.
      1965. Can. Plant Dis. Survey
      45: 111-112.
1320. PETRIE, G.A. & T.C. VANTERPOOL.
      1966. Can. Plant Dis. Survey
      46: 117-120.
1321. PETRIE, G.A. & T.C. VANTERPOOL.
      1968. Can. Plant Dis. Survey
      48: 25-27.

1322. PETRIE, G.A. & T.C. VANTERPOOL.
      1968. Can. J. Bot. 46: 869-871.
1323. PETRIE, G.A. & T.C. VANTERPOOL.
      1970. Can. Plant Dis. Survey
      50: 106-107.
1324. PETRIE, G.A. & T.C. VANTERPOOL.
      1974. Can. Plant Dis. Survey
      54: 37-42.
1325. PETRIE, G.A. & T.C. VANTERPOOL.
      1974. Can. Plant Dis. Survey
      54: 119-123.
1326. PETRIE, G.A. & T.C. VANTERPOOL.
      1978. Can. Plant Dis. Survey
      58: 69-72.
1327. PETRIE, G.A. & T.C. VANTERPOOL.
      1978. Can. Plant Dis. Survey
      58: 77-79.
1328. PIELOU, E.C. & R.E. FOSTER. 1962.
      Can. J. Bot. 40: 1176-1179.
1329. PIENING, L.J. 1961. Can. Plant Dis.
      Survey 41: 299-300.
1330. PIENING, L.J. 1963. Can. J.
      Microbiol. 9: 479-490.
1331. PIENING, L.J. 1965. Can. Plant Dis.
      Survey 45: 116-117.
1332. PIENING, L.J. 1968. Can. J. Plant
      Sci. 48: 623-625.
1333. PIENING, L.J. 1971. Can. Plant Dis.
      Survey 51: 101-104.
1334. PIENING, L.J. 1972. Can. J. Plant
      Sci. 52: 842-843.
1335. PIENING, L.J. & M.L. KAUFMANN. 1969.
      Can. J. Plant Sci. 49: 731-735.
1336. PIENING, L.J., D. DEW & R. EDWARDS.
      1967. Can. Plant Dis. Survey
      47: 108-109.
1337. PIENING, L.J., R. EDWARDS & D.
      WALKER. 1969. Can. Plant Dis.
      Survey 49: 95-97.
1338. PIENING, L.J., T.G. ATKINSON, J.S.
      HORRICKS, R.J. LEDINGHAM, J.T.
      MILLS & R.D. TINLINE. 1976.
      Can. Plant Dis. Survey 56:
      41-45.
1339. PILAT, A. 1940. Studia Botanica
      Cechica 3(1): 1-4.
1340. PILLEY, P.G. & R.A. TRIESELMANN.
      1968. Can. Forestry Service,
      Sault Ste. Marie. Report O-X-80.
1341. PILLEY, P.G. & R.A. TRIESELMANN.
      1969. Can. Forestry Service,
      Sault Ste Marie. Report O-X-108.
1342. PIROZYNSKI, K.A. 1973. Fungi
      Canadenses No. 10, National
      Mycological Herbarium,
      Agriculture Canada, Ottawa.
1342a. PIROZYNSKI, K.A. 1973. Mycologia
      65: 761-767.
1343. PIROZYNSKI, K.A. 1974. Can. J. Bot.
      52: 2129-2135.
1344. PIROZYNSKI, K.A. 1974. Fungi
      Canadenses No. 11 National
      Mycological Herbarium,
      Agriculture Canada, Ottawa.
1345. PIROZYNSKI, K.A. 1974. Fungi
      Canadenses No. 12, National
      Mycological Herbarium,
      Agriculture Canada, Ottawa.
1346. PIROZYNSKI, K.A. 1974. Fungi
      Canadenses No. 13, National

      Mycological Herbarium,
      Agriculture Canada, Ottawa.
1347. PIROZYNSKI, K.A. 1974. Fungi
      Canadenses No. 14, National
      Mycological Herbarium,
      Agriculture Canada, Ottawa.
1348. PIROZYNSKI, K.A. 1974. Fungi
      Canadenses No. 15, National
      Mycological Herbarium,
      Agriculture Canada, Ottawa.
1349. PIROZYNSKI, K.A. 1974. Fungi
      Canadenses No. 21 National
      Mycological Herbarium,
      Agriculture Canada, Ottawa.
1350. PIROZYNSKI, K.A. 1974. Fungi
      Canadenses No. 22, National
      Mycological Herbarium,
      Agriculture Canada, Ottawa.
1351. PIROZYNSKI, K.A. 1974. Fungi
      Canadenses No. 23, National
      Mycological Herbarium,
      Agriculture Canada, Ottawa.
1352. PIROZYNSKI, K.A. 1975. Fungi
      Canadenses No. 61 National
      Mycological Herbarium,
      Agriculture Canada, Ottawa.
1353. PIROZYNSKI, K.A. 1975. Fungi
      Canadenses No. 62, National
      Mycological Herbarium,
      Agriculture Canada, Ottawa.
1354. PIROZYNSKI, K.A. 1975. Fungi
      Canadenses No. 72, National
      Mycological Herbarium,
      Agriculture Canada, Ottawa.
1355. PIROZYNSKI, K.A. 1975. Fungi
      Canadenses No. 73, National
      Mycological Herbarium,
      Agriculture Canada, Ottawa.
1356. PIROZYNSKI, K.A. 1975. Fungi
      Canadenses No. 74, National
      Mycological Herbarium,
      Agriculture Canada, Ottawa.
1357. PIROZYNSKI, K.A. 1976. Fungi
      Canadenses No. 81 National
      Mycological Herbarium,
      Agriculture Canada, Ottawa.
1358. PIROZYNSKI, K.A. & R.A. SHOEMAKER.
      1970. Can. J. Bot. 48:
      1321-1328.
1359. PIROZYNSKI, K.A. & R.A. SHOEMAKER.
      1970. Can. J. Bot. 48:
      2199-2203.
1360. PIROZYNSKI, K.A. & R.A. SHOEMAKER.
      1971. Can. J. Bot. 49: 529-541.
1361. PIROZYNSKI, K.A. & R.A. SHOEMAKER.
      1972. Can. J. Bot. 50:
      1163-1164.
1362. PIROZYNSKI, K.A. & J.D. SMITH. 1972.
      Can. Plant Dis. Survey 52:
      153-155.
1363. PITTMAN, U.J. & J.S. HORRICKS. 1972.
      Can. J. Plant Sci. 52: 463-470.
1364. PLATFORD, P.G. & C.C. BERNIER. 1976.
      Can. J. Plant Sci. 56: 51-58.
1365. PLATFORD, R.G. & C.C. BERNIER. 1973.
      Can. Plant Dis. Survey 53: 61.
1366. PLATFORD, R.G. & C.C. BERNIER. 1975.
      Can. Plant Dis. Survey 55: 75-76.
1367. PLATFORD, R.G., C.C. BERNIER & A.C.
      FERGUSON. 1972. Can. Plant
      Dis. Survey 52: 108-109.

1368. PLATFORD, R.G., H.A.H. WALLACE & C.C. BERNIER. 1974. Can. Phytopath. Soc. Proc. 41: 30.

1369. PLATT, H.W., R.A.A. MORRALL & H.E. GRUEN. 1977. Can. J. Bot. 55: 254-259.

1370. POAPST, P.A., B.A. RAMSOOMAIR & C.O. GOURLEY. 1979. Can. J. Bot. 57: 2378-2386.

1371. POLLARD, D.F.W. 1971. Can. J. Forest Res. 1: 262-266.

1372. POLUNIN, N. 1934. J. Ecology 22: 337-395.

1373. POMERLEAU, R. 1937. Annales Assoc. Can.-Française Avance. Sci. 3: 92.

1374. POMERLEAU, R. 1937. Naturaliste Can. 64: 261-289.

1375. POMERLEAU, R. 1939. Annales Assoc. Can.-Française Avance. Sci. 5: 103-104.

1376. POMERLEAU, R. 1942-43. Soc. Qué. Protection Plantes 29: 69-79.

1377. POMERLEAU, R. 1942. Liste annotée des maladies parasitaires des arbres observees dans le Québec. Qué. Min. Terres et Forêts, Bureau de Pathologie Forestière, TF-650. Mimeo.

1378. POMERLEAU, R. 1944. Can. J. Res. 22: 171-189.

1379. POMERLEAU, R. 1961. Forestry Chronicle 43: 356-367.

1380. POMERLEAU, R. 1965. Can. J. Bot. 43: 787-792.

1381. POMERLEAU, R. 1965. Can. J. Bot. 43: 1592-1595.

1382. POMERLEAU, R. 1966. Can. J. Bot. 44: 109-111.

1383. POMERLEAU, R. 1970. Can. J. Bot. 48: 2043-2057.

1384. POMERLEAU, R. 1979. Phytoprotection 60 (Suppl.): 3-9.

1385. POMERLEAU, R. 1980. Flore des champignons au Québec et régions limitrophes. Les Editions la Presse, Montréal.

1386. POMERLEAU, R. & J. BARD. 1965. Phytoprotection 46: 24-28.

1387. POMERLEAU, R. & J. BARD. 1969. Phytoprotection 50: 32-37.

1388. POMERLEAU, R. & J. BARD. 1969. Phytoprotection 50: 38-45.

1389. POMERLEAU, R. & J. BRUNEL. 1938. Naturaliste Can. 65: 5-12, 98-102, 138-140.

1390. POMERLEAU, R. & J. BRUNEL. 1939. Naturaliste Can. 66: 28-32, 90-94, 123-128, 195-196, 223-228.

1391. POMERLEAU, R. & J. BRUNEL. 1940. Naturaliste Can. 67: 24-30, 91-96, 229-232.

1392. POMERLEAU, R. & W.B. COOKE. 1964. Mycologia 56: 618-626.

1393. POMERLEAU, R. & A.R. MEHRAN. 1966. Naturaliste Can. 93: 577-582.

1394. POMERLEAU, R. & D.E. ETHERIDGE. 1961. Can. Phytopath. Soc. Proc. 28: 13.

1395. POMERLEAU, R. & D.E. ETHERIDGE. 1961. Mycologia 53: 155-170.

1396. PONTEFRACT, R.D. & J.J. MILLER. 1962. Can. J. Microbiol. 8: 573-584.

1397. POVILAITIS, B., R.J. HASLAM & P. ROBINSON. 1964. Can. J. Plant Sci. 44: 126-132.

1398. POWELL, J.M. 1966. Plant Disease Rep. 50: 144.

1399. POWELL, J.M. 1970. Can. Plant Dis. Survey 50: 130-135.

1400. POWELL, J.M. 1971. Can. J. Bot. 49: 2123-2127.

1401. POWELL, J.M. 1971. Can. Plant Dis. Survey 51: 83-85.

1402. POWELL, J.M. 1971. Can. Plant Dis. Survey 51: 86-87.

1403. POWELL, J.M. 1971. Phytoprotection 52: 45-51.

1404. POWELL, J.M. 1971. Phytoprotection 52: 104-111.

1405. POWELL, J.M. 1974. Can. J. Bot. 52: 659-667.

1406. POWELL, J.M. 1975. Plant Disease Rep. 59: 32-34.

1407. POWELL, J.M. & Y. HIRATSUKA. 1969. Can. J. Bot. 47: 1961-1963.

1408. POWELL, J.M. & Y. HIRATSUKA. 1973. Can. Plant Dis. Survey 53: 67-71.

1409. POWELL, J.M. & W. MORF. 1966. Can. J. Bot. 44: 1597-1606.

1410. POWELL, J.M. & N.W. WILKINSON. 1973. Plant Disease Rep. 57: 283.

1411. POWELL, P.E. 1973. Mycologia 65: 1356-1370.

1412. PUA, E.C., H.R. KLINCK & R.L. PELLETIER. 1978. Phytoprotection 59: 189-190.

1413. PUNITHALINGAM, E. 1979. Commonw. Mycol. Inst., England, Descriptions of Pathogenic Fungi & Bacteria 635.

1414. PUNITHALINGAM, E. 1979. Commonw. Mycol. Inst., England, Mycol. Paper 142.

1415. PUNJA, Z.K. 1979. Can. J. Plant Path. 1: 107-110.

1416. PUNJA, Z.K. & D.J. ORMROD. 1979. Can. Plant Dis. Survey 59: 22-32.

1417. PUNTER, D. 1967. Forestry Chronicle 43: 161-164.

1418. PUNTER, D. 1967. Plant Disease Rep. 51: 357.

1419. PUNTER, D. & J.D. CAFLEY. 1968. Plant Disease Rep. 52: 692.

1420. PUTT, E.D. 1965. Can. J. Plant Sci. 45: 207-208.

1421. RAMIREZ, C. & J.J. MILLER. 1962. Can. J. Microbiol. 8: 603-608.

1422. RAYMOND, F.L. & J. REID. 1961. Can. J. Bot. 39: 233-251.

1423. REDDY, M.S. & C.L. KRAMER. 1975. Mycotaxon 3: 1-50.

1424. REDHEAD, S.A. 1973. Syesis 6: 221-227.

1425. REDHEAD, S.A. 1974. Mycologia 66: 183-187.

290

1426. REDHEAD, S.A. 1974. Syesis 7: 235-238.
1427. REDHEAD, S.A. 1975. Can. J. Bot. 53: 700-707.
1428. REDHEAD, S.A. 1977. Can. J. Bot. 55: 307-325.
1429. REDHEAD, S.A. 1979. Fungi Canadenses No. 133, National Mycological Herbarium, Agriculture Canada, Ottawa.
1430. REDHEAD, S.A. 1979. Mycologia 71: 1248-1253.
1431. REDHEAD, S.A. 1980. Can. J. Bot. 58: 68-83.
1432. REDHEAD, S.A. 1980. Fungi Canadenses No. 165, National Mycological Herbarium, Agriculture Canada, Ottawa.
1433. REDHEAD, S.A. 1980. Fungi Canadenses No. 166, National Mycological Herbarium, Agriculture Canada, Ottawa.
1434. REDHEAD, S.A. 1980. Fungi Canadenses No. 175, National Mycological Herbarium, Agriculture Canada, Ottawa.
1435. REDHEAD, S.A. 1980. Fungi Canadenses No. 179, National Mycological Herbarium, Agriculture Canada, Ottawa.
1436. REDHEAD, S.A. & J. GINNS. 1980. Can. J. Bot. 58: 731-740.
1437. REDHEAD, S.A. & D. MALLOCH. 1980. Fungi Canadenses No. 189, National Mycological Herbarium, Agriculture Canada, Ottawa.
1438. REDHEAD, S.A. & P.W. PERRIN. 1972. Can. J. Bot. 50: 409-412.
1439. REDHEAD, S.A. & J.A. TRAQUAIR. 1981. Mycotaxon 13: 373-404.
1440. REID, J. & R.F. CAIN. 1960. Can. J. Bot. 38: 945-950.
1441. REID, J. & R.F. CAIN. 1961. Can. J. Bot. 39: 1117-1129.
1442. REID, J. & R.F. CAIN. 1962. Can. J. Bot. 40: 837-841.
1443. REID, J. & R.F. CAIN. 1962. Mycologia 54: 194-200.
1444. REID, J. & R.F. CAIN. 1962. Mycologia 54: 481-497.
1445. REID, J., B.W. DANCE & H.J. WEIR. 1963. Plant Disease Rep. 47: 216-217.
1446. REID, J. & A. FUNK. 1966. Mycologia 58: 417-439.
1447. REID, J. & K.A. PIROZYNSKI. 1966. Can. J. Bot. 44: 351-354.
1448. REID, J. & K.A. PIROZYNSKI. 1966. Can. J. Bot. 44: 645-653.
1449. REID, R.W. & D.M. SHRIMPTON. 1971. Can. J. Bot. 49: 349-351.
1450. REID, R.W., H.S. WHITNEY & J.A. WATSON. 1967. Can. J. Bot. 45: 1115-1126.
1451. REISENER, H.J., W.B. McCONNELL & G.A. LEDINGHAM. 1961. Can. J. Microbiol. 7: 865-868.
1452. REYES, A.A. 1969. Can. Plant Dis. Survey 49: 56-57.
1453. REYES, A.A. 1969. Plant Disease Rep. 53: 223-225.
1454. REYES, A.A. 1977. Plant Disease Rep. 61: 1067-1070.
1455. REYES, A.A. 1979. Can. Plant Dis. Survey 59: 1-2.
1456. REYES, A.A. 1980. Plant Disease 64: 392-393.
1457. REYES, A.A. & C.A. WARNER. 1967. Can. Plant Dis. Survey 47: 116.
1458. REYES, A.A., G.H. COMLY, L.F. MAINPRIZE, D.A. PALLET, C.A. WARNER & W.A. WILLOWS. 1968. Can. Plant Dis. Survey 48: 95-96.
1459. REYES, A.A., J.G. METCALF, J.T. WARNER & L.W. MATHESON. 1977. Can. Plant Dis. Survey 57: 13-14.
1460. REYES, A.A., J.R. CHARD, A. HIKICHI, W.E. KAYLER, K.L. PRIEST, J.R. RAINFORTH, I.D. SMITH & W.A. WILLOWS. 1968. Can. Plant Dis. Survey 48: 20-24.
1461. REYES, A.A., R.W. DANIELS, E.N. ESTABROOKS, C.C. FILMAN, L.F. MAINPRIZE, W.M. RUTHERFORD, C.A. WARNER & H.M. WEBSTER. 1968. Can. Plant Dis. Survey 48: 53-55.
1462. RICE, P.F. 1970. Can. J. Bot. 48: 719-735.
1463. RICHARD, C. 1977. Phytoprotection 58: 37-38.
1464. RICHARD, C. & C. GAGNON. 1975. Can. Plant Dis. Survey 55: 45-47.
1465. RICHARD, C. & C. GAGNON. 1976. Can. Plant Dis. Survey 56: 82-84.
1466. RICHARD, C. & C. GAGNON. 1976. Phytoprotection 57: 105-106.
1467. RICHARD, C. & C. GAGNON. 1977. Can. Plant Dis. Survey 57: 15-17.
1468. RICHARD, C. & J.-A. FORTIN. 1973. Can. J. Bot. 51: 2247-2248.
1469. RICHARD, C. & J.-A. FORTIN. 1974. Phytoprotection 55: 67-88.
1470. RICHARD, C. & M. O'C. GIUBORD. 1980. Can. J. Plant Sci. 60: 265-266.
1471. RICHARD, C. & J.-G. MARTIN. 1975. Phytoprotection 56: 167.
1472. RICHARD, C. & R. MICHAUD. 1979. Can. J. Plant Path. 1: 59-60.
1473. RICHARD, C. & C. WILLEMOT. 1979. Phytoprotection 60: 163-164.
1474. RICHARD, C., J.-A. FORTIN & A. FORTIN. 1971. Can. J. Forest Res. 1: 246-251.
1475. RICHARD, C., J. SURPRENANT & C. GAGNON. 1979. Can. Plant Dis. Survey 59: 48-50.
1476. RICHARDSON, K.S. & B.J. VAN DER KAMP. 1972. Can. J. Forest Res. 2: 313-316.
1477. RICHARDSON, L.T. 1960. Plant Disease Rep. 44: 104-108.
1478. RICHARDSON, L.T. 1972. Can. J. Plant Sci. 52: 949-953.
1479. RICHARDSON, L.T. & E.J. BOND. 1978. Can. J. Plant Sci. 58: 267-268.
1480. ROBINSON, D.B. 1961. Can. J. Plant Sci. 41: 487-492.

1481. ROBINSON, R.C. 1962. Can. J. Bot. 40: 609-614.

1482. ROBINSON-JEFFREY, R.C. 1963. Can. Phytopath. Soc. Proc. 30: 17.

1483. ROBINSON-JEFFREY, R.C. & R.W. DAVIDSON. 1968. Can. J. Bot. 46: 1523-1527.

1484. ROBINSON-JEFFREY, R.C. & A.H.H. GRINCHENKO. Can. J. Bot. 42: 527-532.

1485. ROBINSON-JEFFREY, R.C. & A.A. LOHMAN. 1963. Can. J. Bot. 41: 1371-1375.

1486. ROFF, J.W. 1973. Can. J. Forest Res. 3: 582-585.

1487. ROSS, R.G. 1964. Can. Plant Dis. Survey 44: 109-112.

1488. ROSS, R.G. & C.A. EAVES. 1971. Can. Plant Dis. Survey 51: 145-147.

1489. ROSS, R.G. & C.O. GOURLEY. 1969. Can. Plant Dis. Survey 49: 33-35.

1490. ROSS, R.G. & C.O. GOURLEY. 1973. Can. Plant Dis. Survey 53: 1-4.

1491. ROSS, R.G. & S.A. HAMLIN. 1961. Can. J. Plant Sci. 41: 499-502.

1492. ROSS, R.G. & C.L. LOCKHART. 1960. Can. Plant Dis. Survey 40: 10-14.

1493. ROSS, R.G. & D.K.R. STEWART. 1960. Can. J. Plant Sci. 40: 117-122.

1494. ROSSMAN, A.Y. 1977. Mycologia 69: 355-391.

1495. ROSSMAN, A.Y. 1979. Mycotaxon 8: 485-558.

1496. ROUATT, J.W., E.A. PETERSON & H. KATZNELSON. 1963. Can. J. Microbiol. 9: 227-236.

1497. ROYLE, D.J. & C.J. HICKMAN. 1963. Plant Disease Rep. 47: 266-268.

1498. RUSSELL, R.C. 1960. Can. Phytopath. Soc. Proc. 27: 15.

1499. RYVARDEN, L. 1972. Can. J. Bot. 50: 2073-2076.

1500. RYVARDEN, L. 1976. The Polyporaceae of North Europe. Vol. 1. Fungiflora, Oslo. 214 pp.

1501. RYVARDEN, L. & R.L. Gilbertson. 1984. Mycotaxon 19: 137-144.

1502. SACCARDO, P.A. 1912. Ann. Mycol. 10: 311-322.

1503. SACKSTON, W.E. 1960. Can. J. Bot. 38: 883-889.

1504. SACKSTON, W.E. 1960. Can. Plant Dis. Survey 40: 44.

1505. SACKSTON, W.E. 1960. Plant Disease Rep. 44: 664-668.

1506. SACKSTON, W.E. 1962. Can. J. Bot. 40: 1449-1458.

1507. SACKSTON, W.E. 1970. Can. J. Plant Sci. 50: 155-157.

1508. SACKSTON, W.E. & J.W. SHEPPARD. 1972. Can. Plant Dis. Survey 52: 72-74.

1509. SACKSTON, W.E. & B.M. CHERNICK. 1960. Can. J. Plant Sci. 40: 690-699.

1510. SAKSENA, H.K. & O. VAARTAJA. 1960. Can. J. Bot. 38: 931-943.

1511. SAKSENA, H.K. & O. VAARTAJA. 1961. Can. J. Bot. 39: 627-647.

1512. SALLANS, B.J. & R.D. TINLINE. 1965. Can. J. Plant Sci. 45: 343-351.

1513. SALLANS, B.J. & R.D. TINLINE. 1969. Can. J. Plant Sci. 49: 197-201.

1514. SALT, J.R. & C.A. ROTH. 1980. Can. Field-Nat. 94: 196.

1515. SAMBORSKI, D.J. 1963. Can. Plant Dis. Survey 43: 183-187.

1516. SAMBORSKI, D.J. 1965. Can. Plant Dis. Survey 45: 30-32.

1517. SAMBORSKI, D.J. 1966. Can. Plant Dis. Survey 46: 33-35.

1518. SAMBORSKI, D.J. 1967. Can. Plant Dis. Survey 47: 3-4.

1519. SAMBORSKI, D.J. 1968. Can. Plant Dis. Survey 48: 6-8.

1520. SAMBORSKI, D.J. 1968. Can. Plant Dis. Survey 48: 107-109.

1521. SAMBORSKI, D.J. 1969. Can. Plant Dis. Survey 49: 80-82.

1522. SAMBORSKI, D.J. 1971. Can. Plant Dis. Survey 51: 17-19.

1523. SAMBORSKI, D.J. 1972. Can. Plant Dis. Survey 52: 8-10.

1524. SAMBORSKI, D.J. 1972. Can. Plant Dis. Survey 52: 168-170.

1525. SAMBORSKI, D.J. 1974. Can. Plant Dis. Survey 54: 8-10.

1526. SAMBORSKI, D.J. 1975. Can. Plant Dis. Survey 55: 58-60.

1527. SAMBORSKI, D.J. 1976. Can. Plant Dis. Survey 56: 12-14.

1528. SAMBORSKI, D.J. 1976. Can. Plant Dis. Survey 56: 123-125.

1529. SAMBORSKI, D.J. 1978. Can. Plant Dis. Survey 58: 53-54.

1530. SAMBORSKI, D.J. 1979. Can. Plant Dis. Survey 59: 67-68.

1531. SAMBORSKI, D.J. 1980. Can. J. Plant Path. 2: 246-248.

1532. SAMBORSKI, D.J. & R.I.H. MCKENZIE. 1972. Can. Plant Dis. Survey 52: 173-174.

1533. SAMBORSKI, D.J., G.J. GREEN & G. FLEISCHMANN. 1963. Can. Plant Dis. Survey 43: 1-22.

1534. SAMPSON, G. & K.S. CLOUGH. 1979. Can. Plant Dis. Survey 59: 51-52.

1535. SAMSON, R.A. & J. MOUCHACCA. 1975. Can. J. Bot. 53: 1634-1639.

1536. SANDERSON, K.E. 1960. Can. J. Plant Sci. 40: 345-352.

1537. SANTERRE, J. 1961. Can. Plant Dis. Survey 41: 321-324.

1538. SANTERRE, J. 1961. Soc. Qué. Protection Plantes 43: 56-60.

1539. SANTERRE, J. 1963. Phytoprotection 44: 69-74.

1540. SANTERRE, J. 1966. Can. J. Plant Sci. 46: 647-652.

1541. SANTERRE, J. 1969. Can. J. Plant Sci. 49: 83-86.

1542. SAVILE, D.B.O. 1952. Can. J. Bot. 30: 410-435.

1543. SAVILE, D.B.O. 1961. Can. Field-Nat. 75: 69-71.

1544. SAVILE, D.B.O. 1961. Can. J. Bot. 39: 909-942.

1545. SAVILE, D.B.O. 1961. Mycologia 53: 31-52.
1546. SAVILE, D.B.O. 1962. Can. J. Bot. 40: 1385-1398.
1547. SAVILE, D.B.O. 1962. Mycologia 54: 321-322.
1548. SAVILE, D.B.O. 1964. Mycologia 56: 240-248.
1549. SAVILE, D.B.O. 1964. Mycologia 56: 452-453.
1550. SAVILE, D.B.O. 1965. Can. J. Bot. 43: 231-238.
1551. SAVILE, D.B.O. 1965. Can. J. Bot. 43: 571-596.
1552. SAVILE, D.B.O. 1966. Can. J. Bot. 44: 1151-1170.
1553. SAVILE, D.B.O. 1967. Can. J. Bot. 45: 1093-1103.
1554. SAVILE, D.B.O. 1968. Can. J. Bot. 46: 461-471.
1555. SAVILE, D.B.O. 1968. Can. J. Bot. 46: 631-642.
1556. SAVILE, D.B.O. 1969. Can. Plant Dis. Survey 49: 29.
1557. SAVILE, D.B.O. 1970. Can. J. Bot. 48: 1553-1566.
1558. SAVILE, D.B.O. 1970. Can. J. Bot. 48: 1567-1584.
1559. SAVILE, D.B.O. 1972. Can. J. Bot. 50: 2579-2596.
1560. SAVILE, D.B.O. 1973. Can. J. Bot. 51: 2347-2370.
1561. SAVILE, D.B.O. 1973. Rept. Tottori Mycol. Inst. (Japan) 10: 225-241.
1562. SAVILE, D.B.O. 1974. Can. J. Bot. 52: 341-343.
1563. SAVILE, D.B.O. 1974. Can. J. Bot. 52: 1501-1507.
1564. SAVILE, D.B.O. 1974. Fungi Canadenses No. 18, National Mycological Herbarium, Agriculture Canada, Ottawa.
1565. SAVILE, D.B.O. 1974. Fungi Canadenses No. 19, National Mycological Herbarium, Agriculture Canada, Ottawa.
1566. SAVILE, D.B.O. 1974. Fungi Canadenses No. 41 National Mycological Herbarium, Agriculture Canada, Ottawa.
1567. SAVILE, D.B.O. 1974. Fungi Canadenses No. 46, National Mycological Herbarium, Agriculture Canada, Ottawa. S380.
1568. SAVILE, D.B.O. 1974. Fungi Canadenses No. 47, National Mycological Herbarium, Agriculture Canada, Ottawa.
1569. SAVILE, D.B.O. 1974. Fungi Canadenses No. 54, National Mycological Herbarium, Agriculture Canada, Ottawa.
1570. SAVILE, D.B.O. 1974. Fungi Canadenses No. 56, National Mycological Herbarium, Agriculture Canada, Ottawa.
1571. SAVILE, D.B.O. 1975. Fungi Canadenses No. 64, National Mycological Herbarium, Agriculture Canada, Ottawa.
1572. SAVILE, D.B.O. 1975. Fungi Canadenses No. 75, National Mycological Herbarium, Agriculture Canada, Ottawa.
1573. SAVILE, D.B.O. 1975. Fungi Canadenses No. 77, National Mycological Herbarium, Agriculture Canada, Ottawa.
1574. SAVILE, D.B.O. 1975. Fungi Canadenses No. 78, National Mycological Herbarium, Agriculture Canada, Ottawa.
1575. SAVILE, D.B.O. 1975. Fungi Canadenses No. 79, National Mycological Herbarium, Agriculture Canada, Ottawa.
1576. SAVILE, D.B.O. 1975. Fungi Canadenses No. 80, National Mycological Herbarium, Agriculture Canada, Ottawa.
1577. SAVILE, D.B.O. 1975. Mycologia 67: 273-279.
1578. SAVILE, D.B.O. 1976. Can. J. Bot. 54: 971-975.
1579. SAVILE, D.B.O. 1976. Can. J. Bot. 54: 1690-1696.
1580. SAVILE, D.B.O. 1976. Can. J. Bot. 54: 1977-1978.
1581. SAVILE, D.B.O. 1977. Fungi Canadenses No. 110, National Mycological Herbarium, Agriculture Canada, Ottawa.
1582. SAVILE, D.B.O. 1978. Fungi Canadenses No. 111 National Mycological Herbarium, Agriculture Canada, Ottawa.
1583. SAVILE, D.B.O. 1978. Fungi Canadenses No. 112, National Mycological Herbarium, Agriculture Canada, Ottawa.
1584. SAVILE, D.B.O. 1979. Fungi Canadenses No. 141 National Mycological Herbarium, Agriculture Canada, Ottawa.
1585. SAVILE, D.B.O. 1983. Fungi Canadenses No. 250, National Mycological Herbarium, Agriculture Canada, Ottawa.
1586. SAVILE, D.B.O. & J.A. PARMELEE. 1964. Can. J. Bot. 42: 699-722.
1587. SAVILE, D.B.O. & J.A. PARMELEE. 1974. Fungi Canadenses No. 24, National Mycological Herbarium, Agriculture Canada, Ottawa.
1588. SAVILE, D.B.O. & J.A. PARMELEE. 1974. Fungi Canadenses No. 25, National Mycological Herbarium, Agriculture Canada, Ottawa.
1589. SAVILE, D.B.O. & J.A. PARMELEE. 1974. Fungi Canadenses No. 26, National Mycological Herbarium, Agriculture Canada, Ottawa.
1590. SCHUMACHER, T. 1978. Norw. J. Bot. 25: 145-155.
1591. SCHUMACHER, T. 1979. Mycotaxon 8: 125-126.
1592. SEAMAN, W.L. 1963. Can. Plant Dis. Survey 43: 208-209.
1593. SEAMAN, W.L. 1967. Can. Plant Dis. Survey 47: 79-80.

1594. SEAMAN, W.L. 1970. Can. Plant Dis. Survey 50: 1-45.
1595. SEAMAN, W.L. 1970. Can. Plant Dis. Survey 50: 48-57.
1596. SEAMAN, W.L. & R.A. SHOEMAKER. 1964. Plant Disease Rep. 48: 69.
1597. SEAMAN, W.L. & V.R. WALLEN. 1962. Can. Phytopath. Soc. Proc. 29: 16.
1598. SEAMAN, W.L., R.J. LEDINGHAM & F.R. HARPER. 1974. Can. Phytopath. Soc. Proc. 41: 31.
1599. SEAMAN, W.L., R.A. SHOEMAKER & E.A. PETERSON. 1965. Can. J. Bot. 43: 1461-1469.
1600. SEDUN, F. 1980. Phytoprotection 61: 117.
1601. SEIFERT, K.A. 1979. Mycotaxon 10: 233-240.
1602. SEMENIUK, G. & A.W. HENRY. 1960. Can. J. Plant Sci. 40: 288-294.
1603. SHAFFER, R.L. 1972. Mycologia 64: 1008-1053.
1604. SHEPPARD, J.W. 1979. Can. Plant Dis. Survey 59: 60.
1605. SHEPPARD, J.W. & J.N. NEEDHAM. 1980. Can. J. Plant Path. 2: 159-162.
1606. SHEPPARD, J.W. & J.F. PETERSON. 1976. Can. J. Plant Sci. 56: 157-160.
1607. SHEPPARD, J.W. & M.A. VISWANATHAN. 1974. Can. Plant Dis. Survey 54: 57-60.
1608. SHERWOOD, M.A. 1977. Mycotaxon 5: 1-277.
1609. SHERWOOD, M.A. 1980. Occasional Papers Farlow Herbarium No. 15, Harvard Univ., Cambridge, Mass.
1610. SHIELDS, J.K. 1969. Mycologia 61: 1165-1168.
1611. SHOEMAKER, R.A. 1962. Can. J. Bot. 40: 809-836.
1612. SHOEMAKER, R.A. 1963. Can. J. Bot. 41: 1419-1423.
1613. SHOEMAKER, R.A. 1964. Can. J. Bot. 42: 411-421.
1614. SHOEMAKER, R.A. 1964. Can. J. Bot. 42: 1297-1301.
1615. SHOEMAKER, R.A. 1965. Can. J. Bot. 43: 631-639.
1616. SHOEMAKER, R.A. 1966. Can. J. Bot. 44: 255-258.
1617. SHOEMAKER, R.A. 1966. Can. J. Bot. 44: 1451-1456.
1618. SHOEMAKER, R.A. 1968. Can. J. Bot. 46: 1143-1150.
1619. SHOEMAKER, R.A. 1974. Fungi Canadenses No. 37, National Mycological Herbarium, Agriculture Canada, Ottawa.
1620. SHOEMAKER, R.A. 1976. Can. J. Bot. 54: 2365-2404.
1621. SHOEMAKER, R.A. & O. ERIKSSON. 1967. Can. J. Bot. 45: 1605-1608.
1622. SHOEMAKER, R.A. & E.G. KOKKO. 1975. Fungi Canadenses No. 65, National Mycological Herbarium, Agriculture Canada, Ottawa.
1623. SHOEMAKER, R.A. & E.G. KOKKO. 1977. Fungi Canadenses No. 100, National Mycological Herbarium, Agriculture Canada, Ottawa.
1624. SHOEMAKER, R.A. & E.G. KOKKO. 1977. Fungi Canadenses No. 101 National Mycological Herbarium, Agriculture Canada, Ottawa.
1625. SHOEMAKER, R.A. & E.G. KOKKO. 1977. Fungi Canadenses No. 103, National Mycological Herbarium, Agriculture Canada, Ottawa.
1626. SHOEMAKER, R.A. & E.G. KOKKO. 1977. Fungi Canadenses No. 104, National Mycological Herbarium, Agriculture Canada, Ottawa.
1627. SHOEMAKER, R.A. & E.G. KOKKO. 1977. Fungi Canadenses No. 105, National Mycological Herbarium, Agriculture Canada, Ottawa.
1628. SHOEMAKER, R.A. & E.G. KOKKO. 1977. Fungi Canadenses No. 106, National Mycological Herbarium, Agriculture Canada, Ottawa.
1629. SHOEMAKER, R.A. & E. MULLER. 1964. Can. J. Bot. 42: 403-410.
1630. SHOEMAKER, R.A., K.N. EGGER & E.G. KOKKO. 1980. Fungi Canadenses No. 190, National Mycological Herbarium, Agriculture Canada, Ottawa.
1631. SHOEMAKER, R.A., P.M. LECLAIR & J.D. SMITH. 1974. Can. J. Bot. 52: 2415-2421.
1632. SHRIMPTON, D.M. 1973. Can. J. Bot. 51: 1155-1160.
1633. SHRIMPTON, D.M. & J.A. WATSON. 1971. Can. J. Bot. 49: 373-375.
1634. SIEGLE, H. 1967. Can. J. Bot. 45: 147-154.
1635. SIGLER, L. & J.W. CARMICHAEL. 1976. Mycotaxon 4: 349-488.
1636. SIGLER, L., A. TSUNEDA & J. CARMICHAEL. 1981. Mycotaxon 12: 449-467.
1637. SILVERBERG, B.A. & J.F. MORGAN-JONES. 1974. Can. J. Bot. 52: 1993-1995.
1638. SIMARD, J. 1962. Can. Plant Dis. Survey 42: 103-106.
1639. SIMARD, J., R. CRETE & T. SIMARD. 1960. Can. Plant Dis. Survey 40: 72-74.
1640. SIMARD, J., R. CRETE & T. SIMARD. 1961. Can. Plant Dis. Survey 41: 353-356.
1641. SIMARD, J., R. CRETE & T. SIMARD. 1962. Can. Plant Dis. Survey 42: 216-219.
1642. SIMARD, T. 1960. Can. Plant Dis. Survey 40: 104-106.
1643. SIMARD, T. 1961. Can. Plant Dis. Survey 41: 310-313.
1644. SIMARD, T. 1962. Can. Plant Dis. Survey 42: 220-223.
1645. SIMARD, T. 1965. Phytoprotection 46: 35-39.
1646. SIMARD, T. 1966. Phytoprotection 47: 61-65.
1647. SIMARD, T. & R. CRETE. 1965. Can. Plant Dis. Survey 45: 113-115.

1648. SIMARD, T. & J. SIMARD. 1961. Can.
Plant Dis. Survey 41: 314-316.
1649. SIMARD, T. & J. SIMARD. 1962. Can.
Plant Dis. Survey 42: 224-228.
1650. SIMARD, T. & J. SIMARD. 1963.
Phytoprotection 44: 47-49.
1651. SIMARD, T., R. CRETE & J. SIMARD.
1965. Can. Plant Dis. Survey
45: 96-99.
1652. SIMARD, T., R. CRETE & L. TARTIER.
1966. Can. Plant Dis. Survey
46: 129-130.
1653. SIMARD, T., R. CRETE & L. TARTIER.
1968. Can. Plant Dis. Survey
48: 124-127.
1654. SIMARD, T., R. CRETE & L. TARTIER.
1968. Phytoprotection 49:
49-54.
1655. SIMMONDS, P.M. 1960. Can. J. Plant
Sci. 40: 139-145.
1656. SIMMONS, E.G. 1967. Mycologia 59:
67-92.
1657. SIMMONS, E.G. 1971. Mycologia 63:
380-386.
1658. SINGER, R. & A.H. SMITH. 1946.
Mycologia 38: 500-523.
1659. SINGH, P. 1969. Can. Forestry
Service, St. John's. Report
N-X-31.
1660. SINGH, P. 1974. Can. Forestry
Service, St. John's. Report
N-X-123.
1661. SINGH, P. & G.C. CAREW. 1973. Can.
Forestry Service, St. John's.
Report N-X-87.
1662. SINGH, P. & G.C. CAREW. 1980. Can.
Plant Dis. Survey 60: 21-22.
1663. SINGH, P. & J. RICHARDSON. 1973.
Forestry Chronicle 49: 180-182.
1664. SINGH, P., C.E. DORWORTH & D.D.
SKILLING. 1980. Plant Disease
64: 1117-1118.
1665. SINHA, R.N. & H.A.H. WALLACE. 1965.
Can. J. Plant Sci. 45: 48-59.
1666. SIVANESAN, A. 1971. Commonw. Mycol.
Inst., England, Mycol. Paper 127.
1667. SKOROPAD, W.P. 1963. Can. Phytopath.
Soc. Proc. 30: 18.
1668. SKOROPAD, W.P. 1966. Can. J. Plant
Sci. 46: 243-247.
1669. SKOROPAD, W.P. & J.P. TEWARI. 1977.
Can. J. Plant Sci. 57:
1001-1003.
1670. SLYKHUIS, J.T. & D.J.S. BARR. 1978.
Phytopathology 68: 639-643.
1671. SMALL, L.W. & R.-L. PELLETIER. 1970.
Phytoprotection 51: 151.
1672. SMERLIS, E. 1961. Forestry Chronicle
37: 109-115.
1673. SMERLIS, E. 1962. Can. J. Bot. 40:
351-359.
1674. SMERLIS, E. 1966. Can. J. Bot. 44:
563-565.
1675. SMERLIS, E. 1967. Can. J. Bot. 45:
1715-1717.
1676. SMERLIS, E. 1967. Plant Disease
Rep. 51: 584-585.
1677. SMERLIS, E. 1967. Plant Disease
Rep. 51: 678-679.
1678. SMERLIS, E. 1968. Can. J. Bot. 46:
597-599.
1679. SMERLIS, E. 1968. Can. J. Bot. 46:
1329-1330.
1680. SMERLIS, E. 1968. Plant Disease
Rep. 52: 167-168.
1681. SMERLIS, E. 1968. Plant Disease
Rep. 52: 403-404.
1682. SMERLIS, E. 1968. Plant Disease
Rep. 52: 738-739.
1683. SMERLIS, E. 1969. Can. J. Bot. 47:
213-214.
1684. SMERLIS, E. 1969. Plant Disease
Rep. 53: 807-810.
1685. SMERLIS, E. 1969. Plant Disease
Rep. 53: 979-983.
1686. SMERLIS, E. 1970. Can. J. Bot. 48:
1613-1615.
1687. SMERLIS, E. 1970. Can. J. Bot. 48:
1899-1901.
1688. SMERLIS, E. 1970. Phytoprotection
51: 47-51.
1689. SMERLIS, E. 1971. Phytoprotection
52: 28-31.
1690. SMERLIS, E. 1973. Can. J. Forest
Res. 3: 7-16.
1691. SMERLIS, E. & M. SAINT-LAURENT.
1966. Can. J. Bot. 44: 361.
1692. SMERLIS, E. & M. SAINT-LAURENT.
1966. Plant Disease Rep. 50:
356-357.
1693. SMITH, A.H. & L.R. HESLER. 1946. J.
Elisha Mitchell Sci. Soc. 62:
177-200.
1694. SMITH, J.D. 1965. Can. Plant Dis.
Survey 45: 118-119.
1695. SMITH, J.D. 1966. Can. Plant Dis.
Survey 46: 123-125.
1696. SMITH, J.D. 1967. Can. Plant Dis.
Survey 47: 112-115.
1697. SMITH, J.D. 1968. Can. J. Plant
Sci. 48: 329-331.
1698. SMITH, J.D. 1969. Can. J. Plant
Sci. 49: 381-383.
1699. SMITH, J.D. 1969. Can. Plant Dis.
Survey 49: 8-13.
1700. SMITH, J.D. 1969. Can. Plant Dis.
Survey 49: 140-141.
1701. SMITH, J.D. 1970. Can. J. Plant
Sci. 50: 746-747.
1702. SMITH, J.D. 1970. Can. Plant Dis.
Survey 50: 95-98.
1703. SMITH, J.D. 1971. Can. J. Bot. 49:
377-381.
1704. SMITH, J.D. 1971. Plant Disease
Rep. 55: 63-67.
1705. SMITH, J.D. 1972. Can. Plant Dis.
Survey 52: 25-29.
1706. SMITH, J.D. 1972. Can. Plant Dis.
Survey 52: 39-41.
1707. SMITH, J.D. & J.G.N. DAVIDSON. 1979.
Can. J. Bot. 57: 2122-2139.
1708. SMITH, J.D. & C.R. ELLIOTT. 1970.
Can. Plant Dis. Survey 50:
84-87.
1709. SMITH, J.D. & R.P. KNOWLES. 1967.
Can. J. Plant Sci. 47: 679-681.
1710. SMITH, J.D. & R.P. KNOWLES. 1973.
Can. J. Plant Sci. 53: 93-99.

1711. SMITH, J.D. & R.P. KNOWLES. 1974. Can. Plant Dis. Survey 54: 108-110.
1712. SMITH, J.D. & E.A. MAGINNES. 1966. Can. Plant Dis. Survey 46: 92-94.
1713. SMITH, J.D. & E.A. MAGINNES. 1969. Can. Plant Dis. Survey 49: 43-45.
1714. SMITH, J.D. & R.A. SHOEMAKER. 1974. Can. J. Bot. 52: 2061-2074.
1715. SMITH, J.D., C.R. ELLIOTT & R.A. SHOEMAKER. 1968. Can. Plant Dis. Survey 48: 115-119.
1716. SMITH, R.B. 1971. Can. J. Forest Res. 1: 35-42.
1717. SMITH, R.B. 1977. Can. J. Forest Res. 7: 632-640.
1718. SMITH, R.B. & H.M. CRAIG. 1968. Forestry Chronicle 44: 37-44.
1719. SMITH, R.B. & H.M. CRAIG. 1970. Forestry Chronicle 46: 217-220.
1720. SMITH, R.B. & E.F. WASS. 1976. Can. J. Forest Res. 6: 225-228.
1721. SMITH, R.B., H.M. CRAIG & D. CHU. 1970. Can. J. Bot. 48: 1541-1551.
1722. SMITH, R.S. 1970. Can. J. Bot. 48: 1731-1739.
1723. SMITH, R.S. & A. OFOSU-ASIEDU. 1972. Can. J. Forest Res. 2: 16-26.
1724. SOLHEIM, W.G. & G.B. CUMMINS. 1979. Mycotaxon 8: 395-401.
1725. SOPER, R.S. 1963. Can. J. Bot. 41: 875-878.
1726. SOSULSKI, F. & C.C. BERNIER. 1975. Can. Plant Dis. Survey 55: 155-157.
1727. SOWELL, G. & R.P. KORF. 1960. Mycologia 52: 934-945.
1728. SPRAGUE, R. 1960. Mycologia 52: 357-377.
1729. SPRAGUE, R. 1960. Mycologia 52: 698-718.
1730. SPRAGUE, R. 1962. Mycologia 54: 44-61.
1731. SPRAGUE, R. 1962. Mycologia 54: 593-610.
1732. STATES, J.S. 1972. Bull. Torrey Bot. Club 99: 250-251.
1733. STELFOX, D. 1966. Can. Plant Dis. Survey 46: 146.
1734. STELFOX, D. & A.W. HENRY. 1978. Can. Plant Dis. Survey 58: 87-91.
1735. STELFOX, D. & J.R. WILLIAMS. 1980. Can. Plant Dis. Survey 60: 35-36.
1736. STELFOX, D., J.R. WILLIAMS, U. SOEHNGEN & R.C. TOPPING. 1978. Plant Disease Rep. 62: 576-579.
1737. STERNER, T.E. 1975. Plant Disease Rep. 59: 638-640.
1738. STIELL, W.M. 1980. Forestry Chronicle 56: 21-27.
1739. STILLWELL, M.A. 1955. Forestry Chronicle 31: 74-83.
1740. STILLWELL, M.A. 1962. Can. Phytopath. Soc. Proc. 29: 17.
1741. STILLWELL, M.A. 1964. Can. J. Bot. 42: 495-496.
1742. STILLWELL, M.A. & D.J. KELLY. 1964. Forestry Chronicle 40: 482-487.
1743. STOLK, A.C. & R. SAMSON. 1972. Centraalbureau voor Schimmelcultures, Baarn, Netherlands, Studies in Mycol. 2.
1744. STRABY, A.E. & R.A. SHOEMAKER. 1968. Can. Plant Dis. Survey 48: 152.
1745. SURPRENANT, J. & C. RICHARD. 1978. Phytoprotection 59: 187.
1746. SURPRENANT, J. & C. RICHARD. 1979. Phytoprotection 60: 163.
1747. SURPRENANT, J., C. RICHARD, M. O'C. GUIBORD & C. GAGNON. 1980. Phytoprotection 61: 1-8.
1748. SUTHERLAND, J.R. 1967. Phytoprotection 48: 58-67.
1749. SUTHERLAND, J.R. 1979. Can. J. Forest Res. 9: 129-132.
1750. SUTHERLAND, J.R. & L.J. SLUGGETT. 1973. Can. J. Forest Res. 3: 299-303.
1751. SUTTON, B.C. 1963. Commonw. Mycol. Inst., England, Mycol. Paper 88.
1752. SUTTON, B.C. 1964. Commonw. Mycol. Inst., England, Mycol. Paper 97.
1753. SUTTON, B.C. 1967. Can. J. Bot. 45: 1249-1263.
1754. SUTTON, B.C. 1967. Can. J. Bot. 45: 1777-1781.
1755. SUTTON, B.C. 1969. Can. J. Bot. 47: 609-616.
1756. SUTTON, B.C. 1969. Can. J. Bot. 47: 853-858.
1757. SUTTON, B.C. 1969. Can. J. Bot. 47: 2083-2094.
1758. SUTTON, B.C. 1970. Can. J. Bot. 48: 471-477.
1759. SUTTON, B.C. 1970. Trans. Brit. Mycol. Soc. 54: 255-264.
1760. SUTTON, B.C. 1973. Commonw. Mycol. Inst., England, Mycol. Paper 132.
1761. SUTTON, B.C. 1975. Ceska Mykol. 29: 97-104.
1762. SUTTON, B.C. 1975. Commonw. Mycol. Inst., England, Mycol. Paper 138.
1763. SUTTON, B.C. 1980. The Coelomycetes. Commonw. Mycol. Inst., England.
1764. SUTTON, B.C. & R.L.C. CHAO. 1970. Trans. Brit. Mycol. Soc. 55: 37-44.
1765. SUTTON, B.C. & A. FUNK. 1975. Can. J. Bot. 53: 521-526.
1766. SUTTON, B.C. & J.J. LAWRENCE. 1969. Plant Disease Rep. 53: 101-102.
1767. SUTTON, B.C. & F.G. POLLACK. 1974. Mycopath. Mycol. appl. 52: 331-351.
1768. SUTTON, B.C. & D.K. SANDHU. 1969. Can. J. Bot. 47: 745-749.
1769. SUTTON, J.C. 1972. Can. J. Plant Sci. 52: 1073-1042.
1770. SUTTON, J.C. 1973. Can. Plant Dis. Survey 53: 157-160.
1771. SUTTON, J.C., W. BALIKO & H.S. FUNNELL. 1976. Can. J. Plant Sci. 56: 7-12.
1772. SUTTON, J.C., W. BALIKO & H.J. LIU. 1980. Can. J. Plant Sci. 60: 453-461.

1773. SUTTON, J.C., A. BOOTSMA & T.J. GILLESPIE. 1972. Can. Plant Dis. Survey 52: 89-92.

1774. SUTTON, J.C., C.J. SWANTON & T.J. GILLESPIE. 1978. Can. J. Bot. 56: 2460-2469.

1775. SUTTON, M.D. & V.R. WALLEN. 1962. Can. Plant Dis. Survey 42: 258-259.

1776. SUTTON, M.D. & V.R. WALLEN. 1963. Can. Plant Dis. Survey 43: 206-207.

1777. SUTTON, M.D. & V.R. WALLEN. 1966. Can. Plant Dis. Survey 46: 143.

1778. SVEJDA, F.J. & A.T. BOLTON. 1980. Can. J. Plant Path. 2: 23-25.

1779 TABER, R.A. & T.C. VANTERPOOL. 1963. Can. Phytopath. Soc. Proc. 30: 19.

1780. TABER, W.A. 1960. Can. J. Microbiol. 6: 503-514.

1781. TABER, W.A. & L.C. VINING. 1960. Can. J. Microbiol. 6: 355-365.

1782. TAKAI, S. & E.S. KONDO. 1979. Can. J. Bot. 57: 341-352.

1783. TAKAI, S., E.S. KONDO & J.B. THOMAS. 1979. Can. J. Bot. 57: 353-359.

1784. TARTIER, L.M. & A.L. DEVAUX. 1977. Phytoprotection 58: 115-120.

1785. TARTIER, L.M. & A.L. DEVAUX. 1978. Phytoprotection 59: 186.

1786. TARTIER, L.M. & T. SIMARD. 1968. Phytoprotection 49: 136.

1787. TARTIER, L.M. & T. SIMARD. 1970. Phytoprotection 51: 150.

1788. TARTIER, L.M., R. CRETE & E.J. HOGUE. 1976. Phytoprotection 57: 116-123.

1789. TAYLOR, L.D. 1970. Can. J. Bot. 48: 81-83.

1790. TEKAUZ, A. 1976. Can. Plant Dis. Survey 56: 36-40.

1791. TEKAUZ, A. 1978. Can. Plant Dis. Survey 58: 9-11.

1792. TEKAUZ, A. & K.W. BUCHANNON. 1977. Can. J. Plant Sci. 57: 389-395.

1793. TEKAUZ, A. & A.W. CHIKO. 1980. Can. J. Plant Path. 2: 152-158.

1794. TEKAUZ, A. & J.T. MILLS. 1974. Can. J. Plant Sci. 54: 731-734.

1795. TEWARI, J.P. & W.P. SKOROPAD. 1976. Can. J. Plant Sci. 56: 781-785.

1796. TEWARI, J.P. & W.P. SKOROPAD. 1977. Can. J. Bot. 55: 2348-2357.

1797. TEWARI, J.P. & W.P. SKOROPAD. 1977. Can. Plant Dis. Survey 57: 37-41.

1798. TEWARI, J.P. & W.P. SKOROPAD. 1979. Can. J. Plant Sci. 59: 1-6.

1799. TEWARI, V.P. 1963. Mycologia 55: 595-607.

1800. THIBODEAU, P.O. & M. SIMARD. 1979. Phytoprotection 60: 170.

1801. THOMAS, G.P., D.E. ETHERIDGE & G. PAUL. 1960. Can. J. Bot. 38: 459-466.

1802. THOMAS, P.L. 1974. Can. J. Plant Sci. 54: 453-456.

1803. THOMAS, P.L. 1974. Can. Plant Dis. Survey 54: 124-128.

1804. THOMAS, P.L. 1978. Can. Plant Dis. Survey 58: 92-94.

1805. THOMAS, P.L. 1984. Can. J. Plant Path. 6: 78-80.

1806. THOMPSON, H.S. 1961. Can. J. Plant Sci. 41: 227-230.

1807. THOMPSON, H.S. 1961. Can. J. Plant Sci. 41: 268-271.

1808. THOMPSON, H.S. 1961. Can. J. Plant Sci. 41: 503-506.

1809. THOMPSON, H.S. & F.J. SVEJDA. 1965. Can. J. Plant Sci. 45: 258-263.

1810. THOMPSON, L.S. & C.B. WILLIS. 1968. Plant Disease Rep. 52: 213-214.

1811. THORPE, H.J. & W.R. JARVIS. 1978. Can. Plant Dis. Survey 58: 107.

1812. TIBELL, L. 1975. Symbol Bot. Upsal. 21(2):1-128.

1813. TILL, B.B. 1968. Can. Plant Dis. Survey 48: 37.

1814. TINLINE, R.D. 1977. Can. J. Bot. 55: 30-34.

1815. TOMS, H.N.W. 1961. Can. Plant Dis. Survey 41: 274.

1816. TOMS, H.N.W. 1964. Can. Plant Dis. Survey 44: 143-225.

1817. TOMS, H.N.W. 1966. Can. Plant Dis. Survey 46: 112-114.

1818. TOMS, H.N.W. 1968. Can. Plant Dis. Survey 48: 28-31.

1819. TOOLE, B. & Z.A. PATRICK. 1977. Can. Phytopath. Soc. Proc. 44: 46.

1820. TRAQUAIR, J.A. 1980. Can. J. Plant Path. 2: 105-115.

1821. TRAQUAIR, J.A. & J.F. AMMIRATI. 1979. Can. Phytopath. Soc. Proc. 46: 72.

1822. TRAQUAIR, J.A. & Y. HIRATSUKA. 1979. Can. Phytopath. Soc. Proc. 46: 72.

1823. TRAQUAIR, J.A. & L.L. KENNEDY. 1974. Can. J. Bot. 52: 1875-1881.

1824. TRAQUAIR, J.A. & E.G. KOKKO. 1980. Can. J. Bot. 58: 2454-2458.

1825. TRAQUAIR, J.A. & W.E. McKEEN. 1980. Mycologia 72: 378-394.

1826. TRIBE, H.T. 1960. Can. J. Microbiol. 6: 309-316.

1827. TRIESELMANN, R.A. & P. NAPHAN. 1970. Can. Forestry Service, Sault Ste Marie. Report O-X-136.

1828. TRIESELMANN, R.A., F.A. BRICAULT & C.A. BARNES. 1974. Can. Forestry Service, Sault Ste Marie. Report O-X-206.

1829. TSUKAMOTO, J.Y. 1965. Can. J. Plant Sci. 45: 197-198.

1830. TSUNEDA, A. & Y. HIRATSUKA. 1979. Can. J. Plant Path. 1: 31-36.

1831. TSUNEDA, A. & W.P. SKOROPAD. 1977. Can. J. Bot. 55: 1276-1281.

1832. TSUNEDA, A. & W.P. SKOROPAD. 1978. Can. J. Bot. 56: 1333-1340.

1833. TSUNEDA, A. & W.P. SKOROPAD. 1978. Can. J. Bot. 56: 1341-1345.

1834. TSUNEDA, A., Y. HIRATSUKA & P.J. MARUYAMA. 1980. Can. J. Bot. 58: 1154-1159.

1835. TU, C.C. & J.W. KIMBROUGH. 1978. Bot. Gaz. 139: 454-466.

1836. TU, J.C. & W.R. JARVIS. 1979. Can. J. Plant Path. 1: 12-16.

1837. TU, J.C. & M.E. McNAUGHTON. 1980. Can. J. Plant Sci. 60: 585-589.

1838. TULLOCH, A.P. & G.A. LEDINGHAM. 1960. Can. J. Microbiol. 6: 425-434.

1839. TULLOCH, A.P. & G.A. LEDINGHAM. 1962. Can. J. Microbiol. 8: 379-387.

1840. TUREL, F.L. 1969. Can. J. Bot. 47: 821-823.

1841. TYRRELL, D. & D.M. MACLEOD. 1975. Can. J. Bot. 53: 1188-1191.

1842. UPADHYAY, H.P. & W.B. KENDRICK. 1975. Mycologia 67: 798-805.

1843. UTKHEDE, R.S. & J.E. RAHE. 1978. Can. J. Plant Sci. 58: 819-822.

1844. UTKHEDE, R.S. & J.E. RAHE. 1980. Can. J. Plant Path. 2: 19-22.

1845. UTKHEDE, R.S., J.E. RAHE & D.J. ORMROD. 1978. Plant Disease Rep. 62: 1030-1034.

1846. VAARTAJA, O. 1975. Can. Plant Dis. Survey 55: 101-102.

1847. VAARTAJA, O., R.E. PITBLADO, R.I. BUZZELL & L.G. CRAWFORD. 1979. Can. J. Plant Sci. 59: 307-311.

1848. VAARTAJA, O., J. WILNER, W.H. CRAM, P.J. SALISBURY, A.W. CROOKSHANKS & G.A. MORGAN. 1964. Plant Disease Rep. 48: 12-15.

1849. VAARTNOU, H. 1970. Can. Plant Dis. Survey 50: 138-141.

1850. VAARTNOU, H. & I. TEWARI. 1972. Plant Disease Rep. 56: 633-635.

1851. VAARTNOU, H. & I. TEWARI. 1972. Plant Disease Rep. 56: 676-677.

1852. VAN ADRICHEM, M.C.J. & J.E. BOSHER. 1962. Can. J. Plant Sci. 42: 365-367.

1853. VAN ADRICHEM, M.C.J. & J.E. BOSHER. 1962. Can. Plant Dis. Survey 42: 118-121.

1854. VAN DER KAMP, B.J. 1975. Can. J. Forest Res. 5: 61-67.

1855. VAN DER WESTHUIZEN, G.C.A. 1963. Can. J. Bot. 41: 1487-1499.

1856. VAN SICKLE, G.A. 1969. Plant Disease Rep. 53: 369-371.

1857. VAN SICKLE, G.A. 1973. Plant Disease Rep. 57: 608-611.

1858. VAN SICKLE, G.A. 1973. Plant Disease Rep. 57: 765-766.

1859. VAN SICKLE, G.A. 1974. Can. J. Forest Res. 4: 138-140.

1860. VAN SICKLE, G.A. 1974. Plant Disease Rep. 58: 872-874.

1861. VAN SICKLE, G.A. 1975. Can. J. Bot. 53: 8-17.

1862. VAN SICKLE, G.A. 1977. Can. J. Bot. 55: 745-751.

1863. VAN SICKLE, G.A. & W.R. NEWELL. 1968. Plant Disease Rep. 52: 455-458.

1864. VANTERPOOL, T.C. 1960. Can. Plant Dis. Survey 40: 59-60.

1865. VANTERPOOL, T.C. 1960. Can. Plant Dis. Survey 40: 60-61.

1866. VANTERPOOL, T.C. 1960. Plant Disease Rep. 44: 362-363.

1867. VANTERPOOL, T.C. 1961. Can. Phytopath. Soc. Proc. 28: 15-16.

1868. VANTERPOOL, T.C. 1961. Can. Plant Dis. Survey 41: 372-373.

1869. VANTERPOOL, T.C. 1962. Can. Plant Dis. Survey 42: 214-215.

1870. VANTERPOOL, T.C. 1963. Can. Phytopath. Soc. Proc. 30: 19.

1871. VANTERPOOL, T.C. 1963. Can. Plant Dis. Survey 43: 212-214.

1872. VANTERPOOL, T.C. 1968. Can. Phytopath. Soc. Proc. 35: 20.

1873. VANTERPOOL, T.C. 1974. Can. J. Bot. 52: 1205-1208.

1874. VANTERPOOL, T.C. & M.J. KUO. 1961. Can. Phytopath. Soc. Proc. 28: 16.

1875. VANTERPOOL, T.C. & R. MACRAE. 1951. Can. J. Bot. 29: 147-157.

1876. VERMA, P.R. & R.A.A. MORRALL. 1971. Can. Phytopath. Soc. Proc. 38: 22.

1877. VERMA, P.R. & R.A.A. MORRALL. 1975. Can. Phytopath. Soc. Proc. 42: 20.

1878. VERMA, P.R. & G.A. PETRIE. 1980. Can. J. Plant Sci. 60: 267-271.

1879. VERMA, P.R., R.A.A. MORRALL & R.D. TINLINE. 1974. Can. J. Bot. 52: 1757-1764.

1880. VON ARX, J.A. 1979. Cross-Reference Names for Pleomorphic Fungi Discussion. In: B. Kendrick ed., The Whole Fungus, Vol. 1: 40. Nat. Museums of Canada, Ottawa.

1881. WAKEFIELD, E.M. 1960. Mycologia 52: 919-933.

1882. WALL, R.E. 1964. Can. Plant Dis. Survey 44: 238.

1883. WALL, R.E. 1966. Can. Plant Dis. Survey 46: 95.

1884. WALL, R.E. 1971. Can. J. Forest Res. 1: 141-146.

1885. WALL, R.E. 1974. Can. Plant Dis. Survey 54: 116-118.

1886. WALL, R.E. & L.P. MAGASI. 1976. Can. J. Forest Res. 6: 448-452.

1887. WALLACE, H.A.H. 1960. Can. Plant Dis. Survey 40: 64-65.

1888. WALLACE, H.A.H. 1964. Can. Plant Dis. Survey 44: 268-278.

1889. WALLACE, H.A.H. 1965. Can. Plant Dis. Survey 45: 120-123.

1890. WALLACE, H.A.H. 1965. Can. Plant Dis. Survey 45: 124-126.

1891. WALLACE, H.A.H. 1969. Can. Plant Dis. Survey 49: 49-53.

1892. WALLACE, H.A.H. 1971. Can. Plant Dis. Survey 51: 3-8.

1893. WALLACE, H.A.H. & J.T. MILLS. 1968. Can. Plant Dis. Survey 48: 141-149.

1894. WALLACE, H.A.H. & J.T. MILLS. 1970. Can. Plant Dis. Survey 50: 74-79.

1895. WALLACE, H.A.H. & R.N. SINHA. 1962. Can. J. Plant Sci. 42: 130-141.

1896. WALLACE, H.A.H., R.N. SINHA & J.T.
MILLS. 1976. Can. J. Bot. 54:
1332-1343.
1897. WALLACE, H.A.H., R.N. SINHA, G.E.
LALIBERTE, B.M. FRASER, P.L.
SHOLBERG & W.E. MUIR. 1979.
Can. J. Plant Sci. 59: 991-999.
1898. WALLEN, V.R. 1960. Can. Plant Dis.
Survey 40: 98.
1899. WALLEN, V.R. 1960. Plant Disease
Rep. 44: 596.
1900. WALLEN, V.R. 1961. Can. Plant Dis.
Survey 41: 365.
1901. WALLEN, V.R. 1962. Can. Phytopath.
Soc. Proc. 29: 18.
1902. WALLEN, V.R. 1962. Can. Plant Dis.
Survey 42: 261.
1903. WALLEN, V.R. 1964. Can. Plant Dis.
Survey 44: 241.
1904. WALLEN, V.R. 1965. Can. J. Plant
Sci. 45: 27-33.
1905. WALLEN, V.R. 1969. Can. Plant Dis.
Survey 49: 27-28.
1906. WALLEN, V.R. 1976. Can. Plant Dis.
Survey 56: 109.
1907. WALLEN, V.R. 1979. Can. Plant Dis.
Survey 59: 69.
1908. WALLEN, V.R. & A.B. EDNIE. 1972.
Can. Plant Dis. Survey 52:
42-44.
1909. WALLEN, V.R. & W.L. SEAMAN. 1963.
Can. J. Bot. 41: 13-21.
1910. WALLEN, V.R. & M.D. SUTTON. 1962.
Can. Plant Dis. Survey 42:
111-114.
1911. WALLEN, V.R. & M.D. SUTTON. 1967.
Can. Plant Dis. Survey 47: 116.
1912. WALLIS, G.W. 1962. Can. Phytopath.
Soc. Proc. 29: 18.
1913. WALLIS, G.W. 1976. Can. J. Forest
Res. 6: 229-232.
1914. WALLIS, G.W. & D.J. MORRISON. 1975.
Forestry Chronicle 51: 203-207.
1915. WALLIS, G.W. & G. REYNOLDS. 1965.
Can. J. Bot. 43: 2-9.
1916. WALLIS, G.W. & G. REYNOLDS. 1970.
Forestry Chronicle 46: 221-224.
1917. WALLIS, G.W. & G. REYNOLDS. 1975.
Can. J. Forest Res. 5: 741-742.
1918. WANG, C.J.K. 1964. Can. J. Bot.
42: 1011-1016.
1919. WARD, E.W.B. & N. COLOTELO. 1960.
Can. J. Microbiol. 6: 545-556.
1920. WATLING, R. 1978. Naturalist
(London) 103: 39-57.
1921. WATLING, R. & O.K. MILLER. 1971.
Can. J. Bot. 49: 1687-1690.
1922. WATSON, A.K. & J.E. MITIMORE. 1975.
Can. J. Bot. 53: 2458-1461.
1923. WATSON, A.K., R.J. COPEMAN & A.J.
RENNEY. 1974. Can. J. Bot.
52: 2639-2640.
1924. WEAVER, G.M. 1963. Can. J. Plant
Sci. 43: 365-369.
1925. WEHMEYER, L.E. 1940. Can. J. Res.
18: 543-545.
1926. WEIR, L.C. 1963. Forestry Chronicle
39: 205-211.
1927. WEIR, L.C. & A.L.S. JOHNSON. 1967.
Phytoprotection 48: 74-77.

1928. WELLS, S.A. & W.P. SKOROPAD. 1963.
Can. J. Plant Sci. 43: 184-187.
1929. WENER, H.M. & R.F. CAIN. 1970. Can.
J. Bot. 48: 325-327.
1930. WENSLEY, R.N. 1964. Can. J. Bot.
42: 841-857.
1931. WENSLEY, R.N. 1970. Can. J. Plant
Sci. 50: 339-343.
1932. WENSLEY, R.N. & C.D. McKEEN. 1962.
Can. J. Microbiol. 8: 57-64.
1933. WENSLEY, R.N. & C.D. MCKEEN. 1962.
Can. J. Microbiol. 8: 818-819.
1934. WENSLEY, R.N. & C.D. MCKEEN. 1963.
Can. J. Microbiol. 9: 237-249.
1935. WERESUB, L.K. 1961. Can. J. Bot.
39: 1453-1495.
1936. WERESUB, L.K. 1971. Can. J. Bot.
49: 2059-2060.
1937. WERESUB, L.K. 1974. Fungi Canadenses
No. 27, National Mycological
Herbarium, Agriculture Canada,
Ottawa.
1938. WERESUB, L.K. 1974. Fungi Canadenses
No. 45, National Mycological
Herbarium, Agriculture Canada,
Ottawa.
1939. WERESUB, L.K. 1976. Fungi Canadenses
No. 87, National Mycological
Herbarium, Agriculture Canada,
Ottawa.
1940. WERESUB, L.K. 1977. Fungi Canadenses
No. 91 National Mycological
Herbarium, Agriculture Canada,
Ottawa.
1941. WERESUB, L.K. & S. GIBSON. 1960.
Can. J. Bot. 38: 833-867.
1942. WERESUB, L.K. & W.I. ILLMAN. 1980.
Can. J. Bot. 58: 137-146.
1943. WERESUB, L.K. & P.M. LECLAIR. 1971.
Can. J. Bot. 49: 2203-2213.
1944. WHITE, G.A. & A.N. STARRATT. 1967.
Can. J. Bot. 45: 2087-2090.
1945. WHITNEY, H.S. 1964. Can. J. Bot.
42: 1397-1404.
1946. WHITNEY, H.S. & A. FUNK. 1977. Can.
J. Bot. 55: 888-891.
1947. WHITNEY, H.S., J. REID & K.A.
PIROZYNSKI. 1975. Can. J. Bot.
53: 3051-3063.
1948. WHITNEY, N.J. & C.G. MORTIMORE.
1961. Can. J. Plant Sci. 41:
854-861.
1949. WHITNEY, N.J. & C.G. MORTIMORE.
1962. Can. J. Plant Sci. 42:
302-307.
1950. WHITNEY, N.J. & C.R. MULHOLLAND.
1979. Can. Phytopath. Soc.
Proc. 46: 76.
1951. WHITNEY, R.D. 1961. Forestry
Chronicle 43:401-411.
1952. WHITNEY, R.D. 1962. Can. J. Bot.
40: 1631-1658.
1953. WHITNEY, R.D. 1966. Can. J. Bot.
44: 1333-1343.
1954. WHITNEY, R.D. 1966. Can. J. Bot.
44: 1711-1716.
1955. WHITNEY, R.D. 1973. Forestry
Chronicle 49: 176-179.
1956. WHITNEY, R.D. 1974. Can. Phytopath.
Soc. Proc. 41: 33.

1957. WHITNEY, R.D. 1976. Can. Phytopath. Soc. Proc. 43: 35.
1958. WHITNEY, R.D. 1979. Can. Phytopath. Soc. Proc. 46: 76.
1959. WHITNEY, R.D. & W.P. BOHAYCHUK. 1969. Can. J. Bot. 47: 1489-1491.
1960. WHITNEY, R.D. & W.P. BOHAYCHUK. 1971. Can. J. Bot. 49: 699-703.
1961. WHITNEY, R.D. & W.P. BOHAYCHUK. 1976. Can. J. Bot. 54: 2597-2602.
1962. WHITNEY, R.D. & W.P. BOHAYCHUK. 1976. Can. J. Forest Res. 6: 129-131.
1963. WHITNEY, R.D. & L.G. BRACE. 1979. Forestry Chronicle 55: 8-12.
1964. WHITNEY, R.D. & D.T. MYREN. 1978. Can. J. Forest Res. 8: 17-22.
1965. WHITNEY, R.D. & H. VAN GROENEWOUD. 1964. Forestry Chronicle 40: 308-312.
1966. WHITNEY, R.D., D.T. MYREN & W.E. BRITNELL. 1978. Can. J. Forest Res. 8: 348-351.
1967. WICKERHAM, L.J. 1960. Mycologia 52: 171-183.
1968. WICKERHAM, L.J. 1964. Mycologia 56: 253-266.
1969. WIDDEN, P. & D. PARKINSON. 1979. Can. J. Bot. 57: 2408-2417.
1970. WILDMAN, H.G. & D. PARKINSON. 1979. Can. J. Bot. 57: 2800-2811.
1971. WILLIAMS, J.R. & D. STELFOX. 1979. Plant Disease Rep. 63: 395-399.
1972. WILLIAMS, J.R. & D. STELFOX. 1980. Can. J. Plant Path. 2: 169-172.
1973. WILLIS, C.B. 1965. Can. J. Plant Sci. 45: 369-373.
1974. WILLIS, C.B. 1965. Can. Plant Dis. Survey 45: 8-11.
1975. WILLIS, C.B. 1966. Can. Plant Dis. Survey 46: 83-84.
1976. WILLIS, C.B. 1966. Can. Plant Dis. Survey 46: 96-98.
1977. WISBEY, B.D., R.J. COPEMAN & T.A. BLACK. 1977. Can. J. Plant Sci. 57: 235-241.
1978. WOOLIAMS, G.E. 1966. Can. J. Plant Sci. 46: 661-670.
1979. WOOLLIAMS, G.E. 1966. Can. Plant Dis. Survey 46: 5-6.
1980. WOOLLIAMS, G.E. 1967. Can. J. Plant Sci. 47: 61-63.
1981. WOOLLIAMS, G.E., L.G. DENBY & A.S.F. HANSON. 1962. Can. J. Plant Sci. 42: 515-520.
1982. WRIGHT, D.S.C. & W.E. SACKSTON. 1973. Can. J. Plant Sci. 53: 391-393.
1983. WRIGHT, E.F. & R.F. CAIN. 1961. Can. J. Bot. 39: 1215-1230.
1984. WRIGHT, N.S. 1968. Can. Plant Dis. Survey 48: 77-81.
1985. YATES, A.R. & W.E. FERGUSON. 1963. Can. J. Bot. 41: 1599-1601.
1986. ZAFAR, S.I. & N. COLOTELO. 1969. Can. J. Bot. 47: 505-512.
1987. ZALASKY, H. 1964. Can. J. Bot. 42: 385-391.
1988. ZALASKY, H. 1964. Can. J. Bot. 42: 1049-1055.
1989. ZALASKY, H. 1964. Can. J. Bot. 42: 1586-1588.
1990. ZALASKY, H. 1965. Can. J. Bot. 43: 625-626.
1991. ZALASKY, H. 1965. Can. J. Bot. 43: 1157-1162.
1992. ZALASKY, H. 1965. Plant Disease Rep. 49: 50.
1993. ZALASKY, H. 1968. Can. J. Bot. 46: 57-60.
1994. ZALASKY, H. 1968. Can. J. Bot. 46: 1383-1387.
1995. ZALASKY, H. 1974. Can. J. Bot. 52: 11-13.
1996. ZALASKY, H. 1978. Phytoprotection 59: 43-50.
1997. ZALASKY, H. & C.G. RILEY. 1963. Can. J. Bot. 41: 459-465.
1998. ZALASKY, H., O.K. FENN & C.H. LINDQUIST. 1968. Plant Disease Rep. 52: 829-833.
1999. ZILLER, W.G. 1960. Can. J. Bot. 38: 869-871.
2000. ZILLER, W.G. 1961. Plant Disease Rep. 45: 90-94.
2001. ZILLER, W.G. 1961. Plant Disease Rep. 45: 327-329.
2002. ZILLER, W.G. 1965. Can. J. Bot. 43: 217-230.
2003. ZILLER, W.G. 1968. Can. J. Bot. 46: 1377-1381.
2004. ZILLER, W.G. 1969. Can. J. Bot. 47: 261-262.
2005. ZILLER, W.G. 1969. Plant Disease Rep. 53: 237-239.
2006. ZILLER, W.G. 1970. Can. J. Bot. 48: 1313-1319.
2007. ZILLER, W.G. 1970. Can. J. Bot. 48: 1471-1476.
2008. ZILLER, W.G. 1974. The Tree Rusts of Western Canada. Can. Forestry Service, Victoria. Publ. 1329.
2009. ZILLER, W.G. & A. FUNK. 1973. Can. J. Bot. 51: 1959-1963.
2010. ZILLER, W.G. & D. STIRLING. 1961. Forestry Chronicle 37: 331-338.
2011. ZIMMER, D.E. & J.A. HOES. 1974. Plant Disease Rep. 58: 311-313.
2012. ZIMMER, R.C. 1974. Can. Plant Dis. Survey 54: 55-56.
2013. ZIMMER, R.C. 1978. Plant Disease Rep. 62: 471-473.

300

INDICES OF SCIENTIFIC NAMES OF FUNGI AND
OTHER CAUSAL AGENTS

Note: Names in Roman type are main headings
in the text. Synonyms are in italics. To
save space the author combinations Berkeley
& Curtis appear as B. & C., Bourdot & Galzin
appear as B. & G., Ellis & Everhart appear
as E. & E., and Hoehnel & Litschauer appear
as H. & L.

I. Genus - species index
Abortiporus borealis (Fr.) Singer 160
Absidia sp. 98, 242, 244, 259, 262
   corymbifera (Cohn) Sacc. & Trott. 244
   glauca Hagem 45, 244
   lichtheimii (Lucet & Cost.) Lendner 244
   orchidis (Vuill.) Hagem 45, 244
   ramosa (Lindt) Lendner 57, 244
Acanthorhynchus vaccinii Shear 257
Acanthostigma clintonii (Peck) Sacc. 178
   parasiticum (Hartig) Sacc. 1, 146
Acanthostigmella pellucida Barr 235
Acarosporina microspora (Davidson & Lorenz)
      Sherwood 13, 252
Acremoniella sp. 108
   atra (Corda) Sacc. 98, 252
Acremonium sp. 66, 136, 178
   apii (Smith & Ramsay) W. Gams 37
   boreale J.D. Smith & J. Davidson 13, 24,
      25, 33, 45, 47, 57, 60, 71, 81, 83, 87,
      89, 97, 108, 117, 119, 125, 132, 134,
      138, 144, 176, 178, 203, 216, 223, 236,
      242, 244, 252
   charticola (Lindau) W. Gams 178
   kiliense Grutz 129
   strictum W. Gams 45, 255, 262
   tsugae W. Gams 247
Acrodictys excentrica Sutton 179
   fuliginosa Sutton 182
   globulosa (Toth) M.B. Ellis 27, 178
Acrospermum cuneolum Dearn. & House 13, 216
Acrostalagmus sp. 13, 48
Acrostaphylus sp. 48
Acrotheca sp. 216
   dearnessiana Sacc. 42
Actidium sp. 162
   nitidum (Ellis) Zogg 114
Actinomucor sp. 98
   elegans (Eidam) Benj. & Hessel. 98
   repens Schost. 98
Actinopelte dryina (Sacc.) Hoehnel 206
Adelopus balsamicola (Peck) Theissen 8
Aecidium sp. 1, 62, 122
   ageratinae Savile 91
   album Clinton 260
   caladii Schw. 41
   columbiense E. & E. 108
   compositarum Mart. 43, 116
   hydnoideum B. & C. 84
   physalidis Burr. 145
   ranunculacearum DC. 36
   violae Schum. 261
Aegerita sp. 178
Agaricus fulvotomentosus Peck 15
Aglaospora anomia (Fr.) Lamb. 211
   profusa (Fr.) de Not. 211
Agrocybe acericola (Peck) Singer 92
   firma (Peck) Singer 48
Alatospora acuminata Ingold 13

Albatrellus peckianus (Cooke) Niemela 92,
   239
Albugo sp. 57
   bliti (Biv.-Bern.) Kuntze 33
   candida (Pers.) Kuntze 38, 57, 63, 83,
      121, 211, 226, 236
   cruciferarum S.F. Gray 38, 57, 63, 83,
      121, 211, 226, 236
   occidentalis G.W. Wilson 232
   tragopogonis (DC.) S.F. Gray 33, 241
Aleurocorticium acerinum (Pers.:Fr.)
      Lemke 253
   alliaceum (Quèl.) Lemke 15, 93, 99,
      139, 204, 253
   candidum (Schw.) Lemke 15, 204
   dryinum (Pers.) Lemke 15
   griseo-canum (Bres.) Lemke 15, 78, 99,
      114, 129, 139, 165, 181, 204, 217, 234,
      237, 240, 253
   incrustans Lemke 4, 15, 39, 108, 197,
      235, 237
   macrosporum (Bres.) Lemke 181
   maculatum Jacks. & Lemke 76, 193
   microsporum Jacks. & Lemke 50, 68, 99,
      139, 204, 240, 253
   nivosum (Berk. ex H. & L.) Lemke 114
   pachysterigmatum Jacks. & Lemke 237
Aleurocystidiellum subcruentatum (B. & C.)
      Lemke 1, 146
Aleurodiscus sp. 1, 13
   abietis Jacks. & Lemke 1
   acerinus (Pers.:Fr.) Hoehnel
      var. alliaceus (Quèl.) B. & G. 15
   amorphus (Pers.:Fr.) Schroet. 1, 146,
      196, 247
   aurantius (Pers.:Fr.) Schroet. 247, 255
   botryosus Burt 13, 236
   canadensis Skolko 1, 13, 48, 146, 192,
      203, 236
   cerussatus (Bres.) H. & L. 13, 34, 178,
      216, 252
   farlowii Burt 196, 247
   fennicus Laurila 146
   grantii Lloyd 1, 247
   lapponicus Litsch. 178
   laurentianus Jacks. & Lemke 1, 146
   lividocoeruleus (Karst.) Lemke 1, 114,
      146, 161, 196, 236
   macrocystidiatus Lemke 39
   minnsiae Jacks. 168, 249
   oakesii (B. & C.) H. & L. 13, 139, 216
   penicillatus Burt 1, 146, 161, 196, 236,
      247
   piceinus Lyon & Lemke 146
   pini Jacks. 168
   roseus (Fr.) H. & L. 18, 30, 184
   spiniger Rogers & Lemke 117, 196, 247
   subcruentatus (B. & C.) Burt 146
   tsugae Yasuda 236
   weirii Burt 196, 247
Allantoporthe decedens (Fr.) Barr 76
   tessella (Pers.:Fr.) Petrak 216
Allescheria terrestris Apinis 1, 146, 148,
   161
Allescheriella crocea (Mont.) Hughes 13
Allophylaria pusiola (Karst.) Nannf. 176
Alternaria sp. 1, 13, 27, 36, 48, 57, 63,
      69, 79, 80, 82, 84, 92, 98, 106, 108,
      114, 116, 117, 120, 125, 127, 129, 132,
      137, 141, 143, 146, 161, 178, 192, 213,
      221, 223, 234, 242, 244, 259, 261, 262

persoonii (Nits.) Hoehnel 30, 78, 130,
193, 194, 218, 230, 231
Libertella sp. 6, 18, 30, 52, 75, 94, 118,
168, 184
acerina Westend. 18
betulina Desm. 52
faginea Desm. 94
Licrostroma subgiganteum (Berk.) Lemke 18
Ligniera pilorum Fron & Gaillat 58, 60, 63,
88, 124, 176, 232, 245
Limacinia sp. 235
alaskensis Sacc. & Scalia 4, 151, 198,
237
arctica (Woronichin) Barr 88
moniliformis (Fraser) Barr
var. quinqueseptata Barr 198
Lindtneria leucobryophila (Henn.) Jul. 30
Linocarpon umbelliferarum Barr 107
Linodochium hyalinum (Lib.) Hoehnel 168
Linospora brunellae E. & E. 191
capreae (DC.:Fr.) Fuckel 218
insularis Johans. 218
tetraspora G. Thompson 184
Lirula abietis-concoloris (Mayr ex Dearn.)
Darker 6
brevispora Ziller 153
macrospora (Hartig) Darker 153
mirabilis (Darker) Darker 6
nervata (Darker) Darker 6
punctata (Darker) Darker 6
Lopadostoma turgidum (Pers.:Fr.) Traverso 94
Lopharia cinerascens (Schw.) Cunn. 211, 253
Lophidium compressum (Pers.:Fr.) Sacc.
var. microscopicum Karst. 52
Lophiostoma caespitosum Fuckel 184
pileatum (Tode:Fr.) Fuckel 18
thujae E. & E. 237
winteri (Sacc.) Rabenh. 261
Lophiotricha viridicoma (Cooke & Peck)
Kauff. 18, 94, 184
Lophium crenatum (Pers.) Sacc. 125
mytilinum (Pers.) Fr. 6, 154, 168, 198,
249
Lophodermella sp. 168
arcuata (Darker) Darker 168
concolor (Dearn.) Darker 168
montivaga Petrak 168
Lophodermium sp. 6, 154, 168, 249
arundinaceum (Schrad.:Fr.) Chev. 81, 121
var. alpinum Rehm 121
aucupariae (Schleich.) Darker 35, 231
autumnale Darker 6
chamaecyparisii Sharai & Hara 115
cladophilum (Lèv.) Rehm 257
consociatum (Grev.) Darker 6
decorum Darker 6
exaridum (Cooke & Peck) Sacc. 116
filiforme Darker 153
hysterioides (Pers.) Sacc. 35
juniperi (Grev.) Darker 70, 115
juniperinum (Fr.) de Not. 70, 115
lacerum Darker 6
laricinum Duby 118
macrosporum (Hartig) Rehm 153
maculare (Fr.) de Not. 256
melaleucum (Fr.:Fr.) de Not. 256
var. melaleucum 256
molitoris Minter 168
nitens Darker 168
orbiculare (Ehr.) Sacc. 70

piceae (Fuckel) Hoehnel 6, 154
pinastri (Schrad.) Chev. 168
pyrolae Parmelee 202
rubiicola Earle 213
septatum (Tehon) Terrier 154
sphaeroides (Alb. & Schw.:Fr.) Duby 120
thujae Davis 237
tumidum (Fr.) Rehm 35, 231
typhinum (Fr.) Lambotte 252
uncinatum Darker 6
versicolor (Wallr.:Fr.) Rehm 218
Lophomerum sp. 6, 154
autumnale (Darker) Magasi 6
darkeri Ouellette 154
septatum (Tehon) Ouellette 154
Lophophacidium hyperboreum Lagerb. 7, 154,
168, 237
Low-temperature basidiomycete 25, 133, 134,
176, 242
Lulworthia sp. 231
halima (Diehl & Mounce) Cribb & Cribb 264
medusa (E. & E.) Cribb & Cribb 168, 194,
198, 237
Lycogala epidendrum (L.) Fr. 154
flavofuscum (Ehr.) Rost. 18
Lycoperdon perlatum Pers. 7, 154, 249
pusillum Pers. 218
pyriforme Schaeff. 7, 18, 94, 249
subincarnatum Peck 94
umbrinum Pers. 18
Macrodiaporthe everhartii (Ellis) Barr 18
Macrodiplodiopsis desmazierii (Mont.)
Petrak 175
Macrophoma sp. 75, 118, 240, 249
candollei (Berk. & Br.) Berl. & Vogl. 61
magnifructa (Peck) Sacc. 237
pellucida Bubak & Dearn. 226
sapinea (Fr.) Petrak 119, 158, 172,
238, 251
smilacina (Peck) Berl. & Vogl.
var. smilacina 177, 226, 255
var. smilacis (Ellis & Martin) Bissett
177, 226
smilacis (E. & E.) Bubak 226
thujana Cooke & Massee 237
tiliacea Peck 240
tumefaciens Shear 181
Macrophomina phaseoli (Maubl.) Ashby 98
phaseolina (Tassi) Goid. 98
Macrosphaerella macrospora (Kleb.) Jorst.
112
Macrosporium sp. 77, 133, 242
fallax Bubak & Dearn. 123
florigenum Ellis & Dearn. 43
mycophilum Bubak & Dearn. 38
Macrotyphula fistulosa (Fr.) R. Petersen 30
*Mamiania coryli (Batsch:Fr.) Ces. & de
Not.* 76
Mamianiella coryli (Batsch:Fr.) Hoehnel 76
var. coryli 76
var. spiralis (Peck) Barr 52, 76
Marasmiellus candidus (Bolt.) Singer 18
filopes (Peck) Redhead 7, 154, 168
paludosus Redhead 70, 88, 102
Marasmius sp. 168, 176
androsaceus (Fr.) Fr. 30, 154
candidus (Bolt.) Fr. 18
capillaris Morgan 205
epidryas Kuehner 85
foetidus (Sow.:Fr.) Fr. 7

infestans (Mont.) de Bary 127, 227
lateralis Tucker & Milbrath 70
megasperma Drechs. 59, 71, 83, 133, 232,
    243
    var. sojae Hildebrand 104
nicotianae Breda de Haarn
    var. nicotianae 215, 225
parasitica Dast. 127
syringae Kleb. 131
Pichia pini (Holst) Phaff 170
Piggotia sp. 100
    coryli (Desm.) Sutton 77
    negundinis Ellis & Dearn. 20
Pileolaria brevipes Berk. & Rav. 209
Pilidium acerinum Kunze 54
Pilobolus crystallinus (Wiggers) Tode 212
Piloderma bicolor (Peck) Jul. 9, 54, 156,
    170, 186, 199, 238, 250
    byssinum (Karst.) Jul. 156
Piloporia albobrunnea (Rom.) Ginns 156, 170
Piptoporus betulinus (Bull.:Fr.) Karst. 54
Pirex concentricus (Cooke & Ellis) Hjort. &
    Ryv. 31, 77, 195
Pistillaria sp. 199
    alnicola Peck 27
    setipes Grev. 189
    spathulata Corner 68
    typhuloides (Peck) Burt 89
Pithoascus intermedius (Emmons & Dodge) Arx
    257
Pithomyces chartarum (B. & C.) M.B. Ellis
    61, 254
Pithya sp. 115
    cupressina (Batsch:Fr.) Fuckel 80, 115,
    238
    vulgaris Fuckel 9, 199
Placosphaeria sp. 106
    onobrychidis (DC.) Sacc. 106
    punctiformis (Fuckel) Sacc. 101
Placuntium andromedae (Pers.:Fr.) Hoehnel 36
Plagiosphaera umbelliferarum (Barr) Barr 107
Plagiostoma acerophilum (Dearn. & House)
    Barr 20
    alneum (Fr.) Arx var. alneum 31
        var. betulinum Barr 54
    campylostylum (Auersw.) Barr
        var. campylostylum 54
    lugubre (Karst.) Bolay 191
    populi Cash & Water. 179
Plasmodiophora brassicae Wor. 59, 207
Plasmopara carlottae Savile 74
    halstedii (Farlow) Berl. & de Toni 106
    latifolii Savile 89
    ribicola Schroet. ex J.J. Davis 210
    viticola (B. & C.) Berl. & de Toni 141,
    262
Platychora alni (Peck) Petrak 31
Platygloea arrhytidiae Olive 170
Platyspora pentamera (Karst.) Wehm. 113
    permunda (Cooke) Wehm. 116
Platystomum sp. 52
Plectania coccinea (Scop.:Fr.) Fuckel 96
Plectosphaera clarae-bonae (Speg.)
    Theissen 257
Pleiochaeta setosa (Kirchn.) Hughes 81
Plenodomus sp. 133, 246
    lingam (Tode:Fr.) Hoehnel 58, 59, 70,
    226, 236
    meliloti Dearn. & Sanf. 125, 133, 134,
    243, 246

Pleomassaria carpini (Fuckel) Sacc. 67
    siparia (Berk. & Br.) Sacc. 54
Pleosphaerulina canadensis Bubak & Dearn.
    226
Pleospora sp. 20, 23, 31, 80, 81, 156, 176,
    195, 219
    ambigua (Berl. & Bres.) Wehm. 35, 108
    betae (Berl.) Nevodovsky 48
    helvetica Niessl 41, 63, 107, 229
    herbarum (Pers.) Rabenh. 26, 43, 48, 97,
    112, 117, 121, 124, 127, 131, 133, 141,
    143, 144, 145, 174, 175, 207, 224, 225,
    232, 238, 243, 247, 260
    hispida Niessl 85
    laricina Rehm 70
    maritima Rehm 243
    moravica (Petrak) Wehm. 191
    phaeospora (Duby) Ces. & de Not. 63
    straminis Sacc. & Speg. 66
    triglochinicola Webster 243
Pleuroceras helveticum (Rehm) Barr 219
    pleurostyla (Auersw.) Barr 219
    populi G. Thompson 184
    tenella (E. & E.) Barr 20
Pleurocybella porrigens (Fr.) Singer 9,
    170, 199, 238
Pleurophomella sp. 172
    spermatiospora Hoehnel 188
Pleurophomopsis sp. 172
    salicicola Petrak 219
Pleurostoma candollei Tul. 156
Pleurostromella sp. 119, 156
Pleurothecium sp. 31
    recurvatum (Morgan) Hoehnel 20, 54, 95
Pleurotus sp. 9, 54, 95, 131, 186, 238
    albolanatus (Peck) Kauff. 199
    atrocaeruleus (Fr.) Kummer 29
        var. griseus Peck. 170
    candidissimus (B. & C.) Sacc. 14, 180
    dryinus (Pers.:Fr.) Quèl. 17, 131
    lignatilis (Pers.:Fr.) Gill. 95
    mastrucatus Fr. 167
    mitis (Pers.:Fr.) Quèl. 8, 199
    ostreatus (Jacq.:Fr.) Kummer 9, 20, 31,
    54, 95, 100, 156, 186, 219, 240, 254
    petaloides (Bull.:Fr.) Quèl. 249
    porrigens (Pers.:Fr.) Gill. 9, 170,
    199, 238
    salignus (Schrad.:Fr.) Quèl. 186
    sapidus (Schulzer) Sacc. 23, 54, 95, 186,
    254
    serotinus (Pers.:Fr.) Kummer 8, 19, 30,
    39, 53, 95, 155, 194, 199, 250, 254
    silvanus Sacc. 187
    spathulatus (Fr.) Peck 29
    strigosus (B. & C.) Singer 9, 20, 54, 95
    subareolatus Peck 20, 39, 187, 231
    tessulatus (Bull.:Fr.) Gill. 253
    ulmarius (Bull.:Fr.) Kummer 17, 94,
    100, 183, 253
    unguicularis (Fr.) Kummer 187
Plicatura aurea (Fr.) Parm. 157
    crispa (Pers.:Fr.) Rea 20, 31, 54, 95,
    156, 187, 195, 200, 240
    faginea (Schrad.) Karst. 20, 31, 95
    nivea (Fr.) Karst. 20, 31, 54, 187, 219,
    240
Pluteus admirabilis (Peck) Peck 20, 250
    atromarginatus (Konr.) Kuehner 170
    cervinus (Schaeff.: Fr.) Kummer 54

Trichurus spiralis Hasselb. 257
Tricladium splendens Ingold 22, 206
Trimmatostroma sp. 56, 67, 77, 85, 189,
     220, 239
  betulinum (Corda) Hughes 56
  salicis Corda 220
Triphragmium clavellosum Berk. 38
Triposporium sp. 239
  elegans Corda 159
Tritirachium sp. 189
  hydnicola (Peck) Hughes 159
Trochila ilicis (Chev.) Rehm 112
Trogia alni (Peck) Peck 31
  crispa (Pers.:Fr.) Fr. 20, 31, 54, 95,
     156, 187, 195, 200, 240
Troposporella fumosa Karst. 12, 189, 196
Truncatella sp. 170, 201, 260
  angustata (Pers.) Hughes 99, 239, 251, 257
  hartigii (Tubeuf) Stey. 173
  truncata (Lèv.) Stey. 251, 257
Truncocolumella rubra Zeller 251
Tryblidiopsis sp. 159
  picea Vel. 119, 159
  pinastri (Pers.:Fr.) Karst. 12, 159, 173
Tubakia dryina (Sacc.) Sutton 206
Tubercularia sp. 23, 56, 96, 101, 159, 196,
     220, 221, 239, 241, 254
  nigricans (Bull.) Link:Fr. 254
  ulmea Carter 22, 64, 101, 159, 221, 254
  vulgaris Tode:Fr. 22, 23, 30, 32, 35, 64,
     71, 75, 77, 78, 86, 95, 96, 111, 113,
     116, 122, 124, 130, 131, 136, 185, 189,
     194, 196, 202, 203, 207, 209, 210, 211,
     212, 213, 218, 220, 221, 231, 232, 240,
     241, 253, 254
Tuberculina persicina Sacc. 12
Tubeufia paludosa (Crouan & Crouan) Rossman
     189
Tubulicrinis accedens (B. & G.) Donk 12, 159
  angustus (Rogers & Weresub) Donk 159, 173
  calothrix (Pat.) Donk 159, 173
  chaetophorus (Hoehnel) Donk 159, 239
  glebulosus (Bres.) Donk 12, 22, 32, 56,
     159, 173, 196, 201, 251
  globisporus Larsen & Hjort. 12
  gracillimus (E. & E.) Cunn. 56, 159,
     173, 201, 239, 251
  hamatus (Jacks.) Donk 12, 159
  inornatus (Jacks. & Rogers) Donk 173
  juniperinus (B. & G.) Donk 159
  medius (B. & G.) Oberw. 159
  propinquus (B. & G.) Donk 159
  subulatus (B. & G.) Donk 12, 159, 173
Tuburcinia clintoniae Komarov 233
  trientalis Berk. & Br. 241
Tulasnella fuscoviolacea Bres. 173
  violacea (Johan-Olsen) Juel 12
  violea (Quèl.) B. & G. 32, 56, 108, 173,
     189, 220
Tylospora asterophora (Bon.) Donk 159
  fibrillosa (Burt) Donk 159, 173
Tympanis sp. 10, 12, 158, 159, 172, 173,
     200, 251
  abietina Groves 12, 119, 159, 173, 251
  acericola Groves 22
  alnea (Pers.) Fr. 32, 56, 116, 119, 131,
     189, 220
     var. hysterioides Rehm 32

confusa Nyl. 12, 119, 159, 173, 251
conspersa Fr. 231
fasciculata Schw. 76, 259
grovesii Ouellette & Piroz. 259
hansbroughiana Groves 12, 119, 159, 173,
     251
heteromorpha Ouellette & Piroz. 189, 220
hypopodia Nyl. 12, 119, 159, 173, 239, 251
hysterioides Rehm 32
laricina (Fuckel) Sacc. 12, 119, 159,
     173, 201, 239, 251
mutata (Fuckel) Rehm 56
neopithya Ouellette & Piroz. 173
piceina Groves 12, 119, 159, 173, 239,
     251
pinastri Tul. 12, 159, 173
pithya (Karst.) Karst. 12, 119, 159, 173,
     239
prunicola Groves 196, 251
pseudoalnea Ouellette & Piroz. 32
pseudotsugae Groves 197
pulchella Ouellette & Piroz. 22
rhabdospora B. & C. 119, 259
salicina Groves 220
saligna Tode:Fr. 220
spermatiospora (Nyl.) Nyl. 188, 189, 196
truncatula (Pers.:Fr.) Rehm 12, 22, 32,
     119, 159, 173, 251
tsugae Groves 12, 119, 159, 173, 251
Typhula sp. 25, 97, 99, 176, 177, 189, 245,
     247, 251, 252
  abietina (Fuckel) Corner 251
  athyrii Remsberg 44
  erythropus Fr. 22, 32, 214, 255
  incarnata Lasch:Fr. 25, 97
  ishikariensis Imai
     var. canadensis J.D. Smith & Arsvoll
        25, 177
     var. ishikariensis 25
  itoana Imai 97
  latissima Remsberg 252
  setipes (Grev.) Berthier 189
  spathulata (Corner) Berthier 68
  trifolii Rostr. 133
  umbrina Remsberg 59
  viburni Remsberg 259
Tyromyces albellus (Peck) Bond. & Singer
     56, 160
  aneirina (Sommerf.) Bond. & Singer 189
  balsameus (Peck) Murr. 12, 159, 201, 239,
     251
  borealis (Fr.) Imaz. 12, 56
  caesius (Schrad.:Fr.) Murr. 12, 22, 56,
     96, 131, 140, 160, 173, 189, 201, 239,
     251
  canadensis Overh. ex Lowe 160
  chioneus (Fr.) Karst. 12, 22, 32, 56, 96,
     160, 189, 196, 206, 220
  fragilis (Fr.) Donk 12, 173, 201, 251
  galactinus (Berk.) Lowe 22, 56
  guttulatus (Peck) Murr. 12, 160, 196,
     201, 251
  immitis (Peck) Bond. 160
  kmetii (Bres.) Bond. & Singer 56
  kravtzevianus Bond. & Parm. 155
  lapponicus (Rom.) Lowe 160, 173
  leucospongia (Cooke & Harkn.) Bond. &
     Singer 12, 173, 201

352

354

II. Species index

aberrans Peck (Puccinia) 226
abieti-capraearum Tub. (Melampsora) 7, 218
abieti-chamaenerii Kleb. (Pucciniastrum)
    9, 89
abieticola (B. & G.) Eriksson
    (Hyphodontia) 153
abieticola Overh. nom. invalid.
    (Polyporus) 13
abieticola (Zeller & Good.) Morelet &
    Gremmen (Grovesiella) 5
abieticola Zeller & Good.
    (Scleroderris) 5
abietina (Bull.) Fr. (Lenzites) 118
abietina (E. & E.) Groves (Pragmopora) 9
abietina (Fuckel) Corner (Typhula) 251
abietina Groves (Tympanis) 12, 119, 159,
    173, 251
abietina (Lagerb.) Morelet (Gremmeniella)
    5, 152, 166
abietina (Peck) Hughes (Verticicladiella)
    160, 201
abietina (Pers.:Fr.) Pouzar
    (Columnocystis) 3, 150, 163, 248
abietina Petrak (Cryptosporiopsis) 8,
    119, 155, 170, 199, 237, 238, 248
abietina (Prill. & Delacr.) Arx & Mueller
    (Botryosphaeria) 2
abietinellum (Dearn.) J. Reid & Cain
    (Nothophacidium) 7
abietinum (Bull.:Fr.) Karst.
    (Gloeophyllum) 118
abietinum (E. & E.) Sutton (Seiridium) 10
abietinum (Hartig) Prill. & Delacr.
    (Fusicoccum) 4
abietinum Peck (Gelatinosporium) 4,
    166, 248
abietinum (Peck) DiCosmo (Foveostroma) 4,
    166, 237, 248
abietinum (Pers.:Fr.) Fr. (Stereum) 3,
    150, 163, 248
abietinum Rostr. (Myxosporium) 248
abietinus (Pers.) Schroet. (Dacrymyces)
    165
abietinus Pers.:Fr. (Polyporus) 11, 39,
    119, 159, 172, 196, 201, 239, 251
abietinus (Pers.:Fr.) Donk
    (Hirschioporus) 11, 39, 56, 119, 159,
    172, 196, 201, 239, 251
abietinus (Pers.:Fr.) Ryv. (Trichaptum)
    11, 39, 56, 119, 159, 172, 196, 201,
    239, 251
abietis Barr (Xenomeris) 40, 201, 251
abietis (W.B. Cke. & Shaw) Hughes ined.
    (Hyphosoma) 1, 247
abietis (Crouan) Groves & Wells
    (Retinocyclus) 10, 157, 171, 200, 250
abietis Dearn. (Bifusella) 6
abietis Dearn. (Dimerosporium) 4
abietis Dearn. (Monochaetia) 7
abietis (Dearn.) Darker (Isthmiella) 6
abietis (Dearn.) J. Reid & Cain
    (Phacidium) 8, 155, 170, 199
abietis (Dearn.) Shoem. (Epipolaeum) 4
abietis Fr. (Valsa) 12, 15, 119, 160,
    201, 239, 251
abietis (Fr.) Link (Heyderia) 152, 167
abietis Jacks. & Lemke (Aleurodiscus) 1

abietis Karsten em. DiCosmo
    (Corniculariella) 150
abietis Mang. & Har. (Rhizosphaera) 10
abietis Naumov (Ascocalyx) 2, 117
abietis Naumov (Rhizothyrium) 10, 157, 250
abietis (Naumov) Seaver (Godronia) 2
abietis (Pers.) Rehm (Cenangium) 148,
    162
abietis Petrak (Rhizocalyx) 10, 157, 250
abietis J. Reid & Piroz. (Pseudoscypha) 9
abietis (Rostr.) Mueller (Delphinella) 4
abietis Sacc. (Cytospora) 3, 69, 70
abietis Sutton (Calcarisporium) 2
abietis (Weir) K. Harrison (Hericium) 5,
    152, 249
abietis Whitney, J. Reid & Piroz.
    (Darkera) 4
abietis Whitney, J. Reid & Piroz.
    (Tiarosporella) 11
abietis-canadensis Arthur (Melampsora)
    184, 218, 249
abietis-concoloris (Mayr ex Dearn.) Darker
    (Lirula) 6
abietis-concoloris Mayr ex Dearn.
    (Hypodermella) 6
acadiensis A.H. Smith (Psilocybe) 89
accedens (B. & G.) Donk (Tubulicrinis)
    12, 159
acericola E. & E. (Phyllosticta) 20
acericola Griffin (Ceratocystis) 14, 203
acericola Groves (Tympanis) 22
acericola Groves & M. Elliott (Ciboria)
    14, 99, 253
acericola (Peck) Sacc. (Pezicula) 19
acericola (Peck) Sacc. (Pholiota) 92
acericola (Peck) Singer (Agrocybe) 92
acerina (Hartig) Deighton
    (Mycocentrospora) 261
acerina (Hartig) Newhall (Centrospora)
    261
acerina Lév. (Melasmia) 18
acerina Oud. (Discula) 18
acerina (Pass.) Sutton (Diplodina) 16,
    21, 136
acerina (Peck) Rehm (Dermea) 15, 21
acerina (Peck) Sacc. (Diaporthe) 16
acerina (Pers.:Fr.) Lemke (Dendrothele) 253
acerina (Westend.) Arx (Discella) 16
acerina Westend. (Libertella) 18
acerina Whet. & Buchw. (Ciboria) 14, 27,
    180, 253
acerinum Kunze (Pilidium) 54
acerinum Peck (Sphaeronaema) 21
acerinum Peck (Stegonsporium) 21
acerinum (Peck) Dearn.
    (Cylindrosporium) 21
acerinum (Pers.) Fr. (Rhytisma) 18, 20
acerinum (Pers.:Fr.) Lemke
    (Aleurocorticium) 253
acerinum J. Reid & Cain (Cryptodiaporthe)
    15
acerinus (Pers.:Fr.) Hoehnel
    (Aleurodiscus)
    var. alliaceus (Quél.) B. & G. 15
aceris Dearn. & Barth. (Cercosporella)
    18
aceris (Dearn. & Barth.) Redhead & White
    (Mycopappus) 18
aceris (Lib.) Berk. & Br. (Septoria) 21

363

buccina (Pers.) Kennedy (Guepiniopsis) 5,
167
bulbigera (Fuckel) Savile (Isariopsis) 221
bulgarioides (Rabenh.) Karst.
(Chlorosplenium) 157
bulgarioides (Rabenh.) Karst.
(Rutstroemia) 157
bullata Berk. (Ustilago) 61, 87, 110
bullata (Berk.) Tul. (Taphrina) 203, 229
bullata (Hoffm.:Fr.) Fr. (Diatrype) 93,
217
bupleuri Rud. (Puccinia) 61
burtii Rom. (Peniophora) 170
burtii (Rom.) Parm. (Phanerochaete) 170
burtii Zeller (Exobasidium) 208
buxi (DC.:Fr.) Berk. & Br. (Volutella) 61
byssina (Karst.) Parm. (Athelia) 156
byssina (Pers.) Rom. (Poria) 156, 171
byssinum (Karst.) Jul. (Piloderma) 156
byssoides (Pers.:Fr.) Bres.
(Peniophora) 1, 146
byssoides (Pers.:Fr.) Eriksson
(Amphinema) 1, 48, 146, 161, 178, 236
cactorum (Leb. & Cohn) Schroet.
(Phytophthora) 37, 76, 77, 127, 131,
134, 156, 170, 195, 203
caerulescens (Desm. & Mont.) Tul.
(Taphrina) 206
caesio-cinerea (H. & L.) Luck-Allen
(Basidiodendron) 39, 162
caesio-cinerea (H. & L.) Rogers
(Sebacina) 39, 162
caesiocinerea (Svrcek) M.J. Larsen
(Tomentella) 189
caesiocinereum Svrcek (Tomentellastrum)
189
caesium C.G. & F. Nees (Gonytrichum) 29
caesius Schrad.:Fr. (Polyporus) 12, 22,
56, 96, 131, 160, 173, 189, 201, 239,
251
caesius (Schrad.:Fr.) Murr. (Tyromyces)
12, 22, 56, 96, 131, 140, 160, 173,
189, 201, 239, 251
caespitica (Karst.) Karst. (Mollisia) 18
caespitosum Fuckel (Lophiostoma) 184
cainii Olchowecki & J. Reid
(Ceratocystis) 148
cajanderi Karst. (Fomes) 4, 118, 151,
166, 182, 193, 198, 248
cajanderi (Karst.) Kot. & Pouzar
(Fomitopsis) 4, 118, 151, 166, 182,
193, 198, 248
cakile Savile (Peronospora) 61
caladii Schw. (Aecidium) 41
calamagrostidis E. & E. (Cylindrosporium)
62
calamagrostidis (Lib.) Sacc. (Septoria) 25
f.sp. koeleriae (Cocc. & Mor.) Sprague
116
calcea (Pers.) Bres. (Sebacina) 151
calcea (Pers.) Wells (Exidiopsis) 151
calcicola (B. & G.) M.J. Larsen
(Tomentella) 159, 188
calcitrapae DC. (Puccinia) 72
caliciiformis Fr. (Crinula) 150
calicioides (E. & E.) Fitzp.
(Caliciopsis) 179
callae Cast. (Rhizoctonia) 171
callista (B. & C.) Sacc. (Fracchiaea) 75
calochorti Peck (Puccinia) 62

calothrix (Pat.) Donk (Tubulicrinis) 159,
173
calothrix (Pat.) Rogers & Jacks.
(Peniophora) 173
calthae Hennebert & M. Elliott
(Botryotinia) 62
calthae Link (Puccinia) 63
calthae (Phill.) Massee (Pseudopeziza) 62
calthicola Schroet. (Puccinia) 63
calthicola Whetz. (Verpatinia) 63, 190
calva (Fr.) Fr. (Mucronella) 7, 154, 169
calyciformis (Willd.:Fr.) Dharne
(Lachnellula) 6, 167
calyciformis (Willd.:Fr.) Rehm
(Dasyscyphus) 6, 167
calyculus (Fr.) Phill. (Hymenoscyphus) 17
cambivora (Petri) Buisman (Phytophthora)
131, 195, 203
cambrensis M.B. Ellis (Corynespora) 75
campanella (Batsch) Fr. (Omphalia) 13,
173
campanella (Batsch:Fr.) Kuehner & Maire
(Xeromphalina) 13, 70, 160, 173, 201,
239, 252
campanulae Lév. & Kickx (Coleosporium) 63
camptospermum (Peck) Maubl. (Toxosporium)
11
campylostyla Auersw. (Gnomonia) 54
campylostylum (Auersw.) Barr (Plagiostoma)
var. campylostylum 54
canadense Bubak & Dearn. (Asteroma) 239
canadense Bubak & Dearn. (Coryneum) 204
canadense (Burt) Eriksson & Weresub
(Amylocorticium) 161
canadense Ellis & Dearn. (Pseudohelotium)
40
canadense E. & E. (Chlorosplenium) 240
canadense Hughes (Trichocladium) 11, 22,
56, 96, 113, 159, 189
canadense Morgan-Jones (Dinemasporium) 181
canadense Savile (Exobasidium) 208
canadensis Arthur (Puccinia) 261
canadensis Barr (Venturia) 215
canadensis Batra (Raffaelea) 200
canadensis Bolton (Marssonina) 98
canadensis Bubak & Dearn. (Phaeoseptoria)
226
canadensis Bubak & Dearn. (Phloeospora) 21
canadensis Bubak & Dearn. (Phomopsis) 141
canadensis Bubak & Dearn.
(Pleosphaerulina) 226
canadensis Burt (Hypochnus) 189, 239
canadensis W.B. Cke. (Solenia) 96
canadensis E. & E. (Dothiorella) 182
canadensis E. & E. (Nectria) 253
canadensis (E. & E.) Berl. & Vogl.
(Calonectria) 253
canadensis (E. & E.) H. Riedl
(Peltosphaeria) 169
canadensis Hughes (Spadicoides) 188
canadensis Kendr. (Phialocephala) 19
canadensis Morton & Sm. (Scopulariopsis)
48
canadensis Nag Raj (Thozetella) 221
canadensis Overh. ex Lowe (Tyromyces) 160
canadensis Overh. nom. invalid
(Polyporus) 151
canadensis Peck (Septoria) 76
canadensis Skolko (Aleurodiscus) 1, 13,
48, 146, 192, 203, 236

fascicularis (Alb. & Schw.:Fr.) Wallr.
(Phibalis) 186
fasciculata Pers. (Solenia) 10, 55
fasciculata Schw. (Tympanis) 76, 259
fasciculata (Schw.) B. & C. (Cyphella)
30, 194
fasciculata (Tode) Petrak
(Cryptosporiopsis) 67, 139
fasciculatus (Pers.) Agerer (Rectipilus)
10, 55
fasciculatus (Schw.) Earle (Merismodes)
30, 194
fastigiata (Lagerb., Lundb. & Melin)
Conant (Phialophora) 9, 19, 53, 119,
156, 170, 186
faulliana Mix (Taphrina) 178
faullii Darker (Bifusella) 6
faullii Darker (Coniothyrium) 3
faullii Darker (Leptosphaeria) 3, 6
faullii (Darker) Darker (Isthmiella) 6,
153
favacea (Fr.) Ces. & de Not.
(Diatrypella) 50
felicis Piroz. & Shoem. (Vestigium) 239
femoralis (Peck) Petrak (Ophiovalsa) 30
femoralis (Peck) Sacc. (Cryptospora) 30
fenestrans (Duby) Petrak (Sydowiella) 89
fenestrata (Berk. & Br.) Schroet.
(Fenestella) 28, 50
fennicus Laurila (Aleurodiscus) 146
fergussonii Berk. & Br. (Puccinia) 261
ferox Long & Baxter (Poria) 39
ferrea (Pers.) B. & G. (Poria) 19, 31,
39, 53, 75, 77, 78, 95, 102, 108, 125,
146, 186, 194, 199, 205, 214, 218, 231,
238, 250, 254
ferreus (Pers.) B. & G. (Phellinus) 19,
31, 35, 39, 53, 75, 77, 78, 95, 102,
108, 125, 146, 155, 186, 194, 199, 205,
214, 218, 231, 238, 250, 254
ferruginea (Fuckel) Deighton
(Mycovellosiella) 42
ferruginea (Pers.) Pat. (Tomentella) 56,
172, 189, 239
ferruginea (Pers.:Fr.) Karst. (Sillia) 77
ferruginea (Sow.:Fr.) Berk. (Pachnocybe)
155
ferrugineofusca Karst. (Poria) 8, 155,
199, 238, 250
ferrugineofuscus (Karst.) B. & G.
(Phellinus) 8, 155, 199, 238, 250
ferrugineum (Pers.) Hughes (Nematogonium)
218
ferruginosa (Fr.) Sacc. (Caldesiella)
56, 189
ferruginosa (Schrad.:Fr.) Karst.
(Poria) 8, 19, 31, 35, 53, 78, 95,
155, 186, 194, 205, 218
ferruginosum Fr.:Fr. (Cenangium) 148, 162
ferruginosus Pat. (Asterodon) 2, 48, 147,
161, 197, 247
ferruginosus (Schrad.:Fr.) Pat.
(Phellinus) 8, 19, 31, 35, 53, 78, 95,
155, 186, 194, 205, 218
fertilis (Nits.) Sacc. (Valsella) 220
festucae Sprague (Phaeoseptoria) 97
festucae (Wegelin) L. Holm (Didymella)
61, 97, 145
fibrillosa (Burt) Donk (Tylospora) 159,
173

fibrillosus Burt (Hypochnus) 159
fibrillosus Karst. (Polyporus) 10, 157,
187, 200, 250
fibrillosus (Karst.) B. & G. (Phaeolus)
157
fibrosa (B. & C.) M.J. Larsen
(Kneiffiella) 172, 189, 200, 220, 239
fibrosa (B. & C.) M.J. Larsen
(Tomentellina) 159, 172, 189, 200,
220, 239
fibulatum Tu & Kimbr., nom. invalid
(Ceratobasidium) 162
ficariae de Bary (Peronospora) 206
ficariae (Cornu & Roze) Fisch. v. Waldh.
(Entyloma) 206
filamentosa (Pat.) Rogers
(Pellicularia) 46, 105, 110, 126,
128, 142, 228, 247
filia (Bres.) Liberta (Trechispora) 172
filicinum (Bourd.) M.P. Chr. (Xenasma)
173, 178, 201
filicinum (Desm.) Starb. (Mycosphaerella)
81
filicinus (Fr.) Nits. (Rhopographus) 201
filiforme Darker (Lophodermium) 153
filiformis (Alb. & Schw.) Fr. (Xylaria) 40
filipendulae Thuemen (Cylindrosporium) 232
filopes (Peck) Redhead (Marasmiellus) 7,
154, 168
filum (Biv.-Bern.:Fr.) Berk. (Darluca)
11, 172, 176
filum (Biv.-Bern.:Fr.) Sutton
(Sphaerellopsis) 11, 172, 176
fimbriata M.P. Chr. (Tomentella) 56, 239
fimbriata Dearn. & House (Venturia) 23,
223, 225
fimbriata Ellis & Halst. (Ceratocystis)
49, 179
fimbriata (Fr.) Quél. (Odontia) 21, 32,
188, 195, 206, 209
fimbriata (Pers.:Fr.) Donk
(Stromatoscypha) 55, 96, 188, 206
fimbriata Schw. (Thelephora) 158
fimbriatum (Fr.) Eriksson (Steccherinum)
21, 32, 188, 195, 206, 209
fimbriatum (Fr.) Fr. (Porotheleum) 55,
188, 206
fimbriatus (Dearn. & House) Barr
(Wentiomyces) 23, 223, 225
fimicola (Rob.) Ces. & de Not. (Sordaria)
99, 257
firma Peck (Naucoria) 48
firma (Peck) Singer (Agrocybe) 48
fischeri Cruchet & Mayor (Puccinia) 222
fischeri (Karst.) Kukk. (Anthracoidea) 66
fischeri (Karst.) Liro (Cintractia) 66
fistulosa (Fr.) R. Petersen
(Macrotyphula) 30
fistulosus (Fr.) Corner (Clavariadelphus)
var. contortus (Fr.) Corner 27
flagellifera (E. & E.) Cicc. (Uniseta)
74, 257
flammans (Fr.) Kummer (Pholiota) 20, 186
flava Farlow (Taphrina) 55
flavescens (Bon.) Rogers
(Botryobasidium) 160
flavescens Estey (Stilbella) 26, 64, 83,
127, 131, 133, 144, 228, 243
flavido-alba Cooke (Peniophora) 95

graminella Sacc. (Stagonospora) 116
graminum Nisikado & Itaka
   (Cephalosporium) 245
gramineum Rabenh. ex Schlecht.
   (Helminthosporium) 109
graminicola (Ces.) G.W. Wilson
   (Colletotrichum) 45, 60, 62, 81, 97,
   108, 109, 144, 224, 245
graminicola Subram. (Pythium) 110, 246
graminis (Busgen) de Wild. (Physoderma)
   175
graminis DC.:Fr. (Erysiphe) 25, 45, 47,
      60, 81, 87, 108, 109, 121, 134, 144,
      176, 196, 202, 224, 245
   f.sp. hordei Marchal 109
   f.sp. tritici Marchal 245
graminis Desm. (Torula) 67
graminis (Desm.) Crane & Schokn. (Rutola)
   67
graminis Fuckel (Scolecotrichum) 60,
   87, 109, 176
graminis (Fuckel) Deighton
   (Cercosporidium) 37, 42, 45, 60, 81,
   87, 97, 109, 144, 176
graminis (Fuckel) Hoehnel (Passalora)
   37, 42, 45, 60, 81, 87, 97, 104, 144,
   176
graminis Ledingham (Polymyxa) 61, 88,
   110, 224, 246
graminis Ledingham (Rhizophydium) 33, 61,
   63, 86, 87, 88, 99, 110, 124, 175, 177,
   190, 224, 246, 263
graminis McAlp. (Hendersonia) 88
graminis (McAlp.) Sacc. & D. Sacc.
   (Wojnowicia) 247
graminis (Pers.:Fr.) Nits. (Phyllachora)
   25, 61, 88, 145, 231
graminis Pers. (Puccinia) 25, 33, 35, 37,
      42, 46, 47, 60, 72, 82, 87, 88, 97,
      110, 128, 135, 145, 176, 224, 244,
      246
   f.sp. agrostidis Eriks. 25
   f.sp. avenae Eriks. & Henn. 46, 47
   f.sp. secalis Eriks. & Henn. 24, 88,
      110, 224
   f.sp. tritici Eriks. 47, 88, 110, 246
   ssp. graminicola Urban 25
   ssp. graminis 246
   ssp. graminis var. graminis 47, 88,
      110, 246
   ssp. graminis var. stakmanii Guyot,
      Mass. & Saccas ex Urban 24, 46, 47,
      88, 110, 224
   var. tritici Eriks. & Henn. 246
graminis Sacc. (Ophiobolus) 25, 109
graminis (Sacc.) Arx & Olivier
   (Gaeumannomyces) 25, 109, 245
   var. tritici Walker 25, 245
graminum (Lib.) Lév. (Dinemasporium) 81
grandinioides B. & G. (Bourdotia) 179
graniforme (Sow.) Ferd. & Winge
   (Cenococcum) 2, 148, 162, 240
grantii Funk (Groveiella) 5
grantii Lloyd (Aleurodiscus) 1, 247
granuliformis Ellis & Holw. (Cercospora)
   261
granulispora Ellis & Gall. ex E. & E.
   (Puccinia) 26
granulosa Fr. (Grandinia) 13, 160, 251
granulosa (Pers.:Fr.) Laurila (Vararia)
   13, 119, 160, 173, 201, 251

grayanum Sacc. & Ev. (Ciliciopodium) 253
gregaria Sacc. (Physalospora) 235
gregata (Allington & Cham.) W. Gams
   (Phialophora) 104
gregatum Allington & Cham.
   (Cephalosporium) 104
grenfelliana Savile (Puccinia) 74
greschikii (Bres.) B. & G. (Peniophora)
   4, 166
greschikii (Bres.) Eriksson (Fibricium)
   28, 34, 237
grindeliae Peck (Puccinia) 90
grisea (Pers.) Bourd. & L. Maire
   (Exidiopsis) 16, 28, 182, 204, 217
grisea (Pers.) Bres. (Sebacina) 16
grisea (Unger) de Bary (Peronospora) 258
griseifolia A.H. Smith (Psathyrella) 31
grisellum (B. & G.) Liberta (Xenasma) 160
griseo-cana (Bres.) B. & G. (Dendrothele)
   15, 78, 99, 114, 129, 139, 165, 181,
   204, 217, 234, 237, 240, 253
griseo-canum (Bres.) Lemke
   (Aleurocorticium) 15, 78, 99, 114,
   129, 139, 165, 181, 204, 217, 234, 237,
   240, 253
griseo-lanatum Funk (Gelatinosporium) 4,
   118, 248
griseofulvum Dierckx (Penicillium) 58
griseopergamacea M.J. Larsen
   (Pseudotomentella) 205
griseoumbrina Litsch. (Tomentella) 119
griseoviolacea Litsch. (Tomentella) 11,
   240, 251
griseum Robak (Oidiodendron) 257
griseus (Krain.) Waks. & Henrici
   (Streptomyces) 247
griseus Peck (Polyporus) 162
griseus (Peck) Bond. & Singer
   (Boletopsis) 162
grohii Groves (Claviceps) 66
grossulariae (Wallr.) Lév. (Microsphaera)
   221
grossulariae (Auersw. & Fleischh.) Barr
   (Antennularia) 210
grossulariae (Auersw. & Fleischh.) Barr
   (Protoventuria) 210
grossulariae (Auersw. & Fleischh.) Sacc.
   (Venturia) 210
grossulariae Sacc. (Phyllosticta) 210
grovei M.B. Ellis (Spadicoides) 11, 96,
   205
grovesii D. Wells (Ascocybe) 248
grovesii Ouellette & Piroz. (Tympanis) 259
grovesii J. Reid & Piroz. (Dermea) 150,
   151
grumiformis (Karst.) Barr (Gibbera) 41
grumosa Sydow & Holw. (Puccinia) 264
guttata (Wallr.:Fr.) Lév. (Phyllactinia)
   20, 31, 47, 54, 67, 68, 69, 76, 77, 78,
   95, 100, 108, 113, 124, 144, 205, 219,
   231, 262
guttatum (Schroet.) Hylander, Jorst. &
   Nannf. (Pucciniastrum) 101
guttulata (Starb.) Wehm. (Apiognomonia)
   24
guttulatus Peck (Polyporus) 12, 160,
   196, 201, 251
guttulatus (Peck) Murr. (Tyromyces) 12,
   160, 196, 201, 251

luteus (Zukal) C.R. Benjamin
(Talaromyces) 113
lychnidis Desm. (Septoria) 126
lychnidis (Fr.) E. & E. (Phyllosticta)
126, 225
lycopersici Bruschi (Fusarium) 127
lycopersici Cooke (Phoma) 127, 227
lycopersici Kleb. (Didymella) 127, 227
lycopersici (Plowr.) Brun. (Ascochyta)
127, 227
lycopersici Speg. (Septoria) 127
lycopodii (Peck) House (Mycosphaerella)
128
lycopodina (Karst.) Arx (Pseudomassaria)
128
lycopodina (Mont.) Sacc. (Leptosphaeria)
128
lyngei (Lind.) Nannf. (Hysteropezizella)
40, 97
macounii (Burt) Eriksson & Boidin
(Clavulicium) 3, 150
macounii E. & E. (Diatrype) 16
macrocarpa (Corda) Hughes
(Sterigmatobotrys) 11, 158
macrocarpum Preuss (Cladosporium) 14, 64,
86, 99, 106, 107, 119, 129, 150, 163,
180, 193, 212, 216, 236, 252
macrocephala (Kohlm.) Meyers & Moore
(Cirrenalia) 163, 197
macrocystidiatus Lemke (Aleurodiscus) 39
macropoda Pass. (Septoria) 176
var. septulata (Gonz. Frag.) Sprague
176
macropus Corda (Dactylosporium) 111
macrosperma Magn. (Uredinopsis) 12, 201
macrospora (Bres.) Lemke (Dendrothele) 181
macrospora E. & E. (Diplodina) 75
macrospora (E. & E.) Wells (Exidiopsis)
4, 28, 151, 198
macrospora Fres. (Ramularia) 63
macrospora (Hartig) Darker (Lirula) 153
macrospora Kleb. (Didymellina) 112
macrospora (Kleb.) Jorst.
(Macrosphaerella) 112
macrospora Peck (Uncinula) 254
macrospora (Peck) Kanouse (Rutstroemia)
49, 93
macrospora (Sacc.) Thirum., Shaw & Naras.
(Sclerophthora) 61, 88, 145, 225, 245,
246, 263
*macrospora (Wollenw.) Ouellette
(Nectria)* 7
macrospora Ziller & Funk (Sarcotrochila)
171
*macrosporum (Bres.) Lemke
(Aleurocorticium)* 181
*macrosporum (Hartig) Rehm
(Lophodermium)* 153
macrosporum (Sacc.) Hoehnel
(Microthyrium) 61
macrostoma Mont. (Phoma) 130
macrotheca (Rostr.) Mueller
(Wettsteinina) 67, 90
maculans (Desm.) Ces. & de Not.
(Leptosphaeria) 58, 59, 70, 226, 236
maculare (Fr.) de Not. (Lophodermium) 256
macularis (Fr.) Mueller & Arx (Venturia)
187, 190

macularis (Wallr.:Fr.) P. Magn.
(Sphaerotheca) 24, 57, 59, 74, 75, 89,
97, 99, 103, 111, 146, 191, 209, 212,
214, 221, 235, 258, 261
maculata Atk. (Fabraea) 34, 78, 80, 230
maculata Cooke & Harkn. (Diplodia) 39
maculata (Jacks. & Lemke) Lemke
(Dendrothele) 15, 76, 193
maculatum (Atk.) Jorst. (Diplocarpon)
207
*maculatum Jacks. & Lemke
(Aleurocorticium)* 76, 193
maculatum Lév. (Entomosporium) 34, 78,
230
maculicola (Rom. & Sacc.) Sutton
(Phaeoramularia) 185
*maculiformans Guba & Zeller
(Pestalotia)* 40
maculiformans (Guba & Zeller) Stey.
(Pestalotiopsis) 40
maculiforme Nees (Circinotrichum) 253
maculiformis (Desm.) Wint. (Venturia) 89
maculiformis (Fr.) Schroet.
(Mycosphaerella) 53, 218, 231
madiae (Sydow) Arthur (Coleosporium) 128
magica Pass. (Urocystis) 27
magnifica Griffin (Ceratocystis) 3
magnifructa (Peck) Sacc. (Macrophoma) 237
*magnispora (E. & E.) Wehm.
(Cryptodiaporthe)* 18
magnum Ellis (Gelatinosporium) 51
magnusiana Koern. (Puccinia) 36, 145
major (Beyma) C. Moreau (Ceratocystis)
14, 163
major Nyl. ex Koerb. (Stenocybe) 11
malacotricha (Niessl) Trav. (Coniochaeta)
163
*mali (Allesch.) Wollenw.
(Cylindrocarpon)* 129, 202
mali Dearn. & Foster (Coniosporium) 129
mali Roberts (Alternaria) 129
mali Roberts (Phomopsis) 131
malicola B. & C. (Trametes) 13, 39
malicola (B. & C.) Donk (Antrodia) 13, 39
malicola A.H. Smith (Pholiota) 95
malicolus (Lloyd) Ginns (Spongipellis)
101, 188, 219
malicorticis Cordley (Gloeosporium) 129
malicorticis (Cordley) Nannf.
(Cryptosporiopsis) 129, 130, 202
malicorticis Jacks. (Neofabraea) 130
malicorticis (Jacks.) Nannf. (Pezicula)
129, 130, 202
malorum Berk. (Sphaeropsis) 131, 231, 262
malorum (Kidd & Beaum.) McCollach
(Phialophora) 194, 203
malorum Peck (Sphaeropsis) 131
malvacearum Bert. ex Mont. (Puccinia) 33,
131
malvacearum (Westend.) Grove (Phomopsis)
131
mammatum (Wahl.) J.H. Miller (Hypoxylon)
17, 29, 52, 94, 183, 204, 217
manitobanum J.J. Davis (Cercospora) 86
manitobanum (J.J. Davis) Sutton
(Cercosporidium) 86
manitobaensis Sutton (Menispora) 52
manshurica (Naum.) Sydow ex Gaeum.
(Peronospora) 104
mansonii (Castell.) Schol-Schwarz
(Rhinocladiella) 59, 157, 200

microscopica (Karst.) Hughes
(Cheiromycella) 163
microscopicum Desm. (Microthyrium) 7
microsora Koern. ex Fuckel (Puccinia) 67
microsora Sacc. (Cercospora) 240
microsperma (Johnston) Sutton (Diplodina)
28, 181, 193, 216, 217
*microsperma (Peck) Petrak
(Cylindrosporella) 48*
microspermum Berk. & Br. (Oidium) 154
microspermum (Peck) Sutton (Asteroma) 48
microspora (Davidson & Lorenz) Sherwood
(Acarosporina) 13, 252
microspora (Jacks. & Lemke) Lemke
(Dendrothele) 50, 68, 99, 139, 204,
240, 253
microspora (Karst.) Gilbn. (Athelia) 147
*microspora (Karst.) H. & L.
(Tomentella) 11, 32, 56, 189, 200, 239*
*microsporum Davidson & Lorenz
(Schizoxylon) 13, 252*
*microsporum Jacks. & Lemke
(Aleurocorticium) 50, 68, 99, 139,
204, 240, 253*
microsporum Karst. (Anthostoma) 92
var. exudans Peck 27
*microsporum (Karst.) ex B. & G.
(Corticium) 2*
microsporum (Unger) Schroet. (Entyloma)
206
*microsticta Berk. & Br. (Coryneum) 212,
257*
*microsticta (Berk. & Br.) Grove
(Coryneopsis) 212*
microstigma Sacc. (Cercospora) 66
milfordensis E.B. Jones (Halonectria)
167, 198
millardetii (Racib.) Meeker (Seuratia)
10, 200
millefolii Fuckel (Puccinia) 23, 42
millefolii (Fuckel) Niessl
(Leptosphaeria) 23
milliaria (Fr.) Sacc. (Eutypa) 28
minax (J.J. Davis) Deighton
(Mycovellosiella) 229
miniatum Bon. (Coleosporium) 212
*miniatus (Scop.:Fr.) Fr. (Hygrophorus)
18*
minima (B. & C.) Underw. & Earle
(Phyllosticta) 20, 231
minima Olchowecki & J. Reid
(Ceratocystis) 163
minima Tul. (Calosphaeria) 192
minimum Laub. (Cryptosporium) 212
minimus J.J. Davis (Uromyces) 136
miniopsis (Ellis) Seaver (Erinellina) 16
minitans W.A. Campbell (Coniothyrium) 59,
107
minnsiae Jacks. (Aleurodiscus) 168, 249
minnsiae (Jacks.) Donk (Laeticorticium)
6, 168, 198, 249
minor (Barr) Barr (Pseudomassaria) 85
minor Bubak (Ophiobolus) 125
minor (Cooke) Rehm (Hysteropatella) 52, 94
minor (Hedgc.) Hunt (Ceratocystis) 3,
148, 163
minor Jagger (Sclerotinia) 83, 107, 228
minor (Karst.) Johans. (Mycosphaerella)
36, 41, 42, 89, 125, 139, 202, 257
minor Peck (Dacrymyces) 4, 76

minor Schroet. (Uromyces) 243
minor Sherwood (Stictis) 188
minor Tul. (Fenestella) 28
minus Shear (Gloeosporium) 256
minusculoides (Pilat) Bond. (Tyromyces) 56
*minusculoides (Pilat) Lowe (Polyporus)
56*
minussensis Thuemen (Puccinia) 117
minuta Barr (Venturia) 220
minuta Olive (Guepiniopsis) 167
minuta Peck (Pezicula) 259
minuta (Siem.) Hunt (Ceratocystis) 3,
148, 163
minutella Bubak & Dearn. (Phyllosticta) 20
minutissima Arthur (Puccinia) 67, 83
minutum Eriksson (Xenasma) 11, 158
minutum (Eriksson) Oberw.
(Sphaerobasidium) 11, 158
mirabilis Bubak (Kabatia)
var. oblongifoliae Conners 125
mirabilis Darker (Hypodermella) 6
mirabilis (Darker) Darker (Lirula) 6
mirabilis (Niessl) Hoehnel (Wettsteinina)
42, 75
*mirabilis (Peck) Magn. (Uredinopsis)
12, 138*
mirabilissima (Peck) Nannf.
(Cumminsiella) 129
mirificum Eriksson (Repetobasidium) 157
misandrae Kukk. (Anthracoidea) 66
misella Niessl (Gnomonia) 138
mitis (Fr.) Singer (Panellus) 8, 199
*mitis (Pers.:Fr.) Quél. (Pleurotus) 8,
199*
miurae Sydow (Uromyces) 101
mixta Fuckel (Puccinia) 26
miyabeana Fukushi (Physalospora) 219
modesta (Desm.) Munk ex L. Holm
(Nodulosphaeria) 107, 206
modesta Hoehnel (Nectria) 7, 19
modonia Tul. (Melanconis) 68
modonia (Tul.) Hoehnel (Pseudovalsa) 68
molitoris Minter (Lophodermium) 168
molliforme B. & G. (Corticium) 172
mollis (Fr.) Kot. & Pouzar (Tyromyces)
12, 160, 173, 201, 251
mollis (Fr.) Staude (Crepidotus) 15, 28,
180, 216, 230
*mollis Pers.:Fr. (Polyporus) 12, 160,
173, 201, 251*
mollis (Sommerf.) Donk (Datronia) 15, 28,
50, 108, 151, 181, 253
*mollis (Sommerf.) Fr. (Trametes) 15,
28, 50, 108, 181, 253*
molliuscula (Schw.) Cash (Dermatea) 50
molliuscula (Schw.) Cash (Dermea) 50, 51
mollusca (Fr.) Pouzar (Leucogyrophana) 6,
153, 168, 198, 237, 249, 253
*mollusca (Pers.:Fr.) Cooke (Poria) 159,
189*
mollusca (Pers.:Fr.) Liberta
(Trechispora) 11, 32, 56, 159, 189,
200, 239, 251
molluscus Fr. (Merulius) 6, 168
molnarii Funk (Nitschkia) 199, 250
molybdaea B. & G. (Tomentella) 11, 172
monesis Ziller (Chrysomyxa) 136, 149
moniliforme Sheldon (Fusarium) 263
var. subglutinans Wollenw. & Rein. 263

olivacea (Wallr.) M.B. Ellis
(Corynespora) 240
olivaceo-album (B. & G.) Jul.
(Confertobasidium) 3, 150, 163, 197,
236
olivaceum Bonord. (Coniothyrium) 130,
256
olivaceum (Schw.) Banker (Hydnochaete)
17, 94, 204
olivaceus Fuckel (Retinocyclus) 157, 171,
250
olivascens (B. & C.) B. & G. (Tomentella)
22, 251
olivascens (B. & C.) Ginns & Weresub
(Leucogyrophana) 184
omnivorum (Shear) Duggar (Phymatotrichum)
10
oncostoma (Duby) Fuckel (Diaporthe) 211
onobrychidis (DC.) J. Mueller (Diachora)
106
onobrychidis (DC.:Fr.) Hoehnel
(Diachorella) 106
f. onobrychidis 106
onobrychidis (DC.) Sacc.
(Placosphaeria) 106
ontariensis (Rehm) Hoehnel (Gorgoniceps)
166
onusta Karst. (Trechispora) 11, 32,
200, 239, 251
opacum (Corda) Hughes (Trichocladium) 22,
99
operculatus B. & C. (Panus) 220
orbicula Peck & Clint. (Puccinia) 191
orbiculare (Berk. & Mont.) Arx
(Colletotrichum) 73, 79, 80
orbiculare (Ehr.) Sacc. (Lophodermium) 70
orbiculare Grev. (Radulum) 2, 27, 49,
92, 162, 179, 192, 230, 247
orchidis (Vuill.) Hagem (Absidia) 45, 244
oreades (Bolt.) Fr. (Marasmius) 176
oregonense Murr. (Ganoderma) 4, 152, 198,
248
oregonensis Goodding (Didymosphaeria) 28
oreophilum Lindsey & Gilbn.
(Steccherinum) 188
orientalis Funk (Caliciopsis) 248
ornata Arthur & Holw. (Puccinia) 215
ornatula Holw. (Puccinia) 261
orobi Sacc. (Ascochyta) 138
orobi Winter (Uromyces) 120
orthoceras Appel & Wollenw. (Fusarium)
var. gladioli McCulloch 103
orthosporum (Sacc.) Wollenw.
(Cylindrocarpon) 181, 216
orthosporum Caldwell (Rhynchosporium) 82,
143
oryzae (Berk. & Br.) Petch (Nigrospora)
245, 263
oryzae Hudson (Khuskia) 245, 263
oryzae Went & Prins.-G. (Rhizopus) 59
osmorrhizae Peck (Septoria) 139
osmorrhizae (Peck) House
(Phloeospora) 139
osmundae Magn. (Uredinopsis) 12, 139
osoyoosensis Savile (Puccinia) 223, 262
osseus Kalch. (Polyporus) 8, 155, 199
ossiformis Olchowecki & J. Reid
(Ceratocystis) 3
ostracoderma Korf (Peziza) 98, 133, 242

ostrea (Blume & Nees:Fr.) Fr. (Stereum)
11, 21, 32, 39, 55, 96, 158, 188, 200,
206, 251
ostreatus (Jacq.:Fr.) Kummer (Pleurotus)
9, 20, 31, 54, 95, 100, 156, 186, 219,
240, 254
ostryae (Dearn.) Wehm. (Melanconis) 139
ostryae de Not. (Gnomonia) 139
ostryigena Ellis & Dearn. (Hendersonia)
139
oudemansii Sacc. (Septoria) 176
oudemansii Tranz. (Puccinia) 141
ovatum (Mérat) Hughes (Stegonsporium) 21
overholtsii (Ryv. & Gilbn.) Ginns
(Flaviporus) 151
ovina (Pers.) Ces. & de Not.
(Lasiosphaeria) 18, 253
owensii Lloyd (Irpex) 31, 77, 195
owensii Sprague (Phyllosticta) 81
oxalidis Diet. & Ellis (Puccinia) 140
oxyacanthae (DC.) de Bary (Podosphaera)
195
oxycocci auct. (Sclerotinia) 257
oxycocci Shear (Sporonema) 257
oxycocci Wor. (Sclerotinia) 257
oxycocci (Wor.) Honey (Monilinia) 256, 257
oxyriae Fuckel (Puccinia) 140
oxyriae Rostr. (Coleroa) 140
oxyriae (Rostr.) Sacc. (Venturia) 140
oxyriae Savile (Mycosphaerella) 140
oxysporum Bomm., Rouss. & Sacc.
(Septogloeum) 24, 25, 62, 87
oxysporum Schlecht. (Fusarium) 4, 45, 47,
58, 71, 80, 100, 109, 117, 123, 127,
130, 132, 134, 144, 152, 166, 174,
193, 198, 227, 242, 245, 256, 263
f.sp. apii (Nels. & Sherb.) Snyder &
Hansen 37
f.sp. asparagi Cohen 42
f.sp. callistephi (Beach) Snyder &
Hansen 62
f.sp. cepae (Hanz.) Snyder & Hansen 26
f.sp. conglutinans (Wollenw.) Snyder &
Hansen 37, 58
f.sp. dianthi (Prill. & Del.) Snyder &
Hansen 84
f.sp. fabae Yu & Fang 260
f.sp. gladioli (Massey) Snyder &
Hansen 103
f.sp. lilii Imle 122
f.sp. lini (Bolley) Snyder & Hansen 123
f.sp. lycopersici (Sacc.) Snyder &
Hansen 127
f.sp. medicaginis (Weimer) Snyder &
Hansen 132
f.sp. melonis (Leach & Currence) Snyder
& Hansen 79
f.sp. narcissi Snyder & Hansen 137
f.sp. pini (Hartig) Snyder & Hansen 198
f.sp. pisi (Hall) Snyder & Hansen 174
f.sp. radicis-lycopersici Jarvis &
Shoem. 127
f.sp. spinaciae (Sherb.) Snyder &
Hansen 232
f.sp. tuberosi Snyder & Hansen 227
var. redolens (Wollenw.) Gordon 93,
123, 132, 134, 198, 227
oxystoma Rehm (Valsa) 28
oxystoma (Rehm) Urban (Cryptodiaporthe) 28

quercuum (Berk.) Miyabe ex Shirai
(Cronartium) 164, 204
quercuum (Schw.) Sacc. (Botryosphaeria)
92, 203
quisqualis Ces. (Ampelomyces) 57, 79
rabiei (Pass.) Labrousse (Ascochyta) 71
racemosa (Burt) Rogers & Jacks. (Vararia)
13, 119, 173, 239
racemula (Cooke & Peck) Barr
(Ditopellopsis) 89
radiata Fr. (Phlebia) 9, 19, 31, 54, 95,
156, 170, 186, 194, 199, 205, 219, 250
radiata (L.) Pers.:Fr. (Stictis) 21, 39,
55, 101, 219
radiatum Peck (Stereum) 179
radiatum (Peck) Parm. (Boreostereum) 147,
179
radiatus Sow.:Fr. (Polyporus) 18, 29,
52, 76, 78, 183, 194
radiatus (Sow.:Fr.) Karst. (Inonotus) 18,
29, 52, 76, 78, 183, 194
radicale Schwartz & Cook (Olpidium) 73,
79, 80
radicata Fr. (Ditiola) 4, 165
radicata Weir (Sparassis) 200
radicicola Vanterpool & Led. (Lagena) 88,
245
radicicola Wollenw. (Cylindrocarpon)
80, 98, 122, 132, 141, 197, 242
radicina Meier, Drechs. & Eddy
(Alternaria) 82, 141
*radicinum (Meier, Drechs. & Eddy) Neerg.
(Stemphylium)* 82, 141
radiosa (Lib.) Baldacci & Cif.
(Pollaccia) 187, 190
radiosum (Lib.) Lind (Fusicladium) 187
radiosum (Pers.) Fr. (Corticium) 13, 160
radula (Fr.) Cooke sensu Bres. (Poria)
29
radula (Fr.) Donk (Hyphoderma) 27, 75,
147, 192, 197, 216, 247
radula (Fr.) Nobles (Basidioradulum) 2,
27, 49, 75, 92, 147, 162, 179, 192,
197, 216, 230, 247
raduloides (Karst.) Donk (Sistotrema) 10,
158, 188, 251
*raduloides (Karst.) Rogers
(Trechispora)* 10, 158, 188
raetica (Mueller) Petrak (Xenomeris) 41
rallum (Jacks.) Hjort. & Ryv.
(Subulicystidium) 11
rallum (Jacks.) Liberta (Xenasma) 22,
160, 201
ramanniana (Moeller) Linnem.
(Mortierella) 169
var. ramanniana 221
ramannianus Moeller (Mucor) 169
ramosa (Lindt) Lendner (Absidia) 57, 244
ramosissima (B. & C.) Wakef. (Tomentella)
32, 56, 159, 172, 189, 206, 239, 240
ramosum (Bull.) Letellier (Hericium)
17, 51, 94, 100, 152, 183, 253
ramulicola Desm. (Diplodia) 91
ranunculacearum DC. (Aecidium) 36
ranunculi (Fr.) Karst. (Fabraea) 206
ranunculi (Fr.) Schuepp (Leptotrochila)
206
ranunculi (Karst.) Lind (Mycosphaerella)
206

raphani Groves & Skolko (Alternaria) 57,
121, 123, 132, 206, 236
raphani Kendr. (Aphanomyces) 206
raptor Moore & Meyers (Zalerion) 173
raveneliana (Thuemen & Rehm) Barr
(Amphiporthe) 203
ravenelii Berk. (Merulius) 7
recedens (O. Massal.) Jones & Weimer
(Stagonospora) 243
recedens Sydow (Puccinia) 224
recisa (S.F. Gray) Fr. (Exidia) 16, 129,
193, 204, 217
recondita in part (Puccinia) 87, 112,
121, 196
recondita Rob. ex Desm. (Puccinia) 46, 224
f.sp. tritici 246
recondita s. l. (Puccinia) 23, 25, 33,
36, 38, 61, 72, 73, 83, 88, 108, 110,
111, 176, 206, 235, 244, 246
recurvatum (Morgan) Hoehnel
(Pleurothecium) 20, 54, 95
redolens Wollenw. (Fusarium) 93, 198,
227
reflexus Bainier (Rhizopus) 137
regularis Hesler & Sm. (Crepidotus) 3
reiliana (Kuehn) Clinton (Sphacelotheca)
263
repanda (Fr.) Nits. (Nummularia) 19, 205
repanda Pers. (Peziza) 185
repandus Overh. (Fomes) 199, 238, 250
repandus (Overh.) Gilbn. (Phellinus) 199,
238, 250
repens Bernard (Rhizoctonia) 171
repens de Bary (Eurotium) 58
repens (Cooke & Ellis) Kendr.
(Phialocephala) 156, 186
repens (Corda) Sacc. (Aspergillus) 244
repens Schost. (Actinomucor) 98
resinaceum Boud. (Ganoderma) 16, 253
*resinae (Ehr.:Fr.) Hoehnel
(Pycnidiella)* 13, 160, 174
resinae (Ehr.:Fr.) Karst. (Zythia) 13
resinae (Ehr.:Fr.) Mudd (Biatorella) 13,
160, 162, 197, 248
resinae (Fr.) Fr. (Sorocybe) 200, 251
resinae (Fr.) M.B. Ellis (Alysidium)
var. microsporum Sutton 178
var. resinae 1, 178
resinae Parbery (Amorphotheca) 146
resinaria (Cooke & Phill.) Rehm
(Lachnellula) 153
resinaria Phill. (Lachnella) 6
resinicola Baranyay & Funk (Helotium) 5,
152, 198, 249
resinosum (Jacks. & Deard.) K. Martin &
Gilbn. (Hyphoderma) 153, 198
resinosus Schrad.:Fr. (Polyporus) 6,
52, 100, 130, 153, 167, 198, 237, 240,
249, 253
resinosum (Schrad.:Fr.) Karst.
(Ischnoderma) 6, 52, 100, 130, 153,
167, 198, 237, 240, 249, 253
reticulata E. & E. (Ramularia) 139
reticulata (Fr.) Cooke (Poria) 14, 163,
179
reticulata (Fr.) Domanski (Ceriporia) 14,
163, 179
reticulata Liro (Ustilago) 178
reticulospora (Greis & Greis-D.) C. & M.
Moreau (Gelasinospora) 256

roseum Link:Fr. em. Snyder & Hansen
(Fusarium) 58, 63, 67, 80, 121,
123, 130, 132, 242, 263
var. acuminatum ined. 132
f.sp. cerealis (Cooke) Snyder & Hansen
263
roseus (Alb. & Schw.:Fr.) Karst.
(Fomes) 4, 50, 118, 151, 166, 182,
198, 237, 248
roseus (Fr.) H. & L. (Aleurodiscus) 18,
30, 184
rosicola B.H. Davis ex Deighton
(Mycosphaerella) 212
rosicola Pass. (Cercospora) 211
rotula (Scop.:Fr.) Fr. (Marasmius) 52, 94
roumegueri (Sacc.) Sacc. (Myxosporium) 75
roumeguerii (Bres.) Donk (Phlebia) 254
roumeguerii (Bres.) Jul. & Stalp.
(Phlebiopsis) 254
rousseauana Sacc. & Bomm. (Fabraea) 62
rubefaciens Johans. (Puccinia) 101
rubella (Bon.) Nannf. (Ramularia) 215
rubellus (Pers.:Fr.) Sacc. (Ophiobolus)
19, 38, 42, 57, 72, 81, 101, 105, 106,
107, 112, 123, 127, 141, 212, 227, 235
ruber (Konig, Spieck. & Brem.) Thom &
Church (Aspergillus) 244
rubescens Fr. (Cytospora) 30, 78, 193,
218, 230, 231
rubi (Fr.) Rehm (Pyrenopeziza) 214
rubi (Lib.) Niessl (Pezicula) 214
rubi Rabenh. ex Schroet. (Peronospora) 214
rubi (Rehm) Winter (Gnomonia) 213
rubi Roark (Mycosphaerella) 213
rubi Westend. (Septoria) 213, 214
rubi (Westend.) Sacc. (Hendersonia) 214
rubi-idaei (DC.) Karst. (Phragmidium) 214
rubi-odorati Diet. (Phragmidium) 214
rubicola Ellis & Dearn. (Coccomyces) 213
rubicola (Ellis & Dearn.) Sherwood
(Karstenia) 213
rubicola (E. & E.) Mueller (Coleroa) 213
rubicola (E. & E.) Theissen (Stigmatea)
213
rubiginosa (Bres.) R. Maire (Tomentella)
159, 241
rubiginosa (Dickson:Fr.) Lév.
(Hymenochaete) 5, 108
rubiginosa Sappa & Mosca (Rhizoctonia) 171
rubiginosum (Pers.:Fr.) Fr. (Hypoxylon)
17, 29, 52, 76, 94, 100, 139, 183, 205,
217
rubiicola Earle (Lophodermium) 213
rubra Peck (Phyllosticta) 78
rubra Zeller (Truncocolumella) 251
rubricosa (Fr.) Sacc. (Valsaria) 173
rubricosum (Dearn. & Barth.) Nannf.
(Mastigosporium) 25, 62, 81
rubrocinctum Magn. (Synchytrium) 223
rubromarginata (Fr.:Fr.) Kummer (Mycena)
7, 154
rubrum Konig, Spieck. & Bremer (Eurotium)
58
rudbeckiae Arthur & Holw. (Uromyces) 214
rude (Karst.) Jul. (Fibricium) 4, 28, 34,
166, 237
rudis (Ehr.) Hughes (Peyronelia) 155
rudis Fr. (Panus) 8, 19, 30, 53, 95, 155,
185, 238

rudis (Sacc.) Hughes (Taeniolella) 11,
158, 239
rufa (Fr.) Boidin (Peniophora) 185, 218
rufa (Fr.) Karst. (Cryptochaete) 185,
218
rufa (Fr.) M.P. Chr. (Phlebia) 9, 20, 31,
186, 194, 205, 219
rufa (Pers.:Fr.) Fr. (Hypocrea) 29, 94,
167
rufilabrum (B. & C.) Sacc. (Hypoderma) 17
rufo-olivacea (Alb. & Schw.:Fr.) Korf
(Velutarina) 56, 239
rufofusca (Weberb.) Sacc. (Ciboria) 3, 197
rufomaculans Peck (Ramularia) 177
rufum (Fr.:Fr.) Nannf. (Dacryonaema) 165
rufus Fr. (Merulius) 20
rugipes (Peck) Ramam. & Korf
(Chlorociboria) 93
rugosa Butin (Ascodichaena) 92
rugosodisca Peck (Omphalia) 5
rugosum (Pers.:Fr.) Fr. (Stereum) 5,
17, 23, 29, 51, 94, 130, 152, 183, 193,
198
rugosum (Pers.:Fr.) Pouzar
(Haematostereum) 5, 17, 23, 29, 51,
94, 130, 152, 183, 193, 198
rumicis (Desm.) Wint. (Venturia) 215
rupestris Kukk. (Anthracoidea) 66
russellii Clinton (Microsphaera) 140
ruttneri Litsch. (Tomentella) 11, 101,
172, 189, 241
sabina (de Not.) Hoehnel
(Eutryblidiella) 115
sabina (de Not.) Petrini, Samuels &
Mueller (Holmiella) 115
sabinae (Fuckel) Dennis (Chloroscypha) 114
saccata (Darker) Darker (Bifusella) 162
sacchari (Butler) W. Gams (Fusarium) 45,
109, 123, 245, 263
sacchari Jenkins (Taphrina) 21
saccharina Fr. (Exidia) 151, 166, 198, 230
saccharinum E. & E. (Gloeosporium) 14
sagmatospora Wright & Cain (Ceratocystis)
148, 163, 197, 236
salicella (Berk. & Br.) Sacc.
(Rhabdospora) 219
salicella (Fr.) Petrak (Cryptodiaporthe)
28, 75, 181, 193, 216, 217
salicicola (Bres.) Magn. (Marssonina) 218
salicicola (Fr.) Sacc. (Septoria) 219
salicicola (Johnson) Sacc. & Trav.
(Diplodina) 217
salicicola Petrak (Pleurophomopsis) 219
salicina (Curr.) Wehm. (Cryptodiaporthe)
181, 216
salicina E. & E. (Cercospora) 216
salicina (Fr.) Burt (Cytidia) 28, 50,
181, 216
salicina Groves (Tympanis) 220
salicina Lév. (Melampsora) 218
salicina (Pers.:Fr.) Fr. (Valsa) 190,
220
salicina (Vest.) Liro (Ramulaspera) 219
salicinum Grove (Camarosporium) 216
salicinum (Peck) Dearn. (Cylindrosporium)
216
salicinum (Pers.:Fr.) Fr. (Rhytisma) 219
salicinus (Fr.) Kummer (Pluteus) 219
salicinus Peck (Panus) 30, 185

408

stilbospora (Corda) Hughes (Taeniolella)
  220
stilbostoma (Fr.) Tul. (Melanconis) 52
stillatus Fr. (Dacrymyces) 4, 151, 165
stillwellii Funk (Gelatinosporium) 248
stipae Arthur (Puccinia) 90, 128, 229, 233
stipata (Fr.) Quél. (Odontia) 7
stipitatus (B. & G.) Neuh. (Dacrymyces)
  165
stipitatus Kauff. (Crepidotus) 93
stipticus (Bull.:Fr.) Fr. (Panus) 30,
  53, 95, 100, 185, 199, 205, 250
stipticus (Bull.:Fr.) Karst. (Panellus)
  30, 53, 95, 100, 185, 199
stipticus (Pers.) Fr. (Polyporus) 201
stipticus (Pers.:Fr.) Kot. & Pouzar
  (Tyromyces) 12, 22, 160, 189, 201, 239
stolonifer (Ehr.:Fr.) Vuill. (Rhizopus)
  99, 107, 195
stolonifera E. & E. (Ramularia) 76
stomatica Ellis & J.J. Davis (Cercospora)
  229
stomaticola (Bauml.) Sutton
  (Pseudoseptoria) 24, 82, 97, 110, 116,
  121, 145, 176, 246
straminis Sacc. & Speg. (Pleospora) 66
streptopi Dearn. & Barth. (Cercospora) 85
streptopodis Peck (Septoria) 85, 233
striatus Schroet. (Uromyces) 92, 133, 243
stricta (Fr.) Quél. (Ramaria) 20
  var. concolor Corner 200
strictum W. Gams (Acremonium) 45, 255, 262
strigosozonata (Schw.) W.B. Cke.
  (Phaeophlebia) 187, 195, 254
strigosozonata (Schw.) Lloyd (Phlebia)
  31, 187, 195
strigosozonata (Schw.) Talbot
  (Punctularia) 31, 157, 187, 195, 219,
  254
strigosum (Pers.:Fr.) Sacc.
  (Dinemasporium) 81
strigosus B. & C. (Panus) 9, 20, 54, 95
strigosus (B. & C.) Singer (Pleurotus) 9,
  20, 54, 95
strigosus (Fr.) Karst. (Gloiodon) 100, 182
striiformis auct. (Ustilago) 25, 62,
  82, 87, 88, 108, 145, 177, 202
striiformis Westend. (Puccinia) 24, 61,
  87, 88, 110, 176, 246
strobi J. Reid & Cain (Coccomyces) 163
strobi Sydow (Phomopsis) 170
strobicola Hilitz. (Leptostroma) 168
strobilicola (Rehm) Sacc. (Tapesia) 172
strobiligena (Desm.) Sacc. (Delphinella)
  248
strobilina (Alb. & Schw.:Fr.) Seaver
  (Chlorociboria) 157
strobilina (Desm.) Died. (Discella) 158
strobilina (Fr.) Sacc. (Phialea) 156
strobilina Lib. (Discosia) 165
strobilinoides Peck (Mycena) 169
strobilinum Bomm., Rouss. & Sacc.
  (Camarosporium) 2, 148
strobilinus Preuss (Sirococcus) 158,
  172, 200, 251
struthiopteridis Stoermer ex Diet.
  (Uredinopsis) 12, 132
stygia (Berk. & Br.) Mueller (Buellia) 179
stygia (B. & C.) Hafellner (Dactylospora)
  181

stygia (B. & C.) Massee (Karschia) 181
stylobates (Fr.) Kummer (Mycena) 205
stylospora E. & E. (Pseudovalsa) 20
stylosporum (E. & E.) Wehm. (Prosthecium)
  20
suaveolens (L.:Fr.) Fr. (Trametes) 189,
  220
suavissimus Fr. (Lentinus) 217
subabruptum (B. & G.) Eriksson & Ryv.
  (Cystostereum) 15, 216
subabruptus (B. & G.) Jul. (Crustomyces)
  15, 216
subacida (Peck) Sacc. (Poria) 9, 20, 31,
  54, 119, 171, 187, 195, 200, 205, 219,
  238, 250
subalutacea (Karst.) Eriksson
  (Hyphodontia) 6, 153, 167, 198
subalutacea (Karst.) H. & L.
  (Peniophora) 6, 153, 167
subareolatus Peck (Pleurotus) 20, 39,
  187, 231
subbadipes Huijsman (Galerina) 66, 252
subcaerulea (Peck) Sacc. (Mycena) 18, 205
subcartilagineus Overh. nom. invalid.
  (Polyporus) 8, 155, 169, 185
subchartaceus (Murr.) Bond. & Singer
  (Hirschioporus) 189
subchartaceus (Murr.) Overh.
  (Polyporus) 56, 189
subchartaceum (Murr.) Ryv. (Trichaptum)
  56, 189
subclavigera Litsch. (Tomentella) 239
subcoronata (H. & L.) Rogers
  (Pellicularia) 147
subcoronatum (H. & L.) Donk
  (Botryobasidium) 49, 147, 162, 197, 236
subcorticalis (Fuckel) Munk (Coniochaeta)
  14
subcorticinum Wint. (Phragmidium) 212
subcruentatum (B. & C.) Lemke
  (Aleurocystidiellum) 1, 146
subcruentatus (B. & C.) Burt
  (Aleurodiscus) 146
subcutanea Dearn. (Venturia) 220
subdendrophora Redhead (Campanella) 143
suberumpens E. & E. (Anthostomella) 252
subferrugineus Burt (Hypochnus) 188, 206
subfimbriata (Rom.) Ginns (Junghuhnia) 29
subfusca (Karst.) H. & L. (Tomentella) 96
subgiganteum (Berk.) Lemke (Licrostroma)
  18
subiculosa (Peck) Cooke (Poria) 6, 198,
  237
subiculosus (Peck) Eriksson & Strid.
  (Inonotus) 6, 198, 237
subincarnata (Peck) Keller (Skeletocutis)
  11, 158, 172, 200, 238, 251
subincarnata (Peck) Murr. (Poria) 11,
  158, 172, 200, 238, 251
subincarnata (Peck) Sacc. (Mycena) 169
subincarnatum Peck (Lycoperdon) 94
subincarnatum (Peck) Pouzar
  (Amylocorticium) 146, 161
subinclusa (Koern.) Bref. (Anthracoidea)
  66
subinvisible Liberta (Sphaerobasidium) 158
sublaevis (Bres.) Jul. (Ceraceomyces) 2
sublilacina (Ellis & Holw.) Wakef.
  (Tomentella) 11, 22, 32, 56, 96, 159,
  172, 189, 195, 200, 239, 251

Addenda

Page
24 *Puccinia graminis* f. sp. *secalis*
should read *P. graminis* ssp.
graminis var. *stakmanii (P.
graminis* f. sp. *secalis).*

25 *Puccinia graminis* f. sp. *agrostidis*
should read *P. graminis* ssp.
graminicola (P. graminis f. sp.
agrostidis).

47 *Puccinia graminis* f. sp. *avenae*
should read *P. graminis* ssp.
graminis var. *stakmanii (P.
graminis* f. sp. *avenae).*

47 *Puccinia graminis* f. sp. *tritici*
should read *P. graminis* ssp.
graminis var. *graminis (P. graminis*
f. sp. *tritici).*

66 After *Orphanomyces arcticus* insert
(Ustilago a.).

88 *Puccinia graminis* f. sp. *secalis*
should read *P. graminis* ssp.
graminis var. *stakmanii (P.
graminis* f. sp. *secalis).*

97 *Hendersonia culmicola* should read
*Stagonospora culmicola (Hendersonia
c.).*

181 Insert *Cytospora nivea* (holomorph
Leucostoma n.): on 7 Ont [1340].

194 After *Monilinia fructicola* add
(Sclerotinia f.).

199, 238, 250 *Paxillus panuoides* should
read *Tapinella panuoides (Paxillus
p.).*

244 Insert *Aspergillus clavatus:* on 6 in
storage Man [1896].

Two important papers dealing with wood decay
fungi were overlooked and the oversight was
not discovered in time to include the
records in: Basham, J.T. & Z.J.R.
Morawski. 1964. Can. Forestry Service,
Ottawa, Publ. 1072, and Engelhardt, N.T.,
R.E. Foster & H.M. Craig. 1961. Studies in
Forest Pathology 23: 1-20. Can. Dept.
Forestry,. Ottawa.

Since the manuscript was completed in
October 1984 the circumscription of several
pertinent genera has changed. These
revisions could not be integrated in the
manuscript.

J. Ginns, October 1986